HYDROPOWER IN THE NEW MILLENNIUM

PROCEEDINGS OF THE 4TH INTERNATIONAL CONFERENCE ON HYDROPOWER DEVELOPMENT
HYDROPOWER'01 / BERGEN / NORWAY / 20 - 22 JUNE 2001

Hydropower in the New Millennium

Edited by

Bjørn Honningsvåg
Lyse Energi AS, Hydro Generation Division, Stavanger, Norway

Grethe Holm Midttømme
Norwegian University of Science and Technology,
Department of Hydraulic and Environmental Engineering, Trondheim, Norway

Kjell Repp
Norwegian Water Resources and Energy Directorate,
Hydrology Department, Oslo, Norway

Kjetil Arne Vaskinn
Statkraft Grøner AS, Trondheim, Norway

Trond Westeren
Norconsult AS, Sandvika, Norway

Taylor & Francis
Taylor & Francis Group

LONDON AND NEW YORK

Cover photo
Hønefossen dam, Norway, during the 1995 flood (photo: G. H. Midttømme)

Published by: Taylor & Francis
    2 Park Square, Milton Park, Abingdon, Oxon, OX14 4RN
    270 Madison Ave, New York NY 10016

Transferred to Digital Printing 2007

ISBN 90 5809 195 3

Publisher's Note
The publisher has gone to great lengths to ensure the quality of this reprint but points out that some imperfections in the original may be apparent

Table of contents

Hydropower as Environment Sustainable Generation

Development of Hydropower Technology and Design

Hydropower in the New Millennium, Honningsvåg et al (eds),
© 2001 Taylor & Francis, ISBN 90 5809 195 3

Preface

These Proceedings contain Papers presented at the International Conference on Hydropower in the New Millennium in Bergen, Norway, 20 to 22 June 2001.

This Conference is the 4th in a series covering a wide variety of topics in the field of hydropower. The topics of the conference are carefully selected to focus on issues expected to be of increasing importance in the years to come.

The power sector is currently being liberalised in both industrialised and developing countries, introducing new concepts for development of hydro power projects as a consequence. Liberalisation of the power markets has affected the operation of existing power plants, as well as the design of new projects, both in hydro dominated systems and mixed systems. New market concepts have also been developed as a result of these processes. A sharper focus is being placed on environmental considerations, for instance CO_2-emissions and greenhouse effects. This has led to more emphasis on sustainable power generation. It is therefore important for hydropower professionals to focus on benefits of hydro generation as a renewable and sustainable source of energy. Development and operation of modern power systems require sophisticated technology. A continuous research and development in this field is therefore crucial to maintain hydropower as a competitive and environmentally well- accepted form of power generation. All these elements are covered by the Conference.

The papers are selected on the basis of a general invitation, except for a few specially invited lecturers. During the evaluation and classification of the submitted papers, the Editorial Committee has been supplemented with Dr. Odd Guttormsen and Dr. Knut Sand. Their contribution is greatly acknowledged.

The Editors wish to thank all contributors, who have made this outstanding collection of papers possible. We hope that this documentation will be useful to all participants, and inspire to further progress on the topics covered by the conference.

All papers have been typed by the authors in accordance with instructions given by the Publisher. The Editors are therefore not responsible for misprints or errors in the text. The opinions expressed are those of the authors, and not necessarily those of the Editors.

Stavanger, 6 April 2001

Bjørn Honningsvåg   Kjetil Arne Vaskinn
Grethe Holm Midttømme   Trond Westeren
Kjell Repp

Hydropower in the New Millennium, Honningsvåg et al (eds),
© 2001 Taylor & Francis, ISBN 90 5809 195 3

Organization

International Centre
for Hydropower

CONFERENCE ORGANIZER

International Centre for Hydropower
Klaebuveien 153
N-7465 Trondheim
NORWAY
Telephone: +47 73 59 07 80
Telefax: +47 73 59 07 81
Internet: www.ich.no

NATIONAL ORGANIZING COMMITTEE

Mr E. Torblaa (Chairman), Statkraft SF
Mrs G. Bangsund, Norwegian Electricity Association (EBL)
Mr I. Haga, Haga Consulting
Mr R. Haugen, Eurokraft Norway AS
Mr B. Honningsvåg, Lyse Energi AS
Mr T. S. Jørgensen, International Centre for Hydropower (ICH)
Mr T. Konow, Norwegian National Committee on Large Dams (NNCOLD)
Mr T. Ljøgodt, Norwegian Water Resources and Energy Directorate (NVE)
Mr P. Mellquist, Glommen's & Laagen's Water Management Association (GLB)
Mrs G. H. Midttømme, Norwegian University of Science and Technology (NTNU)
Mr L. Ramstad, Tafjord Kraft
Mr K. Repp, Norwegian Water Resources and Energy Directorate (NVE)
Mr N. Skjeldal, Bergenshalvøens Komm. Kraftselskap AS (BKK)
Mr K. Sollid, Trondheim Energy Company (TEV)
Mr M. Sørbye, GE Energy (Norway) AS
Mr K. A. Vaskinn, Statkraft Grøner as
Mr T. Westeren, Norconsult AS

INTERNATIONAL ADVISORY COMMITTEE

Mr. S. Alam, France
Prof. K. Baba, Tokyo Science University, Japan
Prof. K. Cederwall, Royal Institute of Technology, Sweden
Mr. M. Marumo, Lesotho Highlands Dev. Authority, Lesotho
Prof. B. Pelikan, University of Natural Resources, Austria
Prof. B. Petry, IHE Delft, The Netherlands
Mr. P.M.S. Pradhan, Nepal Electricity Authority, Nepal
Mr. M. Roluti, US Bureau of Reclamation, USA

Hydropower in the New Millennium, Honningsvåg et al (eds),
© *2001 Taylor & Francis, ISBN 90 5809 195 3*

Acknowledgements

We gratefully acknowledge the financial support from the following organizations
and companies to Hydropower'01:

Bergenshalvøens Kommunale Kraftselskap AS (BKK)
Norwegian Electricity Association (EBL)
GE Hydro
Glommen's and Laagen's Water Management Association (GLB)
Hafslund ASA
Jotun
NCC International
Norsk Energiverk Forsikring AS (NEFO)
Norwegian Water Resources and Energy Directorate (NVE)
Norconsult AS
Norplan AS
Nord Pool
Nord Pool Consulting
Otteraaens Brugseierforening
Otra Kraft
SINTEF Energy Research
Statkraft
Tafjord Kraft
Vannkraft Øst
Veidekke International AS

BKK

Norconsult

EBL
Energibedriftenes landsforening

NORPLAN A.S
CONSULTING ENGINEERS AND PLANNERS

GE Hydro

NORD POOL
C O N S U L T I N G

NORD POOL
THE NORDIC POWER EXCHANGE

HAFSLUND

OTTERAAENS
BRUGSEIERFORENING

JOTUN

OTRA KRAFT

NCC

SINTEF
Energy Research

NORSK ENERGIVERK FORSIKRING AS
NEFO

TAFJORD

V
VANNKRAFT
ØST

Statkraft

Norwegian
Water Resources and
Energy Directorate
N V E

VEIDEKKE

New trends in the development of hydropower projects

BOOT Experiment in Nepal: Recent Practices

Ramesh C.Arya
Water & Energy Commission Secretariat (WECS)
GPO Box 4360, Kathmandu, Nepal

ABSTRACT: Though rich in hydropower resources, very little of the same has been harnessed so far in Nepal. It is mainly due to the capital crunch most of the developing and especially the least developed countries are facing. It is compounded with the small size of the country where the harnessed energy may not all be utilized.

More than bureaucratic approach, business acumen may be useful for finding market for the hydropower. It is expected that the developers with rich experience in hydropower development may be awarded the BOOT contract for the gainful utilization of the resources for national development But in the recent years, there is a growing feeling that the BOOT approach may go against the national interests.

The paper analysis some of the BOOT contracts in the light of tied conditions and examines the same in the greater national interest.

1 INTRODUCTION

Nepal is known for a potential of 84,000 MW of hydropower potential, 34,000 MW of it being economically exploitable.

The State and later the State Owned Enterprises developed hydropower in the country. Bilateral and multi-lateral assistance followed later. In the recent days, private sector has shown interest to build, own, operate hydropower plants in the country. The same plants could later be transferred under the BOOT policy to a competent authority to continue the initiatives taken by the private sector. It is evident from the recent trend worldwide that in the development of infrastructure services, states would seek BOOT initiatives.

2. INVESTMENT IN HYDROPOWER

Nepalese public sector took the first initiatives in electricity generation from petroleum and hydro-resources. The 5-KW Pharping Hydro plant of 1911 and 640-kw Sundarijal Hydro plant of 1934 were the first two ventures in the Kathmandu valley. The 1,600-KW Morang Hydroelectric Company was established in 1939 and commissioned in July 1943 to power Biratnagar Jute Mills in the eastern Nepal. Padma Sundar Malla engineered the project with turbines designed and built in Zurich and generators from New York. The 500-KW plant set up in Birgunj in 1949 ran on diesel generators.

Autocratic rule of the Ranas ended in 1951. After nearly 12 years, a 300-KW plant was established in 1964 in Dharan. Later, a 300-KW Bageshwari Electric Supply Company started in Nepalgunj, a township, in the western region. Both of them were public sector ventures that received loan from Nepal Industrial Development Corporation (NIDC), a government undertaking.

The first external assistance for hydropower development came from USSR with the establishment of 2.4 MW Panauti power plant. The American and British governments financed for diesel plants in Kathmandu, and a number of sub-stations and double circuit transmission lines linking the capital with Birgunj in the south. India offered 1 MW and 21 MW hydropower projects followed by the one from China in 1972. In the later years, Nepal developed its Kulekhani and Marsyangdi projects under multilateral assistance with a number of strings attached such as grace period varying from 23 to 40 years, normal service charge not exceeding 1%, a commitment charge of about 0.5% etc.

In sharp contract to that in the USA and Britain, the power demand in the region was growing at double-digit rate. There were no donors for power sector development. But government did not make serious attempt to build its power generation capacity. It kept on negotiating for comparatively longer period with the World Bank for its ambitious Arun III project which was supposed to generate more than 400 MW. In course of negotiations, the environmental issues were seriously, including the preservation of aquatic life in the Himalayas, were raised by a number of NGOs. The World Bank ultimately dropped the proposal. In between, there were little local initiatives in the development of energy projects in the country with its own resources and expertise.

3. PROMOTING THE PRIVATE SECTOR

The multilateral organizations later started complaining "that despite pouring in billions of dollars there were hardly any trickle down effect to the real poor people. There was no impact on poverty alleviation as manifested by the start 45% of the population wallowing below the poverty line. Its power was in shambles with acute load shedding and extremely high tariff. With the belated realization of the impact of the strings of conditionalities it happily put its signature on, the donors had total grip on the country's macroeconomic activities ands the regular prescription in the form of structural adjustment programs" (Pun, 1999).

The government started promoting rural electrification. But the sector was not a paying sector, as the households were mostly domestic consumers. In some pockets of the kingdom, the population even denied paying the normal charges. Presently, 15% of the population have access to electricity in Nepal.

4. BOOT IN NEPAL

The Hydropower policy and the subsequent acts and rules recognised the need for private sector involvement in the sector. Under the Electricity Act, one could acquire a license to generate, transmit and distribute electricity from a medium sized project for a period of 50 years.

In the 1990s, Nepal entered into BOOT agreements for two projects: Khimti Hydropower Project and Bhote Kosi Project.

Butwal Power Company Limited (BPC) did the ground work for the Khimti Project. This company was established in 1966 to develop the electricity sector with Nepal Electricity Authority (NEA), NIDC and United Mission to Nepal (UMN) as shareholders. In the hydropower sector, the company had first built Tinau Hydropower Plant (1 MW), an underground hydropower plant with 2.4 km tunnel and underground powerhouse. Later, it built two run-of –the river projects at Andhikhola (5.1 MW) and Jhimruk (12 MW).

Soon after the Jhimruk Project was initiated, Ministry of Water Resources requested BPC to undertake a medium sized project to meet the interim power shortage that was envisaged to occur in the system in the near future. At that time, there was not a single medium sized project ready for implementation. BPC turned to the Norwegian Agency for Development Co-operation (NORAD) for total financial assistance similar to Jhimruk but unfortunately the project was too big (base cost at that time was US$ 75 million). However, NORAD was prepared to help financing a bankable feasibility study.

In the 1991-95 period, in collaboration with Nor Power, BPC updated the feasibility study made by NEA and started the promotion of Khimti I Hydropower project. The company carried out the following activities while financial negotiations were being followed: a) Preparatory works for Khimti Project by using its own resources, and, b) 22-km access road to Khimti I, financed by His Majesty's Government of Nepal (HMGN).

In the mean time, Norwegian Electricity Board (NVF), which was wholly owned by the Norwegian Government, was restructured into three companies. Statkraft, as one of the newly formed companies, would be responsible for the entire hydropower Plant operations owned by the Norwegian Government.

As Norway had almost reached its rated capacity, it was necessary for Statkraft to look overseas for new business operations. Following a chance meeting of Statkraft and BPC in 1992, the two companies signed a letter of intent for the development of the Khimti Hydropower Project. BPC was to carry out all the development activities in Nepal and Statkraft would be responsible for developing the financial package for the Project.

Based on the Jhimruk model, BPC made a cost estimate of 75 million US Dollars. However, with the "introduction of the multilateral lending agencies and the foreign lawyers" the project had to face a host of issues far beyond those involved in Jhimruk. After its fact-finding mission, the Asian Development Bank (ADB) revised the cost upward to US$ 125 million. (Harwood, 1998)

Himal Power Limited (HPL) was established to develop, build, own and operate the Khimti I hydropower project through private sector under a 50-year license from the HMGN. HPL has equity capital of US$34.1 million and the quasi equity (income participation certificate and the subordinate loans from international banks) of US$9 million, that is convertible to shares. In the HPC, Statkraft has the highest equity participation with US$25.1 million, followed by BPC, Alston Power, GE Energy and Nordic Development Fund (NDF) with the participation ranging from US$ 2-5 million each. ADB, IFC, NORAD and Eksportfinans together have offered loans amounting to US$92.6 million and the quasi-equity of US$3 million each from ADB and IFC. The overall loan to equity ratio stands at 70:30.

The Upper Bhote Kosi is the second of the privately financed hydro projects in Nepal This too is a run-of-river project with an average energy output of 246 GWh/year. Bhote Kosi Power Company Private Limited had directly negotiated on concession matters. The original concession was taken by Harza of the United States, who subsequently introduced Panda Energy International as US-based IPP company. Panda led the project but the bulk of the equity (70 per cent) is held by a Canadian Investment Corporation MCMIC. The local partner is Himal Power Corporation.

a. Sales and Transfer

A number of small hydropower units (with installed capacity if up to 5 MW) have been selling energy at pre-decided rates to NEA. In the case of medium and large-scale projects, the electricity price is decided under the Power Purchase Agreement (PPA).

Under the BOOT arrangement of the two power sectors, the IPPs produce energy and sell to the NEA. The latter has to develop its grid connection nearest to the power stations built by the developers. NEA pays for the energy received at that point. The transmission and distribution losses and administrative expenses for marketing the energy are born by the NEA.

Under the PPA, the IPPs are required to deliver a fixed quantity of energy annually. As the rivers are snow fed ones, the quantity of water is higher during the summer when snow starts melting and during the monsoon when there is plenty of rain. The period is designated as the wet season. The remaining is the dry season. There is a monthly break up of the supply which is low during the dry season and high

during the wet season. In case of Khimti, for example, the annual supply of 350 GWH is divided into 104 GWH during the dry period (November to April) and 246 GWH during the remaining six months. During the dry period, the energy supply whether from the plants owned by the NEA or the private producers are at the lowest. During the period, if the IPPs happen to produce in excess to the minimum committed under the PPA, the energy is sold to NEA under the take or pay conditions.

The excess energy of the dry period also has a premium value. There is no such binding for buying the excess energy during the wet period. NEA also has some hydro power plants and the energy supply during the period exceeds demand in the normal conditions.

The supply made from the two IPPs for a period of 12 months is shown below. According to the PPA signed by the two parties, the supply for months are taken according to Nepali calendar in case of Khimti and Gregorian calendar in case of the Bhote Kosi. Baisakh, the fist month of the Nepali calendar pertains to a period April 15 to May 14. Similarly January is Pauch 15 to Magh 14.

Monthly energy sales from Khimti and Bhotekosi (in MWH)

KHIMTI PROJECT

Baisakh	Jestha	Asadh	Srawan	Bharda	Aswin
18.00	41.22	41.22	42.56	41.22	39.89
Kartik	Mansir	Pous	Magh	Falgun	Chitra
39.89	27.00	16.50	15.00	13.00	14.50

BHOTEKOSI PROJECT

Jan	Feb	Mar	Apr	May	June
16.6	13.0	13.7	15.3	21.4	23.7
July	Aug	Sept	Oct	Nov	Dec
25.0	24.9	24.2	24.9	23.2	20.2

Finance Dept, NEA

b. Rate of Sales

Khimti sells its energy at 5.94 cents per KWh of energy. The rate applies for 350 GWH of energy bought according to the fixed schedule. The similar arrangement is made with the Bhotekosi for 246 MWH. For the annual commitment, the energy is charged at 6 cents per KWh, escalated at 3 percent for the first 15 years and indexed to the US consumer price index thereafter..

During the dry periods, Khimti's monthly commitment lies between 13-27 GWH and same during the wet season it varies between 39-42 GWH. During the dry and wet seasons, the excess supply is charged at 8.5 cents and 4.2 cents respectively.

Though NEA buys energy at higher rates during

the dry season, it does not in built arrangement to charge any premium to its thousands of clients accordingly.

The power Purchase agreement with Khimti is valid for 20 years. That for Bhote Kosi is for 25 years.

c. The Transfer Arrangement

Khimti works under a 50-year license. After the period, HMGN owns the whole establishment. The license granted to Khimti. As spelt out in the licensing arrangement, the government owns the whole establishment. Bhote Kosi is similarly transferred after 40 years.

At the end of the PPA period (20 and 25 years), NEA gets 50% ownership over the project. The PPA may be further renewed beyond the period.

5. IMPROVING THE BARGAINING POWER

A number of projects undertaken by the NEA would be completed shortly after. Modi came in operation in 2000. Kali Gandaki A and Middle Marsyangdi will operate in 2001 and 2004 respectively. Chilime will be completed in 2002. NEA is also licensed for Kulekhani III and Upper Karnali (300 MW). Kulekhani I and II are already operational and in the fiscal year 1999/00 these two units had record generation of 250 GWh and 123 GWh respectively (NEA.2000).

The state will be in comfortable position for some years as far as the energy supply position is concerned. According to the load forecast of the Nepal Electricity Authority (NEA), the country needs 1967, 2110 and 2300 GWH of energy with a peak of 449, 482, 525 and 575 MW in 2002, 2003, 2004 and 2005 respectively. The growth would thus be 10.05, 7.5, 8.92 and 8.76 percent in the succeeding years.

6. HESITANT INVESTORS

The comfortable position of NEA will put pressure on the IPP to come to more realistic terms. The BOOT process begins with building the project and owning and operating it. The building begins with a survey. The companies licensed to survey are also allowed to build at a later period. But the seriousness must begin from the licensing stage itself. The recent case is of EurOrient Investment Group, an American company.

EurOrient was chosen by the government in late June from amongst three similar interested companies. The other companies applying for the survey

licence included the Canada-based ASTQ Holdings Company. Detailed design works and feasibility studies of the run-of-river hydel project, which will be one of Nepal's biggest once completed, had been finished in 1993. The World Bank, which had supported it for 10 years, pulled out in 1995. At that time, the project's cost then had been estimated at US $1 billion.

EurOrient Group was said to be joining forces with two other multinational companies - ABB and Privam company, "USA's pioneer" power developers. The companies have already developed bigger hydel projects in the US, Europe, China and India.

In February 2001, the company was in the news as it failed to deposit a sum of NRs 42 million (at the rate of NRs 100 per kilowatt) or "performance assurance" required for getting a survey license of the 402 MW Arun III. The deposit sought as guarantee money was 0.05 per cent of Arun III's total estimated cost. The locals of Sankhuwasabha are of course very much interested to get the power project started, as it will lead to development in the region. Situated in the eastern hills, Sankhuwasabha is a remote district without road accesses and other necessary infrastructures (TKP, 2001).

A government owned national media even called this as an attempt " to pocket the Arun III Hydropower Project's survey license without paying the guarantee amount. The company later started knocking the doors of influential power centres including the Prime Minister's Office (PMO) and claims to have received positive assurances from the Ministry of Water Resources. It has now been maintaining that it would open its purse only while applying for the generation license of the 402 MW Project.

The Department of Electricity Development, while inviting proposals from Independent Power Developers (IPD) for 11 hydropower projects' survey licenses last year, had included the performance guarantee as mandatory in the Terms of Reference. Other five other companies, whose proposals for other hydropower projects' survey licenses were accepted together with EurOrient's, have already paid their respective performance guarantee and have even, received the licenses.

Section 24 of the nation's Electricity Regulations 1994 requires the power developers to pay much less amount as royalty (compared to the performance guarantee) while obtaining survey license or generation license. If it were only for the Regulation, EurOrient, for instance, would have to pay only Rupees 60,000 Rupees (15 per cent of the royalty of generation license fixed at the rate of Rupee one per kilowatt) for the survey license. Officials at the Department of Electricity Development point at past records hinting at lack of seriousness on the part of power developers. (TRN, 2001)

The government had invited 17 proposals for different hydropower projects some two years ago. Three IPDs then had walked away with the survey licenses of Middle Bhote Kosi, Kali Gandaki II, and Seti III Hydropower Projects - only to vanish in thin air. The 750 MW West Seti Hydropower Project has been the case with the survey license. Clearly, IPDs like these were sitting tight with their licenses until the validity of the document expired.

These lessons taught the government that all companies are not serious. The Department of Electricity Development has issued around 60 survey licenses to different IPDs over the years - of which currently only 34 are valid.

With regard to the largest project, the Arun III, the government has said it will not give guarantee to purchase the power generated. This onus will remain on the applicants themselves (Spotlight, 2001).

7. OBSERVATIONS

The process has severely impacted the way in which the hydro projects are implemented and the rate of implementation. The private financing favour low risk projects that are not capital intensive and have short construction times and quick returns. Despite the concern for global warming, the private sector is building over 40 megawatts of fossil-fuel thermal plant for each new megawatt of hydro. Unfortunately Nepal cannot go for the fossil fuel based for its long-term power needs due to the high cost of the fuel.

According to a report prepared by the World Bank, Khimti as a private sector power project "has born the brunt of developing the agreements on which other IPP hydro power projects in Nepal will be based"(Head, 2000). Bhote Kosi was the second IPP that made a PPA. The government has not made any further agreements wit the IPPs The following observations in the context are noteworthy.

7.1 State Participation

The economic life of the hydropower projects is well above 50 years. The license for the project is valid for 40 to 50 years. Even the Power Purchase Agreements are valid for a period of 20-25 years. The promoters have acquired term loans in their private capacity. Thus they are required to pay higher interest rates of about 11 per cent and the payback period is normally 11which is relatively short as hydropower projects are accepted as capita intensive

projects. There is no attempt to have private-public sector partnership and sharing of well-defined loan burdens by the two partners. Under such an arrangement, at least the loan made available to the public sector undertaking would have been made available for a longer period and at lower interest rates. This would have helped reduce the tariff for the energy services. Several projects have followed this pattern. As the private sector will be using the project for almost free after the 11th year, 50 years of license for the project is very favourable to the investor.

In Theun Hinboun (210 MW) and Nam Theun II (900 MW) projects there was a close collaboration of the Laos Government. Birecick (672 MW) of Turkey is under public-private partnership. Ita (1450 MW) in Brazil was started in the public sector. Later it accepted limited recourse private financing.

7.2 High Tariff

The government is under pressure from multilateral agencies to increase the electricity tariff. Under similar pressure in the past, the government had twice revised electricity tariff within 12 months. The purchase from IPPs at about 6 cents /KWh to NEA is causing more financial burden on the latter. An effort for a third hike in the recent months caused much public debate and resistance in the parliament.

A commission formed to review and suggest on the Proposal for the Power Hike came up with a number of suggestions. Among others, it suggested the following (MOWR, 2001):

- Compulsory provision for Nepalese investors in projects lesser than 300 MW;
- Reassess the installed capacity and reduce the spill energy;
- Preference to storage based projects;
- Stop one-to-one negotiation and invite competitive rates on energy purchase; and,
- Reduce the project expenditure by refinancing on the external loan on Khimti and Bhote Kosi Projects from the local market.

The public is feeling that the IPPs are doing not really beneficial to the nation. A number of cases have come to light where the IPP have blown up their cost and expenditure. In India, Enron USA and India had signed agreement to build Dhabol Power Plant. After the renegotiation, the power tariff was reduced from IRs 2.4 per kWh to IRs1.89 per kWh. The agreements with IPPs in Pakistan have also not been taken positively by the country. An IPP has invested US$375 million and in one year the company earned US$380 million (Thapa, 1999). Pakistan's state Water and Power Development Authority (WAPDA) has asked the power regularity authority to cut electricity prices to boost demand, despite IMF's pressure for a raising of tariffs. As WAPDA argued, the further tariff increase (by 11.9%) "will ruin the industrial s well as agricultural sector, increasing unemployment and political unrest". On the other hand if "rationalization of tariffs" are allowed, the demand was likely to increase by 1,500 MW. The authority and the government are locked in a bitter disputes with the IPPs over tariffs. The latter are likely to add another 1,600 MW of power to Pakistan's capacity (Asian Power, 1999). The Nepal situation is no better. A number of industries in the industrial belts of the eastern and central Nepal have closed down. High electricity tariff is one of the reasons for the closure.

Due to the tendency of paying back the high interest loan within a very short period, the tariff agreed by the parties are relatively higher.

7.3 Impact of Devaluation

There is a provision that the promoter gets the payment for energy sold in the domestic market in convertible currency at a rate with provisions for adjusting the escalation caused due the increase in consumer price index in the USA. As the income from the sale of services is receivable in the local currency, payment in hard currency may be a tough business. Nepalese currency has devalued from NRs 20 to a dollar to NRs 75 over a short period of just ten years. This depreciation will result on high-rise in domestic services provided under the external loans payable in hard currency. Even without the escalation clause, the generation of remittable resource from the sales in domestic market is a challenge in the light of increasing devaluation.

7.4 Resource from Local Banks

In the projects, the capital availability in the local market was completely undermined. Specially, after the liberalization the industrialization trend in the country has slowed down due to the reduced compatibility in the international market. There are very little avenues left for productive investment. In the past few years, investment is mostly directed at services and financial sectors such as hotels, airlines and commercial banks. Whenever subscriptions have been invited in these sectors these have always been oversubscribed, despite the fact that some of the projects may not pay dividend even in the near future. One deluxe hotel, which has been so over-

subscribed, would not be able to pay dividend for 25 years even if the business runs at its full capacity and at the publicised tariff. Lack of investment opportunities is reflected through the continuously decreasing rate of interests on saving deposits which a few years back was as high as 11-12 percent and now it has come down to 3-4 percent per year.

7.5 Multipurpose Projects

Projects should have been considered for their multipurpose (power and non-power) uses. The Kathmandu Valley, foe example, is facing acute shortage of drinking water, and the government has been negotiating with several multilateral agencies. It has recently come into an agreement for Melamchi Drinking Water Project at 460 million US dollars. The source for the regular supply of 170 million litres of water per day needs a 26.5 km long tunnel. Khimti Hydro project, on the other hand, is located 100 km east of Kathmandu and it used a total of 10-km long tunnels of different sections. Some expert's opinion on its multiple objective development has been overlooked. Services availed from single purpose projects would normally be costlier. The Melamchi water, when availed, will cost four times the present rate and energy from Khimti and Bhote Kosi, bought at source, costs as much as the present domestic rate. On the other hand, the Casecnan (150 MW) multipurpose storage project in the Philippines attracts almost half of its revenue from the sale of water.

7.5 Penalty for Delay

The private promoters need to meet the stipulated deadlines especially as the projects involve loans under relatively high rate of interest. Also the work completed should be reliable and dependable. The parties to the production and delivery are equally responsible for maximum utilization of the goods produced. But such a binding should be applicable to the producer and receiver of the goods and services. It was not so balanced in the case of Khimti, but this was duly considered in case of the second project.

The Khimti Project had imposed penalty clause as much as US$20,000 per day to its subcontractors There was a penalty cause on NEA for failing to provide the transmission line from the designated location for the energy generated by the Khimti Project. But no such clause prevailed for the default if any on the part of Khimti. Fortunately, the subcontractor and both the parties (Khimti Project and the NEA) completed their respective tasks exactly on time. Due to a section to this effect, the contractors of the Bhote Kosi Project had to pay a penalty of as much as US$8,000 per day.

7.6 Disruption in Services
Once regular electric supply of electricity is made available, it should be regularly supplied, as clearly spelt out in the agreement and any irregularity in the supply should be duly compensated to the user of the service. There was a minor breakdown in the Khimti supply immediately after the formal inauguration. The agreement does not provision to this effect.

7.7 Reservoir Projects
Without storage many of the unique attributes that hydro brings to a system are lost, in particular the ability to store and redistribute energy within the load curve, and the ancillary benefits of frequency control and rapid response reserve generation. Unfortunately such a project had not been negotiated seriously by the government. It has set up one storage hydro-project basically to meet the peak time demands.

8. THE BOOT POLICY
The government did gain some experience from the dealings with development of IPPs in the hydropower sector. It has also received requests from the private sector in undertaking the construction of toll roads and several infrastructure development projects. The government recently announced its Build, Own, Operate and Transfer (BOOT) Policy, 2000. It is in process of formal enactment.

The following are the salient features of the Policy:
- Invitation of proposals on a competitive basis on areas identified by the government;
- Direct proposal from the private investor;
- The Project agreement can be for up to 30 years, extendable for another 5 years;
- Public / Private Partnership in "very important sectors" with a maximum of 25 % from the public sector;
- Lease provisions for Land and other infrastructures
- HMG will not take guarantee for external/internal loans;
- Provisions for imposing tolls and taxes on the infrastructure development;

- HMG may prescribe the limit on availing resources from internal/external sources;
- Foreign currency repatriation by purchasing the convertible currency in the open market;
- Foreign investment in equity or shares generally up to 70%;
- Corporate tax not to exceed 20%;
- Depreciation benefits accelerated by 33%;
- Loss of a year carried on to the following 7 years;
- Import of machineries and equipment at 1% custom duty and the spares at a maximum of 20%;
- Re-export of used machinery at a duty not less than 10%;
- HMG takes the ownership of the property after the termination of the Project Agreement period.
- Establishment of a Joint Venture Infrastructure Finance Corporation with an authorised capital of NRs10 billion and paid up capital of NRs500 million (US$ 1=NRs 75).

9. CONCLUSION

For a least developed country endowed with rich hydropower resources, involvement of the IPPs as partners to its development is inevitable. However, as a public service venture, it should not work in complete isolation with the imported capital only. The import is tied with exchange risks that may result in big liabilities to the country. It should utilise the capital and human resources available in the domestic market.

There should be a foolproof mechanism to access the actual cost investment and reasonable profit to the investors.

It is expected that the IPPs would honour the desire of the people where they have their investment risks and would amend some of the provisions in "good faith" clause of their Project Agreements and Power Purchase Agreements for a longer continuation of their profit interests from business in Nepal.

Worldwide, there is very little experience from hydropower related BOOT projects. The host countries and the promoters need to act together in good faith and be willing to correct the mistakes done in course of launching hydro projects which are basically investment in infrastructure projects in many developing countries.

REFERENCE

Asian Power, 1999, *News and Views, p 6.* Vol.7 No. 4., May 1999

Head, Chris.2000. *World Bank Discussion Paper No 420 Financing Of Hydropower Projects*, World Bank, Washington

Harwood, F Peter. 1998 **The Experience Of A Nepal Company Associated With Hydropower Development In Nepal: The Khimti Experience.** Report of The Seminar on Legal Aspects of Water Resources Development for Promoting Private Sector Participation, Nepal Engineers Association, April 1998, Kathmandu.

MOWR.2001. Press Release. *Ministry of Water Resources, His Majesty's Government,* August 30, 2001

NEA.2000. *Fiscal Year 1999/00 A Year in Review,* Nepal Electricity Authority, Kathmandu

Pun, Santa Bahadur. 1999. The Evolving Role of the Public and Private Institutions in the Nepalese Power Sector, *WECS Annual Report-1999,* Water and Energy Commission Secretariat, Kathmandu.

Spotlight. 2001. *Interview* with Bishwanath Sapkota, Secretary, Water Resources, Vol. 10, No 30 February 9-15, 2001.Kathmandu

Thapa, Anand B. 1999 Private Hydropower in Nepal and Power Tariff in *WECS Bulletin,* Vol. 10, No. 1, Water and Energy Commission Secretariat (WECS), Kathmandu

TKP. 2001. News in *The Kathmandu Post,* Kathmandu, February 11, 2001

TRN. 2001 News in *The Rising Nepal,* Kathmandu February 08, 2001 ***

Hydropower in the New Millennium, Honningsvåg et al (eds), © 2001 Taylor & Francis, ISBN 90 5809 195 3

Technical, economic and environmental feasibilities of Nemunas hydro energy utilization

J.Burneikis & D.Streimikiene
Lithuanian Energy Institute, Kaunas, Lithuania

P.Punys
Lithuanian Hydro Power Association, Kaunas, Lithuania

ABSTRACT: Hydropower resources in Lithuania are evaluated as about 2.7 billion KWh/year. The technical hydropower resources of the biggest Lithuanian river Nemunas (937 km length, 98220 km^2 river basin, average annual flow – 634 m^3/s) makes up 1.47 billion kWh/year or 55.6% of all technical resources. Therefore it is very important to investigate the feasibility of these hydropower resources utilization, because large HPP are more efficient and significant comparing with small (< 10 MW). However the environmental requirements for large HPP are stricter. Revival of small HPP have already begun in Lithuania by rehabilitation of abandoned HPP and by erecting small HPP on existing water ponds. The article deals with the technical, economic and environmental feasibilities of the 2 large hydro power plants construction on the middle of the River Nemunas: Birstonas and Alytus HPP. In the case of Ignalina NPP closure the attractiveness of these HPP projects would significantly increase. The cascade of operating Kaunas HPP and proposed 2 new HPP is able to produce about 1 TWh/year.

1 NATIONAL ENERGY STRATEGY

The main source of electricity generation in Lithuania is the Ignalina NPP. During the latter years it generated 80-85% of the total electricity production and with the lowest costs. Total installed capacity of power plants in Lithuania exceeds the present domestic requirement almost 3 times.

Upon comprehensive assessment of technical, economic and political factors, by requirement EU, the following strategy (Lithuanian Ministry of Economy 1999) for further operation of the Ignalina NPP is proposed: in line with the Nuclear Safety Account Grant Agreement, unit 1 of the Ignalina NPP will be closed down by the year 2005, taking into consideration the terms and conditions of long term and considerable financial assistance from the EU, G-7 countries and other states.

Due to the age difference between unit 1 and unit 2, the issue pertaining to the conditions and precise final date of the decommissioning of unit 2 shall be solved in the updated National Energy strategy to be prepared in the year 2004, when more detailed information on the operation of unit 2 will be available. The total negative consequences of the decommissioning of unit 1 on the national economy preliminarily estimated by using various macro-economic models may make up to 40 billion ion Lt.

Upon decommissioning of unit 1 of the Ignalina NPP, the existing capacities will meet the national demand up to the year 2020 in all cases of internal growth only if the Lithuanian thermal power plant (TPP) is maintained and refurbished, and the part of its units upgraded to the combined cycle gas turbine technology. In this event the balance of generation and demand in the year 2020 would be positive, ensuring export potential of 3-5 TWh. Modernization of existing combined heat and power plants and construction of new ones would further increase this potential.

Performed technical-economic analysis shows that should new capacities be required, CCGT CHP, small CHP with gas fired internal combustion engines or gas turbines and a new combined cycle gas turbine (CCGT) would be cheapest source of electricity generation after the refurbishment of available thermal power plants. Taking into consideration the situation in fossil fuel market and requirement to limit CO_2 emissions, the construction of the cascade of hydro power plants (HPP) on the middle track of the river Nemunas may be justified (Lithuanian Ministry of Economy 1999) .The cascade of operating (40 year) Kaunas HPP and 2 new HPP: Birstonas and Alytus would be able to generate about 1 billion kWh/year with the total installed capacity about 250 MW. However, environmental, land ownership, monument protection and other requirements would restrict the possibilities to construct these HPP.

It is also expedient to keep the Kruonis hydro power pump storage (HPPS) not only in a regime of daily regulation with Kaunas HPP but also in regime of weekly regulation, however its role in Lithuanian

Table 1. Energy and electricity balance sheet in 1999.

Indices	Energy		Indices	Electricity	
	Ktne	%		GWh	%
Energy production	8088	100	Gross production	16789	100
Indigenous energy	880	10.86	Ignalina NPP	13942	83
Crude oil	233	2.87	Thermal PP	1974	11.8
Solid fuel	611	7.54	HPP	325	1.9
Hydro	36	0.44	Kruonis HPPS	548	3.3
Nuclear	2570	32.96	Import (+)	4182	24.9
Oil and products	2696	33.28	Export (-)	9341	55.6
Natural gas	1754	21.65	Gross consumption.	11630	69.3
Coal	116	1.43	Own-uses	1668	9.9
Imported fuels	7208	89.0	Kruonis HPPS consumption	771	4.6
Electricity generation.	3110	38	Losses in network.	1778	10.6
Heat consumption.	1413	17.4	Net consumption.	7413	63.7
			Energy sector	792	6.8
			Final consumption.	6621	56.9
			Industry	3192	48.2
			Transport	95	1.4
			Residential sector	1570	23.7
			Agriculture	459	6.9

power sector will depend on the course of implementation of other international projects (the Baltic Ring, electricity transmission line to Poland, etc.).

The Lithuanian TPP will, within the next 10 years, serve as reliable source of half-peak energy, capacity reserve and in the future as a source of basic energy too, neglecting into pollution environment and CO_2 emissions. The final choice of generating source replacing unit 1 of the Ignalina NPP will determine a revised least cost analysis and evaluation of other political, economic, environmental and social factors.

2 ENERGY AND ELECTRICITY BALANCE SHEET

Lithuania (the state in Baltoskandian region, covers 65200 squire km, population 3.7 million) has very limited own energy resources. Local fuels: oil, peat, firewood, waste and hydro energy share was only about 10% in the 1999 fuel mix (Lithuanian Ministry of Economy 2000). Fossil fuels share in the primary energy balance was 67.6% and another 32.4% was the share of nuclear fuel. There is Ignalina NPP operating in Lithuania, which installed capacity is 3 GW. Hydropower in primary energy balance sheet makes up to about 0.5%.

Electricity generation in 1999 was 16789 GWh. Ignalina NPP produced 83% of all electricity generated in Lithuania, thermal power plants - 11.8%, hydro PP - 1.9 %, Kruonis HPPS – 3.3 %. However Kruonis HPPS consumed 771 GWh/year or 4.6% of electricity. The total hydro energy generation in gross electricity production was 5.2% and there were 1.3 % of losses in electricity production. The primary and electricity balance sheet is shown in Table 1.

It is obviously, that Lithuania must use local and renewable energy sources and at first to improve utilization of effective hydro energy resources (by EU requirement renewable energy should make up to 12% in energy balance in 2010).

3 HYDRO AND POWER PLANTS OPERATING IN LITHUANIA

The total installed capacity of the Lithuanian power plants is 6537 MW; including Ignalina NPP with 3000 MW installed capacity and the Lithuanian TPP with 1800 MW capacity. Other CHP have installed capacity of about 828 MW. As one can see from the table the highest share of electricity production in Lithuania is attributed to Ignalina NPP (more than 76% of total electricity production in 1999). The share of hydro was about 5% in 1999 including electricity produced at Kruonis HPPS.

Table 2 presents installed power capacities and electricity output available in Lithuania in 1999.

Table 2. Installed power capacities and electricity output in 1999.

Power plant	Installed capacity, MW	Electricity production, GWh
Lithuanian TPP	1800	1695
Vilnius CHP	384	709
Kaunas CHP	178	236
Mazeikiai CHP	194	433
Klaipeda CHP	11	33
Ignalina NPP	3000	13554
Hydro plants (pure hydro +HPPS)	909	895
Other PP	61	74
Total	6537	17631

There are 2 big hydro power plants in Lithuania. They are: Kaunas HPP with 100.4 MW installed capacity and average 350 million kWh/year electricity production and Kruonis HPPS (800 MW installed capacity). There are also more than 30 small HPP operating currently in Lithuania, which total installed capacity about is 9 MW and electricity production about 30 million kWh/year (Burneikis 1995). They were built in various places on existing water ponds or by refurbishing old small HPP. The water flow in these small HPP is from 0.5 to 10 m³/s; the head of water fluctuates from 3 to 32 m, capacity from 10 to 2460 KW. They were built in period 1955-1999.

The main features of energy sector restructuring are decentralization, demonopolisation and privatization. This requires considerable legislative, institutional, organizational changes (Burneikis & Punys 1999, Burneikis & Streimikiene 1997). The Lithuanian electricity and gas sectors are on the edge of reforms. The government is going to approve the plan to separate electricity production, transmission and distribution, to establish the single buyer and liberalize the market. Energy market liberalization would create more favorable conditions for hydropower development in Lithuania if external costs electricity production by evaluated. The large hydropower projects as Birstonas HPP and Alytus HPP can by actual in new conditions, especially with closure of Ignalina NPP.

4 EVALUATION OF HYDRO POWER RESOURCES IN LITHUANIA

The total potential hydro energy resources of all 477 rivers, which have catchment's area of more than 50 km² or the length of more than 20 km, are evaluated as 5.13 billion kWh/year (Burneikis 1999a). The 2 biggest Lithuanian rivers – Nemunas and Neris have the greatest share of potential hydro energy resources - 60%. There are 40 rivers called as middle rivers, with the share of potential hydro energy resources more than 18%. The rest 435 rivers are considered as small rivers and makes up to 12.8% of all hydropower resources in Lithuania. Potential hydropower production of all Lithuanian rivers is estimated as 79 thousand kWh km²/year. Hydropower resources of Lithuanian rivers are presented in Table 3.

As one can see from the Table 3 the River Ne-

Table 3. Hydropower resources of Lithuanian rivers.

Hydro power resources	Number of rivers	Energy indicators		% from total resources
		Capacity, MW	Production, billion kWh/year	
1. The main resources (>100MW)				
Nemunas	1	240	2.10	41
Neris	1	106	0.93	18.2
Total	2	346	3.03	59.2
2. Local resources of other rivers				
Middle	40	165	1.45	28.2
Small	435	74	0.65	12.6
All	475	239	2.10	40.8
Total	477	585	5.13	100

munas has the greatest share of total hydro energy resources, even 41% and is the most important in future hydro energy development. Large HPP can be built on this river taking into account all environmental requirements and restrictions especially those related to land flooding. The perspectives of hydro energy utilization are evaluated in the following chapter.

Only a part of potential hydro resources can be used technically for energy production. This share fluctuates from 25% to 60%. For the big rivers this share is greater.

So, technically harnessable hydro energy resources can be estimated as about 2.647 billion kWh/year. They consist of:
• river Nemunas 1.471 billion kWh/year or 55.6%;
• river Neris 0.653 billion kWh/year or 24.7%;
• both rivers 2.124 billion kWh/year or 80.3%;
• middle rivers 0.361 billion kWh/year or 12.6%;
• small rivers 0.162 billion kWh/year or 6.1%;
• all middle and small rivers 0.523 billion kWh/year or 18.7% of total resources.

The share of utilization of technically harnessable hydro energy resources in Lithuania is about 15%. Economically harnessable hydro resources makes up to 30% of potential hydro resources and amounts to about 1.5 billion kWh/year, including the share of middle and small rivers (about 0.5 billion KWh/year). The most economically feasible are Birstonas and Alytus HPPs on the middle of the River Nemunas. Kaunas HPP was built on the River Nemunas about 40 years ago. The cascade of those 3 HPPs could produce about 1 billion KWh/year.

The utilization of all economically hydro energy resources in country could supply about 15 % of the total current electricity demand.

5 THE PERSPECTIVES OF HYDRO ENERGY UTILIZATION

The problem of hydropower resource utilization relatively consists from 2 parts:
• The problems related to utilization of the main hydro energy resources of the River Nemunas and River Neris or large HPP;
• The problems associated with the utilization the local hydro resources of all other Lithuanian rivers or small HPP.

This problem is characterized in Table 4.

Definition "small HPP" is relative, in order to define independent power producer. Small HPP are especially interesting for the private sector. Legal advantage are granted only to a sHPP, which installed capacity is <2 MW. Environmental advantages are granted to HPP, which installed capacity is < 500 kW.

Table 4. The problem of hydro power resources utilization in Lithuania

Characteristics	Problem	
	Small HPP	Large HPP
Category	<10 MW	>10 MW
Rivers	All rivers, excluding Nemunas and Neris	Nemunas, Neris
Investors	Private sector	Private sector with guarantied support Government
Time horizon	Today	In perspective
Level of standardization	Typical projects, standard energetic units, construction from wheels	Individual projects, unique energy units, own basis of construction
Necessary activities	In 2 stages: I stage-reconstruction of abandoned sHPP, on existing water ponds. Duration about 5 years II stage –construction of new water ponds and construction of sHPP on suitable sites	Preparation of feasibility studies of hydro energy resources of Nemunas and new schemes. Project organization - Company Birstonas HPP.
Economic effect	Profitable investments based on the financial cash flows	Most efficient development of energy sector using multi criterion optimization concept.
The main constrains	The lack of support from the Government, lack of legal framework, environmental and bureaucratic constrains	The installed capacity of power plants exceeds the required about 3 times, externalities are not taken into account, the lack of capital.

HPP construction as well as construction of other infrastructure projects requires high capital investments (1000-3000 USD/kW), but due to a cheep electricity production the pay-back period is 10 years, depending upon concrete conditions. The technical policies are needed in order to guaranty long term and stable legal, administrative, financial and environment for the investments pay back. In Lithuania regulated are only small HPP on existing water ponds (Burneikis & Punys 2000). The construction of sHPP on the new sites or large HPP is still not regulated.

Taking into account all circumstances the most effective and a most important would be large HPP on the River Nemunas. Preliminary indices of HPP on the middle of the River Nemunas show their technical-economical effectiveness by 10-12 times lower electricity production costs and also lower external costs of electricity production.

Small HPP have already become an attractive business for private investors. There are more than 30 sHPP operating currently in Lithuania, from which more than 20 were built in 1990-2000 year period. It is possible to construct 131 sHPP with total capacity 16.4 MW and more than 60 million KWh/year on the existing water ponds. The part of those sHPP is under construction now.

In the next stage the construction sHPP on the middle and small rivers will be initiated selecting the suitable new sites. The construction of these sHPP would be approximately 1/3 time more expensive comparing with those constructed on already existing water ponds. The duration of the second stage would take a longer time. The long-term hydro energy development target is to cover with hydropower about 15% of current electricity demand in Lithuania.

6 PROPOSED HYDRO POWER PLANTS ON THE RIVER NEMUNAS

The River Nemunas is a low land transboundary river and the largest one in Lithuania, with catchment's area of 97924 km^2. It originates in Belarus and falls into Kursiu lagoon (Baltic Sea). Although the river's length is some 940 km, the source of it lies only 240 m above sea level. The largest part of the Nemunas catchment's (47.5%) belongs to Lithuania, covering some 75% of its territory and the remaining part to Belarus (46.4%) and Poland (6.1%). The length throughout Lithuania is 359 km; 462 km throughout Belarus and 166 km forms the border with Russia (Kaliningrad district). The slope of the river is very gentle: from an average of 0.3 m/km in the upper reaches, the slope reduces to 0.25-0.20m/km in its middle stretch and further decreases to less than 0.12 m/km in the lower reach. The catchments area and seasons conditions govern the hydrological properties of the Nemunas River by its accentuated high flow in spring and the minimum discharge in summer. Their ratio does not usually exceed 20.

The studies covering feasibilities of the Nemunas middle reach (upstream Kaunas city beyond the state border into the Republic of Belarus) water resources utilization for the power generation dates back to the beginning of the last century. All studies, taking into account the favorable topographical conditions of this section (relatively narrow valleys, opportunities to have diversion type plant) emphasized the priory of HPP construction.

The Master plan of the whole Nemunas was developed to exploit its water resources for power and navigation purposes in 60s. The latter foresaw to develop the waterway connecting Baltic Sea and Black

14

Sea, but this plan failed. A cascade of 8 low-head plants of the run-of- river type was planed on the Nemunas River of which 5 were considered to be built in Lithuania in that time. The main updated data of the proposed projects in the territory of Lithuania is presented in (Burneikis & Punys 2000). All 5 projects of HPP create a continuous chain of run-of-river HPP, having earth fill dams. At present, the river power potential is exploited only in the middle reach of River Nemunas by Kaunas HPP, which was commissioned in 1960. It is a run-of-river plant with power house and overflow spillway being an integral part of earth fill dam. Its power production (100.4 MW and average 350 million kWh/year) covers 2.5% of the Lithuanian electricity demand.

The location of Jurbarkas (lower reach), Alytus and Druskininkai HPP projects is definitely determined and they will be located in the upstream to the big towns. The Druskininkai dam will be located upstream to the town Druskininkai. It will impound a reservoir, which stretches beyond the state border, into Republic of Belarus. Only location of Birstonas HPP project, its type and size have to by considered in detail. The project HPP is planned to be built upstream to the existing Kaunas HPP and upper to the Birstonas resort town. One of the proposed options is developed by the Moscow Hydroenergoproject and also consider a run-of-river type plant associated with relatively higher head of the water (35 m) and higher installed capacity (650 MW). The proposed dam will impound relatively large accumulation reservoir which surface area would be about of 134 km^2. The very sensitive area, Punios pine forest would be inundated and reservoir backwater would reach Druskininkai town. The public opinion regarding this project was very negative and consequently it was rejected at that time. An optimal decision was taken in (Burneikis et al. 2000) to divide this large hydropower project into 2 smaller ones, with smaller reservoirs at Birstonas and Alytus, with surface area 25 km^2 and 45 km^2 each, respectively (Burneikis 1999b).

There was also an option to construct the plants of the channel derivation type plant upstream to the Birstonas site. The plant would comprise: a dam located at Nemaniunai (342 km from the mouth rivers), a derivation channel of 4.5 km long and a power house at the mouth of River Verkne, a small tributary of the River Nemunas. Similarly, a dam could be located at Punia (349 km from the mouth), a powerhouse at Margarava (some 327 km from the mouth of Nemunas) with a derivation channel of 3 km long. The latter would have the main advantage - the reservoir would be located upstream to a unique Punios pine forest. However, their disadvantages such as lower energy production, higher investment costs and increased restrictions associated with perspective navigation are crucial. In any case Birstonas run-of-river plant would have an earthfill dam with power-house incorporated in its central part and the overflow spillway. The powerhouse would include 3 Kaplan type turbines, each of 25 MW capacity and totaling to 75 MW. Rated head water would be 17 m and turbine-installed discharge of 180 m^3/s is to be expected. The expected annual average power output will total to some of 300 million kWh/year. To protect above-mentioned Punios pine forest the protection embankment and the ground water level management system would be necessary to install. The total benefit of proposed 2 new hydro plants (Birstonas and Alytus) would be both direct and indirect. The direct benefit would be the installed capacity and electricity production. Indirect benefit would be the supply of water, flood control, reliable and improved navigation, fishery improvements, increased employment opportunities, further development of tourism and recreation etc. The expected investments in both projects (total 150 MW, 630 million kWh/year) are about 300 million USD.

7 THE COMPETETIVENESS OF THE BIRSTONAS HPP PROJECT

The concept of ecological sustainability requires evaluation of damage for environment by using all energy resources. In the case of such environment source as clean air (which has value but hasn't price) we face the market failure. Therefore, utilization of environment is not represented in the cost of energy production and in the sale prices of goods and services. If additional costs incurred by utilizing the environment during the energy generation were included in the price of goods, the market would distribute the resources, including the natural ones, efficiently.

Word Bank proposes such classification of energy projects, which are to be supported:

Type I: national benefits more than national costs and plus global environmental effect;

Type II: national benefits less than national costs, but national benefits plus global environmental benefit more than national costs. All these types of projects are supported.

The environmental damage caused by thermal PP depends on the fossil fuel type, type and capacity of equipment and pollution abatement technologies. TPP, burning high sulphur HFO (as in Lithuania) produces about 4 Lithuanian ct of economic damage per 1 kWh generated. In the case of orimulssion (has place in Lithuanian TPP) - 5 ctLt/KWh, natural gas – 1.5 ctLt/KWh. In average- 2-3 ctLt/KWh. Of course this data are very raff. Other evaluations of the damage are also known (Out 2000). So, Lithuanian TPPs generates about 2 billion TWh/year and creates damage of 40-60 million Lt/year.

We can compare 2 possible projects: Birstonas HPP on Nemunas river and TPP of condense mode using economic criterion supplemented by ecological

15

Table 5. Comparison Birstonas HPP project and TPP by economic-ecological criterion.

The type Of PP	The cost of 1 kW, Lt/kW	Installed capacity C, KW	Electricity production E, million kWh/year	Capital investments K_a, million Lt	Annual exploitation costs O&M+FC, million Lt	Annual economic damage D, million Lt	Total costs of electricity production, CP million Lt
HPP	8000	72000	302	576	4.53	-	621.3
TPP	2000	79200	317	158.4	47.55	9.51	729.0

one (Burneikis 1999b). We will compare total costs of electricity production of both projects using normative pay back period of 10 years. For the comparison of economic efficiency of medium HPP - Birstonas HPP and TPP the total production costs per unit of electricity will be compared. A simple levelized cost methodology, using a uniform real discount rate, is used here in order to make this research as transparent as possible, while still retaining reasonable accuracy in presenting relative costs of different alternatives.

The total production cost per unit of electricity CP is computed as a sum of the capital cost on an annualized basis K_a, annual operations and maintenance costs O&M (fixed and variable) and fuel costs FC:

$$CP/kWh = (K_a + O\&M + FC)/kWh \qquad (1)$$

Annualized capital costs are expressed by following formula:

$$K_a/kWh = K_{kW} * a/T \qquad (2)$$

where K_{kW}- total capital cost (including interest during construction) per kW; a - annuity factor = $r/[1-(1+r)^{-n}]$; r - discount rate; n - financial lifetime (years); T - hours of plant operation per year = c_f •8760 and c_f - plant capacity factor

Operations and maintenance costs are computed using the following formula:

$$O\&M/kWh = O\&M_a/T \qquad (3)$$

where: $O\&M_a$ - annual O&M costs per klW, T - hours of plant operation per year.

Fuel costs are expressed by the following formula:

$$FC/kWh = HR * FP \qquad (4)$$

where HR - heat rate (fuel/kWh) = GJ/kWh=$1/(\eta*3600)$; η - plant efficiency; FP - fuel price (ECU/GJ).

In order to fully incorporate environmental aspects into our analysis, we include annual economic damage caused by air pollution.

So total electricity production costs consist of capital investments, annual operating and maintenance costs, fuel costs and economic damage due to atmospheric pollution. We selected two PP as representatives, which have the similar installed capacity and produce the same about of electricity per year. Such representatives are prospective Birstonas HPP

with total installed capacity of 72 MW and electricity production of 302 million kWh/year and TPP with capacity of 79.2 MW and electricity production of 317 million kWh/year. Data representing the costs of those power plants is presented in Table 5 (Lithuanian Ministry of Economy 2000). So we have the following equation:

$$CP_1/kWh < (CP_2 + D)/kWh \qquad (5)$$

where CP_1 - total costs of electricity production at HPP and CP_2+D - Total costs of electricity production at TPP.

We can make a conclusion, that Birstonas HPP project taking into account incomparable lower exploitation costs and that it does not cause atmospheric pollution is more profitable than alternative TPP during the normative payback period (10 years) according to rough economic-environmental criterion. The share of economic damage in the total costs of TPP is significant, i.e. 13 proc. and the share of exploitation costs –65%.

8 CONCLUSIONS

1. Though the nature is not well disposed towards utilization of hydropower in Lithuania and in the most favorable case only about 15% of total electricity all hydro resources can cover demand, nevertheless it is possibility to promote hydropower utilization.

2. There are good conditions to construct small HPP on the biggest water ponds and on the new suitable sites on the middle and small rivers. This process has already been started in Lithuania.

3. The economically efficient and significant from the energetic point of view are the projects of 2 large HPP on the middle of The River Nemunas - Birstonas and Alytus HPP The total installed capacity of these HPP is about 150 MW and annual electricity production is 632 million kWh/year. Capital investments required about 900 million Lt.

4. In the case of Ignalina NPP closure the probability to implement these HPP projects would be high. The cascade of operating Kaunas HPP and 2 prospective HPP (Birstonas and Alytus) would produce about 1 TWh/year.

5. For the construction of Birstonas or Alytus HPP the BOT financing model should be applied. It is also very important to ensure the participation of the Government in HPP construction in order to reduce political risks and promote privatization process.

9 REFERENCES

Lithuanian Ministry of Economy. 1999. National energy strategy, Vilnius.

Lithuanian Ministry of Economy. 2000. Energy in Lithuania – 99. Kaunas: Lithuanian Energy Institute.

Burneikis J. 1995. Hydropower resources and their exploitation possibilities in Lithuania. *Hydroenergia-95; Proc. of international conference, Milan, 19-20 September 1995.*

Burneikis J. & Punys P. 1999. Hydropower: general framework for legislation and authorization procedures in Lithuania. *Renewable energy in agriculture; Proc. of international conference, Raudondvaris 16-17 September 1999.*

Burneikis J, & Streimikiene D. 1997. Evaluation of hydro energy recourses in Lithuania. *Hydroenergia-97; Proc. of the 5-th International conference, Dublin, October 1997.*

Burneikis J. 1999. The outlook for hydro energy in Lithuania. *Hydropower into the next century – III; Proc. of international conference, Gmuden, October 1999.*

Burneikis J. & Punys P. 2000. Evaluation economic and environmental criteria for large-scale hydropower schemes. *Water Management Engineering vol. 10 (32) (in Lithuanian, summary in English).*

Burneikis J., Streimikiene D. and Punys P. 2000. Possibilities for financing construction Birstonas hydro plant on the Nemunas river. *Hydro -2000 Proceedings of international conference, Bern, October 2000.*

Burneikis J. 1999. Evaluation of hydro energy projects using economic-ecological criterion. *Renewable energy in agriculture; Proceedings of International conference, Raudondvaris, September 1999.*

Out E. 2000. Hydropower planning and the competitiveness of hydro. *Hydropower into the next century-III; Proceedings of International conference, Gmuden, October 2000.*

Hydropower in the New Millennium, Honningsvåg et al (eds), © 2001 Taylor & Francis, ISBN 90 5809 195 3

Development of B.O.O hydroelectric project in China

P.Dong
China International Water&Elecric Corporation (CWE), BeiJing, China

Z.Yuchang & P.Hong
Northeast Investigation Design and Research Institute (NIDRI), ChangChun, JinLin Province, China

L.Guanghua
SongLiao Water Resource Committee, ChangChun, JinLin Province, China

ABSTRACT: The JiangKou project in China is a B.O.O hydroelectric project financed by several domestic investors. The 139 m high dam has double-curved arch concrete section with spillway structure on a rock substratum foundation. The adjacent underground hydroelectric plant is equipped with three units giving the total installed power of 300MW. The concrete and excavation volumes involved are of the order of $570,000 m^3$ and $63,000 m^3$ respectively. The work began in July 1999 and commissioning is scheduled for July 2003. The independent consultant, Northeast Investigation Design and Research Institute, is in charge of guaranteeing the shareholder of the operation company and local government that the design, construction and commissioning of the project will take place in accordance with a predetermined programme and cost and will offer the expected performance and characteristics.

1 INSTRUCTION

A recent project gave Northeast Investigation Design and Research Institute (NIDRI) the opportunity to actively participate in the management of technical, financial, and contractual risks associated with the construction of a large hydropower project in China.

Recently the government of ChongQing, the new municipal city of west area of China, has found it very difficult to keep up with the growth in the demand for energy, especially electric power. This hinders economic growth, and the local government has constantly been looking for a way of speeding up the construction of power stations.

Traditionally, the public sectors are responsible for finance development and own the large hydropower projects, and the level of private funds that can be invested in this field is very low in China.

Since there is an abundance of unharnessed hydropower resource in the west part of China, like many other provinces in the country with strong economic growth, the local authorities decided to give especially funded project specific Chinese companies concession to design, build, and commission given hydropower project, and operate them for a certain period. In exchange for the concession, the local government undertook to buy power from the concessionaires at agreed price.

In July 1999 therefore, after two years preparation and negotiation, the local government awarded the concession for the JiangKou Hydroelectric Power Project to the "JiangKou Development Company"

(JKDC) a Chinese state registered trading company founded by local investors.

Since the project is nearly funded by private investor, it was vital for the shareholders to know that an independent body was checking the design and construction work carried out by the company benefiting from the fund.

Northeast Investigation Design and Research Institute (NIDRI) has been chosen as the independent consultant and is responsible for guaranteeing private investors the design, construction, and commissioning of a large hydro project on the JiangKou, will comply with a determined schedule and cost, and will attain the expected performance and quality level. This task implies regularly updated assessment of the risks associated with execution of the project (quality, compliance with schedule, quantities, and costs) in order to warn the investors, their bankers, and the granter of the concession of any deviation from the forecast, and to undertake corrective action. The assessment is particularly important in JiangKou since the project is a B.O.O contract whereby all parties take a substantial financial risk.

2 CONTRACTUAL FEATURE

This concession was granted under a contract called "Implementation Contract" under the terms of which the local government awarded the concessionaire the JiangKou site for development in accordance with the descriptions and definitions in the Final Design, together with the right to turbine the available flow and owned it.

The company undertook to design, finance, built, and operate the project for preset maximum total price and within a preset maximum time, in exchange for which the authorities undertook to buy the energy produced (in accordance with an Energy Sale Agreement) from the company at a unit price enabling the shareholders to recover the capital they had invested and earn additional remuneration on it. Many other contracts were signed at the same time as the concession contract: Energy Sales Agreement, Insurance Agreement, Water Use Agreement, Share Agreement, Operation Contract, Construction Contract, and Independent Consultant Contract.

The Implementation Contract is a turkey contract for the design, construction, and commissioning of the project for lump-sum price with a determined time.

The JKDC has the right and obligation to check the detailed studies, manufacture's supplies, the method and schedule followed by contractor, and more generally, to ensure that the civil works and erection works carried out at the site are in compliance with the specified regulation, and comply with the schedule and costs defined in the concession contract.

Of course the implementation is also of interests for the other shareholders in the concessionaire, as well as the banks involved in the financing of the project, and it is therefore a decisive factor in the management of the risks for design, poor performance, and cost and deadline overruns.

Local authority is one of the shareholders, this benefits the work of immigration and land imposes.

The JKDC requires independent consultant Contract chiefly involvement safeguarding their interests: first as shareholder in the concessionaire, as a representative of the public interest; and as the future owner at end of concession period, when responsibility for project is ended.

3 TECHNICAL FEATURE

The JingKou project located at SiChuang province, west of China and is intended to harness the last step of the FuRong River, the secondary branch of Wu River and tertiary branch of YangZe River. It consists of a 139m high, 392m crest long concrete double-curved arch dam and a underground power house fitted with three 100MW generating sets. The catchment at the dam site is more than 7740 km^2 in area and has a mean yearly inflow of 479,000Mm^3. The peak design and construction-stage floods for dam and spillway are 17,000m^3/s and 2400m^3/s respectively.

The geological structure at the site is simple: it consists of a stack of sub-horizontal limestone beds

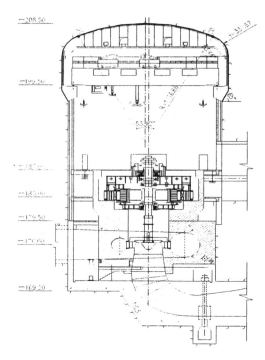

Figure 1. Undergroundpowerhouse typical section

with fewer plains of weakness and cracks. The rock has similar density, modulus of elasticity and shear strength to concrete placed.

The dam consists of a section at left abutment for retaining, a section for spillway structure in riverbeds, another retaining section at right abutment. Five high discharge orifices (12by14.5m) and four middle discharge orifices (6by7m) can provide discharge capacity of 14,582m^3/s.

The spillway has a 160m long, 35~80m wide stilling basin for dissipation, a 10m high secondary dam downstream keep water head in stilling basin.

Power conduit contains power channel, power intake tower and power tunnel. The underground powerhouse (figure1) is an 87m long, 16.5m wide reinforced concrete structure. Its mean annual output is 1071 GWh and the rated discharge of each set is 300m3/s.

The construction began in July 1999 and the detailed construction and commissioning of the project has to be completed with 48 months, i.e. July 2003.

4 THE RISKS TO BE TAKEN BY THE INDEPENDENT CONSULTANT

Given the characteristic of the specific contractual context of the JiangKou project and its special technical feature, the risks that the independent consultant had to manage on behalf of the Concession Company are basically of four types.

These are: The risk of overrunning the contractual completion time for the reasons that cannot be assimilated to cases of Force Majeure:

The risk of exceeding the foreseen quantities of work and materials not offset by possible reduction elsewhere;

The risk of non-compliance with the contractual performance target;

Design risk., design risk can of course have an effect on the first three types of risk, but it can also have a significant effect on the operation of the project

The risk of overrunning the construction time is a major risk that has to be extremely strictly managed in the interest of the shareholder of the Concession Company, the lender, and the local government.

5 THE ROLE OF THE INDEPENDENT CONSULTANT IN THE MANAGEMENT OF THE CONTRACTS

5.1 On the work

5.1.1 Dam Foundation Excavation
The excavation for dam foundation was too consecutive in the Basic Design phase: 10m deep excavation beneath the fresh rock in dam foundation. The fact is that the rock bed is nearly entire and fewer slip planes is found, the fresh, which excavated underground 5~20m, is sufficient enough for support the dam (figure2) according to the independent consultant. This change result in reducing volume of rock excavation of 400,000 m³ and consequently of concrete backfill 300,000m³ also saving investment of 130,000,000 RMB

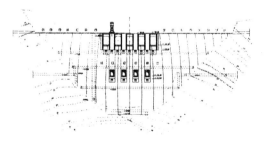

Figure 2. Dam typical section

5.1.2 Stilling Basin
The stilling basin downstream the dam as well as secondary dam is for dissipation, secondary dam in-

tended to provide water pool protecting the dam toe from erosion.

The dam site located the last step on the river, which is very near to the entrance of the main river (2km), Wu river. More than 30m high standing water would keep at dam site in the wet season, the fact was neglected actually in the Basic Design. The dimensions of stilling basin and secondary dam as well as the quantities of excavation at both abutments were reduced advised by independent consultant, this item only would saves 60,000,000 RMB.

5.1.3 Upperstream Cofferdam
There are two cofferdams (embankment and RCC) at upperstream during construction period of dam proposed by contractor.

This was considered unreasonable by independent consultant after carefully analysis of hydrographic data and calculation of work schedule. It can be proved that the stream flow can be diverted by right bank diverting tunnel (12by14 m) in dry season. A combination of diverting tunnel and breach of the dam will provide the discharge at the second rain reason (April to October 2001) at the moment of dam reaching given elevation. Some middle orifices with control gates can also undertake the task in the following rain seasons. RCC cofferdam then was cancelled and 30,000,000 RMB and several months of time were saved.

5.2 On the performance

During the construction and commissioning period, the independent consultant controlled and approved the detailed design document, construction method, quality control program prepared by the contractors.

The adaptations to the design that the design contractor suggested for practical reasons or changes made necessary because of unforeseen condition has been checked and approved by independent consultant who then controlled the need and the suitability of these adaptation and their possible repercussions on other parts of the structure.

Detailed quality control of civil work and HEM works during the construction phrase has been carried out by contractors. The independent consultant discussed, controlled and approved the quality control program defined quality control rules and checked on site that this quality control is properly performed. The performance has been guaranteed. A resident team of specialist has been maintained on the construction site and deviation from these rules has been subjects to special reports and warnings. In this case, the contractor submitted deviation reports to the approval of the independent consultant, explaining the origin and the reasons of the nonconformities and proposed remedial actions.

5.3 On the time schedule

The generation of first unit is scheduled at the end of 2002 and full commissioning of the project in July 2003. The project is now under construction respecting fully the original schedule. The co-operation between all the involved parties from Owner to Contractors and the co-ordination by independent consultant resulted in a timely progress. There is no doubt that is actual performance will reach the expected one. Further more, the design adaptations, which reduce large of work quantities, will fully ensure the foreseen work program and progress and even ahead of the contractual schedule.

5.4 On the cost

The independent consultant advises the Owner in the cost of management of the construction contract by approving the payment certificates established according to the actual progress of the works and unit price contract, since the work quantity might be change due to design adaptations.

The cost saving have made about 220,000,000RMB (about US$26,500,000) due to the suggestions of independent consultant. The expected investment cost of the project has been set as 10,000RMB/KW, and currently is reduced to 8000 RMB/KW, which including the construction, engineering, working capital, estimated escalation, financial expenses and interest at present. It has to be noted that further cost might be expected to save, and final cost is about 6000RMB/KW.

This result has been obtained by a proper management of the construction contract with a permanent concern of saving cost, time and quantities for each component of the project when these were possible.

6 CONCLUSION

The development of JiangKou project on B.O.O basis have designated an independent consultant to assist the concessionaire companies in the management of the construction contract and consequently to manage a number of risks weighing over the shareholders of the company.

The risk of contractual completion time overruns for reasons that cannot be assimilated to cases of Force Majeure, the risk of excess quantities not offset by possible reduction elsewhere, and the risk of non-achievement of contractual performance level are three important types of risk in the example considered here. Of course design risk is not absent either, but in this case it was managed directly by the independent consultant.

These risks, which in the past were assumed by the public authorities, can be considerable and can even result in abandonment of the project. The shareholders of concession companies and their bankers therefore have every interest in appointing an independent consultant to manage a certain number of those risks, particularly those associated with natural condition and, more generally, design, as was the case for the JiangKou project.

The experiences already made shows that proper risk assessment and allocation before construction and risk management during construction is a key element for achieving a successful completion of such a hydropower project developed on a B.O.O basis like JiangKou project in China.

Hydropower in the New Millennium, Honningsvåg et al (eds), © 2001 Taylor & Francis, ISBN 90 5809 195 3

Economic risk- and sensitivity analyses for hydro-power projects

M.Sc. Tor Gjermundsen & Dr.Ing. Lars Jenssen
Statkraft Grøner, Trondheim

ABSTRACT: This paper deals with uncertainty and economical risks in hydropower projects. It discusses the different items that influence the economics of a hydropower project and the uncertainty each item (ie. dam, power house, etc) is burdened with. Different methods for assessing the uncertainty associated with each item, and their influence on the overall project uncertainty are discussed. The use of Monte Carlo simulation for computing risk is illustrated with an example.

1 INTRODUCTION

Before a developer decides to realise a hydropower-project, he wants to have a total overview of the economy of the project. This includes the expected costs and benefits and also the sensitivity; ie. the economic risk and uncertainty.

The developer therefore wants tools / methods to verify the risks and how to reduce the risk to an acceptable level.

This paper describes the main components that influence on the total profit, different methods / programs to calculate and illustrate the uncertainty and one example using a Monte Carlo simulation.

2 IDENTIFICATION OF COMPONENTS THAT INFLUENCE THE OVERALL PROJECT ECONOMICS

The main components that influence the profit from a powerplant are:
– The benefit obtained from the energy production
– Investment costs
– Cost of operation, maintenance and refurbishment
  Multipurpose aspects can also be of importance.
The main components are shown in Fig. 1, 2 and 3.

MAIN ELEMENTS

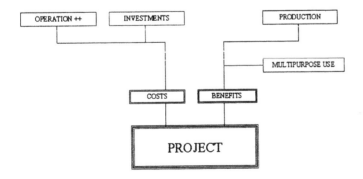

Figure 1. Project structure

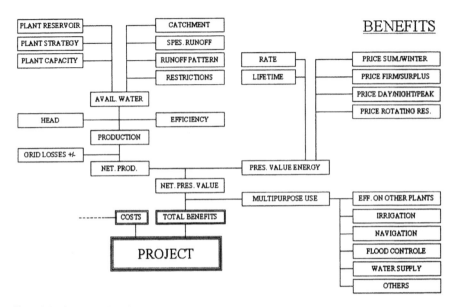

Figure 2. Main structure benefits

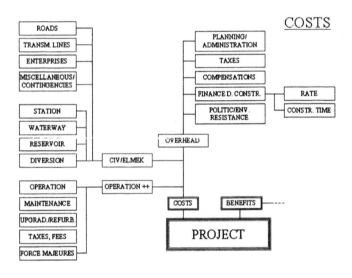

Figure 3: Main structure costs

3 COMPONENT BENEFITS

Even if the main income from the project will be derived from selling electricity, it is also of interest to check whether the owner can obtain other benefits (in cash or in other ways). Such uses may include irrigation, flood control, water supply and navigation. If the project is a "run-of-river" plant in a rural area, the possibility of using the "free" power (electric or direct) during the night and low-load periods should be taken into consideration. Relevant keywords include heating, cooling, mechanical work, drying and pumping,

The following components ought to be considered:

Energy production

- The plant capacity etc.(reservoir, installation, operation)
- Water discharge/available volume (catchment area, runoff, restrictions)

- Gross head
- Total efficiency (including losses in the water-way)
- Grid losses
- Delivery reliability
- Power value / present value (energy prices in the future; firm – surplus?, peak power?, grid connection?, alternative energy sources?, political aspects?, free markets?, rotating reserve? present value factor; function of rates and life time)

3.1 Multipurpose aspects

- Water supply, irrigation, navigation, flood control etc.

4 COMPONENT COSTS

The costs can be divided into two main parts:
- Investment costs
- Operation / maintenance / refurbishment costs

Investment costs

The investments costs can be divided in civil engineering/electromechanical costs and overhead costs:
- Civil engineering / electromechanical costs (diversion, reservoir, waterways and power station, roads, transmission lines, enterprises, miscellaneous and contingencies)
- Overhead costs (planning and administration cost during construction, taxes, compensations, financing during construction, political / environmental resistance (despite of legal permission))

Operation / maintenance / refurbishing / force majeure-costs

The main costs are listed below:
- Operating and maintenance costs (include the normal costs of operating the plant. If the river has a heavy sediment load, special attention should be paid to the implications)
- Taxes and fees (the expected stability of the tax load, possible fees in the future)
- Upgrading / refurbishment / enlargement (especially where the plant will be operating under particularly wearing conditions)
- Force majeure (seismic activity, unstable political system (nationalisation), landslides, etc.)

5 EVALUATION OF THE RISKS ASSOCIATED WITH BENEFITS AND COSTS

Various methods are available to provide an overview of the risks involved:

- Add an item for "unforeseen costs"
- Sensitivity analysis / Star diagram
- Standard deviation as a measure of risk

The mention methods above are assumed to be quite common and therefor not described further .

Step-by-step calculations

Detailed cost calculation is a time-consuming and expensive process that involves dividing the project into a large number of cost items, and estimating the expected value and standard deviation of each and the project as a whole. This may not be feasible in the early stages of a project.

It is important to focus early on cost items and conditions that strongly influence the risks involved in the project. This is the background for the "step-by-step" procedure that was introduced by Lichtenberg (1978). The main steps in this procedure are:
- The project cost is divided into a number of independent items.
- The expected value and variance of the cost (or revenue) is calculated for each item.
- In order to reduce the uncertainty, the item with the largest uncertainty, i.e. the largest variance, is divided further into a number of independent sub-items.
- The expected value and variance of the cost (or revenue) is calculated for each sub-item.

The procedure is iterated until the overall uncertainty in the cost/revenue estimate is regarded as satisfactory, or until further improvement is impossible.

Monte-Carlo simulation

The economic outcome of any project depends on a large number of factors. If the factors are interdependent, the methods mentioned above may not be suitable. In such cases, a Monte-Carlo simulation may be used instead. Using Monte-Carlo simulation makes it possible to take interdependencies into account. The simulation provides a complete frequency distribution of the project outcome, rather than only the standard deviation. This may be important when the distribution is highly skewed.

Before the simulation can be carried out, the analyser must make a model of the project. This model must include the relevant cost and revenue items, and their interdependencies. Each item is described in terms of its expected value and probability distribution.

The simulation involves selecting a single value for each item and calculating the resulting project value using the selected values. This is repeated a large number of times and the results are used to calculate the expected project value, the standard deviation and the distribution of the outcomes.

Model building and simulation are normally done with the aid of a computer program, and several such programs exist. Building a model and running a simulation is relatively easy with a modern, user-friendly computer program. The real problem is to assess the expected value and probability distribution of each item, and to evaluate their interdependencies.

Input to the model:

All cost and income items considered relevant to the project economy are included in the Monte-Carlo model. The main model structure is shown in Fig. 4.

Each cost item are represented by a "node" (filled circle). The statistics of each item (node) is described by the expected value, a upper- and lower bound, and the type of probability density function to represent the node, ie. log - normal. If the component is a tunnel for instance, a certain amount of anchoring / lining is foreseen; the expected value. The lower value will be if the rock quality shows to be very good and nearly no support are necessary. The upper bound can represent a nearly full lining .An example of node input is shown in Fif. 5. The figure shows a skewed probability distribution that is typical for cost items. The graphical user interface makes it easy to construct a simulation model, but the user should be familiar with statistic methods.

Then all nodes are completed, the calculations can be done.

Results:

The results of the simulation may be presented in various ways:
- S-curves (Fig. 6)
- Histograms (Fig. 7)
- Overview of the influence of each component.

It is also important to remember that this method can be used at different stages of a project. At an early stage Monte-Carlo simulation may be used for a rough estimate of the project risk, and to identify the main sources of uncertainty. Work can then be done on these components to reduce the uncertainty and/or improve the project.

As the project develops further, the model is refined and extended. Different items will dominate uncertainty at different stages of the project. By continuously refining and updating the model it is possible to monitor the project uncertainty, and to focus risk management efforts on the most important sources of uncertainty at each stage.

The example in chapter 6 demonstrates how a Monte-Carlo simulation can be used to assess risk in hydropower project.

Overall project profit

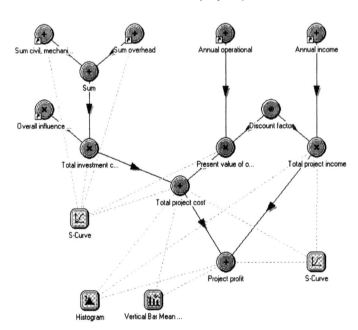

Figure 4. Simplified structure

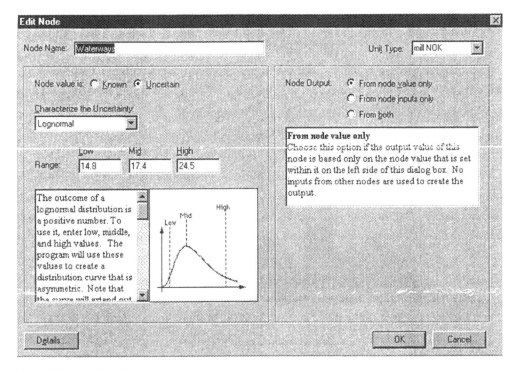

Figure 5. Example of node input

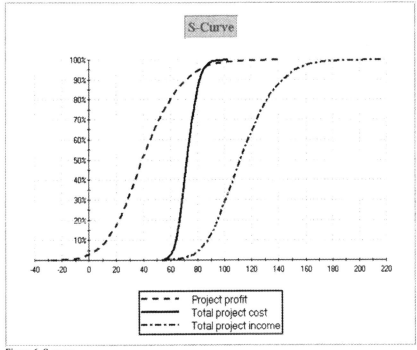

Figure 6. S-curve

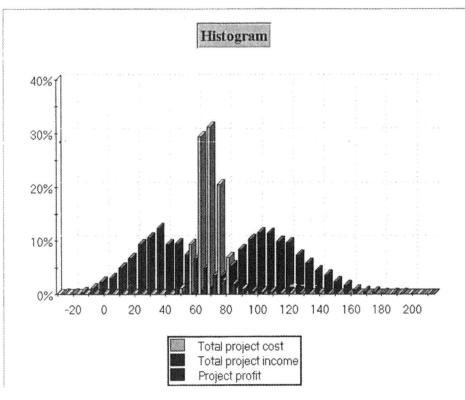

Figure 7. Histogram Project cost, project income and project profit

6   MONTE CARLO SIMULATION. EXAMPLE

In the example uncertainty assessment is illustrated by means of a Monte-Carlo simulation using the computer program "Definitive Scenario".

The example concerns a hydropower project with an estimated production of 39 GWh / year with a total investment of 68 mill. NOK. It consists of a tunnel, a penstock, an outdoor power station and some roads, transmission lines etc.

The same procedure as described in chapter 5 is used.

Fig. 4 shows the main structure (this is again organized in a sub structure, not shown here)

Fig. 5 shows an example of a node input (low-, mid- and high values are put in).

The main input and results are shown in table 1.

Table 1.

	Mill. NOK
Total investment	68
Total cost	73
Total income	114
Project profit	41

(1 $=8.40 NOK).

The results are illustrated by means of S-Curves in Fig. 6. Graphs for the project profit, the project cost and the total project income are drawn. For instance is the probability of a net benefit of 40 mill. NOK 50 % and the chances of loosing money are 3 %.

Fig. 7 shows histograms for the project profit, the total project cost and the total project income The graphs illustrate that the highest uncertainty is on the income side.

7   CONCLUSIONS

The authors have the opinion that a Monte Carlo simulation is well suited to illustrate the economic risk in a hydropower project and how to reduce it to an acceptable level.

LIST OF REFERENCES

Lichtenberg, S. 1978 *Projektplanlægning – i en forandelig verden (Project planning in a changing world)* Polyteknisk Forlag

Jenssen, L., Mauring, K., Gjermundsen, T. 2000 *Economic risk – and sensitivity analyses for small scale hydropower projects* International Energy Agency (IEA)

Hydropower in the New Millennium, Honningsvåg et al (eds), © 2001 Taylor & Francis, ISBN 90 5809 195 3

New trends in the development of hydropower projects in India

D.V. Khera
Member (Hydro), Central Electricity Authority and ExOfficio Additional Secretary to Govt. of India

Major Singh
Director, Hydro Planning & Investigation Division, Central Electricity Authority

(Abstract)
India is endowed with an enormous economically exploitable hydro potential of 84,000 MW at 60% load factor. Only 26% of this has been developed/ under development and cleared by Central Electricity Authority (CEA). Further, share of hydro capacity to total, has steadily declined from 50.6 % on 31.03.63 to 24.3 % at present. This is undesirable for economic operation of a power system. Also, India is facing a severe peaking power shortage, for which hydro power is ideally suited. This necessitates to give new impetus to development of hydro power.

Prior to Independence, most of hydro projects in India were in private sector. Thereafter, priority shifted to Central and State Sectors. Like many other countries, electricity sector in India has also undergone major modifications including institutional reorganization, introduction of new rules, privatization of the sector. To bring additionality to resources for capacity addition, Government of India in 1991 formulated a policy to encourage greater participation by independent power producers (IPPs). Additional incentives were given for hydro development. Many IPPs concluded memorandum of understandings (MOUs). Guidelines were issued for setting mega power projects in November, 1995 (revised in November, 1998). International Competitive Bidding (ICB) route was brought to bring more transparency in selection of IPPs. "Policy on Hydro Power Development" was formulated in August, 1998.

1 INTRODUCTION

1.1 During last century, India has achieved many important milestones in hydro power development. A mini hydro electric scheme, Sidrapong of 2x65 KW near Darjeeling in West Bengal marked the beginning of supply of power for general public good. This was followed by Shivasamudram hydro scheme (6x720 KW) in Mysore State transmitting power at 33 KV over a double circuit transmission line of about 150 Km to Kolar Gold Fields in June, 1902. After that several hydro schemes were installed mainly for serving the need for urban population and industrial demands. Most impressive scheme implemented in the period prior to First World War in 1914 was 48 MW Khopoli of Tata Electric Company (TEC).

1.2 Between two World Wars, many pioneering hydro stations were taken up. 72 MW Bhivpuri and 110 MW Bhira stations were set up by TEC in 1921-24 and 1927. Tamilnadu launched its Pykara scheme in the Nilgiries with an initial capacity of 18.750 MW utilising a fall of 1000 m, one of the highest head in the World then. Uttar Pradesh developed the falls along the Ganga Canal with a chain of 7 power houses (aggregating to

---

* Member (Hydro), Central Electricity Authority and Ex- Officio Additional Secretary to Government of India, 401, Sewa Bhawan (N), R.K.Puram, New Delhi-110066.

** Deputy Director, Central Electricity Authority, 410, Sewa Bhawan (N), R.K.Puram, New Delhi-110066
18.900 MW) from 1927 to 1937. Punjab commissioned 48 MW Jogindernagar Station in 1932. Kerala commissioned Pallivasal Station (15 MW) during 1940-1942.

1.3   Pace of hydro power development up to Independence was rather tardy with a total installation of 508 MW in a period of about 50 years. Most of the installations till 1947 were single purpose. There after, Central and State Governments initiated programmes of development of multipurpose river valley projects, in which hydro power development became one of the primary goals and it made rapid strides. On an average, about 455 MW of new hydro capacity was installed every year during last 53 years. Aggregate hydro capacity in India which was merely 508 MW in 1947 at 12 stations with 51 units has now increased to 24637 MW (48.5 times) at the end of December, 2000 at 231 stations with 666 generating units (excluding schemes up to 3 MW capacity).

1.4   Prior to Independence, most of hydro projects in India were in private sector. Thereafter, priority shifted to Central and State Sectors. Like many other countries, electricity sector in India has also undergone major modifications including institutional reorganization, introduction of new rules and privatization of the sector. To bring additionality to resources for capacity addition, Government of India in 1991 formulated a policy to encourage greater participation by independent power producers (IPPs). Additional incentives were given for hydro power development. Many IPPs concluded memorandum of understandings (MOUs). Guidelines were issued for setting mega power projects in November, 1995 (revised in November, 1998). International Competitive Bidding (ICB) route was brought to bring more transparency in selection of IPPs. "Policy on Hydro Power Development" was formulated in August, 1998. A National Perspective Plan of 12 to 15 years for R&M and Life Extension of hydro power plants has been prepared. 80 hydro schemes have been identified for implementation for RM&U works.

1.5   The paper deals with the hydro power potential, present installation, share of hydro to total installation and new trends in hydro power development in India.

2   ASSESSMENT OF HYDRO POTENTIAL

2.1   India is endowed with enormous economically exploitable and viable hydro potential of 84,044 MW (equivalent to 1,50,000 MW installed capacity) at 60% plant load factor (excluding contribution from small hydro schemes) from 845 major/ medium schemes with generation of 442 billion units per annum. With seasonal energy generation, this energy amounts to 600 billion units. Out of this, about 65,500 MW(78%) is located in the Himalayan Region comprising the Ganga, the Indus and Brahamputra basins. Brahamputra basin has the lion's share of about 42 % of the potential. Regionwise, maximum potential is available in North Eastern Region (31,857 MW) followed by Northern Region (30,155 MW). State-wise, the largest potential is available in Arunachal Pradesh (26,750 MW) followed by Himachal Pradesh (11,647 MW). In addition to this, 56 attractive pumped storage schemes (excluding in operation & construction) have been identified with an installation of about 93,920 MW.

2.2   In addition, small hydro power potential assessed in terms of installed capacity is 6782 MW in 1512 schemes including 356 MW (171 schemes) in operation, 322 MW (86 schemes) under construction and 6103 MW (1255 schemes) investigated/ under investigation.

3   STATUS OF HYDRO POTENTIAL DEVELOPMENT

3.1   Out of hydro power potential assessed at 84,044 MW at 60% load factor in India, only 16.6 % has been harnessed so far and 6.4 % is under various stages of development. 77% of this still remains to be exploited. Region-wise details are given in Table– 1.

Table – 1 : Region – wise status of hydro power potential

Region	Potential at 60% Load Factor (MW) (As on 31.12.2000)				
	Assessed	Developed	Under Development	CEA Cleared	Total
Northern	30155	4560.73	2543.83	1208.75	8313.31
Western	5679	1845.33	1500.22	234.45	3580.00
Southern	10763	5778.75	649.83	90.57	6519.15
Eastern	5590	1369.28	338.78	353.33	2061.39
North- Eastern	31857	388.50	309.72	417.75	1115.97
Total	84044	13942.59 (16.59 %)	5342.38 (6.36%)	2304.85 (2.74%)	21589.82 (25.69%)

4   PRESENT INSTALLATION

4.1   As on 31.12.2000, India has 231 hydel stations (Capacity over 3 MW) housing 666 generating units (Total installation 24637 MW). Region-wise break is given in Table –2.

Table – 2 : Region – wise of Hydro Installed Capacity in India

Sl. No.	Region	Number of PHs	Number of Units	Capacity (MW)
1	Northern	71	208	8352.55
2	Western	41	100	4211.80
3	Southern	80	242	9184.85
4	Eastern	26	81	2246.90
5	North Eastern	13	35	640.70
	Total All India	231	666	24636.80

4.2   High Head Hydro Plants

Suruliar is the highest head hydro electric station (980 m) commissioned in India. Other plants with head over 500 m are Kodayar PH – I (947.62 m), Bhaba (887.20 m), Pykara (867.46 m), Kundah PH –II (713.23 m), Sabarigiri (710 m), Idukki (660 m), Kuttiadi (643.29 m), Pallivasal (570 m) and Lower Lagyap (530.50 m).

4.3   Pumped Storage Schemes

Seven pumped storage schemes (PSSs) with total installation of 1554 MW have been commissioned in India. These are Nagarjunasagar (7x100 MW), Paithon (1x12 MW), Kadamparai (4x100 MW), Kadana (4x60 MW), Panchet (1x40 MW), Ujjani (1x12 MW) and Bhira (1x150 MW). Four PSSs with aggregate installed capacity of 3250 MW viz. Ghatgarh (2x125 MW), Sardarsarovar (6x200 MW), Srisailam (6x150 MW) and Purulia (4x225 MW) are under implementation.

4.4   Underground Hydro Electric Stations

Fourteen underground hydro power stations (total capacity 5010 MW) have been commissioned in India so far. The stations are Maithon (3x20 MW), Koyna Stage I & II (4x65 MW+ 4x75 MW), Chhibro (4x60 MW), Koyna Stage – III (4x80 MW), Idukki

(6x130 MW), Vaitarna (1x60 MW). Tillari (1x60 MW), Pench (2x80 MW), Kadamparai (4x100 MW), Bhaba (3x40 MW), Varahi (2x115 MW), Chamera Stage – I (3x180 MW), Uri (4x120 MW) and Koyna Stage – IV (4x250 MW). Fourteen underground hydro stations with total capacity 8580 MW are under execution.

5 SANCTIONED/ ON – GOING HYDRO ELECTRIC SCHEMES

5.1   There are 59 sanctioned/ on- going hydro electric schemes with aggregate installed capacity of 15136.45 MW in India. Sector- wise details are given in Table - 3.

Table - 3 : On - going Hydro Schemes in India

Sl. No.	Sector	Number of Projects	Installed capacity (MW)
1	Central	12	5670.00
2	State	41	7934.45
3	Private	6	1532.00
	Total	**59**	**15136.45**

6   SHARE OF HYDRO POWER

6.1   Share of hydro installation to total had been varying in India. In the initial years, hydro power initiated industrial growth. By 1915, hydro capacity was 103.5 MW (96.77 %) out of total capacity of107 MW. At the end of First World War, hydro capacity was 103.5MW (79.61%) out of 130 MW. Between two World Wars, thermal plants were set up mainly in and around urban areas. At the beginning of Second World War, capacity was 1069 MW with hydro 442 MW (41.35%). Share of hydro was 37.31% in 1947, which declined to 31.72% in 1953. Then it increased to 50.62% in 1962-63. Thereafter, it experienced a steady decline. At present it is only 24.27% as shown in Table - 4.

Table - 4 : Share of Hydro Capacity in Total Capacity in India

Year	Total Capacity (MW)	Hydro Capacity MW (Including Station Capacity up to 3 MW)	Hydro Capacity % of Total
1947	1361.76	508.13	37.31
1950	1712.52	559.29	32.66
1956	2886.14	1061.44	36.78
1960-61	4653.05	1916.66	41.19
1962-63	5801.19	2936.35	50.62
1965-66	9020.02	4123.74	45.72
1969-70	14102.45	6134.70	43.50
1973-74	16663.56	6965.30	41.80
1977-78	23668.71	10020.22	42.34
1980-81	30213.68	11791.22	39.03
1984-85	42584.72	14460.02	33.96
1989-90	63636.34	18307.63	28.77
1992-93	72319.46	19568.76	27.06
1996-97	85019.31	21644.80	25.46
1999-00	96682.17	23816.01	24.34
2000-01(Upto 12/2000)	100136.37	24723.51	24.69

6.2   Increased share of hydro upto 1963 was mainly due to top priority accorded to agricultural sector thus paving the way for accelerated development of river valley multipurpose projects for irrigation, incidentally with power benefits. From Seventies, share of hydro continued to decline due to a shift of priority from agricultural to industrial sector, triggering emphasis on coal based thermal projects with shorter gestation period.

7.   GOVERNMENT OF INDIA's (GOI) NEW POWER POLICY

7.1   Hydro power development in India has experienced sectoral changes. Prior to India's Independence, most of the hydro projects were in private sector. First Five Year Plan (1951-52 to 1955-56) included a number of major multi-purpose river valley projects, shifting priority from Private to Central and State Sectors.

7.2   To bring additionality to resources for capacity addition in power sector, GOI in 1991 formulated a policy to encourage greater participation by the private entrepreneurs, both domestic as well as foreign in electric power generation, supply and distribution. Like many other countries, electricity sector in India has also undergone major modifications including institutional reorganization, introduction of new rules, privatization of the sector. Legislations governing the Electricity Sector (Indian Electricity Act, 1910 and Indian (Supply) Act, 1948) were amended in October, 1991 to new legal, administrative and financial environment for private enterprises. Additional incentives have also been given to install hydroelectric projects in private sector by issuing revised tariff notification for hydro projects in January, 1995 (amended in January, 1997).

7.3   Like all other developed and developing countries, the initial project solicitation in India was also through the Memorandum of Understanding (MOU) route. A number of MOUs were concluded between the Independent Power Promoters (IPPs) and the concerned State Governments/ State Electricity Boards (SEBs).

7.4   CEA had been issuing "In Principle Clearance" (IPC) (only an administrative clearance) till 31st March, 1997 on the pre-feasibility reports submitted by IPPs to help them to get different financial approvals from Indian/ Inter- national Financial Institutions (IFI) to enable them to prepare detailed project report (DPR) to put up to CEA for TEC under the Electricity (Supply) Act, 1948. Procedure for obtaining IPC of CEA for IPP power projects and outlines of pre-feasibility report to be submitted to CEA were circulated in August, 1995. IPC were accorded by CEA to 13 hydroelectric projects for which MOU were signed. Some DPRs were submitted to CEA for TEC.

7.5   GOI in January, 1996 minimized the number of major clearances from State/ Central Government and other agencies required by CEA while according its TEC to the projects proposed by a private sector registered power company after due compliance of Section 29 (2) of Electricity (Supply) Act, 1948. Clearances now required by CEA are :

Sl. No.	Item	Agency giving Clearance
1	State Government and State Electricity Board Clearance	State Electricity Board/ State Government
2	Water Utilization	Ministry of Water Resources, only if the project is inter- state/ inter-country
3	R & R of displaced families by land acquisition	State Government and Ministry of Environment & Forest (MOEF)
4	Environment & Forest Clearance	MOEF

7.6　Five hydro projects (1516 MW) were accorded techno-economic approval by CEA. Of these final financial package have been approved for 2 projects (486 MW).

Sl No	Name of the Project/ State	Capacity (MW)	Estimated Cost (Rs Cr)	Promoter	Status
I	Cleared by CEA and Financial Package (FP) also Cleared by CEA				
1.	Maheshwar, M.P	10x40 = 400	Rs 1673.00 crores (Completion)	Shri Maheshwar Hydel Power Corp.	-
2	Malana , H.P.	2x43 = 86	Rs 341.911 crores (Completion)	M/s Malana Power Company Limited	-
II	Cleared by CEA but Financial Package Yet to be Finalised				
2.	Baspa St.II. H.P.	3x100 = 300	Rs 949.23 (12/93 price level)	M/s Jaiprakash Hydro Power LTd.	IPP yet to submit FP
3.	Vishnuprayag , U.P.	4x100 = 400	Rs 1614.66 crores (Completion)	M/s Jaiprakash Industries Ltd.	IPP yet to submit FP
4.	Srinagar, U.P.	4 x 82.5 = 330	Rs 1699.12 crores (Completion)	M/s Duncan Industries Ltd	IPP yet to submit FP
	Total	1516			

7.7　CEA formulated guidelines for formulation of DPRs for hydro projects in August, 1993 (revised in June, 1995) to help IPP to properly formulate DPRs for obtaining TEC.

7.8　In-spite of various incentives, response from IPP for development of hydro power has not been very encouraging. No new major/ medium hydro power project has been taken up after the introduction of bidding process for selection of IPP.

8.　POLICY ON HYDRO DEVELOPMENT

8.1　With an objective for faster development of hydro potential, GOI has announced a "Policy on hydro power development" in August, 1998 with following main features :

➢ Additional budgetary support for ongoing and new hydroelectric projects under central Public Sector Undertakings.
➢ Creation of Power Development Fund by levying cess @ 5p/kWh on electricity generated in the country.
➢ Basin-wise Development of Hydro Potential.
➢ Advance Action for Capacity Addition.
➢ Emphasis on Survey & Investigations.
➢ Quick resolution of Inter-State issues.
➢ Renovation, Modernization & Uprating of existing hydro stations.
➢ Promoting Small & Mini Hydel Projects.
➢ Simplified Procedures for Transfer of Clearances by CEA.
➢ Rationalization of Hydro Tariff by allowing premium on sale rate during peak period.
➢ Realistic estimates of completion cost considering geological surprises.
➢ Promoting Hydel Projects with Joint Ventures.
➢ Selection of Developer through MOU / Bidding route and Techno Economic Clearance of CEA.
➢ Govt. Support for Land Acquisition, R&R, Catchment Area Treatment, etc.

Some of the above aspects are discussed below.

8.2 Renovation, Modernization & Uprating of existing hydro stations

In view of renovation, modernization and uprating (RM&U) of the old hydro generating units, which have outlived their useful life & also relatively newer machines which are suffering from generic problems, GOI set up a National Committee in 1987 to formulate strategy on R&M of hydro power plants. Accordingly, 55 hydro schemes (aggregate installation of 9653 MW in 186 generating units) were identified for implementation for RM&U works. Of these, 25 schemes (total installation of 5791 MW) have already been completed. Works on 21 schemes are under different stages of implementation.

GOI in its "Policy on Hydro Power Development", declared in 1998 have laid stress on the need for R&M of hydro power plants. Accordingly, a Standing Committee has been set up by GOI to identify new hydro RM&U schemes to be taken up for execution under Phase –II. A National Perspective Plan of 12 to 15 years for R&M and Life Extension of hydro power plants has been prepared. 80 new hydro schemes (aggregate installed capacity of 12329 MW) have been identified for implementation for RM&U works under Phase – II Programme.

8.3 Modalities for Simplified Transfer of TEC of Hydro Electric Projects

Modalities for transfer of TEC of hydro electric schemes already cleared by CEA have been issued by CEA in January, 1999 (revised in October, 1999) for schemes without any change in scheme features & cost estimates and for schemes envisaging change in scheme features and/or cost estimates. TEC for four hydel scheme, viz, Srinagar, Parbati, Indira Sagar and Omkareshwar, earlier accorded in State Sector, have been transferred to the new Developers.

8.4 Realistic Estimates of Completion Cost Considering Geological Surprises

GOI has made provisions for increase in the project cost arising due to geological surprises, which are not anticipated at the time of preparation of DPR. In such cases, developer would be allowed to submit his proposal for an enhanced cost to Government. An expert Committee would be constituted that would evaluate and recommend cost increase for acceptance.

8.5 Projects - Through Joint Ventures

GOI has envisaged greater emphasis to take up schemes through the joint ventures (independent legal entities registered under the Companies Act and acting as an independent developer) between the Public Sector Undertakings/State Electricity Boards and the domestic and foreign private enterprises.

8.6 Selection of Developer Through MOU/ Bidding Route and TEC of CEA

GOI in January, 1995 have circulated guidelines for competitive bidding route for private power projects. Guidelines include system planning and the selection of the project, project preparation, solicitation documents (Request for qualifications (RFQ), Request for Proposals (RFP), Power Purchase Agreement (PPA), Implementation Agreement between developer and State Government).

GOI in November, 1998 has fixed Rs 1000 crores in relation to hydro electric generating stations prepared by a Generating Company selected through a process of competitive bidding by competent Government or Governments and Rs 250 crores for other hydro-electric schemes, as sum of capital expenditure exceeding which the scheme shall be submitted to Central Electricity Authority for its concurrence. However, all hydro electric schemes utilizing waters of inter-state rivers shall be submitted to CEA for its concurrence.

8.7   Government of India's Policy for Setting up Mega Power Projects

GOI in November, 1995 issued guidelines for setting up of mega power projects of capacity 1000 MW or more supplying power to more than one State. These were revised in November, 1998 providing additional concessions such as duty free import of capital equipment, deemed export benefits to the domestic bidders and liberal income tax holiday. Standing Independent Group (SIG) was constituted by GOI in November, 1997 as an initial apex body to establish parameters for negotiation of mega power projects and to oversee their implementation. These projects would be offered to developers only after all clearances/ land have been obtained so that the project can be started soon. Power Trading Corporation (PTC) would purchase power from the mega power projects and sell to State Electricity Boards.

9   CONCLUSION

9.1   Though the Country during last one hundred years has made spectacular growth in hydro power sector ranking 9th in the World in terms of total hydro installed capacity, the exploitation of hydro power potential is not commensurable with the quantum of potential assessed. Hydro power development which was dominent in sixties, is now exhibiting a declining trend in recent times.

9.2   There is a large scope for development of hydro power potential in India. Hydro schemes in operation and under construction account only 23% of the total assessed hydro power potential. Bulk of potential of the order of 77% still remains to be tapped. Accelerated hydro development in India is also required as the hydro- thermal mix at present is 24:76, against the economically assessed 40:60.

9.3   Government of India in its policy to encourage greater participation by independent power producers has give additional incentives for hydro. Guidelines have been issued for setting mega power projects. International Competitive Bidding route was brought to bring more transparency in selection of Independent Power Promoters.

9.4   National Committee set up in 1987 identified 55 hydro power schemes (aggregate installed capacity 9653 MW) for implementation of RM&U. Of these, 25 schemes (5791 MW) have already been completed. Another 21 schemes are under implementation. GOI in its "Policy on Hydro Power Development", declared in 1998 have laid stress on the need for R&M of hydro power plants. A National Perspective Plan of 12 to 15 years for R&M and Life Extension of hydro power plants has been prepared. 80 new hydro schemes (aggregate installed capacity of 12329 MW) have been identified for implementation for RM&U works.

Hydropower in the New Millennium, Honningsvåg et al (eds), © 2001 Taylor & Francis, ISBN 90 5809 195 3

New trends in the development of hydropower projects: experiences of a utility in a transitioning economy

Jeevan Raj Shrestha
Nepal Electricity Authority, Kathmandu, Nepal

ABSTRACT: Nepal faces the situation that it inherits one of the world's largest hydropower resources but exploitation has been infinitesimal. Nepal Electricity Authority is the key player in the power sector- with consumers that account for only 15 percent of the population. NEA has relied on loans from Development Banks for major financing. The Banks have ensured their returns by incorporating covenants in loan agreements that create healthy revenue streams. But this has created a tariff that ranks as one of the highest in the region.

The early 1990s saw policies that promoted private sector investment. For Nepal, these were identified as one-way avenues in a period of resource crunch. The investors meanwhile discovered that Nepal was high-risk where they should seek safer conditions. This paper intends to record events, options and activities that illustrate efforts made by a small utility in a transitioning economy to survive in controversial economic environs.

1 EARLY DAYS OF POWER TO THE PEOPLE

Nepal's water resources are its foremost natural endowment. Nepal also ranks among the world's least developed nations but it is faced with the ironic situation that although it inherits one of the world's largest hydropower resource, exploitation of this resource has been infinitesimal. This is a major means which could move the country out of the ranks of the underdeveloped to those of the rapidly developing countries. This is the means that must be planned so that the harnessing of its potential brings about economic growth. This is also the means which could solve the nation's chronic trade deficit. But the unfortunate fact is that this potential goes to waste year after year. Presently, only 15 percent of Nepal's populace benefits from an electrical supply and export to India is limited to a diminutive 50MW.

Early repercussions to the finding that Nepal had 83,000 MW of hydroelectric potential in its myriad rivers was followed by immediate political overtones that the power could be delivered to the people as a cheap form of energy. The realisation that hydropower is the country's indigenous energy resource which should be exploited and developed not only for domestic consumption but also for export to neighbouring countries gained momentum soon after the dawning of democracy and the opening of the country to the world in the 1950s. Hydropower gradually gained recognition as the panacea for the nation's ailing economy.

In its early phases, the development of hydropower in Nepal has inevitably been associated with gift packages from friendly governments. The fact that much of the early projects were undertaken on a turnkey basis by the donor countries kept the cost aspects in less transparent red- tape. Self-sustainability and economics of operation did not take precedence. With donor financing providing majority of the development costs and the government covering all the risks and providing the guarantees, this phase of development was devoted to forge ahead and harness the potential of the rivers to provide cheap power to the people and lay the infrastructure for industrial growth. For a decade or two, it was obvious that electricity was supplied without assessing its real cost. It also progressively dawned that electricity was not a cheap form of energy as was envisaged.

The need to obtain economic returns from the country's largest resource was soon realised and led to the formation of a number of Corporations and Boards with statutory provisions to operate and maintain generation plants and transmit and distribute electricity over definite areas of the country along commercial lines. Also, with the greater need to use loan financing in power projects, it was now necessary to form a self-sustainable organisation that could pay back debts to its Lenders. Nepal Electricity Authority (NEA) was created in August 1985 with the responsibility to undertake all planning, construction, operations and maintenance of electrical services such as generation and transmission and distribution throughout the nation, at reasonable cost. NEA was also the amalgamation of major government power related organisations in the country. In a situation where the power sector was emerging as the major force in the country's economy, yet another reason for the creation of NEA was to obtain better co-ordination of power sector activities through a single institution instead of number separate entities. Presently, NEA operates as an autonomous entity under an amended statutory Act that requires NEA to apply for license for generation, transmission and distribution activities and therefore function like any other enterprise in the power sector. NEA holds the distinction of being the organisation that leads in all forms of hydropower studies and applications.

2   NEA'S CONVENTIONAL APPROACH

After its formation, NEA adopted a policy to use the nation's plentiful resource as the prime mode of electricity generation. This was the outcome of intense debate on options to introduce gas turbines to inject cost effectiveness into the highly lumpy nature of NEA's load characteristics. This hydro-based strategy has been generally accepted by the Ministry of Water Resources and the National Planning Commission, acceding to the fact that hydropower remains the nation's major economic resource and therefore needs to be promoted. The exercise within NEA was to maximise the efficiency and application of hydropower for commercial ends.

The design of hydropower projects to achieve a feasible configuration and also to match the system characteristics at the receiving end has continued to pose an unwieldy glitch in system design. Nepal's electrical load comprises of a large percent of domestic load that consists mainly of the lighting load that peaks during the evening hours and fades off rapidly as the evening deepens. The peak demand created by the evening peak is the factor that penalises the economics of the generation system calling for capacity additions that is required only for a narrow slot in time (in the range of three or four hours) and contributing to a non-productive low load-factor characteristics in the system. Such a peak maximises in the winter when the cold weather also injects a heating component to the evening utilisation.

In a scenario where NEA's tariff rates are already one of the highest in the region, a major concern in generation facility design has thus been to keep costs down and therefore the compulsion to use optimised run-of-river (ROR) project design. This done, the project performance confronts another snag that does not assist in the matchmaking process. This is created by the hydrology of the rivers. River flow, and therefore the generated energy from the project, is a maximum during the wet season. But the anomaly is that the energy demand during this season is ironically at ebb, thereby threatening to introduce seasonal surplus in hydro-based design which has been optimised to deliver maximum capacity with wet season peak in focus. Whereas adequate storage capacity would have offered the regulatory mechanism needed in the system, cost consciousness has mandated designs with varying capacities of daily pondage.

NEA approach to hydropower generation planning has been simple and conventional. Hydro generation implies heavy capital expenses the magnitude of which is barely possible to be met from NEA internal resources or that of the Government. Invariably, the only option is to approach development Banks to provide credit financing in the softest terms available. The credit agreement is routed through the Government's Ministry of Finance who impose an interest of twelve percent for their services, but cover NEA from foreign exchange risks over the loan payback period. Besides a host of credit effectiveness covenants to ensure getting back their returns from the organisation, the Banks also impose the recruitment of an international consultant to design the project and oversee its construction to the phase of project commissioning – adding considerably to the project cost. The construction works are then tendered out for international bidding in a number of packages that would be attractive to the bidders and thus aid in a cost cutting. Invariably, NEA's technical staff functions work much in the sidelines and benefit only from technology transfer from mainstream activities. Nevertheless, NEA claims a staff force with considerable expertise second to none in the know-how of hydropower design based on the young and fragile constitution of the Himalayas.

Over the years, NEA has adopted several strategies in the management of hydropower projects. The first is the need to incorporate transparency in its transactions. This arises basically from the large, up-front capital cost of hydropower generation projects. This relatively high capital cost of hydropower projects and the public expenditure during the project phase has always created immense interest in the areas of consultancy, contracting and the electro-mechanical equipment supply. Regarded as a "dying" technology, the industry associated with hydropower generates immense competition to win contracts to stay alive in business. Transparency in such conditions is critical to avoid situations leading to calculated misuse and misinterpretation. Another factor is the need to appreciate and work in compliance with the universal awareness for the preservation of the environment and the human values which may be disturbed by the project implementation. Rehabilitation, acquisition, submergence, the eco-balance and right-of-way issues and their relation to local inhabitants are factors that are gaining importance in a society that is gaining conscious of its humanitarian rights. The lack of transparency and aberrations in adhering to the principles of environment have been blamed for the tragic death of the Arun III hydropower project which was claimed to be a "show-case" hydropower project of the past century.

3 TRENDS TOWARDS PRIVATE POWER

The early 1990s saw the dawn of the democratic constitution in the country and the quick introduction of government policies for a free economy open to foreign investment. It was also accepted that public sector financing for power sector projects the world over was fast dwindling and times were ahead to face a "resource crunch". Wary of survival in such inhospitable circumstances, NEA and the Government soon found themselves thrown from the fire to the frying pan when the World Bank turned down their apparently committed support for the Arun III Hydropower Project which was to see the country through the 1990s as a mainstay generation project.

Although similar policies in the developed world were aimed towards deregulation to provide consumers the opportunities of choice and initiate competition among suppliers, for transitioning economies like Nepal, these were identified as one-way avenues towards private finance. More pessimistic quarters even identified the constriction of public financing to the power sector as a ploy by the hydropower industries in the industrialised nations to grab the fast growing Asian market by compelling the use of private financing which they controlled.

Thus, it was no surprise that a new Hydropower Development Policy was soon adopted in 1993, the Nepalese government bent on promoting private sector investment in the development of the country's nascent hydropower. It is a matter of interest that the policy introduces itself as originating from the Arun III fiasco and virtually points to the private sector as the remedy to cure the serious generation planning flaw created by the fiasco.

Although the swing towards the private sector had now been initiated in many industrialised countries and reforms to stimulate their participation has been completed in a few, the experience of private promoters in developing hydropower projects along BOT/BOOT lines was scarce even in terms of international terms. Risks that were site specific and the long gestation period of the hydro projects were major factors that deterred private sector interests in hydropower. At home, in Nepal, the take-it-lying attitude that the Hydropower Development Policy laid out was designed to tone down the risks and attract venture groups to test the concessions offered.

The 60MW Khimti I Hydropower Project involving a Promoter with Norwegian connections was the first private sector to enter the BOT/BOOT fray. The Promoter was also an affiliate of a reputable power company with long standing experience in Nepal. They had obviously envisioned a strategy similar to the one taken along aid-oriented modalities they had adopted in earlier projects. However, the BOT/BOOT approach based on non-recourse financing was markedly unique and did not come as a regular exercise. Soon after their entering into a loan agreement, it was evident that the Promoters has made major miscalculations on the lending terms for such projects and were forced to approach NEA for a major amendment soon after entering into a Power Purchase Agreement (PPA) with NEA. It was evident that the Lenders such as the Asian Development Bank (ADB) and the International Finance Corporation (IFC) who supported the private sector policy were bent on compelling the independent power producer (IPP) to entering into a Project Agreement (PE) and PPA that ensured that they got back their investment. They obvious saw as murky the prospects of hydropower investment in the geologically risky young mountains of the Himalayas in the control of an even unpredictable political leadership.

The exercise for NEA was even more confounding, facing the new concepts in the capacity of a corner stone of an edifice on which all risks factors of the game rested. The very high-strung drama of power purchase, played in an arena in which NEA professed walking a tight rope with a "learning curve", turned out to be a biased affair. Even the Government seemingly offered a deaf ear to NEA's apprehensions and responded by voicing their strong intent to encourage the private sector as their priority concern. The standoff that ensued during the PPA negotiations had to be moderated by ADB under the banners that an attempt had to be made to divert a major mishap in policy application and save the private sector interest in hydropower investment which was in its fledging stage in Asia.

There was no doubt that the private sector promoters also played against heavy odds in venturing into hydropower generation in a country which was tagged with high risk factors. Their stamina to be on their high horses needs to be complemented but the end result of their "not-cheap price" dampened the aspirations anybody had of the IPP. The IPP contribution to the promotion of hydropower in the country was, to a certain extent, camouflaged by the need to sell all their production to the national utility, NEA. Operating at one of the highest tariff rates in the region, the cost of power purchase from the IPPs accounted for a large percent of NEA's cash flow and created a situation where private power was identified as the principal cause for tariff escalations – a very unpopular step by any standard to a populace wary of the high electricity tariff.

Despite the drawbacks, the interest that accumulated in private sector interest was impressive. This was evident when response to the solicitation called by the Ministry of Water Resources for the issuance of survey and generation licenses of mid-sized hydropower projects. The tragic part was that, although the Government had the correct view of promoting private sector through a process of solicitation, it had overlooked the gravity of its role to create and assure a market for the commodity the IPPs would produce. Except for the very optimistic anticipation of an industrial boom in the country that would send the electricity load forecasts rocketing, it was evident that the major market for the IPP hydropower did not lie within the country. Obviously, the prospects lay in the demand pattern of its neighbouring countries. Where the excellent relations existing between Nepal and its immediate neighbours could have been a fertile ground for creating a market for the generated hydropower, there were no indications that the Government had succeeded in this direction, or even to ratify a Power Trade Agreement it had already signed with the Indian Government. It was also improbable that the small IPPs could succeed in this Herculean task of activating Indian bureaucracy to a timely PPA although some of them did venture.

Minus the market for the IPP generation, the easy shift of the power gear was, therefore, for the IPPs to approach NEA to buy all their generated power. Deprived of the opportunity to develop its own generation at a cost less than that it has to pay the IPP, NEA's relations with the IPPs were oftentimes sour, and sometimes even strained. For NEA, the PPAs that it had signed with IPPs for terms ranging from 20 to 25 years began to reveal omens that threatened its profitable existence. Of foremost impact to NEA was the realisation that the US Dollars, in which the PPA had been agreed, soon turned out dubiously unstable and constantly rising in contradiction to the relative stable constitution predicted during the signing of the PPA -- and that too with NEA squarely facing the exchange risk. Secondly, with its own generation projects coming on-line and its supply capability turning healthy, the "take-or-pay" contracts with the IPPs loomed ahead with the prospects of adding to the generation surplus during the wet seasons when generation from its own ROR projects were of growing concern in a low load factor circumstances.

4 CONCLUSION

For Nepal Electricity Authority, adopting a generation plan based primarily on hydropower has been a challenge that involves both new trends in the power sector financing and management and also on the engineering challenge of matching the river hydrology in generation design with the demand characteristics. Besides the challenges which can get somewhat frustrating at times, NEA is confident that its confidence in a resource that has the premium qualities of a "clean" energy will ultimately pay off as the environment consciousness permeates through a greater fabric of society in a region that is transitioning into a more industrialised economy.

Recent trends in operation of hydropower plants

Hydropower in the New Millennium, Honningsvåg et al (eds), © 2001 Taylor & Francis, ISBN 90 5809 195 3

Real-time inflow forecasting for GilgelGibe reservoir, Ethiopia

G.G.Amenu
Arba Minch Water Technology Institute, Arba Minch, Ethiopia

Å.Killingtveit
Norwegian University of Science and Technology, Trondheim, Norway

ABSTRACT: Ethiopia is naturally endowed with quite a substantial amount of water resources potential. However, due to the erratic nature of its distribution and occurrence with time and in space, the country is also a frequent victim of terrible calamites: drought on one hand and flood on the other. One of the direct victims of these adverse situations is the power production in the country. In the past years power shortages have been occurred several times during dry seasons, while floods passing reservoirs have caused considerable damages during wet seasons. This stems mainly from the fact that there were no efficient hydrological forecasting and water management systems developed for operation of reservoirs in the country. In this study a real-time inflow forecasting model was calibrated and set up for one of the power generating reservoirs in Ethiopia: the Gilgel-Gibe Reservoir. The model will assist the project operators in making a realistic prediction of the reservoir inflow and in optimizing the reservoir water usage.

1 INTRODUCTION

Ethiopia is the owner of quite a substantial amount of water resources potential and is often called the water tower of East Africa. Despite the vast water resources potential available, the country has made no significant progress in development of the sector, the main reasons being luck of professionals in the sector and the country's poor economy.

Though Ethiopia is known with abundant water resources, due to the erratic nature of its distribution and occurrence with time and in space, the country is also a frequent victim of terrible calamities: drought on one hand and flood on the other.

One of the direct victims of these adverse situations is the hydropower production in the country. In the past years, power shortage and, hence, rationing have been occurred several times during dry seasons, while floods passing reservoirs have been caused considerable damages during wet seasons. This fact mainly stems from luck of efficient hydrological forecasting and water management system for operation of reservoirs in the country.

A hydrological forecast is nothing but a prior estimate of the future state of hydrological phenomena (WMO, 1983). The ability of providing reliable forecasts of flows for short periods in to the future is of great value in reservoir operation (Killingtveit and Sɟlthun, 1995). If information on the nature of the inflow is determinable in advance, then the reservoir can be operated by some decision rule to maximize production and minimize downstream flood damage,

through reducing or avoiding spillage and unnecessary releases.

2 OBJECTIVE OF THE STUDY

Reservoir operation forms an important part of water resources management and this is the main challenge facing the Ethiopian reservoirs and power systems today, and seeks an appropriate solution.

The main objective of the present study is, hence, to develop a system, which may assist reservoir operators in efficient utilisation of the reservoir storages in the country. In this study main focus has been given to developing real-time inflow prediction model for the Gilgel-Gibe Reservoir.

3 DESCRIPTION OF THE STUDY AREA

The Gilgel-Gibe Project is situated in the south-western part of Ethiopia, in Oromia Regional State (see figure 1). The project is purely a hydropower scheme, with an installed capacity of 200 MW, aimed to increase energy and power supply to the national grid. The reservoir has a live storage capacity of 657 Mil m^3.

The Gilgel-Gibe River is a right hand tributary of one of the 8 major river basins in Ethiopia, the Omo-Gibe River Basin. The catchment area of the Gilgel-Gibe Basin is about 5125 km^2 at its confluence with the Great-Gibe River and about 4225 km^2 at the dam

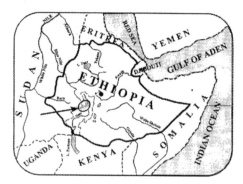

Figure 1. Location of the study area.

site. The Basin is generally characterised by high relief hills and mountains with an average elevation of about 1700 m above mean sea level. The basin is largely comprises of cultivated land.

In general terms, the Gilgel-Gibe Basin is characterised by wet climate with an average annual rainfall of about 1550 mm and average temperature of 19^0c. The seasonal rainfall distribution takes a unimodal pattern with maximum during summer and minimum during winter, influenced by the intertropical convergence zone (ITCZ).

4 HYDROLOGICAL STUDIES

Satisfactory solutions to water resources problems require reliable data on hydrological variables. Therefore, before using any hydrologic data for real practical work, it is very important to make sure that the data are consistent, correct, sufficient, and complete (WMO, 1983).

In this study, the necessary hydro-meteorological data have been collected from the Ministry of Water Resources (MWR) and the National Meteorological Services Agency (NMSA), and data processing has been made to them to the extent required for input to the reservoir inflow modelling (section 5).

18-28 years daily rainfall data for fifteen stations; 10-28 years daily runoff data for four stations; and 21 years monthly climatological data (maximum temperature, minimum temperature, relative humidity, sunshine hours, and wind speed) for one station were available during this study (see figure 2).

The necessary data processing and analysis carried out includes:

- Missing data estimation
- Data quality checking
- Data extension (for runoff)
- Areal determination (for rainfall)
- Variability studies
- Evapotranspiration computation

Missing rainfall data was estimated using Arithmetic Mean and Normal-Ratio techniques, while missing runoff data was infilled by developing correlations between nearby hydrometric stations.

Data consistency checking for both rainfall and runoff was carried out using the double-mass curve analysis technique. The analysis showed that the data are more or less consistent. Moreover, error detection in runoff data was performed by plotting daily differential time-series (changes from day to day) of the flow data.

The Thiessen Polygon technique was applied in determining the areal rainfall over the drainage basin. For the purpose of rainfall-runoff modelling at the dam site, only four rainfall stations located inside the basin were selected for estimation of the areal rainfall.

At the dam site (Deneba hydrometric station) only 10 years runoff data is available. The data is extended based on data at the nearby hydrometric station (Asendabo station) by developing a regression relation between the two stations.

Temporal variability studies of the hydro-meteorological elements indicated that there is a considerable short-term (day to day and month to month) variation, while the long-term (year to year) variations are less significant. The spatial variability of the elements is greatly influenced by topographical variations in the basin.

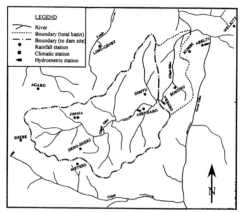

Figure 2. Location of hydro-meteorological stations.

The potential evapotranspiration is computed with the help of Penman method. The data inputs used for the computations were the one recorded at Jimma meteorological station. Accordingly, the annual potential evapotranspiration for the study area was estimated as 1385 mm.

5 RESERVOIR INFLOW MODELLING

5.1 *Model selection and description*

A large number of runoff forecasting models have been developed so far, and the choice of any particular case depends to a large extent on the objective of the study, the facilities and data available, and the hydrologic characteristics of the basin (Linsley et al. 1988). Generally speaking, reservoir inflow forecasting techniques fall in to three categories: time-series models, regression models, and conceptual models.

For the purpose of real-time inflow prediction to the Gilgel-Gibe reservoir, the HBV model – a conceptual precipitation-runoff model – developed at the Swedish Meteorological and Hydrological Institute (SMHI), was selected.

The HBV model, like many other precipitation-runoff models, is based on conceptual representation of a few main components of the hydrologic cycle. With the model, runoff from a catchment is computed from meteorological data like precipitation, air temperature, and potential evapotranspiration.

The standard version of the HBV model uses four main storage components, of which Snow Routine is one. As the area under investigation is located in the tropical climate with no snow experience, the snow routine part of the model has not been used for this particular study. The structure of the model used for the current study, therefore, comprises mainly three storage components as shown in figure 3.

The soil moisture routine receives precipitation (PPt) and potential evapotranspiration (PET) as input and computes the storage of water in soil moisture, actual evapotranspiration (AET), and the so-called net runoff generating precipitation as output to the upper zone of the runoff response routine.

The runoff response routine transforms the net precipitation produced in the soil moisture routine in to runoff. The runoff response function in the HBV model consists of two linear tanks or reservoirs called upper zone and lower zone arranged as shown in figure 3. This routine also includes the effect of direct precipitation on and evaporation from water bodies (lakes and reservoirs) in the catchment. The two linear reservoirs delay the runoff in time and the model can obtain both a quick response for high

flows and slow response for low flows, as normally seen in observed hydrographs.

The upper zone conceptually represents the quick runoff components, both from overland flow and from groundwater drained through more superficial channels, interflow, whereas, the lower zone conceptually represents the ground water and lake/reservoir storage that contributes to base flow in the catchment.

For detail knowledge of the HBV model reference can be made to Hydrology by ! .Killingtveit and N.R.S⌡lthun, 1995.

5.2 Model calibration and results

Model calibration is nothing but determining the value of free parameters in the model that gives the best possible correspondence between observed and simulated runoff of a catchment (Fleming, 1979). In the HBV model, when snow processes are not involved, as in this study, there are about nine parameters to be determined through model calibration.

Calibration of the HBV model for the Gilgel-Gibe catchment was performed using five complete years (1981-85) of daily observations of hydro-meteorological data. Data of four rainfall stations (Jimma, Dedo-Sheki, Asendabo, and Dimtu), one climatic station (Jimma), and one hydrometric station (Deneba) were used for calibration of the model. The aim of the calibration was to obtain model parameters from observed conditions, which will be used during the actual inflow prediction to the Gilgel-Gibe Reservoir.

The calibration of the model utilised a manual trial-and error procedure (see figure 4), whereby relevant model parameter values were changed until an acceptable agreement between observed and simulated flows was obtained. The model was calibrated by maximizing the R^2-value and, at the same time, by visual inspection of the observed and simulated hydrograph match.

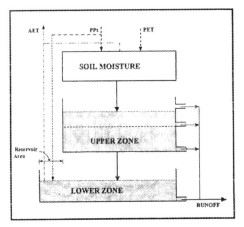

Figure 3. Structure of the HBV model (as applied to the study area).

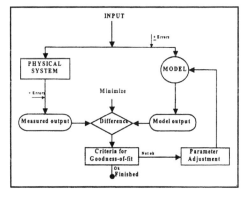

Figure 4. Model calibration process.

A typical hydrograph match obtained during the calibration period is given in figure 5, while the optimum values of model parameters and the R^2-values attained for the study area are presented in tables 1 and 2, respectively.

Table 1. Optimum values of HBV model parameters for Gilgel-Gibe Basin (obtained through model calibration).

Model parameter	Symbol	Value
Rainfall correction	PKORR	1.0
Field capacity in soil moisture zone	FC	350
Threshold value for PET	LP	200
Parameter in soil moisture zone	β	3.3
Fast drainage coefficient for upper zone	KUZ2	0.13
Slow drainage coefficient for upper zone	KUZ1	0.05
Threshold level for fast drainage in UZ	UZ1	25
Percolation for upper to lower zone	PERC	0.25
Drainage coefficient for lower zone	KLZ	0.002

Table 2. R^2- values obtained for the calibration period.

Year	1981	1982	1983	1984	1985
R^2	0.88	0.85	0.90	0.90	0.91

As a final stage in preparing the model for practical application, the verification of the model was carried out using five years independent set of data (1986-90). Table 3 summarises the R^2-values obtained during the verification period.

Table 3. R^2- values obtained for the verification period.

Year	1986	1987	1988	1989	1990
R^2	0.84	0.83	0.91	0.90	0.93

6 CONCLUSIONS AND RECOMMENDATIONS

The results of the HBV model calibration for the Gilgel-Gibe catchment suggests that the model is strongly recommended for use in inflow prediction to the Gilgel-Gibe Reservoir for operational purposes. By utilizing the real-time data such as precipitation, potential evapotranspiration, and runoff (for model updating), the forecast model will provide a realistic inflow prediction for the reservoir.

The accuracy of the forecast model is highly dependent on the quality of the input data. The author, therefore, recommends an implementation of an appropriate quality control to the input data before use in the forecast model. This requires a close collaboration between field data collectors and hydrological forecasters.

As the developed forecasting model can cope up only with short-term prediction of inflow, in order to get correct picture of long-term future inflows, an additional long-term forecasting model needs to be incorporated. This could well improve the optimal operation of the reservoir.

If the forecast indicates water shortages (warns of drought) at any time of the year, then supply of energy must be curtailed in advance to mitigate probable drought loss. The amount to be reduced should be decided by the decision makers based on specified values of rationing. On the other hand, if the forecast indicates water surplus (warns of spill) a pre-release strategy for surplus power production should be considered. In doing so, the project operator can maximize the usage of the reservoir storage while keeping spillage and deficit to the minimum.

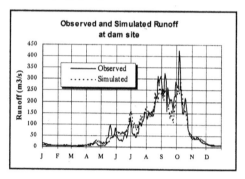

Figure 5. Typical hydrograph match obtained during calibration of the model.

7 REFERENCES

ASCE, Task committee on hydrology handbook, 1996. *Hydrology handbook.* 2nd ed. 784pp.

Baolin, W. 1993. *The calibration of hydrologic models.* ASCE, Engineering hydrology proceedings, San Francisco, California, Edited by Chin Y. K.: pp. 1109–1113.

ENEL-Electroconsult 1997. *Gilgel-Gibe hydroelectric project feasibility study report.* Ethiopian electric light and power authority (EELPA), Addis Ababa.

Ethiopian electric light and power authority (EELPA). *Reservoir and energy management annual reports.* Addis Ababa.

Fjeld, M. & Aam, S. 1980. *An implementation of estimation techniques to a hydrological model for prediction of runoff to a hydroelectrical station.* IEEE Trans. Automatic Control, AC-25(2): pp. 151-163.

Fleming, G. 1975. *Computer simulation technology in hydrology.* Elsevier, New York, 333 pp.

Harlin, J. 1991. Development of a process oriented calibration scheme for the HBV hydrological model. *Nordic hydrology,* 22(1): pp. 15 – 36.

Killingtveit, ! . and Sælthun, N.R. 1995. *Hydrology.* Hydropower development series, Vol. 7, Norwegian institute of technology,Trondheim, 213 pp.

Kürkkünen, K., Sirvi. , H., & Vakkilainen, P. 1998. A comparison of inflow forecasting techniques. *Nordic Hydrological conference,* Vol. II, NHP Report No. 44: pp. 557 – 566.

Kraijenoff, D. A. *(ed.)* 1986. *River flow modelling and forecasting.* Dordrecht, Holland: D. Reidel , 372 pp.

Lindstr. m, G., 1997. A simple automatic calibration routine for the HBV model. *Nordic hydrology,* 28(3): pp. 153 – 168.

Linsley, R. K., Kohler, M. A., & Paulhus, J. L. H. 1988. *Hydrology for Engineers.* Singapore: McGraw-Hill, 492 pp.

Reihan, A., Jevrejeva, S., Kovalenko, O., and P'rh, A. 1998. Modelling of river runoff in to the gulf of Finland. *Nordic Hydrological conference*, Vol. II, NHP Report No. 44: pp. 586 – 595.

Richard Woodroofe & Associates 1996. *Omo-Gibe River Basin integrated development master plan study final report.* Vol. VI – Water resources surveys and inventories, Ministry of water resources, Addis Ababa.

Roald, L. A. 1988. *Quality control of hydrometric data.* Hydrology center, Publ. No. 17, Christchurch, 87 pp.

Shaw, E. M. 1994. *Hydrology in practice.* Van Norstrand Reinholdt(UK), Wokingham. 3rd ed, 569 pp.

World meteorological organisation (WMO) 1983. *Data acquisition and processing.* Guide to hydrological practices, WMO, Vol-I, No.168.

World Meteorological Organisation (WMO) 1983. *Analysis, forecasting and other applications.* Guide to hydrological practices, WMO, Vol-II, No.168.

Yang, X., Parent, E., Michel, C., & Roche, P. 1995. Comparison of real-time reservoir operation techniques. *Journal of Water Resources Planning and Management*, ASCE, 121(5): PP. 345 – 351.

Yonas G. 1998. *Hydrological studies and modeliing in the upper Awash River, Ethiopia.* M.Sc. Thesis, D1-1998-11, Norwegian university of science and technology, Trondheim, 61pp.

Yung, Y. S., Gerald, J. N., and Eric, A. M., 1993. *Real-time forecast and its use for Bradley lake project.* ASCE, Engineering hydrology proceedings, San Francisco, California, Edited by Chin Y. K.: pp. 772 – 777.

Hydropower in the New Millennium, Honningsvåg et al (eds), © 2001 Taylor & Francis, ISBN 90 5809 195 3

Unit Commitment in Hydro Power Operation Scheduling

Michael Belsnes, Olav Bjarte Fosso & Jarand Røynstrand
SINTEF Energy Research

Terje Gjengedal & Eivind Valhovd
Statkraft SF

ABSTRACT: The paper describes the basic principles and results of a method for making detailed hydro unit schedules in a cascaded reservoir system. In this implementation, a commercial program for hydro operation scheduling is further developed to explicit account for the start/stop cost of units. Within the presented case, there is a mix of buffer and storage reservoirs and there are also several units in some plants. The multiple unit case is a special challenge due to the coupling in the unit effective head calculations.

1 INTRODUCTION

Use of discrete variables in short term hydropower scheduling has been investigated for some time. But only recently the use of discrete variables has become an alternative to heuristics and other methods like Dynamic Programming and Lagrange Relaxation.

The reason why we need to include the unit commitment in the hydropower scheduling is the increasing variation in spot price for electric power in the Nordic power market. The power producers need tools for scheduling that account for the costs of starting and stopping units. The need arise when producers want to optimize income and to obtain coverage for costs from granted system services.

Short-term scheduling of hydropower is a challenging task in cascaded reservoir systems. The scheduling must interface with the boundary conditions received from the mid-term scheduling, fulfill discharge constraints within each time interval as well as maintain couplings as ramping constraints and reservoir balances between the intervals. The output of the scheduling problem should be a proper unit commitment sequence with a power production plan on each generating unit. All this results in a large-scale optimization problem with a mix of discrete and continuous decision variables. Proper system modeling and solution strategy is of most importance to be able to solve such a problem. This paper describes our modeling approach, the implementation and discusses the performance on a hydro system. The paper focuses on the modeling of hydropower plants rather than describing the whole model in detail.

2 SYSTEM MODELING

2.1 System topology

In a cascaded reservoir system, there is a need to model all the individual reservoirs as well as the different discharge paths between the reservoirs. Such discharge elements may be power production plants, bypass gates and spill paths. This is illustrated in Figure 1.

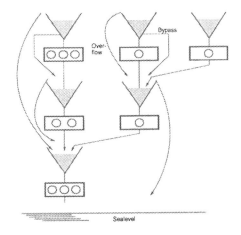

Figure 1: Example topology

Each reservoir will normally have several discharge options and each path may have its separate downstream destination. Several plants can be defined for each upstream reservoir.

2.2 *Reservoirs*

The reservoirs are the main connecting nodes of the watercourse. All plants and discharge gates must be associated with a reservoir.

For each reservoir and time increment of the study period, a reservoir balance equations as given below is set up:

$$- X_i(t-1) + X_i(t) - \sum_{j=1}^{n_u} q_j^u(t-\tau_j) + \sum_{j=1}^{n_d} q_j^d(t) = 0 \quad (1)$$

where:

$X_i(t-1)$ Reservoir content of reservoir 'i' in the end of time interval 't-1'. (Start volume of 't')

$X_i(t)$ Reservoir content in the end of interval 't'

$q_j^u(t-\tau_j)$ Inflow from upstream sources in time interval 't-τ_j' where 'τ_j' is the time delay. These sources can be regular inflows, discharge through plants or through gates.

$q_j^d(t)$ Discharge from the reservoir. This can include a hydropower production plant, and/or a number of bypass gates and a reservoir overflow.

n_d Number of downstream elements

n_u Number of upstream elements

All reservoirs must be given a resource cost at the end of the optimization period. Such a description is necessary since the optimization is based on a multistage formulation where the use of the available resources over the study period will be optimized.

These options for defining the endpoint description can be given:

- Specified reservoir volume in the end of the study period
- Range of feasible reservoir volumes
- An incremental cost of water as a function of the endpoint reservoir volume

The actual method to use depends on how sophisticated the program used for mid-term planning is. Some programs will only supply the reservoir trajectories while others give incremental cost of water. In the current implementation it is possible to use either independent water values for each reservoir or water values that is dependent of the reservoir levels in other reservoirs.

2.3 *Plants*

A hydro production plant description is a main element in the formulation of the optimization problem. The reason for using the plant as a main element rather than the individual units is the hydraulic coupling within the plant. The production on each unit can influence on each other due to common tunnels/penstocks or because of the tailrace effects.

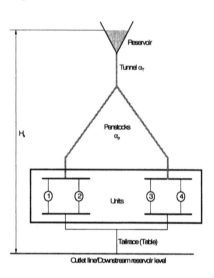

Figure 2: Possible plant topology

The figure above illustrates the internal plant configuration.

2.3.1 *Waterways*

The plant is always connected upstream to a reservoir. The waterways description consist of the main tunnel and/or a number of pressure shafts and is used to calculate the losses from the intake reservoir to the turbines. The main tunnel consists of several segments in the main tunnel with individual loss factors.

There can be several pressure shafts with individual loss factors and the units, like in figure 2, can share common penstocks or be connected to individual penstocks.

This detailed description of the waterways enables an accurate head loss calculation.

Figure 2 also illustrates the problem of including explicit modeling of start/stop in hydropower optimization. The decisions about which unit to run becomes more difficult when the units influence on one another, which is the case here. The unit 2 would normally not be used together with unit 1 if unit 3 or unit 4 were available because of the loss in the penstock tunnel.

In this implementation, these aspects are solved in a two-step approach by using simplified description in the initial iterations when the unit commitments are made. After the unit commitment is made, the accurate modeling of the loss can be included.

2.3.2 Units
Each unit has it own set of efficiency curves for the turbine so that the connection between head and efficiency can be modeled for each individual unit. In this way it is possible to obtain an accurate modeling of the characteristic diagram for the hydropower turbine. Individual generator efficiency curves can also be applied. Minimum and maximum of the turbine can either be fixed values, or be functions of the plant head.

2.3.3 Head
Calculation of head includes the up- as well as downstream reservoir levels. Especially for run of river systems, the tailrace is modeled. This means that rising of the backwater can be accounted for at high water flows.

2.3.4 Constraints
Different types of constraints connected to operation of the hydropower plants can be added to the model:
- Schedules
- Minimum production
- Maximum production
- Ramping on production

2.3.5 Modeling with discrete variables
Using a discrete model for the plant is optional. This make it is possible to model some hydropower plants with continuous and some with discrete variables in the same case. The use of discrete variables is most helpful in the case of strong hydraulic connections in the watercourse. Hydropower plants connected to reservoir with some storage capacity can be handled by other means. This makes it possible to model realistic systems using discrete variables only on the critical hydropower plants. Another consequence of this flexibility is the possibility to use a iteration logic where discrete variables are added for one and one plant in each iteration.

The following equations for the hydropower plants are used in the model. Similar modeling is used for starting and stopping pumps. The equations are valid for the one unit case.

First the startup costs are subtracted from the objective function by adding the following term:

$$\sum_t C_{su} su_t \qquad (2)$$

Calculate the production on the plant:

$$\sum_j \alpha_{j,t} q_{j,t} + \delta_t \gamma_t - p_t = 0 \qquad (3)$$

Ensure $\delta = 1$ when p > 0:

$$p_t - m_t \cdot \delta_t \leq 0, \text{ and } m \geq p_{max} \qquad (4)$$

Ensure $p \geq p_{min}$, when $\delta = 1$:

$$p_t - p_{min} \cdot \delta_t \geq 0 \qquad (5)$$

Calculate start variable su

$$\delta_t - \delta_{t-1} - su_t \leq 0 \qquad (6)$$

Constraints on discharge per segment

$$0 \leq q_{j,t} \leq q_{max\,j,t} \qquad (7)$$

Definition of δ :

$$\delta \in \{0,1\} \qquad (8)$$

su   start up variable
C_{su} is the start up cost
j   is the segment index
t   is the time index
α_j is the incline of the segment
δ_t is the plant running indicator:
     $\delta_t = 0$, plant is standing
     $\delta_t = 1$, plant is running
γ_t is the constant term on the PQ kurve
p_t is sum production
p_{min} is minimum production
p_{max} is maximum production
$q_{j,t}$ is discharge per segment
$q_{max,j}$ is maximum discharge per segment

2.4 Gates
Gates can be defined connecting any two reservoirs in the watercourse. The downstream destination for the gate does not have to be the same as the downstream destination for the plants connected to the same reservoir.

3 SOLVING STRATEGY
The optimization approach is based on successive linear programming. A branch and bound technique is used for handling integer variables. The commercial optimization package CPLEX from the company ILOG CPLEX Optimization Inc. is used as a kernel in the calculations.

     The overall solution consists of iterations (main iterations) using one of two modeling modes. The goal for the first mode (full description) is to find an initial solution that can function as an linearization point for a detailed description. In the second mode (incremental description) a more detailed modeling is used linearizing around the efficiency curves for the committed units.

3.1 Main iterations

The main iterations make it possible to use the results from the last iteration to perform a refinement in modeling from iteration to iteration.

The main motivation for using an iterative approach is that some of the nonlinearities and constraints will depend on the power production/discharge decisions. These decisions are unknown and cannot be taken into account in the first main iteration. An optimization model of the system is built for the entire study period based on the available information. This model couples all time intervals and takes into account the endpoint reservoir constraints. The solution of this problem is the optimal decisions based on the current approximation of the system description.

The sequence of main iterations refines the system description based on the discharge profiles and the reservoir trajectories from the previous iteration. The refinement involves new linearized descriptions of:
- The unit efficiency curves
- The reservoir level - volume relation
- The gate discharges

Major nonlinearities are represented by multiple segments in the linearization process.

The constraint set is also updated in the outer loop. This involves constraints as ramping of reservoir volumes and releases.

After each main iteration that consists of the model building and the solution, a feasible and close to optimal solution will be available unless there are any constraints that are excluded due to the strategy of adding the constraints.

Each main iteration can either be based on a full or an incremental description.

3.2 Full description

In this mode the optimal decisions for production level, gate discharge are computed taking head optimization and constraints into account. Normally two main full iterations are used before it is changed to the incremental. The second full iteration is needed for the reservoir level/plant head optimization. When solving the full description the entire valid production area of the hydropower plants is used. As linear programming require convexity in the modeling the whole working area of the plant or unit cannot be modeled in detail. Losses in the tunnels and penstocks are accounted for in a simplified way. In the full description the discrete nature of the start/stop is

accounted for by introducing binary variables for the unit commitment. This model needs to be solved by a Branch and Bound technique.

3.3 Incremental description

After the unit commitment is made, an incremental formulation is used. Based on the online units and the actual discharge, incremental formulations are established around the current operating point. In these incremental descriptions, the accurate loss models are included. This is possible because the committed units and the tentative production levels already are determined. The number of incremental iterations is normally two to three to obtain satisfactory convergence.

4 TEST SYSTEM

The study is based on a river system with 14 reservoirs and 8 power production plants. A part of this system is shown in the figure 3.

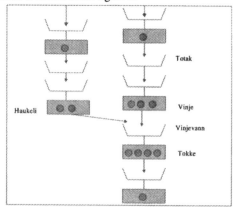

Figure 3: Test system

Studies are made for a price profile over 24 hours on the Norwegian electricity market. The price profile is shown in Figure 4.

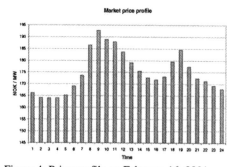

Figure 4: Price profile on February 16, 2001.

There are two high price periods during the day. The last peak has shorter duration and makes the focus on start/stop costs important. The highest price is 193 NOK/MWh (hour 9) and the lowest is 163 NOK/MWh (during the night).

Initially an optimization was performed where costs associated with start/stop was ignored. Each reservoir was given an individual resource cost description (water value). Using water values adds flexibility to the optimization and the discharge from each reservoir will be a function of price level and profile.

The power plant named Vinje got a production profile where it was running from hour 8 to 13 and for a single hour in the afternoon when the price is high. Ignoring the costs involved with start/stop, short periods of operation for marginal changes in market price can occur.

The power plant named Tokke gets a production profile where it runs from hour 8 to 15, then it stops 3 hours but is running again from hour 6 to 9.

Next an optimization is performed where the cost of start/stop is included. For demonstration purposes, it is assumed a cost of 500 NOK for each start/stop cycle of a unit.

The two cases of the plant Vinje are summarized in the figure 5.

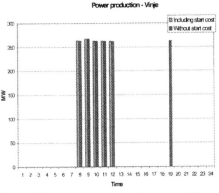

Figure 5: Power production of the plant Vinje

The most important change in the operation of the Vinje plant is that no startup is made for the last price peak. The margin between the market price and the water value in this case was not high enough to make it profitable. The price difference was only 5 NOK/MWH and a unit production of 87 MW gave less income than 500 NOK. There is three units in the plant with a total production of 261 MW.

The two cases of the plant Tokke are summarized in figure 6.

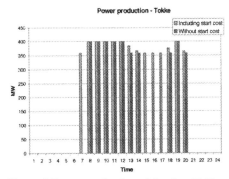

Figure 6: Power production of the plant Tokke

A significant change is observed with start costs included. Now, the plant also runs from 3 pm to 5 pm. The margin between the water value and the market price is low enough to make this profitable. For the periods with high prices, all the units run at maximum output. For the other periods, the plants run according to its marginal cost. The marginal cost is defined by the incremental unit efficiency and the water values. This plant has four units. There are two units on each penstock. In some cases when the difference between water value and price is low, the unit head loss may make it profitable to only run one unit on each penstock. In this case though, all four units were used in the same time intervals.

Figure 7 shows the reservoir trajectories of the reservoir Vinjevann for the two cases.

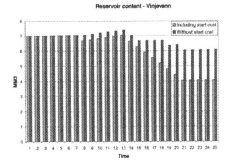

Figure 7: Reservoir trajectory of Vinjevann

Vinjevann is a rather small reservoir that shows a significant change during a day. In the given case where only 24 hours were studied, the reservoir gets a significant draw down in the case with start/stop. As the discharge capacity downstream of Vinjevann is smaller than upstream, it is important for the resource utilization to use the buffer capabilities.

These two cases illustrate well the impact and importance of including start/stop costs. When the discrete nature of these decisions is ignored, it is possible too get plans with to frequent start and stop of units. Such plans will increase the operation cost, and result in less adaptable scheduling plans.

5 COMPUTATIONAL ASPECTS

This work has shown that Branch and Bound techniques can be useful in short-term scheduling of hydropower systems. However, such techniques can be quite time consuming if the number of discrete variables becomes too large. The time horizon of the study and the number of hydro units with explicit modeling of start/stop will be critical for the performance. Most river systems in Norway have less that 15 production units. For such systems it is no problem to study periods of 2-3 days with hourly time resolution without more than a few minutes computation time.

However, it is experienced that some characteristics with the load profile and the market description will impact the computation time.

When using Branch and Bound, the algorithm may find the optimum rather fast, only to spend a long time verifying that the solution is indeed the optimal one. Especially when there are many different solutions with almost the same value of the objective function. This is the case when the price profile is flat, or almost flat.

It is also found that the solution time spent on a specified load profile is higher than to solve a case giving the same production profile against a market price profile. The computation time, on the system and market price profile shown in this paper, is very low.

6 DISCUSSION/CONCLUSION

Introduction of discrete variables in the hydro scheduling is an important extension to the methods based on traditional linear programming. This capability will reduce the operation cost and make unit plans that are more practical to implement. However, since the techniques at this stage are infeasible for really large system, it is important to apply a hierarchy for the decisions. For coordination of multiple river systems with a week's time horizon, it is necessary to use the traditional techniques based on successive linear programming with only continuous variables. The main result from this stage will be how much the different river systems should contribute to the overall plan. The next stage is then the detailed study of how to implement each sub system requirement.

Other principles have also been tested for the hydro unit commitment decisions. Dynamic programming (DP) has been used in combination with the successive linear techniques. The DP was then used

for the post-processing of the plant productions with the sensitivity signals/marginal costs from the linear programming as decision support. This technique was fast and gave almost the same results as the Branch and Bound in cases with some flexibility of the reservoir storages. However, for the buffer reservoirs it was difficult to get satisfactory results.

The Branch and Bound technique was found to be superior on small systems due to it's robustness and systematic approach for finding the optimal solution while taking all flow constraints into account.

The future work of this research project will be to combine the different available techniques in order to solve the multi river coordination problem.

ACKNOWLEDGEMENTS

The writing of this article was supported by Statkraft SF and based on research results financed by EBL and the Norwegian Companies.

REFERENCES

Bellman, R. 1957. Dynamic programming. Princeton, New Jersey: Princeton University Press.

Belsnes, M. M., Røynstrand, J., Fosso, B. O. & Huse, E. S. 2001. Planlegging i serievassdrag med hensyn til start/stopp problematikk. Technical Report A5355, SINTEF Energy Research, Trondheim, Norway.

Flatabø, N., A. Johannesen, E. Olaussen, S. Nyland, K. Hornnes & A. Haugstad 1988. EFI's models for hydro scheduling. Technical Report 3483, SINTEF Energy Research, Trondheim, Norway.

Gjelsvik, A. & S. W. Wallace 1996. Methods for stochastic medium-term scheduling in hydrodominated power systems. Technical Report A4438, SINTEF Energy Research, Trondheim, Norway.

Stage, S. & Y. Larsson 1961. Incremental cost of water power. AIEEE Transactions (Power Apparatus and Systems). 361.

Williams, H. P. 1993. Model Building in Mathematical Programming, 3rd ed. John Wiley & Sons LTD.

Hydropower in the New Millennium, Honningsvåg et al (eds), © 2001 Taylor & Francis, ISBN 90 5809 195 3

Assessment of secondary energy in hydroelectric systems

M.A.Cicogna & S.Soares
Universidade Estadual de Campinas – UNICAMP, Campinas, São Paulo, Brazil

ABSTRACT: This paper presents a study concerning the availability of electric energy in hydroelectric power systems. The study aims to estimate the amount of firm and secondary energy in hydro systems, considering the Brazilian hydro system as a case study. The study consists of determining monthly hydroelectric generation duration curves by simulation over the historical inflows, using three different operation policies. The study was performed in a progressive way in order to highlight the relation between firm and secondary energies. The case-study system is composed of seven large hydro plants located on the Brazilian Southeastern region, with 12572 MW of installed capacity, corresponding to 18% of the total Brazilian power system capacity. Results indicate an expressive amount of secondary energy in hydro systems and a slightly increasing relation with the installed capacity, varying around 50% of the systems firm energy.

1 INTRODUCTION

The Brazilian power system is strongly dependent on hydroelectric generation (94%) and it will keep this characteristic in the future because only a quarter of the estimated hydroelectric potential was already explored (Eletrobrás).

Hydroelectric energy has its availability as a function of the inflow condition. In contrast, thermoelectric energy availability depends only on factors like maintenance scheduling. On the other hand, hydropower generation is strongly dependent of hydrological factors, like the amount of inflow. Other hydro generation dependence is the reservoir operation policy.

In order to measure the energy availability of a hydroelectric plant, a common procedure is based on the concepts of firm energy and secondary energy. The firm energy of a hydroelectric system is the maximum energy that can be continuously produced by the system if the inflows occur as in the historical records. The secondary energy is the excess of energy, considering the low limit determined by the firm energy, that can be produced in the periods of favorable inflows (Fortunato, 1990) and (Cicogna, 1998).

The expansion planning of the Brazilian power system has been based on the concept of firm energy for decades. The increase on energy demand has been attended by the increase on firm energy of the system. This means that secondary energy has been also increased with the system capacity, although

this amount of energy has not been receiving the proper commercial attention.

The present paper aims to determine the different assessment of energy in hydroelectric systems. The study is based on results of a simulation model that uses operation policies based on optimization. The objective is to determine duration curves for the monthly generation, which provides the amount of available energy and the frequency that it occurs.

The paper begins with a look at the initial concepts necessary to determinate the availability of energy on hydroelectric systems. In the following section, the procedure that calculates firm and secondary energies is shown. The study was performed in a progressive way with the installed power capacity. The case-study system is composed of seven large hydro plants localized on the Brazilian Southeastern region, corresponding to 18% of the total Brazilian power system capacity. Finally, the monthly generation duration curves are analyzed in terms of their minimum, maximum and average energies. Some alternatives for exploiting the secondary energy are presented.

2 INITIAL CONCEPTS

Focusing on a hydroelectric plant, the availability of energy depends on the historical inflow values and the policy adopted for the operation of that plant. For instance, if the reservoir is kept full, which means that the operation policy decides to release the inflow each month, the operation is the simple conversion of inflow on power. The assessment of energy will be similar to the assessment of inflow, limited by the upper bound of flow that can be converted on power, which is a characteristic of the turbines. Besides, if the inflow is somewhat greater than this upper bound, the police will produce a considerable amount of spillage. On the other hand, if the operation policy decides to use the reservoir to reduce the chances of spillage, the availability of energy will be greater than in the first policy.

The energy availability measurement, in the Brazilian power system, has been based on the concepts of firm energy and secondary energy (Fortunato, 1990).

- The Firm Energy of a hydroelectric system is the maximum energy that can be continuously produced, considering the repetition of the historical inflow records.
- The Secondary Energy of a hydroelectric system is the excess of energy, with relation to the firm energy, that can be generated in the periods of favorable inflows. The secondary energy is commonly calculated as the difference between the average generation and the firm energy.

The concept of firm energy is associated with an operation policy that decides to generate the largest constant amount of hydropower during the simulation. The secondary energy is calculated as the energy represented by the spillage, which happens in the periods of high inflows.

The next topic will show the procedure for assessment of the hydroelectric systems energy availability. The influence of the generation policy and the topological configuration of the plants will be discussed as well.

3 PROCEDURE

The procedure that calculates the availability of energy was based on simulations performed in a hydroelectric test system composed of some of the most important Brazilian hydroelectric plants, which have accumulation reservoir. The inflows that were assumed in the simulations were those found in the historical inflow records. There are three different policies to consider:

- Deterministic Optimization (DO): this policy utilizes a deterministic optimization model that considers individual characteristics of each plant. This model requires the future inflow sequence as input data. As it knows the future inflows, the results of its utilization can be interpreted as an upper bound for other policies. Besides, the amount of energy determined with the utilization of this policy is the largest value of available energy for a specific set of plants. The methodology is based on a capacitated nonlinear network flow algo-

rithm specially designed for hydrothermal scheduling problems (Rosenthal, 1981) and (Oliveira, 1995).

- Stochastic Dynamic Programming (SDP): this policy represents the set of plants as composite reservoir. The main idea is to aggregate the set of plants in to one simple reservoir that receives, stores and releases energy instead of water (Arvanitidis, 1970). The procedure transforms hydraulic variables into energetic ones. The next step is to calibrate a lag-one periodic autoregressive model to the historical sequence of inflows. This stochastic model is the input to a stochastic dynamic programming algorithm that determines the optimal operation policy to the composite reservoir. Both the aggregation of the plants into composite reservoir and the desegregation of the composite generation decision are made assuming parallel reservoir operation. The SDP methodology has been used as the Brazilian power system's operation policy for last two decades (Terry, 1980).
- Deterministic Optimization with Forecasted Inflows (DOFI): this operation policy is based on the idea of feeding the inflow input to the deterministic optimization model with results of a lagone periodic auto-regressive inflow forecasting model. The operation decision policy decides to release the optimal amounts of the first month. This forecasting-optimization process is repeated at each month of the simulation in order to avoid error propagation.

In order to measure the energy production of the three policies, simulations of the hydroelectric test system were made over the historical inflow record. This registry is monthly-based and it collects the inflow occurrences that have been measured since 1931 until 1996.

The main information extracted from these simulations is the monthly hydroelectric generation. An insightful way of visualizing the amount of energy that each policy has drawn of the system, is to sort the generation and plot the result in a duration curve, as shown in the Figure 1.

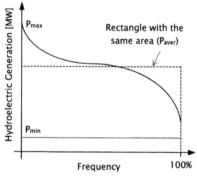

Figure 1. Duration curve scheme.

The scheme of Figure 1 permits the visualization of the principal elements of the energy availability, as the maximum, minimum and average energy that can be drawn from the hydro system.

4 CASE-STUDY

The study was performed on a progressive way for a system composed of seven large hydro plants with accumulation reservoir. This system is located on the Brazilian Southeastern region and has 12572 MW of installed capacity, corresponding to 18% of the Brazilian total system capacity. The spatial arrangement of these plants is shown in the scheme of Figure 2.

Figure 2. Case-study system.

The installed capacity and the reservoir's operational volume are described in Table 1. Note that the major accumulation reservoirs are located in the upstream plants (Furnas, Emborcação and Itumbiara), and the plant with the major installed capacity is the last one (Ilha Solteira, 24 turbines with 3240 MW of total installed capacity).

The progressive composition of the hydroelectric systems is shown in Table 2. The aim is to determine the energy availability when the hydro generation system receives increments of installed capacity.

Table 1. Hydroelectric plant characteristics.

| Plant Name | | Capacity | Volume |
		MW	hm³
1	Furnas	1312	17217
2	Marimbondo	1488	5280
3	Água Vermelha	1380	5169
4	Emborcação	1192	12521
5	Itumbiara	2280	12454
6	São Simão	1680	5540
7	Ilha Solteira	3240	5516

Table 2. Composition of the hydroelectric test subsystems.

| Set # | Plants | Capacity |
		MW
1	1	1312
2	1 and 2	2800
3	1, 2 and 3	4180
7	1 to 7	12572

5 RESULTS

Using the progressive configurations of Table 2, simulations were performed for each operation policy already described. For each simulation process, generation duration curves were determined by sorting the monthly hydroelectric generations.

The studies were made under a CASE system specially implemented to the hydrothermal scheduling problem. This system permits to configure, execute and analyze a wide range of models, like simulation, optimization and forecasting methodologies applied to hydrothermal power systems (Vinhal, 1998).

Table 4 summarizes the results of the DO, DOFI and SDP operation policies applied to each system configuration of Table 2. In each column, the number under the policy name represents the number of plants in the configured system.

Looking only to the numerical results, it is important to note that DO has the largest values for minimum and average generation among all operation policies. This fact happens because the DO policy knows de future inflow occurrences. In this way, its results are interpreted as an upper bound to other operational policies.

Note the difference between the minimum generation and the firm energy for the operation policies. This difference is due to the procedure that determines the firm energy, which does not have an embedded optimization rule on its decision process. As could be verified in Table 4, the DO policy provides de largest minimum generation value among the three policies, which is even larger than the firm energy of the hydroelectric system with seven plants.

Table 3 shows the secondary energy calculated using the firm energy and average generation of the simulations. As expected, the DO policy raises an upper bound to the amount of secondary energy available in the tested systems.

Table 3. Secondary energy

| Policies | E_{sec} | | | |
	1	2	3	7
DO	198	538	909	2972
DOFI	193	512	821	2651
SDP	163	418	721	2563

Values in MW

Table 4. Numeric summary of the simulation results.

	DO				DOFI				SDP			
	1	2	3	7	1	2	3	7	1	2	3	7
E_{firm}	523	1120	1857	5160	523	1120	1857	5160	523	1120	1857	5160
P_{mim}	406	994	1689	5794	153	390	677	2621	136	616	637	1837
P_{max}	1089	2572	3980	11001	1299	2765	4149	12086	1299	2765	4133	12387
P_{aver}	721	1658	2766	8132	716	1632	2678	7811	686	1538	2578	7723

Values in MW

Figures 3, 4 and 5 show de hydroelectric genera-tion duration curves obtained with the pure determi-nistic optimization policy, the operation that com-bines forecasting and optimization, and the stochastic dynamic programming policy.

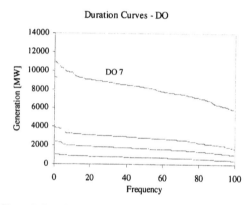

Figure 3. Duration curves obtained with the deterministic op-timization policy applied to each system configuration.

Analyzing the curves, one can notice that the larger installed capacity, the steepest the curve's slope. This behavior suggests that the secondary en-ergy increases with the addition of installed capac-ity.

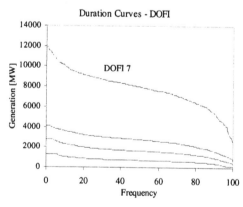

Figure 4. Duration curves obtained with the deterministic op-timization with forecasted inflow policy applied to each system configuration.

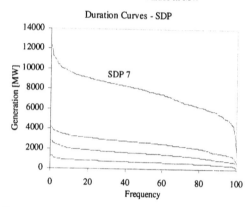

Figure 5. Duration curves obtained with the stochastic dynamic programming policy applied to each system configuration.

Figure 6 shows the duration curves of the three operation policies for the system configuration that collects the seven plants.

The duration curves show insightful information about the frequency of the system's energy avail-ability. This probabilistic information is quite im-portant for planning purposes that aim to implement economic exploitation of hydro energy. For instance, in Figure 6 it can be verified that in almost 30% of the historical registry, the seven plants are able to generate more than 8500 MW, for all three policies. This amount of energy is 65% greater than the firm energy of the seven-plant system.

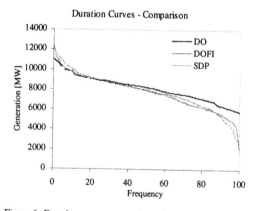

Figure 6. Duration curves comparison obtained with the three policies applied to the seven-plant system configuration.

The DO's duration curve is situated above the other two curves most of the time, mainly for those intervals where the inflow registry represents small amounts of water (inflows where the associated generation is under 8000 MW). Another aspect is the curve slope of the DO policy, which is smaller than the other two policies. This result was expected because the deterministic model knows the future inflow occurrences and can transfer energy between favorable and dry inflow periods. Therefore, the total generation of the deterministic model is more constant than the other policies, which do not know the future inflow and so are real decision-making policies.

Another conclusion that can be drawn from the analysis of the numerical results is that the DOFI policy has its energy availability greater than the SDP policy. This fact occurs in all system configurations, which shows the superiority of this operation policy in comparison with SDP, the technique adopted by the Brazilian power system.

The proportion between firm energy and installed capacity for each configured system is shown in Table 5. These results confirm a known characteristic of the Brazilian hydroelectric system, whose firm energy is approximately 45% of the installed capacity (Eletrobrás). Besides, the average generation is approximately 60% of the installed capacity, which shows the relationship between the energy and the natural hydrologic availability, as shown in Table 6.

Table 5. Firm energy x installed capacity.

E_{firm}/P_{inst}			
1	2	3	7
39,86%	40,00%	44,43%	44,58%

Table 6. Average generation x installed capacity.

Policies	P_{aver}/P_{inst}			
	1	2	3	7
DO	54,95%	59,21%	66,17%	65,50%
DOFI	54,57%	58,29%	64,07%	63,63%
SDP	52,29%	54,93%	61,67%	61,31%

The last comparison that can be extracted from the numerical results, is the relation between the amounts of secondary and firm energies, as shown in Table 7. Comparing the behavior of these energy measurements and the increment of installed capacity, one can conclude, as expected, that the firm and average energy increase with the increment of installed capacity.

Table 7. Secondary energy x firm energy.

Policies	E_{sec}/E_{firm}			
	1	2	3	7
DO	37,86%	48,04%	48,95%	57,60%
DOFI	36,90%	45,71%	44,21%	51,38%
SDP	31,17%	37,32%	38,83%	49,67%

For the seven-plant system, the secondary energy is approximately 50% of the firm energy, or 20% of the installed capacity. If we extend this result to the whole Brazilian hydroelectric system, which capacity is 60000 MW, the amount of secondary energy is approximately 12000 MW. For comparison, the largest hydro plant in operation nowadays is the Itaipu plant, located in the Brazilian Southeastern region. This plant has 12600 MW of installed capacity and, therefore, the amount of secondary energy is equivalent to the Itaipu's maximum generation.

6 SECONDARY ENERGY UTILIZATION

The economic exploitation of the secondary energy is fundamental to the rational use of hydroelectric resources. Two options for implementing an economic utilization of these resources are:

6.1 Secondary markets

This utilization is based on the idea of supply consumers' energy demand that do not have constant load necessity. For instance, industries which energy demand is influenced by the weather seasons. The strategy is to divide the duration curve in horizontal slices with different frequencies of load interruption. Each slice's height represents energy availability with an associated probability. This energy can be offered to the market with decreasing prices in response to increasing availability (Figure 7).

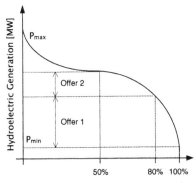

Figure 7. Scheme of the secondary energy economic exploration to supply consumers' energy demand, which do not have constant load necessity.

One advantage of this approach is that this economic utilization of the secondary energy does not require new investments on the generation and transmission systems. Figure 7 illustrates the creation of two energy offers that are situated above the ensured energy. This energy availability has 20% and 50% of shortage probability, respectively.

Using the real simulation result of the DOFI operation policy and keeping the values of shortage probability of the above scheme, Figure 8 illustrates the creation of two energy offers of approximately 6200 MW and 8000 MW, respectively.

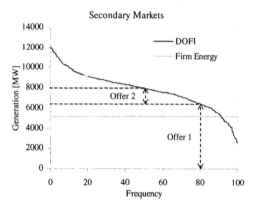

Figure 8. Secondary energy economic exploitation, using the suggestion of secondary markets.

6.2 Non-hydraulic complementary energy

This utilization is based on the idea of transform secondary energy in firm energy by the construction of a non-hydraulic complementary generation system. This approach increases the firm energy of the system using non-hydraulic generation when the hydro system suffers with periods of critical inflow. Figure 9 illustrates the process of increasing the firm energy of a generator system.

The non-hydraulic complementary energy has its usage probability determined by the duration curve. This thermal energy availability information can be used to determine the installed capacity of the thermoelectric system and its usage factor.

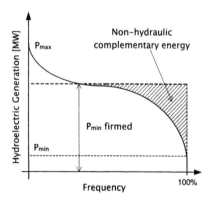

Figure 9. Scheme of the secondary energy economic exploration for non-hydraulic power complementation.

7 CONCLUSIONS

This paper has presented a methodology to assess the energy availability of hydroelectric systems. The study was based on monthly generation duration curves obtained from simulation of a hydroelectric test system for three different operational policies. The simulation was performed in a progressive way in order to highlight the relation between firm and secondary energies. The case-study system is composed of seven large hydro plants located on the Brazilian Southeastern region, with 12572 MW of installed capacity, corresponding to 18% of the total Brazilian power system capacity. The results show an expressive potential of secondary energy that increases with the expansion of the installed capacity. The secondary energy was estimated in 50% of the firm energy, or 20% of the installed capacity.

Any national plan for the economic exploitation of the Brazilian hydroelectric potential should not ignore this huge amount of secondary energy, mainly because its usage does not require new investments on the generation and transmission systems.

8 ACKNOWLEDGMENT

This work has been developed under a research project supported by FAPESP and CNPq.

REFERENCES

ARVANITIDIS, N.V., ROSING, J. 1970. Composite Representation of Multireservoir Hydroelectric Power System – IEEE Transaction on PAS, Vol. PAS-89, n. 2, pp. 319-328 February 1970.

ARVANITIDIS, N.V., ROSING, J. 1970. Optimal Operation of Multireservoir System Using a Composite Representation – IEEE Transaction on PAS, Vol. PAS-89, n. 2, pp. 327-335 February 1970.

CICOGNA, M.A., SOARES, S. 1998. Methodology for Determining the Secondary Energy of Hydroelectric System – in Portuguese. XII CBA Uberlândia, MG, Vol.3, n. 11-15, pp. 1005-1009 September 1998.

ELETROBRÁS, Brazilian Electric System's web site http://www.eletrobras.gov.br.

FORTUNATO, L.A. et al. 1990. Indroduction to the Expansion Planning and Operation of Electric Power Systems – in Portuguese –Universidade Federal Fluminense EDUFF, Niterói 1990.

OLIVEIRA, G.G., SOARES, S. 1995. A Second-Order Network Flow Algorithm for Hydrothermal Scheduling - IEEE Transactions on Power Systems, Vol.10, n.3, pp. 1635 – 1641.

ROSENTHAL, R. A 1981. Nonlinear Network Flow Algorithm for Maximization of Benefits in Hydroelectric Power Systems – Operation Research, vol. 29, n. 4, July-August.

VINHAL, C. N. 1998. A CASE System to the Long-Term Hydrothermal Scheduling of Hydroelectric Systems – in Portuguese. – FEEC/UNICAMP, December 1998.

Hydropower in the New Millennium, Honningsvåg et al (eds), © 2001 Taylor & Francis, ISBN 90 5809 195 3

Benefits of power exchange between hydropower and thermal power systems

P.B.Eriksen

Eltra1, Fredericia, Denmark

ABSTRACT: This paper investigates the consequences of future increased transmission capacity between the hydro-based power production system in Scandinavia and the thermal-based power production system in Europe with main focus on impacts on market prices and exchanges between the two systems. One important consequence is a significant reduction of the spot price volatility of power in Scandinavia. Reduced volatility, and thus stable prices, is beneficial to both investors and consumers.

In addition to this, the paper deals with reduction of CO_2 emissions from the integrated hydro-thermal production system with presentation of two alternative reduction strategies.

The paper is based on model simulations with the EMPS model.

1 BASIS FOR ANALYSIS

1.1 Introduction

Owing to technological differences the Scandinavian hydropower-dominated production system and the European thermal power-based generation system gain mutual benefits from integrated operation.

Due to its superior regulation capability hydropower benefits from exporting "expensive" power during peak hours and importing "inexpensive" power during off-peak periods. Furthermore, the hydropower system improves security of supply in dry years. Similarly, the thermal-based system benefits from a steady operation resulting in improved utilisation of fuel and reduced wear and tear of machines. The need for new thermal capacity is also reduced.

One important parameter, which may limit the integrated operation of the hydro-based system in Scandinavia and the thermal-based system in Europe, is the capacity of the interconnecting transmission lines. This paper investigates the consequences of future increased transmission capacity between the two generation systems – in particular the impacts on exchange and market prices.

Furthermore, the paper deals with reduction of CO_2 emissions from the integrated hydro-thermal production system and the following two alternative reduction strategies will be presented:
- a unilateral national CO_2 emission quota in one country (Denmark)
- an international harmonised approach involving a common CO_2 emission tariff

A comparison will be drawn between CO_2 reductions and distribution of national reductions of the two alternatives. Furthermore, the impacts on power spot prices are estimated and changes in profit for hydropower and thermal power producers are calculated.

1.2 The EMPS model

This paper is based on model simulations with the EMPS model (EFI's Multi-area Power Scheduling Model), which is an integrated model for market-based economic optimisation of hydro-thermal production systems – with the main focus on hydropower. The EMPS model was developed by Sintef Energy Research, Trondheim.

The EMPS model aims at optimal use of hydro resources in relation to uncertain future inflows, thermal generation, power demand and spot type transactions within or between areas.

In the model's strategy evaluation part regional decision tables in the form of incremental water values are computed for each of the areas in the system, using stochastic dynamic programming. A heuristic approach is used to treat the interaction between areas.

---

[1] Eltra is the independent transmission system operator in the western part of Denmark and is responsible for transmission and supply.

In the simulation part, optimal operational decisions are evaluated for a number of hydrological years (typically between 30 and 60). Hydro and thermal production are determined for each time step in a market clearing process based on water values for each aggregate regional sub-system. The aggregate hydropower production within each of these sub-systems is distributed among the available plants, using a rule-based reservoir drawdown model containing a description of all plants and reservoirs. See the listed references for a detailed description of the model.

1.3 *Model area and data input*

The model area includes Scandinavia (Denmark, Norway, Sweden, Finland) and Germany, the Netherlands and Poland, see figure 1.

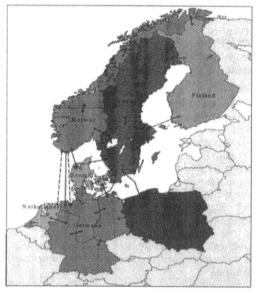

Figure 1. Model area with transmission lines (in 2005)

The data included in the model represent 2005 and have been gathered in co-operation with the independent system operators in Scandinavia.
The following fuel prices (2005) are anticipated:
- Coal: DKK 12.00 per GJ
- Fuel-oil: DKK 17.00 per GJ
- Natural gas: Spot price DKK 24.00 per GJ;
               in addition to this "take or pay
               contracts" at diversified prices form
               part of the fuel supply to thermal
               power plants.
- Light oil: DKK 24.00 per GJ
Price level is DKK, year 2000.
    The EMPS model is a chronological energy model with a time step of one week. The week has been subdivided into three accumulated load periods:

- weekly peak hours (weekdays during daytime:
  5 x 14 hours = 70 hours)
- off-peak hours (weekdays during night:
  5 x 10 hours = 50 hours)
- weekends (48 hours)
By using a time series of 41 hydrological years (1950-90) the stochastic variation of inflows to hydro storage and hydro plants in Norway, Sweden and Finland is simulated.
    It is anticipated that all transmission lines have 90 per cent availability.

2 INCREASED TRANSMISSION CAPACITY BETWEEN THE SCANDINAVIAN HYDRO-BASED SYSTEM AND THE EUROPEAN THERMAL-BASED SYSTEM

Today, transmission between the hydro-based system in Scandinavia and the thermal-based system in Europe is achieved via transit through Denmark (capacity max. 1800 MW) and by using d.c. links from Sweden to Poland (600 MW) and from Sweden to Germany (600 MW), respectively (see figure 1).
    To increase the interconnecting capacity the following two alternatives have been analysed with the EMPS model:
- one new 600 MW transmission line between Norway and Germany
- three new 600 MW transmission lines between Norway and Europe
Figure 2 shows the estimated load duration curves for the utilisation of the new transmission capacity: 600 MW and 3 x 600 MW = 1800 MW. The hours of both peak and off-peak period are depicted. The time record of simulation covers 41 hydrological years.

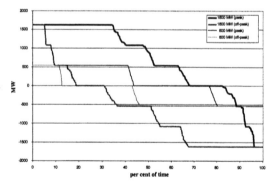

Figure 2. Peak and off-peak load duration curves (41 hydrological years) at transmission capacity of 600 MW and 3 x 600 MW between Norway and Europe (positive numbers represent export from Norway and negative numbers represent import to Norway.

The figure shows that in both cases energy is mainly transmitted from the hydro system to the thermal system during peak hours and then returned during off-peak hours (pump-storage scheduling).

The figure also indicates that, during dry years, import to the hydro-based system takes place in both peak and off-peak periods. Similarly, import to the thermal-based system takes place all day long during wet years.

Figure 2 shows that a transfer capacity of 600 MW will be fully utilised and thus produce bottleneck in about half the period. At 1800 MW the calculation shows fully utilisation and thus bottleneck in approx. 30 per cent of the period. Obviously, there seems to be a need for a transmission capacity of at least 1800 MW.

Figure 3 shows a direct comparison of estimated spot prices in Norway (south) and Germany (north) in case of the new transmission capacity of 600 MW and 3 x 600 MW, respectively. The curves represent the accumulated distribution functions of "spot price".

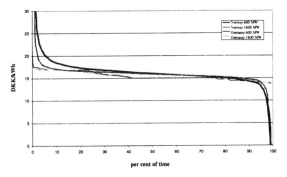

Figure 3. Estimated spot price in Norway (south) and Germany (north) at a transmission capacity of 600 MW and 3 x 600 MW during peak and off-peak periods.

The mean spot price and its standard deviation in Norway and Germany are depicted in figure 4.

From figure 4 it appears that the standard price deviation, and thus the volatility of the spot market, drops significantly (33 per cent) in Norway when increasing the transmission capacity. Moreover, in Norway a small reduction in the mean spot price is observed. The main physical explanation is that increased transmission capacity between the two power systems reduces the duration of periods with high prices – due to draught and periods with low prices in wet years.

In Europe (Germany), however, only small and insignificant reductions in price volatility are observed and the mean spot price remains unchanged. This is due to low transmission capacities compared with the market volumes in Europe.

This means that in particular the hydro-based system in Scandinavia will benefit from an extension of the interconnection. Reduced volatility and thus stable prices are beneficial to both investors and consumers.

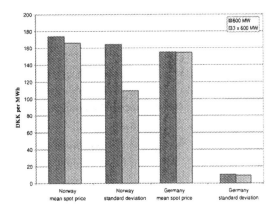

Figure 4. Mean and standard deviation of the spot price in Norway (south) and Germany (north) at a transmission capacity of 600 MW and 3 x 600 MW (mean of 41 hydrological years).

Regarding environmental consequences the analysis shows only minor and insignificant changes in CO_2, SO_2 and NO_x emissions (from thermal power production) when increasing the transmission capacity from 600 MW to 1800 MW – due to modest net exchange via the suggested new transmission lines.

3 REDUCTION OF CO_2 EMISSIONS FROM THE INTEGRATED HYDRO-THERMAL PRODUCTION SYSTEM

Reductions of CO_2 emissions from the integrated hydro-thermal production system have been studied through simulations with the EMPS model. The following two strategies have been investigated:
– A national Danish strategy including a unilateral Danish CO_2 tariff or an equivalent national CO_2 quota
– An international strategy involving a harmonised CO_2 tariff (equivalent to a system of allocation and quota trading)

Figure 5 shows the estimated Danish emissions (million tons of CO_2; 1000 tons of SO_2 and NO_x) and the net export (TWh) in case of a unilateral Danish CO_2 tariff of DKK 0, 40.00, 60.00 and 100.00 per ton of CO_2. It appears that a quota of e.g. 20 million tons of CO_2 corresponds to a CO_2 shadow price of DKK 60 per ton. At this level, the Danish net export to Germany and Scandinavia is estimated to be 7-8 TWh (the average of 41 hydrological years). Similarly, the SO_2 and NO_x emissions amount to 17,000 and 25,000 tons, respectively.

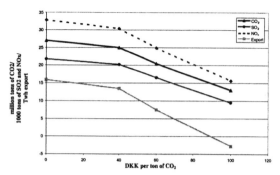

Figure 5. Emission and net export from Danish power plants as a function of a CO_2 shadow price.

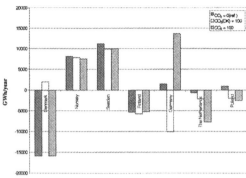

Figure 7. Import/export in one reference case and two alternatives (Import: positive numbers; Export: negative numbers).

Figure 6 shows the amount of CO_2 emission from power production in the model area in one reference case and two alternatives:
- A reference case without any CO_2 tariffs
- A unilateral CO_2 tariff of DKK 100.00 per ton in Denmark
- A harmonised CO_2 tariff of DKK 100.00 per ton in the entire model area

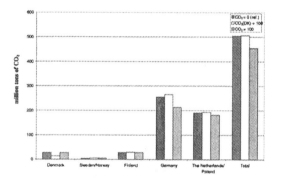

Figure 6. Amount of CO_2 emissions in one reference case and two alternatives.

Figure 7 outlines the import and export balance per country. In general, the assumed data set (2005) shows energy deficits in Norway and Sweden and the model calculations result in imports to Norway and Sweden amounting to 8 TWh and 10 TWh, respectively, on an average yearly basis. Increased demand and insufficient expansion of the production capacity cause the energy deficits in Norway. In Sweden the scheduled closure of nuclear plants (Barseback's reactor one and two) results in large import.

From figure 6 and 7 it appears that a unilateral Danish CO_2 tariff of DKK 100.00 per ton reduces the Danish CO_2 emission (from 27 million tons to 13 million tons). However, the total CO_2 emission from all countries indicates a slightly increasing trend

(from 504 million tons to 506 million tons) due to the increase in Germany and Poland.

With a common CO_2 tariff of DKK 100.00 per ton in all countries the total CO_2 emission is reduced from 504 million tons (reference case) to 454 million tons corresponding to a 10 per cent reduction. The reduction is mainly achieved by:
- scrapping of a part of the old coal-burning units in Germany and Poland and conversion of a part of the new coal-burning units to gas-burning units
- increased utilisation of gas-burning units in the Netherlands

Figure 8 shows the yearly average supply and demand of power covering the entire model area. The two supply curves correspond to the reference case without CO_2 tariff and the case with a harmonised CO_2 tariff of DKK 100.00 per ton in the entire model area.

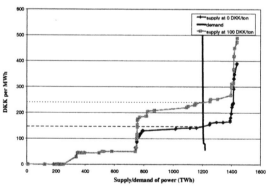

Figure 8. Supply and demand in the power market. Two tariff cases: DKK 0 and 100.00 per ton of CO_2 in the entire model area.

Both supply curves are based on the following production technologies:
  0-250 TWh: hydropower and wind power
  250-750 TWh: CHP (combined heat and

power) and nuclear power

750-1400 TWh: coal-, oil- and gas-burning
thermal power

1400- TWh: gas turbines

However, the two curves differ significantly in the interval above 750 TWh. This is due to different emission factors of CO_2 when burning coal, oil and gas and also due to differences in plant efficiency. Thus, if a tariff is charged a part of the old coal-burning plants are taken out of operation and replaced by oil- and gas-burning units. In addition, conversion of new coal-burning units to gas-burning units is taking place.

Figure 8 indicates that the overall average spot price increases from DKK 145 per MWh to DKK 240 per MWh. In both cases the area between the surplus curve and the dotted horizontal curve represents the producer surplus. As it appears the producers of wind, nuclear and hydropower (0-750 TWh) make large profits when the price increases.

Alternatively, the environmental effects of a harmonised CO_2 tariff could be achieved by implementing an international system of allocation and trading of CO_2 quotas. The two regulation systems are, in principle, equivalent when the tariff on one ton of CO_2 equals the market price of one ton of CO_2 quota.

Figure 9 illustrates the two mentioned regulation principles for CO_2: quota and tariff. Theoretically, it is possible to achieve the same CO_2 reduction by implementing the quota shown or by charging the stated tariff. When considering the financial position of a company, however, there is a considerable difference. In case of a tariff the polluter must pay the indicated additional tariff complying cost.

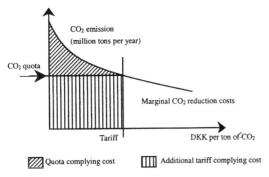

Figure 9. Two regulation alternatives: Quota and tariff.

Europe. The volatility of the spot market in Norway is reduced significantly by increased transmission capacity to Europe. By increasing the scheduled new transmission capacity from 600 MW to 3 x 600 MW the standard deviation of the market price is reduced by 33 per cent and the mean spot price in Norway is expected to decrease. In Europe, however, only a small and insignificant reduction of the price volatility is observed and the mean spot price remains unchanged. Thus it can be concluded that in particular the hydro-system in Scandinavia benefits from an extension of the interconnection. Reduced volatility and thus stable prices are beneficial to both investors and consumers.

Concerning the reduction of CO_2 emissions from the integrated hydro-thermal production system, the analysis shows only insignificant changes in CO_2, SO_2 and NO_x emissions (from thermal power production) when increasing the transmission capacity from 600 MW to 1800 MW. This is due to modest net exchange via the proposed new transmission lines.

When looking at the two reduction strategies, it is obvious that a unilateral Danish CO_2 tariff of DKK 100.00 per ton reduces the Danish CO_2 emission. However, the total CO_2 emission from all countries shows a slightly increasing trend due to an increase in Germany and Poland.

With a common CO_2 tariff of DKK 100.00 per ton in all countries, the total CO_2 emission is reduced by 10 per cent. The reduction is mainly achieved by scrapping a part of the old coal-burning units and by conversion of new coal-burning units to gas-burning units in Europe. Furthermore, the overall average spot price increases from DKK 145.00 per MWh to DKK 240.00 per MWh resulting in large profits to producers of wind, nuclear and hydropower.

REFERENCES

Botnen, O., Johannesen, A., Haugstad, A., Kroken, S. & Frøystein, O. 1992. *Modeling of hydropower scheduling in a national/international context*. Lillehammer, Norway

Http://www.energy.sintef.no/produkt/hydro-thermal/emps.htm

4 SUMMARY AND CONCLUSION

As it appears from the above, it will have consequences for market prices and exchange when increasing the transmission capacity between the hydro-based power production system in Scandinavia and the thermal-based system in

Hydropower in the New Millennium, Honningsvåg et al (eds), © 2001 Taylor & Francis, ISBN 90 5809 195 3

Optimal Control of Adjustable Speed Hydro Machines

T. Gjengedal

Statkraft SF, Høvik, Norway

ABSTRACT: This paper presents the Adjustable Speed Hydro (ASH) machine concept and the control aspects related to some of its applications. The ASH-machine benefits are partly related to the power plant itself and partly to the a better utilization of the power system. The fast control abilities of the ASH may increase the transfer capacity by a quick and precise regulation of rotor currents and thereby active and reactive power supplied from the machine. Hence, the ASH machine can give a considerable contribution to the network both during and after serious conditions in the power system.

1 INTRODUCTION

In established power networks, the tendency of increasing power consumption results in a need for new transmission corridors. This is normally not easily obtainable due to environmental and economical constraints. Therefore, the established network must often come into focus for alternative ways of increasing the power transmission capacity. At the same time, this must not violate the stability constraints associated with the network.

Optimised utilisation of power networks is partly the superior ambition of introducing privatisation, deregulation and competition in power markets worldwide. Competition has resulted in renewed utilisation of assets for generation, transmission and distribution with the aim of enhancing profitability and increasing owners value. This optimal utilisation can not be accepted without maintaining a secure operation of the network. Utilising the power system in a more optimal manner must also be followed by a strategy for maintaining the transient behaviour of the network in order to ensure that the stability is maintained within certain limits. Hence, efforts to improve control of existing transmission systems, as well as increasing the transmission network capacity and effectiveness, are becoming increasingly important. At the same time, the concerns for continued high levels of systems reliability, security and competitiveness will create uncertainties for system operators and planners. [1]

Also in the Norwegian power system, these constraints are being more apparent as the power consumption is increasing. From being an energy rated, regulated power system, it is now de-regulated and in the transition to be rated according to power demand. Increased power consumption in general combined with new HVDC interconnections being set into operation between Norway and Germany/Holland makes a paradigm shift with respect to the pattern of power flow within the Norwegian power system. This also puts more pressure on the transmission lines in accommodating the highly varying demand of power flow and on average, the main grid is loaded closer to its transmission capacity. As a result of these constraints appearing as bottlenecks in the power network, the Norwegian power system is now very often being separated into a number of local price areas.

The aim of this paper is to discuss the application of an Adjustable Speed Hydro (ASH) machine and how it can be controlled to give a better utilization of the resources and the power system.

2 BENEFITS OF ASH

The hydro power plants has until now - both turbine generators and pump storage plants been based on the application of single speed synchronous machines. A synchronous machine, being operated either in pump or turbine mode, will have its dedicated, constant mechanical speed; the synchronous speed set by the system frequency and the number of pole pairs. In hydro plants the overall best efficiency is in general obtained by the 'optimal' combination of speed, head and water flow, giving the «best point». Every change from this best point will give a reduction of the efficiency and thereby reducing the profitability for the owner of the power plant. Hence, in a single speed machine, the number of free parameters available for adjusting to the best efficiency are given by water discharge and head. In power stations with substantial variations in head, considerable changes in load and variable trickle of water, it is desirable to also change the machine speed in order to operate in an optimal way and thereby improve the operating flexibility and hence the revenue for the owner.

During the last decade, advances in power electronics have opened a new world in power generation and system control. Adjustable speed machines have been available for many years, but most applications were in the form of motor drives and were used to meet unique operating constraints imposed by mechanical loads. Although the theory of adjustable speed machines was known and understood, such machines were not applied to hydro generation. However, developments in power electronics now makes it possible to manufacture solid state power conversion and excitation equipment in sufficiently large ratings required for Adjustable Speed Hydro (ASH).

By regulating the machine speed with e.g. ±15% of nominal speed, an ASH generator may obtain optimal efficiency for a number of combinations of head and water flow. Adjustable speed machines allow mechanical speed to be set at the maximum efficiency point at all times. Adjustable speed machines have a broader operating range in the generating mode as compared to single speed machines. The broader operating range, together with fast acting power response gives this class of machine a better load following capability and decreased frequency fluctuations on the grid system.

Hydro plant operators will be able to fine tune turbine efficiency and track variations in head as reservoir levels or available flow changes. In pumped storage plants, adjustable speed machines will allow operators to adjust for optimum efficiency in both pumping and generating modes.

Application of ASH technology will provide benefits to hydro plant designers, plant owners and operators, and to the power system as well. Use of adjustable speed machines in conventional and pumped storage hydro plants will provide the plant operators with new operating choices and increased productivity. Major benefits for conventional hydro plants include

improved efficiency and increased operating hours, and increased capacity when required either due to system or market reasons. The ASH technology is particularly beneficial in pump stations, obtaining an optimal efficiency both in pump and turbine operation. Normally, the pump aggregate is constructed as to have its optimal working point in pump operation. This results in a relatively poor efficiency in generation mode. Hence, the profitability is increased correspondingly by installing an ASH unit.

ASH provides great operating flexibility for units that have to accommodate extreme head variations. By adjusting the speed – increasing and decreasing the speed with increased and decreased heads respectively – the hydraulically optimum operating point can be maintained over a larger head range and less than optimum, but acceptable operating characteristics can be achieved at extreme high or low operating heads

With an adjustable speed machine it is possible to overcome constraints imposed by the mechanical and hydraulically constraints and realize significant additional operating benefits on the electrical side. Thus, hydro electric plants with adjustable speed machines offer the generator owner and system operator an expanded range of operating capabilities while meeting traditional design requirements.

Adjustable Speed Hydro (ASH) technology benefits cover several engineering design disciplines, making assessment of the benefits difficult. Potential plant operating benefits achievable with the adjustable speed generating units primary include: increased efficiency, adjustable pumping load, improved load following capability, operation at reduced cavitation and vibration, fast starting time and improved stability by control of frequency and active power flow, voltage and reactive power support, reserve capacity, reduced dynamic braking and possible energy recovery.

When implementing ASH technology, very important concerns for the generating company are related to both plant and system performance and the improved economic benefits. Therefore, assessments which can quantify both the added plant-, operating and market - as well as added system - economic benefits must be made. Important issues are quantifying the benefits of undertaking investment selection which incorporates not only the added energy and MW-capacity, operating flexibility in steady state performance, but also the dynamic performance as impacted by power system operation. All investment decisions are made in order to increase the investors revenues. Therefore, investing in new technologies such as ASH, requires an added value as compared to traditional technology. It is clear, however, that in order to perform this credible cost-benefit analysis, upon which the decisions related to where the implementation of the ASH devices can be made, one needs a) to identify prospective locations where ASH devices are likely to increase the plant-, operating - and hence the market access and owner benefits and revenues significantly, and b) to evaluate their added value due to improved system performance under both normal and emergency situations.

3 PRINCIPLES OF THE ASH MACHINE CONCEPT

A synchronous machine, being operated either in pump or turbine mode, will have its dedicated, constant mechanical speed; the synchronous speed. The only way of changing the active power production is to change the mechanical torque delivered by the turbine.

In power stations with considerable variations in head and load as well as variable trickle of water, it is desirable to make a change in machine speed in order to operate in an optimal way and thereby increase the income for the owner. By varying the speed within a certain range around nominal speed, an ASH machine may obtain optimal efficiency for a number of combinations of head and water flow.

By injecting definite currents with variable frequency into the rotor (ref. figure 1), the generation of active and reactive power from the ASH machine is controlled continuously and rapidly. The active power from the generator can be controlled independently from the turbine output both in pump and generator mode by means of the active part of the rotor current.

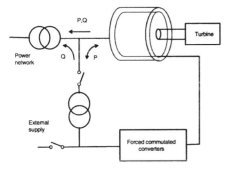

Figure 1: Example of an internal arrangement within an ASH machine unit

After large perturbations in the transmission grid, the frequency will normally oscillate with the speed of the synchronous generators. By changing the frequency of the rotor current, an ASH machine can keep the frequency of the current delivered to the transmission system at its reference value. Depending on the nature of the disturbance and the ASH machine's location in the network, it can either absorb or deliver active power to the network very quickly by transforming rotational energy to electric energy and vice versa.

These currents are fed from a converter into the doubly fed wound rotor over some dedicated slip rings. The stator frequency is maintained at 50Hz by means of the rotor currents, which are compensating for the difference between actual speed and nominal speed.

In a conventional hydro generator, the gate opening controls the active power generated. This regulation has a time constant in the order of seconds, given by the shape of the production.

For an ASH generator, both the active and reactive power production can be changed very rapidly, about the order from 10 to 30 ms. This is possible due to the ability to change the rotor current frequency very fast. By this, the active power supplied or drawn from the generator terminals may change quickly without affecting the network frequency.

4 MODEL DESCRIPTION

The ASH machine model is based on the standard equations for a two-axis reference system. For studies of transient stability, the converter model can be simplified. In this paper, the dynamics of the converter is neglected, assuming that the converter is able to control the rotor current to the reference values at all times.

The rotor current regulators of the ASH machine model are built up based on phasor domain analysis, where the active and reactive power delivered to the network are kept track of by means of the two rotor current phasors $IR\alpha$ and $IR\beta$. This orientation system is selected based on the alignment to the d-axis component of the stator . This is explained in detail in the following.

In the ordinary dq-reference frame, the ordinary machine equations are given by the following expressions :

$$v_{sd} = r_s i_{sd} + \frac{1}{\omega_n} \frac{d\psi_{sd}}{dt} - \frac{\omega}{\omega_n} \psi_{sq} \qquad (1)$$

$$v_{sq} = r_s i_{sq} + \frac{1}{\omega_n} \frac{d\psi_{sq}}{dt} + \frac{\omega}{\omega_n} \psi_{sd} \qquad (2)$$

$$v_{rd} = r_r i_{rd} + \frac{1}{\omega_n} \frac{d\psi_{rd}}{dt} - \frac{\omega}{\omega_n} \psi_{rq} \qquad (3)$$

$$v_{rq} = r_r i_{rq} + \frac{1}{\omega_n} \frac{d\psi_{rq}}{dt} + \frac{\omega}{\omega_n} \psi_{rd} \qquad (4)$$

Active and reactive power in stator are expressed by

$$p = \mathrm{Re}(u \cdot i^*) = (u_{sd} i_{sd} + u_{sq} i_{sq}) \qquad (5)$$

$$q = \mathrm{Im}(u \cdot i^*) = (u_{sq} i_{sd} - u_{sd} i_{sq}) \qquad (6)$$

By assuming stationary conditions, the flux derivatives are equal to zero. The stator resistance is usually very small and as an approximation, it can be ignored. Furthermore, by introducing a new reference frame aligned along the stator flux vector (annotation α-β), the following applies :

$$\Psi_{s\alpha} = \Psi_s \; , \Psi_{s\beta} = (x_s i_{s\beta} + x_h i_{r\beta}) = 0 \qquad (7)$$

$$v_{s\alpha} = r_s i_{s\alpha} \approx 0 \qquad (8)$$

$$v_{s\beta} = r_s i_{s\beta} + \frac{\omega}{\omega_n} \psi_{s\beta} \approx \frac{\omega}{\omega_n} \psi_{s\beta} \qquad (9)$$

Active and reactive power in stator can now be expressed by

$$p_s = u_s i_{s\beta} \; , \; q = u_s i_{s\alpha} \qquad (10)$$

Based on equation (7), the active power in stator can be expressed as a linear function of the rotor current component irβ, i.e.

$$p_s = u_s \left(-\frac{x_h}{x_s} \right) \cdot i_{r\beta} \qquad (11)$$

Correspondingly the expression for reactive power is deduced to

$$q_s = u_s \left(\frac{u_s - i_{r\alpha} \cdot \omega \cdot x_h}{(x_h + x_{s\sigma}) \cdot \omega} \right) \qquad (12)$$

where $x_{s\sigma}$ is the stator leakage reactance.

Hence, the stator active and reactive power are individually controlled by means of the two rotor current components IRα and IRβ, without any considerable loss of accuracy. The converter is supposed to be a Voltage Source Inverter (VSI), controlling the rotor voltage components to obtain the correct rotor currents.

The rotor current regulators of the ASH machine model are built up based on phasor domain analysis, where the active and reactive power delivered to the network are kept track of by means of the two rotor current phasors IRα and IRβ. This orientation system is selected based on the alignment to the d-axis component of the stator flux.

The relatively large speed range of an ASH machine makes it necessary to use accurate models of the hydraulic system. A non-linear model of the turbine assuming inelastic water column is suggested for transient stability studies in [5]. In [6] a more detailed model based on these recommendations is suggested. In [7] and [8], two control strategies are presented. The model presented in our paper is implemented with the power master strategy, which allows the active power output to be controlled independently of mechanical speed. Figure 2 shows a block schematic

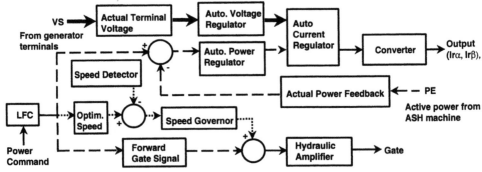

Figure 2: Regulation structure within the ASH machine model
Long dash : Signals associated with active power control. Short dash : Speed control. Bold :
Voltage control

of the regulation algorithm implemented in the model.

The control signals for active and reactive power based on terminal voltage (VS) and active power (PE) measurements are given as input to the auto current regulator.

To control the speed, a common PID controller with a first order filter is used. The speed control can not be so fast that it impairs the response of the active power output. This means that the gain, both proportional and integral have to be set at low values to stabilise the system.

ASH-machine represents generator G2, respectively. The cases described in this paper are a comparison of the phase angle of generator G1 as related to that of generator G3 by using synchronous versus ASH machine for generator G2. This is done to illustrate the impact of the ASH machine with respect to the stability margin.

The question is: How could these properties of the ASH machine be utilised in ordinary transmission networks? In order to illustrate this, investigations have been carried to different case studies as described below..

The two 700 MVA generators are producing near the maximum power output. This is done intentionally

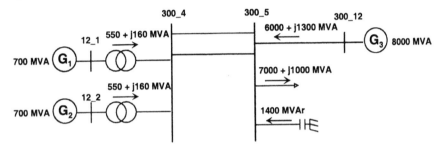

Figure 3: Power system model

5 POWER SYSTEM MODEL

The network model used in this paper is given schematic in figure 3. This is a simplified model of a transmission system used to show the principle behaviour of the ASH machine as related to conventional machines. The two generators G1 and G2 are rated 700 MVA and they are connected to the load point (300_5) via transmission lines. Generator G3 represents the auxiliary network.

A fifth order synchronous machine model is used to represent the generators applied in the network model. The synchronous machines are associated with a general conventional governor, which is measuring the speed deviation, while the ASH machine takes into account the deviation of power production related to the power reference.

The ASH machine or the respective synchronous machines is not equipped with Power System Stabilisers (PSS) since this study concerned about transient stability.

6 SIMULATION RESULTS

In the following, attention will be paid to some simulation cases where a synchronous machine or an

to illustrate the impact of the ASH machine on the stability margins.

The initial load flow is the same for both cases as given in figure 3.

6.1 Case 1: Maximum fault clearing time for synchronous generators

For the conventional machine case, the fault period is successively increased until the network falls out of synchronism. This is obtained for a fault period equal to 250ms. The simulation results presented below apply to a fault period marginally less than the limit found (i.e. 240ms). A corresponding simulation is arranged for generator G2 being an ASH machine. In figure 4, the phase angle difference between the generators G1 and G3 is shown for generator G2 being a synchronous machine and an ASH machine, respectively.

With only conventional machines included in the network, the difference in pole angle increases continuously within the fault period (240ms). When the fault is removed, the angle difference is –0.26 radians, increasing to –0.4 radians before the angle difference turns in positive direction. In this phase, the most critical amplitude occurs, being equal to +1.6 radians.

By introducing the ASH machine for generator G2,

Synchronous machine

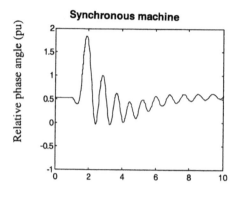

ASH machine

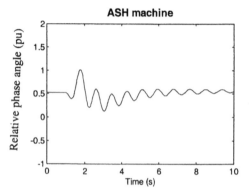

Figure 4: Difference in pole angle between generators G1 and G3 with generator G2 being a conventional machine (upper curve) and an ASH machine (lower curve).

the maximum angle difference reduces to 1.0 radians. The angle difference is turned in positive direction near instantaneously. When the fault is removed (240ms), the difference is near zero radians. The most critical amplitude occurs in this first swing, being effectively damped in the following oscillations. This result indicates that the ASH machine increases the stability margin and that more active power could be transferred without coming out of phase in serious fault situations.

6.2 Case 2: Maximum fault clearing time for ASH machine

As to examine the stability margin of the ASH machine in this network, the fault period is successively increased until the generators get out of phase. The corresponding maximum trip time is found to be 500ms. . The respective curve for the difference angle between generators G1 and G3 is shown in figure 5. The large maximum amplitude is due to the increased fault period. Also in this case, the most criti-

cal amplitude occurs in the first swing, being effectively damped in the following oscillations.

ASH machine

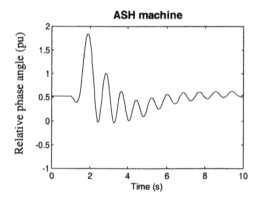

Figure 5: Difference in pole angle between generators G1 and G3 with G2 being an ASH machine. Maximum tripping time is 500 msec.

7 CONCLUSION

In this paper, a direct comparison has been done between an ASH machine and a conventional synchronous machine. The simulation results show that the ASH machine is able to maintain the torque equilibrium even during serious failures. Using a fixed active power reference, the rotor current controller fine-tunes the current to give the correct power output from the machine. In transient situations, this is obtained by extracting power from the rotor (rotating energy).

The simulations have shown that by replacing conventional machines with ASH machines, an improved utilisation of the transmission system may be obtained. Restrictions to power transfer due to stability problems in the network may be circumvented by installing this promising technology.

8 REFERENCES

[1] T. Gjengedal, J.O. Gjerde, R. Fløo: *"Ancillary Services and Transmission System Operation in a Deregulated Market"*, ICEE '96, Beijing 1996

[2] N.G. Hingorani, *"Flexible AC transmission"*, IEEE Spectrum Vol. 30, April 1993, pp. 40. - 45.

[3] E.V. Larsen, J.J. Sanchez-Gasca, J.H. Chow, *"Concepts for design of FACTS controllers to damp power swings"*, Paper 94 SM 532-2 PWRS, presented at the 1994 IEEE/PES Summer Meeting, San Francisco, 1994.

[4] R. Fløo, J. O. Gjerde, T. Gjengedal, *"Selection of optimal TCSC damping control*

parameters and placement by use of eigenvalue and sensitivity analysis methods", Technical note, Norwegian University of Science and Technology, Trondheim, 1995.

[5] Kundur, "Power System Stability and Control", McGraw-Hill, 1993, pp.387-393.

[6] B.H. Bakken, "Technical and Economic Aspects of Operation of Thermal and Hydro Power Systems", Dr.ing.thesis, Norwegian University of Science and Technology, Trondheim, 1997.

[7] T.Kuwabara, A.Shibuya, H. Furuta, E.Kita, K.Mitsuhashi, "Design and Dynamic Response Characteristics of 400MW Adjustable Speed Pumped Storage Unit for Ohkawachi Power Station", IEEE Transactions on Energy

[7] Conversion, Vol. 11, No. 2, June 1996. Goto, Shibuya, Inoue, Ishizaki, Tezuka, "Power System Stabilising Control by Adjustable Speed Pumped Storage Power Station using Stabilising Signals", Cigre Symposium Tokyo 1995, Paper No. 510-01.

[8] J.O.Gjerde, R.Sporild, T.Gjengedal, R.Flølo, A.Elstrøm, "Enhancement of Power System Stability in a Deregulated Environment - ASD-Machines a Promising Alternative", PowerCon'98 Proceedings pp. 1393-1397, Beijing, August 1998.

73

Kulekhani reservoir: our operational experience

Rajendra Prasad Hada
Nepal Electricity Authority, Nepal

ABSTRACT: Kulekhani reservoir supplies the power discharge to Nepal's only peaking cascade hydro- power system with installed capacity of 92 MW which is about 30 % of system load. The catchment was supposed to be fed by sediment free inflow of dry season's average of 2.1 cumecs with annual rainfall of 1,500 mm, however in 1993, the catchment was struck by unprecedented storm rainfall causing huge sediment inflow endangering the life of reservoir. The papers briefs about the sediment handling technique in Nepal's only man-made reservoir.

1 INTRODUCTION

Nepal's only man-made 7 Km long reservoir has been created by construction of 406 meter long, 114 meter high rock fill dam in Kulekhani River at Kulekhani Gaon (Village). The reservoir is locally named as Indra Sarovar. This reservoir collects the water during the rainy season and utilizes during the dry season through cascade system of Kulekhani No. 1 Power Station (60 MW) and Kulekhani No. 2 Power Station (32 MW). These two power stations with total installed capacity of 92 MW shares about 30% hydropower system capacity in Nepal and contributes greatly to the peak power demand of the electricity supply in the country. The location of the reservoir and power stations are shown in Figure 1. Location Map and the salient features of the existing structures are detailed in Annex – A (1 – Kulekhani No. 1 Power Station, 2 – Sediment Mitigating Structures, 3 – Kulekhani No. 2 Power Station).

2 RESERVOIR OPERATION :

The natural run off of the Kulekhani river at the dam site is as small as 2.1 cubic meter per second (cumecs) in the dry season while it is 6.2 cubic meter per second in the wet season. Thus the runoff in the wet season is being stored in the reservoir for the power generation in the dry season. The standard storage curve is shown in Figure 2. Reservoir Capacity Curve.

2.1 *Normal Operation :*

The impounding of reservoir started from October 23, 1981, since then, the reservoir has been operating as per design standard rule curve for Eleven (11) hydrological year. The standard Reservoir Level and the designed generation for each month is shown in Table 1. The total designed energy generation per year is 165,500 MWh from Kulekhani No. 1.

Table 1. Standard Level and Generation

Month	Level m.	Generation (MWh)
January	1517.5	21,500
February	1510.8	20,000
March	1501.9	20,000
April	1491.1	20,500
May	1476.0	20,500
June	1480.3	13,500
July	1495.9	7,000
August	1507.1	7,000
September	1515.9	7,000
October	1523.6	7,000
November	1530.0	7,000
December	1524.1	14,500
Total		165,500

During this period, the minimum and maximum reservoir level has been recorded as shown in Table 2 which shows the reservoir experienced maximum level rise of about 8 meter in one day (within 24 hr.) in July 17, 1989. However no danger occurred to the infrastructure in the Kulekhani Reservoir.

The spilling operation through ungated spillway as well as gated spillway was within the standard normal rules and no technical challenges have been encountered during this period. However the Dam

Table 2. Yearly Extreme Reservoir Level

Year	Minimum Reservoir Level		Maximum Reservoir Level		Remarks
	Level (EL.)	Date	Level (EL.)	Date	
1983	1,483.60	Jun. 26	1,530.07	Nov. 5	
1984	1,504.67	Jun. 14	1,526.94	Nov. 25	
1985	1,513.41	May. 18	1,530.20	Oct. 10	
1986	1,501.50	Jun. 23	1,530.21	Sep. 27	
1987	1,483.82	Jul. 02	1,529.31	Nov. 1	
1988	1,488.01	Jun. 10	1,518.95	Sep. 24	
1989	1,478.54	Jun. 30	1,506.33	Oct. 14	
1990	1,484.45	May. 12	1,530.21	Sep. 25	
1991	1,484.11	May. 21	1,517.85	Oct. 29	
1992	1,481.14	Apr. 29	1,502.66	Nov. 02	
1993	1,479.43	Apr. 14	1,530.08	Nov. 16	1993 Flood
1994	1,489.40	May. 30	1,520.69	Nov. 14	
1995	1,484.50	Apr. 21	1,530.47	Nov. 15	
1996	1,493.03	Jun. 22	1,528.25	Oct. 27	
1997	1,472.50	May. 15	1,517.52	Sep. 14	S.I. Construction
1998	1,486.71	Jun. 20	1,530.05	Aug. 30	
1999	1,483.50	Jun. 10	1,530.45	Jul. 30	
2000	1,484.30	Jun. 12	1,530.46	Sep. 22	

Break Study has been conducted by Norwegian Consultant in 1993 (April), to access the maximum level of inundation downstream and its disaster management plan for evacuation. The recommendation are indicated in Annex – B.

2.2 Extreme Passage of Flood :

On 19th and 20th July 1993, an unprecedented storm and flood hit Nepal including the Kulekhani Project area. An extreme 1 day rainfall of 540 mm was observed on July 19, 1993 at Tistung located in the catchment area of Kulekhani Dam.

In general, the flood operation team, stationed at Dam house starts functioning only after the reservoir level reaches EL. 1525.00. However, On July 19th 1993, the maintenance team was maintaining the one of the gate by lifting full (i.e. 11.0 m open) and they stopped the work for that day at 6.00 PM and at that time the reservoir level was 1498.63 m.

The storm on that night was such a furious that the reservoir water level rose upto 1522.05 even though one of the gate was fully opened.

Average basin rainfall on this day was estimated at 395 mm. The return period for this rainfall corresponds to 1000 years for rainfall dates of 1968 to 1990. However, the same storm rainfall over the Kulekhani Water shed corresponds to about 90 years return period, for further flood analysis.
The peak flood was calculated to be 1,340 cubic meters at 23 hr. on July 19th by unit hydro-graph approach.

No structural damages was noted in reservoir area, however the flood caused whole reservoir surface coloured indicating enormous amount of sediment inflow. This 1993 flood washed way the penstock line of about 100 meter in Jurikhet River and completely damaged the headwork for Kulekhani No. 2 Power Stations which results the generation from Kulekhani Power Stations were stopped for 5.0 months (July 19, 1993 – December 23/93).

As the total length of reservoir was coloured during the flood. The sedimentation survey was conducted in December 1993 indicates about 5 million cubic meter of sediment inflowed into one season. To sustain the life of reservoir, a Master Plan Study on Sediment Control in Kulekhani Watershed was conducted in November 1994.

THIS REPORTS REVEALS :

a. 32 m Rise of river bed in front of existing bell mouth intake indicating only 12 meter depth is remaining.

b. 1993 flood deposited 4.8 Million Cubic meter (Mm^3) of sediment into the reservoir while 10.5 Mm^3 of sediment came into the reservoir in subsequent year.

c. The loss of 5.1 Mm3 of dead volume out of 12.0 Mm3, thus only 6.9 Mm3 of dead volume is remaining in front of intake EL. at 1471.0 m.

d. If no counter measure for sediment control is carried out, the average sediment inflow to dead volume is remain only 21 years and if flood of 1993 scale comes, that would be only 12 years,.

THIS REPORT RECOMMENDS

1 Sloping Intake to reduce the risk of intake closure. The sloping intake will allow for power generation even after the dead storage is full of sediment and the life of the storage will be more than 50 years, as the intake level can be raised by inserting the stoplog.

2. To mitigate the sediment deposition in Kulekhani Reservoir, the Check Dams on upstream of reservoir as Check Dam No. 5 in Palung Khola and Check Dam No. D-0 in Thado Khola.
Further to this study, the Japanese International Cooperation Agency (JICA) in March 1997 conducted the study on the Disaster Prevention Plan for severally affected areas by 1993 disaster in the Central Development Region of Nepal including Kulekhani Watershed area. This report recommends following three approaches to sustain the life of Kulekhani reservoir. The basic concept is indicated in Annex – B.

a. Sloping Intake Approach
b. Sand Resources Development Approach
c. Integrated Watershed Management Approach

2.3 *Abnormal Reservoir Operation during Sloping Intake Construction :*

The Normal Operation of Reservoir is between EL. 1476 (Dry Season) and EL. 1531.00. (During Peak Flood) The bottom elevation of existing bell mouth intake was in EL. 1471.0 m. In order to convert existing intake into sloping intake, the reservoir level has to be drawn down upto EL. 1471 and keep constant till the bottom block of the Sloping Intake was Constructed.

The construction activities of the Sloping Intake comprises of the following:
- excavation
- Anchor Bars
- Top Structures
- Bottom Structures
- Cofferdam
- Closing Existing Inlet Structure
- Demolition of existing intake structure
  - Coffer Wall inside inlet structure
  - Installation of Drainage pipes and valves
  - Installation steel liner
  - Concrete work
- Valve Closing
- Installation of Trash racks

The construction of sloping intake has been executed from top to the bottom as the reservoir water level is lowered as per the Reservoir Lowering schedule during the dry season (November to June). Construction of the bottom portion needs the abnormal operation of the reservoir. The reservoir level has been kept within allowable range by generating power from the released water by drainage pipes and valves. The general Artist view of sloping intake and installed drainage facilities during construction has been shown in Figure 3. Sloping Intake

2.4 *Post Sloping Intake Operation:*

Upon the completion of Sloping Intake, the lowest power draw down level (sill) is EL. 1480.069 m. During non flooding period, the ungated spillway level at EL. 1530.0 m with 65 m long has been temporarily heighten by 50 cm using sand bags (Photo – 1) to store about one million cubic meter of water. This height does not affect the behavior of rock fill dam. Though the bed level of reservoir has changed and its storage capacity has been decreased due to sediment, the energy generation is as per system requirement and there is no significant changes in power generation as could be indicated from Table 3.

Table 3.   Actual Generation (Long term)

Year	Generation (MWh)	Remarks
1996/97	167,981	Before S.I.
1997/98	121,571	During Construction
1998/99	195,737	After S.I.
99/2000	225,000	after S.I.

It has been noted that due to sedimentation as well as raising of intake level from the EL. 1471 to 1480, the usable capacity of the reservoir has been reduced, however with the installation of Telemetering System comprises of three (3) rain gauge stations in its catchment area and reservoir level transmits in hourly basis, upto 10 mm), the utilization of Kulekhani reservoir water is efficiently used. This has been verified by last year's generation which is more than 165,500 MWh. The revised capacity has been shown in Table 4. Reservoir Storage Volume and indicated in Figure 2.

Table 4.   Reservoir Storage Volume

Elevation (meter)	Original (Mil. m^3)	Revised (Mil. m^3)
1427	0	0
1430	0.045	0
1440	0.525	0
1450	1.775	0
1460	4.205	0.086
1470	8.465	1.522
1480	14.855	5.167
1490	23.345	11.150
1500	34.065	19.330
1510	47.735	30.412
1520	64.915	45.283
1530	85.285	63.225

3 CONCLUSION:

The Central Nepal Grid System Consists of 1,370 Km long 132 Kv, 332 Km long 66 Kv and 1,506 Km long 33 Kv Transmission line with total installed capacity of 390 MW.

The role of Kulekhani Power Stations are vital as this is the only peaking hydropower system in the grid which is able to supply power during the dry season from December to May in each year. To sustain the reservoir life for power generation, the following approaches are now being utilized:

- To handle the sedimentation of the reservoir, the sloping intake approach has been implemented.

Based on the annual sedimentation survey of the reservoir, the insertion of stoplog is being decided with the following condition:

a. If dead volume below EL. 1480 > 5.0 Million cubic meter.
No stoplog is inserted for that season.

b. If dead volume below EL. 1480 < 5.0 Million cubic meter.
Stoplog is/are inserted to make dead volume equal to 5 million cubic meter for the normal flood.

c. If the flood of 1993 scale is observed, then the Stoplog is/are inserted to make dead volume equal to 12 million cubic meter.

- To mitigate the reservoir sedimentation, at the upstream reaches of the reservoir, the Check Dams No. 1 and No. 5 have been constructed so that the deposited sand are being removed and being utilized as construction material upto Kathmandu valley.

- Watershed Management is being implemented by the Department of Soil Conservation through a project named Bagmati Integrated Watershed management Project financed by European Community.

ANNEX - A
Salient Feature

1 KULEKHANI NO. 1 POWER STATION

Reservoir
- Catchment Area :	126 Sq.Km.
- Total Length :	7.0 Km.
- Maximum Reservoir Water Level :	1,530.0 above mean sea level
- Minimum Operating Level :	1,483.0 above mean seal level

Dam and Spillway
- Type :	Zoned Rockfill with inclined Core
- Dam Height :	114.0 meter
- Crest Length :	406.0 meter
- Crest Width :	10.0 meter
- Embankment Volume :	4.4 Million Cubic meter
- Two Radial Gates (11.5 X 9.0 meter) (with Sill Level at EL. 1519.0	
- Ungated Spillway at EL. 1530 masl	65.0 meter long

Power Intake Structure
- Bell mouth Intake converted to Intake :	98.0 meter
- 2.5 m. Circular Tunnel :	6233 m. long
- 3.0 m. Circular Surge Tank :	92 m. high
- Penstock :	1,356 m. long with diameter Varying diameter from 2 to 1.5 m.

- Underground Power house dimension	
- Height	30.4 meter
- Width	15.5 meter
- Length	49.0 meter
- Power Output	

- Installed Capacity	30X2 Megawatt meter
- Turbine Type	Pelton
- Transmission Line	
- Transmission Voltage	66.0 Kilo Volt
- Length	42.0 Kilo Meter

Stoplog
- Dimension :	0.8 m thick X 1.5 high X 3.2 m wide
- Weight :	10.0 ton/piece
- Number of pieces :	25.0 upto EL. 1501.7 m

Stoplog Handling Devices
- Lifting Beam Assembly	Equiped with two hooks jointly connected through a linkage mechanism to catch and release automatically the lifting bracket of each concrete stoplog
- Winch and Control Room	Motorized low speed type with lowering and raising speed not exceeding 1 m/sec.

2 SEDIMENT MITIGATING STRUCTURES

Check Dam No. 1 (at Upstream Edge of Reservoir)
- Dam Type :	Concrete Gravity
- Dam Height :	14.0 meter
- Dam Crest Length :	98.5 meter
- Width of Dam Crest :	2.0 meter
- Crest Elevation :	1,533.4 masl
- Foundation Elevation :	1,514.5 masl

Check Dam No. 5 (Palung River)
- Dam Type :	Concrete Gravity
- Dam Height :	9.0 meter
- Dam Crest Length :	89.9 meter
- Crest Elevation :	1,748 masl

3 KULEKHANI NO. 2 POWER STATION

Intake Structure :
- Diversion Weir :	36 m. long 12 m. high concrete weir
- Embedded Headpond :	1800 cubic meter storage with side spillway at EL. 911.0 m
- Embedded Mandu Intake :	0.2 cumecs
- Pump Rapti Intake :	0.2 cumecs

Waterways :
- Headrace Tunnel :	5848 m. long X 2.5 diameter circular
- Surge Tank	3 m. diameter, 44 m. high with upper chamber (45 m long X 5.5 m high X 5m wide)
- Penstock Line	843 m long with varying diameter from 2.1 m to 1.5 m.

Powerhouse and Generation :
- Power Output :	
- Installed Capacity :	2 X 16 MW
- Turbine Type :	Francis
- Transmission Line :	
- Transmission Voltage	132.0 Kilo Volt
- Length	42.0 Kilo Meter

ANNEX - B

1 Recommendation of Dam Break Study

The effects of a possible dam break may be mitigated by assessing the flood hazards and vulnerability of the downstream areas, and preparing proper plans for evacuation of people living in the vulnerable areas.

The Following should be initiated in addition to the installation of warning system:
– Assessment of Vulnerability shall be done to enable the preparation of scenarios of possible flood emergencies
– Evacuation Plans
  It will be necessary to identity the possible escape routes, check their state of repair and remove any obstacles to the rapid movement of evacuees into the refugees zone.
– Training for evacuation
  Training for evacuation shall be made through simulated emergency exercise which shall help to reveal any deficiencies in the plan and remedy them. Leaflet and other printed information shall be distributed to the local authorities and general public.
– Microzoning Studies and Physical Planning
  Criteria shall be formulated for the selection of safer area for important installations and industrial works.
– Alternative ways to supply power in the system
  A study shall be undertaken to establish ways of providing the National Grid with electricity in case of a prolonged power cut.

2 Recommendation of JICA Study

a. Sloping Intake Approach
The main objective of the construction of sloping intake is to prevent its clogging due to sediment. Since the location of the intake in the reservoir is EL. 1471 m. and the sediment level has reached upto EL. 1460 m. The countermeasure which have been constructed, will make it possible to raise the intake water level, then sediment is prevented from entering the water way of the hydropower structures. The basic principle is as shown in Figure

b. Sand Resource Development Approach
Considering the remarkable amount of remaining sediment in along the river course as well as on the slope, sediment inflow into the reservoir will continue and the accumulation will affect the function of reservoir gradually. In order mitigate such sediment inflow, the sand resources development approach introduces the excavation of sand from the effective

storage of the reservoir and to transport and sell the excavated material as construction material in the local vicinity and in Kathmandu Market. The main objective of sand resources approach is to sustain the function of Kulekhani reservoir of which main role is to guarantee peak power operation in dry season.

c. Integrated Watershed Management Approach.
Integrated watershed management is a very complicated approach which are difficult to be defined clearly. In case of Kulekhani watershed, however clear objectives which are difficult to be defined clearly. In the case of Kulekhani watershed, however clear objectives can be identified as to mitigate sediment yield in the watershed and to encourage people to participate in the watershed management activities including income generating activities. The Department of Soil Conservation has already implementing a project called Bagmati Integrated Watershed Management Project under the financing of European Union.

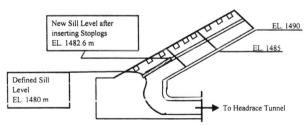

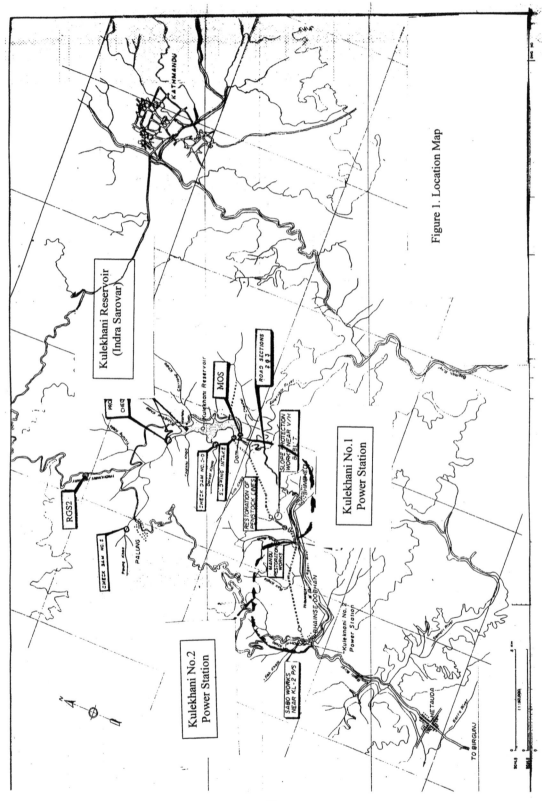

Figure 1. Location Map

80

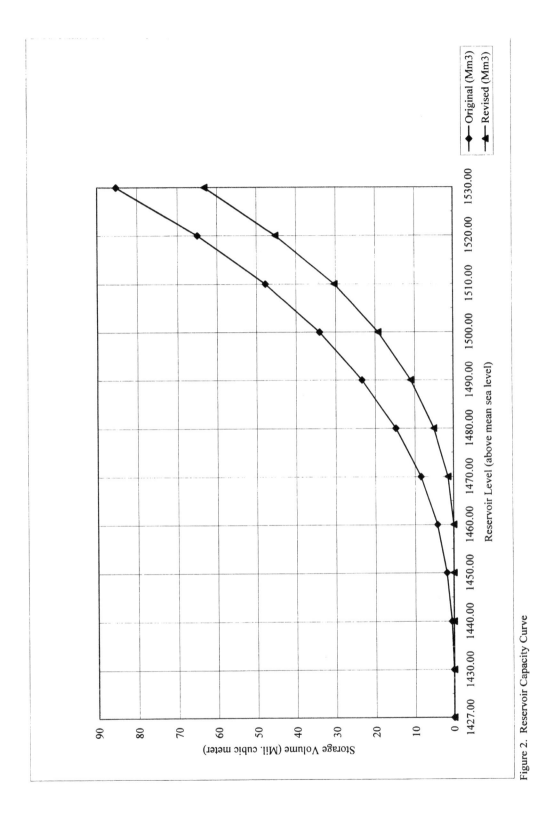

Figure 2. Reservoir Capacity Curve

81

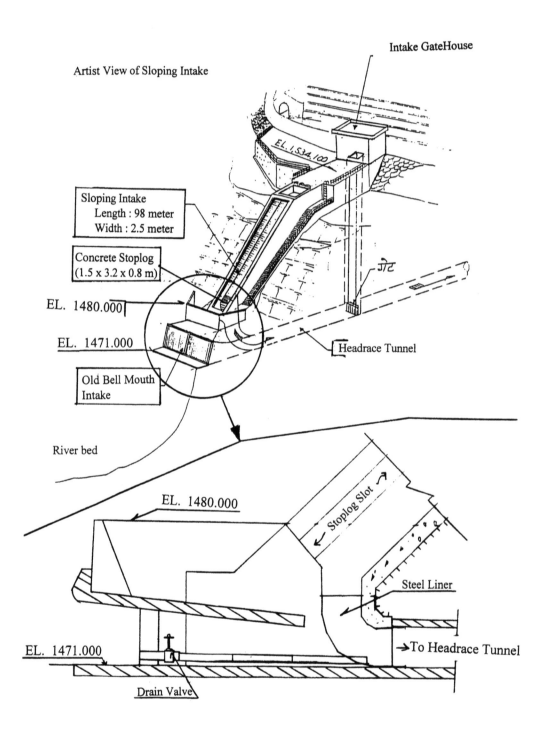

Figure 3   Sloping Intake with Constructional Drainage System

Photo - 1. Placing of Sand Bag in Ungated Spillway
to raise reservoir water by 50 cm from EL. 1530.0
(Full Level)

Scenario Techniques for Long-Term Hydrothermal Scheduling

L.Martinez & S.Soares

Unicamp – Electrical Engineering Faculty, P.O. Box 6101-13083-970 Campinas, SP, Brazil

ABSTRACT: This paper presents a performance analysis of different scenario techniques in the long-term hydrothermal scheduling. In these approaches, scenarios are used to represent the uncertainty of the stochastic parameters of the problem and decision-making policies are established based on deterministic optimization models. The stochastic parameters of the hydrothermal scheduling problem, represented here by scenarios are the water inflows. Two techniques were used: the unique scenario technique, represented by the most probable future inflow sequence, and the multiple scenario technique, where various possible future scenarios are considered, rather than been limited to the most probable. Different forecasting models were used for the determination of the most probable scenario. For the case of multiple scenarios, historic inflow data were used. The approaches were compared using simulations based on inflow records. The hydroelectric system studied is composed of the hydro plants on the Grande River, in the Southeast Brazilian system, with 7188 MW of total installed power capacity. The results obtained using scenario techniques considering a single reservoir were compared with those of the stochastic dynamic programming approach. The conclusions indicate that a unique scenario with appropriate water inflow forecasting and terminal conditions should be more efficient than the multiple scenarios technique or the stochastic dynamic programming approach.

1 INTRODUCTION

Hydrothermal scheduling is a complex problem due to various aspects of modeling, including randomness of inflows, interconnection of hydro plants in cascade and nonlinear functions for hydro production and thermal costs, as well as operational constraints on electrical transmission. In Brazil the fact that most electricity is produced by hydro generation increases the complexity of the problem.

Due its complexity, the problem is usually divided into long, mid and short term planning so that each problem emphasizes certain details while simplifying or even ignoring others. Long-term hydrothermal scheduling involves the operation of a system for a period of a few years ahead, considered in monthly steps. For this planning horizon, the transmission system can be neglected, and the main modeling aspect to be considered is randomness of inflow.

Stochastic Dynamic Programming (SDP) has been used extensively for the optimization of hydrothermal scheduling problems. This approach, however, is limited by the so-called "curse of dimensionality", since the computational burden increases exponentially with the number of state variables included in the problem. Various approaches have been suggested to overcome the problem of dimensionality in SDP, including the aggregation of the

hydroelectric system through a composite representation (Arvanitidis & Rosing 1970a,b), (Duran et al. 1985), (Cruz & Soares 1996), (Turgeon & Charbonneau 1998) and the use of stochastic dual dynamic programming based on Bender's decomposition (Pereira & Pinto 1985), (Rotting & Gjelsvik 1992).

Alternative to SDP have also been proposed, including approaches based on deterministic optimization of the problem, with inflows provided by a forecasting model (Hanscom et al. 1980), (Bissonnette et al. 1986), (Oliveira & Soares 1990). In this case, the randomness of inflow is represented by the most probable inflow sequence provided by a forecasting model. The deterministic model permits the representation of the hydro system in detail, with the consideration of each hydro plant individually, including its operational constraints and nonlinear production characteristics. The stochastic model considered for the representation of inflows in this model can be quite general, and based on any methodology and appropriate for each specific hydro plant.

Other approaches have recently been suggested, such as the use of analysis scenarios for the treatment of the uncertain in stochastic problems (Rockaffelar & Wets 1991), (Dembo 1991), (Mulvey et al. 1995), (Escudero et al. 1996). In this case, various possible scenarios of future inflows are considered to represent the stochasticity of the problems, rather than only the most probable one. The deterministic optimization model thus solves the problem for a variety of different scenarios and the solutions obtained for each scenario are combined to yield a single solution for the underlying stochastic problem.

The present paper presents a performance analysis of two different scenarios techniques for long-term hydrothermal scheduling. The first one corresponds to a unique scenario technique, where only the most probable inflow sequence is considered in the optimization process; two different inflow forecasting models are used for determination of the unique scenario: Average Long Term inflow and a lag one Autoregressive Periodic Model. The second technique involves multiple scenarios, with the randomness of the problem represented by a set of possible scenarios. In this case, historic records of inflow were used for the representation of the set of scenarios.

The comparison of the two techniques was performed using simulations based on historical inflows. The effectiveness of the two approaches was measured as a function of mean and standard deviation values for hydro generation and operational costs during the planning period.

The test system involved 10 hydroelectric plants and 5 reservoirs in a cascade on the Grande River, in the Southeast Brazilian system, with 7188 MW of total installed capacity. Results obtained by the scenario techniques considering a system composed of a single hydro plant were compared those obtained by the SDP approach, since in such a system the manipulations necessary to implement the SDP for multiple hydro plant system are unnecessary.

The present paper is structured as follows: Section II presents the formulation of the long-term hydrothermal scheduling problem. Section III presents a brief outline of scenario techniques in general as well as the specific techniques implemented in this paper. Section IV presents the comparison of the approaches and Section V presents the conclusions of the study.

2 PROBLEM FORMULATION

The deterministic version for long-term hydrothermal scheduling can be formulated by the following nonlinear programming problem:

$$\min \sum_{t=1}^{T-1} \psi_t (H_t, D_t) + V(x_T) \tag{1}$$

subject to:

$$H_t = \sum_{i=1}^{N} k_i [\phi(x_{it}) - \theta(q_{it} + s_{it}) - hp_i] q_{it} \tag{2}$$

$$x_{i(t+1)} - x_{it} - \sum_{j \in \Omega_i} (q_{jt} + s_{jt}) + q_{it} + s_{it} = y_{it} \quad \forall i \tag{3}$$

$$\underline{x}_{it} \leq x_{it} \leq \bar{x}_{it} \quad \forall i \tag{4}$$

$$\underline{q}_{it} \leq q_{it} \leq \bar{q}_{it} \quad \forall i \tag{5}$$

$$s_{it} \geq 0 \quad \forall i \tag{6}$$

$$x_{i1} \; given \tag{7}$$

$$\forall t, \; t = 1, 2, ..., T-1$$

The objective function (1) is composed of the operational costs ψ_t during the planning period and the future costs $V(x_T)$ associated with final storage in the reservoir. The function ψ_t represents the costs from complementary non-hydraulic sources such as thermoelectric generation, imports from neighboring systems, and load shortage and is obtained by the prior optimization of all sources available. As a consequence, ψ_t is a convex decreasing function of the total hydro generation H_t and depends on the system load demand D_t. The function $V(x_T)$ is a terminal condition, which represents future operational costs as a function of reservoir storage. This term is essential for the equilibrium between the use of water during the planning period and its use afterwards.

Hydro generation at plant i and stage t is a nonlinear function, represented by equation (1), where x_{it} is the water storage in the reservoir, q_{it} the water discharge through the turbines and s_{it} the water spillage from the reservoir. The constant k_i is the product of water density, gravity acceleration and average turbine/generator efficiency; ϕ_i is the forebay elevation as a function of water storage; θ_i is the tailrace elevation as a function of total water release, and; hp_i is the average penstock head loss in the reservoir.

Equality constraints in (3) represent the water balance in the reservoirs, where Ω_i is the set of plants immediately upstream from plant i, and y_{it} is the incremental water inflow at plant i in stage t. Other terms such as evaporation and infiltration have not been considered for the sake of simplicity. Lower and upper bounds on the variables, expressed by constraints (4)-(6) are imposed by the physical operational constraints on hydro plants, as well as the constraints associated with multiple uses of water.

In the present paper, the deterministic solution of the optimization problem is obtained using a nonlinear network flow algorithm, especially developed for hydrothermal scheduling (Oliveira & Soares 1990).

3 SCENARIOS

A scenario can be defined as some kind of limited representation of information about uncertain elements or about how such information may evolve (Rockaffelar & Wets 1991). In this way, scenario sets are formed by different possible description of the stochastic elements of the system, although not necessarily the most probable ones, according to stochastic available information.

The fundamental idea in scenario analysis is to combine the solutions obtained for each different scenario to form a solution for the underlying stochastic problem. In the case of dynamic systems certain approaches differ in regard to anticipation of solution obtained, i.e., whether to assume knowledge about future stochastic variable or not.

The nonanticipativity principle establishes that *"if two different scenarios s and s' are identical up to stage t on the base of the information available about them at stage t, then the solution under the scenarios s and s' must be identical up to stage t"* (Rockaffelar & Wets 1991).

In according with this principle, a policy cannot require two different courses of action at the time t in relation to the two scenarios s and s' if there is no way to tell at time t which of the two scenarios one happens to be following. This condition guarantees that the solution obtained by the model at each stage does not depend on information that is not yet available. This approach was proposed by Escudero et al. (1996) for the optimization of hydrothermal scheduling problems.

On other hand, the scenario technique proposed by Dembo (1991) assumes that is difficult to predict or model the dynamic behavior of the probabilities associated with a scenario set. The approach thus considers a rolling planning horizon mode with the model solved periodically to readjust decision-making policies over time. The decisions for the first stage are the only ones taken, and the stochastic parameters are reevaluated according to changes occurring in the system.

The following the scenario techniques considered in this paper are presented.

3.1 Unique scenarios

Deterministic optimization with unique scenario takes into consideration the most probable future forecast for the uncertain data of the problem. In the present paper, two different inflow forecast models were used to generate the inflow of the unique scenario. The first considers the Long Term Average (ALT) inflow, which represents the monthly average of inflow for each hydro plant of the system. This is a simple unique scenario technique, whose intention is to plan system operation assuming that future hy-

drological conditions will be equivalent to the historical average.

The second forecast model used to generate the scenario is the lag-one Parametric Autoregressive model PAR (1). This model considers that the inflow in stage t only depends of inflow from the previous stage $t-1$, based on the methodology proposed by Box & Jenkins (1994). The PAR (1) model is also used in SDP to provide the conditional probability density function of the inflows.

In the unique scenario technique, optimization determines an optimal decision based on the current forecast of future values for each stage of the planning period, and this decision is utilized until a new forecast becomes available, which is again based on the latest information available to the system.

3.2 Multiple Scenarios

The basic idea of deterministic optimization with multiple scenarios is to establish a decision-making policy that considers the information available under various scenarios without being limited to any one of them. The fundamental question in scenario analysis is how to combine the solutions obtained under different scenarios to form a single reasonable solution for the underlying stochastic original.

In the present paper, historical inflow records form the basis for scenario set. The decision-making policy is obtained by a two-stage model. In the first stage, the deterministic model is used to find a solution for each of the different scenarios constituting the scenario set. This stage can be viewed as a sampling of the solution space of the underlying stochastic model. In the second stage, the solution obtained under each of the scenarios is supplied for a coordinator model that aims find the solution with the best fit for the behavior of the system under uncertainty. The coordinator model is flexible in that it can include additional penalty terms or constraints, depending on the context (Dembo 1991).

In order to facilitate the presentation of the solution procedure, certain simplifications will be introduced into the model (1)-(7), so that the objective function (1) is represented by the polynomial function f and all bounds on the variable considered to be just non-negativity conditions, thus obtaining the following representation of the problem:

$$\min_{x} f(x)$$
$$\text{s.to}: \begin{cases} Ax = b \\ x \geq 0 \end{cases} \tag{8}$$

where the vector x represents the decision-making variables of the system, the linear equality constraint represents the water balance in the reservoirs, and the non-negativity constraint represents the upper and lower bounds on the variables.

Let x^s be the solution obtained by the deterministic model under the scenario s belonging to the scenario set S, i.e.,

$$x^s \cong \min_x f(x,s)$$

$$s.to: \begin{cases} Ax = b^s \\ x \geq 0 \end{cases} \qquad (9)$$

The objective of the coordinator model is to track the scenario solutions as closely as possible. For example, one possible coordination model could be written as follows (Dembo 1991):

$$\min_x \sum_{s \in S} p_s \left\| f(x) - f(x^s) \right\| + \sum_{s \in S} p_s \left\| Ax - b^s \right\|$$

$$s.to : x \geq 0 \qquad (10)$$

where, p_s is the probability associated with the scenario s.

In the objective function of the problem (10), the term first attempts to find a solution, which is as close as possible to that obtained under the various scenarios, while the second attempts maintain the feasibility of the solution.

The coordinator model adopted in this paper is a simplified model which attempts find a solution for the first time interval which is as close as possible to the optimal decision obtained for each scenario; this can be written as follows:

$$\min_x \sum_{s \in S} p_s \left(x_1 - x_1^s \right)^2 \qquad (11)$$

For each stage of the planning period, the solutions obtained by the deterministic model under each scenario are provided to the coordinator model, which will then find a solution for the first time interval, in a rolling planning horizon.

The task in a multiple scenario approach is a multiple of that required for a single scenario deterministic problem; however, the structure of this approach is favorable to parallel computation, so that the entire process can be expected to require a small multiple of the time required for the solution of a single deterministic problem (Dembo 1991).

In the following section, a comparison of the results obtained using the unique and multiple scenario techniques is made. In the case of the system composed of a single plant, the results of the scenario techniques are also compared to those of the SDP approach.

4 RESULTS

The hydroelectric system selected for study consists of the hydroelectric plants on Grande River, located as in Figure 1. The main operational characteristics of the hydro plants are shown in Table 1.

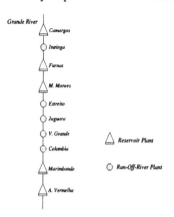

Figure 1. Grande River Cascade

Table 1. Hydro plant characteristics

Hydro Plants	Installed Capacity (MW)	Storage Capacity (hm^3)	Discharge Min/Max (hm^3/s)
Camargos	48	672	84/757
Itutinga	54	-	84/641
Furnas	1312	17217	515/4446
M. Moraes	478	2500	649/3490
Estreito	1104	-	622/5329
Jaguará	616	-	670/4110
V. Grande	380	-	723/4163
P. Colômbia	328	-	807/5224
Marimbondo	1488	5280	1159/7737
Água Vermelha	1380	5169	1317/7679

The operational costs ψ_t are, in general, obtained by the optimal dispatch from available non-hydraulic sources. For the non-hydraulic aspects of the Brazilian system, the operational cost can be estimated by the following quadratic function:

$$\psi_t = 0.5 \left(D_t - H_t \right)^2 \qquad (12)$$

The load demand D_t, treated as constant during the planning period, was considered to be equal to the total installed hydro capacity, assuring a balanced hydrothermal system.

The terminal condition adopted for the deterministic model attempted to maintain the storage reservoirs of the hydro plants as full as possible at beginning of the dry season, i.e. April in Southeast Brazil. This decision is supported by the fact that the optimal determinist solutions based on historical inflow records indicates that the storage of a hydro plant should almost always be at maximum level at the beginning of the dry season. According with the terminal condition adopted in the deterministic model, the optimization horizon should be variable throughout the planning period, so that the final stage will always be April.

Scenario techniques are compared through simulations for the period from May 1931 to April 1990, considering a rolling planning horizon, where for each stage of the planning horizon, the inflow scenario at each hydro plant is brought up to date according the scenario approach adopted. The following scenario techniques were considered:

− Unique scenario ATL;
− Unique scenario PAR (1) and
− Multiple scenarios (historical).

The PAR (1) model was applied to the actual historical inflow records after normalization of the series by subtracting the expected value and dividing by the standard deviation. As a result, the PAR (1) model is represented by:

$$\frac{y_{t(r,m)} - \mu_m}{\sigma_m} = \phi_m \frac{(y_{t(r,m)-1} - \mu_{m-1})}{\sigma_{m-1}} + a_{t(r,m)} \qquad (13)$$

where, $y_{t(r,m)}$ is the inflow at time $t(r,m) = 12(r-1) + m$, with r being the year and m the month; μ_m is the expected value of the inflow for month m, σ_m the standard deviation of the inflow for month m, ϕ_m the autocorrelation coefficient of the normalized series, and $a_{t(r,m)}$ a sequence of uncorrelated random variables with distribution $N(0, \sigma_m^-)$.

A number of estimates for the ϕ_m periodic function have been suggested by statisticians. In this paper, the Maximum Likelihood Estimate method (Box & Jenkins 94) is used for autocorrelation function in the PAR (1) model.

For the first study, a single operation of the Furnas hydro plant was selected. In this case, the performance of the scenario techniques was also compared to the SDP approach, in which, the same PAR (1) model was adopted to provide the conditional probability density functions of the inflows. Table 2 presents the values of the mean and standard deviation of hydroelectric generation and operational costs for the various simulations realized.

Table 2. Statistics of simulations for Furnas hydro plant.

Scenarios	Hydro Generation		Operating Cost	
	Mean	Std. Dev.	Mean	Std. Dev.
ALT	722.6	237.5	2.0184×10^5	1.3523×10^5
PAR(1)	725.5	248.9	2.0291×10^5	1.4274×10^5
Multiple	721.9	226.1	1.9963×10^5	1.2651×10^5
SDP	719.6	198.2	1.9475×10^5	1.2094×10^5

According to these results, higher average hydroelectric generation was obtained with the unique scenario technique, specially for the unique scenario using the PAR (1) model. However, the standard deviation was also higher, indicating greater fluctuation in hydro generation, and leading to higher final

operating costs, since this evolve a convex and increasing function.

Although the multiple scenario approach resulted in lower operational costs than other scenario techniques, the difference in the behavior of the solution obtained using this approach in relation to that of unique scenario using ALT is small; hence, it can be concluded that the two approaches produce similar results. In the case of SDP, however, the results revealed less fluctuation in hydro generation during the planning period, thus leading to lower final operational costs than the scenario techniques in general.

Figure 2 shows the trajectories of the water storage for the period of 1950 to 1960 in the reservoir of the Furnas hydro plant as a function of the approaches simulated. The optimal trajectory obtained by assuming that the exact inflows are known for the planning period is also presented in the figure (Optimal Solution). The historical inflow records in this period are presented in Figure 3.

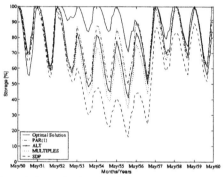

Figure 2. Storage Trajectories of Furnas from 1950 to 1960.

A look at the trajectories shows that larger differences were verified for the critical period of 1952 to 1956 when water was scarcer, while in years with average streamflow, such as 1950/1951, 1951/1952 and 1959/1960, the difference between the approaches was reduced. This can be explained by the fact that although inflow is critical and cannot be correctly estimated by forecasting models, the terminal condition considered in the deterministic model of the scenario techniques is responsible for maintaining the higher level of storage in the reservoir.

This higher level of storage leads to higher values for water head, thus increasing the productivity of the plants and therefore improving their efficiency. This is an interesting feature of the approaches based on deterministic model, since dry periods, when higher operational costs are registered, are the more important in operational planning.

A higher level of storage during the critical period was verified with the use of unique scenario technique. The differences in behavior verified for the different scenario techniques suggest that forecaster models may have a great deal of influence on deter-

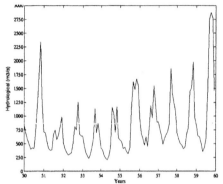

Figure 3. Historical Inflow Records of Furnas from 1950 to 1960.

ministic optimization. Possible benefits associated with the unique scenario technique may results from the use of models that are more efficient than the lag-one parametric autoregressive model in the representation of the actual stochastic process of the inflow.

For the second study, the system comprising a cascade of 10 hydroelectric plants and 5 reservoirs was used. Table 4 shows the values of the mean and standard deviations for hydro generation and operating costs obtained in the simulation of this system using the scenario techniques. In this case, the scenario techniques were not compared with the SDP approach, since in this latter the dimensionality problem require modeling manipulation and the idea of the present paper is just to focus the effectiveness of the different approaches in coping with the randomness of inflow.

Table 3. Statistics of simulation for the Cascade of the Grande River.

Scenarios	Hydro Generation		Operating Cost	
	Mean	Std. Dev.	Mean	Std. Dev.
ALT	4676.3	1352.8	4.0679×10^6	3.8566×10^6
PAR(1)	4678.0	1282.4	3.9109×10^6	3.6106×10^6
Multiples	4676.3	1324.6	4.0303×10^6	3.8050×10^6

For the multiple hydro plant system, higher average hydroelectric generation was again obtained for the unique scenario technique using the PAR(1) model. The standard deviation obtained for this case was also lower, thus leading to lower final operating costs. These results reveal similar behavior for the unique scenario ALT and multiple scenario techniques, which shows that deterministic optimization considering the monthly average of the historical inflow can furnish decision-making policy similar to that politic obtained by considering the average of the deterministic solutions obtained from historical records.

5 CONCLUSION

This paper has compared two scenario techniques for the treatment of the randomness involved in the long-term hydroelectric scheduling problem: the unique scenario technique, where the randomness of inflow is represented by the most probable future scenario of inflow provided by a forecasting model, and the multiple scenario technique, where various possible future scenarios of inflow are considered, not only the most probable. Two different forecasting models were used in connection with the unique scenario technique: the average long term inflow and a lag-one parametric autoregressive model. For the case of multiple scenarios, the scenario set was represented by historic inflow records.

The system selected for testing constituted of a hydro plant in cascade on the Grande River, in the Southeast Brazilian system. The results obtained considering the system as composed of a single hydro plant were compared to those obtained with the SDP approach. The comparisons were made through simulation based on historical inflow records.

The results showed that the unique scenario technique led to a somewhat higher average for hydroelectric generation in all simulations performed, although for a single hydro plant case higher standard deviations were not compensated for by enough of an increasing in average generation, leading to higher final operating costs than those obtained by other approaches simulated.

However, the scenario techniques are especially efficient during dry streamflow periods, and these are critical for operation planning purposes. This efficiency is due to the terminal condition considered by the deterministic optimization model. Moreover, other possible benefits associated with scenario techniques may be result from a detailed representation of the system in the case of multiple hydro plant systems.

Although the unique scenario technique was found to be efficient, the use of models which are more efficient than the lag-one parametric autoregressive model to represent the actual stochastic process of inflows should improve the performance of this approach even more so that to will surpass that of stochastic dynamic programming, since this latter cannot cope with such representations, due to the "curse of dimensionality". The unique scenario technique also surpass the performance of the multiple scenario technique, since the performance of the latter was similar to deterministic optimization when the monthly average of historical inflow was considered.

REFERENCE

Aranitids N. V. & Rosing J. 1970a. Composite representation of multireservoir hydroelectric power system, *IEEE Transaction on Power Apparatus and Systems* vol. PAS-89, no. 2: 319-326.

Aranitids N. V. & Rosing J. 1970b. Optimal operation of multireservoir system using a composite representation, *IEEE Transaction on Power Apparatus and Systems* vol. PAS-89, no. 2: 327-335.

Bissonenette V. & Lanford L. & Côté G. 1986. A hydrothermal scheduling model for Hydro-Quebec production system, *IEEE Transaction on Power Systems* vol. PWRS-1, no. 2: 204-210.

Box G. E. P. & Jenkins G. 1994. *Times Series Analysis*, Prentice-Hall, Inc., Englewood Cliffs, New Jersey.

Cruz G. Jr. & Soares S. 1996. Non-uniform composite representation of hydroelectric systems for long-term hydrothermal scheduling, *IEEE Transaction on Power Systems* vol. 11, no. 2: 701-707.

Dembo R. S. 1991 Scenario optimization, *Annals of Operations Research*: 63-80.

Duran H. & Pueh C. & Diaz J. & Sanchez G. 1985. Optimal operation of multireservoir system using an aggregation-decomposition approach, *IEEE Transaction on Power Apparatus and Systems* vol. 104, no. 8: 2086-2092.

Escudero L. F. & Fuente J. L. & Garcia C. & Pietro F. J. 1996. Hydropower generation management under uncertainty via scenario analysis and parallel computation, *IEEE Transaction on Power Systems* vol. 11, no. 2: 683-689.

Hanscom M. L. & Lafond L. & Lasdon L. S. & Pronovost G. 1980. Modelling and resolution of the deterministic midterm energy production problem for Hydro-Quebec system, *Management Science* vol. 26: 659-688.

Mulvey J. M. & Vanderbei R. J. & Azenios S. 1995. Robust optimization of large-scale systems, *Operation Research 43*: 264-281.

Oliveira G. G. & Soares S. 1990. A second-order network flow algorithm for hydrothermal scheduling, *IEEE Transaction on Power Systems* vol. 10, no. 3: 1635-1641.

Pereira M. V. & Pinto L. M. V. G. 1985. Stochastic optimization of a multireservoir hydroelectric system: a decomposition approach, *Water Resources Research* vol. 21, no. 6: 779-792.

Rockaffelar R. T. & Wets R. J.-B. S. 1995. Scenario and policy aggregation in optimization under uncertainty, *Mathematics of Operation Research vol. 16*: 119-147.

Rφtting T. A. & Gjelsvik A. 1992. Stochastic dual programming for seasonal scheduling in the Norwegian power systems, *IEEE Transaction on Power Systems* vol. 7: 273-279.

Turgeon A. & Charbonneau R. 1998. An aggregation-disaggregating approach to long-term reservoir management, *Water Resources Research* vol. 34, no. 12: 3585-3593.

Hydropower in the New Millennium, Honningsvåg et al (eds), © 2001 Taylor & Francis, ISBN 90 5809 195 3

Southern Africa power electricity trade development

Paper by M. Marumo
Chief executive Lesotho Highlands development authority

OVERVIEW

Regional electricity in Southern Africa trade has a history dating back to the early 1950s when a 220kv interconnection was built from the Katanga Province in Democratic Republc of Congo (DRC) to Zambian Copperbelt. In more recent years, network interconnections and associated exchange arrangements have multiplied, particularly following the creation of Southern African Power Pool (SAPP), by 1997, the volume of electricity traded in Southern Africa reached the order of 1,000 MW and 7,700 Gwh with an estimated annual trading value in excess of US$125 million. By 1998 this increased to about 1,500 MW and 11,500 Gwh at an estimated value of US$140million. It is estimated that regional trade will continue to grow with estimated firm power contracts exceeding 1500MW supplemented by another 300 – 500MW of peaking and/or interruptible contracts.

Cooperation in the energy sector is gradually developing in Southern Africa, led by a growing trade in electricity among Southern African Development Community (SADC) countries. The region is rich in potential energy supplies, with massive rivers in the north and reserves of fossil fuels in the south, but has yet to develop its resources to the fullest extent.

The first significant step towards fostering a regional electricity market was made in 1995 when the 12 member states of the SADC (since expanded to 14) formed the Southern African Power Pool (SAPP) and signed an agreement. The members comprise of the following:-

Angola, Botswana, the Democratic Republic of Congo, Lesotho, Malawi, Mauritius, Mozambique, Namibia, the Sychelles, South Africa, Swaziland, Tanzania, Zambia and Zimbabwe.

The objective of the Southern African Power Pool is to provide reliable and economic electric supply to the consumers of each SAPP member consistent with reasonable utilisation of natural resources and effect on the environment. Basis of SAPP initially were on cooperation, recently there has been a shift to introduce competition amongst members necessitated by a need for economic growth based on competitive and affordable prices for electricity. There are agreed basic principles under which SAPP operates.

The co-ordination of and the co-operation in the planning and operation of the various systems to minimize costs while maintaining reliability and, the full recovery of costs and the equitable sharing of the resulting benefits is paramount. Among the benefits that are in process of being achieved are the reduction in required generating capacity, reduction in fuel costs and improved use of hydro electric energy. In addition, all the members have the right and obligation, regardless of size or type of organization, to own or otherwise provide the facilities required to provide its electricity service requirements.

SAPP STRUCTURE

The Management structure of SAPP as approved, consists of the SADC Government Ministers and officials. SAPP Executive Committee; and Management Committee which has three Sub-Committees:-

Operating Sub-Committee, Planning Sub-Committee and Environmental Sub-Committee and finally the Coordination Center.

SADC Energy Ministers and Officials:

The Southern African Development Community (SADC) Government Ministers and Officials are responsible for policy matters which are normally under their control in terms of the national administrative and legislative mechanisms that regulate the relations between the Government and its respective power utility.

The Executive Committee convenes regarding requests for membership by non-SADC countries and major policy issues that may arise from the operations of the SADC Energy Ministries.

Executive Committee:

The Executive Committee is composed of the Chief Executives of only those Member Electricity Supply Enterprises who generate, wholesale and retail power to end-use customers. Independent Power Producers are not eligible to participate in the Executive Committee. The Committee acts as the Board of the Pool. Chief Executives report to the SADC Energy Ministers through the Technical Administrative Unit (TAU) based in Angola which acts as the Secretariat for the Executive Committee. A country having more than one utility meeting these requirements designates one utility to represent it on the Executive Committee. For example; in the case of Lesotho, Lesotho Electricity Corporation (LEC) is the member and Lesotho Highlands Development Authority (LHDA) has been granted observer status. LHDA qualifies to participate as an Independent Power Producer.

Management Committee:

The Management Committee oversees the administration of the Pool and ensures that the objectives of the Pool, are met. The Management Committee makes recommendations to the Executive Committee. Independent Power Producers are not eligible to participate in the Management Committee.

Planning Sub-Committee:

The Planning Sub-Committee reports to the Management Committee and is responsible for:

> Establishing and updating common planning and reliability standards which have an impact on SAPP. In addition, and based on individual Member's plans, an overall Pool Plan is developed every two years which,

highlights the benefits and opportunities for cost savings that can be derived by the Members from the co-ordination of activities.

Subsequently, the planners take into account the forecasted demand and energy consumption in each Member's system, including Demand Side Management. Planning also forecasts anticipated sales and purchases by each Member, including those with Electricity Supply Enterprises or Independent Power Producers or Non-Members of SAPP.

The Planners also use the characteristics, location and commissioning dates of the new generating units and new transmission facilities of 110kV and above which are planned in each Member's system. Also in the equation are the characteristics, location and commissioning dates of the new telecommunication, telecontrol and supervisory facilities which are planned in each Member's system, when such facilities have a significant impact on the operation of the interconnected system.

Lastly, the planners identify and record new generation, transmission, telecommunication or telecontrol facilities to be installed in the systems of Members and Non-Members and evaluate software and other tools which will enhance the value of planning activities such as load forecasting, the determination of planning or reliability standards, cost-benefit analysis or system studies.

Operating Sub-Committee

The Operating Sub-Committee reports to the Management Committee and is responsible for operating and other duties. The Operating Sub-Committee consists of representatives of Members which are signatories to the SAPP Agreement between Operating Members. The committee is composed of a maximum of two representatives per Member and these representatives are of sufficient seniority in their own organisation to make all relevant decisions. The main representative is also a participant in the Management Committee.

The duties of the Operating Sub-Committee are in accordance with the Agreement between Operating Members.

Environmental Sub-Committee:

The Environmental Sub-Committee reports to the Management Committee and is responsible for alerting and advising the Management Committee concerning environmental matters. Under the direction of the Management Committee, the Environmental Sub-Committee keeps abreast of world and regional matters relating to air quality, water quality, land use and other environmental issues. Where Governments have in place related Environmental Organisations, this Committee liaises with them to assist one another on specific issues. The Sub-Committee presents all findings and recommendations to the Management Committee, the Planning and Operating Sub-Committees and carries out other functions and activities as assigned or approved by the Management Committee.

Coordination Centre

A Co-ordination Centre has been created since there was a demonstrated need for monitoring the entire SAPP member agreement. This independent entity that reports to the Management Committee is responsible for the following to ensure orderly and regulated operation of the Pool.

The centre monitors continuously the transactions between Operating members and between members and non-members.

In addition, time correction procedures, inadventent power flows, control performance criteria, technical support, protection performance and other technical tasks are under the jurisdiction of the coordination centre.

MODELLING ELECTRICITY TRADE IN SOUTHERN AFRICA

Since the purpose of the SAPP Memorandum of Understanding is to establish a framework under which the signatories pronounce their clear intention to enhance regional power co-operation through the establishment and operation of the aforementioned Southern African Power Pool, trading within SAPP presents opportunities and along with those, certain modelling requirements.

There are twenty articles including three appendices that govern the SAPP agreement. Contained in this document is the total cost of doing business associated with the procedures set forth.

To determine a competitive market's impact on the region a new regional price-forecasting model was constructed. The model is the basis upon which the Southern African Power Pool simulates its scenarios for electricity trade among members.

The modelling system serves as a good basis because of its comprehensive, integrated characteristics. It incorporates four submodels: one each on demand, supply, finance, and rates. It simulates the interaction of customer demand, system generation, total revenue requirements, and customer rates. By providing varying sets of assumptions, it deals with the uncertainty of such factors as economic growth, construction costs, and fossil fuel prices.

For the purposes of joint planning, the long-term model is utilized, enabling regional approach for development of least cost expansion and utilization plans.

The model minimizes the total expansion and production costs in an integrated manner, which includes transmission. The inclusion of transmission constraints in the expansion model is particularly important, since, shortages in transmission capacity have been hindering the substitution of clean, cheap hydropower from the north for the more expensive, polluting thermal power in the south of the region.

The model operates under varying constraints from each individual member country and the constraints include:-

- Supply/demand equations, which insure that user specified demands are met for each hour modeled for each of the day types in each of the seasons:
- Capacity constraints, which make sure that each plant's generation does not exceed current rated capacity:
- Reliability constraints, which make sure that a proper reserve margin is maintained between installed or purchased capacity and peak demand:

- Country autonomy constraints, which make sure that domestic capacity is always greater than a pre-determined fraction of domestic peak demand.

Three commodities are traded in the model:

- Spot power to meet demand, obtained in the open market, with no guarantee of availability;
- Firm power to meet demand, available up to a prescribed maximum guaranteed by the seller;
- Firm capacity, which the buyer may use to satisfy reserve requirements, and which sets the upper limit on firm power imports.

NET EFFECT OF THE LONG TERM MODEL

The model presents a business case that addresses regional/local politics, drought, economics, engineering, geography etc. since it is constructed to reflect the advantages of regional, rather than country-by-country, generation and transmission capacity planning.

It addresses:

- Lower Reserve Requirements: As individual generators represent a smaller fraction of the total system load, their unplanned outages are less likely to result in an overall generation outage. Thus, more diverse generation sources result in lower reserve requirements. Joint planning for utilities will increase generation diversity, thereby resulting in lower reserve requirements than would occur under separate planning.
- Load Diversity: Not all utilities experience peak load conditions at the same time of day due to the different characteristics of the customers they serve. Similarly, they experience annual peak demand on different days. Therefore, the chronological sum of the individual utility loads provides a peak that is lower than the sum of the individual peak demands. Since generation capacity must be capable of handling the peak demand during the year, separate planning will result in larger generation requirements than will joint planning.
- Economies of Scale: Generally, it requires less capital to construct one large facility than is required to build an equivalent capacity with several smaller units. Similarly, multiple units at single site are cheaper to build than the same units at numerous different sites. These economies of scale result from common use of facilities, such as fuel handling, transformers, and transmission lines. Joint planning allows these economies to be captured more frequently than separate planning does by allowing utilities to share a jointly planned unit.
- More Available Options: Joint planning may allow a utility to utilize generation options for both energy and capacity requirements that are otherwise unavailable when planning is done separately. Thus a utility with little or no hydro sites available will not have to build a more expensive type of generation.

In addition to reflecting these advantages, the model takes into account the extraordinary uncertainty regarding demand growth in the SAPP region, as well as uncertainty on the supply side ie. the impact of drought, and line or unit failure.

Long-run expansion decisions must consider alternative growth and supply scenarios. It is almost a certainty that an expansion plan based on most likely growth and supply scenarios will not be the preferred option if its performance is measured against all scenarios. Flexible capacity expansion scenarios; ones where the cost of over or under estimating demand/supply are not catastrophic to the region are always preferred.

An added feature of the modelling in the region is to allow each SAPP participant to decide on the maximum level of dependence on imports expressed as a domestic generation reserve margin – domestic energy production capacity divided by peak demand. This depends on each country's need for security and autonomy.

CASE FOR COOPERATION – PENNSYLVANIA, NEW JERSEY & MARYLAND (PJM) VS SAPP

There is a parallel between SAPP's activities now and those of the Pennsylvania, New Jersey, Delaware, Maryland, Virginia and District of Columbia in North America where PJM Interconnection is responsible for the operation of the centrally dispatched system.

In addition to ensuring the reliable supply of energy from generating units to wholesale customer, PJM administers bid-based wholesale markets in which participants buy and sell electric products including energy and generation capacity. PJM coordinates the operation of more than 540 generating units in its control area, providing approximately 58,000 megawatts of generating capacity over a grid of more than 8,000 miles of high voltage transmission circuits. PJM facilitates the delivery of electric power to over 23 million customers in Pennsylvania, New Jersey, Delaware, Maryland, Virginia, and District of Columbia.

With more than 140 Members representing every segment of the electric power industry, PJM's markets have become some of the most liquid and active electric markets in the United States. PJM is recognized throughout the world as a leader in the use of advanced technology to support robust, non-discriminatory, and competitive electric markets.

PJM has evolved over seven decades as the commercial and regulatory environment has changed, becoming an Independent System Operator in 1998. Today, the PJM control area is the largest of nearly 150 control areas in North America, and the fourth largest centrally dispatched system in the world.

In the Southern Africa, the issues of Commercial and regulatory environments are only emerging and so much remains to be done.

Committees and User Groups for PJM

Committees and User Groups are central to PJM's structure. They provide a forum to respond to ongoing and emerging issues efficiently, proactively, and fairly. Committees and User Groups enable PJM to balance the concerns of power marketers and traders, utilities, regulators, environmental groups, and electric consumers. They are the source of the collaborative solutions on which PJM's success depends.

PJM Committees continuously consider and recommend changes to the PJM Operating Agreement, the Open Access Transmission Tariff, and the Reliability Assurance Agreement.

As opposed to SAPP structure, PJM structure is much broader and inclusive of all parties affected in the Electricity Supply Industry. The structure is composed of the following Groups:-

Board of Managers
Office of the Interconnection
Members Committee
Planning Committee
Operating Committee
Energy market Committee
Finance Committee
Tariff Advisory Committee
Audit Advisory Committee
Alternate Dispute Resolution Committee
Liaison Committee
User Groups
Reliability Committee
Transmission Owners' Agreement Administrative Committee

THE BENEFITS OF POWER POOLING

For both scenarios, the benefits of Power pooling are increased reliability, the assurance of an adequate supply of electricity, the security to withstand disturbances, and reduced operating costs.

Reliability means providing customers with the amount of electricity they need when they need it. Reliability exists when the electric system is in balance – supply exactly matches consumption – or can be rapidly brought back into balance following any disruption, without having any transmission system overloads. A more stable supply of reserves is achieved through pooling than could be achieved by individual members acting alone.

Adequacy is the assurance of sufficient supply of electricity to meet customers' needs. Adequacy exist when the system has sufficient generation and purchased power available to meet the load and reserve requirements. Through pooling, each energy producer can reduce plant investment through lower reserve requirements than would be necessary if operating by itself.

Security is the operation of a power system to withstand sudden large and unanticipated disturbances. Security exists when actual or contingency flows on all key transmission facilities are within safe and reliable limits.

Economy in the operation of a system, composed of many generators, requires that the units be dispatched in order of increasing marginal cost regardless of ownership or location relative to pool load. In this way, the operating costs to all customers are minimized. The pooling of generation resources, dispatched by least-cost scheduling, results in the optimal use of resources for all participants.

JOINT MAINTENANCE SCHEDULING

Allows facility owners to perform necessary maintenance on generating units and transmission lines in coordination with others. A coordinated maintenance plan helps to assure pool reserve capability and optimal use of generation and transmission resources.

These compare closely with SAPP advantages and benefits referred to earlier.

The importance of PJM has been noted for three reasons:-

- Its operational excellence has brought certainty on a number of issues, notably reliability and planning.
- The inclusive stakeholder process has made it a place for collective problem solving. The stakeholders have come to understand the need and are willing to work together and adopt rules as the market develops.
- Independence has been a critical aspect of their success because it has brought equity and allowed all issues to be addressed in a fair way.

While SAAP has gone a large way in addressing first issue; a lot more remains to be done on the other two.

PROPOSED CHANGE IN THE ESI IN SOUTHERN AFRICA IS AN IMPERATIVE

The current SAPP Management structure is sub-optimal when compared with PJM structure and can not meet the needs of a modern electricity market. From such a structure because of stranded costs from individual utilities, electricity has become unnecessarily expensive. There has been poor incentives for efficiency improvements amongst suppliers and this has had a detrimental effect on national and regional growth as well as on customers.

There are factors that will remain as drivers that precipitate change in the Electricity Supply Industry and it is imperative that they are addressed by SAPP structure and these include among others:

- a need to maximise financial and economic returns through fiscal revenue and debt reduction.
- economic efficiency through applying allocation efficiency e.g. in respect of the next investment in generation capacity.
- widened resource availability and technological change via competitive imports from Southern African Power Pool.
- the need for the improved customer service and choice

There are, however, advantages that exist in the current structures of SAPP which should not be lost and should be protected.

With efforts of creating the African Union it may be possible to restructure the entire Electricity Supply Industry to one where supply will be available at low prices with electricity vendors selling their services to customers who will have a choice.

Such a structure should have separate generating companies which will supply this electricity to a power pool to be known as the South African Power Exchange. all the transmissions should be handled by a new transmission electricity company. The retailers would comprise Regional Electricity

DIAGRAM

PROBABLE NEW DISPENSATION

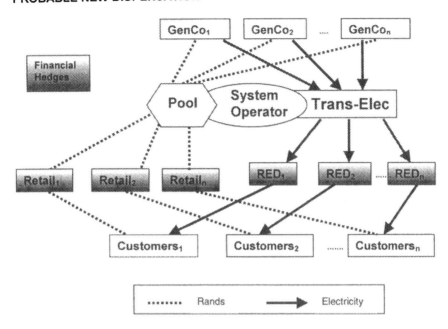

Distributors (REDs) who would buy electricity from the pool and distribute / sell it to their customers.

Those parts of the business that form a monopoly, in other words the wires will require to be neatly separated out and regulated while the rest of the business will remain competitive. Retailers will buy where they can and will sell electricity as an attractive package to a customer.

THE GRID AND SAPP

The past decade has seen a major shift in attitude throughout the region, with most countries accepting the view that regional integration of the ESI is the most economic alternative for growth and development. Hence intra-regional electricity trade has increased with new interconnections being planned and constructed. With the exception of Angola, Malawi and Tanzania all the mainland SADC countries are interconnected with the regional high voltage grid which forms the backbone of the SAPP. Power exchanges in the region are increasingly being done under SAPP principles and pricing schedules. In this respect some supply contracts are being revised to reflect the SAPP agreements.

The co-ordination Centre is established in Harare Zimbabwe. It will not be a control centre, but merely an entity responsible for the monitoring of operations, transactions, time corrections procedures and inadvertent power flows and returns. It will also provide routine reports and has a mandate to investigate disturbances on the SAPP network.

SAPP membership is restricted to national electricity utilities. In March 1998 HCB (Cahora Bassa) was granted temporary membership of the various committees and sub-committees, but not in the SAPP's most senior body, the Executive Committee. At the same meeting moratorium was declared prohibiting any other new members from joining the SAPP for the next two years while the SAPP considers the role of IPPs and Independent Transmission Companies (ITCs).

CHANGE DRIVERS AND RESPONSES

The power sector is undergoing reform both at national and regional level to overcome the following constraints:

- inefficiencies;
- technical weaknesses (particularly in the distribution area);
- poor financial performance of ESI participants (return on investment < cost of financing) and strong limitation on public sector financing;
- electricity prices being used as political instruments;
- unsustainable subsidies to various consumers groups; and
- an ESI structure that does not allow for efficiency through competition.

Statutory and regulatory frameworks are being overhauled in most of the countries. Utilities are being commercialised, and legislation is being amended to allow new entrants in the market. Private operators, especially IPPs, are being under active consideration in most countries. Other priorities include third party access to transmission systems and electricity tariff reform. Reforms being discussed include:

- changes in the role of the public and private sector;
- industry structure and composition;
- new institutional arrangements for electrification; and
- regulation; particularly licensing and price regulation.

FUTURE TRADING CONDITIONS

The scope for competition in generation and supply differs from country to country in southern Africa. Most countries in the region have a very limited number of sources of supply, and these sources are often not comparable (it is difficult to compare a base load coal fired station against a mid-merit or peaking hydro supply). Furthermore, in some countries where competition could be possible, e.g. the DRCongo with several independent producers, the real possibilities are hampered by the lack of an integrated national transmission grid.

At present, only South Africa with a large number of thermal stations holds the potential for real competition between national generators. However, in practical terms this is limited as Eskom holds most of the capacity. In the future, the existence of IPPs might create scope for competition in Zimbabwe, Mozambique and Zambia, while the situation in Namibia and the DRCongo remains unclear.
The private sector is increasingly being looked to for providing new generation capacity and, to a lesser extent, transmission systems. Whether IPPs, and thereby increased competition, will actually materialise will depend on the attractiveness of the power sector as an area for investment. The fiscal frameworks are generally improving, or allow a fair scope for negotiation. This situation is most likely best in Mozambique, Namibia and Zambia, and least attractive in South Africa. In terms of political and commercial risks, SADC countries are still perceived as moderate to high risk. Uncertainty is, however,

further created by the general lack of cost reflective pricing for electricity and weak and/or underdeveloped regulatory frameworks.

So far little competition has taken place in the selection of private developers of new generation plant. Generally, selection of private participants has been by negotiation with a preferred developer. This approach of negotiated deals has resulted in some of the deals being very costly to the country. With the opening up of the power market, it is expected that competition will prevail, both from the supplier and customer point of view. This should contribute to increased efficiency and a lower real cost of supply, as has happened in South America for instance. Similar developments in southern Africa should benefit both individual countries and the regional power sector as a whole.

At present, whereas a few of the new Acts or draft Acts are clear on third party access to the transmission system, there is silence on direct transactions between IPPs and large consumers, although no prohibition against such arrangement appears to exist. This could imply that each case could be viewed on its own merits and demerits – i.e. direct transactions could be negotiated with Government and the public utility.

Despite nominal tariff increases in most southern African countries during 1998/99, tariff levels measured in US$ terms have decreased in most of the countries over the last 12 months due to the depreciation of the South African Rand and many of the other regional currencies, e.g. the Zimbabwe Dollar. With tariffs already being below long-run marginal cost in most of the countries, this has worsened the financial situation of the power utilities. Financial constraints are today a major obstacle to the realisation of new generation expansion projects, but also to payment for import of power from neighbouring countries, despite such imports often carrying a lower cost than internal supply options. The fact that most import contracts are based on US$ prices has resulted in many of these increasing substantially in local currency terms.

Electricity prices vary considerably for the existing export/import contracts in place between SAPP utilities.

The prices are lowest for firm 'take-or-pay' contracts and for non-firm energy trading, while the prices for traditional maximum-demand contracts are generally higher. However, the present level of trading prices clearly reflects the prevailing situation of surplus capacity. As this capacity is exhausted in the medium-term, prices for power traded through the SAPP will increase.

Regional trading prices will gradually increase towards 2007-09 when new generation capacity is required in key markets and for the region as a whole. At this point, the trading prices is expected to reach a level of about 3.0 – 4.0 Usc/kWh in real terms (measured in 1998 US$). This is a doubling of regional trading prices in real terms over a 10 year period.

Regional power trade in the short term is likely to make more active use of existing surplus capacity through innovative arrangements, resulting in a combination of short and long term contracts. The longer term trading conditions will largely depend on ESI market reforms and SAPP developments. Further enhancement will be provided from the SAPP's moves towards pooled operation. Finalisation of a common pricing methodology for all transmission services (wheeling, import/export and nationally) will provide additional impetus.

However, there may be a number of risks associated with regional trading. There are technical risks associated with an increased complexity in system planning and operations. The possibility of 'imported' system disturbances may also arise. Institutional risks would be posed by an uncertain regulatory environment, while currency (and payment) uncertainties may add financial risks. Further risks may arise from self-sufficiency concerns rooted in political or supply reliability considerations.

In summary, the key challenges for future regional power trade are the following:-

- Increasing the number of players through reforms and the introduction of IPPs (agree on how to handle the broadening of SAPP membership to also include IPPs and privately owned transmission companies);
- Continued strengthening of the regional grid;
- Resolution of wheeling issues through third party networks;
- Agreement on the currency of payments;
- Stimulation of short term trade and innovative solutions; and
- Establishment of dispute resolution mechanisms.

- **Summary**

In summary, the challenge for the region is to get the economies moving. The way forward is to take advantage of Africa's enormous hydropower potential in the north and the enormous electricity demand in the south. This coupled with the regional approach to trading electricity would greatly assist in the economic development of the region. Lowering the cost of energy has enormous benefits and may result in a Power for Jobs policy taken up by the respective country leadership.

Integrated tool for maintenance and refurbishment planning of hydropower plants

B.Mo, E.Solvang, J.Heggset, D.E.Nordgård & A.Haugstad
SINTEF Energy Research

ABSTRACT: Optimal investment in maintenance and refurbishment of hydropower plants is very complicated. This paper describes a new model currently under development. The model consists of several modules, each solving separate tasks and passing on necessary information to the other modules. There is a module for computation of losses from disrupted production caused by forced outages, a module for computation of production losses during maintenance, a module for specification of unit (i.e. turbine, generator and control system) failure probabilities as functions of time, a module for presentation and evaluation of qualitative criteria and finally a module for optimal timing of projects. The whole concept is adapted to decision makers who own many production units, although parts of the system may also be useful for smaller utilities.

1 INTRODUCTION

Due to increasing average age of existing hydropower plants in Norway there is a growing need for maintenance/upgrading and refurbishment in the Norwegian system. In this paper we will not distinguish between maintenance and refurbishment and will consequently use the term maintenance meaning maintenance, upgrading and refurbishment. Market based operation of the plants, i.e. more starts and stops, also increases the need for maintenance.

Optimal maintenance planning for hydropower systems is complicated for a number of reasons:

- The lifetime of the different components and the consequences (quantitative and qualitative) of failures are not known.
- Plant efficiency as a function of time of use is unknown.
- It is difficult to estimate probabilities of unit failure due to limited statistics or the applicability of available statistics.
- Loss of sales income due to plant failures or scheduled maintenance is dependent both on future market prices and inflow which are uncertain.
- In a hydropower system with several plants along a watercourse optimal maintenance for these plants may be related.
- There may be limited personnel and financial resources for maintenance in a given period.
- The goal of maintenance planning is a multi-criteria decision problem since both profit and HES (Health, Environment and Safety) are important factors.

During the last few years, several tools and methods for decision support in connection to maintenance of hydropower plants in Norway have been developed. These include:

- Methods and information needed to describe unit efficiency as a function of time.
- Methods and information needed to estimate probabilities of unit failures.
- Methods used to handle qualitative objectives such as health, environment, safety and negative publicity, Tangen 1996.
- A tool that can be used to compute economic losses of a given unit failure, Tangen et al. 1999.
- A tool that can be used to compute economic value of the production loss due to maintenance.

This paper summarises these methods and focuses on a new tool under development. The goal of this new system is to integrate the different models/tools into one decision support system that can solve the overall maintenance planning problem. Recent work has focused on different parts of the planning problem but there is yet not any tool available that can be used to solve the overall problem. The next chapter will describe the model concept, which consists of separate models that interact with each other.

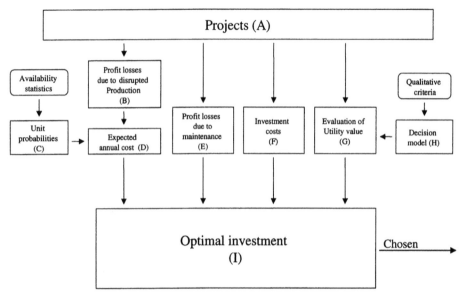

Figure 1. Integrated model for maintenance planning.

2 DESCRIPTION OF INTEGRATED MODEL CONCEPT

Figure 1 illustrates the integrated model concept. As already mentioned, the concept consists of several modules that interact with each other, but that can be used separately.

Initially the decision maker considers a number of possible maintenance projects. These projects can go through a pre-screening process in order to reduce the number of projects. The reduced number of projects are input to the integrated model. In Figure 1 these projects are illustrated by module A 'Projects'. The output from this module is a list of possible maintenance projects. Projects that are obvious should be excluded from the list.

If there is no 'physical' connection between projects and no personnel or financial constraints, projects can be decided independently. In our integrated model these types of dependencies are treated differently. Physical dependencies have to be included in the list of possible projects. For instance if a hydro production system consists of two plants 1 and 2 which are physically connected, i.e. located along the same water course, the list of possible projects should consist of three projects, maintenance of plant 1, maintenance of plant 2 and simultaneous maintenance of both plant 1 and 2. Simultaneous maintenance of both has to be included in the project list. The optimization module (I) accounts for project dependencies due to personnel or financial constraints.

Module B in Figure 1 represents a model whose purpose is to compute the expected value of production loss due to unit failure. The model actually calculates a probability distribution for the production losses, where different price and inflow scenarios give the distribution. Only the expected value is output from the module since the goal of the model concept is to minimize expected costs.

Inputs to module B are unit number and downtime in addition to the physical description of the hydro system, inflow statistics and market description. The model is based on the EOPS (EFI's One area Power market Simulator) model, described by Flatabø et al. 1998 and Haugstad et al. 1997. This is described more fully in the section 4.

Module C represents collection of failure statistics and specification/computation of probabilities of failures for the different units. The statistics are limited and the user often must specify these numbers directly based on expert experience. However, in the future a more probabilistic based approach could be used. The output from this module is probabilities of unit failures as a function of time for all units.

Based on inputs from module B and C, module D calculates the expected annual costs due to disrupted production.

Module E represents a model that computes the value of the production losses due to preventive maintenance. These losses are part of the investment costs and very dependent on the time of year, market prices, storage capacity etc. The losses could be due to lost water or to increased production at lower prices. The EOPS model is also applied for this purpose, but run in a different mode. The model is further described in section 3.

The F module is used to specify investment costs for each project.

The G module is used to evaluate and document the trade off between the qualitative values associated with a project. The outputs from this model are relative weights for each qualitative value, i.e. Health, Environment and Safety or other. The properties of this module are described in more detail in section 6.

As background for this evaluation module H contains the decision model for the trade-off between different qualitative criteria.

The I module is essential in the integrated concept. The module gets input from all the other modules and gives as a result a list of investments. The optimization part of this module is based of Dynamic Programming for a given set of possible maintenance decisions.

3 LOSSES CAUSED BY PREVENTIVE MAINTENANCE

During maintenance parts of the production system may be unavailable. This can result in lost water or profit losses because production has to be moved to periods with lower prices. These losses are an important part of the total maintenance cost. However, it is possible to reduce these costs with good planning. Reservoir levels can be scheduled lower than 'normal' beforehand and maintenance could be timed to periods with expected low prices. Low prices are however generally correlated with high inflows.

The EOPS model was developed for expansion planning and long to medium-term generation scheduling in predominantly hydropower production systems. It is mainly used for local planning, since it

is a single-area model with a single busbar and no grid. The optimal scheduling of hydro-resources is sought in relation to uncertain future inflow and market prices, taking into account specified constraints, contracts, demand and available thermal generation capacity. Both inflow and market price are stochastic variables in the model.

The long-term model consists of two parts:

1. A *strategy evaluation part* computes a decision table in the form of expected incremental water values for an aggregate model of the hydropower system. These calculations are based on use of a stochastic dynamic programming (SDP)-related algorithm.

2. A *simulation part* simulates optimal operational decisions for a number of corresponding inflow and price scenarios. Weekly generation is determined based on the incremental water value table calculated in the model's strategy part. Aggregate hydro generation is for each week distributed among available plants using a rule-based reservoir drawdown model containing a detailed description of the modelled hydro system.

Hydropower is represented in a fairly detailed manner, as indicated in Figure 2, based on use of standard plant/reservoir modules as shown in Figure 3. Flatabø et al. 1998 includes a detailed description of the properties that may be attached to each hydropower module.

The model may include thermal generation capacity, local demand, and other types of contracts for electricity sales or purchase, as indicated in Figure 2. For our analyses, however, only a spot market represents the market. This is modelled using a price forecast consisting of different scenarios for price development. Forecasts for market price are obtained by using models such as the EMPS model from SINTEF Energy Research, Haugstad & Ris-

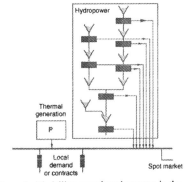

Figure 2. Modelling a producer's system in the EOPS model.

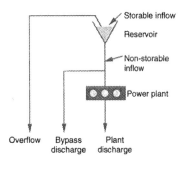

Figure 3. A hydropower system is modelled using standard plant/reservoir modules.

mark 1998. The system owner is considered to be a price taker, as it is assumed that short-term variations in generation do not influence the market price.

The EOPS model can include specified maintenance periods for specified units, but the model does not find the optimal timing of maintenance projects. In the current version of the model this has to be done by running the model for different maintenance timing and comparing the economic results. Production losses due to maintenance are given by the difference between the sales income for a run with no scheduled maintenance and the sales income with maintenance included. These costs may depend on the current state (market price, reservoir levels) of the system. For example, if the current reservoirs levels are much above normal, it may be that maintenance should be delayed.

In the I module forecasted maintenance costs for the current year and for all the other years in the planning period are needed. The current state of the system may effect the system for about three years in the Norwegian hydropower system. For the following years we assume that the production system is stable and that it is only necessary to compute maintenance costs for the third and for the last year in the planning period. This simplification is done in order to reduce computation time. Maintenance costs for the intervening years are found by linearization. Even if forecasted market prices are time varying, this will usually be a reasonable simplification.

We have not included an example of computation of maintenance costs with the module because these calculations and types of results are identical to what is shown in the next section. Only the running mode of the model is different; maintenance is seen in advance and the hydro schedule is adapted to the maintenance plan.

4 LOSSES CAUSED BY DISRUPTED PRODUCTION

A key economic figure for maintenance selection is the potential revenue loss if a project is postponed or cancelled and a major breakdown occurs. Usually, power plants in waterways with low storage capacity (compared to inflow) are more sensitive to halts in operation than power plants in waterways with large reservoirs able to store inflow for longer periods. The EOPS model has been adapted to and applied to estimate the potential energy and revenue losses related to unplanned outages. A more detailed description of this module is found in Tangen et al. 1999.

The EOPS model is designed to simulate system behaviour assuming a *normal* state of operation, including planned outages. The model has been modified to simulate *unplanned* outages, so that the operational strategy is not adapted to the future outage being analysed. In practice this module is the same

as described in the previous section, but the model is run in a different mode.

The timing of an outage has to be decided. One option is to simulate random outages. This would require simulation of a great number of scenarios, and also a decision about the probability and duration of a breakdown. At this stage, however, this is not what we are looking for. Our strategy is to find out how much an outage is likely to cost (in lost revenues) *if* one occurs, and then evaluate the probability of occurrence as a separate operation. In the chosen approach the timing and duration of an outage is specified to the model for each simulation. Since the expected loss of revenues from an outage is likely to depend on the time of year, several analyses are conducted for each plant with outages placed at different times of year. Simultaneous outages of several plants in the same watercourse have not been considered. This is a reasonable simplification, considering the high reliability of a hydropower plant. Expected value of production loss due to a given unit failure is calculated by simulating the production system with and without the specified unit failure for possible price and inflow scenarios.

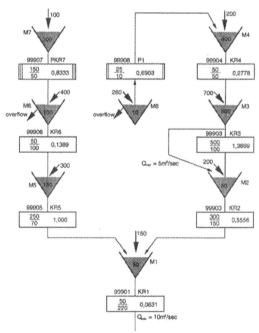

Figure 4. Example of hydro production system.

Figure 4 shows an example of a hydro production system. The system consists of 6 hydropower plants, one pump (P1) and one pump storage (PKR7). The numbers in the figure show expected inflow

(Mm³/year), maximum discharge (m³/s), corresponding production (MW) and maximum storage capacity (Mm³). We have used the EOPS model to calculate the expected losses for disrupted production in plant KR4. This plant consists of one production unit with production capacity of 50MW. We assume that the repair time for a failure will be nine weeks and calculate the expected losses caused by disrupted production for four different parts of the year. The results are shown in Table 1. If we assume that the probability of a unit failure is independent of the time of the year we can estimate the expected loss due to plant failure by taking the average of the four values in Table 1.

The results show that production losses are very dependent on when production is disrupted. In our example, the reservoir above the plant (M4), has relatively large storage capacity compared to expected yearly inflow. In this case plant failure during the winter period is much more costly than for the other periods of the year.

Table 1. Expected losses caused by disrupted production in plant KRV4 for four different parts of the year.

	Week 6-14	Week 21-29	Week 31-39	Week 42-50
Expected losses (kNOK)	7848	65	952	992

The values in Table 1 are the expected values for 60 different inflow and price scenarios, which are assumed to have equal probability. Figure 5 shows simulated losses for each scenario, assuming plant failure from week number 6 to 14, and the expected loss as a straight line. The figure shows that the losses are very dependent on future market price and inflow, i.e. the scenario number. These values correspond to column one in Table 1. The figure shows a simulated profit (about 1.0 mill NOK) for scenario 20 if production is disrupted. This is because this scenario is a extreme low price scenario with normal local inflow. In this case we are able to store the inflow in the reservoir and were lucky to produce later at higher prices. This might happen also in real operation. Scenario number 31 is an extreme high price scenario with inflow a bit above normal resulting in large losses if the production is disrupted.

Since our integrated model does not include risk aversion, only the expected cost of a given unit failure is passed on to the I module. However, it is possible for the user of the B module to check detailed results for a given plant failure. Detailed results include simulated reservoir operation, production, overflow etc. for all price and inflow scenarios. These are the same type of results that are available when the model is used to calculate the production losses for a given maintenance.

5 STATISTICAL MODELLING OF FAILURES

The output from module B, described in the previous section, is the expected loss *if* an outage occurs. We are, however, interested in the expected *annual* costs due to forced outages. Thus, the output from module B must be multiplied with the corresponding failure probability. To achieve this, statistical models for failures in individual components in the power plant are needed. Registration of failures

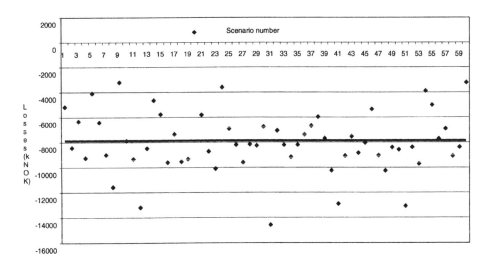

Figure 5. Simulated production losses for different inflow and price scenarios.

in existing power plants will be used as input to the probability distributions. To provide the necessary data basis an extension of the FASIT reliability data collection system (Heggset & Kjølle 2000) has been developed to suite maintenance purposes. This system specifies formats for registration of failures, inventories, maintenance parameters and relevant external conditions for various components. Data from the system will be used to develop probability distributions for failures in a given power plant. However, this requires flexibility in the software so that a user may be able to calibrate the models for specific projects. The calibration will depend on the relevant external conditions, maintenance actions, incidents and operational patterns.

The tool we are developing focuses on the unit level, i.e. the module calculating the losses caused by disrupted production will only consider failures that lead to disrupted or reduced output from the plant. This means that failures with other consequences will be neglected in the model. Due to this we must perform a Failure Modes, Effect and Criticality Analysis (FMECA) for the plant and use only the failures that cause disrupted production in the further analysis.

After mapping all possible failures, the probability distribution for each relevant failure must be estimated. These distributions will be used as a basis for calculating the resulting failure probability distribution for the whole plant. Module C will help the user to choose or estimate the probability distributions that will be used together with the results from module B to calculate expected annual costs due to disrupted production. These costs will be compared with the costs associated with relevant maintenance actions performed to reduce the probability of the failure in question. This optimization will be performed in module I (Figure 1).

6 EVALUATION OF QUALITATIVE CRITERIA

When evaluating projects, also non-economic (qualitative) criteria should be considered. Examples of such criteria may be a project's impact on safety and environment. Several methods have been developed for handling qualitative criteria in a structured manner; among these are the Multi Criteria Decision Making (MCDM) methods. Tangen 1996 describes the theoretical fundament for MCDM and practical implementation of such methods are discussed.

MCDM is made for formalizing the decision process using decision models and value functions to describe a project's impact on predefined criteria.

By using such methods the decision process is improved in several ways. Examples of improvements are:

- Standardized procedures for evaluating the project's qualitative utility value
- Objectivity and consistency when comparing projects
- Establishment of a systematic information basis
- Improved documentation of decisions

MCDM is included in the integrated model in addition to economic analyses in the evaluation of projects. This is illustrated by module (G) in Figure 1.

Based on the identified criteria applicable for the company, a decision model is established. Furthermore an evaluation is performed where the criteria's importance compared to each other is decided.

The resulting model contains numerical weights of the criteria and also scales suitable for each criterion for the projects in question.

When combining the decision model with information from each project, a project specific utility value is computed.

An example of the structure of a decision model is shown in Figure 6.

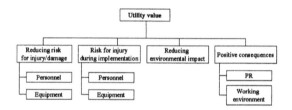

Figure 6. Example of decision model.

Figure 7 shows an example of how results from a MCDM-analysis can be presented.

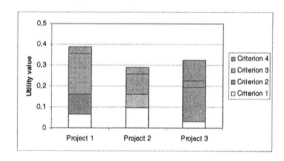

Figure 7. Presentation of utility value results.

7 OPTIMAL PLANNING

The purpose of the essential module (I) of the concept is to calculate optimal maintenance plans. The result is a list of projects to be carried out in the current year and a maintenance schedule for the rest of the years in the planning period. An example of how this could be presented is shown in Figure 8.

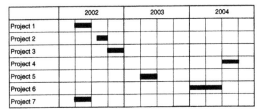

Figure 8. Example of optimal maintenance schedule.

The optimization will be based on dynamic programming and take into account financial and personnel constraints defined by the user of the model. This optimization module has a yearly time resolution.

The objective function is to minimize the total present value of sum relevant costs over the planning period. The relevant costs include:

- Investment costs
- Lost production due to maintenance
- Expected costs of disrupted production
- Decrease in production due to reduced efficiency

Qualitative utility values will not be included in the optimisation but be part of the presentation for the optimal projects.

Input to this module is a list of possible projects with corresponding investment costs and HES values as indicated in Table 2.

Table 2. Example of project description.

	Project description	Investment cost	Public relations	Environment	Safety
		mill NOK	scale 1-4	scale 1-4	scale 1-4
Project 1	New runner plant C	5	1	2	1
Project 2	Turbine maint. plant B	4	1	3	2
Project 3	Generator maint. plant D, unit 2	17	2	4	1
Project 4	Generator maint. plant D unit 1	15	1	3	2
Project 5	New hatchet plant C	3.5	2	2	3
Project 6	New runner plant A	4.5	1	1	2
Project 7	New runner plant D	6	1	2	2

The state variable in the optimisation is a number that describes the state of the whole production system for a given year. Transition from one state to another in a time period implies maintenance in one or more units in that period. Many transitions are not possible, or obviously not optimal to do. Table 3 shows an example of the connection between state variable and project investments. For example, going from system state 1 to system state 2 represents investment in project 2, project 1 is already done. In order to cover all possible combinations the number of rows in the table can be very large even for a limited number of projects. It is therefore important to use relevant information to reduce the number of system states. Expected costs due to production losses during the maintenance period and changes in expected losses due to plant failures are as mentioned before calculated for each possible state transition by the B and E modules.

Table 3. Possible specification of system state in Dynamic Programming approach.

System state	Project number						
	1	2	3	4	5	6	7
1	x						
2		x					
3			x				
4				x			
5					x		
6						x	
7							x
8	x	x					
9		x	x				
10	x			x			
11		x	x	x			
12			x	x	x		
13			x	x			x
14		x	x	x	x		
...							

8 FURTHER WORK

Further development of the integrated concept will be split into two phases:

In phase one we will develop the model as shown in the Figure 1, but the model will not include optimisation of the maintenance schedule. The model only calculates sum profit and qualitative utility values (eg HES) for a given maintenance schedule. The system will be user friendly and the user will be able to calculate and compare sum profits and check personnel and investment constraints for many different maintenance schedules in a relative short time. The results will also include a graph that shows profit and qualitative utility values.

In phase two we will include optimisation of the timing problem. This development phase includes a further development of the interaction between the modules.

This two-phase development will allow us to deliver a useful tool at an early stage. Also, utilities with only a few plants may not need the optimisation of the timing.

The B and E modules in Figure 1 have already been delivered to several utilities and are used in current maintenance planning. These modules only need minor changes in order to fit into the integrated concept.

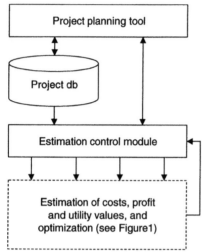

Figure 9. Integration with project planning.

Module A in Figure 1 represents the maintenance projects, which are the basis for the calculation of optimal maintenance plans. Relevant information about the projects is normally managed by the utility's project planning tool and stored in a project database (see Figure 9). The estimation and optimisation model in Figure 1 will be a module, that will support the project planning tool with complex analyses when that is needed. Figure 9 illustrates one solution of integration with the project planning tool and the project database. In addition to the model in Figure 1 we will develop a control module (see Figure 9), whose purpose is to communicate with the project planning tool and the project database as well as control the estimation and optimisation processes, including the data flow.

We believe that the pre-screening process in order to reduce the number of projects for detailed analyses, as mentioned in Section 2, should be handled by the control module. This solution is favorable regarding a 'standard' interface with various project planning tools.

One utility has already established a 'manual' interface between its project planning tool and the modules B and C in Figure 1. The interface in terms of manual routines, specifies how to apply the two modules in maintenance planning. Estimation of profit losses is performed based on i.a. data from the project database, and estimated cost figures are then stored in the same database. All data entries are car-

ried out manually. The control module will improve this interface and the whole planning process. The manual operations will be automated as far as it is appropriate from the users point of view.

Utilities use various kinds of project planning software. We will therefore specify an interface with our control module, which is flexible with regard to this problem.

9 CONCLUSIONS

This paper presents a new tool for maintenance and refurbishment planning that is currently under development. The tool consists of several modules that may be run separately. The concept introduces several improvements to current practice. The model includes evaluation of both economic and qualitative values. Physical connections and financial or personnel limitation which have implications for the project prioritising are included in the model. The model calculates automatically the optimal maintenance schedule and thus reduces the manual work necessary.

The model should give improved maintenance decisions and documentation of why the decisions were taken. It will also provide for a uniform documentation of all possible maintenance projects.

10 REFERENCES

Botnen, O.J., Johannesen, A., Haugstad, A., Kroken S. & Frøystein, O. 1992. Modelling of Hydropower Scheduling in a National/International Context. *Proceedings from Hydropower '92.* Rotterdam: Balkema.

Flatabø, N., Haugstad, A., Mo, B. & Fosso, O.B. 1998. Short-term and Medium-term Generation Scheduling in the Norwegian Hydro System under a Competitive Power Market Structure. *EPSOM'98 (International Conference on Electrical Power System Operation and Management), Switzerland*

Haugstad, A., Mo, B. & Belsnes M. 1997. Evaluating Hydro Expansion or Refurbishment in a Deregulated Electricity Market. *Proceedings from Hydropower'97, Norway, pp. 271-277.* Rotterdam: Balkema.

Haugstad, A. & Rismark, O. 1998. Price Forecasting in an Open Electricity Market based on System Simulation. *Proceedings from EPSOM'98, Switzerland.*

Heggset, J. & Kjølle, G. 2000. Experiences with the FASIT Reliability Data Collection System, *Proceedings from IEEE/PES Winter Meeting 2000, Singapore.*

Tangen, G. 1996. Decision making support applied to hydropower plant upgrading, *Dr. ing. thesis, Norwegian Institute of Science and Technology.*

Tangen. G., Haugstad. A., Green, T. & Teppan, P. 1999. Planning refurbishment based on analysis of Potential Losses, *Uprating and Refurbishing Hydro Power Plants, Berlin, Mai 1999.*

Hydropower in the New Millennium, Honningsvåg et al (eds), © 2001 Taylor & Francis, ISBN 90 5809 195 3

Economie risk assessment for hydro plants using computer software

M.Morgenroth
Acres International, Niagara Falls, Ontario, Canada

F.Welt & H.W.de Meel
Acres Productive Technologies, Niagara Falls, Ontario, Canada

ABSTRACT: In order to decide whether an equipment upgrade, the rehabilitation or the refurbishment of any component of a hydroelectric facility is economically feasible it is essential to consider not only capital costs and benefits, but take into account the risk exposure associated with the aging equipment. Quantifying this risk exposure in terms of a cost stream hinges on a good understanding of the probability of failure as it varies over time and the consequence costs of a failure. Within an integrated, interdependent system of components, the failure probability experiences a coupling between individual components through the effect of one failure on another. Further, the costs associated with risk mitigating interventions is also coupled through the effect of outage concurrence. It is desirable that timing and type of risk mitigation interventions be selected in an optimized manner. This optimization is best accomplished by the use of a transparent and rational process in the form of a computerized algorithm. Building on and briefly recapitulating the methodology used for individual components, the present paper will discuss the methodology to calculate a coupled cost target function for a system of interdependent components. Further, the optimization algorithms used in the HydroVantage software are described. An application of the method is reported to underscore the potential benefits of the computerized method versus other less structured capital planning approaches.

1 CONTEXT

To manage hydroelectric generation assets successfully means to find the best course of action to maximize generation revenue, minimize costs such as maintenance, outages, repair or component replacements, and to maintain the value of the asset over the long term.

Given the fact that more than two-thirds of all hydro stations in North America are 40 years or older, most plant owners and operators need to deal with aging equipment. Effective tools are required to support decisions about the available options of continued operation, retirement, rehabilitation or replacement of equipment. Questions about an old piece of equipment need answer such as:

- What is the probability that it will fail?
- What would happen if it should fail?
- What should be done with the component? - Rehabilitate or replace?
- If required, when should it be rehabilitated or replaced?
- Which projects should be done first within available capital expenditure budgets?

The current approach to capital expenditure planning is mostly based on past and present performance and engineering judgment alone. Neither formal economic life cycle evaluation nor analytical optimization of intervention timing is conducted. This intuitive approach may easily result in intervention priorities being incorrectly assigned, interventions being incorrectly timed, large savings remaining unutilized or unnecessary expenditures and cost penalties being incurred.

Computerized optimization algorithms can help to translate the available technical and cost information into a format that empowers to make an informed decision. It must be noted that engineering judgement is still required - however, it is based on defined, organized and quantified information. Further, with the use of user-friendly decision support software sensitivity analyses on assumptions and estimates can be undertaken rapidly while maintaining consistency in the approach.

2 CALCULATED EQUIPMENT RISK

The risk-based methodology recognizes that exposure to risk is a real cost and has to be accounted for like any capital or Operating and Maintenance (O&M) expense. Various publications exist that deal with the application of this methodology in the hydroelectric field (de Meel et al. 1997, Bhan et al. 1998, de Meel & Donnelly 1998, de Meel & Westermann 1999, Morgenroth et al. 1999, de Meel & Morgenroth 2000, and Westermann et al. 2000). By its statistical nature, it may not affect each and every plant or component in exactly the predicted way, but if sufficient time is permitted to pass and a sufficient number of components are observed, the statistical average will represent reality well[⊕]. As a matter of definition, the risk-cost used throughout this paper is the cost arising from a failure event, directly or as a consequential associated cost, factored by the likelihood of this event occurring.

$$R = p \cdot C \qquad (1)$$

Where R = Risk cost; p = probability of occurrence of failure event; and C = consequence cost that is incurred for this failure event.

Typically an initial period of early, infancy failures is followed by a period where failure is entirely random, i.e. the failure rate is constant. As wear-out of the equipment sets in, the likelihood, or probability of failure, for a component increases typically with the age of the equipment as shown in Figure 1. Most equipment exhibits this typical behavior, but the parameters describing the failure probability curve can only be determined from statistical analysis of historic data or a physics-based. capacity demand approach (Morgenroth et al. 1999).

If the piece of equipment under investigation is similar to the equipment represented by the failure probability curve and its service conditions are also similar, then application of the failure-probability curve for prediction of the future likelihood of a failure will provide a realistic estimate.

However, the actual position of a specific piece of equipment on the failure-probability curve is determined by its representative age, not its calendar age. To obtain the representative age, a condition assessment, typically expressed through a condition index, is required. To support the condition assessment and detect trends in the condition of a component traditional methods which are conducted at discrete time intervals during equipment shutdown are

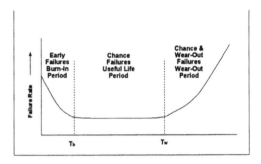

Figure 1. Typical reliability development of a component during its life

supplemented by modern on-line tools, such as partial discharge analysis (PDA) or air gap and vibration monitoring.

In determining the best timing for implementing a risk-mitigating intervention, one must consider the least total costs in today's present worth. This means on the one hand that money spent on an intervention later on costs less. On the other hand, the risk (cost) increases if a risk-mitigating intervention is deferred. Figure 2 shows how this balance constitutes for a single individual component an one-dimensional optimization problem in terms of the intervention timing.

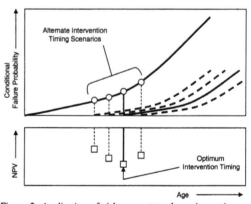

Figure 2. Application of risk concept to determine optimum timing of a risk mitigating intervention

3 FORMULATION FOR AN INTERDEPENDENT SYSTEM

For an interdependent system the situation is more complex than just summing up all costs for the individual components in this system. Coupling effects exist which can increase or decrease the total cost arising during the study duration. These total costs are defined as

$$CTF = \sum_{i=1}^{n} NPV(annualcosts_i) \qquad (2)$$

Where CTF = Cost Target Function, annual costs$_i$ = sum of capital intervention costs, O&M costs, benefits (negative costs) and risk costs in year i, i = counter for year starting at 1 and running to n, the end of the study duration, and NPV( ) = application of a net present value factor for the year, i:

$$NPV(cost_i) = cost_i \cdot \frac{1}{(1+d)^i} \qquad (3)$$

Where d = constant annual discount rate.

Coupling arises from the fact that a failure mode can affect a subsequent consequential failure on the same or other components. Therefore, the total risk cost, which is the sum of the individual risk costs for each component's failure modes, is increased by a factor corresponding to the likelihood of any failure mode causing another failure mode:

$$R_i = (p_i + p_j \cdot p_{ji}) \cdot C_i \qquad (4)$$

Where R_i = risk cost for failure mode i, p_i = probability of failure mode i individually, p_j probability of failure mode j individually, p_{ij} = probability of failure mode j causing failure mode i, and C_i = consequence costs of failure mode i.

Additionally, coupling arises from the fact that if interventions that are planned in the same year, typically some overlap of the outage time to implement an intervention mode exists. For example, if an unit is taken down to implement a generator rewind then possibly little additional outage costs arise to do refurbishment work on the turbine, transformer or, depending on the water passage configuration, even the penstock or intake gates. If individual intervention modes are outage coupled then the outage costs are directly entered into the program replacing the term:

$$\sum_{i=1}^{n} od_i \cdot or_i \qquad (5)$$

Where i = counter that runs from 1 to n, n = total number of intervention modes for all components that concur in a given year, od_i = outage duration to implement intervention mode i in days, and or_i = daily rate for a planned outage caused by intervention mode i.

This approach of prescribing a single cost for a combination of interventions would require detailed scheduling and cost estimating for all combinations of interventions that can possibly concur. Since such a level of detail in data preparation is not practical for a large system, another more global approach to model outage concurrence coupling was introduced into the software. Expressed as a global degree of concurrence the user may enter his concurrence input on a percentage scale between 0%, and 100%. In this scale 0% is equivalent to all outages occurring sequential if implemented in the same year, and

100% to all intervention outages having a complete concurrence with only the intervention requiring the longest outage governing the total outage costs.

4 OPTIMIZATION ALGORITHMS

Optimization is a common task in many areas of technology, business and daily life. Loosely the term is used for the process of finding the best solution to a problem.

In the present context, however, the term optimization is used more specifically. It means to find the minimum value of the scalar cost target function, the total costs over a predefined period of time, which is dependent on a vector of independent values which are the times at which each intervention mode is implemented.

4.1 Stochastic dynamic programming (SDP)

Dynamic Programming is a classic optimization approach, particularly well suited for problems whose dimensionality is low. It has been applied to a wide diversity of topics in Operations Research, such as Water Resource Management and Power Systems Control (Wurbs 1993, Hachem et al. 1997). A similar formulation to the present one is described in Dreyfus & Law 1977, under the title of the "Equipment Replacement" problem.

The system dimensionality is determined here according to the number of interdependent components. When this number of interdependent components within the plant is relatively low, such as two or three, the Dynamic Programming approach can provide a very efficient, robust and complete resolution of the problem. Furthermore, the problem at hands is discrete, nonlinear and fundamentally stochastic, which makes it difficult for other traditional approaches to solve without making a significant number of approximations.

4.1.1 Problem formulation

A number of different intervention modes can be applied to every component within a plant, at various time steps. The formulation allows for finding the best combination of interventions that minimizes the overall cost over the study period. A multiple number of interventions can be applied to the component, as the choice of intervention mode at any given time step constitutes the basic discrete decision variable of the problem.

The stochastic nature of the problem is introduced by calculating the mathematical expectancy of the cost function, defined here as the product of the probability of the future failure or non failure events times their costs. All possible events can be considered in the SDP formulation, i.e., the component can fail once, several times, according to different modes, etc.

The nonlinear characteristic of the problem comes from the fact that the future probabilities of failure depend on the present decision and its cost. This is illustrated in the following paragraph.

4.1.2 Target Function

The stochastic cost function (SCF) can be written at time $T0$, as:

$$SCF = \sum_i p_{faili(T0)} \cdot C_{faili(T0+1)} + (1 - \sum_i p_{faili(T0)}) \cdot C_{nofaili(T0)}$$

(6)

Where p_{faili} = probability of failing in mode i in the current time step, C_{faili} = cost of failing in mode i in the current time step, $C_{nofaili}$ = cost of not failing in mode i in the current time step.

The recursive element of Equation 6 is based on the fact that the cost of not failing in a given time step t is the cost of failing in time step t+1 times its probability, plus the cost of not failing in time step t+1, times its probability, respectively. Mathematically,

$$C_{nofail(T0)} = \sum_i p_{faili(T0+1)} \cdot C_{faili(T0+1)}$$
$$+ (1 - \sum_i p_{faili(T0+1)}) \cdot C_{nofaili(T0)}$$

(7)

4.1.3 State variables

The failure costs are strictly dependent on the age of the component. On the other hand, the O&M, capital and performance benefit costs are dependent on the history of intervention on the component. Although the history of intervention affects the component age through its rejuvenation characteristics, it does not define it completely because failure can also affect the component age (e.g., through complete or partial destruction). There are therefore two state variables that define the components:

– Component age (referred to as "age state")
– History of intervention (referred to "intervention state")

There are quickly many possible age states to the problem, as the component can fail in succession, under different failure modes, etc. However, the number of intervention states is generally quite low (number of defined intervention modes with benefits).

4.1.4 Resolution

A backward Stochastic Dynamic Programming approach has been used, where the problem is optimized at each stage (time step of a year), for every

possible state of the system, starting from the end of the study period. The least cost path to time step 0, for the current age of the component, provides the optimal SCF. The process is illustrated in Figure 3.

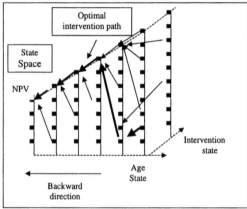

Figure 3: Search for the optimal SCF

4.1.5 Other Considerations

Aggregation of the age state has been found to be very effective at reducing the dimensionality of the problem and increasing performance, without a significant impact on the accuracy of the solution. Without interdependencies, a multiple failure and intervention mode problems can be solved within a fraction of a second, whereas a two interdependent component problem is generally of the order of a few seconds when aggregation of the age state variable is applied. The method provides a global optimal, which can be used to compare against the results of the Downhill Simplex approach.

4.2 Downhill simplex

4.2.1 Description of Method

The downhill simplex method (Press et al. 1992) is suitable for higher dimensional problems and uses only evaluation of the target function itself. No knowledge of its derivatives is required. Its algorithm is self-contained in that it does not require an one-dimensional optimization algorithms to perform.

This method competes mainly with directional methods in that both methods are suitable for multi-dimensional problems, don't require explicit knowledge about the target function derivatives and require data storage proportional to the square of the problem dimensionality. However, the downhill simplex has shown in the present application as well as in the application to an inverse design problem (Favaretto et al. 1998) to be superior in terms of robustness and speed.

The method is geometrically inspired and makes

use of a simplex, i.e. a figure with n+1 corners or vertices for an n-dimensional problem. This figure, starting from an initial guess, reflects, expands or contracts to "roll" down the n-dimensional topography of the solution domain to converge at a minimum

4.2.2 Solution Constraints

The solution domain is bounded by the problem constraints. These constraints limit the implementation timing to neither reach into the past (which can in the non-Orwellian context not be changed) nor go beyond the horizon of the study duration (where it was decided in the problem definition that costs are inconsequential for the present day). However, not to implement an intervention is a valid implementation outcome and is coded with the value zero for the affected intervention mode.

For the cost target function itself no constraints exist in that both positive and negative values are permissible. In the latter case the benefits of a capital plan outweigh its costs.

4.2.3 Solution Strategy

To get started the algorithm needs an initial guess of the geometry and a set of basis vectors for the solution space. Since any multi-dimensional optimization problem is plagued by the nagging question whether the optimum found is truly global or only local it is good practice to use as much information about the problem as can be obtained.

Knowledge that can be obtained at relatively cheap computational cost is the best timing of an intervention for the one-dimensional problem of a single individual intervention mode. Further, it is known that the outage concurrence coupling of intervention modes always yields a less expensive cost target function than an equivalent uncoupled system.

Therefore, this information is used to construct the initial simplex placing its vertices on points in the domain that represent individual optimum timing and outage concurrence points.

4.3 Comparison of optimization methods

Marked differences between the two approaches exist. While the dynamic programming approach deals with the stochastic values of risk costs according to their true nature the downhill simplex method uses a pseudo-deterministic approximation.

What is the advantage of the former approach, i.e. an accurate model of all possible paths through time, for a system of one or two components becomes its downfall for models of higher dimensionality. This is where the approximate nature of the downhill simplex is more advantageous.

For the user it translates into the following profile for the two approaches (see Table 1)

Table 1. Comparison of Downhill Simplex and Stochastic Dynamic Programming optimization approach

	Downhill Simplex	SDP
Stochastically Accurate	No	Yes
Suitable for Individual Components	Yes	Yes
Suitable for Larger Coupled Systems	Yes	No
Able to Optimize Intervention Mode and Timing	Timing only	Yes

5 APPLICATION

The application for the HydroVantage model described above varies broadly. Potential uses include:
- Devise a risk-based capital expenditure (CapEx) plan for a whole plant or individual components
- Review an existing CapEx plan from a risk-based perspective
- Demonstrate viability of a rehabilitation or upgrade project
- Prioritize competing projects
- Determine optimum timing for a viable project
- Determine the optimum intervention mode for a component at risk
- Determine costs of deferring or anticipating an intervention (if overruling reasons exist to do so)

Similarly, the potential user group is diverse:
- Operating or maintenance staff
- Maintenance manager
- Station/plant manager
- Engineering staff
- Asset manager

The following application example was carried out on behalf of the station management to review the 20-year CapEx plan developed by the owners' head office engineering group and asset managers.

5.1 Plant description

The utility-owned plant, built in the mid 1950's, consists of two units with a combined installed capacity of 60 MW which deliver a total of about 280 GWh of electricity per year.

The equipment to be modeled was selected in a screening process outside the actual software using general and specific outage statistics. The model components were selected to include the equipment that posed the greatest risk exposure. The generating equipment including windings, excitation system, bearings, turbine regulation, runner and governor as well as the transformer and parts of the water conveying system of one unit were part of the model. Figure 4 depicts a diagram of the modeled components as well as their failure modes and intervention modes.

5.2 Application of individual component optimization

In an initial analysis step components were analyzed on an individual basis without considering their interdependencies. Table 2 shows a comparison between the existing CapEx plan and that devised from a single component risk-based analysis.

The largest difference between the existing and risk-based plan can be realized by introducing benefit driven interventions into the CapEx schedule. This shows most prominently for the intervention modes "upgrade" on the turbine and remove on the inlet valve. Between these two components a difference of about $ 3.25 M can be realized. However, the changes in timing on the risk driven interventions (exciter replacement, turbine overhaul and governor replacements) are in themselves attractive and a difference of about $ 1.85 M can be realized there. The benefit driven and risk driven amounts are not exactly cumulative, because of an overlap in components and for a comprehensive intervention schedule solution a fully interdependent model needs to be consulted.

5.3 Application of interdependent system optimization

5.3.1 Effect of failure coupling

To investigate the effect of interdependencies of failure modes, or failure coupling, alone the components whose failure modes had a chance of inducing other failure modes were grouped together and modeled. These groups can be identified in Figure 4 as being linked by dashed lines.

It was found that the failure coupling, which can be viewed in some way as a secondary or consequential damage cost associated with a failure mode, is small enough to generally not affect the intervention schedule significantly. Only for one intervention mode, the transformer replacement, was a different optimum intervention time by one year determined. The associated total net present value of all costs over the entire study duration, short NPV, could only be compared directly between individual analysis and failure-coupled analysis where the failure coupling remains within a single component as is the case for the generator and transformer. In Table 3, the difference refers to the difference in NPV between failure coupled and individual intervention mode optimization The small difference relative to

the absolute value of NPV confirms that the failure coupling is in deed minor.

However, this application may be peculiar in that the components which may be subjected to expensive failure modes such as the water conveyances or civil structures are outside the scope of components that were analyzed. To account for these failures their costs are factored into the risk costs of failure modes of the components which are part of the scope, i.e. the costs are treated implicitly. Therefore, it can be asserted that in a model that includes such components explicitly within its scope a strong failure coupling may be present.

5.3.2 Effects of Outage Coupling

A further increase of sophistication in the model is the coupling of intervention modes through their outage concurrence. The recommendations to the Owner's were derived from these results.

Interdependency between intervention modes, or outage coupling, arise from the fact that implementing multiple interventions simultaneously saves on outage time. Typically and also in this application the high value of outage costs relative to costs for material and labor create a strong outage coupling in the simulation.

Unlike failure coupling, which increases only the computational time that is required for each combination in the intervention schedule, the outage coupling increases the number of combinations of intervention modes that form an intervention schedule. This affects computational time much more dramatically than failure coupling through increasing the dimensionality of the optimization problem. For this reason, a step-wise solution strategy was adopted where outage interdependencies were investigated first in pairs, then in groups of three, four, five and finally for all 13 relevant intervention modes.

From the pair-wise model runs, it can be learned that the concurrence of outages results in a strong affinity between intervention modes. Over and above the net gains from an optimization of the individual modes, up to 30% improvement relative to the NPV for single optimization can be gained. For three to five simultaneous intervention modes, this percentage rises to 150%.

For high problem dimensionality, 13-dimensional in the present application, it can never be proven rigorously that a global minimum of the NPV costs has been found because above about five intervention modes, the complete mapping of the solution domain becomes too computationally expensive. Therefore, various optimization start simplices were tried and compared based on intuition and previous lower-dimensional results. With such a solution strategy high confidence in the global validity of the found minimum NPV cost can be assured, in spite of the lack of complete rigor.

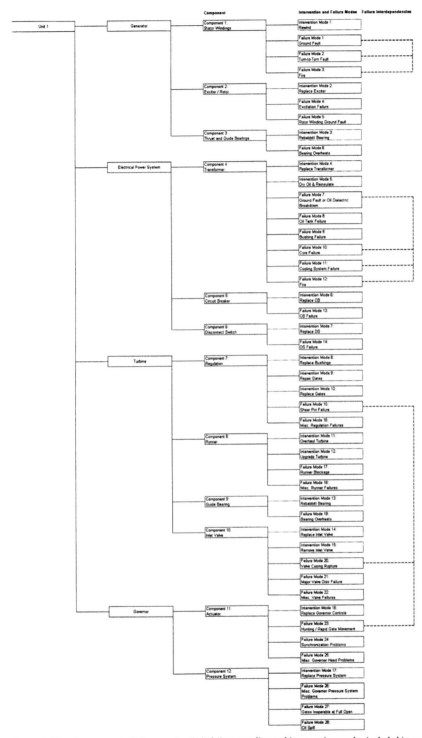

Figure 4. List of components, failure modes, their failure coupling and intervention modes included in analysis

Table 2. Comparison of existing CapEx plan and risk-based one using only individual component analyses

Component	Intervention Mode	Existing 20-year CapEx Year	NPV	Risk-Based Analysis Year	NPV	Difference in NPV
Generator Stator Winding	Rewind	2011	$ 2146 k	2009	$ 2145 k	$ 1 k
Exciter /Rotor	Replace	2001	$ 642 k	2017	$ 509 k	$ 133 k
Thrust and Guide Bearing	Rebabbitt	n/a	n/a	don't	$ 323 k	n/a
Transformer	Replace	2010	$ 829 k	2009	$ 826 k	$ 3 k
	Overhaul	now	$ 1152 k	2004	$ 1123 k	$ 29 k
Circuit Breaker	Replace	don't	$ 166 k	don't	$ 166 k	$ 0
Disconnect Switch	Replace	now	$ 146 k	now	$ 146 k	$ 0
Turbine Regulation	Replace Bushings	n/a	n/a	now	$ 359 k	n/a
	Overhaul Gates	2011	$ 451 k	don't	$ 265 k	$ 186 k
	Replace Gates	2011	$ 292 k	don't	$ 265 k	$ 27 k
Turbine	Overhaul	2011	$ 1316 k	don't	$ 46 k	$ 1270 k
	Upgrade Runner	2011	($ 1504 k)	now	($ 4007 k)	$ 2503 k
Guide Bearing	Re-babbitt	don't	$ 169 k	don't	$ 169 k	$ 0
Inlet Valve	Replace	don't	$ 400 k	2002	$ 141 k	$ 259 k
	Remove	don't	$ 400 k	now	($367 k)	$ 767 k
Governor Actuator	Replace	2008	$ 337 k	don't	$ 151 k	$ 186 k
Governor Pressure System	Replace	2008	$ 304 k	don't	$ 19 k	$ 285 k

Table 3. Comparison of risk-based individual component analysis and only failure coupled analysis

Component	Intervention Mode	Individual Analysis Year	NPV	Failure Coupled Analysis Year	NPV	Difference in NPV
Generator Stator Winding	Rewind	2009	$ 2145 k	2009	$ 2203 k	$ 58 k
Transformer	Replace	2009	$ 826 k	2008	$ 836 k	$ 10 k
Turbine Regulation	Replace Bushings	now	$ 180 k	now	n/a	n/a
	Replace Gates	don't	$ 265 k	don't	n/a	n/a
Turbine	Upgrade Runner	now	($ 4007 k)	now	n/a	n/a
Inlet Valve	Remove	now	($ 367 k)	now	n/a	n/a
Governor Actuator	Replace	don't	$ 151 k	don't	n/a	n/a

Table 4. Comparison of existing CapEx with failure and outage concurrency coupled risk-based analysis

Component	Intervention Mode	CapEx Schedule	Optimum Intervention Schedule
Generator Stator Winding	Rewind	2011	2009
Exciter / Rotor	Replace	2001	2009
Thrust and Guide Bearing	Rebabbitt	don't	don't
Transformer	Replace	2010	2009
Circuit Breaker	Replace	don't	don't
Disconnect Switch	Replace	now	now
Turbine Regulation	Replace Bushings	don't	now
Turbine Upgrade	Runner	2011	now
Guide Bearing	Rebabbitt	don't	don't
Inlet Valve	Remove	don't	now
Governor Actuator	Replace	2008	don't
Governor Pressure System	Replace	2008	don't
Total NPV		$ 3348 k	-($ 518 k)

As shown in Table 4, NPV cost savings between the CapEx intervention schedule and the risk-optimized schedule amount to about $3.87 million or 115% of the total NPV for the existing capital plan.

6 CONCLUSIONS

Traditionally, risk costs are not considered in CapEx planning quantitatively in the form of risk cost stream forecasts. However, to translate engineering concerns over reliability of aging equipment into a financially tangible form the inclusion of risk costs is deemed an appropriate, consistent and transparent approach. The inclusion of these stochastic risk costs allows for a more realistic model on which to base decisions.

To compute risk costs, applicable and adequately researched failure probability curves need to be available which represent the equipment under investigation and for the life stage they are in.

Components can be analyzed in an individual, component-by-component fashion and these results already help to make better decisions than to decide by judgement and intuition alone. However, to reap the full benefit of a risk-based optimization analysis it is far superior to consider interdependencies arising from failure and outage concurrence coupling.

Significant savings are typically found between an existing traditionally prepared CapEx plan and one based on risk principles. For the presented application the difference in total costs reached a level where the revision of the CapEx schedule was worth more than the initial plan itself.

The HydroVantage model is a unique tool for the complex risk analysis of a system of interdependent components of a hydroelectric power plant that supports capital planning decisions. Easy access to the model is insured by providing it as an internet-based user-friendly application.

7 FUTURE WORK

Various improvements and enhancements to the presently offered features are planned as the software matures and its use becomes more wide spread. Some of the more significant are:

A major value of the software stems from the quality of the failure probability and component specific cost data that is provided as "default" values. To maintain the underlying data base a mechanism needs to be created to feed back information that resides with each individual user. Web-enabled polling software is planned for integration with the HydroVantage model.

Presently the model formulation represents explicitly only intervention modes that affect the component reliability. However, a class of interventions acts not on the component itself, but rather mitigate the failure risk by giving early warning of failure or by reducing the consequence costs or outage exposure, such as monitoring equipment and provision for spare parts at site. A more realistic formulation for such intervention modes is currently in progress.

Component interdependencies exist not only through failure and outage concurrence coupling, but also in some cases through benefits coupling. Such a coupling exists when the benefits for an intervention on one component are conditional on intervention on another component. An example for such a benefit coupling would be a runner upgrade It provides for a higher turbine capacity which may be constraint by the generator capacity and conditional upon a generator stator rewind. Formulations that allow for this type of coupling are presently under development.

REFERENCES

Bhan, K., de Meel, H.W., Gibbs, C.J. & Westermann, G.D. 1998. Life cycle management of hydro assets. CEA Report G9720008.

de Meel, H.W., Donnelly, C.R., MacTavish, B.A. & Walsh, H. 1997. The use of an advanced risk-based asset management system for evaluation of rehabilitation options for an aging hydroelectric station. CDSA/CANCOLD Conference.

de Meel, H.W. & Donnelly, C.R. 1998. Managing hydro assets rationality: a risk based approach". Hydro Review Magazine. December 1998.

De Meel, H.W. & Morgenroth, M. 2000. Internet-driven optimization of capital expenditure programs. Hydro 2000 Conference, Bern, Switzerland

de Meel, H.W. & Westermann, G.D. 1999. Hydro plant cost savings using risk management methodologies. Hydropower & Dam. October 1999.

Dreyfus, S. E., and Law, A. M. 1977. The Art and Theory of Dynamic Programming, Academic Press, N.Y.

Favaretto, C.F., Morgenroth, M & Ferreira, V.C.S. 1998. Design optimization methodology applied to a mixer. ENCIT - 7th Brazilian Congress of Engineering and Thermal Sciences, Recife, Brazil.

Hachem, S., Hammadia, A.G., Welt, F., and Breton, M. 1997. Dynamic Models for Real Time Management of Hydro-Plants, Proceedings of HydroPower '97, Trondheim, Norway.

Morgenroth, M. , Donnelly, C.R., Westermann, G.D., Huang, J.H.S. & Lam, T.M. 1999. Quantitative risk assessment using the capacity demand analysis. CDA – 2nd Annual Conference, Sudbury, Ontario, Canada.

Press, W.H., Teukolsky, S.A., Vetterling, W.T. & Flannery, B.P.1992. Numerical Recipes – The Art of Scientific Computing. Cambridge University Press

Rodenburg, R. L. 1995. Replacements, Units, Service Lives, Factors. Prepared for U.S.Department of Energy, U.S. Department of the Interior and U.S. Bureau of Reclamation. PO No. AA-P0-12652-22503.

Westermann, G.D. 1999. Penstock rupture: Assessing the risks, identifying potential problems. Hydro Review Magazine. July 1999.

Westermann, G.D., Bhan, K. & de Meel, H.W. 2000. Generator rewinds: A model to predict optimum timing. HydroVision 2000 Conference.

Wong, C. T. 1990. Determining O&M Costs Over the Life of a Hydro Station, Hydro Review. December 1990.

Wurbs, A.R.. 1993. Reservoir Simulation and Optimization Models, Journal of Water Resources Planning and management, Vol. 119, No. 4, July 1993, pp. 454 – 472.

Hydropower in the New Millennium, Honningsvåg et al (eds), © 2001 Taylor & Francis, ISBN 90 5809 195 3

Study of transient two-phase flow of the turbine start of the Coo-2 hydroelectric storage power station

MA. Motte & A.G.H. Lejeune
University of Liège, Laboratories of Hydro-mechanics, Belgium

ABSTRACT: In order that the Coo-2 storage hydropower station could uphold the electrical network rapidly, the owner plans to modify the turbine starting operation, to let the wheel running in the air and to open the valve and to expulse the plug toward the downstream reservoir with the result of two-phase transient and unsteady flow. The study describes the effects of the outflow on the holding of the turbine, the draft tube and pipes. Three approaches have been achieved: by computation, scale tests and site tests. The paper will present the most characteristic results of those studies.

For instance, we note there is a relationship between the evolution of the air pocket, after opening of the control valve and : - the degree of valve's opening
        - the speed of opening of the control valve.
A more advanced study of void fraction and velocities of bubbles, with double optical probes.

1 INTRODUCTION

The hydroelectric storage hydropower plan of Coo - Trois Ponts of ELECTRABEL located in the Belgian Ardennes, has a power of 1.100 MW about with 6 Francis turbines, 3 of 160 MW and 3 of 205 MW, head varying between 230 to 275 m.

The figure 1 gives a schematic view of the plan.

Figure 1. Schematic view of the Coo-Trois Ponts site.

If in a first time, 3 minutes were needed to start in fully automatic operations to supply power to the trash rack from 0 to 100% , now the starting operation time has been reduced to less of 2 minutes. In view to sustain the primary network in a electric en-

ergy European market, it appeared interesting to study the possibility of very short starting, either of the order of 30 seconds.

The present process consists in starting the wheel-turbine (figure 2) of zero to its nominal rotating speed in water. The project to take under consideration is to operate the turbine as motor or compensa-

Figure 2. Cut in the turbine

tory synchronised on the network permanently but unwatered with compressed air in view to reduce losses. In case of energy demand, it will be a fast opening of the gates (spherical valve and wicket gate) to allow the turbine to shift from motor or compensatory operation to generator and on the same time, to push the air pocket downstream in the draft tube, pipes and the downstream reservoir .

The new-born two-phase flow will be essentially transient and no homogeneous due the presence of air in water.

The present study will try to value the effects of the outflow on the holding of the turbine, the draft tube and on the connecting pipes close to the downstream reservoir.

Three different approaches have been achieved:
- Computation: a modelling of the water unsteady outflow is performed by numerical model of the transient flow with variable time opening.
- Scale tests: hydraulic scale model have been achieved in transparent Plexiglas to identify measure the main characteristics of the water flow and air pocket pattern.
- Site tests : measurements of vibration, pressures, opening of valves, turbine power were performed on the site for various starting time of the turbine.

2 NUMERICAL MODEL

The modelling of the unsteady out-flow is achieved for the site of Coo-2 as sketched in the diagram of the figure 3.
The main features of the site are the following one:
- upstream maximal level: 507,01 m
- downstream minimal level: 233,75 m
- total nominal discharge: 3*96 = 288 m³/s
- upstream tunnel constituted of 3 sections AB, BC, CD of lengths and equivalent diameters as shown in the table below
- downstream tunnel EF
- hydraulic resistance coefficient supposed identical for all sections: f = 0,0096

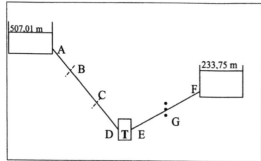

Figure 3

Table 1: measurements of conducts

Sections	Length (m)	Diameter (m)
AB	336	8
BC	630	6.5
CD	184	6.12
EF	296	8.85

- trash rack in G, middle of EF,
- For a ratio between surface/total surface equal to 0,28 (value considered by LEVIN for the majority of French power station) Idel'Cik indicates a coefficient of head loss of the trash rack equal to 0,5.

The computation of unsteady outflow generated by the opening of the valve was performed with the method of characteristics (integration of equations to the partial derivatives of hyperbolic type) that:
- requires the previous knowledge of the celerity of overpressure and depression waves;
- requires the subdivision of the pipes of hydraulic circuit in order to respect the Courant criteria

Taking account of these imperatives, we substituted to the real site of Coo-2, a ideal one that essentially differs some by the following points :
- The total discharge of 3*96 = 288 m³/s is supposed to fully pass by only one turbine and no three,
- The downstream tunnel is supposed to be a single tunnel EF and not to divide in two, near the reservoir downstream,
- The trash rack is supposed at midway between EF rather than to the downstream reservoir,
- the numeric simulation is done with a fluid single-phase, the influence of the air pocket will make the object of the modelling by tests.

The number of sections chosen is:
- 3 for AB then dx = 336/3 = 112 m
- 6 for BC then dx = 630/6 = 105m
- 2 for CD then dx = 184/2 = 92 m
- 2 for EF then dx = 296/2 = 148 m

The chosen time step is equal to 0,1 seconds, that gives the respective celerities:

$Dt = Dx/a$ then $a_{AB} = Dx/Dt = 112/0,1 = 1120$ m/s

$a_{BC} = Dx/Dt = 105/0,1 = 1050$ m/s

$a_{CD} = Dx/Dt = 92/0,1 = 920$ m/s

$a_{EF} = Dx/Dt = 148/0,1 = 1480$ m/s

These values correspond quiet well to the value, supposed constant, adopted by the BECA office (the designer consultant office), to know: a = 1360 m/s.
The time of opening of the valve will be the variable parameter and will allow to proceed of various simulations. It has been chosen from 1 to 10 seconds. In the modelling, the outflow were computed during the 10 seconds after the opening.
Places for which we raised variations of pressure are

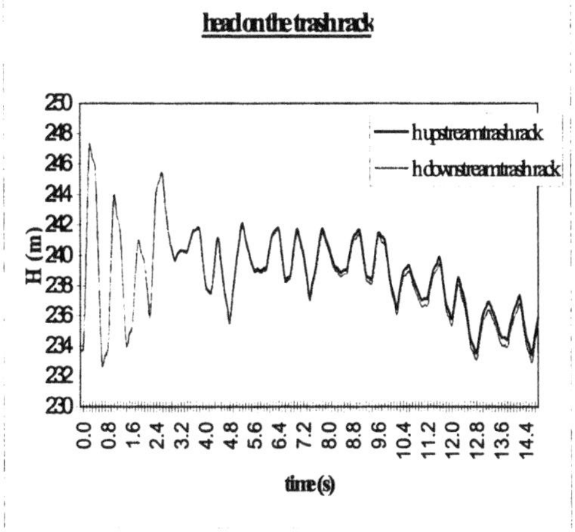

head on the trash rack

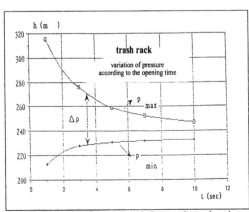

Figure 4. Upstream and downstream head of the trash racks

3.1 First scale model

A first model was achieved in transparent Plexiglas to be able to visualise the outflow of air/water flow. It essentially included:
- a pump centrifugal P simulating the upper basin
- a main stop valve corresponding to the spherical valve
- a control valve representing the gating
- a transparent vertical pipe, of approximate length of 1 m, of interior diameter of 32 mm, including a hole of compressed air injection to its higher part
- a transparent pipe with a slope of 10%, a diameter of 32 mm, simulating the tailrace tunnel and including a trash rack in expanded metal
- the downstream basin .

The manoeuvre of simulation consisted in injecting, control valve closed, a plug of compressed air whose length could vary at will.

The evolution of the air plug, after opening of the flow-regulating valve depends:
- of the degree of opening of the stop valve,
- of the speed of opening of the control valve.

A slow opening of the control valve gives place to an outflow where the plug of air is subdivided in a isolated bubble rosary lodged to the generating superior of the tailrace tunnel.

A fast opening, even though the evolution of the outflow is more difficult to follow visually, seems to indicate the maintenance of the initial shape of the air plug.

A simultaneous and adequate manipulation of stop valves and control valves permits, thanks to the transparency of ducts used, to transport a liquid discharge, the plug of air having split up in isolated bubbles, but captive in the vertical duct.

The passage of the air plug, in fast opening case, to the place of the trash rack, doesn't permit to draw a conclusion about the impact of a possible water-hammer.

It is to underline that the model has not been achieved to the prototype scale. As information that we can have, they are solely qualitative order and no quantitative (to satisfy, for example, to the law of Reynolds similitude would have led to the impracticable velocities in model, if we wanted to use the same fluid transported). It is noted however that the average velocities achieved in the model tunnel correspond to those of the prototype tunnel.

the trash rack and the turbine. Results are summarised on diagrams that follow.

In all diagrams, the heights h corresponds to the piezometric head. The elevation of the turbine is 215,75 m and the alleviation of the trash rack 227,37 m. As example, the figure 4 below shows the piezometric head in function of the time upstream of the trash rack during the 15 seconds after to a 10 seconds opening.

We note that the maximum of head (equal to 247,24 m) has place to the time 0,3 s and the minimum of pressure (equal to 232,7 m) to the time 0,7 s. The behaviour of the trash rack is best summarised on the figure 5 that shows the this pressure in function the opening time of the gating.

The piezometric head upstream of the trash rack for the permanent outfow (toward which unsteady outflow tends after one theoretically infinite time) is 234,50 m.

Figure 5. Summary of pressure variations on the trash rack according to times of opening

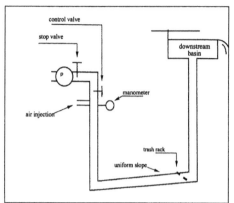

Figure 6: Scale model 1

3.2 *Second scale model*

To approach the existing reality, a second model (figure 7) was achieved, essentially including the 3 turbines with a draft tube and leading on a common collector.

The operating mode is similar the one achieved on the first model, to know:
a) partial regulation of the total discharge by opening of the stop valve
b) closing of the control valve 1 and injection of compressed air
c) opening of the control valve 1
   repetition of the phases b) and c) while changing:
   d) the time of opening of c)
   e) the total discharge, either by
      f) other opening of a)
      g) opening of regulation valves 2 and /or 3

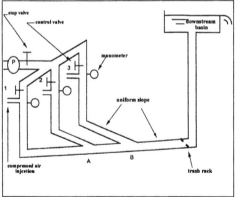

Figure 7 : Scale model 2

The conclusions are :
- the dissociation of the air plug in isolated bub-

bles depends on the speed of opening of the control valve;
- Valves 2 and 3 being opened or closed either, the fast opening of valve 1 doesn't provoke an increase of pressure to the junction A and B of around 80 cm;
- if valves 2 and 3 are closed, the fast opening of valve 1 provokes an increase of pressure of 50 cm to the trash rack;
- if valves 2 and/or 3 are opened, the fast opening of valve 1 doesn't provoke an increase of pressure of 30 cm to the trash rack.
- Whatever is the speed of opening of the control valve, the plug of air never has tendency to go up again toward turbines 2 and 3.
- A simultaneous adjustment of the stop valve and the control valve allows, by visual control, to transport a discharge Q_f while keeping a steady mixture air/water, captive in the draft tube. In spite of the difference of similitude between model and prototype, a prompt measure permitted to establish that the ratio of discharges "valve discharge Q_f / discharge achieved for an identical opening of the stop valve and total opening of the control valve" is
$$0,03 / 4,2 = 0,007$$
- The impact of the air hunt on the working of the two neighbouring turbines is analysed by the next procedure: Valves 2 and /or 3 being opened, they flow a certain discharge Q_2 and Q_3 (dependent of the available loads and losses of load provoked); pressures indicated on manometers 2 and 3 indicate values of the order of 20 m.

After opening of the valve 1, the passage of the air plug to the junction A and B provokes an increase of pressure that is not owed to a possible waterhammer however. We record pressures of the new regime, pressures that don't absorb themselves with time contrary to a waterhammer.
Finally we will notice that terms "slow" and "fast" opening translates the physical process further that the comparison with return time put by the wave of overpressure-depression from the valve until the reservoir downstream. Indeed, a celerity of the order of 1200 m/s, on a circuit of 10 m (model), makes that an opening achieved for example in 0,2 second, cannot be qualified of fast: (t = L/a = 10/1200 = 0,008 <0,2! !)

3.3 *Third scale model*

Currently, a third model is to the survey. Conducts of 100 mm of diameter permit to disregard the tensions of surface. Indeed, according to Hager (1999) both viscosity and surface tension effects can be neglected provide the pipe diameter is larger than 100 mm and the involved are water and air.

It is about the circuit in buckle represented the following figure 8.

Figure 8 : Scale model 3

Three optical probes whose position varies in the horizontal and vertical parts of the conduct permit to characterise the distribution of phases, speeds and diameters of bubbles as well as their number.

The floodgate of control automated by an electric servomotor permits to change the time of floodgate's operation of a systematic and constant way.

Figure 9. Pierced bubble by an optical probe

These tests have for goal to establish a relation between the time of opening of the floodgate and the dissipation of the air pocket in dispersed bubbles as well as to characterise the evolution of the air pocket at the time during its passage of horizontal conduct in vertical conduct.

4 MODELLING BY SITE TESTS

Tests had for goal to verify if a turbine of Coo could answer quickly to a demand of power of the network. When the group is in pumping, it is very easily and quickly relieved.

It remains therefore the problem of its reaction time when the group is to the stop: it needs to a few less than one minute to take the parallel then it reaches its full head in about twenty seconds. If the group is in compensatory turbine, it needs 75 seconds to begin to provide the power. Tests consisted therefore in testing possibilities to accelerate the passage of compensatory turbine to turbine and to measure strains inflicted to the group and the trash rack of the lower basin.

Measures of pressure, of position of floodgates, strains, vibrations and cavitation (signal provided by Voight), have been achieved in the cave and to the downstream basin.

The following conclusions have can be set up:
- Measures show that the passage of compensatory turbine to turbine, without submersion of the wheel doesn't provoke a superior strain to the normal starting. The main fear concerning the passage of the air bubble in trash racks of the intake of the lower basin proved to be vain, being given that this bubble is nearly completely left by the trash rack of cofferdam, the rest escaping by holes situated in the roof of the water hold.
- Variations of pressure and vibrations to the main trash rack are comparable at the time of a starting or a passage of compensatory turbine to turbine, with or without submersion of the wheel.
- The passage of compensatory turbine to turbine has can be achieved in 36 seconds instead of the 75 seconds in the sequence with submersion.
- The time of power supplying could be reduced while turning in compensatory turbine with open spherical floodgate, but in this type of working, the group consumes 18 MW. To this consumption, it is necessary to add the net loss corresponding to the water leak. In these conditions, although the cavitation is practically hopeless, eccentricities are about twice those to 120 MW.
- While opening meadow-dawns and the circle of gating in order to produce 0 MW, the cavitation increases to 40% but the consumption of water is even higher. On the other hand, the power production becomes practically immediate.

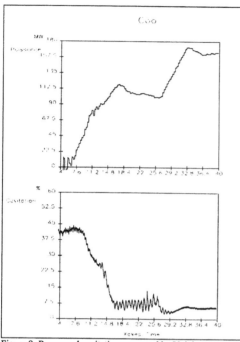

Figure 9: Power and cavitation measured by Voight

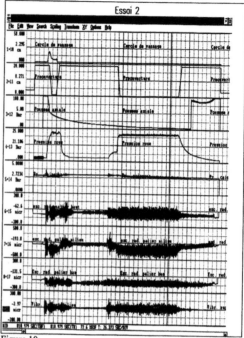

Figure 10

5 CONCLUSIONS

Following these studies, it appeared several important findings that can be summarised as follows:

- The feasibility of the turbine quick start is acquired,
- The present system of floodgate opening is one of brakes to a fast manoeuvre,
- The observed out-flow doesn't have features of a waterhammer,
- The good general agreement between the numeric studies, the study on scale model and measures in situ,
- The detailed information about the behaviour of the air bubble, its out-flow and its dividing thereafter in the draft tube, pipelines and the other works,
- Phenomena of shocks owed to the fragmentation of the air pocket can appear that have not been put in evidence by these studies,
- A complementary survey is necessary in order to analyse in detail the behaviour of the air pocket and its fragmentation.

This study will permit better to understand the transient two-phases flows and the behaviour of the bubbles and the plug, including for the oil industry and to understand problems in the interconnecting piping and for instance when a change of liquid is performed in the pipeline system.

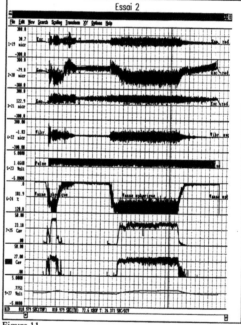

Figure 11.

The survey should permit to determine how much time the grid, the pipe will resist to the air pocket and to these strains.

REFERENCES

CUBIZOLLES, G. 1996. *Étude stéréologique de la topologie des écoulements diphasiques à haute pression*, Thèse de Doctorat, École centrale de Lyon.

HAGER, W.H. 1999. Cavity outflow from a nearly horizontal pipe, *International Journal of Multiphase Flow*, Vol. 25, 349-364.

IDEL'CIK, I.E. 1969. *Mémento des pertes de charge: coefficient des pertes de charge singulières et des pertes de charge par frottement*. Paris : Eyrolles.

LABORELEC - FLUID SYSTEMS, 1997. *Coo: Passage rapide de compensateur à turbine*, Confidential report

LEJEUNE, A. ; MOTTE, M-A., & FONCK, R. 1997. *Modélisation du démarrage en turbinage de la centrale hydroélectrique de Coo-2*.

WYLIE, E.B & STREETER, V.L. 1978. *Fluid Transients*. New York :Mc Graw Hill.

Hydropower in the New Millennium, Honningsvåg et al (eds), © 2001 Taylor & Francis, ISBN 90 5809 195 3

Reversible Pump Turbine Plant for increased grid utilisation

Torbjørn K.Nielsen
GE Hydro

Per Olav Haughom
Sira-Kvina Power Company

1   Introduction

Increased demand for power in Norway as well as the possibility of power export to Central Europe initiated a national research program for finding technical solutions to the new situation. The Norwegian power producing units and systems are initially designed for optimised energy production. The new energy marked made a huge change in the way the power plants were operated, following the price fluctuation in the marked. Problems with sufficient pricing of ancillary services giving enough priority to this kind of production in order to maintain the grid capacity, was, and still is a major obstacle for building for power.

One of the research projects has been looking into the problem of adjusting the existing power plants for the new situation with priority on power production. One of the issues has been utilising reversible pump turbine plants for ancillary services such as spinning reserves, frequency and voltage governing. Reversible pump turbines (RPT's) will very effectively increase the grid capacity.

Tonstad Power Plant owned by Sira-Kvina Power Company has been used as an example for these studies.

The Sira-Kvina power plants, with an installation of 1760 MW and a mean yearly production of 6 TWh, is a substantial production unit in the Norwegian power system. The condition for expanding the power output is good, particularly in Tonstad Power Plant. Here is a reservoir with 450 m head well suited for pumping.

Tonstad Power Plant is one of the major energy producers in the southern part of Norway with installed power of 960 MW on five high head Francis turbines. The power plant's location is very central regarding the DC cables to Denmark, starting from Kristiansand. There are plans for additional cables, both to England and to Germany. The plans involve continuing the DC cables to Feda, a terminal in the neighbourhood of Tonstad, see Figure 1.

In the years to come, it is expected that several new cable connections from the southwest part of Norway to the continent and to England will be established. The power plants at Sira-Kvina and particularly the power plant in Tonstad will be central located in the intersection of the line network for power interchange abroad.

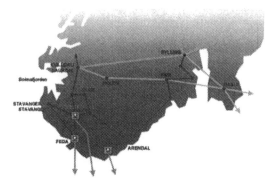

Figure 1: 300 and 420 kV main grid in south west Norway and the existing and future
connections to Central Europe.

However, most of the major power production in this part of Norway is located north of
Stavanger. The present grid, as shown in Figure 1, is mainly built for power
transportation west to east, while the future power transportation will increase in the
north – south direction. Hence, the grid is not sufficient for handling the new demand
for energy transportation in this area.

2   Frequency control – is the present system sufficient to meet the future?

An important requirement for frequency regulation is a maximum stationary frequency
deviation of ± 0.1 Hz. In a transient period, ± 0.5 Hz is accepted. The dynamic
properties of our hydro power plants for meeting the requirements, are set by
dimensioning criteria according to what was optimal before the energy marked was
liberalised

The dynamic properties regarding the primary frequency control is mainly defined by
the dominating time constants for the system, which are penstock time constants and the
time constant of the rotating masses. For most power plants in Norway as well as hydro
power plants in most other countries, the penstock time constant is approximate 1
second and the time constant for rotating masses is about 6 seconds. These time
constants together with the stability, or rather quality requirements, decide the governor
settings. Since the time constants and requirements are similar, the dynamic behaviour
of the hydro power plants becomes very similar.

Regarding the secondary control, i.e. to keep the frequency in accordance with the
requirements in the long run, the system operator will, when ever necessary, demand for
more (or less) power according to what is announced available. In Norway this is a
manual operation. The system operator telephones the energy producer that has the
cheapest offer and asks him to effectuate. The producer then has 15 minutes to get the
additional power on line.

For the time being, the rate of load change in the Norwegian power system is maximum
50 MW/minute. If one assume that the system frequency initially is at 50 Hz and that the

frequency bias is 6000 MW/Hz, the time available before violating the +/- 0.1 Hz requirement is:

6000MW/Hz * 0.1 Hz / 50 MW/minute = 12 minutes.

Adding a future exchange with Continental Europe, it is possible that the rate of load change might be up to 200 MW/min. With the same frequency bias, the time available will only be 3 minutes. With the 15-minute's response, the frequency requirement can be violated even today, and for the future situation the ability of power regulation is far from satisfactory.

The situation has to be met by either increasing the frequency bias or by dedicating large power plants for following the load variation, a variant of Automatic Generation Control (AGC). The solution must of course be acceptable in the free energy marked [Ref.1].

Reversible pump turbines are very suitable for this task.

Then to the transient conditions: Shutdown is first of all a matter of safety. In case of load rejection, the wicket gate must be closed fast in order to avoid run-away-speed, but not so fast that the transient pressure overrides the strength of the penstock.

When start-up from complete stand still, first the inlet valve must open, then the wicket gate is set in a minimum position and the turbine gain speed. The governor takes the speed of rotation up to synchronous speed. If the generator and the grid have the same AC phase, the aggregate can be connected. If not, the phase is adjusted by means of altering the RPM. This means that stability at ideal speed is imperative. Optimising the start-up procedures to make it more time effective has up to now, at least in Norway, not been an issue.

The conclusion is that the typical characteristic times regarding frequency and power regulation for the Norwegian water power system is:

• Start-up from complete stand still:	3 - 10 minutes
• Full load addmitance from idle speed	20 – 30 seconds
• Transient period after a load change	10 seconds
• Full stop:	5 – 10 seconds

The Norwegian power system is perhaps the fastest power regulator in Europe due to the dominance of water power. The question is, however: Is this good enough in the present and future energy market which is very much in need of power?

3   Improvements in dynamic properties

RPTs in the system can effectively improve the dynamic abilities. To illustrate the possibilities, Dinorwig Power Plant in UK is a good example. Some characteristic times are listed:

• Auto-start, synchronise and load to 150MW	125 seconds
• Generate normal step load change 150MW-250MW	6.5 seconds

- Emergency response from spin-generate to generate   150 MW in 8.5 seconds
- Emergency response from 150 MW generate   70 MW in 4 seconds
- Pump to spin-pump   95 seconds
- Generate 200 MW to spin-generator   11 seconds
- Spin-pump to pump   41 seconds

In normal operation Dinorwig can regulate 100 MW in 6.5 seconds which equals 923 MW/min. The emergency response equals 2250 MW/min, while traditional hydro power plants will have trouble with meeting 200 MW/min.

This shows that the technical ability of regulating power and the quality of frequency control can be improved. However, it will imply a certain investment that will not give additional energy production. In the new electricity marked, this is a major problem. All prices are linked to energy delivery, but the problem is to meet the demand for power combined with bottlenecks in the transmission system. The major obstacle for investing in power is the price system.

The case at Tonstad illustrates this issue.

4   Two alternatives for grid reinforcement

When the new licenses and agreements for power exchange to the Continent (3x600 MW), the Norwegian Grid Company Statnett applied for license for a new 420 kV transmission line and for expanding the existing lines. Because of environmental impacts, especial north of Stavanger where the new lines had to cross preserved high mountain areas, there has been considerable resistance against the plans.

An alternative solution was brought in which made the 420 kV line north of Tonstad unnecessary, but would imply a pump power plant with a capacity of 1000 MW at

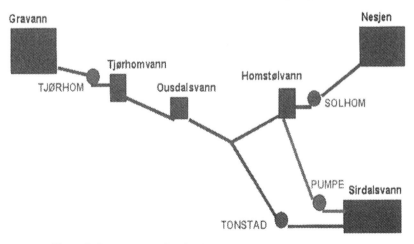

Figure 2: Pump power plant implemented in the Sira-Kvina system

Tonstad. The arrangement, with the new reversible pump power plant between the reservoires Sirdalsvatn and Homstøl is shown in Figure 2.

In order to analyse and illustrate the consequences of the two alternatives, i.e.1) new transmission lines with full capacity and 2) a minor reinforcement south of Tonstad combined with the 1000 MW pump power plant, Statnett and Sira-Kvina Power Company made a comparison which also involved the socio-economic benefits and mutual benefits for the actors in the marked. The result is shown in the Table 1 below with attached comments.

Already in 1995, The Norwegian Water Resources and Energy Directorate (NVE) together with the power company made a technical/economic evaluation of possible additional installations with the intention of increasing the power production in the area. In Tonstad Power Plant a doubling of the installation from 960 MW to 2000 MW was actually evaluated. The price was estimated to 2100NOK/ kW. This will not give the additional benefits that the pump power plant will contribute to.

Table 1: Table for comparing the two alternatives.

Comments	Consequences	Alternative 1	Alternative 2
1	Energy loss in the production system without the pump	321GWh	0
2	Energy loss in the transmission system without the pump	135GWh	0
3	Energy loss when pumping	0	500GWH
4	Water loss in existing plant	137GWh	0
	Sum energy losses	593 GWh	500 GWh

	Capitalised energy loss (10-15 øre/kWh)	1000 mill.NOK	600 mill.NOK
5	Investments	700 mill.NOK	2000 mill.NOK
5	Capitalised running expence	50 mill. NOK	200 mill.NOK
6	Operational costs:		
	Condensator batteries	40 mill.NOK	0
	Reactors	105 mill.NOK	55 mill.NOK
	Dynamic SVC		
	Dynamic phase compensator	20 mill.NOK	20 mill.NOK
	Sum operational costs	165 mill.NOK	75 mill.NOK
	Sum costs	1915 mill.NOK	2875 mill.NOK

	Additional:		
7	Environemental impact	100 mill.NOK	
8	Reduced import in low-load periodes	285 mill.NOK	
9	Power balance (equilibrium)	1000 mill.NOK	0
10	Power of frequency characteristics	12 mill.NOK	
11	Ability to regulate power following the price	60 mill.NOK	
12	Industrial development	50 mill.NOK	
	Total costs	3422 mill NOK	2875 mill.NOK

Comments:

1   Without the pump power plant it will be necessary to squeeze the existing production system in order to obtain more power in peak load periods. This will give extra losses because the turbines must run at a unfavourable operational point. Simulations indicate an extra yearly loss of about 320 GWh/year compared with the pump power plant.

2   A pump power plant at Tonstad will relieve the transmission system to the DC cabel connections. Reduced electrical losses are calculated to 135 GWh/year.

3   Energy loss in the pump power plant is calculated to 500 GWh/year.

4   Without the pump power plant, the existing power plant will be operated with less margins in intake reservoirs and thereby increase the possibility of water losses. Simulations indicate this loss to 130 GWh/year.

5   The investment for new transmission lines if the pump power plant is not built amount to 700 mill.NOK. Investment for building the reversible pump power plant is estimated to 2000 mill.NOK (2x500 MW). Capitalised operational costs are calculated to 50 mill.NOK for the new lines and 200 mill.NOK for the pumping power plant.

6   The requirement for technical installations for secure the grid stability is different for the two alternatives. New lines without the pump power plant will make it necessary with new capacitor banks for 40 mill.NOK and an additional investment in reactors to 50 mill.NOK.

7   Building new transmittion lines will have a significant environmental impact because of the crossing of the vulnerable mountain area. On the other hand, the pump power plant will not have any impact on the nature because all the installations will be underground and there will be no need of additional reservoirs. The environmental impact with new transmission lines compared to the pump power plant is estimated to 100 mill.NOK.

8   The pump power plant will give better possibilities to increase the import in low load periods and thereby give a better energy balance. The difference in energy import with and without the pump power plant is estimated to 285 mill.NOK.

9   If the pump power plant is not built, about 1000 MW must in near future be obtained else where for instance to expand existing power plants. It is assumed that the cost of alternative additional power is 1000 NOK/kW.

10   A reversible pump power plant at Tonstad will give access to regulating power very central in the main network. The value of this compared to new transmission lines is calculated to 12 mill.NOK

11   With access to 1000 MW power with the ability of very fast regulation it is possible to cash in on even short high-price periods. The profit is estimated to 60 mill.NOK

12   Development of new technology regarding pump turbines and eventually
      generators with variable speed will give valuable national competence and
      expertise that could be exploited in increased export. The value of this is
      estimated to 50 mill.NOK.

5   Simulation of the revenue

In the simulations of the revenue we have used a day-and-night price profile with four
different price zones. The profit is of course dependent on the price level, price
difference over day and night and the operation time.

The energy production at the different price zones and alternative hours of pumping
during night is optimised, see Figure 4.

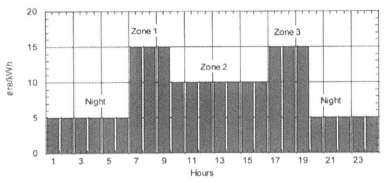

Figure 3: Price profile used in the calculations of the revenue.

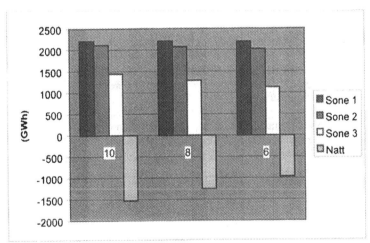

Figure 4:   Energy production and pumping during 24 hours with alternatives 6, 8 and
             10 hours pumping during night.

With the price profile shown in Figure 3, which is according to the best guesses, the profit is in the order of 100 mill.NOK/year. This revenue is by no means sufficient for defending the investment of the pump power plant at Tonstad as an energy producing plant. If the additional benefits are taken into account, the picture is different. The question is: Who pays the bill?

6   Conclusions

Evaluated in a socio-economic perspective, a new pump power plant at Tonstad is obviously beneficial compared to building new transmission lines. However, for Sira-Kvina Power Company it will not be profitable with the current energy prices and the external conditions the power companies has to comply with.

To build a pump power plant of this size has benefits for all the actors in the marked. The main benefit of the pump power plant is to increase the capacity of the transmission lines and preserve the quality of frequency control and the ability of regulating power. The investment cannot be justified unless the savings connected to new lines and technical installations, environmental impact and other benefits for the society as a whole are taken into account.

Pricing of power in today's marked is not very predictable and gives no basis for big investments. With the current price system, price on power will only be achieved when there is a shortage of power in the marked and because of "bottlenecks" in the transmission lines. New power installation, which really is necessary, will only contribute to keep the prices low.

The challenge with the new cables to the Continent is so special and extensive that it is called for unorthodox arrangements in order to find solutions to the perspective of the society's best interest, and optimal for the energy business sector as a whole.

REFERENCES

1   B.H.Bakken,A.Petterteig, A.K.Nystad:"Ramping as an ancillary
         Service",Proc.13th Power System Computation Conference, Trondheim,
         Norway, 1999, pp 395-401.

2.   F.O.Rasmussen: "Simulations of the power production with
         new reversible pump turbines at Tonstad"(in Norwegian), Report 98007,
         GE Hydro, Marsh 1998.

Hydropower in the New Millennium, Honningsvåg et al (eds), © 2001 Taylor & Francis, ISBN 90 5809 195 3

Extension of price forecasts for hydropower scheduling and risk management tools

Ingrid Schjølberg & Birger Mo
Sintef Energy Research, Trondheim, Norway

ABSTRACT: Reliable price forecasts are essential for hydropower scheduling and risk management in the electricity market. This paper presents a method for improving the representation of uncertainty in price forecasts. The basis for the work presented is price forecasts generated by simulation of the Nordic electricity market. The price forecasts are described by a number of price scenarios. These scenarios are limited in number and a larger number is required to give a better price distribution. An example is given to demonstrate the effect of using more scenarios in hydropower scheduling.

1 INTRODUCTION

Forecasting future spot prices is essential for hydropower scheduling and trading in a deregulated power market, where maximising profit and risk management are key issues. It is important that the forecast describes uncertainties in the prices next week in order to reduce risk.

Many Scandinavian producers use the EMPS model (EFI's Multiarea Power Market Simulator) (Botnen et al. 1992) to simulate the Scandinavian electricity market, and to forecast future prices (Haugstad & Rismark 1998). Price forecasts are input to hydropower scheduling and risk management tools (Mo & Gjelsvik 2000). The price forecasts generated by the EMPS model are represented by a limited number of scenarios. Each scenario is assumed to have equal probability and is related to an inflow scenario. The number of price scenarios is limited because each scenario represents a historical inflow year (Wangensteen et al. 1995). The EMPS model will not give a perfect description of the market due to factors like unpredictable psychological effects and varying oil prices. These are elements of uncertainty which are not represented in the EMPS price forecast. Furthermore, experience shows that the EMPS model describe a to small price uncertainty. A more realistic price forecast can be obtained by using a larger number of scenarios with a better price distribution. In hydropower scheduling and risk management tools a good price distribution will improve simulation results.

Two methods for generating a larger number of price scenarios with better price distribution have been considered in this paper. The first uses the EMPS model error, which is defined as the difference between the simulated system price and the observed system price for the same period. A larger number of price scenarios is generated by adding the model error to an existing price forecast. The second method extends the number of price scenarios by directly adding a noise function to existing forecasts.

The main contribution of the present work is to study how the information in the EMPS model error can be applied to achieve a larger number of price scenarios with better price distribution.

The outline of this paper will be the following. In Section 2 a presentation of error models is given and in Section 3 a case study is presented. Section 4 gives the conclusions of this work.

2 MODEL DESCRIPTION

The starting point of this work is a time series describing the difference between the simulated and observed system price (EMPS model error). A function is fitted to this time series and an error model is established. The error model describes the error in EMPS system price prediction.

The error model considered in this work is a linear auto regressive model; a non-linear function is optional. The auto regressive model can be fitted to describe a time series without dominating non-linearities (Sjøberg et al. 1994). The dynamic time-varying model is given by

$$A(q)x_k = C(q)e_k \qquad (1)$$

where x_k is the model input; e_k is the noise term;

$A(q)$ and $C(q)$ are polynomials in the shift operator q^{-1}; k is the week number.

The number of polynomial terms in $A(q)$ is equal to the model order. The parameters in the polynomials can be identified using least-squares estimation methods.

A k-step predictor is given by

$$\hat{x}_{k+1} = A(q)\hat{x}_k + C(q)\varepsilon_k \tag{2}$$

where $\hat{x}_{k+1}$ is the predicted output in week $k+1$; and ε_k is additive white noise to excite the process.

3 CASE STUDY

This section demonstrates methods for extending the number of price scenarios. The section is divided into three parts. In Section 3.1 the EMPS model error for a specified period is established. Error models are applied for generating more price scenarios and this is described in two examples in Section 3.2. In Section 3.3 new price scenarios are applied in hydropower scheduling.

3.1 *EMPS model error*

The EMPS model error is established by calculating the difference in simulated and observed system price. A simulated system price is generated by feeding the model with real time data for a given period. Brennvik & Foshaug (1997) simulated system prices using the EMPS model for the period week 31 in 1994 to week 30 in 1997, totally 156 weeks. The simulated system price is shown in Figure 1, and with the observed system price for the same period.

The error between simulated and observed system price contains information that will be utilized to make an error model. The error between the simulated and observed system price is given by

$$z_k = y_{s_k} - y_{o_k} \tag{3}$$

where z_k is the model error in week k; y_{s_k} is the simulated system price; and y_{o_k} is the observed system price.

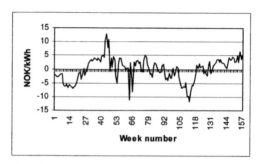

Figure 2. Difference in simulated and observed system price.

The measured error established in Brennvik & Foshaug (1997) is plotted in Figure 2.

The model error (Fig. 2.) is normalised according to

$$x_k = (z_k - \mu)/\sigma \tag{4}$$

where x_k is the normalised error in week k; z_k is the EMPS model error; μ is the mean of the error; and σ is the standard deviation of the error.

The normalised error is fitted to the model described in Equation 1. The obtained error model is given by the function

$$x_k = -0.93x_{k-1} - 0.24\varepsilon_k \tag{5}$$

where x_k is the normalised error in week k; and ε_k is white noise term with mean equal to zero and variance equal to one.

A simulation result using the error model is shown in Figure 3. A good fit is established using a first order model in one step prediction.

3.2 *Extended price forecasts*

In this section the error model is used to extend the number of price scenarios. Two examples to demonstrate this are included at the end of this section.

A price forecast with weekly resolution is generated using the EMPS model. The planning period is 156 weeks, starting in week 1, 1997. Percentiles describing the price forecast are shown in Figure 4.

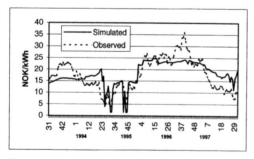

Figure 1. Observed and simulated system price generated by the EMPS model.

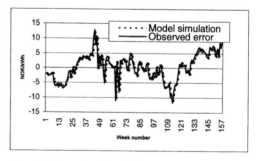

Figure 3. Observed and simulated model error.

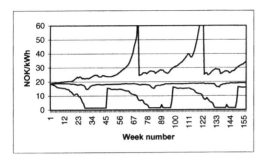

Figure 4. EMPS price forecast in (0, 50, 100) percentiles three years ahead.

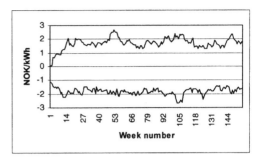

Figure 6. Range of predicted error scenarios.

The forecast is represented by 60 price scenarios. The covariance function between the scenarios in each week is shown in Figure 5.

The covariance function shows that the EMPS model produces scenarios with little uncertainty at the beginning of the planning period and in the winter period (week 46-70, 98-122) (Fig.5). To achieve a better price distribution in the scheduling period the price scenarios should describe a higher uncertainty. The covariance function shows that the variation in price is small in the start of the forecast period and the uncertainty in the price for one week to the next is more than 0.8 (Fig.5). This indicates that the prices are too correlated compared to the variations observed in the market. This confirms the fact that there are factors that are not modelled in the EMPS.

In the following two examples demonstrates how the model of the error between observed and simulated system price is used to add more scenarios to this forecast.

3.2.1 Example 1
In this example the first order error model given in Equation 5 and the price forecast shown in Figure 4 are used to obtain an extended price forecast. A finite number of white noise vectors are generated to excite the error model. This affects the distribution of the model output. Predicted error scenarios are calculated according to Equation 5. The range of predicted error scenarios is shown in Figure 6.

Four error scenarios are randomly selected from this range and added to each price scenario generated by the EMPS model (Fig.5). In this case study the EMPS model generates 60 price scenarios, and adding four error scenarios to each gives totally 240 price scenarios. The range of the scenarios is restricted such that negative prices are not induced.

The range and covariance of the new price forecast are shown in Figures 7-8. The covariance function shows that a larger uncertainty is achieved in the extended price forecast, in the beginning of the planning period and during the winter period.

3.2.2 Example 2
In this example a noise function is directly added to the EMPS price forecast. This is choosing $A(q) = 0$ in the fitting process. The obtained error model is given by the function

$$x_k = 0.6\varepsilon_k \qquad (6)$$

where x_k is the normalised error in week k; and ε_k is white noise term.

Figure 9 shows the new price forecast generated by using the error model given in Equation 6. Figure 10 shows the covariance function of this forecast. The covariance function in the extended forecast has decreased and more uncertainty is added to the fu-

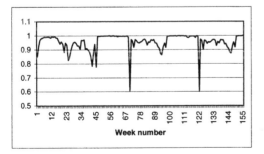

Figure 5. Covariance function in EMPS price forecast one week to another.

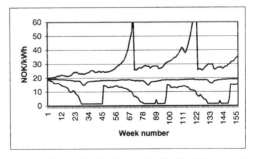

Figure 7. Extended price forecast in (0, 50, 100) percentiles three years ahead.

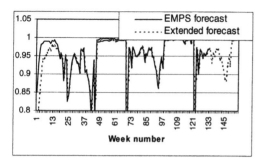

Figure 8. Covariance function in EMPS forecast and extended forecast.

ture prices. The distribution of the extended price forecast is improved.

3.3 *Applying the extended price forecast*

In this section the effect of using a larger number of price scenarios in hydropower scheduling will be demonstrated.

The extended price forecast presented in Example 1 is applied to a hydropower system consisting of 11 reservoirs and 7 hydro plants.

The system is simulated using EOPS model (EFI's One-area Power Market Simulator) (Flatabø et al. 1998). The EOPS model is designed for local sched-

Table 1. Standard deviation of sales income.

Week	1-52		53-104	
Scenarios	60	240	60	240
St.dev. (MNOK)	6.9	6.3	9.5	8.6

uling. The model operates in two phases: Firstly, it calculates a scheduling strategy using dynamic programming. Secondly, it simulates the hydropower system for a given number of price and inflow scenarios. Weekly generation is based on the strategy established.

The hydropower system is simulated twice using two price forecasts. In the first simulation the forecast with 60 scenarios is applied (Fig.3). In the second simulation the forecast with 240 scenarios is used (Fig.7). The scheduling strategy is established using 60 price scenarios, and the same strategy is applied in the second simulation.

The sales income is registered in two periods (week 1-52, 53-104). The distribution of the sales income for the periods is given in Figure 11-14. The distribution is more continuous using 240 price scenarios. The standard deviation of the sales income using 60 and 240 price scenarios is given in Table 1. The deviation is reduced using a larger number of price scenarios.

This demonstrates how the information in the EMPS model error can be applied to achieve a larger

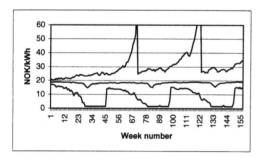

Figure 9. Extended price forecast in (0, 50, 100) percentiles three years ahead.

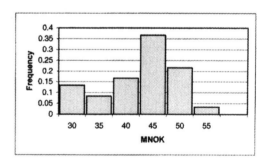

Figure 11. Distribution of sales income week 1-52 using 60 price scenarios.

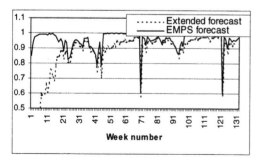

Figure 10. Covariance in EMPS price forecast and extended forecast.

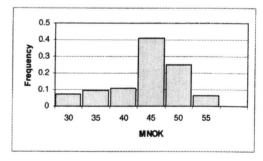

Figure 12. Distribution of sales income week 1-52 using 240 price scenarios.

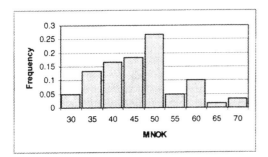

Figure 13. Distribution of sales income week 53-104 using 60 price scenarios.

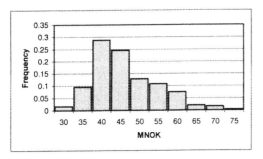

Figure 14. Distribution of sales income week 53-104 using 240 price scenarios.

number of price scenarios. Furthermore, a larger number of price scenarios describes more uncertainty in the beginning of the planning period. This gives a more realistic price forecast and more evenly distributed sales income.

4 CONCLUSIONS

This paper studies how information in the EMPS model error can been used for generating a larger number of price scenarios. This has been done in three steps. Firstly, the EMPS model error in a given planning period is calculated. This gives a description of the factors which are not modelled in the EMPS. Secondly, an error model is established and error scenarios are generated. Finally, the error scenarios were added to a given set of price scenarios, and this resulted in a larger number of price scenarios.

The effect of applying more scenarios and thereby a more realistic price forecast has been demonstrated. Schedules based on a larger number of scenarios with a better distribution result in more evenly distributed sales income.

REFERENCES

Botnen, O.J. & Johannesen, A. & Haugstad, A. & Kroken, S. & Frøystein, O. 1992. Modelling of Hydropower Scheduling in a National/International Context. *Proc. Hydropower Lillehammer, Norway, 1992.* Rotterdam: Balkema.

Brennvik, A. & Foshaug, A. 1997. Langsiktige prisbevegelser i kraftmarkedet. *Hovedoppgave ved NTNU, Institutt for industriell økonomi og teknologiledelse.* In Norwegian.

Flatabø, N., Haugstad, A., Mo, B. & Fosso, O. 1998. Short term and medium term scheduling in the Norwegian hydro system under competitive power market structure. *EP-SOM'98, Zurich, Switzerland.*

Haugstad, A. & Rismark, O. 1998. Price forecasting in an Open Electricity Marked based on System Simulation. *EP-SOM'98, Zurich, Switzerland.*

Mo, B., & Gjelsvik, A. 2000. Integrated risk management of hydropower scheduling and contract management. *IEEE Transactions on Power Systems, PE-161PRS (12-2000).*

Sjøberg, J. & Hjalmarsson, H. & Ljung, L. 1994. Neural Networks in System Identification. *IFAC Symposium on System Identification 1994, Copenhagen, Denmark.*

Wangensteen, I., Mo, B., Haugstad, A. 1995. Hydro Generation Planning in a Deregulated Electricity Market. *Proc. Hydropower into the Next Century, Barcelona, Spain 1995.* Sutton, UK: Aqua-Media International.

Hydropower in the New Millennium, Honningsvåg et al (eds), © 2001 Taylor & Francis, ISBN 90 5809 195 3

Experience from Peaking Operation of a High Head Hydropower Plant in Tanzania

O.Skuncke & O.A.Røvang
NORPLAN A.S, Power and Renewable Energy Division, Drammen, Norway

D.P.Mhaiki
Tanesco, Kihansi Hydro Power Plant, Tanzania

ABSTRACT: Lower Kihansi Hydropower Plant was commissioned in year 1999 / 2000 adding 180 MW to the national power system. Three Pelton units, each with a capacity of 60 MW have been installed in the underground powerhouse in the present phase. Two additional units with the same capacity are foreseen to be installed in a future phase when peaking needs to justify the addition. The paper presents the experience from the first year of operation of Lower Kihansi hydropower plant. In addition to the station's abilities for peaking, it is also characterised by a great flexibility permitting the operation in other stations to be optimised better than previously.

1 GENERAL DESCRIPTION OF THE LOWER KIHANSI HYDROPOWERPROJECT

1.1 Project development

The Lower Kihansi Hydropower Project (LKHP) in Tanzania is located in the Rufiji basin, some 550 km southwest of the capital Dar-es-Salaam. It is owned by the Tanzania Electric Supply Company (TANESCO) which is practically the only entity which generates, transmits and distributes power in the country.

The project was conceived in the *Rufiji Basin Hydropower Study* of 1984, which suggested that the Kihansi river was the most favourable for future hydro power development in the basin.

A feasibility study which was published in 1990, demonstrated the technical feasibility and economic viability of the Lower Kihansi project. Based on this study the World Bank and TANESCO agreed to incorporate the implementation of the Lower Kihansi project into a sector loan package - the Power VI Project Program.

In December 1991, Norplan was selected to carry out a feasibility review of the Lower Kihansi scheme, and if this proved positive, to continue with the final design. Tendering and supervision of construction followed as an extension to NORPLAN's initial contract.

The actual construction phase of the project started in July 1994, and the units were successfully commisioned in December 1999, March 2000 and May 2000.

1.2 Description of the project

The Lower Kihansi Hydropower project includes a 25 m high concrete gravity dam which impounds a small reservoir with a total storage of 1.6 Mm^3, of which 1 Mm^3 can be used for daily regulation. The dam is equipped with sediment flushing gates and an intake structure with trashracks and gates. The intake connects to the headrace tunnel via a circular unlined vertical headrace shaft (25 m^2), some 500 m deep. The tunnel which is unlined, and slopes down

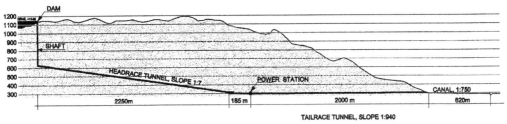

LONGITUDINAL SECTION

Figure 1. Longitudal section of the Lower Kihansi Hydropower plant.

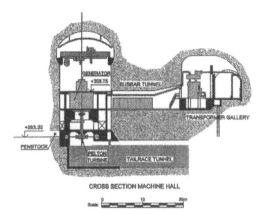

Figure 2. Cross section of the powerhouse at Lower Kihansi Hydropower plant

stream at an inclination of 1:7, is 2200 m long and has a cross-section of 30 m². No surge shaft or chamber has been provided due to the relatively short tunnel length and the low water velocities, combined with the use of Pelton turbines.

The penstock section provides for two branches, one for units 1, 2 and 3, and the other for the future units 4 and 5. The tailrace tunnel which is designed as a free-flow conduit, has a length of 2000 m and a cross-section of 40 m². It slopes gently downstream at an inclination of 1:900, and connects to an 800 m open-cut canal that evacuates the water into the existing Kilombero river system.

The powerhouse cavern is 12.6 m wide, 70 m long and 30 m high including necessary space for the future units 4 and 5. The powerhouse is accessed by a 1900 m access tunnel (40m² in cross-section) and also by a separate cable tunnel which was chosen as an extra security measure for the 220 kV cables passing from the underground power house to the outdoor switchyard.

Three turbine-generator units are installed in the powerhouse including four-jet Pelton turbines with a capacity of 60 MW and synchronous generators with a capacity of 70 MVA. The rotational speed is 600 rpm.

The switchyard at Kihansi comprises SF₆ switchgear, including a 20 MVAr shunt reactor for voltage regulation, and a 220/33 kV transformer for the local supply. For additional reliability of power transmission, two 220 kV lines were built from Kihansi to connect with the TANESCO grid. One line (97 km) runs to the north-west and joins the system at the Iringa substation. The other line (178 km) runs to the north-east and joins the grid at the Kidatu substation.

The project was designed principally as a peaking station, with a daily / weekly reservoir storage capacity and no restrictions on sudden restrictions in downstream flows.

2 PARTICULAR ASPECTS RELATED TO PEAKING

2.1 Load factor

Table 1 shows output, plant factor and firm production for Kihansi with three and five units. Table 2 shows corresponding data for other hydropower stations and for thermal plants.

Already in the present phase the load factor for Kihansi is lower than for other hydropower plants in Tanzania. With all five units installed, Kihansi will have a considerable peaking capacity.

2.2 Structures and equipment, general

The high head (850 m) gives a high energy per cubic meter of water: roughly 5 times more than for Kidatu and 10 times more than for Mtera. This makes it possible to obtain a large capacity, favourable for peaking, keeping a moderate size (and cost) for the intake pond, hydraulic structures and tunnels.

The intake pond has ample capacity for day peaking. It is dimensioned to permit daily peaking in

Table 1. Plant Factor for Lower Kihansi Hydropower station.

	Number of units	Output per unit MW	Plant factor	Firm Prod. GWh
Present	3	60	0.41	650
Future	5	60	0.25	650

Table 2. Load Factor for other hydropower stations and for thermal plants.

	Output MW	Load factor	Firm Prod. GWh
Other hydro	360	0.56	1775
Thermal*	112	0.80	780

* Before commissioning of ITPL plant with 104 MW.

Figure 3. Map over the national grid in Tanzania (Only main lines 220 kV and 132 kV are shown).

all flow conditions with 5 units in operation, with in addition a certain amount of week peaking. This is obtained with a relatively small dam and very limited flooded areas. The reservoir contained no permanent habitation.

The tunnel system, including hydraulic structures at intake and tailrace outlet, have moderate cross-sections for a station of this capacity. The trashrack and intake gate have areas of respectively 50 m^2 and 20 m^2. The vertical shaft has a cross section of 25 m^2 while the cross section of the headrace tunnel is 30 m^2.

The penstocks represent generally an expensive item for a Pelton station. The solution adopted for Kihansi with a deep headrace tunnel permits to have comparatively short penstocks: they have a length of approx. 185 m, while a solution with a shallow tunnel would have required penstocks with a length in the range of 1200 m.

3 GRID DEMAND AND POWER PRODUCTION CAPACITY

3.1 Grid demand

The demand in the grid is in constant growth. In the second half of year 2000, the peak demand was around 400 MW. Figure 4 shows the grid demand during the day for different week days in June 2000. The curve shows a marked peak in the evening and a lower but longer peak during daytime.

The rate of growth of the demand is in the range of 30 to 40 MW per year.

3.2 Production capacity

Before commissioning of Kihansi, power available has been a limiting factor. Table 3 shows the pro-

Table 3. Production capacity at all power stations in Tanzania before commissioning Lower Kihansi.

Power plant(s)	Production capacity [MW]
Kidatu*	150
Mtera**	70
Pangani	68
Hale	12
Nyumba ya Mungo	5
Ubungo gas turbines	112
Small diesel stations	7
TOTAL	424

* The total capacity at Kidatu is 200 MW, but this was reduced to 150 MW due to rehabilitation works on one of the units.
** Capacity is depending on reservoir level

duction capacity practically available from the different stations in the grid system at that time. The figures shown are in many cases not corresponding to the maximum capacity of the plant as different factors can reduce the capacity currently available during a peak.

Considering the spinning reserve requirements, it is necessary to have a capacity in the range of 450 MW to be able to cover a demand of 400 MW.

The possibility to produce enough peak power has been a limiting factor before commissioning of Kihansi, with occasional load shedding at peak hours. The state of some other units was poor, and in reality it has been difficult for TANESCO to cover peak demand of even 350 MW.

After the commissioning of Kihansi, the situation was radically changed. The relative increase in production capacity is so large that the availability of peak power will represent no problem during the first years of operation of Kihansi.

This does however not mean that the energy needs are satisfied in Tanzania at all times. Specially in drought periods, the availability of energy may remain a problem, due to limited storage to cover dry season requirements.

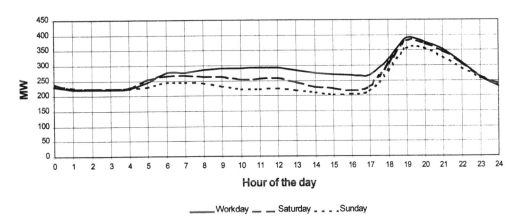

Figure 4. Grid demand during the day for different weekdays (June 2000).

147

The production obtained at Kihansi during the first year of operation shows that the previsions for "firm power" are well satisfied. Firm production was calculated to be 650 GWh. Although 2000 was a dry year, the total production for the year at Kihansi reached 592 GWh, with unit 2 starting generating only in March, and unit 3 in May.

4 REVIEW OF OPERATION CONDITIONS AFTER COMMISSIONING OF KIHANSI

4.1 General

The first hydropower stations in Tanzania were of moderate size. The old station at Pangani Falls was commissioned in 1934 originally with a capacity of 7,5 MW later increased to 17,5 MW. It was followed by Hale (21 MW) and Nyumba ya Mungu (8 MW).

Kidatu was in the 1970s the first of a new generation of stations, followed by Mtera, New Pangani Falls, and now Lower Kihansi. Kidatu with its 200 MW has for a long time been the largest power station in Tanzania. It has represented the backbone both for power and for energy production. This reflects in the operation practice: the frequency has been controlled mainly by Kidatu, and rebuilding of the grid after a blackout has also been started with Kidatu. At first, little was changed in this operation practice when Kihansi was put in operation in the beginning of year 2000.

Kihansi introduces a new situation with an output in the same range as Kidatu. It offers new possibilities to optimise the energy production in the country. Review of operating conditions was made during a workshop arranged with TANESCO operation personnel in autumn 2000. The different aspects influencing the operation of the grid were analysed systematically, present practices for operation were reviewed, and conclusions were drawn concerning future operation.

This review of operation did not include the long-term planning for reservoir operation at Mtera and Nyumba ya Mungu.

4.2 Efficiency considerations

Efficiency represents the most obvious parameter that has to be considered when combining the operation of the units within a station and combining operation of different stations.

In a first step the distribution of load between the units of each station should be determined in order to obtain an optimal combination. For stations with identical units, this will generally be obtained by sharing the load equally between the units. Taking into account the efficiency for the waterways (hydraulic losses), for the turbines and for generators and transformers one obtains a combined efficiency

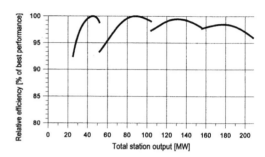

Figure 5. Efficiency curve for Kidatu Hydropower station.

curve for the station. This represents an essential tool for the combination of operation of different stations. Curves for the four larger hydropower stations in Tanzania are shown on Figures 5, 6, 7 and 8

The Francis units have comparatively steep curves giving overall curves with marked tops for each unit, and dips in between. This is particularly pronounced for the stations with 2 units as Mtera and Pangani. Kidatu with its four units has a more regular curve in the area corresponding to three and four units in operation.

Kihansi with its Pelton turbines has a very flat overall efficiency curve: from 30 to 100 MW the efficiency is not varying by more than 1.5%!

The load shall be distributed between the different stations in the grid to follow the demand curve.

The gas turbines at Ubungo will generally be op-

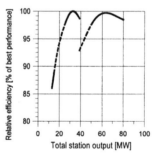

Figure 6. Efficiency curve for Mtera Hydropower station (Dashed line shows the cavitation area).

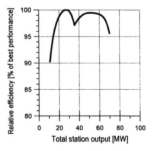

Figure 7. Efficiency curve for Pangani Hydropower station

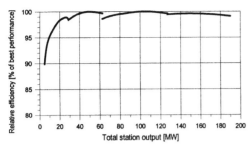

Figure 8. Efficiency curve for Lower Kihansi Hydropower station.

erated as a base load, the different units being operated close to their maximum load (where the efficiency is maximum).

The load variations will be covered mainly by the hydropower stations.

Combination between the different stations should be made so that each station is operated in the areas with good efficiency. A natural pattern for stations with 2 units will be to operate with one unit around best efficiency in off-peak periods, and two units around best efficiency during peak, the length of the peaking period being adjusted in function of the mean flow at the moment. Kidatu should also in principle be operated close to the outputs corresponding to the tops of the efficiency curve. Kihansi offers much more freedom of operation, as the efficiency curve is practically flat.

4.3 Speed droop spinning reserve considerations

Simple efficiency considerations are not sufficient to determine the operation of the different stations: another main requirement that affects operation is to keep a stable frequency in the grid in all conditions.

The limited number of stations of importance in the Tanzanian grid makes that each of them will to some extent have to participate to the frequency regulation. Kidatu has traditionally had the largest part of this regulation with a speed droop of 2% while other stations have had speed droops ranging between 3 and 5%. Operation with a small speed droop makes that the actual load at which the units will be operating may be rather far from the load set point, when the frequency is deviating from 50 Hz. For the units at Kidatu, a frequency deviation of ± 0.2 Hz, which is the maximum value accepted normally for the Tanzanian grid, will correspond to a load deviation of ± 10 MW around the set point.

Choice of operating point has to be made taking these conditions into consideration: as far as possible, the load variation range around the set point should be kept in a favourable area of the efficiency curve, and the maximum should not exceed the capacity of the unit.

Another aspect is the global spinning reserve in

the grid. A generally accepted principle is to have a spinning reserve exceeding the capacity of the largest unit in the grid. This will give the possibility for the remaining units to pick up in the case of unforeseen shut down of one large unit.

4.4 Reactive power balance

Not only the active power has to be in balance between production and consumption (keeping the frequency in the grid), but also the reactive power balance has to be kept to control the voltage level.

The active power balance is determined mainly by producers and consumers in the network. For the reactive power, the transmission system itself influences the balance. The capacitance of the transmission lines produce a considerable amount of reactive power. Reactive power is consumed in the inductance of the transmission lines only the when current flows through them. With a lightly loaded system there will often be an excess of reactive power in the system, leading to a too high voltage. This situation is frequent in TANESCO's transmission system characterised by long transmission lines.

Running the generators underexcited represents one way to consume reactive power. This can however be uneconomic as it may require to run the units at low load (and poor efficiency) in order to consume enough reactive power. A more economic way to consume reactive power is to install reactors. A 20 MVAr reactor was installed at Kihansi in order to reduce the need to use the Kihansi generators to balance reactive power. In order to improve the conditions in the grid further, a 10 MVAr shunt reactor has been planned and is presently under installation at Mbeya. It is scheduled to be commissioned at the end of year 2001.

4.5 Review of present operation practices for Kihansi and the other stations in the grid

Present operation practices have been reviewed in the light of the different aspects mentioned above. Typical load distributions in off-peak and peak conditions are shown in the Tables 4 and 5. The load distributions shown correspond to a comparatively dry period. In the example considered, Mtera is shut down during the off-peak periods to save water.

Figure 9 shows the efficiency curve for Kidatu with the load set point corresponding to the values in the table, and the load and efficiency variation for a frequency $\pm 0,2$ Hz and a speed droop of 2%. Figure 10 shows the corresponding curve for Kihansi. The speed droop used so far for Kihansi was 3%. Figure 11. shows the "Moderate peaking operation" of Kihansi corresponding to these conditions.

In peak conditions, operation is practically optimal. Kidatu and Kihansi, as well as Pangani and Mtera are operating at optimal efficiency; also the

Table 4. Typical off-peak load distribution

Power station	No of units x output MW	Station output MW
Gas turbines	2 x 35	70
Hale	1 x 8	8
Pangani	1 x 19	19
Mtera	-	-
Kidatu	3 x 33	99
Lower Kihansi	2 x 20	40
Nyumba ya Mungo	1 x 4	4
Small diesels	-	-
TOTAL		240

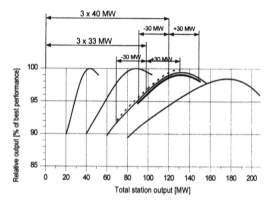

Figure 9. Examples of load and efficiency variations at Kidatu Hydropower station (Load variation around set point for frequency ± 0.2 Hz and speed droop of 2%).

two (out of four) gas turbines in Ubungo are operating at full load, which means at high efficiency. The spinning reserve is in the range of 80 MW which is suitable.

The situation is less favourable in off-peak conditions. Both Kidatu and Kihansi are operated with several units at comparatively low load. The main reasons given for this are the requirement of spinning reserve and the need to consume reactive power.

Operating Kidatu with more units than strictly required from an output consideration and at low load and efficiency is to some degree a habit inherited from the time Kidatu was the main station for the control of the frequency. With Kihansi on the line, this should no longer be necessary to operate in this way. Kihansi offers an important spinning reserve at a very low cost as the efficiency loss to obtain this reserve is minimal, in contrast to the stations with Francis units where spinning reserve can require a loss of efficiency of several %. A way to make better use of the regulating properties at Kihansi is to lower the speed droop, from the value of 3% used formerly to a value of 2%, as for Kidatu.

Still the system has to be seen as a whole, considering also the dynamic behaviour for the different stations. The Francis turbines at Kidatu have a faster reaction time than the Pelton turbines at Kihansi. Kidatu will therefore react quicker in case of sudden increase of the load on the grid, picking up load momentarily. Kihansi will follow with some delay taking over one part of the load from Kidatu. Such conditions are difficult to analyse in theory, and

must be experimented carefully step by step in practice. For each step the behaviour of the system should be analysed carefully. It was decided to try operation with a speed droop of 2% at Kihansi.

Reactive power consideration is a second aspect which can justify operation of units at a lower load than what is suitable from the point of view of efficiency. The necessity for the stations to consume enough reactive power in the periods with light loads on the grid remains one important issue in the Tanzanian grid. Analyse of reactive power balance is made difficult by the facts that the voltage varies throughout the grid, and is depending on the direction of the reactive power flow. Here also, a purely theoretical approach is difficult, and investigations should be made in practice to determine an optimal distribution of active and reactive loads.

The situation will have to be reviewed after the 10 MVAr shunt reactor at Mbeya has been commissioned. The important is to keep in mind that consuming reactive power with the generating units at low load is expensive and should be avoided as far

Table 5. Typical peak load distribution

Power station	No of units x output MW	Station output MW
Gas turbines	2 x 35	70
Hale	1 x 9	9
Pangani	2 x 25	50
Mtera	1 x 30	30
Kidatu	3 x 40	120
Lower Kihansi	2 x 50	100
Nyumba ya Mungo	1 x 4	4
Small diesels	-	7
TOTAL		390

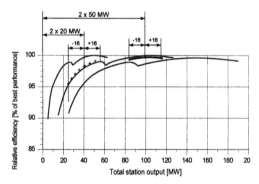

Figure 10. Examples of load and efficiency variations at Lower Kihansi Hydropower station (Load variation around set point for frequency ± 0.2 Hz and speed droop of 3%).

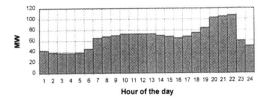

Figure 11. "Moderate peaking operation" of the Lower Kihansi Hydropower station.

as possible. If it appears that the reactive power balance still represents one governing factor for the operation of the units at Kihansi and Kidatu it should be considered to install more reactors at suitable locations in the network.

5 CONCLUSIONS

In addition to its peaking abilities, Lower Kihansi offers the possibility to optimise the operation in other stations of the Tanzanian grid. Kihansi can operate over a very wide range of outputs with a practically constant efficiency, providing spinning reserve as required and permitting to operate other stations close to their point of best efficiency. Optimising the operation at Kidatu, the major energy producer in Tanzania, can bring a sensible gain in production on national basis.

To be able to take fully advantage of the possibilities of Kihansi in this respect it is important that the reactive power balance can be obtained in a suitable way, without having to sacrifice efficiency. This question has to be followed up further and the situation analysed after completion of the installation of the shunt reactor at Mbeya.

As this paper goes to press, the water situation in Tanzania has changed. The flow has increased at Kihansi which is now operated in a typical peaking pattern with an output up to its maximum of 180 MW. It has been possible to stop releases from Mtera to fill up its reservoir, while Kidatu is run only on unregulated tributary inflow between Kidatu and Mtera. In these conditions, Kihansi represents the largest single power producer in the country during the peaks, and its contribution is essential in avoiding load shedding in the coming years.

Hydropower in the New Millennium, Honningsvåg et al (eds), © 2001 Taylor & Francis, ISBN 90 5809 195 3

River Basin Management in India vis-à-vis Perspective HE Development

R.N.Srivastava
Chairman, Central Electricity Authority and Ex-officio Secretary to Govt. of India

D.V. Khera
Member (Hydro), Central Electricity Authority and ExOfficio Additional Secretary to Govt. of India

Major Singh
Director, Hydro Planning & Investigation Division, Central Electricity Authority

Synopsis

The optimum development of water Resources is considered to be key element in the socio economic development of a country. India is endowed with abundant water resources which can be broadly classified in six major river systems. Apart from meeting requirements for drinking water, irrigation and flood control etc., Hydro Power ,a cheap, environment friendly and renewable source of energy forms an important parameter in evolving a frame work for formulation of river basin plans.

The need for optimum development of Hydro Electric Resources has been duly recognised for meeting the ever increasing demand for power requirements in the country. The paper highlights the issues and constraints associated with harnessing of the vast hydro electric potential in various river basins of the country about 75% of which is yet to be tapped. The role of basin-wise Hydro Electric Surveys of India particularly, the large scale development of Multipurpose-River Valley Projects which ushered an era of green revolution in the post independence era have also been discussed.

The bulk of Hydro Electric potential of India is located in river basins of Himalayan region which is ecologically sensitive. There has also been greater environment awareness and emphasis on sustainable development of water resources. An attempt has been made to highlight the imperative to have a re-look into the Hydro Electric potential considering environmental aspects & additional data on consumptive uses of water for various purposes. The detailed studies to firm up the parameters of hydel projects identified in various river basins are necessary to facilitate maximizing benefits and prioritizing execution of projects in the basin as a whole.

1.0   Introduction

1.1   Water resources is the key element & a prime natural resource and a precious asset which is basic requirement for survival of mankind. The ever increasing demand on account of rapid growth of population and urbanisation, causing increase in agriculture and industrial requirements on the other hand the water resource is becoming very scarce but the amount of water available is more or less constant. It is therefore essential to not only to carry out systematic assessment of water resources but also improve the efficiency of planning, development & utilisation of this source to mitigate its adverse effect on socio economic development of the country.

1.2   India is endowed with vast network of rivers which carry an estimated natural flow of about 1900 Cu. Km.(annual). The flows available in the rivers have substantial variation with respect to time & location as most of the river flows are available during South West Monsoon (June to September). There is therefore an urgent need to manage the water resources compatible with socio economic developmental programme & environmental policies. It has been assessed that only about, 37.% of the available surface flow   could only be utilised which can be increased to about 50% by inter basin diversion of water from surplus to deficit/drought affected basins.

----

*   Chairman, Central Electricity Authority and Ex-officio Secretary to Govt. of India.
**   Member (Hydro), Central Electricity Authority and Ex-Officio Additional Secretary to Govt. of India.
***   Director, Hydro Planning & Investigation Division, Central Electricity Authority.

1.3   The objective of water resource management is to make available the right amount of water for particular use at right time and of right quality. According to National Water Policy adopted by Govt. of India in 1987, Planning for water resources should be done by considering basin as a hydrologic unit. According to Policy the priority for water use could be in the order of Drinking Water, Irrigation, Hydro Power, Navigation, industrial and other uses. It is however essential to differentiate between consumptive and non-consumptive use etc. Under non-consumptive use hydropower generation, development of navigation/recreation for which certain rate of supply is essential but not consumed except for evaporation and seepage losses etc.

1.4   At present only about 25% of the hydro electric potential of India has been harnessed. Water resources projects involve construction of storage and/ or diversion structure on the river & water conductor system. The hydro electric potential of most of the ideal and comparatively easy sites have already been harnessed. The remaining sites which are geologically and topographically complex & difficult are yet to be developed.

2.0   River Basin Management

2.1 According to the National Water Policy all the developmental projects should be formulated within the frame work of an over all basin/sub-basin. Basin is defined as a basic unit for water resource planning and is considered as an area covered by a system of a surface and   with water flowing into common terminus. There is an inter relationship between surface water, ground water and the supporting natural environment with the catchment boundary. There cannot be an universal policy and strategy to suit the development of all river basins which are solely governed by local conditions.

2.2 According to Central Water Commission (Govt. of India) river basin can be classified into three groups on the basis of catchment area. The river basin draining catchment area above 20000 sq. km are grouped as major basins, the basins intercepting drainage areas between 20000 sq. km and 2000sq.km are termed as medium basins and those with drainage area less than 2000 sq. km are called minor basins. The run off contribution from major and medium basins is assessed to be over 92% whereas remaining 8% is received from minor basins.

3.0 River Basins of India

3.1   Physiographically India can be divided in three major divisions namely in Himalayas and their associated young fold mountains, the ancient block of Peninsular India and the Indo-Gangetic plains lying between the two. The three regions are vastly different in geological history and in character of their terrain. Out of these three divisions, the Himalayan range comprising Greater Himalayas, the Lesser or Middle Himalayas and the Shivalik range possess vast hydro potential. Greater Himalayas being inaccessible provide little opportunity to harness hydro potential but they do act as reservoirs of water for all the rivers of this range. This leaves the other two ranges viz Lesser Himalayas and Shivalik as potential source for development of Hydro Electric potential.

3.2   For the purpose of hydro electric potential survey, the country has been classified into six major river systems namely Indus, Brahmaputra, Ganga , Central Indian River System, East flowing river system and West flowing river system. For the purpose of studies these river systems have been further divided in to fortynine basins. The details of these river system and basin are given below

A   **Indus**   Indus,Jhelum,Chenab, Sutlej , Ravi & Beas

B   **Ganga**   Upper Ganga ,Upper Yamuna , Lower Yamuna, Chambal , Sarda - Gomati - Ghaghra , Sone , Betwa -Sind , Kosi- Gandak -Mahananda, ,Lower Ganga , Damodar.

C   **Brahmaputra**   Upper Brahmaputra, Teesta, Subansiri, Kameng, Kalang, Dihang-Dibang, Luhit, Lower Brahmaputra & Barak and neighboring river system.

D   **Central India River (CIR)System**   Narmada, Tapti Subernarekha , Barhmini - Baitarni, Mahanadi , Mahi , Sabarmati & Luni - Banas and other rivers.

E   **West Flowing Rivers (WFR) System**   Mindhola - Daman ganga, Vaitarna-Savitri,Vashishta-Tillari, Mandvi - Sharavathi , Varahi -Kuttiyadi, Baypore-Periyar and Pamba -Paraliyar.

F   **East Flowing River (EFR) System**   Rivers between Mahanadi & Godavari, Godavari, rivers between Godavari and Krishna , Krishna, rivers between Krishna & Penner , Penner, Rivers between Penner & Cauvery , Cauvery & rivers between Cauvery and Kanykumari

4.0 Hydro Electric Potential of India

4.1 The advent of planned development of the country in the post independence era found the country almost unprepared with a necessary data pertaining to hydel resources of the country. The first systematic survey to assess the economic hydro electric potential of India was carried out by erstwhile Central Water and Power Commission (Power Wing) Govt. of India during 50's when the country's potential was assessed about 42 million KW at 60% Load Factor with annual energy generation of 222 Twh from a total of 250 nos. Hydro Electric schemes. This survey could not cover some of the areas of this country for which adequate data was not available. However the results of this survey proved to be extremely useful to the power planners and provided basic document for planned development of river valley projects in the country. The studies were carried out with several constraints because of availability of very limited data on river flows and topographical information of major portion of Himalayas. This region of the country which has always been known to possess sizeable hydro electric potential. The basinwise results of first Hydro Electric survey are given below :

i)   **River system**   **Potential at 60% Load Factor (Million kW)**

River system	Potential at 60% Load Factor (Million kW)
Indus	6.58
Ganga	4.82
CI R System	4.30
Brahmaputra	13.42
WFR system	4.35
EFR system	8.65
Total	42.12

Based on this survey the large scale construction of multipurpose projects ushered an era of green revolution as well as industrial development in India.

4.2 With the passage of about quarter century since the first survey, there have been major changes in pattern of developments of water resources. It was therefore, felt that a more detailed and comprehensive hydro electric survey of the country became inevitable. Central Electricity Authority carried out Reassessment of Hydro Electric Potential of the Country during 1978-87. The scope of Reassessment Studies was considerably enlarged as compared to first survey. According to the studies the country 's hydro electric potential was placed at about 84 Million Kw at 60% Load Factor from a total number of 845 HE schemes. These schemes have possible Installed Capacity of about 1,50,000 MW. The basin-wise details of the reassessment studies are given below :

Sl.No	Name of River System	No of Basin Studied	Firm Potential (MW)	Potential at 60%LF	Theoretical potential (MW)	Annual Energy Generation (Gwh)	
						90% dependable year	50%
1	Indus	6	11992.8	19988.0	50712	147751	176868
2	Brahmaputra	9	20951.9	34919.8	146170	267663	335766
3	Ganga	10	6428.8	10714.8	52938	81100	98856
4	Central Indian Rivers	8	1644.2	2740.3	14888	14998	23900
5	West Flowing Rivers	7	3689.4	6149.0	9437	35680	40315
6	East Flowing Rivers	9	5719.0	9531.7	26972	52901	63710
	Total	49	50426.1	84043.6	301117	600093	729235

4.2.1 It could be seen that according to the Reassessment studies, the total theoretical potential is estimated at about 301 Million Kw and economic power potential as about 50.4 Million kw (firm) for about 84 Million Kw at 60% Load Factor . . On an average the economic potential of the country works out to about 16.75 % of the total theoretical Potentials. The river system wise ratio of economic potential to theoretical potential is the highest for Western Flowing River system as 39.1% followed by 23.6% of the Indus and 21.2% for East Flowing River system. The total annual energy potential of these scheme is estimated to be about 600 Twh in 90% and 739 TWh in 50% dependable flow conditions.

4.2.2 According to the Reassessment Studies, out of 845 identified Hydro Electric schemes , 331 schemes envisage storage type development and the remaining 514 schemes are of run - of - river type. At the stage of detailed design for run- of- river. small storage to provide diurnal regulation would however be considered. Among the 331 storage type scheme, 51 nos. envisage large storage by constructing dam with height more than 100 meters. The remaining 280 nos. involve construction of low or medium height dams. The total submergence by the storage schemes would be about 0.77 % of the entire drainage area of the country.

4.2.3 In view of the typical topographical characteristics of our river system, water storage have to play an extremely important role in water management of the country. Considering present status of development of the entire Himalayan region is destined to attract more attention of the water management authorities. The country will thus have to be increasingly dependent on the Himalayan water resources. Unfortunately, the Himalayan Region, possessing power potential of about 64 million kw at 60% Load Factor. representing about 76% of the total power potential of India and feeding the major river system of India which carry annually about 1113086 m.cu.m of the discharge, appears to have suitable topographical features to provide only very limited regulatory facilities. Most of the major schemes which constitute the major part of the submergence area are of multipurpose type providing primarily irrigation and flood protection facilities.

4.3 It is obvious that importance of creation of storages in the Himalayas in view of the peculiar flow conditions and increasing pressure on the utilisation of scarce water resources cannot be over emphasized. The possibility of creating such facilities is already restricted because of the topographical features and other considerations. All attempts must therefore be made to ensure early utilisation of such resources. It is recognised that large civil structure and manmade large reservoirs have some adverse environmental effects. It has also been recognised that impact of many of these factors can be substantially mitigated, if advance protective measures are taken. Therefore due advantage can be taken of the Reassessment Studies carried out by Central Electricity Authority to ensure that these projects which are bound to form basic essential element of modern water management system. It is well known that the distribution of the power potential of the country is extremely uneven , as would be expected in view of the diverse topographic features and other factors. Long term comprehensive planning at National level is therefore imperative for optimum development of country power systems by carrying out Feasibility Studies of Hydro Electric Schemes which are yet to be developed.

5.0 Basin-wise Status of Development of HE Potential of India

5.1 As on Jan.2001, out of total assessed potential 84044 MW at 60% Load Factor, Hydo Electric schemes with aggregated potential of 13942.6 MW(16.59 %) have already been developed & HE projects with total potential of 5342.38 MW at 60% load factor (6.36%)are under development. In addition to this Hydro Electric Schemes with total potential of 2304.85 MW at 60% Load Factor (2.74%) have been cleared by Central Electricity Authority and are awaiting investment sanction . Considering the potential of existing , ongoing and Central Electricity Authority cleared schemes it is evident that 25.69% of the assessed Hydro Electric potential have either been developed or is under various stages of development and balance 74.31% potential is yet to be harnessed .

The river system wise status of HE potential of development of India (as on 1.1.2001) is given below :

Basin	Potential Assessed At 60% LF MW	Potential Developed At 60% LF MW	% Devel- oped	Potential Under Develop ment MW	% Under Develo Ment	% Dev'ed +Under Dev'ment	CEA Cleared Pot. At 60% Lf MW	% CEA Cleared Pot.	% Total
Indus	19988.00	3165.02	15.83	1191.67	5.96	21.80	1014.42	5.08	26.87
Ganga	10715.00	1877.00	17.52	1355.55	12.65	30.17	308.50	2.88	33.05
Central Indian Rivers	2740.00	634.33	23.15	1528.00	55.77	78.92	221.95	8.10	87.02
West Flowing Rivers	6149.00	3667.17	59.64	537.63	8.74	68.38	52.83	0.86	69.24
East Flowing Rivers	9532.00	4061.75	42.61	300.98	3.16	45.77	37.73	0.40	46.17
Brahmaputra Basin	34920.00	537.33	1.54	428.55	1.23	2.77	669.42	1.92	4.68
All India	84044.00	13942.60	16.59	5342.38	6.36	22.95	2304.85	2.74	25.69

6.0 Trends of Hydro Electric Development

The first hydro electric development in India dates back to 1897 but the pace of hydro development remained tardy upto independence (1947) when total installed capacity of Hydro Electric stations was only 508 MW. The hydro capacity addition has since 1947 experienced a steady rise from 508 MW to 24746 MW now. The share of hydro power in the total installed capacity was upto 50.62% in 1963 and thereafter had a steady decline and at present accounts to only 24.6 % of the total installed capacity of power stations. One explanation for increased share of hydro upto 1963 is that country accorded top priority to development of Agriculture sector thus paving up the way for accelerated development of river valley multi purpose projects primarily for irrigation & incidentally with power benefits. In the seventies a shift in the priority from Agriculture to industrial sector became perceptible. This seems to have triggred concentration of coal based power projects. When the required of pace of capacity addition became more intense , the priority was equally accorded to coal & gas based projects. This resulted in marked decline in hydro development in the country.

7.0   Constraints in Development of Hydro Power

The main reasons for slow development of hydro electric potetials of India are as under:

-       Hydro Projects are site specific and even in case of extensive geological investigations using new techniques of survey & investigations, an element of uncertainty remains in the sub surface geology & geological surprises during actual construction are encountered. This problem is particularly faced in the Hydro Electric projects in Himalayan Region where underground tunneling is required.

-       A large number of hydro electric projects having common river system between adjoining states are held up on account of inter states disputes . A few of such projects have even received Techno Economic Clearance (TEC) of Central Electricity Authority but the investment decisions could not be accorded due to inter state aspects.

-       The investment decision for new projects and implementation of many sanctioned projects are held up / delayed on account of environmental and / or forest clearance from Ministry of Environment & Forest (MOEF).

-       The progress of some of the hydro electric projects has been severely affected on account of opposition to construction by Project Affected People (PAP).The resolution of Rehabilitation & Resettlement (R&R) aspects is often time consuming and results in delay for implementation .

8.0   Policy on Hydro Power Development

8.1   With an objective of faster development of hydro potential , Govt. of India (GOI) announced a policy in 1998 incorporating several steps/measures.

8.2   Basin- wise Development of Hydro Electric potential is one of the important step included in the policy. According to the policy further studies to firm up the parameters of projects identified by Central Electricity Authority would be taken up on the basis of development of Hydro Electric Potentail in a basin as whole for maximising benefits and priortising the projects.

8.3   As a long term strategy efforts will be made to ensure that DPRs which are under various stages of processing are finalised and cleared so that start could be made on these projects. Survey & Investigation of potential hydro sites on an advanced scientific basis would be an essential requirement for future development of Hydro Electric Potential.

8.4   Govt. of India recognises the need for evolving an approach to ensure that the available Hydro Electric potential is fully utilised without prejudice to the right or riparian states. A consenuses could evolved amongst the basin states regarding location of Hydro Electric projects, basic parameters involved and mechanism through which these projects would be constructed and operated. It will be responsibility of State Govt. to acquire the land (Govt./pvt./forest) for the project and also negotiate on its own terms with land owners as per policy adopted by respective State Govt. All the issues of Rehabilitation & Resettlement (R&R) associated with projects have to be addressed by State Govt

9.0 Review of Hydro Potential

9.1 The studies for Reassessment of Hydro Electric Potential carried out by CEA in 1987 had a number of limitations such as old and inadequate hydrological data, topo-sheets with contour interval of 20-40 m, limited information for present/future irrigation requirements. More over environmental aspects like submergence were assigned only broad consideration. It has now therefore become essential to review the hydro potential by making use of updated/long term hydrological data, topo-sheets of shorter of contour interval and by making use of detailed information regarding consumptive uses of water. The studies are proposed to be carried out by interaction with Ministry of Environment & Forest(MoEF) by obtaining the views of concerned states so that development of identified schemes are not hampered on account of such factors.

10.0 The Role of States in Water Planning & Management in India

10.1 India is a union of states and most of India's river basins are inter state in nature. States can develop the water resources within their boundaries. It also gives certain power to the Union Government to regulate and develop the interstate rivers. The water development has, therefore, mostly been planned and executed with the State, or part of the basin within the state, as a unit

10.2 According to Constitution of India Electricity is concurrent subject. Apart from state sector the power development is being done in Central as well as Private Sectors Central Electricity Authority (CEA) is statutory organization constituted under Electricity (Supply) Act, 1948. Apart from various functions Central Electricity Authority carries out Techno Economic Appraisal of power projects. At present all the proposals for development of Hydro Electric Projects costing more than Rs.250 crores as well as all schemes involving inter state aspects are required to be examined in CEA for Techno economic clearance (TEC).

11.0 Approach for Basin-wise Development in India

11.1 The concept of basin wise planning was initiated in India in the post independence period by establishment of Damodar Valley Corporation(DVC) by Act of Parliament in 1948. DVC was established for development and management of water and other infrastructural facilities in Damodar basin. The functions of DVC include promotion and operation of irrigation, water supply, drainage, hydro electric and thermal power generation, flood control etc. for overall development Damodar basin. Besides DVC a few other corporations and statutory organisations have been created at Public Sector with a view to develop hydro potential resources & master plans of river basins in States not equipped with adequate financial resources.

12.0 Inter Basin Transfer of Water

12.1 Inter Basin transfer of water is another important consideration of water resource development. Though river basin is considered as a basic unit for water resources development this may not always lead to optimum utilisation. Many basins in India are in the ultimate stage of development, while other basin would already be facing water shortages or drought conditions. Inter Basin transfer of water could thus play an important role to meet such situations. The short water transfer links appears to be quite attractive as implementation of long links are likely to have various problems like consensus among basin state, quantum of surplus water availability, environmental issues etc.

13.0 Conclusion

13.1 The concept of basin level planning & development has been attempted in India immediately after independence by establishment of Damodar Valley Corporation (DVC). Subsequently, other corporations and statutory organisations have also been set up for development of vast hydro electric potential of various rivers and for preparation of perspective master plans for overall development of river basins. The studies for Basin-wise Hydro Electric Survey of India have provided valuable input for taking up development of river valley projects. There is an immediate need to formulate the strategy for optimum development of balance 75% un-tapped Hydro Electric potential of India. The policy instrument / measures regarding Basin-wise Development of Hydro Electric potential, Resolution of Inter State Aspects and Survey & Investigation on advanced scientific basis could play vital role for systematic development of un-harnessed Hydro Electric Potential of India.

Hydropower in the New Millennium, Honningsvåg et al (eds), © 2001 Taylor & Francis, ISBN 90 5809 195 3

Water resourees management in a cascade of hydropower plants

S.Theobald, A.Celan & F.Nestmann
Institute for Water Resources Management, Hydraulic and Rural Engineering (IWK), University of Karlsruhe, Germany

ABSTRACT: The operation of barrages in rivers has to fulfil different sets of requirements since most reservoirs have a multi functional character. The paper describes a software tool for structuring the automated operation of hydropower plants. The program optimizes the parameters of automation and allows the definition of the best method of operation for a cascade of barrages or hydropower plants.

1 INTRODUCTION

The operation of hydropower plants (including the weirs) and the management of water resources has to fulfil the following - often contrary - requirements: safety of navigation, intensive use of hydropower, flood protection, coverage of industrial water demand, supply of water for irrigation and minimization of the number of actuator operations (turbines, weirs). The main goal of the operation should be a constant and steady discharge within small head water level tolerances.

Today, the automated control of a reservoir or a cascade of reservoirs is of increasing importance. The requirements mentioned above have to be fulfilled by the control system with increasing demands in control performance and a high reliability of operation.

On the other hand, there is a complex, highly nonlinear, unsteady and locally distributed hydraulic behaviour in the reach as well as the variable operation conditions of the barrage. The operation conditions depend on water quantity and availability. The main influence on the hydraulic behaviour is determined by the demand driven production of electric energy, the operation of locks, the outflow and inflow of the reach, the precipitation, etc. The controlled variable is the water level (usually the headwater level), the manipulated variable is the discharge at the barrage (see fig. 1). The water level should be controlled in such a way that the most economic and complete exploitation of the water power is achieved. For the sake of reliability each reach is controlled by a local controller. In the case of a cascade of reservoirs, the discharges at the hydropower plants and the reference values of the water level control are given by a central water management system considering global effects, such as power modulation by the needs of energy production, navigation, weather and other requirements.

The control of a barrage is usually realized using industrial process control systems with highly reliable components and partly redundant structures. It is a well known fact that many automated control installations of hydropower plants are suffering from latent stability problems caused by flow or water level oscillations. During operation, the control loop may change from stable to oscillatory and limit cycle behaviour. The oscillations, when encountered, are usually subdued by opening the water level control loop and changing to pure flow control, power control or even manual operation. The manual as well as the automated operation of a cascade of reservoirs may result in extreme flow oscillations - especially

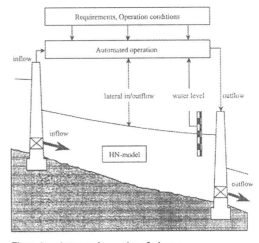

Figure 1: Automated operation of a barrage

at the end of the cascade - caused by the amplification of flow changes downstream from barrage to barrage.

2 TASK AND METHODS

Before an automated control of a barrage can be realized, investigations to evaluate the influence and the effects of the automatization to the reach are necessary. These investigations refer to the interaction between the operation of the hydropower plants, changing the discharges, and the hydraulic situation within the reservoir. Using numerical models, it is possible to:

— analyse the hydraulic behaviour of the river reach
— investigate the existing and new control structures for operating hydropower plants
— evaluate and verify the automated control
— develop strategies to optimize the management within the reach
— teach and train operators with the simulator

The program *KASSMO* (Karlsruher-Staustufen-Simulationsmodell) is used to find a solution to the problems named above (see fig. 2). It has been developed at the Institute for Water Resources Management, Hydraulic and Rural Engineering (IWK) at the University of Karlsruhe especially for numerical simulation of barrage operations. *KASSMO* has been successfully applied in numerous projects in practice. The program consists of four main parts:

Part 1: Reservoir model: Hydrodynamic-Numerical model (HN-model) for the simulation of the hydrodynamic behaviour in the reach

Part 2: Functions to control the regulating units with output of time-dependent discharges at the barrages depending on the state of process and the type of operation strategy

Part 3: Optimization tool for the parameters

Part 4: Visualization tool

The aim of the control design process is to determine the structure and parameters of the automated control of a hydropower plant or a cascade of hydropower plants. The adjustment of the existing control functions is based on the investigations made with the program system *KASSMO*.

3 SIMULATION OF THE HYDRAULIC PROCESS

3.1 Requirements of the reservoir model

The complete numerical method is based on the modelling of the reservoir, because the investigations for the automated control refer to the interactions of the hydraulic behaviour in the reservoir and the variation of discharge at the barrages. To find the proper closed loop flow control, the influence of variations of inflow and outflow on the variation of the water level has to be known. The following requirements for the modelling of the reservoir have to be considered:

— simulation of the reservoir hydraulics (unsteady flow calculation of water level and discharge at certain stations as input and output of the control)
— quality of modelling (comparison between the process in nature and model)
— calculation of characteristic reservoir parameters
— small computational and modelling efforts

3.2 Reservoir modelling

For the modelling of reservoir hydraulics, the geometry of the reservoir as well as the flow control at the barrages have to be taken into account. This includes geometric and topographic characteristics under consideration of flow characterizing cross-sections, parts of retention, branches of the river system and special hydraulic structures. In the model system, those boundary conditions have to be considered, where variations of discharge affect the state of flow through the reservoir. The control stations can be weirs, gates, turbines, lock channels and separate intake or diversion structures e.g intake structures of cooling water for thermal power plants or intake or diversion structures of pumped-storage plants. Inflows or outflows are taken into account by the size of the gradient and the amount of input or output.

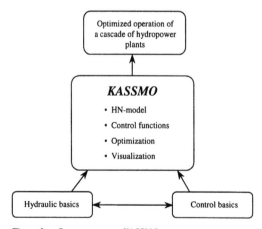

Figure 2: Program system *KASSMO*

3.3 Quality of modelling

Hydrodynamic processes in reservoirs are three-dimensional, complex and unsteady and can be described as a non-linear dynamic system. To minimize the costs of control design, the main task of the engineer is to model the real process as exact as necessary and not as exact as possible. Before practical use, the models have to be evaluated thoroughly. To get a good similarity of the hydraulic process values in nature and model, the numerical model has to describe the real process on the basis of physical equations. Only when the process values achieved by the numerical method correspond sufficiently with those of nature, they can be translated into practice.

3.4 Calculation of hydraulic parameters of the reservoir

With the knowledge of the wave propagation time T_L and the retention time T_R, a classification of the reservoir as well as an appropriate control strategy can be made. The wave propagation time T_L is defined as the period of time between a variation of discharge (e.g. the inflow into a reservoir) and the resulting variation of a reference water level (e.g the upstream water level of a barrage). The retention time T_R is the time interval needed to take in/out the retention volume at given changes in discharge with the retention volume being the volume difference corresponding to two different steady state discharges at the same headwater level. The wave propagation time T_L as well as the retention time T_R depend on the discharge and the upstream water level.

As a basis of the parameter optimization of the automated control, the following hydraulic values of the reservoir have to be known:

- amplitudes of negative and positive surge
- wave propagation of negative and positive surge
- volumes of retention
- times of retention
- impounded surface

Therefore, it has to be taken into account that the values are not constant but dependent on the discharge and the upstream water level. The state of the reservoir varies widely between high flow and low flow, which has to be considered in the optimization process of the control system.

3.5 Choice of the numerical method for the reservoir modelling

The consideration of the requirements mentioned above (3.1 - 3.4) is essential for a wise choice of a reservoir model. For this reason, an one-dimensional unsteady hydrodynamic-numerical model is used to describe the hydraulic process. The basic equations of the model are the classical equations of Saint

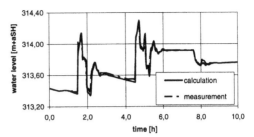

Figure 3: Stage hydrograph at the hydropower plant RADAG. Comparison between measurement and calculation

Venant for 1-D open channel flow (continuity equation and dynamic equation). The geometry of the landscape is described in the form of cross-sections.

At the locations where the hydropower plants and barrages are usually situated, unsteady boundary conditions of discharge, water level and rating curves have to be chosen. Special elements like weirs, gates, storage basins, bridges, siphons and looped or meshed river systems can be considered as well. Time (t) and place (x) are the independent variables in this equation system, whereas the dependent variables are the discharge $Q(x,t)$ and the wetted area $A(x,t)$ or the water depth $y(x,t)$. This equation system is solved by using the Preismann implicit method.

In the following section the application of the HN-model is described, referring to the investigations in the reach of the Rheinkraftwerk Abbruck-Dogern (RADAG). To calibrate the detailed HN-model, field investigations at unsteady discharge were performed. For different waves generated by the hydropower plant, the water levels were measured simultaneously at 12 stations in the reach. These measurements are suitable to perform the verification of the HN-model. Previously the calibration of the model was performed, based on fixed water levels at steady flow conditions. Figure 3 shows a typical result of this investigation with the comparison between the calculation and the measurement of the water level at the hydropower plant RADAG. The values determined by calculation and those determined by measurement corresponded almost exactly. For the calculation and the measurement, the water level, the amplitudes of the waves and the wave propagation time were nearly the same. This proves, that the HN-model is suitable for quantifying the hydraulic behaviour of the reservoir.

4 AUTOMATED CONTROL OF BARRAGES

The increasing requirements on dynamic performance, economic and guaranteed operation have stimulated the development of sophisticated control strategies, allowing fully automated operation. The economic goal of the project engineer is to design and implement the control system with a minimum effort. This effort is mainly determined by three aspects: complexity of the process to be controlled, control requirements given by the user, and operating conditions (e.g. given by laws).

For the implementation and investigation of automated control, the HN-model was extended at the IWK with special tools for different kinds of flow and water level control. The outflow is calculated considering information of the inflow and the water level. The original control algorithms of the real industrial realization platform are integrated into the simulation model. The main extensions are control functions. In order to fulfil the often contrary requirements, the choice of intelligent operation strategies for the barrage depends on the analysis of the state of the hydraulic situation in the reservoir. Therefore an a priori classification of the reservoir is necessary as a ratio of wave propagation time T_L and retention time T_R. For a reservoir with $T_R > T_L$ the standard closed loop level control can be used, because the water level can be fixed to a constant reference value without amplification of the outflow in relation to the inflow. The closed loop level control is defined as a barrage operation where the water level at a reference point - in most cases the upstream water level at the barrage - is fixed to a constant value. The ratio of inflow and outflow of the reservoir are not taken into consideration.

For a reservoir with $T_R < T_L$ the standard closed loop level control leads to an amplification of outflow in relation to the inflow and is therefore an improper operating method. In these reservoirs a variation of the water level can not be avoided. To avoid the amplification of outflow the information about incoming discharge has to be given to the control of the outflow in advance (before T_L). This can be done in two different ways.

The first type is based on a feedforward control to the controller output. The inflow is given time shifted, but directly as a set value to the outflow (feedforward control). An additional cascaded proportional-integral (PI-type) level controller has the task to keep the water level in a given tolerance even in cases when the feedforward control operates with incorrectly measured inflow values as well as incorrect parameters. There is no change of the reference value water level of the level controller. In Germany this type of control of barrages is often called OW/Q-method.

The second alternative is called anticipation-method. In this case the referring water level depends on the variation of inflow. That means a variation of inflow is causing a variation of outflow indirectly via the water level controller by a changed reference value. The principle is the follow-up control.

The advantages and disadvantages of OW/Q-method and anticipation-method in practical use, which

are mainly dependent on practical influences (inaccurate determination of discharges, disturbances at the gauge) are not named explicitly in this report.

The A/Q-method is a combination of the two methods above. It includes the feedforward control to the change of outflow and the change of the reference value (fig. 4). It combines the advantages and reduces the disadvantages of the OW/Q-method and the anticipation method.

The implementation of the control functions in *KASSMO* is realized in cooperation with the University of Applied Sciences of Fulda. The parameters for the control algorithms are chosen depending on the discharge. The following operation strategies are implemented in *KASSMO*:

–   closed loop level control
–   closed loop flow control
–   headwater/discharge-control (OW/Q), which means feedforward control to controller output (discharge)
–   anticipation-method, which means feedforward control to reference value (water level)
–   anticipation/discharge-control (A/Q), which is the combination of the two above
–   headwater/discharge-control with pumped-storage operation
–   anticipation-method with pumped-storage operation
–   central water management system

In the case of a cascade of hydropower plants, a central water management system is integrated to calculate the reference discharges and water levels at

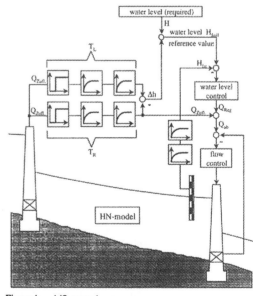

Figure 4:   A/Q-control

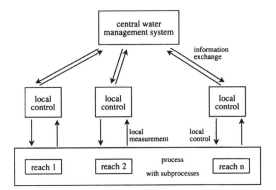

Figure 5: Hierarchical control structure

every barrage. It is useful and necessary for the management of the reservoirs to have enough volume to allow variable upstream water levels. The hierarchical control structure with a central water management system is shown in figure 5.

5 OPTIMIZATION

Advanced control design uses optimization methods to automatically optimize the time-variant parameters of the nonlinear control structure on the basis of detailed simulation runs. Therefore the program system *KASSMO* includes an optimization tool to find the best control parameters. For a given inflow hydrograph, the control parameters are optimized considering defined criteria of the control quality. The control design process is supported by a genetic algorithm which uses calculated performance indices of the simulated automated process as input and produces improved control parameters as output. The final parameter sets are achieved by numerous simulation runs. The following criteria of the control quality are defined:

– deviation of the water level from the reference value
– time shifted forwarding of the outflow in relation to the inflow, amplification of discharges
– number of actuator operations
– number of actuator operations for short

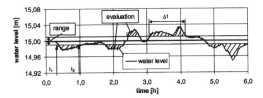

Figure 6: Evaluation of water level

contrary movements of the regulating units (opening, closing)

KASSMO also includes a special tool for optimizing energy production, which is a further important criterion.

5.1 Deviation of the water level from the reference value

One of the most important requirements in the operation of hydropower plants is to keep the water level in a narrow range. The following evaluation considers the magnitude and the duration of the deviation of the water level from the reference value (fig. 6). These considerations are shown in equation 1.

$$S_{OW\text{-}Stand} = \sum_{i=1}^{N} \int_{t_{2i-1}}^{t_{2i}} \left(y_{ist}(t) - y_{soll}(t) \right)^2 dt \sqrt[3]{t_{2i} - t_{2i-1}} \qquad (1)$$

where $S_{OW\text{-}Stand}$ = quality criterion for deviation of water level; t_{2i} and t_{2i-1} = intervals with water level outside of the given range; N = number of intervals; $y_{ist}(t)$ = water level; and $y_{soll}(t)$ = reference value.

5.2 Amplification of discharges

A second very important criterion is to avoid an amplification of the discharge from barrage to barrage. As examples two different kinds of inflow extremes are defined. Within a given time period, extreme (A) is a short, non relevant event with a little peak, whereas extreme (B) lasts longer and has a high peak and is therefore relevant. The non relevant extreme (A) can be neglected. If the outflow, however, surpasses the value of the relevant extreme (B) the amplification of the discharge has to be considered and calculated. The situation is shown in fig. 7. The calculation is performed by equation 2.

$$S_{Qverst.} = \sum_{i=1}^{N} \int_{t_{2i-1}}^{t_{2i}} \left(Q_{ab}(t) - Q_{extr,rel}(t) \right)^2 dt \qquad (2)$$

where $S_{Qverst.}$ = quality criterion for the amplification of discharge; $Q_{ab}(t)$ = outflow; $Q_{extr,\,rel}(t)$ = discharge value for the relevant extremum.

5.3 Number of actuator operations

In order to minimize wear of the mechanical system, the number of actuator operations is to be kept low. The two following criteria consider the number of actuator operations.

$$S_{Stell} = n_{Stell} \qquad (3)$$

where S_{Stell} = quality criterion for the number of all

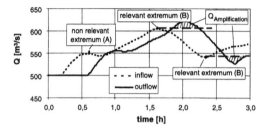

Figure 7: Evaluation of discharge with regard to amplification

actuator movements and n_{Stell} = number of all actuator movements.

$$S_{Umkehr} = n_{Umkehr} \qquad (4)$$

where S_{Umkehr} = quality criterion for the number of short contrary movements of the regulating units and n_{Stell} = number of short contrary movements of the regulating units.

5.4 Total criterion

Finally, all single criteria are summarised to one criterion S_{gesamt}. As different operating conditions focus on different criteria of main interest, it is necessary to emphasize the relevant criteria for the case considered. This is done through weighting coefficients α_i (see equation 5). Small values of a criterion correspond to a good fulfilment.

$$S_{Gesamt} = \alpha_1 S_{OW-Stand} + \alpha_2 S_{Qverst} + \alpha_4 S_{Stell} + \alpha_5 S_{Umkehr} \qquad (5)$$

where S_{gesamt} = total criterion; α_i = weighting coefficients; $S_{OW-Stand}$ = quality criterion for deviation of water level; $S_{Qverst.}$ = quality criterion for the amplification of discharge; S_{Stell} = quality criterion for the number of all actuator movements; and S_{Umkehr} = quality criterion for the number of short contrary movements of the regulating units.

5.5 Production of energy

The main goal for the owners of hydropower plants is to optimize energy production. This is a challenging task for cascades of reservoirs with volumes for water management. A tool is currently being developed to help maximizing profits. It has to combine the following:

-   use of the full capacity of the reservoir within the concession limits
-   high energy production during peak times of energy demand
-   optimal filling and emptying of the volume

6 VISUALIZATION AND TRAINING TOOL

The increasing automation of hydropower plant operation with sophisticated and powerful control systems results in the loss of capability of the operators to manually operate the plant. The processes underlying the automated operation are too complex for the operators to fully comprehend all influences on hydraulics and automation. Nowadays, as the usual operation of a hydropower plant is almost fully automated, the operator's main task is focussed on monitoring the process. He thereby loses more and more knowledge of how to manually operate the plant. If a failure of the automated control (a possibility that cannot be eliminated) coincides with extraordinary hydraulic situation, a high competence in manual operation is required to avoid danger and damage to humans and nature. Since the operator is unchallenged during the normal operation, he does not feel capable to safely master an exceptional hydraulic situation which leads to stress and operational mistakes. This problem can be solved by continuous training with a simulation program. Relevant cases for simulation are for example:

-   sudden change of inflow for different basic discharges
-   gradual change of inflow for different basic discharges
-   sudden increase or decrease of the reference value
-   operation of upstream/downstream locks
-   rapid closing of turbines upstream/downstream combination of locks and passage of ships
-   flood event
-   characteristic inflow hydrograph of the reach

VisuCas, an integral part of *KASSMO*, is such a simulation tool for operators of hydropower plants and complies with didactic requirements (see fig. 8). Ideally, as the plant is operating automatically during normal conditions, the operator can work with the simulation tool for learning and training purposes. With the possibility of interaction, he has an active part in the simulation but in contrast to the operation of the real hydropower plant, he is allowed to make mistakes without risk. To make mistakes is part of the learning process. Like operating a flight simulator, the operator exercises the operation of a barrage and verifies the consequences of his own operations and collects valuable experiences. The main parts of the visualization tool are the following:

-   online presentation of hydraulic values i.e discharges and water levels as hydrographs for monitoring the process
-   possibilities for interactive manipulation of the process
-   real-time simulation

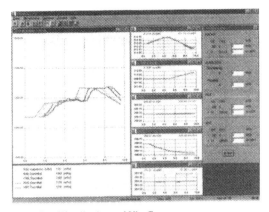

Figure 8: Visualization tool VisuCas

Not only dangerous situations are part of the teaching program, also daily operations can be trained.

7   APPLICATION OF THE SIMULATION TOOL KASSMO

7.1 River Neckar

The investigated reach is the 200 km stretch of the river Neckar between Mannheim and Plochingen with 27 barrages. The owner of the hydropower plants intended to successively implement an automated operation as part of the modernisation process of system. The main requirement of the operation is to maintain the navigability of the river while complying with the strict limitations of headwater level changes and avoiding the amplification of discharge from one barrage to the next. A headwater/discharge control is planned at every barrage. In addition to the models of the single reservoirs, a complete model for the entire navigable reach of the Neckar with all 27 barrages was developed. The simulations showed the feasibility of this project and provided the structure and parameters for the automation.

7.2 Rhine Reach Säckingen

This study was carried out for the hydropower plant Bad Säckingen/Rhine (RKS). It included the installation of an automated control in order to be able to forward the inflow practically unchanged. The reservoir of the RKS also acts as the downstream basin of the pumped-storage power plant Säckingen. Therefore, the automated control system uses a special feedforward control in order to determine the reference water level for filling and emptying the reservoir. According to the management, the automated control has been complying with all requirements since its installation in the beginning of 1997.

7.3 Rhine Reach Albbruck-Dogern

The automated control of the Albbruck-Dogern Rhine hydropower plant (RADAG) has to comply with similar tasks as the one described in the previous chapter, since the volume of the reach is used as the downstream basin of the pumped-storage power plant Waldshut. However, there are two more difficulties in the reach RADAG. There are two inflows (Rhine and Aare) and because of the spatial separation (approx. 3.6 km) of weir and power plant there also are two outflows. The concession gage is located in the head water of the weir RADAG, hence during operation of the power plant alone, a dead time between the change of the manipulated variable (outflow) and the controlled variable (water level) occurs. The dead time depends on the outflow in the channel of the power plant. Seen from the perspective of feedforward control, these circumstances are not easy to handle as the transition from power plant to weir operation is known to be difficult. The solution lies in the consideration of the outflow dependent wave propagation times between the locations of the pumped-storage power plant (input, output) and the weir (reference gage) when applying the feedforward control to the command variable, which is necessary for the operation of the pumped-storage power plant. Since the installation in 1999 the system has been complying with all requirements demanded by the management. The comparison of simulations of the already installed with the restructured and parametered automated control raises expectations for a substantial improvement concerning the required uniformity of the outflow (see fig.9).

7.4 Schluchseewerk AG

The IWK also carried out surveys of the operation modes of the Schluchseewerk AG hydropower plants, consisting of two pumped-storage power plants and the management of the reservoirs of three run-of-river hydropower plants (RADAG, RKS and Ryburg-Schwörstadt) and the storage basin. The reaches of the run-of-river hydropower plants serve as downstream basins of the pumped-storage plants. In order to manage this system, target values for outflow and retention have to be specified for the central control, so that the water level in the Rhine reaches remains within the concession limits. The aim of the management is using the reservoirs in such a way, that the concession limits are not exceeded during the operation of the pumped-storage plants and oscillations of the Rhine outflow do not occur. Furthermore, oscillations of the inflow into the cascade of reservoirs have to be compensated by the operation of the run-of-river hydropower plant at the end of the cascade. The visualization and operation recommendation tool developed by the IWK is currently being installed.

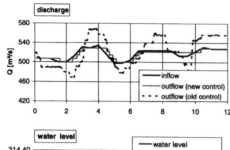

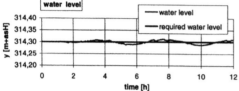

Figure 9: Automated operation RADAG: hydrograph of in
flow, outflow and upstream water level

7.5 River Volga

The river Volga, with a length of 3650 km, is the
longest and also the largest river in Europe. However, the difference in altitude from its sources to its
estuary is only 260 m, with 150 m being taken by 8
hydropower plants. The installed power of these 8
run-of-river hydropower plants of the Volga totals
8400 MW. Up to now, in the context of a German-
Russian joint venture, the IWK developed a GIS-
model for the collection of topographical data and a
HN-model for the simulation of the flow conditions
in the 335 km long reach Tscheboksary. For fear of
increased flood danger and waterlogging, the
hydropower plant at Tscheboksary is presently
operated with a headwater level reduced by 5 m. The
reason is insufficient information on the influence of
the backwater curve at different flows.

First simulations with the newly developed model, however, show the possibility to introduce a
flow dependent headwater level at the hydropower
plant Tscheboksary which increases the power production at the plant discharge by approx. 400 MW
from 1000 MW to 1400 MW. The objective of the
further project work is to model the Volga cascade
using *KASSMO* in order to optimize its operation.

7.6 Bavarian Danube

In the free flowing reach of the river Danube downstream of the hydropower plant Straubing, flow oscillations occurred at low water causing flow reductions which deteriorate the conditions for navigation.
These oscillations are assumed to be the result of
inflow variations due to hydrologic conditions in the
upstream part of the Danube, by the simultaneous
management of the Main-Danube Channel, and by a
not yet optimized operation of the run-of-river

hydropower plants. In a preliminary study, a cascade
of four reaches is currently analysed. The possibility
to reduce the oscillations and thereby improve the
conditions for navigation by changing the mode of
operation of the run-of-river hydropower plants
and/or the management of the Main-Danube Channel is investigated.

8 CONCLUSION

The operation of barrages in rivers has to fulfil different sets of requirements since most reservoirs have
a multi functional character. Today, the automated
control of a reservoir or a cascade of reservoirs is of
increasing importance in order to comply as much as
possible with these different, sometimes contrary
requirements. The interaction between the operation
of the discharge elements and the hydraulic situation
within the reservoir needs to be modelled in order to
be able to design the structure and parameters of an
automated control. The newly developed program
system described in this paper is appropriate for
various processes and requirements and includes an
optimization tool to provide improved control
parameter sets. It combines the basics of hydraulics
and control to develop an automated operation and
training program for cascades of hydropower plants.
The program has been applied successfully in
numerous projects in science and practice. Case
studies of the rivers Neckar, Rhine, and Danube in
Germany as well as the river Volga in Russia show
its practical application.

REFERENCES

Celan, A. 1997. Einsatzmöglichkeiten von Simulationspro-
grammen. Universität Karlsruhe (ed.), *Workshop "Gestaute
Flußsysteme - Dynamik und automatisierte Betriebsführung", veranstaltet vom FA 2.5 des DVWK*

Cuno, B. 1997. Grundlagen der Automatisierung. Universität
Karlsruhe (ed.), *Workshop "Gestaute Flußsysteme - Dyna-
mik und automatisierte Betriebsführung", veranstaltet vom
FA 2.5 des DVWK*

Cuno, B. und Theobald, S. 1997. The relationship between con-
trol requirements, process complexity an modelling effort in the
design process of river control systems. *Proceedings of the 2nd
Symposium on Mathematical Modelling of the International
Society of Mathematics and Computers in Simulation
(IMACS)*

Nestmann, F. & Theobald, S. 1994. Numerisches Modell zur
Steuerung und Regelung einer Staustufenkette an Beispielen
von Rhein und Neckar. *Wasserwirtschaft, 2/94*: 72-78 and
BAW Mitteilungsblatt, Nr. 71: 1-14

Theobald, S. 1999. Numerische Simulation von Staustufenket-
ten mit automatisiertem Betrieb. Universität Karlsruhe,
Institut für Wasserwirtschaft und Kulturtechnik (ed.), *Mitteilungen 201.*

CONTACT INFORMATION

Dr.-Ing. Stephan Theobald
Dipl.-Inform. Andjelko Celan
Prof. Dr.-Ing. Dr. h.c. Franz Nestmann
Institut für Wasserwirtschaft und Kulturtechnik
Universität Karlsruhe
Kaiserstr. 12
D-76128 Karlsruhe

e-mail: iwk@uni-karlsruhe.de
www: http://www.uni-karlsruhe.de/~iwk

Hydropower as environment sustainable generation

Hydropower in the New Millennium, Honningsvåg et al (eds), © 2001 Taylor & Francis, ISBN 90 5809 195 3

Project Construction and Environmental Impact Assessment Approach

B.Berakovič
Faculty of Civil Engineering, University of Zagreb, Zagreb, Croatia

Z. Pletikapič
Elektroprojekt Consulting Engineers, Zagreb, Croatia

Z.Mahmutovič
Elektroprojekt Consulting Engineers, Zagreb, Croalia

B.K.Karleuša
Faculty of Civil Engineering, University of Rijeka, Rijeka, Croatia

ABSTRACT: Decision-making on construction of complex projects is highly demanding process involving comprehensive and integrated consideration of the related issues. The accelerated economic development in the twentieth century has adversely affected the environment and resulted in a more complex valuation of economic activities. This particularly applies to the decision-making on investment projects and development of major infrastructure (e.g. power generation, transportation) where the once "objectively" biased engineer has been replaced with the subjectively biased green movement.

The environmental management policy of a community is directly dependent on available resources and the level of national development. For an investment to be justified and environmentally acceptable, the methods are being looked for a project environmental impact assessment that should result in sustainable development.

The possibility of applying the economic analyses based on neoclassic economic thought to human exploitation of the environment is highlighted. In the decision-making process, properly used economic mechanisms are still a dominant tool. Comprehensive and thorough insight into the environmental and social impact gained through the environmental economics and economic values has the advantage over nonobjective and unmeasurable multi-criteria methods.

The assessment and ranking approach to recommended and comparable concepts has the following advantages: increased objectivity-transparency, measurability, respect of time component (project LCA), and accounting for uncertainty and risks (particularly in the future).

1 INTRODUCTION

Hydro power projects, which today constitute the backbone of the country's energy system, began to be constructed in Croatia on a systematic basis as recently as the second half of the 20th century. Today's 25 hydro power plants in Croatia with a combined capacity of 2075 MW supply 50 to 60 percent of the total annual electricity needs of the country.

However, towards the end of the '80s of last century, the hitherto continuous effort to build new hydroelectric projects was stopped despite plans that had been prepared for construction of additional 60 hydro power plants, totaling about 1400 MW. The reasons for this are to be looked for not only in the establishment of a new economic system or in the collapse of a relatively strong industry and industrial lobby capable of constructing and fully equipping such plants, but also in that the previous projects had practically used up the last available sections of the waterways in Croatia for which the project viability

had only been in utilization of water power. In other words, the remaining potential utilization of waterways in Croatia for power generation either require participation by other potential users to be profitable (i.e. construction of multi-purpose hydraulic systems) or a solution for significant environmental impacts (which, because of demands by local population and the green, significantly increase construction and operation costs).

On top of the above, requirements were introduced for the generally hard-to-measure environmental and societal impacts of construction of such projects, such as impacts on habitats of endangered plant and animal species or on traditional way of life of local population, or agreements with neighboring countries on transboundary pollution in the case of border waterways.

All this leads to the conclusion that it will be hardly possible to continue making decisions on new hydro power projects in Croatia in the old way, protecting only the interests of the power industry and perhaps potential partners interested in (multi-purpose) benefits of these projects. The decision making process now requires (even under current legislation) involvement of local population and a number of environmental and nature conservation institutions as equal partners.

In other words, such partnerships demand redefining of the process of decision making on new hydro power projects, which is based on standard assessment of profitability of their construction and operation from power generation perspective or on cost-benefit analyses (which include all other direct and indirect costs and benefits of project construction) from profit perspective. As licensing, as well as agreeing participation of potential partners in the project, has included general population and politicians and highly specialized experts, the tendency has been growing to make decisions on the basis of "simpler" assessment processes, adjusted to the level of knowledge of the general population and of various professions and disciplines, rather than using economic methods.

It should be noted that all the decision-making procedures elaborated in this paper with regard to the hydropower projects could equally be used for any multipurpose hydropower and other complex projects inserted in space.

2 OVERVIEW OF DECISION-MAKING METHODS

2.1 General

Project construction decision making is essentially a process of selecting the best solution from a number of alternative solutions which lend themselves to system analyses. Thus, it is possible to identify construction objectives, determine constraints for project implementation, select criteria to establish and assess the quality of each alternative, and determine measures for evaluation responsiveness to individual criteria.

In the water resources management system (and in the case of using hydraulic potential of waterways alike) criteria and measures have a triple task of:
- assessing current degree of meeting water resources management objectives,
- assessing possible solutions to select the best for future water management,
- assessing the success of the implementation of the selected solution.

When starting from the basic objective of the water resources management, which can be defined as establishing and maintaining the balance between human wishes and actions on the one hand and natural processes on the other, then each solution must be assessed using as minimum the following two criteria:

- criterion of satisfying reasonable human needs, wishes and actions,
- criterion of establishing and maintaining acceptable changes in the environment.

Both criteria can be broken down into less complex criteria, but with two kinds of consequences:
- simplified assessment of possible alternatives as per individual criteria,
- the integrity of the solution may be lost, which can jeopardize the success of the solution selection process.

Also, for the selected criteria, different measures can be applied to assess the meeting of individual criteria. These can be selected from a set of different physical quantities or can be in the form of assessing the extent to which each solution meets the desired target. In selection of measures, it is necessary to take into account the need for integral understanding and addressing of an issue.

It is important to take into consideration the timeframe within which the best solutions are to be the selected. In the water resources management system, it is common to set the timeframe so as to meet the human needs in a relatively near future (thirty to fifty years). However, time determinants of processes in nature have increasingly been considered. As a result, the sensitivity of the selected criteria and measures or assessment methods becomes an especially significant issue as these need to be adjusted and upgraded related to the new requirements within a particular timeframe. Starting from the above principles of system analysis, a number of methods have been developed which can be used to select the best solution in accordance with the diverse selected criteria. All assessment methods and/or the decision-making methods for selection of the best solution must however start from the postulate that with complex objectives, and with different criteria for assessment and varying time constraints (which is all characteristic of the water resources management system) a absolute best solution cannot be produced, so that the optimum solution should be searched for within a group of non-inferior solutions.

2.2 Multi-criteria decision-making methods

All multi-criteria decision-making methods start from the postulate that, because of the different nature of criteria (numerical and descriptive, more and less important), which hamper objective selection and the finding of best solution), a consistent ranking system should be introduced to assess the meeting of preset criteria by alternative solutions.

In principle, in all methods, the optimum solution from the group of non-inferior solutions is produced through a decisive contribution of the institutions responsible for decision-making and their implementation (client) who provide the defined criteria and preferences (the ranking system), and importance (weight) of each criterion, or at least the defined conditions for teaming up for implementation of the decision-making process. The multi-criteria optimization methods can be classified in five groups:

- Methods for identification of non-inferior solutions (weighing method, multi-criteria simplex method, constraint method)
- Methods with known preferences (multi-attribute benefit method, target programming, ELECTRE, PROMETHEE, AHP method)
- Interactive methods (STEM, SEMOPS)
- Stochastic methods (PROTRADE)
- Methods for "highlighting" a non-inferior solutions subgroup (compromise programming).

In case of solutions with significantly different system configurations and advantages and disadvantages, in order to narrow the room for decision-making and highlight the facts important to make decisions, frequent use is made of multi-criteria ranking of solutions to identify the sequence of solutions assessed under all criteria. The methods ELECTRE, PROMETHEE, ICOR and AHP (of the above) are based on this principle. These methods are also collectively called higher rank methods. Each higher rank method includes two phases:

- Establishing the higher rank relations
- Using this relation as aid to decision maker.

As the above phases can be carried out in several different ways, the methods differ.

ELECTRE (ELimination and (Et) Choice Translating REality) method is based on the assumption that each solution represents a node for which preference is identified according to a set of criteria against other nodes (solutions). For each criterion, the relation between two solutions (nodes) can be presented as a solution preferred over or equivalent to another solution.

On the basis of the relations established among the solutions, a graph is constructed in which nodes represent alternative solutions. The result of numerically processing the graph is a kernel which represents alternative solutions preferred by a given preference structure. For m solutions and n measures of goodness, on the basis of which the best solution is selected, n graphs with m nodes should be constructed. The synthesis of different criteria is done by using a concord index and a discord index so that the dominance relation among the solutions is determined. The development of ELECTRE I was followed by ELECTRE II, III, and IV. ELECTRE I should be used

if only one final solution is desired, where some very good solutions can be rejected in the process. ELECTRE II is used to rank alternatives. The methods ELECTRE III and IV include quasi-criteria with preferential thresholds (degrees of preference are introduced, strong or weak preferences).

PROMETHEE (Preference Ranking Organization METHod for Enrichment Evaluations) is a method that is used to obtain partial (PROMETHEE I) or full ranking of alternatives (PROMETHEE II).

This method is characterized by three segments:

- formulation of the decision maker=s preference, where six possible functions of the preference are considered for each criterion and the one best describing the preference under the criterion is selected; values of preference functions range from 0 to 1 and represent the probability of realization of solution under a particular criterion (e.g. 1 represents strong preference)
- construction of the estimated higher rank relation
- use of higher rank relation.

PROMETHEE methods are capable of accurately describing characteristics of criteria and analyzing stability of solutions through changes in weights and parameters, and presenting results of calculations, which allow a detailed insight into their structure (GAIA program).

PROMETHEE I method produces partial relations, which means that some solutions (under a certain criterion) are comparable and some are not. Partial ranking gives significant information to decision maker about relations between solutions, which in the full ranking under PROMETHEE II may be lost.

ICOR (Iterative COmpromise Ranking) is a method developed on the basis of compromise programming. The basic characteristic of the compromise programming is that a solution for the task (compromise solution) is determined by minimizing departures from an ideal point toward the adopted distance measure, including all criteria. An important parameter is "p", a balancing factor between the total benefit and a maximum individual departure. The minimization of "p" parameter increases a group benefit but reduces an individual benefit under one criterion and vice versa. The result of ICOR method is the alternatives ranking list.

AHP (Analytic Hierarchy Process) method is based on four axioms:

- reciprocity of relationship between alternatives (if a first solution is five times better than a second, then a second is five times worse than a first),
- in comparing two alternatives under a certain criterion neither may be infinitely better than the other,

- hierarchical formulation of a problem is possible,
- it is desirable to include in the process all criteria and alternatives considered by the decision maker

AHP method includes the calculation of weights (importance) of criteria and alternatives. Comparison matrices are formed to define preferences obtained by comparison of pairs (equally important, more important, significantly more important, very significantly more important, extremely more important, and there are compromise mid-values of preferences in between). These preferences are determined for alternatives and criteria. For each criterion in a group of criteria, one alternatives comparison matrix is formed. For determination of preference weights of the criteria, criteria comparison matrix is formed. Columns in matrixes are normalized and criteria weighing vectors and alternatives weighing vectors are determined under all criteria. The result should be a alternatives weights matrix in which alternatives weighing vectors by criterion constitute columns. The result is a alternatives ranking list.

2.3 Economic decision-making methods

The economic methods are, in principle, divided into macro-economic and micro-economic. A large number of non-economists are of the opinion that the majority of the today's problems emerging from the social relations and environmental protection are due to uncontrolled economic growth and assume that the macro-economic analyses are more relevant for resolving of such conflicts. They are, however, not aware that macro-economics mainly resolves the issues related to the economic structures, labor market and their growth.

The economists themselves are using the macro-economic analyses only in making decisions related to e.g. launching of projects, when they attempt to prove that most of the conflicts are those related to concrete business activities. They advocate the approach that the market mechanisms should regulate the relations between the human activities and possibilities for exploitation of environment, through the relation of prices for different products, which should maintain the economic growth within the set limits. More concretely, the economic science takes an approach that the product demand will be regulated by the product price increase resulting from the costs of the production waste recycling, price escalation due to limited production resources and escalation in price of production premises.

The cost-benefit analysis is considered to be the most adequate among the economic decision-making methods for selection among competitive alternative solutions. This method has been developed in order for the project launching decisions to be made while taking account of the interest of the entire community rather than the interest of the investor accountable for the project implementation. Otherwise, the investment feasibility analyses (financial analyses) would be used which consider the project only with regard to the current market prices and relations, with respect to the interest of the project developer rather than overall relations and their changes in time.

The cost-benefit analysis generally enables that all expected, direct and indirect project costs and benefits be included in calculations, since the project affects every member of the community be it a stakeholder of the one affected by the project implementation. Therefore, the method can be used on both micro-economic and macro-economic levels, its implementation being widespread in economic analyses conducted in the public sector and natural resources economics.

The cost-benefit analysis is carried out in following steps:

(i) Project definition and identification

The problem is identified, the constraints are determined (spatial, time-related, available information), the criteria are selected for integrated and comprehensive understanding of the problem solution, and possible alternative solutions defined, along with constraints and criteria.

(ii) Determining of input and output physical quantities relevant for the project implementation

All physical consequences of the project are determined for all the alternative solutions by e.g. determining quantity of material and work necessary for the project construction and operation, space occupancy, quantities of waste emission from the production process, and other input and output physical quantities. Quantification of all remaining, hard-to-measure project impacts on the human community and environment is conducted. The flow of all so quantified values is determined for the project duration.

(iii) Pricing (cost and benefit) of input and output quantities for the project duration intervals

The input and output values quantified under (ii) are priced for alternative solutions and for the duration of the project, so the same measure is used to quantify all the desirable and undesirable impacts of the project on the investor, the community and environmental conservation.

(iv) Project Net Present Value (NPV) calculation

The NPV calculation, based on conducted quantifications, results in evaluation of economic and total feasibility of project based on its total present value adjusted for the initial year of the project's lifetime. All the solutions with NPV greater than zero are desirable for the community and should be considered in selection of the best solution.

(v) Calculation sensitivity analyses for changes in assumed input parameters and alternative solutions ranking

This step includes recalculation by introduction of alternative inputs in order to check impact of some insufficiently investigated solution elements (or those changing with time) on the final result - NPV based ranking of alternative solutions.

The sensitivity ranking analyses are particularly important in determining impact of different project types and alternative solutions on generally adopted human society sustainable development requirements. Namely, selection of the discount rate in the NPV calculation generally determines the conditions under which a community is willing to renounce some gains for the benefit of the future generations, so it is the sensitivity analysis of individual environmentally sound solutions to the discount rate selection that regulates conditions of the solution implementation.

Further, the sensitivity analyses are important for determining of alternatives ranking sensitivity to extent of the impact of natural and artificial environment on the project implementation. E.g. it is not easy to forecast the impact of escalation of the input material price ensuing from exploitation of non-renewable resources, escalation of prices for recycling the materials in the production process, or cost estimate for protection of human life and other values which might be affected by the project.

3 DECISION-MAKING APPROACH COMPARISON

Assuming that a decision-making process on a project construction has involved precise defining of the project construction objectives and constraints, the best solution selection procedure is most affected by the selection of criteria and measures according to which the quality of offered solutions is valuated and the criteria performance measured.

However, the more complex a project the more intensive a trend to subdivide the criteria (and consequently the measures) in order to achieve a more comprehensive and simpler overview of the project. However, this might result is loss of comprehensive approach to the problem and affect successful search for the best solution. Namely, if the solution is selected on the basis of a number of criteria, and there is a number of them whose successful performance is evaluated by comparison with different performance patterns, it is very difficult to determine inter-relations

and individual effects (weights) of these criteria on final selection of the solution. Only when an equal measure is selected, the evaluation of performance of all the criteria becomes comparable and the integral problem approach maintained.

Both assessment methods (multi-criteria and economic) are based on the assumption that the criteria performance system is integral and the equal measures are used. In the former case, the measures are numerical scores and numerically expressed weights of individual criteria, and in latter case these are the monetary units. However, this is where the comparison ends.

The multi-criteria assessment methods apparently have some advantages over the economic methods being:
- tailored to creation of "ad hoc" design teams comprising of experts from various disciplines and other stakeholders, suitable for understanding of varying interests and exploitation of general experience and specific knowledge in the decision-making process,
- suitable for quick decision-making since, generally, no major preparations, investigations and detailing of a problem resolving approach is requested.

On the other hand, the main disadvantages include subjective decision-making present in criteria and determination procedure, and in determining "weights" of the method impact on the solution selection, and during the criteria performance evaluation. Further, and not less important is the disadvantage of these methods when the time-related factors are included in the decision-making procedure for selection of the best solution.

A frequent starting point in selecting the economic assessment methods for solution selection among different alternatives is a totally wrong assumption that economic assessment methods are methods that allow the assessment of the quality of a solution by no other than one B economic criterion (whose measure of performance is money). Or, it is assumed that this economic criterion seeks to find as the best solution the one that generates profit to the client, which means that other interests are neglected or placed in inferior positions.

On the other hand, it is generally accepted that this assessment method allows a transparent assessment and comparison of alternatives.

However, both assumptions may be questioned. The first assumption is questionable because in principle all kinds of criteria can be included in economic assessment methods, at least by identifying costs of substitute elements of the solution which increase the "goodness" of the total solution under a "problems"

criterion. For example, the comparison of solutions by the following criteria of their impact on the environment (which are considered hard to measure):
- physical criterion, which defines critical values of certain variables of the environment below/above which problems occur (water acidity, amount of nitrates in water, etc.)
- criterion of life structure, which determines (non)existence of mixed aquatic and terrestrial forms of life,
- environmental standards,
- concept of environmental balance in a given region (integrated resources),

can be done by some of monetary assessment methods. A similar approach applies to the comparison of alternative solutions by "sociological criteria", such as:
- benefit to the individual (quality of life or meeting his/her vital needs)
- benefit to the society (standard of living, capability of various subsystems and institutions to meet basic
- human needs, a social system as a whole).

The monetary assessment methods applied to environmental and sociological values should not be confused with the methods applied in determining market value of products and services, since their results are suitable only for understanding of total direct and indirect costs and benefits of a project. These methods include:
- Productivity Method
  Estimates economic values for ecosystem products or services that contribute to the production of commercially marketed goods
- Hedonic Pricing Method
  Estimates economic values for ecosystem or environmental services that directly affect market prices of some other good. Most commonly applied to variations in housing prices that reflect the value of local environmental attributes.
- Travel Cost Method
  Estimates economic values associated with ecosystems or sites that are used for recreation. Assumes that the value of a site is reflected in how much people are willing to pay to travel to visit the site.
- Damage Cost Avoided, Replacement Cost, and Substitute Cost Methods
  Estimate economic values based on costs of avoided damages resulting from lost ecosystem services, costs of replacing ecosystem services, or costs of providing substitute services.
- Contingent Valuation Method
  Estimates economic values for virtually any ecosystem or environmental service. The most

widely used method for estimating non-use, or "passive use" values. Asks people to directly state their willingness to pay for specific environmental services.
- Contingent Choice Method
  Estimates economic values for virtually any ecosystem or environmental service. Based on asking people to make tradeoffs among sets of ecosystem or environmental services or characteristics. Does not directly ask for willingness to pay; this is inferred from tradeoffs that include cost as an attribute.
- Benefit Transfer Method
  Estimates economic values by transferring existing benefit estimates from studies already completed for another location or issue.

Briefly, the answers are also searched for by approving additional cost of solutions which either ensure a fair compensation for the values lost or afford improvement of the overall standard of living to population affected by changes, in excess of possible damages caused by the project.

A second assumption is questionable because the impact of including different criteria in economic assessment methods is neglected. Or, it is often necessary to assess the impact of substitute elements on the final result (the final solution), which requires systematic analyses of sensitivity of solutions to changes in individual input parameters. The optimum solution in this case too is usually selected from within the range of possible best solutions.

The economic assessment methods certainly have the following advantages:
- use of objective and transparent system for measurement of the criteria performance for individual solutions, since the monetary value is used as measure recognized by our society as a generally accepted value,
- involvement of time factor in solution selection, and the factors of sustainable development of human society through the discount rate.

Possible disadvantages of these assessment methods include:
- risk that some criteria which are not easily expressed in monetary terms be excluded from the solution selection process,
- risk of one-sided approach to the problem because only interests of the investor as the project developer are taken into consideration,
- risk of adopting an assumption that wrong assumptions used in the decision-making will be corrected by the market as an ultimate regulating mechanism,
- need for data and documentation input and lasting

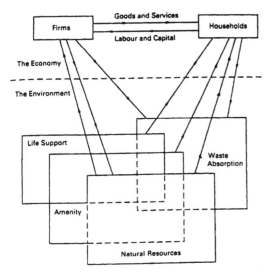

Figure 1   Relations between the environment and economic activities (Source: Figure 9.2; A.P. Thirwall: Growth and Development, 5th edition, Macmillan, London 1993)

procedures in which the alternative solutions are analyzed and evaluated, which are more comprehensive than under multi-criteria assessment methods which considerably increases the costs and time necessary for the decision-making.

These disadvantages may be avoided or controlled if all acceptable environmental standards, along with the heritage and natural resources standards are adopted and their effect on the market conditions defined (Fig. 1).

4 PROPOSED DECISION-MAKING APPROACH

The multi-criteria assessment methods are increasingly advocated by the NGOs, organizations and institutions involved in environmental protection and nature conservation, and some political groups. This is the consequence of lack of knowledge about the possibilities the economic methods offer for the decision-making process on projects which might affect wider human and natural environment on one hand, and apparent "significant" advantages of the multi-criteria methods ("democratic" approach to criteria selection and performance evaluation for each alternative solution, simplicity, speed and comparatively low price of their implementation) on the other.

It is proposed that the economic methods be adequately

used and advanced, and be given a leading role in the decision-making process. Consequently, the users should change their approach in the following:

-   the criteria used for understanding and evaluation of the solution quality should be defined as interdisciplinary, with involvement of all the stakeholders,

-   gaining wider support from different professions for finding of methods and ways of monetary presentation of the project environmental impacts that have so far been considered unmeasurable.

The market should and could not be burdened with the role of the only actor which uses its mechanisms to regulate the environmental protection and social relations. The state institutions, acting pursuant to legislation and administrative regulations, should be a correction mechanism acting from the platform of moral and ethics. In such engagement, the institutions need to be supported by the professional and scientific organizations which are expected to undertake a pro-active approach since the decision-makers need to be persuaded that such conditions will not affect the competitive qualities of the economy and that they are a move in right direction for the entire community.

REFERENCES

B. Beraković: Water Resources Management, Hrvatske Vode journal (1/1993)

M. Common: Environmental and Resource Economics: An Introduction, Longman, London, 1988

A.P. Thirwall: Growth and Development, 5th edition, Macmillan, London 1993

R.J.Johnston: Nature, State and Economy, 2nd edition, John Wiley & Sons, London 1996

Hydropower and Environment; an Experience from Khimti I Hydropower Project, Nepal

Meg B Bishwakarma
Research Engineer, Hydro Lab Private Limited, G.P.O. Box-21093, Kathmandu, Nepal

ABSTRACT: Nepal has a very rich bio-diversity, especially that of the mountainous areas. With a rich background of rivers, diverse flora and fauna, and harsh terrain, the environment is sensitive to development. Protection of the environment has been a global concern. Sustainable use of natural resources for the development of mankind is a primary need for today. Hydropower is a clean energy, however there are growing concerns regarding the overall environment and ecology, which posses a great challenge on planners and designers. Traditionally, the environment and its conservation have been of secondary importance in Nepal. However, with recent changes in the government regulations and also strict requirements of the donor agencies, it has become an important issue. In this paper, the writer tries to share the environmental management experiences from 60 MW Khimti I Hydropower Project in Nepal. This Project has pioneered the consideration of environmental impact minimisation in Nepal.

1 BACKGROUND

1.1 General Introduction

Khimti I Hydropower Project (KHP) is a 'Run-of-the-River' type of hydroelectric plant designed for nominal capacity of 60 MW and annual energy generation of 350 GWh to supply to the national grid of Nepal. The Project is located about 100 KM due east of Kathmandu, the capital city of Nepal as shown in Figure 1. KHP has been built as a Build, Own, Operate, and Transfer (BOOT) agreement with His Majesty's Government of Nepal (HMGN). This is a very first hydropower project in Nepal, which is owned and operated by a private company called Himal Power Limited (HPL) using the international financing from ADB and IFC. Electricity is sold to Nepal Electricity Authority (NEA). The banks and investors will be repaid during the operation. The plant will then be transferred to HMGN after 50 years of operation. Butwal Power Company Limited of Nepal, Statkraft SF, ABB Energy and Kværner Energy of Norway are the main shareholders of HPL.

1.2 Salient features:

Catchment area	358km2
Capacity (installed)	60 MW (5 nos. Pelton Turbines)
Power Production	350 Gwh/year
License	Issued for 50 years by HMGN
Net head	660 meters
Rated flow	2.15 m^3/s
Headrace Tunnel	7885m long. X-section 11.5 m^2

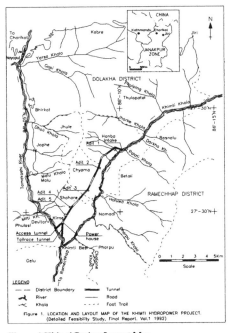

Figure 1. LOCATION AND LAYOUT MAP OF THE KHIMTI HYDROPOWER PROJECT.
(Detailed Feasibility Study, Final Report, Vol.1 1992)

Figure-1 Khimti Project Layout Map

Penstock	Steel lined on a 45° incline. 1000m long. X-section 3.1 m²
Powerhouse	Underground Cavern. 70m long. X-section 120 m²
Access Tunnel	900m long. X-section 24.5 m²
Tailrace Tunnel	1433m long. X-section 15.7 m²

Financing:

Cost	US$ 140 million
Funding	75% bank financing and 25% shareholders

Lenders   Asian Development Bank. International Finance Corporation. Eksportfinans AS (Norway). Nordic Development Fund (Finland). Norwegian Agency for Development Co-operation (NORAD).

Project organisation:

Lenders' Representative & Independent Engineer: Morrison Knudsen International. Inc. USA

Project Manager and Engineer: Khimti Services Consortium (Statkraft Engineering AS. Norway. Butwal Power Company Ltd.. Nepal)

Civil Construction and Engineering Contractor: Civil Construction Consortium (Statkraft Anlegg AS. Norway: Himal Hydro and General Construction Ltd. Nepal)

Equipment Supply and Installation Contractor: Consortium of ABB and Kværner (ABB Kraft and Kværner Energy a.s. Norway)

Local Sub-Contractor: Nepal Hydro & Electric Pvt. Ltd.. Nepal.

1.3 Brief Description of the Project

The headworks facilities are located in Palati site. Headworks consist of a low concrete/ boulder sill structure in the steep river through which fish pass arrangement has been made. Minimum release of 500 litres per second is also maintained through this weir. Water is conducted through a bed load trap. a gravel trap and a coarse trashrack into a two-chambered settling basin. There is a fine trashrack and a roller gate before the water enters into the headrace tunnel. From the downstream end of the headrace tunnel the water is conducted down to the powerhouse through steel lined inclined pressure shaft. The powerhouse is located underground with an access tunnel from Kirne. the camp area. The Outdoor Control Building. Transformers bay. Switchyard facilities and a back up diesel power plant are located at the close proximity of Ac-

cess Tunnel portal in Kirne. The generated power is transferred through cables to the outdoor transformers and then stepped up to 132 kV for supplying to the national grid. Nepal Electricity Authority (NEA) is responsible for the grid operation. The project has been completed and started commercial production from 11 July 2000.

2 ENVIRONMENTAL IMPACT ASSESSMENT (EIA)

The environmental impact assessment was carried out together with the Khimti Khola Hydroelectric Project Feasibility Study during 1992/93. Thereafter a supplementary Environmental Assessment was carried out during 1994. The EIA study aimed to identify the potential effects of the proposed Khimti Project on the existing physical. biological and socio-economic environments of the project area. It addressed the beneficial and adverse impacts and provided measures to mitigate the adverse impacts. The specific objectives of the study were to:

- Identify the type and extent of potential environmental impacts with emphasis on significant beneficial/adverse effects.

- Recommend mitigation and strategies to minimise adverse environmental and socio-economic impacts. and

- Assist planners and decision makers in evaluating the project's feasibility based on its potential environmental impacts

The impact study included the following specific tasks. which were considered to be the potential impact areas due to the construction and operation of the Khimti Project.

- Climate and air quality

- Topography and geology

- Soils. sedimentation and erosion

- Hydrology and ground water

- Water users

- Flora and fauna

- Fishery

- Human settlement

- Health and sanitation

- Gender issues

- Culture and archaeology

- Community development

- Agriculture and forestry

- Safety

The report concluded that if the suggested mitigation measures were properly carried out. the project's adverse effects on the local environment would be small. The access roads should be of benefit to the area. as there had no road access prior to the project. However. the effects of socio-cultural impacts were uncertain. The suggested major mitigation measures were as follows:

- Installation of a fish ladder across the weir

- Minimum release during the low flow periods

- Compensation for land and houses

- Erosion protection measures e.g. tree planting

- Guards and User Groups to prevent illegal hunting. fishing and tree cutting

- Community development programme

3 ENVIRONMENTAL MITIGATION AND MONITORING PLAN (EMMP)

On the basis of EIA study and the supplementary Environmental Assessment. an Environmental Mitigation and Monitoring Plan (EMMP) was prepared for the owner during 1996. The purpose of this plan was to specify in details the impacts. how they could be mitigated. the responsibilities of the various parties. and the methodology for monitoring the impacts. The plan did not set out all the details of all mitigation measures. but identified where the individual measures were specified. EMMP was a part of the document prepared to meet the requirements of the International Lenders. and was incorporated as part of the Project Licence issued by His Majesty's Government of Nepal (HMGN). The EMMP was also included as part of each contract agreements with different contractors. In addition to this. Programme for Environmental Control and Monitoring (PECAM) was prepared as a format to ensure that the effects of construction and operation on the biophysical and social environments be mitigated and monitored.

3.1 *Organisation and responsibilities*

As the owner of the project. the primary responsibility was rested on HPL. HPL implemented its responsibilities through the various contracts. Keeping the objectives of the plan in mind responsibility was put on the different parties involved in the project. The responsible parties were as follows.

Electricity development Centre (EDC): EDC is a Government Department under Ministry of Water Resource. responsible for issuing the licence. They were responsible to ensure that HPL abides by the condition of the Licence and carry out audits as and when necessary.

The Owner, Himal Power Limited (HPL): HPL would manage much of this work through the various contracts. however the final responsibility including the collation of monitoring reports and submission to EDC remained with it.

The Project Manager, Khimti Services Consortium (KSC): KSC was responsible to ensure that the engineering works. mitigation measures and the method of work and its impacts on the environment were complying with the terms of each contract. KSC would report to HPL on a monthly basis.

Khimti Environment and Community Unit (KECU): KECU was established under KSC and was responsible to conduct community related mitigation measures (not related to the construction contracts) and monitoring of the projects impacts and implementation of the mitigation measures.

The Contractors: The Civil and Electro-mechanical contractors were responsible to prepare a Health. Safety and Environmental Plan (including industrial hazards waste disposal sections). an Environmental Management Plan and for Civil Contractor a Revegetation Plan. HPL had to approve the plan and the contractors were to comply with all these plans.

The mitigation plan was prepared categorising the measures as preventive. corrective and compensatory measures and issues were specified for the different construction as well as for the operation phases of the project.

4 PUBLIC RELATION

4.1 *Information Dissemination*

Since the beginning of the project activities. public relations were performed by the project's Public Relations Officer (PRO) together with a Liaison Officer from EDC. These staffs were responsible for land acquisition/resettlement and dealing with the local public and local government authority on behalf of the project.

Prior to the start of construction activities. public hearings were conducted in the major places of project area including the district headquarters in the presence of local and central authorities of the Government. During the construction period, the project's PRO and other senior officials disseminated project information to the Government officials and the local publics. Queries concerning the project development from the public and Government officials were also answered frequently. Various journalists and reporters were received and project's activities were briefed.

4.2 Land acquisition

The project acquired a total of 24 of private land and 53 hectors of public land. The local people were consulted early in the process of land acquisition and were invited to cooperate. A notice concerning land acquisition was issued to the concerned landowners individually. Land valuation committee was formed comprising the Chief District Officer (CDO) of Dolakha District, local representative and the project's representative to decide the land price. Among the 145 landowners, who were affected by the land acquisition, only 3 families were fully displaced, seven families had to partially resettle. The displaced family requested the project to provide money and moved out to a place of their interest.

5 CONSTRUCTION PHASE ENVIRONMENTAL MONITORING

The Khimti Services Consortium (KSC) carried out the overall project monitoring and control. In order to ensure the project's environmental compliance KSC KECU had specialist staff for carrying out day-to-day monitoring works at sites and interface with the contractor's site staff. Such KSC staff included Environmental Co-ordinator, Environmental Monitoring Officer and Health and Safety Monitoring Engineer. Their basic duty was to ensure the implementation of the EMMP and HSE Plan of the project on day-to-day basis. An organisation structure of KECU is given in Figure-2. The scope of work for KECU was defined in the KHP Management Contract.

Figure-2 KECU Organisation Chart

5.1 Methodology for monitoring

The health, safety and environmental monitoring and mitigation guidelines were given in the project documents such as EMMP, Health, Safety and Environmental Plan (HSEP) and the Program for Environmental Control and Monitoring (PECAM). The following arrangements were made for the biophysical as well as the socio-economic environmental monitoring.

- Periodic reviews by the Independent Engineer's Environmental Consultant

- Six monthly and Annual Environmental reviews by the Independent Environmental specialists' team

- Regular environmental monitoring by KECU

The Environmental Action Plans (EAP) were prepared by compiling the observations, comments and recommendations from above reviews and actions were taken accordingly. KSC Environmental Monitoring Officer (EMO) and Safety Monitoring Engineer (SME) who directly reported to Environmental Co-ordinator (EC), carried out the day-to-day health, safety and environmental monitoring at the construction sites. Contractors had their own internal set up to ensure the implementation of project's guidelines. The Civil Contractor had an extensive team of Health, Safety and Environmental matters. Apart from other support staff CCC's environmental team included full time HSE Manager, Safety Officer and Environmental Officer.

Environmental monitoring was done in the form of regular qualitative site inspections of air quality, water quality, slope stability, spoil tip deposition, revegetation, flora and fauna. KSC EMO, SME and the Environmental and Safety Officers from CCC and ESIC arranged regular joint site inspections. Based on the site inspections, discussions at site level with the site Incharge and then also discussed during the weekly Health, Safety and Environmental meetings when actions were set with time limits. Any issues not resolved were referred to weekly Site Meetings. If the issue did not resolve within the given time then matters were forwarded to the Monthly Project Management Meetings, where key personnel of the project parties participated. This meeting had authority to instruct the contractors for the environmental compliance. KSC, KECU had also contractual authority to issue non-conformity report against any environmental infractions. Such arrangements ensured the timely mitigation of health, safety and environmental issues at site during construction.

5.2 Biophysical monitoring

The monitoring parameters and the adopted mitigation measures under biophysical environmental monitoring is presented below.

5.3 Socio-economic Monitoring

The monitoring parameters and the adopted mitigation measures under the socio-economic environmental monitoring is presented below.

Table-1 Biophysical environmental monitoring

Monitoring parameters	Adopted Mitigation measures
Air quality	Vehicle and Engine emission was reduced: by reducing distance between camp and work areas: use of electricity instead of diesel where possible: engines to comply with Nepal standards for exhaust gases and servicing to be done regularly Dust Control was done by watering dusty areas: provision of breathing equipment where necessary and use of gravel in heavily trafficked areas Noise control was done by measuring noise levels with Decibel Meters on a weekly basis. The Ventilations fans were built with insulators in order to dampen the noise level.
Forest use	Formation of forest users groups: ban on tree cutting: liaison with District Forest Office (DFO): protection of endangered species: fire prevention education: ensure availability of kerosene or gas for cooking and other domestic purpose: establish project nurseries for revegetation: forest protection conservation and management
Wildlife	Ban on hunting within the project vicinity: rubbish control for pest management: habitat protection: and education on importance of wildlife conservation
Fisheries:	A ban on fishing within the project area for project employees only: care not to block passage of migration fish during construction period: constructing weir to allow fish to pass
Water Quality	Minimising oil and other toxic chemical spills: runoff control: use of sediment pools. grease traps and settling basins: testing of water quality on quarterly basis.
Hydrology	Release of at least 500 l/s of flow in the Khimti river at all times.
Ground Water	Flow measurement was taken twice in the year. in dry season on March/April and winter season in August/September.
Slope Stability	This was taken care through suitable engineering design. construction and with the use of bioengineering and maintaining the slope of the spoil tips.
Rehabilitation	At the completion of the project. all disturbed areas of the site were returned to its original condition through the implementation of revegetation. reinstate of agriculture land. slope stabilisation with gabion structures.
Soil	Fuel adequately stored & protected from spilling: all exposed area protected against erosion & landslides: all land used by approval of appropriate authority.

5.4 Environmental accidents during the construction period

Environmental accidents were not common to this project. When an incident occurred. it was quickly observed. reported and action taken. One major accident occurred with the spoil tip at Adit 5. which released large amount of spoil and sediment into private land near by Kirne. Immediate action was taken and compensation paid. The next one was a large landslide occurred in the 1999 monsoon at Adit 5. Owners of damaged lands were compensated for two crops and paid to rehabilitate their fields.

5.5 Fish monitoring programme

As stipulated in the EIA. fish-monitoring programme was included in the Civil Engineering and Construction Contract for the construction phase. A total of four sampling locations were fixed and pre-monsoon. monsoon and post-monsoon fish monitoring was carried out. Cast net method was used for the sampling. Water quality parameters such as temperature. turbidity. PH value. etc were measured. The fish species found were namely spotted snow trout. Stone carp. Torrent catfish and Copper mahaser. In order not to disturb the fish population during construction. all project workers were strictly prohibited for fishing activities. suitable fish passage was maintained particularly during the period from May to September and due attention was paid not to disturb the fish nurseries.

After the construction was over. HPL has been continuing the fish-monitoring programme through the consultant. The study so far has not found any new migratory fish species in the Khimti river.

6 COMMUNITY DEVELOPMENT AC ON WEEKLY BASIS BY USING THE TIVITIES

Apart from the environmental monitoring of the project. the KSC-KECU was responsible to carry out all the community development activities in the affected Village Development Committees (VDC) in the project area. Basically KECU had launched its programmes in the VDCs namely Sahare. Chyama and Hanba. KECU's programmes also included the transferred community. The activities are briefly presented below.

6.1 Non Formal Education (NFE)

The Non Formal Education (NFE) was basically focussed on the adult education. The NFE classes were aimed to serve as a forum for creating basic awareness raising along with literacy. All other programs were gradually introduced through these classes. The NFE courses included 7 months basic course, 7 months follow up course and then establishment of Trunk Li-

Table-2 Socio-economic environmental monitoring

Monitoring parameters	Adopted Mitigation measures
Health and sanitation	General clean up and awareness program in the project staff camp area and also in the nearby community; guidelines for handling of waste products and management as per the HSE Plan; building adequate number of toilets; project clinics at each site and water quality testing
Local culture and customs	Protection of the religious sites; respect with local and worker's cultural practices; avoid practice of the alcohol drinking and gambling at all construction sites and local community.
Local employment	During the construction period, job priority was given for effected households and local people in the different activities of the project.
Inflation	As per contracts, the Civil Contractor had to ensure sufficient food supply and basic commodities for project staffs and workers in a reasonable market price.
Waste treatment, disposal and industrial hazards	An industrial Hazards Management Plan was prepared to ensure the control and handling of the waste materials including human, organic, non-organic waste as long as handling and use of all hazardous materials including petroleum compound, concrete additives and admixtures, paints and acids used for construction activities at the project sites and facilities.
Human and domestic organic and non-organic waste	This was controlled through: General clean up and awareness program, guidelines for handling of waste products; trained waste collector; adequate number of waste bins; recycling the waste materials; burning pit or incinerators for burning materials; proper disposal of waste materials; septic tanks and soak pits (located a minimum of 25m away from the water sources)
Toxic and industrial hazards waste	This was done by: Controlling unwanted and deteriorated explosives; controlling unwanted and deteriorated detonators; collection of scrap metals; collection of scrap timbers; proper use of the set and expired cement; spillage of oil, diesel and petrol and chemical hazards; construction of oil water interceptors; construction of plant workshops; construction of drainage channels for water outlets; proper storage system for oil, diesel, petrol and chemical hazards; immediate clean up of spillage area; proper disposal of collected oil and chemical hazards.
Health and safety	The Civil Contractor was responsible for provision and operation of clinical facilities at each construction site. CCC established a fully staffed main Clinic at Kime. An ambulance was kept stand by all the time at site in case of emergency. Regular staff check up and vaccination were launched by the clinics from time

Monitoring parameters	Adopted Mitigation measures
	to time. The clinic was accessible to the staff and their families of all contractors and the local community. In order to ensure the occupational safety of the workers, the project provided Personal Protective Equipment (PPE) such as Hard Hat; Eye glass/ Goggles; Mask; Ear plug/ Ear muff; High visible Jacket/ Overhaul; Gum boot; Safety belts; etc. *Note: Unfortunately a total of 7 work related fatalities occurred. Besides that, there were other 3 fatalities, which ware associated with the project but not work related.*

braries for their continuing education. Summary of the achievement of NFE programme is given below:

Table 3: Summary of NFE programme.

Activities		Total
No of basic NFE classes conducted		49
No of NFE follow-up classes conducted		38
Participants in the basic course	Male	143
	Female	717
	Total	860
NFE participants in follow-up course		619
No. of NFE facilitators trained		49
No. of Trunk libraries established		27

Similarly, Non-formal education classes were arranged for the project workers and their families. A total of 72 person from this community benefited from the course.

6.2 Agriculture Development Programme

Agriculture Development Programmes were considered for awareness raising of the local farmers on improved farming practices. This included the regular advice given by the Agricultural Worker of KECU for improved yield and introduction of new varieties of crops and cropping patterns. Improved varieties of vegetable seeds were distributed and demonstration plots established. 30 different varieties of vegetable seeds were distributed and about 800 kitchen gardeners benefited from this activity. Now, people have

Table-4 Summary of Agricultural Activities

Activities	Total
Improved vegetable seeds distribution. (No of beneficiaries)	4637 farmers
Production of fruit / seedlings. No	13900
Distribution of fruit/ seedling. No	10900
Kitchen Garden Training	29
Organic Pest Management Training	8
Kitchen Garden Demonstration Plot	27
Vegetable seedling Production and distribution	28800

Table-5: Summary of skill development training

Activities	Total	Remarks
Basic Tailoring Training	19	All female
Advance Tailoring Training	4	All female
Knitting Training	45	All female
Poultry farming Training	4	All female
Bee keeping Training	8	All female
Saving Credit Training	31	Male-7. Female-24
Bamboo Furniture Training	15	Male-12. Female-3
Pickle Preparation Training	12	All female
Potato Chips Training	10	All female
Pig farming Training	16	Male-12. Female-4

started to produce different varieties of vegetables for their own use and for selling out to the market.

6.3 Skill Development / Income Generation / Women Groups.

Skill development activities were mainly focussed to women groups for the development of skill and income generation. For this. KECU introduced and provided varieties of trainings to the locals. The following trainings were conducted during the project period.

Altogether 15 women groups were formed. Some of the groups are capable of functioning independently. Besides this, a total of 4 numbers of Saving and Credit groups were also formed. Several women rights awareness raising programmes were organised aiming for improving the condition of local women on education for family planning. gender rights. campaign against excessive alcohol drinking and gambling. Training on "Women Leadership and Development" was also organised.

6.4 Health and Sanitation Programme

Under the health and sanitation programme. KECU specially focused on the construction of toilets and smokeless stove in the project areas. Before constructing these facilities training on preparation and operation of toilet & smokeless stove was given to the interested people. Some of the activities conducted under this programme are presented in the table below.

Table-6: Summary of health & sanitation activities

Activities	Total	Remarks
Toilet Construction (no).	251	3nos training conducted
Smokeless Stove Installation	77	9 males and 9 female trained
Nutrition Training	2/59	All female
Road Drama	8	
Community Heath Awareness Day (CHAD)	3	170 infants health check-up

On CHAD Sarbottam Pitho and Jeevan Jal were distributed to the mother of each infant with orientation on preparation procedure. Awareness raising programme on health and sanitation. common disease. childcare and women rights was also organised.

6.5 Forest development and conservation programme:

In order to raise awareness among the local people regarding the conservation and management of local forest resources. KECU had implemented various activities. These activities were aimed for raising nurseries for afforestation and formation of local forest users' groups.

6.5.1 Nursery development and plantation

KECU established 4 nurseries with fodder and forage seedlings in various locations of the project during the program period. The nurseries were established at Kirne. Bhotechhap (near surge shaft area). Chyama and Hanba areas. These nurseries produced a total of about 120.000 plant seedlings of various species. which were distributed among the local peoples for planting them in the community forest as well as to their private lands. The community forest plantation was organised and managed by KECU. In order to encourage the locals for plantation. the seedlings were distributed free of cost. During the programme period. plantation was done in about 70 hectors of land in the project area. Various species of timber. fodder trees. fruit trees and endangered species of forest plant were planted for increasing the forest density. soil conservation and animal grazing. A preliminary survey indicated that about 65% of the planted seedlings have survived. The above figures do not include the revegetation carried out by the Civil Contractor in the areas directly affected by the construction activities.

6.5.2 Formation of Community Forest User's Groups (CFUG)

The major activity of the forestry component of the KECU programme was to organise the local people and educate them to become capable for conservation of the available local forest resources and use for their benefit. In order to facilitate these processes. various training. discussion. meeting and workshops were organised for the community members. During the project period a total of 13 Forest User's Group were formed. Six of them have been registered with District Forest Office in Charikot and rest of the groups are in the process of registration.

6.6 Observation tours

In order to raise awareness among the local people. various observation tours to other projects were arranged for the community members. As part of this 14 facilitators were taken to Community Development Program Chisapani. Ramechhap launched by SDC (a Swiss Project). Other 15 NFE facilitators were taken to visit Madan Pokhara of Palpa District in the Community Health Development Programme. Similarly a tour for the Forest Users' group executive members (24 Participants) was also organised. Among the total 50% participants were female.

7 TECHNOLOGY TRANSFER AND TRAINING PROGRAMME (TTTP)

Technology transfer and training progremme was mainly focused on providing training to the staff involved in the construction activities on behalf of different contractors and the staff those were to be involved in the operational phase of the project. Separate package of training was designed for both of the schemes. KSC conducted series of training courses in Nepal and also in Norway for the operational staff and the respective contractors provided training to their own staff. These programmes helped to build up capacity for Nepalese staff involved in the project construction and operation.

8 HPL IMPLEMENTED ACTIVITIES

8.1 Improvement of the services within the project area

As part of the mitigation program outlined in the EMMP. HPL provided financial assistance. and KSC provided technical and managerial support to various schools. health-posts and other services sector in the project area. These assistances were aimed for better education and opportunity for the local children. better health care and improved water supply and sanitation facilities to the local areas. Summary of such activities is presented in the table below:

8.2 Jhankre Rural Electrification and Development Programme (JREDP)

In addition to the KECU. HPL also established a separate programme called Jhankre Rural Electrification and Development Programme (JREDP). A 500 kW Jhankre Mini Hydro plant was built for supplying construction power to Khimti project. The aim was to

Table-⁻ Summary of HPL implemented activities

Services	Provided assistance
Khimti Project School	Establishment and operation of a primary level school
Trikuteswor Primary School	Provided land and new building with furniture
Janajagrit Primary School	Doors windows furniture fencing and one teacher for two years
Shankheshower Primary school	Extension of building. roof and furniture
Mahendra Primary School	Construction of a complete new building
Chyamma Health post	Construction of a complete new building and training for staff
Hanwa Health post	Construction of a complete new building and training for staff
Kalika Secondary School	Improvement of science laboratory and black boards
Bhim Primary School	Improvement of roof and furniture
Bhumeswori Primary School	Doors windows furniture fencing and one teacher for about two years
Water supply	Five new schemes constructed
Irrigation scheme	Assistance for maintenance of the canal

hand over the plant to local user's group for long term operation after the construction of Khimti was over. The main objective of the project was to promote broader economic development within the most affected areas of KHP through the medium of rural electrification. Specific objectives were:

– To develop the managerial and technical capabilities of the local population.
– To strengthen the local economy and provide long-term economic sustainability.
– To develop a technical master plan for enterprise and domestic electrification within the JREDP area.
– To implement the technical master plan.
– To create awareness among local people regarding family health and sanitation. co-operatives. etc. through NFE programme.

This programme was mainly focussed in the surrounding area of Jhankre power plant called Thulopatal VDC. Rural Electrification was also carried out in Sahare. Chyama. Betali and Rasnalu VDCs. Apart

Table-8 Summary of JREDP implemented activities

Implemented Activity	Total	Remarks
Rural Electrification (no. of households)	390	KHP affected VDCs
Electricity based enterprise establishment (nos.)	11	Small cottage industries
NFE Programme (classes)	12	Total 275. 90% women
Skill development training (nos)	14	

from the RE work. JREDP also conducted community development activities in Thulopatal VDC. This programme also included orientation tour to the similar projects a strengthening of the local management capacity. The summary of the JREDP implemented activities and achieved outputs are presented below.

HPL has been continuing the JREDP programme during the operation period and planning to hand over the plant to the local community once the local capability is suitably trained and prepared to take over.

9 PROBLEMS ENCOUNTERED

The activities were implemented on the basis of EMMP. Environment/community Programme Long Term Plan and the Jhankre Rural Electrification and Development Programme. All implemented activities were run smoothly with the close co-operation and support from the local community. However, some problems were also experienced during the implementation, which is briefly presented below:

Disturbance in the construction works by the locals: In the early stage of the construction at the headworks area. the local interrupted work several times demanding for the road from a place called Jiri to the site. This part of the road was taken out in order to reduce the over all cost of the project by the government authority. The locals were not happy with the decision. Helping the locals for constructing the tractor able road solved this problem.

Complain against the house cracks: The locals complained that their houses were cracked due to the tunnel construction. The base line survey did not include the house condition close to the tunnel alignment. Therefore. the project had to hire a consultant for the investigation of the cause. The consultant after field investigation submitted a report concluding that the reason of the house crack was earthquake. which occurred some months back. There was no harmful effect on the surface structures due to the tunnel construction.

Complain against the stream leakage: There was also a complaint regarding the reduction of stream discharge. The base line study did not include the stream discharge measurement. However. regular monitoring of stream discharge was carried out during construction. The monitoring results showed that this was not a serious problem and the locals did not complain later when the tunnel was filled up with water.

Demand for electricity: There was a debate among the locals over for identification of those who would re-ceive electricity first. In order to make it more transparent and participatory a management committee including 50% local representation was formed. who decided the priority.

10 CONCLUSION

Khimti I Hydropower Project is the first major hydropower project in Nepal constructed in private funding. which has implemented a dedicated and structured Environmental Mitigation and Monitoring Plan plus a Community Development programme in the project area. The KECU fully implemented the programme required by the project's EMMP and PECAM. As a result of these programmes. the local community has benefited from the project in many ways. These programmes have encouraged and educated the local people for changing their life style and behaviour in such a way that their life standard in general is improved.

Careful attention was paid to protect the surrounding environment. The environmental clauses included in the various contracts and the general understanding of the importance of protecting the environment among the project parties were the key factors of success. Khimti project has been considered as one of the environmentally friendly projects where international funding is mobilised.

11 RECOMMENDATIONS

The implemented programmes have been very popular in the local community. In order to make the programmes sustainable. it is recommended that some of the activities such as awareness raising. income generation. local entrepreneur development. rural electrification. monitoring and strengthening of the forest resources. slope stability and re-vegetation. etc should be carried out during the project operation phase. Monitoring of fishery and minimum release through the weir must be continued. Efforts need to be continued to build up the local capacity for handing over the Jhankre Mini Hydro Power Plant and its rural electrification programme. Proposed JREDP II would be a good programme from the long-term sustainability point of view.

For a hydropower project under study. it is very important to carry out a comprehensive base line study including the stream discharge measurements and the existing house conditions (with respect to cracks) for the comparison during and after construction periods.

REFERENCES:

1. BPC Hydroconsult and Norpower as. 1992/93. Khimti Khola Hydroelectric Project Detailed Feasibility Study.
2. BPC Hydroconsult. 2000. Completion Report Jhankre Rural Electrification and Development Project.
3. BPC Hydroconsult 1998 to 1999. Fish Dynamics in the Khimti Khola Fish Monitoring Report.
4. BPC Hydroconsult. 1995. Khimti Hydropower Project a Socio-economic Benchmark Survey Report.
5. Himal Power Limited. 1996. Khimti I Hydropower Project. Environmental Mitigation and Monitoring Plan.
6. Khimti Services Consortium. 2000. Project Completion Report. Khmti I Hydropower Project

Hydropower in the New Millennium, Honningsvåg et al (eds), © 2001 Taylor & Francis, ISBN 90 5809 195 3

A Decision Support System for Hydropower Peaking Operation

P.Borsányi & Å.Killingtveit
Department of Hydraulic and Environmental Engineering, Norwegian University of Science and Technology, Trondheim, Norway

K.T.Alfredsen
SINTEF Energy Research, Trondheim, Norway

ABSTRACT: This paper describes a modelling system for balancing hydropower economy against environmental impacts using a case study from the river Nidelva, Norway. Several types of models were integrated into a decision support system to analyse and rank the effects on river habitats of different hydro-peaking patterns. The qualification of these patterns is based on both power production finance and environmental effects on fish habitats in the downstream river. The modelling system consists of a hydropower simulation model (nMAG), a dynamic river hydraulics model (BOSS DAMBRK) and a habitat simulation and classification model (HABITAT).

1 INTRODUCTION

Hydropower peaking is a common economical way of producing power during hours when consumption is high. But the environmental consequences of this strategy may be negative if the power plant discharges directly into a river reach. In this case, the swift water level and discharge variations may degrade habitat quality for species in the river ecosystem. Recent research has shown that the risk of stranding fish on the shore is a function of time, duration and rate of peaking, especially during daytime in the winter season (Bradford 1997, Saltveit et al. 2001). It would therefore be useful to adjust the pattern of peaking based on knowledge about physical and biological responses in the river, and by this to strike a better balance between hydropower economy and environmental impacts in the river.

Here a modelling system was established in order to study different peaking methods; it consists of three major modules and their linking framework.

The Norwegian nMag (Killingtveit 2000) model was used for the hydropower simulations. From catchment runoff data, it can calculate power production, discharge, spillage and other hydropower related values in a predefined system.

The BOSS version of the DAMBRK hydraulic model applied here is able to model fluctuating river discharge regimes. It is a 1-D hydrodynamic flood routing model, which can consider the effects of spillway and turbine flow, downstream tail-water elevations, frictional resistance, and lateral inflows

and outflows amongst other features. (BOSS International 2001.)

The habitat modelling part combines hydraulic modelling of the river reach with the habitat selection of different species (Alfredsen 1997). Only one fish species is taken into account here under the assumption that it represents a wide range of effected animals. Preference curves describe the fish response to different hydro-physical parameters (e.g., water velocity or depth) by assigning preference values to hydro-physical parameters (Borsányi 1998). All of these parameters relate to water flow; in this case to the results of the hydrodynamic simulation. The effects of other important aspects such as cover or substrate are neglected here.

The combination of these models results in a decision support tool that gives the functional relationship between power production and the habitat availability. The connecting framework allows data to move seamlessly from one model to the other and demonstrates some of the results.

2 HYDROPEAKING

If power is sold in a system that operates within a regular open market, the price can vary substantially seasonally, daily and hourly. Indeed, the peak-hour price can be a several times that of the off-peak price (Figure 1).

Power consumption is high on weekday mornings and early afternoons, and is low during nighttime or when industries are closed. Some variations

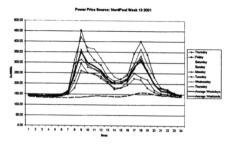

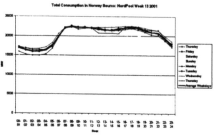

Figure 1: Power price and consumption Norway, on week 13, 2001.

can be seen at the seasonal scale, albeit with a lesser amplitude. More power is used during winter periods than summer (heating, illumination etc.). Since there is no economical way of storing power, production must follow consumption. A suitable hydropower system is capable to follow large variations in consumption by applying peaking operation. So peaking in this context means a frequent variation in production, which is achieved by fluctuating the use of stored water.

If water is released into a natural river, there can be adverse effects on the ecosystem, especially fish and vegetation related to high variations in physical habitat (Saltveit et al. 2001). Both experiments and experience have shown that the quick and large fall of water level causes stranding of fish at particular reaches and increased growth of water vegetation. Stranding is particularly high among young fish for reaches where the substrate is coarse, and the riverbed is moderate in slope in the cross-sectional direction. That is, large areas can dry out in short time periods, and so young fish are often "pool-trapped". To avoid this negative environmental impact, some modified (moderated) peaking patterns should be followed.

Therefore, the Norwegian Research Council and several power producers have initiated a research programme to investigate the impacts on the ecosystem of hydro peaking and to provide tools for analysing the impacts of future hydro-peaking projects (Harby et al. 2001). A decision support tool is needed for this large-scale research programme that is able to evaluate habitat quality and to calculate power production in different cases of peaking. By

linking the three models mentioned above, the resulting system should fulfil these needs.

3 THE MODELLING SYSTEM

The system consists of a hydropower simulation model, a one-dimensional dynamic hydraulic model, a habitat model and the links between them.

3.1 The nMag model

This is the second version of the original ENMAG model, developed in 1984-86 at the Norwegian Hydrotechnical Laboratory. The major difference is that nMag can handle more than one reservoir. The main components of the hydropower system are the:
- reservoir;
- power plant;
- interbasin transfer; and
- control point.

By defining the internal links and providing the necessary and optional data that describe each component, the model can calculate power production, water release, spillage and other values at them. For example in the case of a "Power Plant" component, the necessary data are:
- addresses of turbine water, bypass release, flood spill;
- maximum capacity (m^3/s); and
- energy equivalent (kWh/m^3).

Optional data are:
- nominal head (m);
- intake and tailwater levels (m.a.s.l.);
- head loss coefficient (s^2/m^5);
- efficiency; and
- peaking schedule (figure 2);

Simulations can be organized according to monthly, weekly, daily or hourly time-steps. Runoff data from the sub-catchments and daily/annual price variations should be presented as input. Peaking is defined by hourly ratios of proposed daily loads, while seasonal variations can be described by their relationship to one another (since total runoff cannot be foreseen).

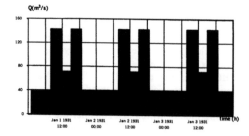

Figure2: Peaking schedule

192

3.2 Boss Dambrk

This programme was designed to predict flood wave propagation in a river channel. The flood source can be, for example, a dam failure with breach development or simply a given inflow hydrograph. At the outlet of the channel, for example, a stage hydrograph can be described to ensure a proper boundary condition at the downstream end of the model. The channel is defined by cross sections (geometry, friction, distances from one another, etc.). Flood routing follows the initialisation of the starting conditions, and results in stage and discharge time series for the cross sections.

Serious limitations of the system are the limited number of graph-points allowed for specifying the inflow hydrograph, the method of describing the cross sections, and the strong sensitivity to backwater effects. Characteristic points of which DAMBRK accepts only 30 specify the inflow hydrograph. Therefore the simulation period is restricted if there are significant variations in the inflow. The programme can handle only symmetric cross-sections, so the actual cross-section must be converted accordingly. Computational problems can often occur where there is a local depression in the channel, so where slopes are negative in the downstream direction between two cross sections.

3.3 Habitat

The first version of the programme was created as the habitat modelling part of the River System Simulator in 1996. It was used to describe channels with respect of their suitability for fish life. The main purpose within RSS was to minimize the negative environmental impacts of regulations or power production on fish production. The version used here can handle various types of input sources such as HEC-2/HEC-RAS, SSIIM, and more. One-, two- and three-dimensional modelling results can be handled. The user interface is a control file for the major module; the user gives various options and switches that influence the programme input, operation and results. These results can be further analysed by the presentation module (Habplot) or other means. The system is able to create classical preference curves (habitat classification curves), WUA (weighted usable area) curves, habitat time series, habitat plots, and –in a preliminary form– habitat patches (indices).

3.4 The linking framework

The three models above need to be "linked" so, that results from hydropower simulations (nMag) are used by DAMBRK that in turn provides inputs to the Habitat programme. In this first prototype version, the emphasis was on defining and testing useful analysis and presentation methods. Less weight was put on automating the computational process. Since all three models are included in the River System Simulator (RSS), in principle it is possible to realize the linking process therein and this work is planned for the next phase.

Because the models operate on different temporal and spatial scales, it will be necessary to develop scripts to transform, for example, nMag output discharges into inflow hydrographs for DAMBRK. At present, transformation and linking procedures are done manually with Excel spreadsheets and scripts.

Average velocities in the cross-sections are calculated from discharge, water level and geometry. When calculating habitat indices, Habitat needs a cross-sectional velocity distribution as input for each sub-section of the river reach. These are calculated from the average velocities that are distributed according to each sub-sections area. (Figure 3)

3.5 The test case

The lower section of Nidelva, which flows through Trondheim, was used as a test case. This river is famous of its sport fishing. Like many other rivers in Norway, the river is also utilized for hydropower production, thus actual flow conditions are strongly influenced by the operation of the power plants.

The river reach extends from the outlet of two hydropower stations, Nedre Leirfoss and Bratsberg (Figure 4), down to the river mouth that flows into the Trondheim Fjord (Trondheimsfjorden). Nedre Leirfoss is a nearly 100 years old and is run with almost continuous minimum flow of 30-40 m^3/s. Bratsberg is a modern power plant with peaking capacity. Its intake is at the large regulating reservoir Sebusjøen, the outlet of the tailwater channel discharges at the same place as Nedre Leirfoss. The capacity of this plant is at most 103 m^3/s, when peaking full capacity is used.

According to measurements, an immediate complete shutdown from full to zero production results in drop in discharge from ~130 m^3/s to ~40 m^3/s within 5-6 minutes at the outlet into the river. The

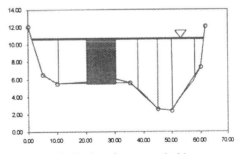

Figure 3: Distribution of average velocities

The Nea-Nidelva watercourse

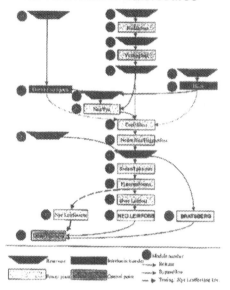

Figure 4: The Nea-Nidelva hydropower system

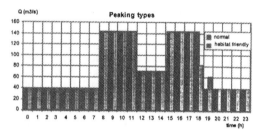

Figure 6:Peaking studies

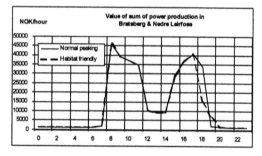

Figure 7. Power production in Bratsberg and Nedre Leirfoss

discharge attenuates in downstream direction, for example about 5 km from the outlet discharge drops within 40 minutes, while another 5 km further downstream it takes more than one hour for the river to reach minimum flow (Figure 5).

Two different peaking schedules were defined in the study as "normal peaking" and "habitat friendly peaking". In the first case, the simulated discharge drop appears abruptly as for usual case as described above. In the second case the power plant is not shutdown immediately; initially there is still some production, but not so low as to damage machinery (cavitation) or be uneconomical (inefficient production). This level is set at 30% of the maximum capacity. Shutdown is completed over two hours in two steps (complete shutdown, see figure 6).

To help compare the economic losses or benefits from the two cases, the value of production is shown in figure 7.

For habitat simulation, experimental preferences for Brown Trout (*Salmo Trutta*) were used assuming that these preferences are valid all over the reach. This assumption is not correct in general, but

it seems to be suitable for the purposes of the experiment.

4 RESULTS

In both cases ("normal" and "habitat friendly") a winter peaking situation of 1 day (24 hours) was simulated. For each habitat index values were calculated along the reach at different time-steps, additionally habitat plots were created. Figure 8 shows an overview of the reach and a habitat plot for one of the time steps. The different grey-tones on the plot show the habitat quality between the cross sections. The quality factor is defined here by depth and velocity, fish habitats are said to be usable, indifferent or avoided.

Figure 9, habitat indexing, shows how habitat suitability, as given above, changes in time. Such indices allow the two cases to be easily compared.

If avoided conditions continue for long time periods, or if usable areas change abruptly to avoided, fish are likely to be negatively affected. If usable areas appear more often during peaking, or if the habitat is changed in a gentle manner, the physical habitat is probably less degraded (Saltveit et al. 2001). By applying "normal" peaking avoided conditions ensue around one hour longer in the evening (18h-19h). Moreover, the "habitat friendly" method results in usable areas first becoming indifferent and then avoidable rather than an abrupt change (Figure 10).

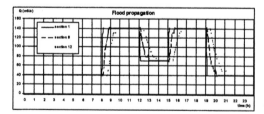

Figure5: Discharge-time series at different sections

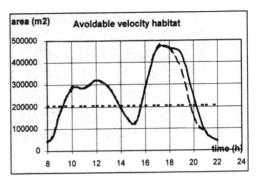

Figure 9: Comparison of variations of velocity habitat in time

habitat friendly peaking

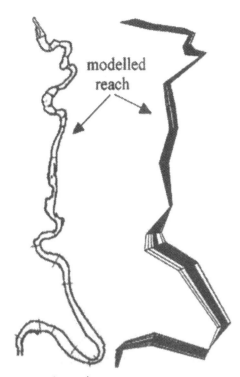

Figure 8: Reach overview

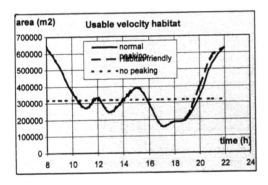

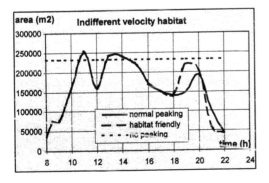

normal peaking

Figure 10: Habitat plots at t=18:00, 19:00 and 20:00 on the simulated day

The production differences for the two simulations was small, but their cost is highly dependent on how daily price variations are predicted.

By comparing the two series of habitat plots, the critical areas (i.e., where stranding potential is probably high) can be identified. Figure 10 shows a part of the simulated reach at three sequential time steps for each case.

5 CONCLUSIONS

The economic and ecological consequences of hydro-peaking must be computed and balanced during operation planning. It can obviously be difficult to balance different strategies, but it may prove helpful to employ a system similar to that described above. Such systems should be tested on a river reach influenced by peaking. At present, this system has only been tested on River Nidelva, Norway, but in principle it can be transferred to any other situation. By relating variations in habitat distribution to "their price" (average power production for either of the peaking methods) could lend some support for decision making.

It has to be noted, that in its present state the system is more experimental than practical. So far it points out not only numerous directions of further research needs, but also many possibilities. The data used here was collected for other purposes, and habitat preferences may not be representative for all species. Extension of the system is possible both towards more sophisticated habitat analysis, and even towards including a runoff-forecasting model.

ACKNOWLEDGEMENT

Special thanks to Annette Semadeni-Davies, who corrected this paper with untiring energy.

REFERENCES

Alfredsen, K. 1997. *A Modelling System For Estimation Of Impacts On Fish Habitat*. Proceedings of the IAHR Conference: Water For a Changing Global Society. San Francisco

Borsányi, P. 1998. *Physical habitat modeling in Nidelva, Norway. Diploma thesis D1–1998–29*. Institutt for vassbygging

BOSS International 2001. *BOSS DAMBRK Reference documentation*

Bradford, M. J. 1997. *An experimental study of stranding of juvenile salmonids on gravel bars and in side-channels during rapid flow decreases*. Regulated Rivers 13: 395-401.

Harby, A., Alfredsen, H.P. Fjeldstad, J.H. Halleraker, J.V. Arnekleiv, P. Borsányi, L.E.W. Flodmark, S.J. Saltveit, S.W. Johansen, K. Clarke, D.A. Scruton 2001. *Ecological impacts of hydro peaking in rivers*. Hydropower 2001 Bergen.

Killingtveit, Å. 2000. *nMag2000, a computer program for hydropower and reservoir operation simulation. User's manual. Version 18/01-2000*. Institutt for vassbygging

Saltveit, S. J., J.H. Halleraker, J.V. Arnekleiv A. Harby 2001 (in press). *Field experiments on stranding in juvenile Atlantic salmon (Salmo Salar) and brown trout (Salmo Trutta) during rapid flow decreases caused by hydropeaking*. Regulated Rivers.

Hydropower in the New Millennium, Honningsvåg et al (eds), © 2001 Taylor & Francis, ISBN 90 5809 195 3

The Norwegian R & D Programme for Environmental Flows

John E.Brittain & Jan Henning L'Abée-Lund
Norwegian Water Resources and Energy Directorate (NVE), Oslo, Norway

ABSTRACT: Instream flows are a strategic element in the management of hydropower development. Norwegian licensing procedures require the setting of specific flows in the reaches affected by dams, as well as water diversions and transfers. Licensing authorities require an objective system that takes account of the needs of the developer, the environment and other present and future users, also with respect to remedial measures. Environmental needs encompass a wide range of aspects, including water quality criteria, groundwater, esthetical and landscape considerations and of course the needs of the flora and fauna. In addition to new schemes, many old licenses are due for renewal and will require an assessment of the adequacy of present flows. The new Water Resources Act emphasises greater flexibility and more emphasis on taking into account ecological, aesthetic and economic considerations, thus increasing the need for site specific knowledge. The five year R & D Programme, which will involve both practical trials and the development of new, user-friendly decision analysis models, aims to provide the licensing authorities with the tools needed for a modern and environmentally based allocation of flows in regulated rivers.

1 INTRODUCTION

Allocation of instream flows is paramount in the licensing procedure of watercourse encroachments. The determination of instream flows is an area management has to make decisions on a daily basis with respect to new licenses, renewal of old licences and in response to the new Water Resources Act. This new law, brought into force on 1 January 2001, opens for a more flexible treatment of instream flows. However, the basis for making decisions on instream flows has frequently been inadequate, resulting in the granting of a trial period for the set instream flows. However, even after the trial period sufficient data has often not been available, resulting in an extension of the trial period, largely on biological grounds. For example, the Alta power station has had trial instream flow since its inception in 1987.

At present Norway is preparing for implementation of the European Union Water Framework Directive. The Directive has clearly defined environmental standards and water quality must be defined according to ecological criteria. Discharge to a large extent determines the nature and development of the freshwater ecosystem. Thus, the Directive will be of importance in the setting of instream flows.

Traditionally the concept of minimum flows has been used as the flow below which the regulator is not allowed to go below. However, not least inter-

nationally, this concept has been shown to be inadequate. Environmentally based flexible instream flows are more in tune with modern sustainable watercourse management.

In order to meet the challenge of sustainable watercourse management and address the problem encountered in the licensing procedures, the Norwegian Water Resources and Energy Directorate (NVE) has instigated a five year R & D programme on environmental instream flows. The objective of the programme is to increase knowledge of the effects of strongly reduced discharge in order to form the basis for development of appropriate methods that will enable management to set ecologically sound instream flows.

2 PREVIOUS NORWEGIAN R & D

The topic of reduced discharge and the setting of instream flows have been addressed in previous R & D programmes. However, much of the work has been theoretical in nature and largely based on existing knowledge, rather that obtaining new knowledge concerning the ecological consequences of a particular instream flow. In particular, two programmes, the Environmental Effects of Hydropower Development (MVU; 1982-88) and Effective Energy System (EFFEN; 1992-96), have considered the question of minimum flows. In addition two other programmes, the Weir Project (1973-83) and the Biotope Adjustment Programme (1985-95) have increased knowledge of remedial measures in regulated rivers and lakes.

The MVU Programme had flexible regulation and requirements for minimum flows as one of its project areas. The aim was to develop methods to improve the basis for decisions regarding the setting of the regulation regime and the effects of different strategies. A number of interesting desk studies were completed (Ziegler, 1986), but practical demonstration projects were not carried out. The EFFEN programme had "Environment" as one of its five programme areas. Here the aim was to increase knowledge of the environmental consequences of hydropower development. Attempts were made to develop an expert method for the setting of minimum flows. The method was tested in four watercourses (Faugli, 1997). However, it proved difficult to put forward specific flows as the user interests were insufficiently documented. The Weir Project and the Biotope Adjustment Programme did not address the question of setting environmental flows, but did increase knowledge of remedial measures, especially weirs, in regulated rivers in association with reduced flows (Eie *et al.*, 1997).

3 INTERNATIONAL PRACTICE

The setting of environmental flows has been the subject of considerable interest internationally and several countries are addressing the problem. There has been a change from the concept of minimum flows below large dams developed in the 1950s and 1960s, followed by instream flows in the 1970s, hydrological and habitat based methods on the 1980s, to today's multidisciplinary catchment based criteria. In Norway the idea of "minimum flows" is still prevalent and there is a clear need to think more in terms of environmental flows. A certain degree of flexibility by setting different flows at different times of the year has been instigated in many instances, although there is necessary to incorporate year to year variations. For example, it may be possible to allocate more water in wet years compared to dry ones.

Methods for the setting of instream flows can be allocated to three main categories with increasing degree of complexity and resource requirements (Dunbar *et al.*, 1998):

- *Look up techniques*. These are the simplest and are frequently based on one or more hydrological indices, such as specific proportion of average discharge. These techniques are widely used, and require a relatively low level of resources. Such methods are undoubtedly of value in an initial screening process or in low conflict situations.
- *Discussion based approaches and hydrological analysis*. There has been an increasing tendency

to use expert opinion in the setting of environmental flows in combination with hydrological time series comparing historical, natural and alternative flow regimes. Such an approach also includes holistic methods that have been particularly well developed in Australia and South Africa where the whole river system including the river channel, riparian zone and groundwater is the focus for field assessment, hydrological modelling and workshops (Arthington, 1998; Arthington & Zalucki, 1998). In certain cases, instead of starting with no discharge and then determinimg what is necessary for specific uses, one starts with the maximum acceptable deviation from the norm, i.e. how much water can one remove without producing significant geomorphological or ecological damage.

- *Biological response modelling*. These techniques are usually the most resource intensive, but they are considered more defensible, although not without their problems (Gore & Mead, 2000). Within this category the Instream Flow Incremental Methodolgy (IFIM) is the most widespread. One of the elements in IFIM is PHABSIM (Physical Habitat Simulation). These techniques are primarily based on physical variables, but are linked to the physical requirements of fish. Considerable efforts are being made to improve these techniques and to include invertebrates. In Norway the River System Simulator, a habitat modelling framework, has been developed for modelling change resulting from hydropower development, with salmonid fish as the target species (Harby *et al.*, submitted).

4 THE NORWEGIAN ENVIRONMENTAL FLOWS PROGRAMME

The are several considerations that must be addressed when setting environmental flows. These include energy production, pollution, ice problems, sediment transport and erosion, aesthetics and biology/ecology. At present the are certain areas where there is an urgent need for improving our practical knowledge of the consequences of reduced flows in regulated watercourses. These will form the focus of the Environmental Flows Programme and they fall into seven main categories:

- *Estimation of low flows*. There is a need to improve the methods for calculating low flows, also in catchments where observational data is not available. There is a need for knowledge on seasonal variations and to incorporate these into the modelling process.
- *Groundwater*. Our knowledge of the interactions between flows in the main channel and groundwaters in the floodplain needs improving. There is the need for predictive models to determine

the consequences of reduced flows for groundwater. The new Water Resources Act also includes groundwaters in addition to surface waters, and all impacts that affect groundwaters, both quantity and quality, must be evaluated in the granting of hydropower licences.

- *Water temperatures.* Water temperature is critical factor for many of the processes taking place in rivers and for the river biota. There is the need to develop improved models for predicting water temperatures, especially in high gradient rivers with reduced discharge. Data is also lacking on the consequences of reduced flows on water temperatures in small rivers.

- *Sediment transport and erosion.* Reduced flows will have consequences for sedimentation and erosion processes, which will in turn affect ecological conditions and the aesthetic appearance of the watercourse. There is a need to improve knowledge of the relationship between reduced flows, groundwater levels and erosion and the importance of increased sedimentation for the frequency of damaging floods. Increased sedimentation can also lead to the filling of interstitial species in stony substrates and subsequent colonisation by aquatic mosses, thereby creating problems for fish spawning and fry habitat. The relationship between reduced flows and sedimentation and erosion is likely to vary between different types of catchment, but this needs to be evaluated.

- *Biological effects.* The effects on the biota are frequently of prime importance in determining environmental flows. However, management lacks the necessary tools to determine environmental flows on the basis of biological criteria. Increased macrophyte growth has been observed in several Norwegian regulated rivers. However, there may be several reasons for this that are not directly related to low flows, and it is necessary to determine the role of hydropower regulation. It is necessary to develop methods to predict the consequences of reduced flows on fish production and biodiversity, as well as assessing the impact of variations in flow from drought conditions to major floods.

- *Aesthetics.* It is necessary to evaluate the effects of variation in flows in relation to aesthetics, as well as attitudes towards natural and impacted geomorphological processes.

- *Remedial measures.* Reduced flows must always been considered with respect to possible remediation in order to maximise biodiversity and the appreciation riverine landscapes. Weirs have been a common measure in Norwegian rivers with reduced flows. However, there are a number of adverse effects, necessitating the evaluation of modified or alternative techniques.

The possibility of combining many of these methods and approaches in a comprehensive management decision support system will also be addressed. The Programme will involve extensive co-operation within NVE and between NVE and those involved in hydropower regulation, including other government agencies, power companies, research institutes and universities. The programme will start during 2001 and has a preliminary 5-year time frame, although the complexity of the topic and the necessity for long-term studies, may necessitate an extension for a further 5 years.

Projects selected for support must fulfil the following criteria:
- Relevance to the needs of management
- Practically useful
- High scientific quality
- Feasible within the constraints of economy and time

The setting of instream flows is extremely complicated topic and in, addition to national R & D, it will be necessary to draw on international experience and expertise. By addressing the specific needs of management, it is hoped that the Norwegian Environmental Flows R & D Programme will make a significant contribution towards a more environmentally appropriate allocation of instream flows in order to maintain the ecological quality of those Norwegian watercourses developed for hydropower production.

5 REFERENCES

Arthington, A.H. 1998. Comparative evaluation of environmental flow techniques: review of holistic methodologies- Land and Water Resources, Australia, Occasional Paper Series 26/98.

Arthington, A.H. & Zalucki, J.M. (eds). 1998.Comparative evaluation of environmental flow assessment techniques: review of methods. Land & Water Resources, Australia, Occasional Paper Series 27/98

Dunbar, M.J., Gustard, A., Acreman, M.C. & Elliott, C.R.N. 1998. Overseas approaches to setting River Flow Objectives. Institute of Hydrology, U.K., R & D Technical report W6-161.

Eie, J.A., Brittain, J.E. & Eie, J.A. 1997. Biotope adjustment measures in Norwegian watercourses. Kraft og Miljø 21. Norges vassdrags- og energiverk.

Faugli, P.E. (ed.) 1997. Fastsettelse av minstevannføring på faglig grunnlag i tidligere regulerte vassdrag. Ekspertmetoden. Report EFFEN Miljø.

Gore, J.A. & Mead, J. 2000. The benefits and dangers of eco-hydrological models to water resource management decisions. In Ecohydrology: a new paradigm. United Nations/UNESCO, Geneva and Cambridge University Press.

Harby, A., Bakken, T.H., Bjerke, P.L., Halleraker, J.H., Heggenes, J., Tjomsland, T. & Vaskinn, K.A. Submitted. Application of the River System Simulator for optimising environmental flow in a Norwegian regulated river. *Environmental Modelling and Software*.

Ziegler, T. (ed.). 1986. Fleksibel manøvrering. MVU rapport A7. Norges Teknisk-Naturvitenskapelige Forskningsråd, Oslo.

Hydropower in the New Millennium, Honningsvåg et al (eds), © 2001 Taylor & Francis, ISBN 90 5809 195 3

Short-term regulation in reservoirs – biological impacts and mitigation measures

E.M.Brodtkorb
Statkraft Grøner, Lysaker, Norway

ABSTRACT: This paper describes potential biological impacts in reservoirs following short-term regulation (hydropower peaking operation). We have focused on effects regarding aquatic vegetation, plankton, zoobenthos and fish. Assessments of potential biological impacts are based on results from full scale peaking operations in lake Vinjevatn (southern Norway) during autumn 1997 and 1998, as well as evaluation of relevant literature. We have focused on making the results applicable for other short-term regulation sites.

INTRODUCTION

Environmental impacts from seasonal hydropower regulation of Norwegian lakes and reservoirs are well know (e.g., Gunnerød & Melquist 1979, Aass &. Borgstrøm. 1987). Apparently, less is known about the impacts following short-term regulation of lakes and reservoirs (Pettersen 1998). So far, research has mainly focused on the effects of such regulation in rivers. Extensive reviews of these effects are given by Cushman (1985) and Hunter (1992).

The aim of the project has been to upgrade the knowledge of biological impacts caused by short-term regulation of lakes and reservoirs. We have used the following definition of short-term regulation: *Short-term regulation is frequent changes in power production in order to cover the variation in power demand*. We have studied the effects of hydropower peaking on different biological factors in Lake Vinjevatn, southern Norway. Aquatic vegetation, plankton, zoobenthos and fish have been examined. The project is part of the multidisciplinary program: "Short-term regulations – Environmental impacts and mitigation measures", funded by a consortium of several institutions.

The environmental impacts caused by short-term regulation falls within the same categories as for seasonal regulation. Normally, impacts following a new regulation will be greater than the impacts due to a change in regulation regime from seasonal to short-term. Whether or not such a change will lead to other impacts and/or increase existing impacts, depends on possible occur of changes in important abiotic factors.

In our study we therefore focus on the following questions:

– What are the main differences in abiotic factors between short-term regulation and seasonal regulation?
– What is the effect of these differences on biotic factors?
– What environmental factors are important to asses when planning short-term regulation?
– What types of basins or reservoirs are suitable for short-term regulation?
– How can we reduce/prevent negative biological effects?

Additional to the results from the field surveys in Lake Vinjevatn, our conclusions are based on general knowledge of effects from seasonal hydro power regulation on abiotic and biotic factors, and experiences from peaking operation of river reservoirs (Hildebrand et al. 1980, Thornton et al. 1990).

The results and conclusions drawn in this paper are not necessarily valid for all types of short-term regulation, but more a review of possible "worst case" scenarios.

STUDY AREA

The Tokke scheme in Telemark in Southern Norway was selected as study area, and Lake Vinjevatn chosen for the field surveys (Fig. 1).

Lake Vinjevatn is one of several reservoirs in the Tokke scheme, developed in the 1950's and 60's. Lake Vinjevatn has a total volume of 21.3 mill. m³

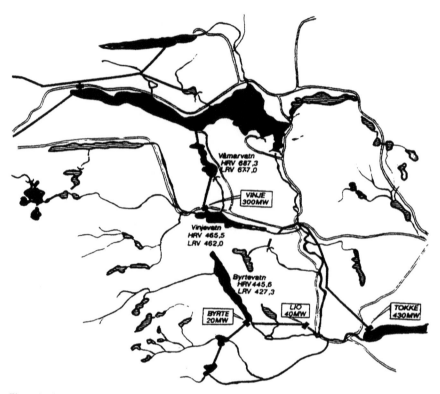

Figure 1. The Tokke scheme in Telemark, southern Norway

and a regulated volume of about 11.2 mill. m³. It is 7.5 km in total length and has a mean depth of 7 m. The regulation height is 3,5 m, but only the upper 1,5 m is used regularly. Lake Vinjevatn is oligotrophic. 12 species of aquatic vegetation are registered. Density of zoobenthos is generally low, dominated by species typical of regulated lakes. Registered species of fish are trout (*Salmo trutta*), char (*Salvelinus alpinus*), three spine stickleback (*Gasterosteus aculeatus*) and minnow (*Phoxinus phoxinus*).

METHODS

During the experimental periods in 1997 and 1998 the power stations at both ends of the reservoir were operated for fourteen days in peak diurnal regulation mode, followed by reference periods with slower water level changes (Fig. 2). In September 1997, the water level fluctuated between the 462.5 and 465.5 contour line. The amplitude of the diurnal fluctuations varied between 1.5 m and 3.0 m. In 1998, the peaked diurnal water level fluctuation was less extreme (2 m) and generally nearer HwL.

During these two trial periods, the effects of the

peaking regulation on abiotic and biotic factors were measured and observed.

In 1997 and 1998 water samples were collected in the reservoir, in all the tributaries and at the reservoir outlet, to determine suspended sediment concentrations. Sediment traps were placed at four locations in the reservoir to measure deposition rates and erosion processes recorded during field inspections.

During the experiments, water temperature was measured with several temperature loggers and termistor strings placed in the lake.

The effects on aquatic vegetation were examined on three sections on exposed and protected shores. Zoobenthos were collected with a 0.025 m² Van Veen grab at four stations, with a total of 15 samples from each station. Samples were taken at depths of 0.5 m, 1 m. and 1.5 m, before and after the diurnal regulation. Density and composition of species were examined.

Plankton was collected prior to, during, and after the experiment in 1998. A total of 14 qualitative and 52 quantitative samples were collected. Density and composition of species were examined.

Stranding of fish was studied at 7 stations during the two most distinct drawdowns in 1997. Number

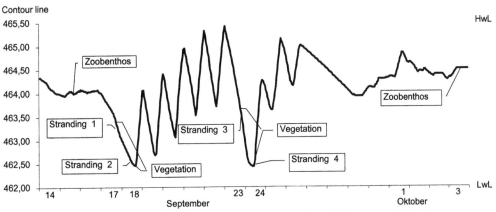

Figure 2. Water level fluctuation in Lake Vinjevatn in 1997. The timing of different field surveys are indicated.

of stranded fish was counted manually. The total area of the fish stations was 11075 m^2, and it was searched four times.

More thorough descriptions of the methods used in these field experiments are given in Bogen & Bønsnes (2001), Kvambekk (2001) Førde et al. (1997, 1999), Johansen & Rørslett (1999) and Færøyvig et al. (1999).

RESULTS AND DISCUSSION

What are the main differences in abiotic factors between short-term regulation and seasonal regulation?
The most important difference between short-term regulation and seasonal regulation is usually a faster and more frequent water level fluctuation. The frequent changes in water level appear in addition to the changes caused by seasonal regulation. This usually leads to a more unstable and "stressed" littoral zone (Thornton et al. 1990).

Short-term regulation might lead to increased erosion, turbidity and siltation in a reservoir or lake. In Lake Vinjevatn, the peaked regulation caused heavy erosion along shorelines and in shallow areas (Bogen & Bønsnes 2001). Suspended sediment concentrations up to 470 mg/l were recorded in the swash zone along the shores. The relatively intensive erosion during the diurnal peaked regulation period, was found to be a result of both increased sediment supply from groundwater erosion and increased wave erosion. Peaked regulation creates suitable conditions for groundwater erosion. The rapid drawdown of the water level results in uncompensated porewater pressure in the sedimentary deposits along the exposed shoreline. At some locations this excess porewater pressure caused sheet slides and groundwater spring sapping erosion (Bogen & Bønsnes 2001).

Short-term regulations might increase water temperature variation during summer and winter season compared with seasonal "long-term" regulation (Kvambekk 2001). In a regulated watercourse with one reservoir, considerable changes in water temperature would not be expected. However, in more complex systems, several factors might affect water temperature. Depending on the temperature differences, additional water from adjacent regulated reservoirs, or natural run off from larger tributaries, might lead to temperature effects in the reservoir (Kvambekk 2001). However, such temperature effects will normally be restricted to smaller mixing zones (both vertically and horizontally) near the inlet and outlet of the reservoir or lake (Kvambekk 2001).

Periodically, increased variation in hydraulic flushing rate and water retention time is expected, especially in smaller reservoirs (Hildebrand et al. 1980, Thornton et al. 1990). In smaller reservoirs, ice conditions might be more unstable due to the above mentioned factors (Kvambekk 2001).

What is the effect of these differences on biotic factors?
Biological impacts due to more rapid and frequent changes in water level and water surface area in the littoral zone are mainly negative. Aquatic vegetation might be additionally exposed to frost and drought, depending on the magnitude of the water level changes and time of the year (Johansen & Rørslett 1999). However, large seasonal water level fluctuations can be expected to be more effective in reducing vegetative growth than frequent small fluctuations (Hildebrand et al. 1980). So if a reservoir previously has been used for seasonal regulation, with regulation heights larger than 4-5 m, vegetation left in the littoral zone is sparse (Rørslett 1985, 1989). In such a system, short-term regulation would have minor consequences for littoral vegetation because the damage is already done.

Littoral benthic invertebrates are more exposed to stranding during rapid water level decreases, because they are unable to follow the drawdown (Hynes & Yaday 1985). Attempts to migrate to deeper water during drawdowns probably result in significant predation mortality. Larger species are often most vulnerable (Kaster & Jacobi 1978). Several studies have shown low macrobenthic diversity in reservoirs with rapidly fluctuating water level (Fillion 1967, Kaster & Jacobi 1978, Hruska 1973). This was also the case for Lake Vinjevatn. After the period with short-term regulation in 1997, density of benthic invertebrates on exposed sites was heavily reduced (Færøyvig et al. 1999).

Stranding of fish closely associated with the littoral and shore zone is expected when water level decreases rapidly. In Lake Vinjevatn, stranding of sticklebacks was most abundant. Few trouts were found stranded. This is probably due to the fact that juvenile trout mainly live in riverine habitats, and thereby avoid the fluctuating water level. Larger fish avoid stranding by migrating to deeper waters. Other factors than water level fluctuation might be important concerning stranding potential in a reservoir. Factors like substrat coarseness, shore line slope, time of day/year and water temperature, are of importance for the potential of stranding in rivers (Cushman 1985), and are also assumed important in reservoirs.

Increased erosion, turbidity and siltation will generally have negative effects on a reservoir or lake ecosystem. Ground water erosion and wave induced erosion, due to short-term regulation, might lead to detachment of whole plants, cleaning of plants, exposure of roots, and siltation (Brookes 1986, Johansen & Rørslett 1999). Increased turbidity, and thereby reduced water transparency, have lead to brief decreases in phyto plankton biomass (Duthie & Osrofsky 1975, Duthie 1979). However, production increases quickly when transmission of light recovers. Increased erosion might negatively affect zoobenthos directly through further deterioration of natural habitats in the littoral zone, or through siltation of habitats in or below this zone.

Generally, potential changes in water temperature due to short-term regulation will have minor effects on the ecosystem. Negative effects are not expected concerning plankton, zoobenthos or fish, due to small and rapid changes in temperature. This mainly because such effects will be of local character.

Periodically increased variation in hydraulic flushing rate and shorter water retention time will normally not affect aquatic vegetation, zoobenthos or fish. Plankton communities however, might be negatively affected if the flushing rate is high and water retention time is low (Færøyvig et al. 1999). A reduced plankton production or biomass will eventually lead to reduced fish production.

Unstable ice conditions during the winter, with broken ice near the shores, will seldom have severe impacts on the ecosystem. However, zoobenthos in the littoral zone, that bury in the ground when water drops, might be negatively affected if lack of ice cover exposes zoobenthos to frost (Paterson & Fernando 1969).

The direct consequences of changes in abiotic factors following short-term regulation in reservoirs, are assumed to be modest for fish populations. However, if short-term regulation leads to a decline in available prey (zooplankton, benthic invertebrates and smaller fish), fish composition and production are assumed to be negatively affected.

What environmental factors are important to asses when planning short-term regulation?
For site-specific assessment of potential impacts from short-term regulations, it is important to gather background information concerning present and historical regulation regime, present biological status and planned short-term regulation regime for the given hydroelectric site.

1   Present and historical regulation regime
Impacts from short-term regulation will depend greatly on the historical and present regulation regimes. Relevant information is listed below:

- Storage capacity of the reservoir
- Regulation height
- Reservoir topography
- Normally used regulation height
- Water retention time
- Maximum useable flow
- Production strategy/Discharge regimes
- Outlet of power station
- Erosion and sediment conditions
- Coupling with other hydroelectric reservoirs

2   Present biological state
General knowledge of the biological state of a reservoir is important when assessing possible impacts of a short-term regulation. In reservoirs with harsh, seasonal regulation, biodiversity and production are usually low. Reservoirs with a more modest regulation, or unregulated lakes, normally have higher diversity and an overall better production. Information about the ecosystem will therefore indicate its vulnerability to short-term regulation. A closer look at the following factors will give valuable information about a system:

- Aquatic vegetation
- Plankton
- Zoobenthos
- Fish

Important information about these factors is species composition, number and density of different species, life histories, special environmental requirements and behavioural adaptations.

3 Planned short-term regulation regime

Introducing short-term regulation in a system will normally change several physical conditions. In order to predict environmental impacts, information about these physical changes is necessary. The most important changes concerning environmental impacts are:

- Water level fluctuations
- Frequency of water level fluctuations
- Water retention time
- Planned production strategy/Discharge regime

What type of basin or reservoir is suitable for short-term regulation?

Based on physical and biological data from the actual site, and theoretical considerations, a table for classifying the environmental suitability of different sites for short-term regulation has been made (Table 1). A suitable lake or reservoir for short-term regulation, as regards biological impacts, should have a large volume. A large volume usually causes less change in water level. Lakes or reservoirs with large regulation heights and mean depths are suitable, because they often have steep shoreline slopes and few shallow shore areas. This reduces the impact of water level fluctuations on the littoral zone. Basically, water retention time should be long, and hydraulic flushing rates low, due to the increased variation following short-term regulation.

However, some variation in hydraulic flushing rates might be favourable if the ecosystem is adjusted to it. The outlet of the power station should preferably be to the sea, because this normally has less negative impacts than outlet in rivers.

Bottom substrate should be coarse to reduce the effect of erosion. The lake or reservoir should not be linked to other lakes or reservoirs. Linkage might lead to impacts throughout the whole system. Bio-

Table 1. A suitability classification system for reservoirs were short term regulation is planned.

	Classification	
	Favourable	**Unfavourable**
Present regulation		
Reservoir volume mill. m³	Large (> 150)	Small (< 150)
Regulation height m	Large (> 10 m)	Small or none (< 10 m)
Reservoir topography	Mean depth large Steep shoreline slopes / few shallow areas	Mean depth little Slack shoreline slopes and large shallow areas
Normally used regulation height	Large	Small or none
Water retention time / Water discharge	Long/small	Short/large
Maximum operating flow m³/s	Small (< 30 m³/s)	Large (> 100 m³/s)
Production strategy / Discharge regimes	Varying	Stable
Outlet of powerstation	Fjord/Sea	River
Substrat stability and erosion	Coarse bottom substrat, stable sediments	Fine bottom substrat, unstable sediments
Coupling with other reservoirs	No	Yes
Tributaries	Small	Large
Biological state		
Aquatic vegetation	Sparse vegetation in the littoral zone	Vegetation in the littoral zone
Plankton	Simple communities, few species dominate	Complex communities, many species
Benthic invertebrates	Simple communities, few species dominate	Complex communities, many species
Fish	Simple communities, few species totaly dependent of the littoral zone	Complex communities, many species dependent of the littoral zone
Planned short-term regulation		
Water level fluctuations	Small	Large
Frequency of water level fluctuations	Slow	Fast
Water retention time	Long, stable	Short, variable
Planned production strategy / Discharge regime	Regulation rest periods in spring and autumn	No rest periods

logically it's an advantage if the reservoir is already heavily regulated, with an ecosystem already strongly impaired, with species adapted to an unstable environment.

How can we reduce/prevent negative effects from short-term regulations?

Several mitigating measures for reducing environmental impacts exist. The most obvious is to minimise water level fluctuations. Water fluctuations should take place within the regulation height normally used, and not necessarily between the whole max and min regulation height. It's also important to minimise water level fluctuation during vulnerable periods for the ecosystem (spring and autumn). Drawdowns should be slow to prevent stranding of benthic invertebrates and juvenile fish. Short-term regulation between heights that reveals large areas of dry land should be avoided to reduce the risk of stranding.

ACKNOWLEDGEMENTS

Thanks to Leif Lillehammer and Finn Gravem for reviewing this manuscript. I am indebted to Per Færøyvig, Gunnar Barstad, Morten Kråbøl Stein Johansen, Leif Lillehammer, Vegard Pettersen, and the people at Vinje power station for their work within the project. Thanks to the Norwegian Research Council, Statkraft SF, the Norwegian Electricity Federation (Enfo) and the Norwegian Water Resources and Energy Administration (NVE) for funding the project.

REFERENCES

Bogen, J. & Bønsnes, T. 2001. Virkninger av effektregulering på erosjon og sedimentasjon i vannkraftmagasiner. Rapp. Nr. 16 *Effektregulering- Miljøvirkninger og konfliktreduserende tiltak:* 1-65. Høvik

Brookes, A. 1986. Response of aquatic vegetation to sedimentation downstream from river channelization works in England and Wales. *Biological Conservatio,* 38: 351-367.

Cushman, R. M. 1985. Review of ecological effects of rapidly varying flows downstream from hydroelectric facilities. *North American Journal of Fisheries Management* 5: 330-339.

Duthie, H. C. 1979. Limnology of subartic canadian lakes and some effects of impoundment. *Arctic and Alpine Research* 11: 145-158.

Duthie, H. C & Ostrofsky, M. L. 1975. Environmental impact of the Churchill Falls hydroelectric power project: a prelimnary assessment. *J. Fish. Res. Board Can.* 32: 117-125.

Filion, D. B. 1967. The abundance and distribution of benthic fauna of three mountain reservoirs on the Kananaskis River in Alberta. *J. Appl. Ecol.* 4: 1-11.

Færøvig, P., G. Barstad, & E.M. Brodtkorb. 1999. Effektregulering-virkninger på bunndyr og plankton i Vinjevatn.. Rapp. Nr 8. *Effektregulering - Miljøvirkninger og konfliktreduserende tiltak:* 1-17. Høvik.

Førde, E. , Magnell, J. P., Kvambekk, B., Bogen, J., Brodtkorb, E., Johansen, S., Nordli, P. E., Pettersen, S. 1997. Forsøkskjøring i Vinjevatn september 1997. Virkninger på fysiske og biologiske forhold – foreløpelige resultater. Rapp. Nr 2, *Effektregulering – miljøvirkninger og konfliktreduserende tiltak:* 1-40. Høvik.

Førde, E. Kvambekk, B., Bogen, J., Brodtkorb, E., Nordli, P. E., Pedersen, T. B. og Magnell, J. P.(red) 1999. Prøvekjøring Vinjevatn august/september 1998. Virkninger på fysiske og biologiske forhold – foreløpelige resultater. Rapp. Nr 7, *Effektregulering – miljøvirkninger og konfliktreduserende tiltak:* 1-25. Høvik.

Gunnerød, T. B. & Mellquist, P. 1979. *Vassdragsreguleringers biologiske virkninger i magasiner og lakselver,* 1-294.

Hildebrand, S.G., R.R. Turner, L.D. Wright, A.T. Szluha, B. Tshantz, & S. Tam. 1980. Analysis of environmental issues related to small scale hydroelectric development.III: Water level fluctuation. *Oak Ridge National Laboratory, ORNL/7453.* 1-132.

Hruska, V. 1973. The changes of benthos in Slapy Reservoir in the years 1960-1961.. In F. Hrbacek and Straskraba, M. (eds.). *Hydrological Studies 2.* Academic Publ. House, Czechoslovak Acad. Sci. Prague.

Hunter, M. A. 1992. Hydropower flow fluctuations and salmonids: a review of the biological effects, mechanical causes, and options for mitigations. *State of Washington. Department of Fisheries. Technical Report nr. 119:* 1-46.

Hynes, H. B. N. & Yaday, U. R. 1985. Three decades of post-impoundment data on the littoral fauna of Llun Tegid, North Wales. *Arch. Hydrobiol* 104: 39-48

Johansen, S.W. & B. Rørslett. 1999. Effektregulering-virkninger på vannvegetasjon av prøvekjøringene i Vinjevatn 1997 og 1998. Rapp Nr. 11. *Effektregulering-Miljøvirkninger og konfliktreduserende tiltak:* 1- 44. Høvik.

Kaster, J. L. & Jacobi, G. Z. 1978. Benthic macroinvertebrates of a fluctuation reservoir. *Freshwater. Biol.* 8: 283-290.

Kvambekk, Å. 2001. Virkninger av effektregulering på is og vanntemperatur i vannkraftmagasiner. Rapp Nr 15 *Effektregulering-Miljøvirkninger og konfliktreduserende tiltak:* 1- 42. Høvik.

Paterson, C. G & Fernando, C. H. 1969. Macroinvertebrate colonization of the marginal zone of a small impoundment in eastern Canada. *Can J. Zool.* 47(5):1229-1238.

Pettersen, V. 1998. Innsamling og systematisering av erfaringsdata. Rapp Nr 3 *Effektregulering-Miljøvirkninger og konfliktreduserende tiltak:* 1- 26. Høvik.

Rørslett, B. 1985. Vannvegetasjon og vassdragsreguleringer. *K. Norske Vidensk. Selskap. Rapp. Bot. Ser. 1985,* 2:109-124.

Rørslett, B. 1989. Forekomst av vegetasjon i regulerte vassdrag. Problemidentifisering og omfang. *NIVA rapport O-88003*

Thornton, K.W., B.L. Kimmel, & F.E. Payne. 1990. *Reservoir limnology: Ecological perspectives.* pp. 33-246. John Wiley & Sons, Inc. New York.

Aass, P. & R. Borgstrøm. 1987. Vassdragsreguleringer. In: Borgstrøm, R. & L.P. Hansen (eds.),. *Fisk i ferskvannøkologi og ressursforvaltning:.* 244-266. Oslo: Landbruksforlaget.

Hydropower in the New Millennium, Honningsvåg et al (eds), © 2001 Taylor & Francis, ISBN 90 5809 195 3

Hydropower development and environmental requirements in Lithuania

J. Burneikis
Lithuanian Energy Institute, Kaunas, Lithuania

P.Punys & G.Zibiene
Lithuanian University of Agriculture, Kaunas, Lithuania

ABSTRACT: This paper describes the environmental requirements relating to the development of hydropower projects. A particular emphasis is given to hydropower and environmental legal frameworks. An overview of national hydropower development tendencies is presented, analysis of the environmental legislation and the EIA procedures is given, and ecological flow regulations currently in force are treated. Only small hydropower (P<10 MW) regulations well established and harmonized in accordance with the similar EU regulatory standards are applied in Lithuania. Efforts of enactment of the comprehensive hydropower legislation including both small and large hydro issues are currently under way. Environmental legislation regarding hydropower development is well founded and clearly defined. Environmental Impact Assessment (EIA) procedures correspond entirely to those required by the EU Directives and other regulations dealing with the implementation of the EIA. A major barrier hindering country's hydropower development is due to the requirements ensuring the life and reproduction of migrating fish. The simple empiric method, based on the hydrology of the river is used in practice to quantify ecological (instream) flow. By comparing it to the more advanced methods used in other countries, in most cases, the calculated minimum flow values are less. Therefore, there is the need to develop the methods based on biological and ecological considerations, especially for the rivers characterized by more concentrated fish habitat.

1 INTRODUCTION

Lithuania is a low-lying country and the fall of watercourses is relatively low. There are no suitable sites for impounding storage reservoirs. Only low head run-of-river projects are developed here. Small rivers are being used now for electricity generation, harnessing the power of large ones is intended in the future.

A general overview of existing hydropower resources and their role in country's energy balance is given in Burneikis et al. (2001b). It has to be emphasised that hydropower share represents nearly 2.5% of the total country's electricity generation. Consequently, Lithuania's power sector is heavily dependent on imported fuels. According to the studies performed hydropower could reduce their share at least by 15%.

There is no officially approved hydropower development program so far in Lithuania. Only its main development trends have been established. The Figure 1 illustrates the major phases under which hydropower development takes place.

In recent years, consideration of "environmental"

factors has become an increasingly important part of project design and development. Today, environmental considerations are a very significant part of hydro project planning or operating. Typically, environmental activities take from 10 to 20% of the planning and design effort on a major project. From a legal perspective, taking into account the prevailing conditions in Lithuania, the ultimate goal is to reconcile the two basic requirements that frame hydropower development: protecting the environment and ensuring everyone's right to economic development.

The overall objectives of the study have been to:
- Overview national hydropower legislative framework,
- Analyse national environmental legislative framework, focusing particularly on the EIA procedure, water regime and fish protection, protected areas,
- Describe the predominant species of the river fish habitat,
- Revise the national methodology for determination of ecological flow and compare it with others common used in practice.

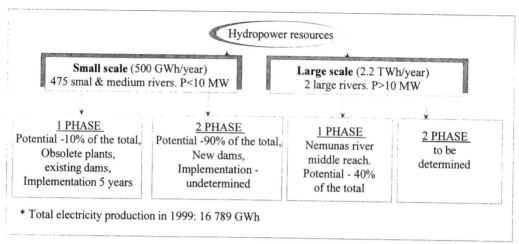

Figure 1. Hydropower resources development phases

2 OVERVIEW OF NATIONAL LEGAL & REGULATORY FRAMEWORKS

2.1 Hydropower legislation

After Lithuania regained its Independence and Independent Power Producer (IPP) status was established, it was necessary to proceed with hydropower sector regulation issues. Notwithstanding this, there is no specific hydropower legislation so far in Lithuania. The plans to establish its comprehensive legal framework, which can deal with both small and large hydropower, are currently under way (Burneikis & Punys 1999a).

To date the laws, decrees and orders issued by the Government, Ministry of Economy (Department of Energy Development), and Ministry of Environment, Ministry of Agriculture, former Ministry of Construction and Urban Development (now the department subordinated to the Ministry of Environment) regulate hydropower.

The administrative act (1996) issued by the Ministry of Environment (MoE) dealing with environmental issues distinguishes between schemes with an installed power up to 10 MW (small abstractions) and schemes with an installed power over 10 MW (large abstractions). This definition regarding the hydropower plant scale has been set in accordance with the ESHA (European Small Hydropower Association) recommendations (EC DG XVII 1995). It has to be noted that existing hydropower legislation in Lithuania deals only with small hydropower plants to be built at existing dams, i.e. the development of the first phase of small hydropower resources (Fig.1). In fact the last large hydropower plant was commissioned in 1959 (the 100 MW Kaunas HPP), except the 800 MW Kruonis pumped-storage plant commissioned in the 1980s

(still uncompleted to design capacity). Notwithstanding this, large hydropower projects (in the scale of local topographic conditions) are being considered in Lithuania.

Existing Law on Energy (1995) is a framework law for all energy (electricity, heat, gas) sectors. It requires a permit from any body to be involved in the energy activities, including the generation of electricity. This Law empowers the utilities to connect to the grid IPP's plants using renewable resources, including hydro, to buy the electricity at agreed buy-back rates. The Minister of Economy by his order, taking into consideration the suggestions of the State Commission for Pricing and Control Energy (a regulatory under supervision of President's administration), fixes energy price to be paid by the state owned utility "Lithuanian Energy" (currently this utility is under restructuring). This order lays down the limit of power capacity for a hydro plant (P<5 MW) which power delivered to the grid has to be purchased. Power purchase agreement (PPA) has to be contracted between IPP and local grid authorities.

A new Law on Electricity harmonized with the EU electricity sector requirements has been recently passed. This law describes regulation for generation, transmission and distribution of electricity and lays down the rules relating to the organization and functioning of the competitive electricity market. It supports consumers to buy the power produced by renewable energy sources.

2.2 Environmental legislation

In the former Soviet times there was in force the environmental legal framework, which was designed for the existing political and economic system. Having restored Lithuania's Independence one of the first legal act was the enactment of the Law on Environmental Protection (LEP). The LEP of 1992 lays down the basic principles of environmental protection. It is a framework law and forms the legal basis for the enactment of all laws and administrative acts that regulate the use of natural resources and protect the environment. It sets up the Ministry of Environment, which initiates the policy on environmental protection. The main objective of this law is to achieve an ecologically sound and healthy environment on which human activities have little negative impact and which can maintain Lithuania's typical landscape as well as its diversity of biological systems. The law foresees environmental impact assessment (EIA), prescribes polluter-pays principle. According to the Law on Taxes of State Owned Natural Resources introduced in 1991 and the further decision of Government of Lithuania there is no charge imposed on water use for small hydropower plants. It has to be noted that the size of a hydroelectric project to be considered as a small is not so far approved by a high level legislation, except earlier mentioned administrative act issued by the MoE.

Water use and protection in Lithuania is regulated by the Water Law (1997), which states that no one can utilize the energy of the streams and rivers without the State's authorization. Following this law a hydropower developer or operator is obliged to comply with established rules of reservoir use and maintenance in order to maintain imposed water levels fluctuations, ecological flow and assure the free passage of migrating fish. In addition, developers of the new schemes must compensate losses caused by the creation of reservoirs. A specific methodology established by MoE exists to calculate the losses for the transformation of ecosystems and physical modification in monetary terms.

Water Law is being now amended taking into account the requirements of the recently adopted the EU Water Framework Directive.

The national river authority is responsible for granting water rights for hydropower developer up to 50 years with possibility to be extended for another period of time. The request for water use is free of charge. Land needed to develop a hydropower site can be purchased or leased out up to 99 years.

2.2.1 Regulatory framework for EIA

The term EIA, following the definition made in the EU Directive 85/337/EEC of 27 June 1985 applies to the identification, description, and assessment of direct and indirect effects of a project on: human beings, fauna and flora; soil, water, air, climate and landscape; the interaction of these factors; and on material assets, and the cultural heritage.

Lithuania like most industrialized countries has a generalized Environmental Impact Assessment (EIA) legislation aimed at all types of development projects. According to the former Law on EIA (1996) two categories of EIA were foreseen: initial and full EIAs. Taking into account the EU Directive 97/11/EC of 3 March 1997 and other requirements relative to environmental impact studies this law has been recently amended (2000). The law entitled "Law on Environmental Impact Assessment of the Proposed Economic Activity" takes now into consideration all elements and aspects usually included in EIAs: from project preparation, screening, scooping to decision by competent authority. A specific legal act published by the MoE (2000) provides well-determined rules for public involvement in the EIA process.

The Figure 2 summarises the requirements for the EIA procedure regarding water management, particularly focussing on hydropower sector. The commentaries given in the brackets are not represented in the Law. Only underlined statements are official.

Depending on a particular project size there are two options: mandatory requirement or screening. The typical guidance is available to perform the screening process. As can be seen from the Figure 2 hydropower is not directly included in the mandatory list for the EIA. However impounding reservoirs or cross-watershed diversions are closely related to hydroelectric projects. The size of a hydroelectric project with capacity P>0.1 MW subject for the screening must be considered as a strong requirement.

As stated in IEA technical report (IEA 2000) the criteria established worldwide in the various legislative frameworks for EIA regarding hydroelectric projects fall into a wide range.

2.2.2 Water regime and fish protection

The specific regulations issued by the MoE or its subordinated institutions are applied for water regime and fish protection.

The maximum draw down height in a reservoir generally varies from 1 to 2 m. A daily draw down value is fixed at 0.1 to 0.2 m. During the spawning period water levels variations have to be smaller to 0.05 to 0.1 m. It has to be noted that above values are highly site specific and therefore they are imposed case-by case.

Following recent changes in fish screening regulation applicable to small hydropower plants the space between slots does not exceed 3 to 3.5 cm wide. A screen may be interpreted as any device that will prevent the entry of fish, whether it is a physical mesh or a so-called "behavioural" screen, which uses a deterrent stimulus (e.g. light). When downstream migrating fish pass through the turbines

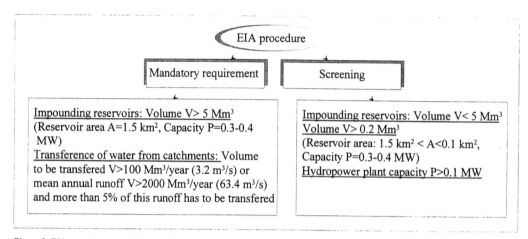

Figure 2. EIA procedure according to the Law on EIA for water management practices

only 10 individuals a day may be injured or event killed. The migrating periods from 1 April to 1 Jun and from 10 September to 20 October are specified. In this case the indicated rate of fish losses is admissible. If fish mortality is greater, the compensation must be paid according to the existing methodology.

The MoE published a list of 24 small dams (1997) at which the fish passes must be constructed. Height of dams varies from 2 to 20 m. Among them the largest hydropower plant in Lithuania Kaunas HPP (P=100 MW) is included. Despite the fact that there are about 430 small reservoirs (area exceeds 5 ha) so far only 5 fish ladders are operating.

Note that some frictions exist between hydropower developers and fisheries bodies (including regulatory agencies) that have justifiable fears that already declining fish stocks might be damaged further.

2.2.3 Protected areas

The establishment of the existing network of protected areas started in 1960, with the formation of 89 nature conservation reserves thought the first State nature reserve with prohibited any economic activity was created in 1937. There is the Law on Protected Areas introduced in 1992. The existing system of protected areas covers 11.7% of the total area in Lithuania. There are 54 landscape, 36 hydrographical and 10 ichtiological reserves.

Taking into account the fact that the first phase of small scale hydropower development approaches to the end the MoE published the list of watercourses where damming or rehabilitation of old dams is prohibited. The list annexed numerates the sites of existing dams or old mills where building small hydropower plants are preferred. I.e. before developing new sites on wild rivers, the

residual potential of regulated rivers has to be exploited. The protected territories stretch along about one third of the total river length suitable for hydropower developments. Notwithstanding this rigorous list, there are potential sites for building large hydropower plants (capacity of 10 to 100 MW) on the largest rivers like the Nemunas and Neris. It is clear that the final decision related to their construction can be made in accordance with the EIA.

3 ECOLOGICAL FLOW REGIME

Anthropogenic activity, which was especially intensive in the 7th – 9th decades, considerably changed the state of surface waters in Lithuania. Recent development of hydraulic engineering – land reclamation, river training, and erection of dams, reservoirs – changed the natural landscape, ecosystems of rivers. An area of 30 460 km² (almost 47% of Lithuania's territory) were drained, more than 70 thousand various hydraulic structures were built. Kaunas hydropower plant, Kruonis pumped storage hydroelectric plant and over 30 small hydropower plants were constructed. Nowadays there are more than 1100 reservoirs with an area over 0.5 ha. A number of them can be used for small hydropower generation (Punys et al. 1999). 20 largest Lithuania's dams are registered in ICOLD's World Register of Large Dams.

To mitigate their impact a determined minimum flow is required to pass from the reservoirs through a dam down to the river or leaving the residual flow in the riverbed.

These determined minimum flows in various countries or even in various parts of the same country and in various times are denominating very different (e.g. residual flow, instream flow, minimum acceptable flow, ecological flow, environmental flow, river flow objective, reserved flow, etc.). Its simple definition is: the minimum flow required to maintain aquatic life.

The imposed values of the ecological flow can be crucial for the development of a hydropower site or operation a hydropower plant. Too large an ecological flow can make an otherwise good project economically unfeasible. It is important to know how an ecological flow will affect reservoir operations, as it is to know how reservoir operations will affect instream habitat. According to the study performed by Strobl et al. (1994) in small hydropower plants in Switzerland, Austria and Germany the losses in electricity production due to ecological flow regulations range between 0 and 65%. The percentage of electricity production losses of the total EDF generation increased from 1.5% to 3.7% when the value of ecological flow depending on the fraction of mean annual flow passed from 1/40 to 1/10 (Veyre 1992). It is clear that the largest losses in power production resulting from ecological flow occur in the schemes of diversion type. Fortunately many of such schemes are rare in Lithuania because of the flat topography.

It is common practice to protect species that are especially vulnerable during specified times intervals. One of the crucial of the biological characteristics in determining ecological flow value is the river fish habitat. Any works undertaken in a riverbed must provide a minimum flow ensuring the life and reproduction of fish. Below the characteristics of the river fish habitat in Lithuania are described.

3.1 River fish habitat

According to the ichthyologists' data, near 50 fish species live in Lithuanian rivers and reservoirs (Virbickas 1986, Bukelskis et al. 1998). Up to 10 – 15 fish species live in the smallest rivers (brooks, rivulets, upstream of the rivers), up to 20 – 25 species – in the middle-sized rivers with length up to 100 km, 25 – 30 fish species – in the rivers with length up to 200 km and in the middle parts of the largest rivers, and 40 – 50 fish species live in the reaches of the largest rivers. Fish species predominating by density and by biomass in various Lithuanian river fish communities are presented in Table 1.

Small rivers (brooks, rivulets) which length is less than 10 km consists of 98% of all rivers in Lithuania, but they cannot be used for hydropower. Only larger rivers which length is over 20 km and catchment area exceeds 50 km² may be considered for hydropower generation. Their number is 477 from which only 42 are undoubtedly attractive for hydropower (Jablonskis et al. 1996). 19 rivers are longer than 100 km and only 9 are longer than 200

km. River basins are small: 62% of them are less than 10 km² and only 22 river basins are over 500 km².

Water of the most brooks, rivulets and streams is cool and clean. Among the middle-sized rivers cool water is in the rivers of the Merkys, Minija, Zeimena, Neris. Water of some larger rivers is cool as well as some – is warm. Sufficient conditions exist for salmonides and rare fish species spawning in almost all rivulets, streams and rivers with cool water.

An important part from 50 fish species living in the rivers is rare and protected. Salmon and sea trout are under protection of Lithuania's Red book. 26 fish species are under protection of the Bern convention, however, most of these species are numerous in the rivers. Endangering species are the Baltic salmon, sea-trout, trout, grayling, vimba etc.

When building hydropower plants a particular attention must be paid to these rivers where precious fish species migrate (so-called *anadromuos* fish). These migrating fish species are very sensitive to the changes of habitat. The results of various investigations prove that migrating fish stocks are decreasing.

The main causes of losses are water pollution, poaching, uncontrolled fishing and dam building. The migrating fish species especially decreased after the hydropower dam was commissioned on the main Lithuania's river the Nemunas near Kaunas in 1959. It is one of the largest hydro schemes without the fish ladder in Lithuania. The efforts to search necessary funds are currently under way to integrate the fish ladder into the existing dam. In order to preserve and restore the stocks of the wild Baltic salmon from endangering the special program named "Salmon 1997–2010" has been initiated.

Table 1. Fish species predominating by density and by biomass in the rivers

Type of river	River length	Predominant fish species
Brooks, rivulets	Up to 10 km	Three-spined stickleback, river trout, European sculpin, brook lamprey
Streams	Up to 50 km and its upper sections	(Predominated fish species in brooks) + dace, gudgeon, European grayling, European roach, Northern pike, perch, burbot
Middle-sized rivers	Up to 200 km and its middle sections	(Predominated fish species in brooks and streams) + chub, bleak, riffle minnow, loach,
Largest rivers	Over 200 km and its lower sections	ide, migrating salmonides (Predominating fish species in middle-sized rivers, less of predominated fish species in brooks and streams) + barbell, roach, asp, bream, vimba

Recent channel dredging of the Klaipėda strait (the Baltic Sea) and reduced level of water pollution resulted in increased access of migrating salmon to return in their breeding grounds of the upper reaches of Lithuania's rivers. About 6 to 8 thousand of salmon, 8 to 11 thousand of sea-trout migrate every year into Lithuania's rivers. Good conditions for salmon spawning are in the rivers Minija, Sventoji, Zeimena, Neris, and Nemunas which hydropower potential is high.

According to the results of records of salmon smolts in the Lithuania's rivers carried out by ichthyologists in 1998 these rivers are grouped as follows:

- Rivers, where salmon lives;
- Rivers, where salmon has not been detected, but can live;
- Rivers, where salmon is disappeared.

However, any program for other fish species restoration still is not proposed in Lithuania, although there are quite a lot of data on diversity, fish density and biomass of rare and protected fish species as sea-trout, vimba, etc.

It has to be noted that the investigations on habitat suitability and habitat requirements still have not been carried out in Lithuania. Lack of these investigations hinders the use of habitat-based methods for ecological flow estimation.

Variety of the rivers communities changes from the upper to lower section of the river as in similar geographical zones. The changes are observed not only in fish biomass and diversity, but also in diversity and biomass of the smallest water organisms (*xerophytes, zoophytes, phytoplankton*), which are as food for the fish. According the typical fish species Lithuania's rivers may be divided in 4 zones: trout zone, grayling zone, barbell zone and bream zone. The same dividing into zones is applied in the USA (4 zones), Germany (7 zones) and other countries. The peculiarities of these biogeographical zones have to be taken into account when building dams.

3.2 Methods for ecological flow estimation

All methods used for ecological flow estimation can be divided in two groups: empirical methods and analytical methods. The latter are more advanced based on hydrological, hydraulic, hydrobiological data, taking into account social and economical factors. Empirical methods are simple ones based only on hydrological data.

The review of the empirical and analytical methods – from simple methods (standard-setting or look-up techniques) to modern methods based on detailed biological response simulation, such as IFIM (Bovee et al. 1998), CASIMIR (Giesecke & Jorde 1998), EVHA (Harby & Alfredsen 1999),

RSS (Souchon et al. 1998) etc. has been done (Zibiene 2000). Results show that the main trends of methods for ecological flow (or instream flow) estimation development in the different countries are as follows:

- Evaluation of habitat taking into account riverine morphology, flow regime and used simulation tools; use of advanced one-, two- and three-dimensional hydraulic - habitat computer models,
- Use of statistical and hybrid models in looking for links between hydraulics and biology,
- Development of energetic and individual fish based models.

The Lithuanian methodology for the ecological flow estimation, currently used in practice, may be referred to as empirical methods (Juozapaitis & Punys 1997). This methodology is based on the hydrology of a stream. According to this methodology an ecological flow Q_e is equal to the mean of minimum flows of 30 days in the dry period (April – October) of 95% or 80% probabilities ($Q_{30min95\%}$ or $Q_{30min80\%}$). These probability values depend on the river flow index φ, which is derived from the annual flow duration curve. Its values theoretically vary from 1 (very regular river regime) to 0 (very irregular river regime) and are mapped. The calculation is as follows:

For the rivers characterised by irregular river regime (i.e. $\varphi \leq 0.65$),

$$Q_e = Q_{30min80\%}$$

For the rivers characterized by regular river regime (i.e. $\varphi > 0.65$),

$$Q_e = Q_{30min95\%}$$

In the case of the derivation project a minimum ecological flow equivalent to 10% of the long-term average flow is set.

The estimated value of ecological flow may be corrected taking into account the possibilities to store the necessary water amount in the reservoir and water uses.

The advantages of this methodology are as follows:

- Possibility to divide the rivers into drying-out rivers (no need to estimate ecological flow) and rivers for which environmental flow has to be estimated as percentage from minimum flow (this part depends on river flow index φ),
- Indirect evaluation of ecological considerations of reservoir ecosystem taking into account the percentage of the surface area which appears on the banks of reservoir when the water level is draw downed,
- Application of methodology does not require

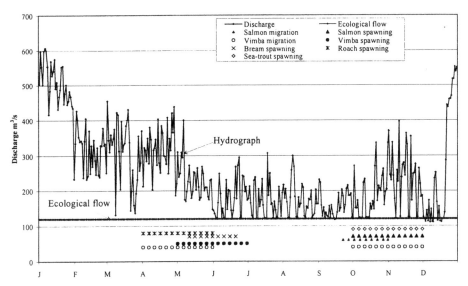

Figure 3. Hydrograph (year 1993) of the Nemunas below Kaunas HPP (P=100 MW), ecological flow and the time of migration and spawning of rare, protected and most valuable species of fish

survey, simulation; therefore the financial needs are very low.

Main disadvantages of this methodology are as follows:

- Methodology is based only on the river hydrology,
- Biological and ecological considerations are taken into account only indirectly.

The rivers like the Nemunas, Neris, Zeimena, Sventoji, Minija, Dubysa, Venta are the migration ways of rare and protected species of fish such as Atlantic salmon (*Salmo salar*), sea-trout (*Salmo trutta trutta*), vimba (*Vimba vimba*). Therefore, the estimation of different ecological flow values during the periods of spawning migration of the above mentioned fish species is suggested. This may be done by comparing the river hydrograph and time of migration and spawning of rare, protected and most valuable species of fish on the same scale (as showed in Fig. 3).

Established values of ecological flow of Lithuanian rivers were compared with values of minimum flows of the same rivers to maintain fish habitat requirements established according to Tennant, Montana, Tesman methods (see Fig. 4). The so-called "wetted perimeter" method have been used for ecological flow estimation of the Nemunas near Kaunas hydropower plant.

The comparison of the results shows that ecological flows established according to the official methodology in Lithuania in most cases are less than flows that meet minimal requirements of fish habitats.

4 CONCLUSIONS

1. Hydropower in Lithuania is regulated by a variety of laws, decrees or orders issued by Government or different Ministries. Only small hydropower (P<10 MW) regulations harmonized with the similar EU regulatory frameworks exist there. Efforts to enact the comprehensive hydropower legislative framework including both small and large hydro are currently under way.

2. Environmental legislation concerning hydropower development is well founded and clearly defined. Environmental Impact Assessment (EIA) procedures correspond entirely to those required by EU Directives and other regulations dealing with the EIA. The list of watercourses and their sections where damming or rehabilitation of old dams is prohibited exists there. Regardless its strong requirement there are quite a few sites for selecting hydropower projects event large ones (P>10 to 100 MW). However, the residual potential of regulated rivers must be exploited at first.

3. When developing new sites a particular attention has to be given to the migrating fish protection. A number of conventions and special programs to protect or restore stocks of endangered specious are made there. This issue constitutes a major barrier hindering country's hydropower development.

4. The simple empiric method, based on the hydrology of the river is used in practice to set up ecological flow value. By comparing it to the

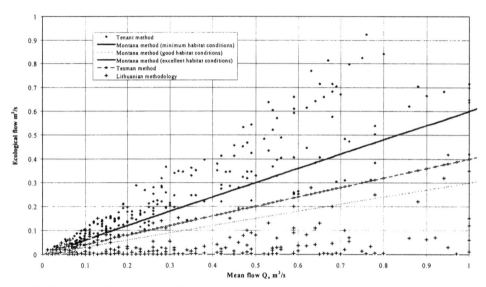

Fig. 10. Comparison of ecological flow (Q_e), estimated by the official Lithuanian methodology and other methods for small-dammed rivers

more advanced methods used in other countries, in most cases, the calculated minimum flow values are less. Therefore, there is the need to develop the methods based on biological and ecological considerations, especially for the rivers characterized by more concentrated aqua habitat.

5. Most hydropower schemes due to the relatively topography are of run-of-river type. So-called river diversions projects are usually exceptional. Therefore the losses resulting from maintaining ecological flow downstream of dams are insignificant.

REFERENCES

Bovee, K. et al. 1998. Stream habitat analysis using the instream flow incremental methodology. *Information and Technology Report USGS/BRD/ITR-1998-0004:* U.S. Department of the Interior, U.S. Geological Survey, Biological Resources Division. Fort Collins, USA.

Bukelskis, E. Kesminas, V. & Repecka, R. 1998. *Fresh water fish.* Vilnius: Dexma (in Lith.).

Burneikis, J. & Punys, P. 1999a. Hydropower: general framework for legislation and authorization procedures in Lithuania. *Renewable energy in agriculture, Proc. intern. conf. Kaunas, 16-17 September 1999:* 179-186.

Burneikis, J., Streimikiene, D. & Punys, P. 2001b. Technical, economic and environmental feasibilities of Nemunas hydro energy utilization *(see these proceedings).*

Giesecke, J. & Jorde, K. 1998. Simulation and assessment of hydraulic habitat in rivers: *Modeling, Testing & Monitoring for Hydro Powerplants – II;. Proc. Conf., October 1998, Aix-En-Provence, France:* The International Journal on Hydropower & Dams:71 – 82.

Harby, A. & Alfredsen, K. 1999. Fish habitat simulation models and integrated assessment tools: *An International Workshop on Sustainable Riverine Fish Habitat. Proceedings. 21-24 April 1999, Papers on CD-ROM.* 1999, Victoria, B.C., Canada.

IEA Hydropower Agreement: Annex III. Hydropower and the Environment. 2000. IEA technical report. Subtask 4. *Survey of existing guidelines, legislative framework and standard procedures for EIA of hydropower projects.*

Jablonskis, J. Punys, P., Tautvydas, A. & Savelskas, V. 1996. *Data book on small hydropower in Lithuania.* Kaunas. (in Lith.).

Juozapaitis, A. & Punys, P. 1997. Evaluation of environmental issues when constructing hydropower plants on existing dams. *HIDROENERGIA 97, Dublin, Sept. 29-Oct. 1, 1997:* 500-507.

Punys, P., Ruplys, B. & Vansevicius, A. 1999. Prospects for installing small hydro at existing dams in Lithuania. *Hydropower into the next century III, Proc. Conf., 18-20 October, 1999, Gmunden, 1999:* 99- 107.

EC DG for Energy XVII, ALTENER, ESHA. 1995. *Small Hydropower. General framework for legislation and authorisation procedures in the European Union.* Brussels.

Souchon, Y. et al. 1998. *Technical Specifications for the Assessment, Classification, Monitoring and Presentation of Ecological Status of Surface Waters. Implementation of The Water Framework Directive. Task: Habitat Quality.* Lion.

Strobl, Th., Maile, W. & Heilmar, Th. 1994. The problem of minimum flow in diversion type hydropower plants. *ESHA Info.* 14: 31-33.

Veyre, G. 1992. Débit réservé – bilan du 1/40 du module – passage au 1/10. *Insertion de Petites Centrales Hydroélectriques dans l'Environnement, Sophia Antipolis, 1et 2 Décembre, 1992:*Ademe :71-81.

Virbickas, J. 1986. *Lithuanian fish.* Vilnius: Mokslas (in Lith.),

Zibiene, G. 2000. *Substantiation of instream flow assessment methodology.* Kaunas. Doct. thesis (in Lith,. summary in Engl.).

Hydropower in the New Millennium, Honningsvåg et al (eds), © 2001 Taylor & Francis, ISBN 90 5809 195 3

Cameron Highlands Hydroelectric Scheme: Landuse Change – Impacts and Issues

F.K.Choy & F.B.Hamzah
TNB Engineers Sdn. Bhd., Kuala Lumpur, Malaysia

ABSTRACT: The Cameron Highlands Hydroelectric Scheme constructed between the late fifties and the early sixties is located in a hilly area in the Peninsular Malaysia. It consists of four run-of-river and one small storage hydro schemes.

The Ringlet reservoir impounded behind a 40-m high concrete buttress dam, forms an integral part of the Cameron Highlands Hydroelectric Scheme. The reservoir feeds water for hydroelectric power generation at the 100 MW Sultan Yussuf Power Station as well as at the 150 MW Sultan Idris II Power Station of the Batang Padang Hydroelectric Scheme located downstream.

The catchment for the Cameron Highlands Scheme has been subjected to increasing land clearing and development activities since the sixties. The construction of the scheme has provided improved access into the interior areas of the catchment in addition to facilities for electricity generation. As the development need and economic growth in the Cameron Highlands area increases, the development area demarcation is enlarged and more land clearing occurs. The development activities, which consist mainly of cultivation and urbanization, have led to widespread soil erosion over land surface, increased sediment load in rivers of the Cameron Highlands catchment and high sedimentation rate in the Ringlet reservoir. The current situation of land clearing and sedimentation problem are a cause for concern. The land use causes a gradual change to the environment. The cultivation and urbanization activities has affected the natural drainage of the Telom and the Bertam river basins of the catchment, including the water and sediment discharge of the basins which subsequently influence the operation and energy generation of the power stations.

This paper gives a brief description of the land use change and sedimentation problem in the Cameron Highlands catchment and presents some views on the land use impact, issues of catchment preservation and mitigation works.

1 INTRODUCTION

The Cameron Highlands Hydroelectric Scheme is situated in the northwest of the state of Pahang, Malaysia. It was constructed in the period between 1957 and 1964. The scheme consists of four small run-of-river and one storage hydro projects and has five power stations.

The main features of the storage project beside the 100 MW underground power station are a 40-m high concrete buttress dam with gated spillways, four side-stream diversion schemes of Sg. Plau'ur, Sg. Kial, Sg. Kodol, and Sg. Telom, some 20 km length of tunnels, the Bertam intake and other appurtenant structures. The location plan and layout plan of the scheme are as shown in Figures 1 and 2 respectively.

The Ringlet reservoir is a man-made lake created upstream of the concrete dam on Sg. Bertam and forms an integral part of the Cameron Highlands Hydroelectric Scheme.

The dam that is presently known as the Sultan Abu Bakar Dam, consists of a mass concrete gravity section on the left bank and a rockfill section on the right bank. It impounds the waters of Sg. Bertam and

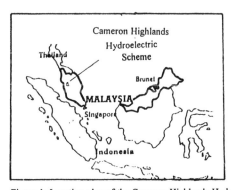

Figure 1: Location plan of the Cameron Highlands Hydroelectric Scheme.

Table 1: Powerstations, installed capacity and annual average energy generated in Cameron Highlnads Hydroelectric Scheme.

Power Station	Installed Capacity (MW)	Rated Head (m)	Annual Average Energy (GWh)
Kampung Raja	0.8	80	6
Kuala Terla	0.5	37	5
Robinson Falls	0.9	235	7
Habu	5.5	91	32
Sultan Yussuf (Jor)	100	546	320
Total	107.7		370

its tributaries and those of Sg. Telom, Sg. Plau'ur, Sg. Kodol and Sg. Kial which have been diverted from the Telom catchment through the Telom tunnel into the Bertam catchment.

The reservoir was impounded in early 1963 and is located about 570 m above the Jor Power Station. It is about 3 km in length with varying width and has a surface area of approximately 60 hectares. The designed gross storage of the reservoir is about 6.7 million cubic meters, of which, 4.7 million cubic meters is usable storage. Water from the Ringlet reservoir is channeled through a tunnel to the Jor power station and then is discharged through a tailrace tunnel into the Jor reservoir of the Batang Padang Hydroelectric Scheme.

At the time of study of the Cameron Highlands Hydroelectric Scheme in 1950's, the measured sediment contents of the rivers were not high. Sedimentation had not been anticipated to become a problem as it is now. According to the project study report (Service, 1956), the Ringlet Reservoir, which has an estimated dead storage of 2.0 million cubic meters, would have a useful life of approximately 80 years.

2 DESCRIPTION OF THE CATCHMENT

2.1 Topographical Characteristics

The Cameron Highlands catchment is situated on the Main Range of the Peninsular Malaysia. Unlike

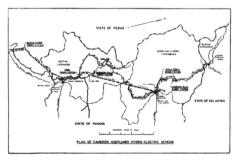

Figure 2: Layout plan of the Cameron Highlands Hydroelectric Scheme

other parts of the Main Range which is narrow and sharply defined, the Cameron Highlands area seems to broaden out into a dissected massif approximately 24 km long from north to south and averaging 6.4 km wide from the east to west. The massif generally consists of an inner zone of hills and valleys surrounded by a chain of mountain peaks. The average elevation of the catchment area is approximately 1180 m with the highest peak, Gunung Brinchang at approximately 2032 m above sea level and many of the other peaks are over 1524 m.

The massif area has developed physiographically as a separate unit lying high above the country to east and west of it, and it has experienced its own isolated cycles of erosion, deposition, rejuvenation, river capture and so forth. The upper reaches of the Sg. Telom and Sg. Bertam lie within the Cameron Highlands massif and exhibit unusual physiographic characteristics. The valley walls are steep and are benched by river terraces that occur at levels ranging from a few meters to several hundred meters above the present riverbeds. Such valleys are especially well developed in the valleys of the Sg. Telom (four distinct terraces at heights of 1158.5 m, 1181.4 m, 1105.2 m and, 1272.9 m) and its tributaries near Kuala Terla and the Sg. Bertam near Tanah Rata.

2.2 Geological Characteristics

Most of the Cameron Highlands area consists of granite. However, there are two sizable masses of schist in the southern half.

The granite is part of the Main Range batholith that forms the backbone of the Peninsular Malaysia and extends unbroken in the northwesterly direction for approximately 480 km from Malacca in the south to Thailand in the north. The schist masses are remnants from the granite erosion. The age of the schist is not certain but it is shown in the 1948 Geological Map of Malaya as being Carboniferous or Permian. The age of the granite itself is more closely known to be of late Mesozoic.

2.3 Catchment Characteristics

The Cameron Highlands catchment that contributes water to the Ringlet Reservoir includes several sub-catchments and has an area of 183 km^2.

The main rivers of the Cameron Highlands

Table 2: Detailed breakdown of the Cameron Highlands' catchment area.

Major Catchment	Sub-Catchment	Area (km^2)	Total (km^2)
Telom	Plau'ur	9.6	
	Kial	22.7	110
	Kodol	1.3	
	Telom	76.4	
Bertam	Bertam	73	73

Description	Catchment		
	Plau'ur	Telom	Bertam
Area of Basin (km^2)	9.6	100.4	73
Length of Basin (m)	5 400	16 000	14 000
Average Width of Basin (m)	4 000	10 000	8 000
Basin Perimeter (m)	16 937.75	50 389.92	61 307.31
Basin Circularity Ratio	0.426	0.498	0.467
Drainage density	4.305	4.052	Upper – 3.945 Lower – 4.587
Slope	0.0657	0.0415	0.0457

catchment, namely the Sg. Telom and Sg. Bertam drain eastwards into the Sg. Pahang and subsequently into the South China Sea in the eastern coast of the Peninsular Malaysia.

3 LANDUSE CHANGES

The construction of the Cameron Highlands/Batang Padang Hydroelectric Scheme in the sixties has provided in addition to the power generation facilities, some reasonable good access into the interior of the Cameron Highlands area. It may have therefore helped to encourage land development and other economic activities growth in the area during and post construction periods. Forest in the Cameron Highlands area is progressively being cleared to make ways for agricultural and construction activities.

Table 4a: Landuse changes in Bertam catchment

Vegetation/Land Use	Bertam Catchment (km^2)		
	1950's	1980's	1990's
Forest	46.5	45.1	43.5
Tea/Orchards	15.2	10.4	6.6
Vegetable/Flower	5.1	7.0	8.1
Urban	-	4.1	4.2
Open/grassland/Scrub forest	5.8	6.0	10.2

Table 4b: Landuse changes in Telom catchment

Vegetation/Land Use	Telom Catchment (km^2)		
	1950's	1980's	1990's
Forest	99.1	90.3	74.1
Tea/Orchards	6.3	6.2	6.2
Vegetable/Flower	5.0	10.7	23.2
Urban	-	0.5	1.1
Open/grassland/Scrub forest	-	2.7	5.8

The Structure Plan (1995 – 2000) prepared by the Local Planning Authority of the Cameron Highlands District Council in 1996 reports that in the period between 1950's and 1990's, agriculture development activities increased from 10% to 34% in the Telom valley and from 28% to 36% in the Bertam valley. The area of farm land in the Cameron Highlands district (including the Bertam Valley downstream of the dam) is reported to have increased from 4,816 hectares in 1991 to 5,125 hectares in 1993. About 40% of the farmland is tea plantation and the other 60% of the area is cultivated with vegetable, flower plants, fruit crops and other cash crops.

In the aspect of urbanization, the Structure Plan reports that the number of houses in the Cameron Highlands area increased from 3860 units in 1980 to 5526 units in 1991, 32% of which is found in the Tanah Rata area, 11% in the Ringlet area and the remaining 57% in other parts of the Cameron Highlands area. Many of the houses are holiday bungalows and apartments catered for holiday-makers and tourists. The number of tourist visiting Cameron Highlands in 1995 has been estimated to be about 320,000 and the number has been projected to reach 830,000 in the year 2005 and 1,340,000 in the year 2010.

About 700 hotel rooms and 185 units of condominium/apartments were completed in the last few years.

4 LANDUSE IMPACT

Extensive deforestation and indiscriminate earth bulldozing in the Cameron Highlands area for agricultural and housing development has resulted in widespread soil erosion over the land surface. The extent of soil erosion occurring in the area particularly that of Cameron Highlands scheme catchment is increasing, the impact of natural vegetation destruction on the environment and the stations' generation/operation is a major cause for concern. It leads to sedimentation of the streams and the Ringlet reservoir from which water is drawn to the power station. Excessive sediment particles drawn into the power station will cause damage and shorten the useful life of the mechanical parts of the generating plant, whereas excessive sediment deposited in the Ringlet reservoir will affect its storage as well as useful life of the reservoir.

There were incidences of water qualities and aquatic lives in the Ringlet reservoir being badly affected in the past when excessive sediment/silt content and high chemical content (e.g. pesticides and fertilizers) resulted in dying of fish and wide spread of water hyacinth. Cases of landslide have also been reported which have led to the destruction of properties and crops as well as loss of lives.

A number of studies were conducted in the past to

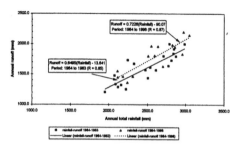

Figure 3: Rainfall-Runoff Correlation for Upper Bertam (at two different periods)

examine the effects of development in the hydro scheme catchment on the hydro power station. The studies (Choy, 1987, 1989, and 1991) have shown that the operation and maintenance as well as energy generation of the hydro stations of the Cameron Highlands scheme are affected by the land development. The severity of sedimentation has led to a study which reported that the carrying capacity of the Telom Tunnel had been reduced to 40% (WLPU, 1986)

A further study of the land use impact (Choy, 2000) and a review of some previous studies recently done, give the following findings:

□ Average annual temperature for urbanized area in the catchment has increased slightly in recent years.

□ There is an increase in the annual mean runoff for the catchment subsequent to the land use change (Figures 3 and 4).

□ Water utilization for generation decreases due to increased infiltration losses in dry days and spilling losses in wet days because of intake closing or increases in peak flow (Figures 5 and 6).

□ The suspended sediment load in the Sg. Telom and Sg. Bertam river have increased by twenty fold and seventeen fold respectively since 1960's (Figures 7 and 8).

Since its operation in 1963, the Ringlet reservoir has lost nearly 53% of its gross storage to sedimentation, which is presently estimated as reaching a volume of about 3.5 – 4.0 million cubic meters. Fig-

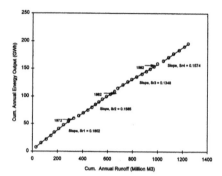

Figure 5: Double Mass Curve (Annual Runoff vs. Annual Energy Output) Robinson Falls Power Station (1964 to 1998)

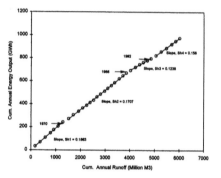

Figure 6: Double Mass Curve (Annual Runoff vs. Annual Energy Output) Habu Power Station (1964 to 1998)

ure 9 shows the storage capacity curve of the Ringlet reservoir. The currently estimated sediment deposition rate in the Ringlet reservoir is in the range between 350 000 to 400 000 cubic meters per year. Figure 10 shows the decline in trap efficiency over time for the Ringlet Reservoir calculated using the Brune's method on an annual basis, which shows the current trapping efficiency of the reservoir to be about 56 % as shown in Figure 10.

Most of the incoming sediment is therefore unlikely to be deposited in the Ringlet reservoir and will be carried down to the Jor station and discharged into the Jor tailrace and reservoir. Presently,

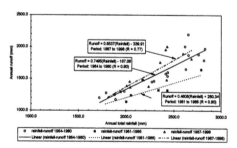

Figure 4: Rainfall-Runoff Correlation for Sg. Telom (at 3 different periods)

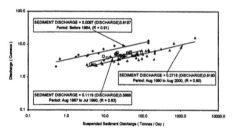

Figure 7: Sediment Rating Curve – Sg. Telom at 49 M.S. (before 1964 and from August 1987 to now)

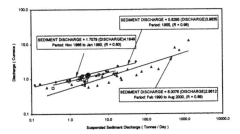

Figure 8: Sediment Rating Curve – Sg. Bertam at Robinson Falls (before 1955 and from November 1986 to now)

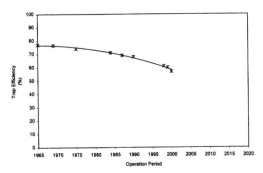

Figure 10: Ringlet Reservoir – Trap Efficiency vs. Operation Period

45.2% of the 3.9 million cubic meters gross storage of Jor reservoir is filled with sediment.

A study in 1999/2000 (SNC-Lavalin, 2000) indicates that the sediment deposition in the Ringlet reservoir has not affected the safety of the dam. However, the substantial reduction in the live storage of the reservoir would largely reduce the capability of the Jor station to generate peaking power for the load requirement of the National Grid and would also reduce the capability of the reservoir to regulate incoming high flow. The former will make the Jor station to be no different from other run-of-river stations and will result in financial loss; the latter will pose a risk of the dam being without flood control capacity, hence leading to frequent spillage and

flooding of the downstream Bertam Valley during monsoon.

5 MITIGATION AND REHABILITATION WORKS

Since the early seventies, mitigation measures (Choy & Fuad Omar, 1990; Choy & Fauzan, 1997) have been carried out periodically at various locations of the Cameron Highlands scheme to minimize the impact of sedimentation on the operation and maintenance of the five hydro stations. These measures include:

☐ Manual removal of sediment accumulating behind the intake weirs of Kampung Raja, Kuala Terla and Habu.
☐ Construction of a small timber silt retention weir on Sg. Ringlet in the late seventies and a concrete silt retention weir downstream of the Habu Power Station in the early nineties.
☐ Pumping and excavation of sediment accumulating at upstream of the Telom intake and Robinson Falls intake and in the Ringlet and Habu silt retention ponds.
☐ Desilting of the Telom tunnel in the seventies and again in the early nineties.
☐ Construction of a new desander in front of the original Telom intake structure to improve the sediment self-scouring efficiency.

As a result of the above mitigation measures, the inflow and accumulation of sediment in the Ringlet reservoir has been checked and reduced by a significant amount as shown in the following Figure 11.

☐ If there is no action at all to check the sediment inflow, the reservoir will theoretically lose its entire storage before or by the end of the year 2000 due to sedimentation.
☐ The desilting works carried out during the period between mid-seventies and early nineties appears to have prevented more than 1.5 million cubic meter of sediment from entering and accumulating in the Ringlet reservoir.

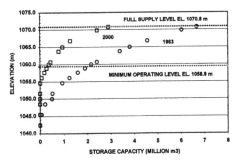

Figure 9: Ringlet Reservoir – Reservoir Capacity Curve (1963 vs. 2000) (Choy & Hamzah, 2000)

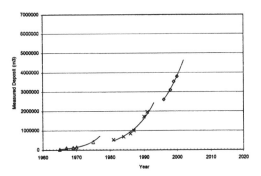

Figure 11: Ringlet Reservoir – Sediment Deposition Rate

□ The completion of the Telom new desander and the Habu silt retention weir and subsequent desilting works has apparently prevented at least another further 1.5 million cubic meter of sediment from accumulating in the Ringlet reservoir during the last few years.

6 LANDUSE AGREEMENT, LEGISLATION AND ISSUES

The 1959 Agreement signed between the State Government of Pahang Darul Makmur and the Tenaga Nasional Berhad (then National Electricity Board) stipulates among others the need for preservation of the catchment and the discouragement of uncontrolled or improper development. To achieve these objectives, the Agreement includes some attachment providing useful guidelines to land development on hill slopes such as:
□ Methods of clearing vegetation
□ Methods of establishing vegetative cover
□ Selection of types of crops to be grown
□ Choice of subsequent cultivation
□ Mechanical methods of planting

The Land Conservation Act 1960 also contains the provision for control of hill land cultivation and control of silt and erosion in development.

The above Agreement and Act are evidently not effective in restraining any uncontrolled development and ensuring the adoption of proper method of cultivation in the Cameron Highlands area. Cash punishment or fine has failed to deter the farmers from indiscriminate leveling of hill land. The land tenure system of issuing Temporary Occupation License (TOL) for agricultural land that is subjected to annual review and renewal, does not help to encourage land conservation practices. The fact that water and land matters are within the jurisdiction of the State Government makes the advisory role of the Tenaga Nasional Berhad ineffective in ensuring proper watershed management and control of catchment developments.

Notwithstanding the requirements of the Land Conservation Act and the Environmental Quality Act, three major road projects in the Cameron Highlands area are at the various stages of planning and implementation. These are:
□ Proposed Cameron Highlands – Kuala Lipis Road.
The road will link the Bertam valley at downstream of the Ringlet Dam to Kuala Lipis and Raub in the east.
□ Proposed Cameron Highlands – Bukit Fraser – Genting Highlands Road.
The road will connect Cameron Highlands to the other two hill resorts of Bukit Fraser and Genting Highlands in the south. The entire length of this road will be situated on high elevation and in hilly areas.

□ Simpang Pulai – Cameron Highlands – Lojing, Gua Musang Road.
This proposed road is presently under construction. It connects Kampung Raja in Cameron Highlands with Simpang Pulai in Perak and Lojing/Gua Musang in Kelantan then further west with Kuala Berang in Terengganu. The road forms a second east-west link across the Peninsular Malaysia.

The construction of the above three roads will certainly worsen the soil erosion and sedimentation problem already prevailing in the Cameron Highlands area.

The Structure Plan of Cameron Highlands also includes provision for expansion of the existing development site boundary and increase in the number of hotel room and apartment units.

7 CONCLUDING REMARK

Cameron Highlands is one of the few hill resorts in Malaysia, which are frequented by Malaysian public and foreign tourists. It is also a main area for producing tea, vegetable and flower for domestic market and export. These two factors perhaps account for the occurrence of growth in the property and agricultural development in the Cameron Highlands area. The incident of soil erosion and sedimentation occurrence is indeed a setback to the requirement by the authority for land and environmental preservation.

The source of sediment yield is mainly at the development sites where soil erosion occurs. The provision of proper silt traps and soil conservation works on the development sites in the Cameron Highlands area is the most logical and cost effective solution to the sedimentation problem in the area. The failure in observance of the Act (by the farmers) and the lack of strict enforcement of the law (by the Authority) has led to the problem of sediment accumulation occurring at the locations of the intakes of the several hydro schemes and the Ringlet reservoir. Substantial cost has been incurred by Tenaga Nasional Berhad to mitigate the impact of sedimentation on the Cameron Highlands scheme.

The desilting works, Telom new desander and Habu silt retention weir completed in the period between the seventies and early nineties has produced positive and fruitful results of reducing the inflow of sediment into the Ringlet reservoir. Nevertheless given that:
□ The mitigating measures are only able to lessen but not stop the sediment accumulation in the Ringlet reservoir.
□ Nearly 53% of the storage capacity of the Ringlet reservoir has already been filled with sediment
□ There is plan to expand the development area in the Cameron Highlands area in the future.
It is prudent to take early action on the following:

- ☐ To construct another silt retention weir on Sg. Ringlet at the Ringlet end of the Ringlet reservoir so as to further reduce the sediment inflow into the reservoir.
- ☐ To examine the feasibility of desilting the reservoir so as to regain portion of the lost storage capacity due to sedimentation.
- ☐ To implement a watershed management plan and to impose the requirements for soil conservation practices in all future development.

8 ACKNOWLEDGMENT

The authors would like to thank the Management of Tenaga Nasional Berhad (TNB) for the permission to publish this paper. Any views expressed in the paper are the authors' and do not represent those of TNB Management.

REFERENCE:

Cameron Highlands District Council. *Rancangan Struktur.* (1996).

Choy, Fook Kun. *Cameron Highlands/Batang Padang Hydroelectric Scheme. Generation Performance Assessment.* Bulletin IEM. (1987)

Choy, Fook Kun. *Cameron Highlands Hydroelectric Scheme – A Study of River Sedimentation Load and Reservoir Sedimentation.* Bulletin IEM. (1989)

Choy, Fook Kun and Omar, Mohamad Fuad. *Cameron Highlands Hydroelectric Scheme – The Sedimentation Problems.* Proceedings of IEM Seminar on Hydropower and Flood Mitigation Project without Dams – Prospect and Issue. (1990.)

Choy, Fook Kun. *Cameron Highlands Hydroelectric Scheme – Effect of Development of the Telom and Upper Bertam Catchment on Water Yield and Energy Generation.* Bulletin IEM. (1991)

Choy, Fook Kun and Hamzah, Fauzan. *Cameron Highlands Hydroelectric Scheme – Performance of the Ringlet Reservoir.* (1997)

Choy, Fook Kun. Master of Philosophy Thesis (Draft) "Land Use Impact on Annual Runaoff and Energy Generation For the Cameron Highlands Stations. (2000)

Service,Harold. Director General of Geological Survey, Federation of Malaya, *Geological Report on Cameron Highlands Stage 1.* (1956)

Gupta, Ram S. *Hydrology and Hydraulic Systems.* Prentice Hall (1989).

Institute of Postgraduate Studies and Research (IPSR), University of Malaya. *'The Development of Hydroelectric Catchment Management Information System for Cameron Highlands-Batang Padang Hydroelectric Scheme* (2000)

SNC-Lavalin. *Cameron Highlands Batang Padang Hydroelectric Rehabilitation Project, Sedimentation Study* (2000)

WLPU Consultants. *Telom Tunnel and Intakes. Preliminary Observation on the Silting Problem.* (1986)

Hydropower in the New Millennium, Honningsvåg et al (eds), © 2001 Taylor & Francis, ISBN 90 5809 195 3

Upgradation of small hydro power plant at Maniyar, India

Devadutta Das
Director & Professor (Hydroelectric), WRDTC, University of Roorkee, India

R. Muraleedharan Nair
Plant Manager, MHEP, Carborundum Universal Limited, India

A.Sivakumar
G.M. (Corporate Technology Group), Carborundum Universal Limited, India

Nayan Sharma
Associate Professor, WRDTC, University of Roorkee, India

ABSTRACT : Carborundum Universal Limited's (CUMI) small Hydroelectric Power Plant at Maniyar was designed to generate maximum power of 12 MW at head of 14.5 meters and discharge of 100 cumecs. During actual running, the plant was not able to generate beyond maximum of 8.0 MW. The discharge was limited to 80 cumecs and the design forebay level with design discharge could not be maintained. Investigations carried out, brought out the inadequacy of design and construction of Water Conductor System. Remedial measures based on analytical modelling were validated by physical modelling. Execution of the improvements on the existing running plant was done with minimum effect on power generation. The paper brings out the methodology and execution of upgradation of existing small Hydroelectric Power Plant.

1 INTRODUCTION

CUMI belongs to Murugappa Group based at Chennai, India, having strong presence in various businesses such as Abrasives, Electrominerals, Refractories, Ceramics, Fertilizers, Sugars, Engineering, Plantation etc, with an annual turnover of above 700 million US dollars.

The company manufactures Electrominerals and Refractories in Kerala, and is principally dependant on power.The manufacturing operations have been supported by pioneering efforts towards establishing power plant based on renewable energy by setting up of run of the river small (12 MW) hydropower plant at Maniyar across Pambar river. (Figure 1)

The project utilizes the surplus water available at Maniyar barrage after meeting the irrigation needs of Pambar Irrigation Project. The power plant has a plant load factor of 30% utilizing the run-off of seasonal rains and diversion of water for irrigation during the dry months. The project included construction of 350 meters long tunnel to carry a maximum of 100 cumes of water from the pond to the forebay. The project consists of 3 nos. of Kaplan turbines of 4 MW capacity each. The plant was commissioned in 1994.

Though the performance of the mechanical and electrical equipment was found to be good, the plant could not develop beyond 8.5 MW. The maximum flow that could be passed through the water conductor system was around 80 cumecs only. Any additional inflow of water resulted in lowering of the forebay level and a net head below the permissible minimum operating head of 12.5 meters. Non utilization of

additional flow of water led to increase in pond level, forcing gates of the barrage to be opened. As a result, water in tailrace level shot up, reducing effective working head and reducing power generation to around 4.0 MW.

Having stabilised the operations for more than 3 years, CUMI started to look at improving the power generation. In this connection, they approached WRDTC (Water Resources Development Training Centre, Roorkee) to study the situation and recommend suitable improvements to improve the power generation.

2 INVESTIGATION INTO THE CAUSES OF POOR PERFORMANCE

Detailed study of the existing operations was made and plant operating records were gone through.

Table 1. shows the values of various parameters of the project in the design and those actually achieved and measured during operation.

The entire length of the water conductor system from the intake up to the intake structure and tail race channel was thoroughly inspected to identify the probable causes responsible for higher than predicted head losses in the water conductor system. (See Figure.2) Inspections revealed the probable causes attributing to high head losses which as mentioned below :

a) Wing wall projections at the intake entrance causing

Figure1. Maniyar Hydro electric Power Plant-Overall View

obstructions to flow and also causing flow separation.

b) Unequal and non-prismatic expansions and contractions of intake to tunnel creating pockets for eddy formation.

c) Occurrence of pronounced flow separation along with wing wall near the entrance to tunnel due to turbulence and vortex formation.

d) Non-stream lined pier noses at the entrance.

e) Tunnel - walls and bed highly undulated and with projections

f) Eddy formation in the forebay due to steep slope from the tunnel exit up to intake structure, irregular

surface of the forebay walls (unlined rock surface) and non-uniform divergence of the side walls.

g) Protrusions in the tailrace near the right side wall impeding smooth flow expansion.

h) Marked indulations in the bed of the tailrace.

i) Backwater effect in the tailrace during floods.

j) Backwater effect in the tailrace during floods.

Head losses in the water conductor system from the intake up to the trash rack in the power dam were calculated using established relationships for different discharge values. In order to check the accuracy of the predicted performance by mathematical analysis, it was decided to conduct a model test before applying the suggested modifications at site to improve the performance.

A geometrically similar physical model was constructed with the scale of 1:15 for horizontal and vertical dimensions. The barrage was also modeled with the similar scale of 1:15. Since the bed in the prototype is of hard rock (granite), the bed of the model was also constructed as a fixed bed with concrete lining. The discharge scale for the model worked out to 1:871.4.

Table 1. Designed and observed parameters of the project

Sl. No.	Parameter	Design value (for 100 cumecs discharge)	Observed value (for 80 cumecs discharge)	Remarks
1	FRL at barrage	34.62 m	34.62 m	
2	Forebay level	34.27 m	33.3 m	
3	Maximum tail water level	19.3 m 23.0 m 18.1 m	19.9 m	With no spill With spill Minimum TWL
4	Average gross head	14.97 m	13.4 m to 14.0 m	
5	Head loss in water conductor system	0.35 m	1.32 m	
6	Net head	14.15 m	12.55 m	
7	Maximum power output	12 MW 4 to 5 MW	8 to 9 MW	With no spill With spill

Table 2. Head Loss in Water Conductor System

| Sl. No. | Segment | | Head Loss for Discharge in meter | | | |
| | | | 80 Cumecs | | 100 Cumecs | |
			Analytical	Actual on model	Analytical	Actual on model
1	A-A to F-F	Intake	0.423	0.40	0.680	0.50
2	F-F to H-F	Tunnel	0.661	0.83	1.041	1.30
3	H-F to I-I	Forebay	0.148	0.12	0.230	0.20
		Total	1.232	1.35	1.951	2.00

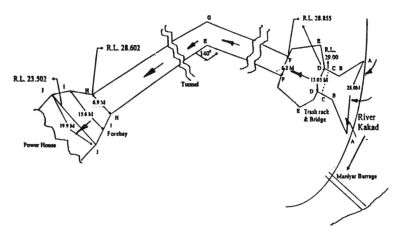

Figure 2. Layout of Water Conductor System

Table.2 shows the head loss as calculated and as measured on the physical model during test for two sets of flow conditions namely 80 cumecs and 100 cumecs.

The head loss as actually measured at site was 1.32 m (refer Table 1) against a measured value of 1.35 m on the physical model which proved the accuracy of the model constructed for the purpose

3. EVALUATION OF IMPROVEMENTS

Since the head loss encountered in the intake channel tunnel and the forebay were much higher than the values projected in the design, it was concluded that improvement in hydraulic performance could be achieved only by modifying the design of the intake channel and effecting remedial measures in the tunnel and forebay. The fact that the structures and machines were installed and were in operation, placed constraints on the modifications and renovations that could be effected. Keeping the foregoing fact in view, the following modifications and remedial measures were considered so that major construction work is not involved and also the work is completed during the lean flow period of about four months so that loss of revenue on account of loss of generation is minimal.

Remedial measures considered are :

a) Modification of the intake channel up to tunnel entry
b) Lining of the tunnel
c) Modification to the forebay
d) Lowering and leveling of the tailrace channel before confluence with the river

3.1 Modification to the Intake Channel

As mentioned earlier, existing intake was a source of head loss due to the expanding and contracting shape which contributed to formation of eddies and flow separation zones.

The hydraulic phenomena observed in the intake channel were :

(i)   Unequal flow distribution along the length of the intake
(ii) Vortex formations near the approach wing walls
(iii) Formation of eddies and back flow in the cavity in front of the tunnel mouth.

To achieve smooth hydrodynamic flow condition and to avoid conditions conducive to formation of eddies and vortex pockets, the following alternative proposals were considered :

Alternative Proposal I :

Providing an alignment resembling S-curve to side walls after trash rack till tunnel mouth to the existing structure

Alternative Proposal II :

Designing the intake tentatively as per Chaturvedi's [3] transition.

Alterntive Proposal III :

Shaping the intake as per Hind's [3] transition.

Proposal II when considered, resulted in increase in the mouth of the intake and increase in trashrack area to accommodate more flow. This proposal was hence dropped out of consideration as it involved demolition of existing intake and trash track and construction of new structures and trash track.

Proposal III was also dropped out of consideration as the intake needed widening of the mouth involving demolition of existing structures.

Proposal I was found to be the best possible solution as it was found feasible to provide a smooth S-curve simply by filling up the pockets. It also did not involve demolition of the existing structures. The

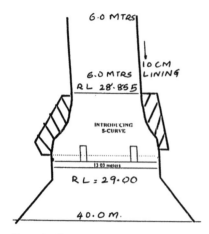

Figure 3. S-curve at the Intake

proposed modification to the intake channel is shown in Figure 3.

3.2 Smoothening of tunnel bed and walls

The tunnel has been constructed through sound granite rock. But while constructing, there were undulations in bed and walls. In order to meet the deadline and not to miss one year of power generation,only 5% of the tunnel was lined during execution of the Project, not realizing its impact on long term basis. The tunnel is a free flowing tunnel.

Due to severe irregularity on the surface, the frictional head loss was predominant. The only feasible solution that could be considered was to provide a 100 mm thick cement concrete lining (1:1.5:3).

3.3 Modifications in the fore bay

Major modifications in the fore bay were ruled out due to high costs involved, which were not commensurate with increased revenue on account of resulting higher energy production. Smoothening of the forebay walls by removing projections in front of turbine no.3 intake by use of jack hammers was therefore considered adequate.

3.4 Modifications to tail race channel

Since the bed configuration of the tailrace channel was undulated up to the confluence with the river, it was proposed to deepen the floor level to RL 16.5 m and provide a concrete bed.

With the velocity of 2 meters per second and with Manning's coefficient (n) as 0.02, the tail water depth was calculated to be 2.6 meters for a discharge of 100 cumecs. Consequently, the tail water level worked out to RL 19.1 m (= RL 16.5 + 2.6) against RL 19.9m. Thus by modifying the tail race channel as mentioned above, it was possible to gain in head by 0.8 m which would contribute to higher generation.

4 RESULTS OF THEORETICAL EVALUATION OF HEAD LOSS AND FROM MODEL TEST

4.1 The head loss was calculated by using established relationships for the modified intake and lined tunnel as mentioned in the foregoing paragraph.The modifications as contemplated in the foregoing paragraph were also implemented on the physical model and tests were conducted.

The results of theoretical calculation and model tests after effecting the modification as outlined above are mentioned in Table 3.

Table 3. Head loss in Water Conductor System after modifcation to intake and lining of tunnel

Sl. No.	Segment	Head loss in meter after modificatio for 100 cumec discharge	
		Calculated	From model studies
1	AA to FF Intake	0.090	0.400
2	FF to HH Tunnel	0.394	0.200
3	HH to II Forebay	0.120	Negligible
	Total	0.604	0.60

4.2 Implementation of modifications suggested

The modifications proposed and confirmed on the model test resulted in a head loss of 0.60 metre upto the power dam with the design discharge of 100 cumecs. Consequently, the fore bay level would be RL 34.02 m when the FRL at the barrage is RL 34.62 m. With the existing structural features, the fore bay level dropped to RL 32.5 m when a discharge of 100 cumecs passed through the water conductor. Hence the modifications suggested proved by model tests indicated increase in head by 1.52 m (= RL 34.02 m – RL 32.5 m).

5 PRACTICAL PROBLEMS FACED DURING THE EXECUTION OF THE PROJECT

5.1 The execution of the project be divided into four phases:

(i) Facilitate survey of the entire site (before analysis by WRDTC)
(ii) Study the extent of lining work to be carried out over the side walls of the tunnel and near its entry and exit (after receipt of report from WRDTC)
(iii) Try out cross polymer lining as an innovative measure to replace tunnel lining work
(iv) Execute civil work as recommended

All the above jobs have to be done with minimum disruption to the actual working of the plant. Otherwise the loss of power generation will be tremendous and also add to the cost of upgradation work.

The chart given below gives the seasonal inflow of water and consequent power generation. As seen from the chart, the maximum power generation is achieved during monsoon months of July, August, September, October, the lean months being January, February, March, April and May

5.2 *Planning and execution*

The project, the first of its nature on running plant was executed through detailed planning and coordinated efforts of all agencies. The major aim is to implement the recommendations with minimal disruption to current operations and generation of power.

The activities are broadly grouped under :

(i) Ordering of assignment on consultancy
(ii) Mathematical and physical modelling
(iii) Planning for implementation
(iv) Execution of the project

The activities were started in November 1997 and completed by May 2000.

For initial proposal to use imaging through

satellite was dropped in view of the cost involved. Instead, survey was done by using a boat to go across the entire pond area and tunnel. During the operation, the plant was shut off. Except for minor disruption of work, no problem was experienced during this part of activity.

For lining purpose, estimation of work on lining of tunnel was required. Since the tunnel had no gate, coffer dam of two meters height , has been constructed at the inlet. During this period, inflow of water into the pond was low and excess water was let down through the barrage. As such there was no problem in completing the coffer dam. The entire tunnel area was divided into number of segments and the surface irregularities were measured both vertically and horizontally and entire mapping was done subsequently to find out the extent of filling to be done. The coffer dam was then dismantled and regular work was started immediately causing minimum disruption of power generation.

5.3 *Trial on Polymer Liner*

Earlier, during the course of analysis of upgradation, an idea has been thrown up to provide a lining made of cross polymer to reduce the cost of upgradation and time spent on the same. The liner is to be kept on both sides of the wall to serve as a substitute for concrete liner. The idea has further developed by Mr. B.N. Sreedhara, Free Lance Structural Engineering Innovator from Bangalore. The cross polymer sheet is hung parallel to the tunnel wall. As water is present on both sides, there is no side thrust and water flow is smooth and streamlined due to flexibility of sheet and low co-efficient of friction. The sheet is expected to last about 4-5 years and will have to be replaced periodically. Its cost will be less than 10% of the concrete lining work and the time taken will be one week as against two months for lining work. (See Figure .5)

The cross polymer liner failed after three days of

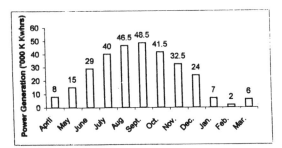

Figure 4.Seasonal Variation in generation of Power

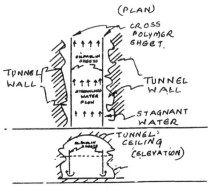

Figure 5. Water flow with cross polymer (sirpaulien) liner

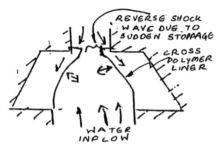

Figure 6.View explaining failure of Polymer Liner

successful operation. The investigation revealed the following :

(i) Whenever the turbines tripped due to grid condition, the back pressure led to heavy pressure of water locked between the wall the liner, against the clamps on to which the sheets are hooked.

(ii) Further at the inlet stage, the area of water flow was reducing and side thrust was felt by the liner.

Based on the above, it was concluded that the cross polymer liner would work under following conditions :

(i) The polymer liner should be placed only in straight portion and at its ends, concrete filling is to be done to prevent rise in pressure of water in between liner and the wall.

(ii) Where cross section changes or direction of flow changes, concrete lined wall only should be used and cross polymer liner should be avoided here. Schematic sketches of proposed method is given in Figure7.

CUMI however did not want to risk putting cross polymer liner and decided to adopt conventional concrete lining of walls.

5.4 Coffer Dam of six Meters height

The actual upgradation work was started in February 2000. The first hurdle was : the Coffer dam which was well in front of intake was to be raised to the height of 6 meters. The conventional method involves huge cost and time. Hydro bags were considered since they were not available in India, the idea was dropped. After design analysis by civil engineers, it was decided to use the existing civil structure for the water intake and build coffer dam around it as shown in Figure8.

Both the thrust and bending moments had been considered while constructing the coffer dam.(See Figure 8.) The bags were placed in front and rear side of the trash rack located in the intake area. The load was shared by the existing concrete structure, trash rack and the sand bags.

Apart from the technical problem, certain social and infrastructural problems affected execution of the project. Annual religious convention used to be held on the river bed down stream for a fortnight and during this period water flow should be avoided. Hence, work at coffer dam had to be done hurriedly to meet the deadline.

The inter-union labour problems during the upgradation work also caused some delay which have been overcome through negotiation. Unseasonal rains created further delay in transportation of material for lining from outside.

5.5 Project Completion

Due to participative and tactful management, the project was completed restricting the over run to 15 days beyond the schedule. It was facilitated by the decision to skip partial lining of the tunnel wall having effect of head loss of 10 cms only. This work is proposed to be taken during next year.

The project being first of its nature, has taken thirty months for completion. However, similar projects in future can be executed within nine months.

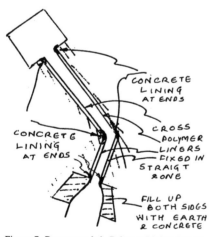

Figure 7. Recommended Polymer Lining System

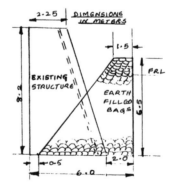

Figure 8. Design of Coffer Dam

6. PERFORMANCE AFTER UPGRADATION

During the first year after upgradation, the project has shown promising results. Values of head loss before and after upgradation are in given in Table 4 below.

Table 4. Head loss in Water Conductor System before and after modifications

Sl. No.	Discharge in (cumecs)	Observed Head Loss (in meters)	
		Before Modifications	After Modifications
1	33	0.178	0.100
2	66	0.711	0.400
3	80	1.232	0.500
4	92	1.900	0.700

Maximum power generation of 10.5 MW was achieved after up gradation as against previous figure of 8.5 MW of power .The model tests showed an improvement in available head by 1.52 Metersat 100 cumecs of flow ,and the actual gain at 92 cumecs of flow was 1.2 Meters.With the completion of 5% OF pending lining work next year and water flow 100 cumecs , next year,the gain as per test will be achieved.

7. CONCLUSIONS AND RECOMMENDATIONS

(i) While designing and constructing mini and small hydroelectric stations, there has been a tendency on the part of the designers and the developers to resort to compromises in order to simplify the schemes and save time and expenditure. The impact of such simplification on long terms is overlooked during the process.

(ii) Location of intake and its alignment vis-à-vis the river flow is very crucial for its hyudraulic performance. It has been observed at the intake site of this project that two distinct stream flows— one from the river and the other from the direction of barrage (located downstream of the intake) are converging in front of the intake mouth creating a vortex. Such a hydraulic condition is not desirable which causes loss in head. Since it was not possible to effect change in the orientation of the intake, it has been left as such with modifications limited to those areas where these were physically possible.

(iii) Existing small hydro power plants can undertake study to determine the potential for upgradation.Based on recommendations for upgradations, it can execute the scheme with marginal disturbance to existing operations.

8. REFERENCES

Das, D. 1999.Report on Upgradation of generation at Maniyar small Hydroelectric station(unpubl.), Water Resources Development Training Centre, University of Roorkee, Roorkee, India.

Albertson, L.M & Kia, A.R 1989. Design of Hydraulic Structures

Ceballos Pacheco, R. Energy loss in Channel Bends, Theory and Application : pp341-347

Varshney, R.S et al. Theory and Design of Irrigation Structures, Nem Chand & Sons, Roorkee, India

Arora, K.R, 1996. Irrigation Water Power and Water Resources Engineering, Standard Publishers & Distributors, New Delhi, India

Hydropower in the New Millennium, Honningsvåg et al (eds), © 2001 Taylor & Francis, ISBN 90 5809 195 3

Mitigation measures for environmental protection – sediments and fisheries in relation to the Kjøsnesfjorden Hydropower Project, Norway

S.E.Haarklau
Natural Resources Division
NORPLAN A.S., Consulting Engineers and Planners, Oslo, Norway

E.T.Pettersen
Power and Renewable Energy Division
NORPLAN A.S., Consulting Engineers and Planners, Oslo, Norway

ABSTRACT: The objective of this paper is to illustrate how timely and thorough environmental studies which are integrated into technical project planning can assist in making hydropower projects environmentally acceptable from a fisheries point of view, even in sensitive areas with important fish stocks.

Major concerns were initially raised about the proposed Kjøsnesfjorden Hydropower Project. In particular, negative impacts on genetically unique Brown Trout (*Salmo trutta*) populations and the most valuable Brown Trout inland fishery in Norway were presented as major arguments against developing this project.

The Environmental Impact Assessment (EIA) has since shown that the probability of significant negative impacts on fish populations and fisheries from developing this project are unlikely. The integrated approach using both environmentalists and engineers in the planning and EIA process facilitated the development of effective mitigation measures. The implementation of these measures could in fact lead to a slight, long-term increase in fish production and improvement in fish quality in Lake Kjøsnesfjorden.

1 INTRODUCTION

1.1 Background

The objective of this paper is to illustrate how timely and thorough environmental studies which are integrated into technical project planning can assist in making hydropower projects environmentally acceptable from a fisheries point of view, even in sensitive areas with important fish stocks.

The proposed Kjøsnesfjorden Hydropower Project (KHP) will affect a lake that supports the most valuable inland, Brown Trout fishery in Norway. The local Brown Trout populations also posses unique local adaptations and therefore are of particular scientific interest and receive special attention from the environmental management authorities as well as environmental NGOs. Before the full EIA was carried out significant negative impacts were expected on the fish populations, fisheries, tourism and recreation dependent upon the fish stocks.

1.2 Hydropower development in Norway

Hydropower is one of Norway's major natural resources. During the past hundred years of major hydropower development Norway has had the highest per capita electricity production in the world, and virtually all of this from hydropower.

Norway has developed top expertise within a range of fields related to hydropower development, including the environmental aspects.

2 AREA AND PROJECT DESCRIPTION

2.1 Location and physical features

The proposed Kjøsnesfjorden Hydropower Project is located in Jølster Municipality, Sogn and Fjordane County in Western Norway (see Figure 1). Parts of the project area are within the Jostedalsbreen Glacier National Park.

The KHP will collect water from 11 steep streams around Lake Kjøsnesfjorden, an arm of Lake Jølstravatnet (see Figure 2). A small lake in the mountains east of Lake Kjøsnesfjorden (Lake Trollavatn) will act as the reservoir for the project and the tail-

Figure 1. Location of the Kjøsnesfjorden Hydropower Project, Sogn and Fjordane County, Western Norway.

race tunnel will return the water to Lake Kjøsnesfjorden. Lakes Kjøsnesfjorden and Jølstravatnet are used as reservoirs for two other hydropower stations, 8 and 17 km downstream the outlet of Lake Jølstravatnet. A few details about Lakes Kjøsnesfjorden and Jølstravatnet are presented in Table 1 below.

Table 1. Physical data for Lake Kjøsnesfjorden and Lake Jølstravatnet (modified from Sægrov 2000a).

Parameter	Lake Kjøsnesfjorden	Lake Jølstravatnet
Area (km^2)	7.7	32.2
Volume (mill. m^3)	472	3076
Max. depth (m)	149	233
Average depth (m)	61	95
Catchment area (km^2)	84	384
Average annual runoff (mill. m^3)	267	927
Water retention time (yr)	1.77	3.32
Average flow (m^3/s)	8.5	29.5
Regulated water levels (m a.m.s.l.)	HRW: 207.35 LRW: 206.10	HRW: 207.35 LRW: 206.10

2.2 Hydrology and sediment transport

The total catchment area for the KHP[1] is about 33.7 km^2, of which 17.2 km^2 consist of glaciated areas and are covered with permanent snow and ice. The catchment is located above 1000 m a.m.s.l., with the highest areas about 1650 m a.m.s.l. There is only minimal human activity (recreation) in the catchment and no permanent settlements. Annual average discharge from the rivers supplying the hydropower project is 4.01 m^3/s and annual average runoff is 126.3 mill. m^3.

As about 51 % of the catchment consists of glaciers, the water contains a significant amount of sediments generated by the glaciers' erosion. The majority of sediments enter the rivers and lakes between late June and October. This is the warmest time of the year when melting of snow and ice increase river flows and sediment transport. Sediment load in glacier fed rivers is extremely variable, and the amount of suspended solids in Norwegian glacier fed rivers has at times been estimated to 800 g/m^3. During the winter and spring months (November to May) river flow and sediment load are very low.

2.3 Terrestrial ecology

The majority of the Kjøsnesfjorden catchment is covered by permanent snow, ice and bare rock. The project area has limited vegetation cover and terrestrial flora and fauna species diversity and abundance are low.

2.4 Aquatic ecology

2.4.1 Water quality

Water quality is generally good, with low or insignificant levels of man-made pollution. Long-range, transboundary air pollution has resulted in "acid rain" and loss of freshwater biodiversity in large parts of Southern Norway. However, the project area has so far not been severely affected by acidification.

Plant nutrient levels are generally low, with average summer values for total phosphorous around 4-6 μg P/l and total nitrogen about 180-200 μg N/l.

Due to the high sediment load from the glacier fed streams turbidity is high during summer and autumn months (June - October) compared to typical Norwegian conditions. Turbidity, measured as the

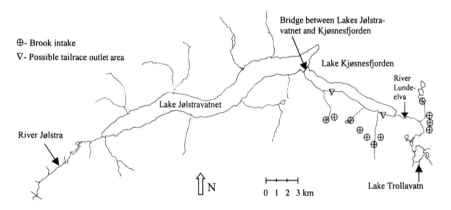

Figure 2. Illustration of Lakes Jølstravatnet, Kjøsnesfjorden and Trollavatn, the largest inlet rivers around these lakes as well as River Jølstra. Brook intakes have been indicated (note: not all streams are included in the illustration).

[1] When discussing features of the proposed Kjøsnesfjorden Hydropower Project we refer to alternative 2, which is the second largest of the four alternatives for this project and the alternative the Developer has prioritized for development.

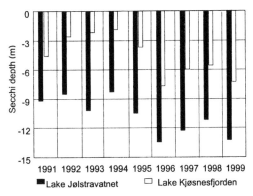

Figure 3. Secchi depth in Lakes Jølstravatnet and Kjøsnesfjorden in August 1991 – 1999 (modified from Sægrov 2000a).

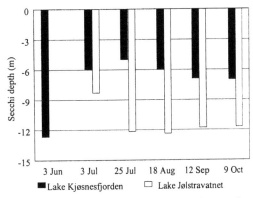

Figure 4. Secchi depth in Lakes Jølstravatnet and Kjøsnesfjorden from June to October 1997 (modified from Sægrov 2000a).

secchi depth[2], is very variable both between and within years (see Figure 3 and 4 below). Construction of a bridge across the strait between Lakes Kjøsnesfjorden and Jølstravatnet in 1969 reduced the length of the free flowing section of the strait from 400 m to 75 m. This resulted in a change in the currents and circulation patterns in the Lakes. As a consequence of the bridge construction Lake Kjøsnesfjorden is now much more affected by the turbid melt water from glaciers than Lake Jølstravatnet.

2.4.2 Aquatic vegetation and plankton

The lake that is proposed to be the future reservoir (Lake Trollavatn), the rivers to be affected as well as Lakes Kjøsnesfjorden and Jølstravatnet mostly lack vascular plants as aquatic vegetation. Primary producers in the Lakes are mainly algae/phytoplankton. Primary production in the glacier fed streams is very low. The zooplankton community in Lakes Kjøsnesfjorden and Jølstravatnet is dominated by *Daphnia galeata* and *Bosmina longispina*. Species such as *Holopedium gibberum* and *Cyclops scutifer* are also common. Other species, like *Polyphemus pediculus* and *Bythotrephes longimanus*, are only found in very low densities but despite these low densities still represent a very important food source for Brown Trout. Seasonal developments in the algae/phytoplankton and zooplankton communities as well as their species composition indicate oligo-

trophic[3] conditions in Lakes Kjøsnesfjorden and Jølstravatnet (Brettum 1989).

2.4.3 Fish ecology

Until 1990 the fish fauna in Lakes Jølstravatnet and Kjøsnesfjorden consisted only of Brown Trout, which is a relatively rare freshwater ecosystem today. In 1990 the European Minnow (*Phoxinus phoxinus*) was discovered in the lakes. This species has been spread tremendously by humans during the last century with negative consequences for many Brown Trout populations. A locally developed method of controlling the Minnow population has so far prevented negative impacts on the Trout population in Lakes Jølstravatnet and Kjøsnesfjorden.

Genetic as well as other studies have shown that Brown Trout populations in Lakes Jølstravatnet and Kjøsnesfjorden consist of several genetically and ecologically distinct populations. Lake Kjøsnesfjorden has a population of lake-spawning Brown Trout. The majority of the population spawns in January-February, which is 3-4 months later than Brown Trout that spawn in the inlet rivers in the area. Lake-spawning is a rare phenomenon among Brown Trout populations that prefer more oxygen rich riverine habitats for spawning.

Lake Jølstravatnet has a Brown Trout population that mainly reproduces in the side streams of the Lake. In addition, a cannibalistic, large-trout population reproduces in River Jølstra and utilizes Lakes Kjøsnesfjorden and Jølstravatnet as its feeding areas as adults. The future reservoir does not have any naturally reproducing fish populations. The affected streams are mostly too steep and fast flowing for any fish populations to survive with the exception of the lowest section of River Lundeelva, which is flatter and provides a habitat for the Brown Trout despite the impact of glacier melt water.

---

[2] The secchi depth is a measure of turbidity/visibility of water and gives an indication how deep light penetrates into the water column. This depth is often measured by lowering a white disc (20-25 cm diameter) down in the water in a shaded area until the disc is no longer visible. The depth where the disc disappears when lowered and appears again when lifted is defined as the secchi depth. The secchi depth often represents a level where 1-5 % of the light at the surface is remaining (Økland 1983a,b, Borgstrøm et al. 1995).

---

[3] Oligotrophic: Low concentrations of plant nutrients.

2.5 Fisheries, tourism and recreation

Fisheries and tourism based on fishing have long traditions dating back to the early 20th century. The fisheries in Lakes Jølstravatnet and Kjøsnesfjorden form the basis for the largest Brown Trout inland fishery in Norway, with annual catches of 11,000 – 25,000 kg. Fish is sold on the local, regional and national market, partly fresh and partly processed. The fisheries provide a significant source of income for local land owners as well as various tourist businesses. Fishing is locally and regionally important as a recreation activity but the economic value of this activity is difficult to determine.

2.6 The Kjøsnesfjorden Hydropower Project

Kjøsnesfjorden Kraftverk AS, the Developer, has applied for a concession to develop the Kjøsnesfjorden Hydropower Project. The Norwegian authorities are in the process of making a decision regarding whether such a concession should granted, and if so, under which conditions.

The Developer consists of Sunnfjord Energi AS (utility), Veidekke ASA (contractor) and local land owners who own the right to utilize the water resources for hydroelectric power production. The Developer has considered four alternatives for the Kjøsnesfjorden Hydropower Project. Alternative 2, the second largest and the one prioritized by the Developer, is presented briefly below.

The KHP is proposed as a high head (796 m) project with an installation of three Pelton turbines providing a total generation capacity of 56 MW. The project has an underground power station and an average annual production of 228 GWh. The reservoir, Lake Trollavatn (987.5 m a.m.s.l.), will cover an area of 1.1 km^2 and have a live storage of 35.1 mill. m^3, which is equivalent to 28 % of the average annual catchment runoff. The highest and lowest regulated water levels are proposed to be 1003 and 970 m a.m.s.l. respectively. A system of tunnels will collect water from 11 smaller streams around Lake Kjøsnesfjorden and divert this water to the power station. The total length of water bearing tunnels is 18.3 km. Project costs are estimated to be about NOK 524 mill. (USD 58 mill.) and construction is expected to take 3 ½ years.

2.6.1 The Environmental Impact Assessment
NORPLAN A.S, Consulting Engineers and Planners of Norway, has been responsible for carrying out an Environmental Impact Assessment (EIA) in parallel with the design of the project. The EIA included 13 specialist studies, of which the freshwater ecology and fisheries study was the largest. This study was sub-contracted to the Norwegian Institute for Nature Research (NINA) and Rådgivende Biologer AS (RB). This paper is mainly based on the NINA/RB

study (Sægrov 2000b,c,d, Sægrov & Johnsen 2000, Sægrov & Urdal 2000, Hindar & Balstad 2000, Hvidsten et al. 2000, Sægrov et al. 2000).

3 MATERIALS AND METHODS

3.1 EIA procedure

NORPLAN applied a structured three step procedure in the EIA, making impact assessment, conclusions and recommendations more objective, easier to understand and possible to trace back if desired. The procedure combines: 1) an objective and neutral description and assessment of the impact zone's value, and 2) an evaluation of the expected magnitude of positive and negative impacts. These are combined to arrive at 3) an overall assessment of impacts.

3.2 Physical and chemical parameters

The future reservoir, Lake Trollavatn, is mostly covered with ice from October/November to July/August and in some years the Lake is never fully free of ice. There is no natural reproduction of fish and primary production is low. It was therefore decided to focus the sampling in Lakes Kjøsnesfjorden and Jølstravatnet, the streams that would be affected by the project and the River Jølstra.

Hydrological data was collected and analyzed both for the purpose of the EIA and for the design of the hydropower project. Due to short time series of site specific data, comparable catchments were used for correlation and reference purposes.

Depth conditions for Lakes Kjøsnesfjorden, Jølstravatnet and Trollavatn had been surveyed previously and provided useful input for the EIA. For the purposes of the EIA, studies were carried out on current patterns in Lake Kjøsnesfjorden to get an overview of current speed and direction at various depths in the Lake. Temperature was measured using loggers at various locations in Lakes Kjøsnesfjorden and Jølstravatnet as well as in the streams around Lake Kjøsnesfjorden that would be affected by the project. Temperature profiles in Lakes Kjøsnesfjorden and Jølstravatnet were measured in order to study the stability of the water column throughout the year.

Water quality samples were collected regularly and analyzed for various parameters, with particular emphasis on plant nutrients (nitrogen and phosphorous). Light penetration/availability in the water column was studied by regularly measuring the secchi depth as well as turbidity (FTU) of collected water samples.

3.3 Algae, plankton and fish ecology studies

Ecological studies have been carried out in Lakes Kjøsnesfjorden and Jølstravatnet since the 1960s, and since the early 1980s studies have been carried every or every second year. At present Lakes Kjøsnesfjorden and Jølstravatnet are probably the best studied Brown Trout lake ecosystems in Norway (e.g. Klemetsen 1966, 1967, Nordland 1981, Sægrov 1985, 1990, 1993, 1995, 1997, 2000a, Hvidsten et al. 2000).

For the purpose of the EIA algae and zooplankton samples were collected in Lakes Kjøsnesfjorden and Jølstravatnet and species composition determined in the laboratory. Volume and density of the various groups/species were then also determined.

Sampling and analysis of a range of fish ecology aspects were carried out, including collection of data on:

☐ Habitat utilization by Brown Trout and European Minnow using multi-mesh size gill nets (5–55 mm) in the littoral and pelagic zones
☐ Migration patterns between Lakes Kjøsnesfjorden and Jølstravatnet using mark-recapture techniques
☐ Genetic variation (including the presence of several sub-populations) through analysis of juvenile samples using standard eletrophoresis techniques (Aebersold et al. 1987, Utter et al. 1987)
☐ Density and size of adults in River Jølstra using diving equipment
☐ Density and size of juveniles in the littoral zone in Lakes Kjøsnesfjorden and Jølstravatnet, as well as streams and rivers flowing into the Lakes and in River Jølstra, using electrofishing equipment
☐ Spawning areas with physical signs of spawning activity
☐ Variation in the size age cohorts based on sampling as well as catch data from fishermen
☐ Catches of Brown Trout based on daily reports as well as other reports from fishermen

Collected and analyzed data were used as the basis for the assessment of potential positive and negative impacts and development of mitigation measures and monitoring activities.

Data quality and quantity are generally considered to be very good. The Brown Trout population and its ecosystem is one of the best studied and understood Brown Trout freshwater ecosystems in Norway.

4 ISSUES OF CONCERN

Hydropower projects worldwide have received a lot of attention as a result of negative impacts on fish populations and fisheries. Before commencement of the KHP EIA, some major issues of concern were raised. One of them was that Kjøsnesfjorden Hydropower Project could affect the largest and most valuable inland fishery for Brown Trout in Norway. This was of particular concern given their unique adaptations and therefore important biodiversity val-

ues. The only way for the Developer to potentially avoid large protests at a later stage would be to thoroughly investigate the implications of the project for fish and fisheries. The project would have to be justified in relation to fish and fisheries by either showing that negative impacts would be small/minimal or the Developer would have to define acceptable mitigation and/or compensation measures.

4.1 Fish ecology and biodiversity

Brown Trout in Lakes Kjøsnesfjorden and Jølstravatnet are unique in several respects and are therefore of great interest from a scientific as well as fisheries management point of view. The following aspects are of particular importance and also indicate a high biodiversity value of the Brown Trout:

☐ Brown Trout in Lake Kjøsnesfjorden posses a special adaptation as they use the littoral zone of the lake as their spawning grounds. This is rare for Brown Trout which usually prefers more oxygen rich conditions provided by running waters/rivers. Timing and location of spawning activities, as well as egg and fry development, are carefully timed with the natural regimes of snow/ice melting and variation in sediment load in Lake Kjøsnesfjorden.
☐ Lakes Kjøsnesfjorden and Jølstravatnet support a population of large Brown Trout that are cannibalistic. This is a population type that receives particular interest from environmental authorities.
☐ The main food source for Brown Trout in Lakes Kjøsnesfjorden and Jølstravatnet is the zooplankton located in the pelagic zone. The majority of Brown Trout populations are dependent upon alternative sources of food, primarily various insects.

Construction of the KHP could result in changes in currents and sediment load in Lake Kjøsnesfjorden, with potentially detrimental effects on trout egg/fry survival. The lake-spawning habitat is marginal with a survival rate of about 60 %, while the survival rate in more oxygen rich riverine habitats is around 85 % (Sægrov 1990, Lura 1995). There were concerns that even a small increase in sediment load could reduce survival and recruitment significantly.

4.2 Interaction between Brown Trout and European Minnow

Introductions of European Minnow into rivers and lakes with Brown Trout population have previously resulted in detrimental effects for Brown Trout (DN 1995, Hesthagen & Sandlund 1997). Once the Minnow was discovered in Lakes Jølstravatnet and Kjøsnesfjorden in 1990 local people as well as the environmental authorities feared negative impacts for the valuable fish population and fisheries in the lakes.

So far the European Minnow have only colonized Lake Kjøsnesfjorden to a limited extent, while the species is found in extremely high densities in several shallow and relatively warm areas in Lake Jølstravatnet. It is not clear whether the low density of Minnow in Lake Kjøsnesfjorden is a result of lower temperature and higher turbidity, or some other cause. In general, Lake Kjøsnesfjorden is deep and has few shallow areas where the Minnow seems to thrive.

Concern was raised that the Kjøsnesfjorden Hydropower Project might change temperature and turbidity conditions in Lake Kjøsnesfjorden and that the Minnow would subsequently colonize Kjøsnesfjorden. This could have negative impacts on the Brown Trout population.

4.3 Fisheries

The Brown Trout populations in Lakes Jølstravatnet and Kjøsnesfjorden support the largest inland fishery for Brown Trout in Norway, with annual catches of 11,000 – 25,000 kg. The majority of the catches, about 90 %, are made by local land owners using gill nets. The fishing is partly for own consumption and partly for sale. Fishing is an economically important activity for these land owners. Sports fishing and recreational fisheries are also popular and important, but the total catches are not comparable in size to the semi-commercial fisheries.

Catches in Lake Jølstravatnet are larger than catches in Lake Kjøsnesfjorden, 4.4 kg/ha and 2.5 kg/ha respectively. As there is migration of fish between the two lakes, it was feared that negative impacts in Lake Kjøsnesfjorden could also negatively affect the yield in Lake Jølstravatnet. The level of impacts would be dependent upon the extent of the migration and importance of fish from Lake Kjøsnesfjorden in the catches in Lake Jølstravatnet.

4.4 Tourism

Fishing opportunities in Lakes Kjøsnesfjorden and Jølstravatnet, as well as River Jølstra, have attracted tourists to the area for almost 100 years. The economic value of the fishing-related tourism is significant for the local land owners who benefit from the sale of fishing licenses as well as renting out accommodation on a small scale. Various other tourist oriented businesses also benefit from the visitors. If recreational fisheries were to be negatively affected by the hydropower project this could have side effects on tourism in the area.

5 EVALUATION OF POTENTIAL IMPACTS

5.1 Reservoir

The proposed reservoir, Lake Trollavatn, does not have natural reproduction of fish and no impacts on fish or fisheries are therefore expected in this lake.

5.2 Affected rivers and streams

The project will reduce the flow in several streams and rivers around Lake Kjøsnesfjorden. The majority of these rivers and streams are too steep and heavily affected by glacier melt water to provide suitable conditions for Brown Trout. Only a short section of one river (River Lundeelva) provides conditions that can support a low density of trout, and this stretch will be affected by an average reduction in river flow of about 65 %. This stretch is of minor importance both as a spawning and feeding area, and impacts on the fish populations and fisheries are negligible.

5.3 Lake Kjøsnesfjorden

5.3.1 Change in sediment load – impacts on egg survival and recruitment

Brown Trout in Lake Kjøsnesfjorden has adapted its life cycle to the natural regime of snow/ice melting and sediment input to the lake. Under natural conditions water in Lake Kjøsnesfjorden is clear from October/November to early June, which is the period when the Brown Trout spawns and the eggs hatch. The fry then come up from the gravel and sand in which the eggs were buried during the winter months.

Drawing the reservoir down during the winter months will cause erosion in the large amount of sediments in the reservoir. Consequently, the hydropower project will cause an increase in turbidity and sediment load in Lake Kjøsnesfjorden during a period when turbidity and sediment load naturally is very low. Sedimentation of clay and silt over trout spawning areas can reduce oxygen availability for the eggs and fry, with a subsequent reduction in the survival rate in this marginal habitat. The project might thus reduce recruitment to the trout population, and cause loss of unique biological diversity.

Based on calculations from comparable rivers that are influenced by glacier melt water, a worst case scenario for sediment influx to Lake Kjøsnesfjorden was evaluated. In the highly unlikely event of the power station running at full capacity (8.1 m^3/s) from the 1st of January to 31st of May, the overall sediment input[4] was estimated to 116 tons. This is based on the assumption that the average clay content in the long-term operation phase of the project will be 1.1 mg/liter (Tvede & Hougsnæs 1993). The total area of Lake Kjøsnesfjorden is 7.7 km^2 and the trout use the littoral zone down to 10 m for spawning. Assuming that the sediments (clay) will be deposited equally over the relatively evenly located spawning grounds, a sediment layer of 0.02

---

[4] Looking only at clay component of the sediments, that is, the component that can be a problem for the eggs and fry.

mm will be deposited. This is only 1/50 of the 1 mm sediment layer that is found on the spawning areas under natural conditions during late spring.

However, experience has shown that during the first years of operation sediment load can be 10-20 times higher (Tvede & Hougsnæs 1993, Bogen et al. 1996). Whether this short term higher sediment load will result in a temporary decrease in trout egg and fry survival is not clear, but is expected to be unlikely. Mitigation measures have been proposed to further reduce the chances of such impacts.

5.3.2 Change in sediment load – impacts on production of plankton and fish

Variation in the fishery yields (measured as kg fish/ha) shows a statistically significant correlation with secchi depth during the 1990s. In years with high turbidity and low secchi depth, availability of light seems to be the limiting factor for primary production. Consequently, density of zooplankton and fish growth/yield are reduced. In years with low turbidity the concentration of phosphorous is probably the limiting factor for primary production and density of zooplankton and fish growth/yield increase. Also, the quality of the fish is higher in years with low turbidity.

High turbidity reduces food availability for Brown Trout, which causes increased competition and cannibalism and therefore indirectly reduced recruitment.

The Kjøsnesfjorden Hydropower Project will store sediment rich melt water during the summer filling of the reservoir. The sediment load and turbidity in Lake Kjøsnesfjorden will therefore be significantly reduced during the summer. This is the main production period and primary production, zooplankton density, fish growth and fish production will all increase compared to the natural conditions. The change will be most pronounced in years where sediment input to Lake Kjøsnesfjorden would otherwise have been large. If the project is able to include the most sediment rich rivers and streams, production in Lake Kjøsnesfjorden might increase significantly and reach levels presently experienced in Lake Jølstravatnet. This is under the assumption that water from the reservoir and power station will not, or only to a limited extent, reach the productive zone in Lake Kjøsnesfjorden during the summer months (see "mitigation measures" below).

Storage of water in the reservoir during summer months will reduced the glacier melt water's dilution of plant nutrients in Lake Kjøsnesfjorden. Subsequently, a small increase in primary as well as fish production can be expected.

The color of the Brown Trout's meat is considered as a sign of the quality, where red fish meat indicates good quality and white fish meat indicates poorer quality. Reduced turbidity and changes in the zooplankton community towards conditions close to those experienced in Lake Jølstravatnet are likely to improve the Brown Trout quality by increasing the proportion of trout with red meat.

The cannibalistic, large-trout population uses Lake Kjøsnesfjorden as a feeding area. As recruitment to the Brown Trout population there is not likely to be negatively affected by the KHP, no impacts are expected on the important large-trout population.

5.3.3 Changes in temperature

Under natural conditions the melt water from the surrounding glaciers lowers the temperature in Lake Kjøsnesfjorden 1-2 °C compared to Lake Jølstravatnet. Storage of cold melt water in the reservoir during the summer will increase the summer temperatures in the surface layers Lake Kjøsnesfjorden. Streams and rivers with reduced river flow will also experience increased temperature.

Water released from the power station during the summer months will be colder than the water from the natural streams around Lake Kjøsnesfjorden. This is caused by removing the warming effect (exchange of energy with the warmer air) that would have occurred if the water was allowed to run in the natural streams instead of through the headrace and tailrace tunnels. No major impacts are expected. If recommended mitigation measures are implemented, a slight increase in surface layer's temperature during the summer months is expected.

In cold winters when Lake Kjøsnesfjorden is covered with ice, water from the power station released at the surface is likely to cause a small increase in the surface layer's temperature. No significant impacts on fish populations are expected.

5.3.4 Changes in interaction between Brown Trout and European Minnow

Comprehensive studies of interactions between Brown Trout and European Minnow during the 1990s have not been able to show any negative impacts on the Brown Trout populations. It is likely that without the locally developed methods for controlling the Minnow population, negative impacts would have been experienced. At present it seems unlikely that the Kjøsnesfjorden project will change the interactions between Brown Trout and European Minnow populations significantly. However, in light of the importance of the Kjøsnesfjorden Brown Trout population, mitigation measures for dealing with such impacts have been recommended as a contingency plan.

5.3.5 Impacts on fisheries

Impacts on fisheries are clearly linked to the impacts on the Brown Trout populations, but it is important to maintain a distinction between i) the populations as unique elements of biodiversity and ii) the economic value of the fisheries. It is possible to have negative impacts on one or more of the Brown Trout sub-populations and even loose valuable biodiversity, without necessarily reducing the economic output of the overall Brown Trout fisheries in Lakes Kjøsnesfjorden and Jølstravatnet.

From the discussion of impacts above it can be seen that long term negative impacts on the Brown Trout populations are not likely to occur. Thus, the basis for fisheries is not likely to be affected by the operation of the Kjøsnesfjorden Hydropower Project. In fact, there might be a slight increase in fish production and improvement of fish quality in Lake Kjøsnesfjorden in the long term if proposed mitigation measures are implemented.

There is, however, some uncertainty about the erosion in the reservoir and sediment input to Lake Kjøsnesfjorden in the short term, that is, immediately after starting the operation of the KHP. If project start up and the initial reservoir draw down(s) are in May/June the first year(s) of operation, there might be a temporary increase in turbidity in the productive zone of Lake Kjøsnesfjorden. Spring blooms of phytoplankton can, without any mitigation measures, be negatively affected and fish production be reduced temporarily.

5.4 Lake Jølstervatn

Productivity and fish catches in Lake Jølstravatnet (32.2 km^2) are larger than in Lake Kjøsnesfjorden (7.7 km^2). The total catch in Lake Kjøsnesfjorden is on average only about 12 % of the total catch in Lake Jølstravatnet. Lower productivity in Lake Kjøsnesfjorden is mainly the result of higher turbidity and sediment load from mid-June to October/November in Kjøsnesfjorden compared to the less glacier-influenced Jølstravatnet.

Lake Kjøsnesfjorden acts as a recruitment area for Lake Jølstravatnet. Fish migration studies in the EIA have shown that fish from Lake Kjøsnesfjorden only to a limited extent contribute to the catches in Lake Jølstravatnet. About 10-15 % of the adult fish population migrate from Lake Kjøsnesfjorden to Lake Jølstravatnet and contribute to 2-3 % of the fish catch in Lake Jølstravatnet. Consequently, even in the unlikely event of negative impacts on the Brown Trout population in Lake Kjøsnesfjorden no significant impacts are expected on the fisheries in Lake Jølstravatnet. The limited importance of Lake Kjøsnesfjorden as a recruitment area for Lake Jølstravatnet also indicates that an increase in fish production in Lake Kjøsnesfjorden can only result in a minimal increase in catches in Lake Jølstravatnet.

Even though fish production and overall catches in Lake Kjøsnesfjorden are not likely to be negatively affected, small negative impacts can be experienced for some types of recreational fishing.

Where the rivers and streams around Kjøsnesfjorden meet the Lake, sport fishing is popular. This is due to the high density of fish around the river outlets, in particular during rainfall periods when river flows increase. Reduced flow in three significant rivers around the Lake is likely to result in smaller catches for sports fishing at these outlets.

5.5 River Jølstra

River Jølstra, the outlet river from Lakes Jølstravatnet and Kjøsnesfjorden, provides breeding grounds for the cannibalistic, large-trout population. As discussed above, impacts on this population are expected to be minimal or none. Therefore no impacts are expected on the fisheries in River Jølstra.

6 POTENTIAL MITIGATION MEASURES

Environmentalists and engineers discussed a number of mitigation measures during the technical planning and EIA phase to avoid or reduce potential negative impacts. Some of the measures are discussed below.

6.1 Dam and reservoir

Lake Trollavatn will be used as the reservoir for the Kjøsnesfjorden Hydropower Project and this lake has previously received large amounts of sediments from the surrounding glaciers. Regulating the water level in the lake will cause erosion in the sediments and some of these sediments will be transported through the power station to Lake Kjøsnesfjorden.

Initial plans for the Kjøsnesfjorden project proposed the highest (HRW) and lowest (LRW) regulated levels to be 1000 and 960 m a.m.s.l. respectively. In order to reduce erosion in the flatter and deeper areas of the reservoir (close to LRW), it was suggested to increase LRW to 970 and thereby reduce the sediment input to Lake Kjøsnesfjorden, and possibly also turbine wear and tear by sediments. The live storage would be maintained by raising HRW to 1003 m a.m.s.l. Raising HRW by three meters will only inundate a limited additional area (8 ha) covered by bare rock and snow. However, this additional area in located within the Jostedalsbreen Glacier National Park. Environmental authorities will therefore need to consider the feasibility of this mitigation measure.

6.2 Tailrace tunnel outlet

Various options for the location of the outlet of the tailrace tunnel have been considered. First of all the outlet will be dependent upon the location of the power station. Two locations for the power station have been considered, one at the east end of Lake Kjøsnesfjorden and one 6 km further west. Considering the distribution of spawning grounds and current patterns in Lake Kjøsnesfjorden, these two alternatives do not appear to be different in terms of their likely impacts on fish populations.

Various measures have been considered in order to reduce the sediment input to the productive zone in Lake Kjøsnesfjorden. It is believed that an underwater outlet from the tailrace tunnel is a feasible solution. By locating the outlet at least 20 m below the surface of Lake Kjøsnesfjorden, it is believed that water with sediments from the tailrace tunnel will remain under the thermocline during the period of thermal lake stratification (June to October/November). This means that turbid water from the reservoir/power station will remain in the cold and unproductive deep sections (hypolimnion) of Lake Kjøsnesfjorden and not enter the warmer and productive upper zone (epilimnion) of the Lake.

In terms of egg and fry survival the estimates of sediment load and deposition of fine sediments (clay) over the spawning areas (see above) indicate that negative impacts are not likely to occur. The underwater outlet will reduce the risk of a decrease in recruitment even further.

Compared to the pre-project situation, an underwater outlet will improve turbidity and temperature conditions in the productive zone in Lake Kjøsnesfjorden in the period of thermal lake stratification. This is likely to increase primary production and zooplankton density. The fish yield in Lake Kjøsnesfjorden may then increase from the present average level of 2.5 kg/ha to levels comparable to the average yield in Lake Jølstravatnet (4.4 kg/ha).

6.3 *Brook intakes*

The number and location of brook intakes are important for the turbidity conditions in Lake Kjøsnesfjorden. The more sediment rich glacier melt water that can be diverted to the reservoir and through the power station to the hypolimnion, the larger will be the decrease in turbidity in the productive epilimnion during summer months be. An unusual situation exists where the more water that is diverted for power production, the more positive the impacts for fisheries in terms of production. However, this has to be weighed against other environmental impacts, for instance changes in landscape characteristics due to reduced river flows.

6.4 *Operational procedures*

When the hydropower project enters into operation and starts drawing down the reservoir, there is likely to be severe erosion in existing sediments in Lake Trollavatn as well as washing out of sediments in the tunnels generated by tunnel construction. Even though long term negative impacts are unlikely, it will be important to avoid water with a high sediment load in the initial operation period entering the productive zone of Lake Kjøsnesfjorden. Start up and the first draw down(s) of the reservoir should therefore be done during a period of thermal lake stratification, that is, between mid-June and October/November.

An underwater outlet at 20 m depth may not ensure a significant reduction in the turbidity in the productive zone in Lake Kjøsnesfjorden during periods when the water column in the lake is not thermally stratified (November-June). May and early June is an important period in terms of primary production as phytoplanktons have a spring bloom. Turbidity in Lake Kjøsnesfjorden during this period is naturally low as the major snow and ice melt period and transport of sediments into the Lake does not start until the second half of June.

It has therefore been recommended that operation of the power plant in May and June be done using water directly from the brook intakes with lower sediment load instead of water from the reservoir that is likely to contain a higher sediment load.

6.5 *Monitoring as part of mitigation*

Close monitoring of turbidity conditions, in particular just before, during and after start up of operation of the hydropower project, is very important to verify whether the conclusions of the EIA are correct and in order to suggest corrective measures if needed. The intensity of monitoring can be reduced as sediment and turbidity conditions in the reservoir as well as Lake Kjøsnesfjorden stabilize.

Should water be drawn from the reservoir during May/June the first years of operation, it is important that this is closely followed up with measurements of turbidity and secchi depth. This monitoring can be basis for adjusting operation of the power plant if necessary. If it is not possible to show a negative impact of using the reservoir for power production in this period, monitoring might act as a tool for justifying higher power production than would otherwise be recommended.

7 CONCLUSION

Major concerns were raised before commencement of the Environmental Impact Assessment (EIA) for the proposed Kjøsnesfjorden Hydropower Project. In particular, negative impacts on genetically unique Brown Trout populations and the most valuable Brown Trout inland fishery in Norway were presented as major arguments against developing this project.

A thorough EIA, including detailed studies of the Brown Trout populations and the freshwater ecosystem, carried out in parallel with technical planning have shown that the probability of significant negative impacts on fish populations and fisheries are unlikely. In fact, there could be a slight long term increase in fish production and improvement in fish quality in Lake Kjøsnesfjorden if proposed mitigation measures are implemented.

Currently, the Brown Trout populations in the lakes and rivers that will be affected are among the best studied of their kind in Norway and conclusions and recommendations in the EIA are therefore well documented.

REFERENCES

Aebersold, P.B., Winans, G.A., Teel, D.J., Milner, G.B. & Utter, F.M. 1987. *Manual for starch gel electrophoresis: A method for the detection of genetic variation*. NOOA Technical Report, NMFS 61, Seattle.

Bogen, J., Bønsnes, T.E., Elster, M. & Olsen, H.C. 1996. *Erosjon i Storglomvatn – magasinet, Svartisen Kraftverk.* NVE rapport nr. 37. Norwegian Water Resources and Energy Directorate (NVE), Oslo. (in Norwegian)

Borgstrøm, R., Jonsson, B. & L'Abée-Lund, J.H. 1995. *Ferskvannsfisk. Økologi, kultivering og utnytting*. The Research Council of Norway (NFR), Oslo. (in Norwegian)

Brettum, P. 1989. *Alger som indicator på vannkvalitet. Planteplankton*. NIVA-rapport nr. 2344. Norwegian Institute for Water Research (NIVA), Oslo. (in Norwegian)

DN 1995. *Spredning av ferskvannsorganismer*. Seminarreferat. DN-notat 1995-4. Directorate of Nature Management (DN), Trondheim. (in Norwegian)

Hesthagen, T. & Sandlund, O.T. 1997. Endringer i utbredelse av ørekyte i Norge: årsaker og effekter. NINA Fagrapport 03. Norwegian Institute for Nature Research (NINA), Trondheim. (in Norwegian)

Hindar, K. & Balstad, T. 2000. Genetisk variasjon og stammetilhørighet hos Jølstraaure. In Sægrov, H. (ed.). *Konsekvensutgreiing Kjøsnesfjorden Kraftverk - Fiskebiologiske undersøkingar*. Rådgivende Biologer AS, Bergen. pp. 43-48. (in Norwegian)

Hvidsten, N.A, Sægrov, H., Jensen, A.J. & Johnsen, G.H. 2000. *Konsekvensutgreiing Kjøsnesfjorden Kraftverk - Delutgreiing fisk og fiske*. – NINA Oppdragsmelding 629. Norwegian Institute for Nature Research (NINA), Trondheim. (in Norwegian)

Klemetsen, A. 1966. *Ørreten i Jølstervann. Ernæring, vekst og beskatning*. M.Sc. thesis. University of Oslo. (in Norwegian)

Klemetsen, A. 1967. On the feeding habits of the population of brown trout (*Salmo trutta* L.) in Jølstervann, West Norway, with special reference to the utilization of planktonic crusraceans. *Nytt magasin for Zoologi* 15: 50-67.

Lura, H. 1995. *Domesticated female Atlantic salmon in the wild: spawning success and contribution to local populations*. Dr. scient thesis. University of Bergen.

Nordland, J. 1981. *10-års verna vassdrag i Vest-Norge. Jølstravassdraget*. Fiskerikonsulenten i Vest-Norge, Bergen. (in Norwegian)

Sægrov, H. 1985. *Optimal storleik for innsjøgytande aurehoer, Salmo trutta L., i Kjøsnesfjorden (Jølstravatnet), Vest-Norge*. M.Sc. thesis. Department of Zoology, University of Bergen. (in Norwegian)

Sægrov, H. 1990. Er innsjøgyting hos aure undervurdert? In Vassdragsregulantenes Forening. *Fiskesymposiet 1990*. Kompendium. pp. 99-113. (in Norwegian)

Sægrov, H. 1993. *Aure og ørekyt i Jølstravatnet – Kjøsnesfjorden*. Department of Zoology, University of Bergen. (in Norwegian)

Sægrov, H. 1995. Prøvefiske og næringsfiske i Jølstravatnet og Kjøsnesfjorden i 1995. Rådgivende Biologer AS, Bergen. (in Norwegian)

Sægrov, H. 1997. *Prøvefiske og næringsfiske i Jølstravatnet og Kjøsnesfjorden i 1996*. Rådgivende Biologer AS, Bergen. (in Norwegian)

Sægrov, H. (ed.) 2000a. *Konsekvensutgreiing Kjøsnesfjorden Kraftverk – Fiskebiologiske undersøkingar*. Rådgivende Biologer AS, Bergen. (in Norwegian)

Sægrov, H. 2000b. Vandring frå Kjøsnesfjorden til Jølstravatnet. In Sægrov, H. (ed.). *Konsekvensutgreiing Kjøsnesfjorden Kraftverk - Fiskebiologiske undersøkingar*. Rådgivende Biologer AS, Bergen. pp. 59-74. (in Norwegian)

Sægrov, H. 2000c. Variasjon i årsklassestyrke av aure. In Sægrov, H. (ed.). *Konsekvensutgreiing Kjøsnesfjorden Kraftverk - Fiskebiologiske undersøkingar*. Rådgivende Biologer AS, Bergen. pp. 75-89. (in Norwegian)

Sægrov, H. 2000d. Næringsfiske i Kjøsnesfjorden og Jølstravatnet. In Sægrov, H. (ed.). *Konsekvensutgreiing Kjøsnesfjorden Kraftverk - Fiskebiologiske undersøkingar*. Rådgivende Biologer AS, Bergen. pp. 90-98. (in Norwegian)

Sægrov, H. & Johnsen, G.H. 2000. Lokalitetsbeskrivelse og produksjonsgrunnlag. In Sægrov, H. (ed.). *Konsekvensutgreiing Kjøsnesfjorden Kraftverk - Fiskebiologiske undersøkingar*. Rådgivende Biologer AS, Bergen. pp. 5-31. (in Norwegian)

Sægrov, H. & Urdal, K. 2000. Habitatbruk hos aure og ørekyte. In Sægrov, H. (ed.). *Konsekvensutgreiing Kjøsnesfjorden Kraftverk - Fiskebiologiske undersøkingar*. Rådgivende Biologer AS, Bergen. pp. 5-31. (in Norwegian)

Sægrov, H., Hellen, B.A. & Kålås, S. 2000. Gytebestandar og gytelokalitetar. In Sægrov, H (ed.) *Konsekvensutgreiing Kjøsnesfjorden Kraftverk - Fiskebiologiske undersøkingar*. Rådgivende Biologer AS, Bergen. pp. 5-31. (in Norwegian)

Tvede, A.M. & Hougsnæs, R. 1993. *Suspensjonstransport i Vetlefjordelva og i overføringen, Mel kraftverk, 1992*. HM-notat nr. 16/93. Norwegian Water Resources and Energy Directorate (NVE), Oslo. (in Norwegian)

Utter, F., Aebersold, P. & Winans, G. 1987. Interpreting genetic variation detected by electrophoresis. In Ryman, N. & Utter, F. (eds.) *Population Genetics and Fishery Management*. University of Washington Press, Seattle. pp. 21-45

Økland, J. 1983a. *Miljø og prosesser i innsjø og elv. Ferskvannets verden 1*. Universitetsforlaget/Scandinavian University Press, Oslo. (in Norwegian)

Økland, J. 1983b. *Planter og dyr – Økologisk oversikt. Ferskvannets verden 2*. Universitetsforlaget//Scandinavian University Press, Oslo. (in Norwegian)

Hydropower in the New Millennium, Honningsvåg et al (eds), © 2001 Taylor & Francis, ISBN 90 5809 195 3

Influence of streamflow variability on the design of small hydropower plants

Z.Hailu & Prof. Dr.-Ing. H.-B.Horlacher
Dresden University of Technology, Dresden, Germany

ABSTRACT: *Small-scale hydropower plants are associated with less environmental impact than large hydropower plants. However, they are of less reliability in rivers in semi-arid regions which are characterised by high seasonal variability of flow. The search for an optimal small hydropower development level which satisfies both criteria of environmental compatibility and supply reliability in these regions is, therefore, tied up with a big challenge facing water resources engineers and environmentalists in general.*

A case study is made based on a small hydropower site in Ethiopia to show the influence of streamflow variability on run-of-the river as well as on small-scale storage options. Backing up of the energy supply appeared to be an important measure which could ensure reliability of supply while at the same time allowing to minimise adverse environmental impacts. The paper concludes by demonstrating the economic attractiveness of a diesel/hydro hybrid system for both types of developments.

1 INTRODUCTION

The paper considers a case study of the Ropi hydropower site in the Bilate basin in Ethiopia to show the effect of seasonal variability of river flow on the magnitude and reliability of power supply.

The Bilate basin is found in the Abaya-Chamo lakes basin and its geographic location is south west of Addis Ababa, Figure 1. Within this basin, the Ropi hydropower site is situated at 7°08'N,38°07'E on the upper part of the Bilate river.

Both cases of storage and diversion types of devel-

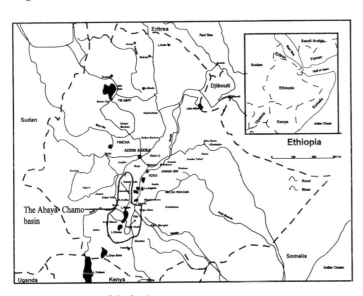

Figure 1 Location of the basin

opments are considered to show that operation at firm yield (say at 90% reliability) leaves a greater portion of the energy potential of the river unutilized. Suggestions are made for the increased utilization of the energy potential of the river by increasing the installation capacity.

2 NATURE AND EFFECT OF STREAMFLOW VARIABILITY

Two gauging stations with more than twenty years of service are found upstream and downstream of the hydropower site. The data of the two gauges showed a very strong correlation and provided the database for estimating the flow-duration at the hydropower site. Figure 2 is a month-by-month flow-duration curve for the Ropi site based on daily flow data. It shows how excess flow is concentrated during few months of the year while there is very limited flow during the remaining period. This is a typical characteristic of mountain rivers in tropical regions with high runoff factors. Most of the runoff occurs during the wet season leaving an insiginificant amount of base flow during the dry season.

Even without deducting the minimum downstream requirements, the dependable flow (90% exceedance) seems to be very low except for the few wet months of the year. This fact baffles designers of small hydropower plants in the choice of optimum design flow. If a reliable supply is to be assured throughout the year, the design discharge must be set so low that it would lead to under utilization of the power potential of the river. Increase in the design flow would, on the other hand, imply low guarantee of supply at the installation level.

Often, the provision of storage is proposed as a solution to curb the above dilemma. But, such a solution can only be approved only when the associated impacts are tolerably low. Issues revolving around impacts caused by reservoirs vary from context to context. For this reason, particular facts of the Bilate

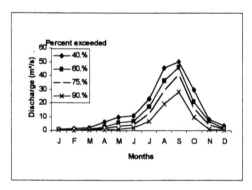

Figure 2 Flow-duration curve at the site

basin are discussed in a little more detail before proposing a general solution to the dilemma.

3 SMALL-SCALE STORAGE DEVELOPMENT AT THE ROPI SITE

Storage reservoirs represent the most effective means of time-redistribution of water according to socio-economic requirements. In serving such purposes, reservoirs must be in compliance with environmental, and social criteria of development. Otherwise, the human and environmental impact of water control and storage may outweigh any benefits to be gained from it.

Reservoirs often inundate arable and pasture land as well as places of historical or cultural value to communities. Water impoundment has often deprived people from having access to traditional common property resources. Flora and fauna also become submerged leading to the loss of bio-diversity.

A sad experience of negligence to observe questions revolving around environmental issues has already occurred in the upper part of the Bilate basin. Despite the completion of the construction of a dam in 1990, it could not go into operation because of the unwillingness of upstream dwellers to move to downstream settling zones. The people rejected offers of compensation because they were unwilling to see their indigenous village submerged by the filling of the reservoir. They won the legal controversy and managed to protect their village. The experience is an outstanding lesson for the need to take the interests of the population very seriously before undertaking the construction of dams.

3.1 Environmental compatibility Vs reliability

An ideal dam site may be defined as a site which has little negative environmental impacts and which provides high reliability of supply of the design flow with minimum requirement of foundation treatment and minimum volume of dam body. However, such sites hardly exist, since the satisfaction of one criterion usually causes the violation of another criterion. In particular, the criterion for environmental compatibility stands in conflict with the criterion for reliable supply.

To be environmentally compatible, a reservoir has to be as small as possible to minimize the negative impacts. In contrast to this, a reservoir has to be big enough to guarantee reliability of supply of the forecasted load. Compromising these two criteria is a challenge facing water resources engineers in general. The challenge assumes greater proportions in semi-arid regions which are characterized by very high seasonal variability of river flow. Reservoir reliability analysis for dams in such areas shows the need for substantial storage volume to regulate the seasonal variation. Moreover, the requirement to allow sufficient dead storage for sediment deposition minimizes the usable live storage. Thus, the need for a still greater storage volume has to be accommodated.

In a large scale storage, the episodes for shortages occur at the end of relatively long wet and dry periods, and the fluctuations of flow during most individual years are largely inconsequential since they cannot induce either spills or shortages. On the other hand, for small scale storage, the uncertainity in long-term fluctuations is irrelevant since the reservoir fills up every year and a water shortage also can occur every year. The probability of occurrence of either of the two episodes depends exclusively on the conditions in that one year only, Klemes (1983). Therefore, the preparation and use of a reliability curve for the selection of the capacity of small scale dams is highly recommended.

3.1.1 Steps in establishing the reliability curve

The reliability of a reservoir is defined as the probability that it will deliver the expected demand throughout its lifetime without incurring deficiency. In this sense lifetime is taken as the economic life, which is usually between 50 to 100 years. An operation study is made for the period of the economic life on assumed future inflow to determine the reservoir reliability. The study may analyze only a selected critical period of very low flow, but modern practice favors the use of a long synthetic record. In the first case the study can only define the capacity required during the selected drought. With the synthetic data it is possible to estimate the reliability of reservoirs of various capacities, Linsley(1979).

The sequent-peak algorithm suits itself to operation study based on long synthetic data. As a first step, values of the cumulative sum of net inflow i.e. inflow minus withdrawals are calculated. The first peak and the sequent peak are then identified. The required storage for the interval is taken as the difference between the initial peak and the lowest trough in the interval. The process is repeated for all cases in the period under study and the largest value of required storage is determined. This is done for many scenarios of inflow. The different storage values are then analyzed by probabilistic methods to establish the reliability curve. Once the reliability curve is fixed, it can be used to estimate the reliable release corresponding to an acceptable storage volume and to a predetermined value of reliability. In so doing, the topographical and environmental limits at the dam site are considered to determine the acceptable storage volume. Topographical features of the Ropi reservoir site are obtained from an elevation-volume curve established by the Water Resources Commission, WRC(1974).

3.1.2 Reliability for Ropi reservoir

The reliability at Ropi reservoir is estimated by stochastically generating 100 traces equal in length to the adopted project life (50 years.). Each trace may then be said to represent one possible example of what might occur during the project lifetime, and all traces are equally likely representatives of this future periods. The Thomas-Fiering model which considers the month to month correlational structure of mean monthly flows is used for the generation of stochastic data, Equation 1, Mutreja(1986). The month to month correlational structure is represented by the correlation coefficient of the flows of consecutive months r_j and the slope of the correlation equation B_j. Independent random numbers of a uniform density function (t_p) are generated in the interval $(0,1)$ which are then transformed into random normal variate using the Box and Muller's method, Press(1992).

$$Q_{p,j+1} = MQ_{j+1} + B_j(Q_{p,j} - MQ_j) +$$
$$t_p S_{j+1}(1 - r_j^2)^{\frac{1}{2}} \qquad (1)$$

where: MQ_{j+1} = Mean monthly flow of month j+1
$\quad MQ_j$ = Mean monthly flow of month j
$\quad Q_{p,j+1}$ = Mean monthly flow of year p and month j+1
$\quad Q_{p,j}$ = Mean monthly of year p and month j
$\quad S_{j+1}$ = Standard deviation of flow for month j+1

The storage required to deliver a specified demand is calculated for each trace by the sequent peak algorithm, and the resulting values of storage are fitted to Gumble frequency distribution. The result is a family of reliability curves which indicate the probability of satisfying demand throughout the project life for a selected reservoir capacity, Figure 3. In constructing the curves, a portion of the reservoir volume is left for the dead storage which is necessary to prevent the dam from filling up with sediment. In the absence of adequate data to estimate sediment yield, a 30% allowance is made based on recommendation from local experience, Hagos (1984).

The family of curves provide a useful basis for determining a compatible reservoir level. A hydropower engineer and an environmentalist can examine the curves to ascertain an acceptable storage level from their own perspective. The hydropower engineer may prefer to choose storage values from the upper right part of the curves as far as the topography permits. On the other hand, the environmentalist would prefer to look at the lower left part of the curves in order to minimize environmental impacts. The curves provide these experts with a summarized information which can help them in their search for a compromising solution.

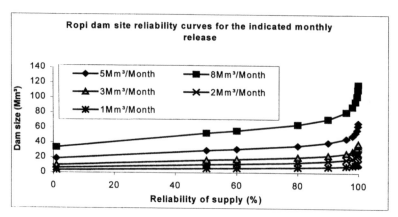

Figure 3 Ropi dam – reliability curve

3.2 Under utilization of potential in storage type hydropower plants

Storage type small hydropower plants in stand alone operation must be designed with sufficient reliability for continuous supply. A design discharge corresponding to high reliability is very low and thus the installation capacity is low too. But the natural inflow is not always found at the extreme low level and it is at times possible to release higher flow than the design discharge for more power production. By designing the power conduit and the turbines to entertain more flow than the design discharge, it is possible to produce several times more energy in the operation period of the plant. To give an image of the situation, the optimal release policy of the Ropi dam is carried out with a discrete dynamic programming technique for one random trace of the generated data series.

The adopted objective function is the maximization of energy production and the decision variable is the release at any particular month. The amount of release lies between the upper limit of conduit capacity and the lower limit of reliable reservoir yield which determine the maximum and minimum power production potential (P_{min}, P_{max}) respectively. These limits are used to estabilish the constraints in the dynamic programming algorithm. Equation 1 shows the constraints for both cases where P_t is the power produced at a particular month. The lower and upper bound of power production limits correspond to a reliable release of 3Mm³/month and a maximum pipe capacity of 9Mm³/month respectively.

$$\left. \begin{array}{l} P_{min} - P_t(S_t, S_{t+1}, Q_t) \leq 0 \\ P_t(S_t, S_{t+1}, Q_t) - P_{max} \leq 0 \end{array} \right\} \qquad (2)$$

P_t - actual power production for the given state
P_{min} – lower bound on power production

P_{max} - upper bound of power production
S_t - Storage at current month
S_{t+1} - Storage at coming month
Q_t - Release at current month

The following formulation by Mimikou(1987) expresses the objective function in its non-linear form. The monthly energy production B_t is estimated as a function of the average release rate through the turbines Q_t and the average head H over a month's period. As shown in Equation 3, the average head is expressed as a function of the storage S_t at the beginning and S_{t+1} at the end of a particular month. K is a factor depending on the units in use.

$$B_t = kQ_t H \left(\frac{S_t + S_{t+1}}{2} \right) \Delta t \qquad (3)$$

The computational procedure of the dynamic programming is based on the original work of Bellman(1957). The sequential decision technique is designed such that it steps from stage to stage and determines the optimal volume of water that should be released based on the natural inflow and water balance conditions of the reservoir, Equation (4). The procedure produces the optimal energy production curve on a monthly basis. The advantage lies in the flexibility of operation of the power plant which enables the maximum utilization of the storage for energy generation.

$$S_{t+1} = S_t - Q_t - E_t(S_t, S_{t+1}) + R_t \qquad (4)$$

where
   R_t – natural inflow to the reservoir
   E_t – Average Evapotranspiration

Such a curve for the Ropi dam is shown in Figure 4. The optimal monthly energy production for arbitrary

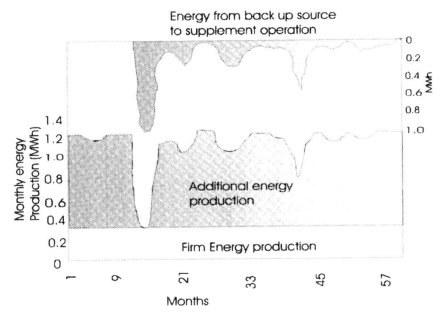

Figure 4 Monthly optimal energy production

five years period shows that there is a considerable increase in the total energy production as a result of setting the installation capacity of the power plant higher than the firm capacity.

However, shortage occurs at times of extreme low flow during which the power plant's operation would be restricted to the reliable reservoir yield. In order to maintain a constant supply at a higher production level, the provision of a standby supply source seems mandatory. A diesel generator or any other appropriate source can be used as a supplement to the storage dam during such extreme conditions.

4 RUN-OF-THE RIVER PLANT AT ROPI SITE

Run-of-the river plants are designed to operate on the naturally available flow with no artificial regulation. The extent of availability of natural flow is determined from the flow-duration curve as discussed in a previous section. A more elaborate information of the power potential can be given in a power-duration curve. A power-duration curve is similar in feature to a flow-duration curve except that it contains additional information on the available head at the site. Figure 5 shows a power-duration curve for the Ropi site based on monthly flow data. The curve provides a useful basis for determining the relationship between installation capacity and its degree of dependability. A design limited to 90% exceedance, for instance, leaves a greater portion of the energy generation potential unutilized. In contrast, design

for higher flows, say 75% exceedance, would result in higher energy utilization as shown in the power-duration curve. However, the later requires substantial back up energy than the former. These facts point to the need for an optimum choice of installation capacity. In general, a way has to be found to increase the energy production while at the same time maintaining the reliability of supply. As mentioned before, one solution could be to supplement the hydropower production with another source of energy such as diesel generation. The role of the diesel generator being the contribution to the attractiveness of small scale hydropower utilization.

4.1 Diesel backed hydropower generation

Figure 6 shows the energy production costs for diversion type hydropower, for diesel generation and for a combination of the two at the Ropi site. The production costs of the hydropower plant are smaller than the diesel generator. But, the diesel generator can supply at its guaranteed installation capacity as long as there is no shortage in diesel supply. The hydropower plant on the other hand can be operated at installation capacity level only during excess flow periods. Thus the energy production falls down to lower levels during drier seasons.

It is apparent that the combined operation of a diesel and a run-off-river plant gives a slightly higher production cost than a purely hydropower supply. But, the combined system provides more insurance for covering the peak demand. In fact, the difference in

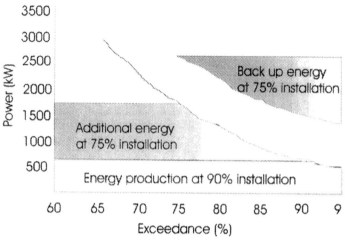

Figure 5 Power-duration curve of Ropi site

production cost between the pure hydropower option and the combined system is the price that must be paid to ensure reliability of supply. It is necessary to limit the contribution of the diesel generation both from economical and environmental aspects. Normally, the contribution of the diesel generator increases with increase in installation capacity. In the considered case, the total energy production by the diesel unit is less than 20% of the overall production. However, the diesel generator plays a big role in improving the reliability and the maximized utilization of the hydropower supply.

4.2 Combined operation of renewable energy sources

Diesel generator is considered as a back up source because it is suitable for quick start up and shut down according to the demand on back up energy. However, its use has to be gradually reduced in favor of environmental friendly energy sources. Because renewable resources are abundant (and varied) around the world, renewable hybrid energy systems offer reliable, economically competitive, and envi-

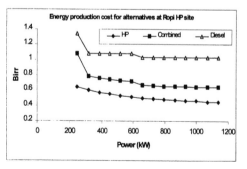

Figure 6 Energy production cost (1USD≅ 8Birr)

ronmentally friendly power for many currently non-electrified villages throughout the developing world, Flowers et. al(1994). Renewable energy/diesel hybrid systems, that may use a combination of photovoltaics, wind, mini hydropower, biomass, batteries, and diesel do offer practical alternatives for remote power generation, Morris(1998). But, thorough research is necessary in this area to achieve the smooth operation of renewable energy systems. Research must also be geared towards reducing the relatively high unit cost of production from solar energy, Horlacher (1994). This is so because solar energy is the most promising one among the renewable energy sources in tropical regions.

4.3 Supply to Grid....a possible compromise

Legistlation in most countries has been or is in the process of being amended to permit small producers connection to the public grid at the same time obliging the large utilities to buy power from small producers, Biermann(1995). This trend plays an important role in promoting rural electrification. The law that permitted private level power production in Ethiopia has attracted the attention of investors. The process of power development at private level could be facilitated if a reliable buyer of the producible energy can be created through grid connection.

The strategy of supply to the grid has a mutual benefit to all concerned bodies. First and foremost, the public will benefit from assured energy supply. Secondly, the burden on national electric utility will partly be shared by the private power developer. Thirdly, the developer will have the incentive of operating the power plant at full production level to maximise the energy production and thereby his profit. Last but not least, the variable energy production from a diversion type or from a storage type hydropower plant can be admissible since the grid can compensate for periodic deficits in production from excess production at other supply points.

CONCLUSION AND RECOMMENDATION:

The paper shows that the variability of stream flow is one of the major constraints of small hydropower design in semi-arid regions. Even though the study is not considered to be sufficient to make an absolute generalisation, an adequate insight has been provided as to how the operation of small hydropower plants could be highly restricted unless necessary measures are taken either to regulate the river flow or to supplement the power supply of the hydropower plants. Flow regulation by means of small reservoirs can prove to be a solution only so long as care is taken to keep the reservoir size within environmentally tolerable limits. The paper recommends the use of reservoir reliability curves for the choice of optimum and environmentally compatible reservoir size.

Storage type hydropower plants with environmentally acceptable storage level can be made to deliver economically attractive amount of energy if their operation is backed up by other energy sources. The same applies for unregulated run-of-the river type of developments. At present, the most suitable form of back up source is diesel generation. However, the paper recommends the provision of back up from viable renewable energy sources as the best strategy for future. This, of course, requires focusing extra attention on making energy accessible to the rural poor. Provision of adequate energy in general and renewable energy in particular is the only solution that could prevent the threat of environmental degradation that is hanging over most developing countries.

REFERENCE:

Bellman, R. 1957. Dynamic Programming. Princeton University Press, Princeton, N. J.

Bierman, E.; 1995, Small hydropower – An option with a future, GTZ, Eschborn, Germany

Flowers L.J. et. al. , 1994, Village Power hybrid systems development in the united states, In Proceedings of first Philippine wind power international conference 8-10. Arlington,USA

Hagos M. Ph.D. 1984 Dissertation,Multi-purpose water resources study for areas with data scarcity, Case study, Arsi Region, Ethiopia, Royal Institute of Technology Stockholm , Sweden

Hailu Z. & H.-B. Horlacher, 2000, Kleinwassernutzung in Entwicklungsländern bei Flüssen mit großen Ab flußschwankungen, Drittes Anwenderforum Kleinwasserkrafwerke, Passau, Germany

Horlacher H.-B.,1994, Potentiale, Kosten und Nutzungsgrenzen regenertiver Energiequellen zur Stromerzeugung in Deutschland, Wasserwirtschaft 84,Germany

Kuester J. L., Mize J. H., 1973,Optimization techniques with Fortran, McGraw Hill Book company

Klemes V., August 1983,Hydrological uncertainity in Water- management context Sceintific Procedures Applied to the Planning, Design and Management of Water Rsources Systems; (Proceedings of the Hamburg Symposium) IAHS pulb. No.147

Linsley R.K., 1979,Water Resources Engineering McGraw Hill company

Mayer-L. G.;1998,Probleme der Energieversorgung Gebiete in ländlcher Africa, Weltforum Verlag, München,

Morris E. , 1998, Analysis of Renewable energy retrofit options to existing diesel mini-grids APEC Publication number 98, New York

Mimikou M. and Kanelopoulou S., January 1987, Optimal firm hydro energy Journal of Water Power & Dam construction

Mutreja K N,1986, Applied Hydrology, Tata McGraw-Hill Publishing company , New Delhi

Press W.H. and et. al,1992, Numerical Recipes in Fortran Cambridge University Press

Water Resouces Commission (WRC) ,1974,Ethiopia: preliminary water resources potential studies in the Bilate basin

Hydropower in the New Millennium, Honningsvåg et al (eds), © 2001 Taylor & Francis, ISBN 90 5809 195 3

Ecological impacts of hydro peaking in rivers

A.Harby, K.T.Alfredsen, H.P.Fjeldstad & J.H.Halleraker
SINTEF Energy Research, Trondheim, Norway

J.V.Arnekleiv & P.Borsányi
The Norwegian University of Science and Technology, Trondheim, Norway

L.E.W.Flodmark & S.J.Saltveit
The University of Oslo, Oslo, Norway

S.W.Johansen
Norwegian Institute for Water Research, Oslo, Norway

T.Vehanen & A.Huusko
Finnish Game and Fisheries Research Institute, Paltamo, Finland

K.Clarke & D.A.Scruton
Fisheries and Oceans, Environmental Sciences Section, St. John's, Newfoundland, Canada

ABSTRACT: The deregulated Norwegian energy market will probably lead to increased use of hydro peaking. Sudden reductions in river flow may cause high mortality of juvenile salmonids through stranding. Much higher incidence of stranding was found in daylight during winter conditions compared to at summer temperatures. Diving observations and radio telemetry show that juvenile salmon and trout move rapidly into newly water-covered areas when discharge is increasing. Salmon parr does not show more movements during frequent flow variations than for stable conditions. Young brown trout seek cover during frequent and rapid flow changes. We found no significant increased stress level in young brown trout subject to repeated peaking operations. In rivers with almost daily peaking operations, the distribution and composition of invertebrates and water vegetation are severely disturbed only in the dewatered zones. New simulation modelling techniques including GIS-analysis, are being developed further to ensure a holistic integration of results.

1 INTRODUCTION

It is well known that rapid fluctuations in river flow lead to a dramatic change in the habitat for the freshwater biota. Previous studies in American rivers have documented numerous biological impacts of hydropower induced flow fluctuations, and some mitigation requirements are suggested by Hunter (1992) and various frameworks are summarised by Morrison & Smokorowski (2000). In France, Valentin et al (1996) found severe impacts from hydro peaking only when the residual flow at non-peaking hours were very low. Lauters et al (1996) found severe impacts when the morphodynamic conditions of the influenced river were poor, leading to little heterogeneity in habitats. Further, scientific studies of biological impacts of hydro peaking are also documented from Norway. Hvidsten (1985) reported large losses of Young-of-the-Year (YoY) Atlantic salmon and brown trout in river Nidelva in Mid-Norway, and suggest that the poor recruitment of trout in the anadromous part of the river was due to hydro peaking. Arnekleiv et al. (1994) documented significant cross section zonation of macroinvertebrates linked to the dewatering zones of river Nidelva after a period of hydro peaking. This river

has a rich bottom fauna, but the fauna in the shallow zone is very sparse. They linked the reduced diversity to frequent water fluctuations caused by hydro peaking of the river.

The recent deregulation of the Norwegian energy market will most probably lead to increased use of hydro peaking production, utilising the full potential of the hydropower system for daytime power production and running a low production or power import at night time. The effects of such a strategy on the aquatic ecosystem are not well understood, and the Norwegian research council and several power producers have initiated a research program to investigate the impacts on the ecosystem from hydro peaking and to provide tools for analysing the impacts of future hydro peaking projects.

The multidisciplinary project lead by SINTEF has been initiated to study the effects of hydro peaking on aquatic ecosystems. The project was started in 1997 and will be completed by the end of 2001 to reach the following main objectives:
(a) To quantify the ecological consequences of rapid changes in fluvial water levels on fish, benthos and water vegetation,
(b) To develop management tools and guidelines to decrease the harmful effects of hydro peaking.

To reach these goals, the following subprojects are defined:

1 Field experimental studies carried out in two Norwegian rivers and in laboratory to quantify stranding of juvenile Atlantic salmon and brown trout under various climatic and hydraulic conditions.
2 Short-term temporal variation in composition and distribution of available physical habitats for juvenile salmonids and benthos analysed at study sites downstream peaking hydropower plants. Both habitat- and bioenergetic modelling for fish will be carried out.
3 Fish behaviour and shelter type selection during rapid water level variation investigated in laboratory and subsequently studied in natural situations using high precision telemetry.
4 Water vegetation (moss and algae) and river benthos monitored and linked to modelled hydraulic conditions in hydro peaking rivers. Time series analysis to show the impact from the variable hydraulic conditions on the composition and distribution of water vegetation.
5 Energy consumption, growth and stress for juvenile fish will be studied to quantify the energetic costs for fish due to rapid and frequent changes in water levels.

All the subprojects will be integrated and results will be analysed together, leading to an overall development of new methods and simulation models to assess environmental impacts of hydro peaking.

2 STUDY SITES

Four regulated rivers in Norway and one regulated river in Canada were selected for field studies. These are river Nidelva in Trondheim and river Surna, both in Mid-Norway, the River Dale close to Bergen in western Norway, Mandal River system in southern Norway and the West Salmon River in southern Newfoundland, Canada. The typical range of flows during studies in the test rivers are 30 to 110 m^3s^{-1} (Nidelva), 5 to 30 m^3s^{-1} (Dale), 1.3 to 30 (80) m^3s^{-1} (Smeland in Mandalselv), 1.5 to 10 m^3s^{-1} (Laudal in Mandalselv), 15 to 33 m^3s^{-1} (Surna) and 1.3 to 7 m^3s^{-1} (West Salmon River). All the test rivers are most of the time operated with frequent and rapid changes in flow, mainly due to hydro peaking.

For the experimental field stranding studies, large enclosures were built in the drawdown zone of river Nidelva at "Trekanten" and in river Dale.

The Trekanten area in the river Nidelva was also chosen for physical habitat studies.

Four flumes in the laboratory of The Finnish Game and Fisheries Research Institute were used to study fish behaviour during fluctuating flow.

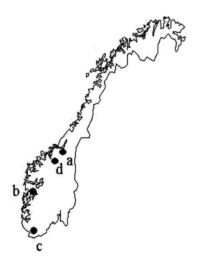

Figure 1. Norwegian study sites: a) River Nidelva, b) River Dale, c) Mandal River system, d) Surna river.

The 6 km bypass section of Laudal power plant was chosen for fish behaviour studies in the river Mandalselv. The flow was manually varied between 1.5 and 10 m^3s^{-1} by adjusting a gate. The area around the Trekanten site in river Nidelva was also used for behaviour studies using radio telemetry.

Two river reaches of approximately 100m each, situated respectively 500m and 800m downstream the outlet of Smeland power station in the Mandal river system, were chosen for studies of water vegetation. The Smeland power station is normally operated with peaking flow at 30 m^3s^{-1} during daytime and no water at night time, leading to about 1.3 m^3s^{-1} discharge in the river due to minimum flow release and natural runoff.

To study special situations with a high level of control on the environment, we have established an indoor channel in the SINTEF laboratory. The channel is 20 x 4 m, water flow can be varied with maximum 300 ls^{-1} and water velocities vary from 0 to 0.5 ms^{-1}. The channel has a deep gutter on the right bank and is gradually shallower towards the left bank, representing the half of a natural river with substrate and some moss from the river, natural river stones, gravel and sand.

Figure 1 show the location of the Norwegian study sites.

3 METHODS, MODELS AND MATERIALS

3.1 *Stranding of juvenile salmon and trout*

In the river Nidelva the enclosure fenced off an area of 75 m^2 within one big enclosure. The enclosure covered the area left dry, and fish leaving the area during flow reduction were trapped in a net bag, which remains under water at minimum flow.

In the river Dale, three parallel enclosures were built with a total area of ca 90 m^2. The site in river Dale has a coarser substrate than the river Nidelva site. Fish leaving the enclosed area being dewatered during flow reduction were trapped in a plastic half-pipe, which remains under water at minimum flow.

In both rivers, the enclosures were stocked with a known number of group tagged fish, varying from approximate 1 to 2 fish per m^2, depending on fish size. Wild brown trout and/or Atlantic salmon were caught by electro shocking, and then acclimated for more than 24 hours in perforated boxes in the river before they were gently transferred into the enclosure. The fish were given time to hide and establish territories within the enclosures for 6 - 120 hours before the stranding experiment started.

Experiments were conducted with different speed and shape of drawdown at different water temperatures, air temperatures, cover and time of day. After dewatering, fish found in the net bag/half-pipe were counted and fish left in the substrate were sought after. A more detailed description of the stranding experiments is given by Saltveit et al. (2001).

3.2 Physical habitat for fish and benthos

Approximately 9600 m2 of river bed at Trekanten were topographically mapped using a theodolite with electronic distance meter ("total station"). Local water velocities were measured in 116 points at two different flow situations. The "Viking stick" was used to measure velocities by wading. The Viking stick is a rod with 4 Ott C2 current meters connected to an Aanderaa pulse counter and an Aanderaa data logger. An Aanderaa pressure sensor with air pressure equalizer is also connected to the Viking stick. The substrate was classified according to size (in meters) using visual observations, and borders between areas of uniform substrate were mapped during the total station survey.

The three-dimensional hydraulic model SSIIM (Olsen, 2000 and Olsen & Stokseth, 1995) was used to model water depth and water velocities at different flows. Fish preference data were derived from snorkelling observations of brown trout and Atlantic salmon under summer, autumn and winter conditions at the same site under various discharge (30, 50, 70 and 90 m^{3}s^{-1}). Exact positions and detailed hydraulic characteristics of the microhabitat for each fish observations were recorded. The Habitat model (Alfredsen, 1999) in the River System Simulator (Alfredsen et al, 1995) was used to combine SSIIM results with fish preference data to make habitat analysis (Borsányi, 1998).

Invertebrate composition and biomass were measured throughout a year at Trekanten in Nidelva.

Samples were taken prior to, and after hydropeaking episodes. The exact position of each sample could be linked to modeled and observed data of water velocity and water depth. Drifting invertebrates were sampled prior to, during and after hydropeaking episodes to quantify the amount of available fish food organisms related to river flow.

3.3 Fish behaviour

Fish behaviour and cover type selection were studied in artificial flumes described by Vehanen et al (2000).

In the river Nidelva, 20 salmon parr were caught by electro shocking and tagged with ATS micro tags implanted in the abdominal cavity, They were acclimated 3 days before 10 fish were released to the enclosures described in chapter 3.1. The other 10 fish were released in the river and their positions were tracked four times per day for 14 days with rapid and frequent changes in flow in the range of 30 to 110 m^{3}s^{-1}. The modelling methods for physical habitat studies (see chapter 3.2) were also used for telemetry studies.

In the river Mandalselv, 16 salmon parr were caught by electro shocking and tagged with ATS transmitters implanted in the abdominal cavity. They were acclimated 4-5 days before tracking of fish positions 4-12 times per day for 12 days with rapid and frequent changes in flow in the range of 1.5 to 10 m^{3}s^{-1}. The SSIIM and Habitat models were used to describe the physical mesohabitat. Additionally, the CASIMIR (Jorde, 1996) was also applied to the river reach to quantify physical mesohabitat conditions.

On the West Salmon River, Newfoundland, Canada, experimental flow manipulation studies to simulate hydro peaking flows has been conducted in the summer and fall of 1999, the fall of 2000, and in the winter of 2001. Juvenile Atlantic salmon (*Salmo salar*) and brook trout (*Salvelinus fontinalis*), from 10 to 15 cm length, were captured by electro shocking, surgically implanted with Lotek or ATS radio transmitters, and immediately returned to the river. After two days acclimation, habitat selection and movement, including time budgets, of experimental fish were determined. The number of fish studied varied between 9 and 16 in each of the study periods. Flows were varied between 1.3 and 7.0 m^{3}s^{-1} on a daily basis over the diel cycle, with both evening and morning peak flows being simulated. Fish position was determined from 4 to 12 times per day, more frequently during flow transition. Time budgets of individual fish were also determined by continuously recording fixed telemetry stations. The RHABSIM (Payne 1998), River 2-D (Waddle et al, 1996), and SSIIM models were used to describe physical microhabitats in one, two, and three dimensions, respectively. Telemetry derived positions are also being used to contrast observed fish position with that predicted by habitat-hydraulic models in a 'validation' exercise.

3.4 Water vegetation

To quantify the physical conditions at the Smeland sites, the river topography was mapped using a total station. Local water velocities were measured at two different flow situations. The Viking stick (see chapter 3.2) was used to measure velocities by wading. The SSIIM model was applied to the reaches.

Photographs of 30 x 40 cm of river bottom were taken 4 times per year in 1998 and 1999 in totally 473 points, and they were used to analyse abundance and cover percentages of macroscopic visible water vegetation. In addition, we took qualitative and quantitative samples of green algae biomass. The exact position of each photograph was known and could then be linked to modelled and observed values of water velocity and water depth. Water temperature and discharge were also recorded for the sampling period.

3.5 Stress physiology

The laboratory channel at SINTEF (3.8 x 21 m) was used to study stress physiology in one-year-old brown trout subjected to rapid and frequent changes in flow (60 – 190 ls⁻¹). The experiments were done without dewatering of the substrate, and hence a bank full channel despite the flow variations. Plasma levels of cortisol and glucose provided indicators of a physiological acute stress response. Control experiments were carried out on fish exposed to a constantly high discharge.

To check the acclimation to laboratory conditions, plasma cortisol and glucose levels were measured in juvenile trout after catching wild fish in river Nidelva and at several time steps after introduction to the laboratory.

3.6 Model development and integration of results

In the river Nidelva, several studies are being carried out for research purposes. The studies were not designed to be integrated in a total assessment of hydro peaking impacts, but integration of results can be done as an example. A more holistic approach will be carried out in river Surna for 2001-2002. Some additional models are also being developed (Borsányi et al 2001).

4 RESULTS

4.1 Stranding of juvenile salmon and trout

Figure 2 shows the shapes of dewatering measured at the lowest end of the stranding enclosure, 1.3 km

downstream of the power station outlet in river Nidelva. Normal shut down procedures of the power station implies dewatering of this site within ca 60 minutes. However, for the experiments we also tested various stepwise and slow shut down procedures, giving flow reductions from 110 to 30 m³s⁻¹ within ca 240 and ca 300 minutes. The slowest dewatering episodes gives an average dewatering speed of 0.2 cm per minute at the site, while normal dewatering gives up to an average of 1.3 cm per minute in our experiments.

Our controlled field experiments showed that sudden reductions in river flow caused high mortality of juvenile salmonids through stranding. Figure 3 shows in general, a far higher incidence of stranded Atlantic salmon during daytime mid-winter conditions (2.0 - 4.3 °C) compared to nighttime or daytime spring (1.7 – 3.5 °C) conditions. Notice the differences in stranding rates between day and night and between winter and spring conditions. The same tendency was also found in laboratory experiments and additional field experiments with brown trout (YoY). Under summer and autumn conditions at water temperatures above 9 °C the stranding of both Atlantic salmon and brown trout was considerable lower than at cold water, except for slow dewatering rates. At this water temperatures the worst stranding episodes occured at dewatering during nighttime (Saltveit et al., 2001).

Figure 3 also shows that stranding almost was eliminated by a moderate stepwise shut down procedure (14 – 19 cmh⁻¹) of the power station, but the use of artificial cover raised the stranding rates (not showed here).

4.2 Physical habitat for fish and benthos

Figures 4 shows the results from simulations of physical habitat for young Atlantic salmon under summer conditions at the Trekanten area of Nidelva. The results are based on preliminary studies of fish habitat use and hydraulic simulations from Borsányi (1998).

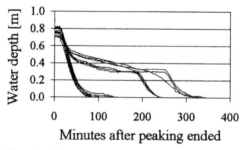

Figure 2. 13 drawdown episodes from stranding experiments in the enclosures of river Nidelva in the spring of 1999.

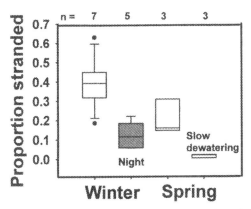

Figure 3. Proportion of stranded wild Atlantic salmon (one-year-old) derived from field experiments in river Nidelva under winter (November-December 1998) and spring (April-May 1999) conditions. The boxes represents the inner 50 per cent of the data, while the vertical lines represent the median values. The grey box is night experiments, while the rest is daytime dewatering experiments. Whiskers are 10th and 90th percentiles while dots are individual outliers outside whiskers.

After periods of almost daily peaking operations, the distribution and composition of macroinvertebrates was severely disturbed in the dewatered zones, and a clear zonation of the fauna was established. Most species of *Ephmeroptera, Plecoptera, Trichoptera* and *Simuliidae* were heavily affected, while *Chironomids, Oligochaetes* and *Tipulids* were less affected. The macroinvertebrates re-colonized the dewatered zone after about one month with a stable high flow. The permanent wetted area had a high biomass and a high diversity of macroinvertebrates at any time, also during hydropeaking periods.

4.3 Fish behaviour

Results from Vehanen et al (2000) showed that velocity shelter is important for juvenile trout when flow is increased rapidly in small flumes. More fish were flushed out during the first increase in flow than later increases during repeated peaking experiments. Juvenile trout can also recover from flushing

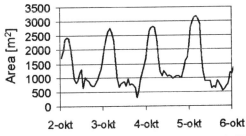

Figure 4. Time series from October 2, 1999 to October 6, 1999, of how the total area of simulated suitable velocity habitat conditions for salmon juveniles varies through several peaking operations at Trekanten in Nidelva.

and re-establish their position when flow is reduced after a peaking operation.

Radio telemetry experiments with salmon parr show no difference in number of movements during rapid flow changes compared to stable conditions in cold water in river Mandalselv (Harby et al 2001). Berland & Nickelsen (1999) found the same results at summer temperatures during radio telemetry studies in river Nidelva.

Harby et al (2001) found that fish living in a reach with varied habitat conditions changed position twice as many times as fish from a large pool when both groups were subjected to rapid and frequent flow changes.

On the West Salmon River, as discharge was increased during peaking flow experiments, the mean column velocity and depth of holding positions of both Atlantic salmon and brook trout juveniles increased. Salmon were observed to behaviourally adapt to flow changes through more contact with the substrate and use of pectoral and pelvic fins to hold position. Both species used instream cover to a greater extent at both low and high flow extremes. At the flow increments used in the study, all fish were able to maintain their general position in the stream and none were displaced from the study reach.

4.4 Water vegetation

River moss (dominated by *Nardia compressa*) and green algae (*Microspora palustris, Micorspora palustris* var. *minor, Binucleria tectorum* and *Zygogonium sp3*) can grow at any depth within the zone being dewatered and also in the permanent wetted area. Green algae (*Klebshormidium flaccidum*) growing on gravel and stones were only found in the zone being dewatered during hours without power production. Red algae (*Batrachospermum sp.*) were only found within the permanent wetted area, while macrophytes (*Juncus supinus*) preferred the permanent wetted area. We found no correlation between local water velocity and the abundance or composition of water vegetation.

4.5 Stress physiology

Flodmark et al. (2001) found a significantly elevated plasma cortisol level in juvenile brown trout after one day of cyclical flow fluctuations compared with undisturbed fish. On the fourth and seventh day of fluctuating flow, no elevation in plasma cortisol above control levels was observed. No secondary stress response measured as elevated blood glucose was detected.

Arnekleiv et al (in prep.) found an elevated blood plasma cortisol level in juvenile trout immediately after introduction to the laboratory, but the cortisol

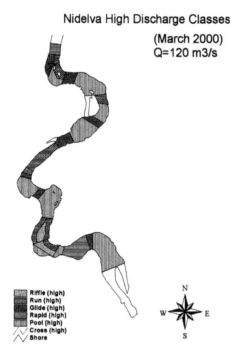

Nidelva High Discharge Classes

(March 2000)
Q=120 m3/s

Riffle (high)
Run (high)
Glide (high)
Rapid (high)
Pool (high)
Cross (high)
Shore

N
W ✶ E
S

Figure 5. Results from classification of approximately 4 km of the river Nidelva.

level was reduced to a natural level after 24-72 hours of acclimation in the laboratory.

4.6 *Model development and integration of results*

A classification system based on hydraulic properties of the river is under development, and a preliminary application has been used for river Nidelva. The system divides the river into meso habitat classes based on the slope, surface pattern, depth and water velocity of the actual part of the river. The results have been applied to a digital map of the areas by the use of GIS. Figure 5 show results from this classification at two different flow situations.

Results from all the other studies in the research project will be linked to meso habitat classes and transferred to all the identified meso habitat objects on figure 5.

5 DISCUSSION

The results from the stranding experiments show clearly that dewatering during daytime in cold water (temperatures less than 4.5 °C) will increase the stranding risk for juvenile salmon and trout. Hydropower operations should be done without flow reductions during the time of daylight to avoid this risk at cold water. However, a reduced dewatering

rate slower than 19 cm per hour, can improve the conditions. Hence, long shut down procedures of the turbines during daytime, decreased stranding of Atlantic salmon (7-9 cm) drastically under spring conditions.

The decreased stranding risk during night time dewatering at cold water is probably mainly due to a lower fish activity during the cold season and a substrate seeking behaviour especially during daytime. Stranding was lower at night, probably due to a predominant night active behaviour. In autumn, the stranding was in general low compared to mid-winter, and the lowest stranding was found during daytime at water temperatures higher than 9 °C. Temperature, season and light conditions have pronounced effect on stranding of juvenile salmonids (Saltveit et al. 2001). It is possible to reduce stranding by taking into account these ecological considerations during hydro peaking operations and to adapt speed of the shut down procedures to avoid rapid dewatering of river reaches with preferable habitat for juvenile salmonids.

The physical habitat use for juvenile salmon and trout varies throughout the seasons. At the investigated site in river Nidelva, we found high densities of juvenile salmon and trout during summer conditions even in the shallow areas. The juvenile fish moved rapidly into newly wetted areas during hydro peaking, and seemed to follow the water edge up and down, and they found suitable habitat conditions in this zone. During non-peaking hours, the habitat conditions varied incrementally, allowing the fish to find suitable habitat conditions close to their previous position.

Longer periods of regular hydro peaking reduced the diversity and patchiness of the macro invertebrate fauna in the zone being wetted during peaking, in favour of specialist species. However, no impacts on benthos were found in the permanent wetted zone. As long as the permanent wetted zone is not too small, the impact on fish food will be very moderate as Valentin et al (1995) found.

Experimental studies in small flumes showed that juvenile trout were able to recover and re-gain their previous position at low flow after being flushed downstream during high flow. However, some of the smallest fish did never recover and the long term effects of frequent flow fluctuations may be severe when the high flow is strong enough to flush fish away from their shelter (Vehanen et al, 2000).

Telemetry studies of individual salmon parr showed that habitat type and variation is important for how much impact flow fluctuations will have on their movements. When the habitat conditions stay relatively stable during flow fluctuations like in large pools, salmon parr will move less. However, we found no significant change in number of movements in salmon parr being subject to flow fluctuations compared with stable flow conditions. This indicates that flow fluctuations alone is not controlling fish movements.

River moss and green algae does not seem to have any preference for growing in permanent or partly wetted area in rivers with regular peaking operations. Some of the other water vegetation has a strong preference for either the permanent wetted area or the partly wetted area, leading to increased patchiness and reduced local diversity in water vegetation. Water velocity and shear stress can vary a lot within small time steps in rivers with peaking operations. Unexpectedly, we found no correlation between maximum water velocity and water vegetation abundance or composition. This is probably due to a stabilisation of the river bed and rapid re-growth of vegetation if parts of it are torn off during peaking.

Juvenile trout does not seem to gain evaluated stress levels from repeated flow fluctuations at sites without dewatered substrate, leading to the conclusion that hydro peaking does not affect fish stress level when peaking is done frequently in rivers with high minimum flow (Flodmark et al, 2001). These findings can be seen in the light of the lack of flow induced movements achieved in the telemetry studies. With few exceptions the trout was not observed to leave the substrate under or after the flow reductions. Observations under field stranding experiments support this (Saltveit et al, 2001), and Bradford et al. (1995) reported that fish did not leave their consealment in the substrate until their backs or tail became exposed to air under dewatering. Typically fish often wait until it is almost too late before it leaves its microhabitat even under complete dewatering (Halleraker et al, in prep.).

However, peaking operations that are carried out on an irregular basis with many days of stable flow between each peaking, is probably more harmful for the stress level. More investigations over longer time periods should be conducted with additional studies of fish energy budget.

The preliminary integration of subprojects and simulation models looks promising, but the method must be applied at a broader scale and validated in a suitable river system. This river system must be regulated with traditional hydro operations today that are transferred into peaking operations for the future.

6 ACKNOWLEGDMENTS

The Norwegian Research Council, the authorities and several power producers have sponsored the project.

The authors would like to thank the following scientists and students for their contribution to the project: E. Ahtari, T.H. Bakken, P.L. Bjerke, T.C. Daae, T. Forseth, J. Heggenes, N.A. Hvidsten, K. Jorde, Å. Killingtveit, B. Kjelsberg, N.R.B. Olsen, B. Kohler, A. Mäki-Petäys, T. Ott, C. Pennel, D. Perry, E. Stephenson and K.A. Vaskinn.

REFERENCES

Alfredsen, K. 1999. An object oriented framework for application development and integration in hydroinformatics. Doktor ingeniør thesis. Institute for hydraulic and environmental engineering, Norwegian University of Science and Technology, Trondheim.

Alfredsen, K. 1997. A Modelling System for Estimation of Impacts on Fish Habitat. Proceedings, *XXVII IAHR Congress Water for a Changing Global Community, San Francisco, USA, ASCE*

Alfredsen K., Bakken T.H. and Killingtveit (editors) 1995. User's Manual. SINTEF NHL report.

Arnekleiv, J.V., Urke, H.A, Kristensen, T., Halleraker, J.H. & Flodmark, L.E.W. in prep. The recovery time and stress response in wild, juvenile brown trout (Salmo trutta L.) subjected to electroshock, temperature-shock, handling and flow fluctuations. To be submitted to Journal of Fish Biology.

Arnekleiv, J. V., Koksvik, J. I., Hvidsten, N. A. & Jensen, J. A. 1994. Virkninger av Bratsberg-reguleringen (Bratsberg kraftverk) på bunndyr og fisk i Nidelva, Trondheim (1982 – 1986). Vitenskapsmuseet, *Rapport Zoologisk Serie* 1994-8. (In Norwegian)

Berland, G. & Nickelsen, T. 2000. Atferdsrespons hos stor parr av laks ved effektkjøring av kraftverk. Hovedoppgave ved Høgskolen i Telemark, studieretning for Naturforvaltnig. (In Norwegian)

Borsányi, P., Killingtveit, Å. & Alfredsen, K.T. 2001. A decision support system for hydropower peaking operation. Proceedings of *Hydropower 01*.

Borsányi, P. 1998. *Physical habitat modelling in Nidelva, Norway.* Diploma thesis, Department of Hydraulic and Environmental Engineering, NTNU.

Bradford, M. J., Taylor, G. C., Allan, A. & Higgins, P. S. 1995. An experimental study of the stranding of juvenile Coho Salmon and Rainbow Trout during rapid flow decreases under winter condition. *North American Journal of Fisheries Management:*, vol 15, 473-479.

Bradford, M. J. 1997. An experimental study of stranding of juvenile salmonids on gravel bars and in side channels during rapid flow decreases. *Regulated Rivers: Research & Management*, vol 13, 295-401.

Flodmark, L.E.W., Urke, H.A., Halleraker, J.H., Arnekleiv, J.V., Vøllestad, L.A. & Poléo, A.B.S. 2001. Physiological stress responses in juvenile brown trout (*Salmo trutta* L.) subjected to a fluctuating flow regime in an artificial stream. Submitted to *Can. J. Fish. Aquat. Sci.*

Halleraker, J. H., Saltveit, S. J., Arnekleiv, J. V., Fjeldstad, H.-P. & Harby A. in prep. Factors influencing stranding of wild juvenile brown trout during rapid and frequent flow decreases in an artificial stream. To be submitted to Journal of Regulated Rivers.

Harby, A., Clarke, K., Pennell, C., Scruton,D.A., Stephenson, E., Ott, T., Haugland, S. & Heggenes, J. 2001. Habitat use and movement patterns of Atlantic salmon parr (*Salmo salar*) during rapid variations in discharge. The Fourth Conference of Fish Telemetry in Europe. Trondheim, 26-30 June 2001.

Heggenes, J., Krog, O.M.W., Lindås, O.R., Dokk, J.G. & Bremnes, T. 1993. Homeostatic behavioural responses in a changing environment: brown trout (*Salmo trutta*) become nocturnal during winter. *Journal of Animal Ecology* 62: 295-308.

Hunter, M. A. 1992. Hydropower flow fluctuations and salmonids: A review of the biological effects, mechanical
causes and options for mitigation. State of Washington,
Department of Fisheries, *Technical Report* No. 119.

Hvidsten, N. A. 1985. Mortality of pre-smolt Atlantic salmon,
Salmo Salar L., and brown trout, *Salmo Trutta L.*, caused
by fluctuating water levels in the regulated River Nidelva,
central Norway. *Journal of Fish Biology* 27: 711-718.

Jorde, K. 1996. Ecological evaluation of instream flow regulations based on temporal and spatial variability of bottom
shear stress and hydraulic habitat quality. *Proceedings of
the 2nd International Symposium on Habitat Hydraulics*,
Quebec, Canada. B: 163-186.

Lauters, F., Lavandier, P., Lim, P., Sabaton, C. & Belaud, A.
1996. Influence of hydropeaking on invertebrates and their
relationship with fish feeding habitats in a Pyrenean river.
Regulated Rivers, 12: 155-169.

Morrison, H. A. & Smokorowski, K. E. 2000. The application
of various frameworks and models for assessing the effects
of hydropeaking on the productivity of aquatic ecosystems.
Canadian technical report of fisheries and aquatic sciences
no. 2322.

Olsen, N.R.B. 2000. A three-dimensional numerical model of
sediment movements in water intakes with multiblock option. Version 1.1 and 2.0 for OS/2 and Windows. User's
Manual. Trondheim, Norway.

Olsen, N.R.B. & Stokseth, S. 1995. Three-dimensional Numerical Modelling of Water Flow in a River with Large
Bed Roughness. IAHR Journal of Hydraulic Research, Vol
33, No. 4.

Payne, T.R. & Associates. 1998. RHABSIM 2.1 for DOS.
User's manual.

Saltveit, S.J., Halleraker, J.H., Arnekleiv, J.V & Harby, A.
2001. Field experiments on stranding in juvenile Atlantic
salmon (*Salmo salar*) and brown trout (*Salmo trutta*) during
rapid flow decreases caused by hydropeaking. *Submitted to
Regulated Rivers.*

Valentin, S., Wasson, J.G. & Philippe, M. 1995. Effects of hydropower peaking on epilithon and invertebrate community
trophic structure. *Regulated Rivers, 10: 105-119.*

Valentin, S., Lauters, F., Sabaton, C., Breil, P. & Souchon, Y.
1996. Modelling temporal variations of physical habitat for
brown trout (*salmo trutta*) in hydropeaking situations.
Regulated Rivers, vol 12, numbers 2 & 3, pp 317-330.

Vehanen, T., Bjerke, P. L., Heggenes, J., Huusko, A. & Mäki-
Petäys, A. 2000. Effect of fluctuating flow and temperature
on cover type selection and behaviour by juvenile brown
trout in artificial flumes. *Journal of fish biology* 56: 923-
937.

Waddle, T., .Steffler, P., Ghanem, A., Katopodis, C. & Locke,
A. 1996. Comparison of one and two-dimensional hydro-
dynamic models for a small habitat stream. In: *Leclerc, M.,
Cote, Y., Valentin, S., Capra, H. .& Bodreault, Y. (eds.)
EcoHydraulics2000, Quebec City, Canada.*

Hydropower in the New Millennium, Honningsvåg et al (eds), © 2001 Taylor & Francis, ISBN 90 5809 195 3

Impact of climatic change on hydropower investment

G.P.Harrison & H.W.Whittington
Dept. Electronics & Electrical Engineering, University of Edinburgh, Edinburgh, UK

ABSTRACT: The increased use of renewable energy is critical to reducing emissions of greenhouse gases in order to limit climatic change. Hydropower is currently the major renewable source contributing to electricity supply, and its future contribution is anticipated to increase significantly. However, the successful expansion of hydropower is dependent on the availability of the resource and the perceptions of those financing it. Global warming and changes in precipitation patterns will alter the timing and magnitude of river flows. This will affect the ability of hydropower stations to harness the resource, and may reduce production, implying lower revenues and poorer returns. Electricity industry liberalisation implies that, increasingly, commercial considerations will drive investment decision-making. As such, investors will be concerned with processes, such as climatic change, that have the potential to alter investment performance. This paper examines the potential impact of climatic change on hydropower investment. It introduces a methodology for quantifying changes in investment performance, and presents preliminary results from a case study. These inform discussion of the implications for future hydropower provision and our ability to limit the extent of climatic change.

1 INTRODUCTION

Climatic change is expected to be the outcome of increases in atmospheric concentrations of "greenhouse" gases resulting from human activities (Houghton et al., 1990). The emissions are caused, in part, by fossil-fuelled electricity generation, and as world energy demand is expected to at least triple by the end of the twenty-first century (Nakicenovic et al., 1998), emissions and hence concentrations are expected to rise considerably. The impact of climatic change could be significant especially if less developed countries expand their electricity supply systems using fossil fuels.

In an attempt to control greenhouse gas concentrations and slow down the greenhouse process, governments are aiming to cut or stabilise emissions relative to 1990 levels. To achieve this target, the energy sector will have to change the way it operates: it could reduce its reliance on fossil fuels, use more renewable energy, and practice greater energy efficiency. Together with other means, such measures should allow the climate to reach and stabilise at a new equilibrium level.

Over the next century or so, during which this new set of equilibrium conditions will be reached, generating plant could be expected to be replaced twice (the design life of the electro-mechanical equipment in a power station is rarely greater than

50 years). Increasing demand and the move to deregulated electricity systems means that private investment is likely to be used to fund new and replacement capacity. This, in turn, means that the perceptions of current and future investors will play a major role in whether emission cuts are achieved.

2 CLIMATE CHANGE

Many greenhouse gases, including carbon dioxide (CO_2), occur naturally and keep the earth warm by trapping heat in the atmosphere. However, since the Industrial Revolution, man-made sources of CO_2 have added greatly to atmospheric concentrations. In particular, transportation and the burning of fossil fuels for electricity generation are frequently cited as major sources.

Enhanced levels of greenhouse gas concentrations are predicted to cause a significant rise in temperature over the next century, with rates of increase anticipated to be greater than at any time in the past. The current consensus is that under present rates of economic and population growth, global mean temperatures will rise by around 3°C by the end of the next century, although there is considerable uncertainty surrounding the degree of climate sensitivity. Figure 1 shows that throughout the twentieth century, temperatures have been rising and that the rate

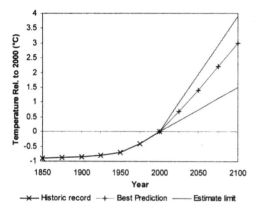

Figure 1. Historic and future temperature rise (adapted from Houghton et al., 1990)

of increase is accelerating. The rise in temperature is expected to be accompanied by increases in global mean precipitation levels of up to 15% (Houghton et al., 1990).

Many predictions of future climate are based on the output of complex numerical General Circulation Models (GCMs) which simulate physical processes in the atmosphere and oceans. Although GCMs differ in the detail of their methodologies, most agree on the general temperature trend (Gates et al. 1990, Wood et al. 1997).

There are many potential impacts of climatic change including: loss of land due to sea level rise, damage from increased levels of storm activity, and threats to bio-diversity (Houghton et al., 1990).

Under the Kyoto Protocol (UNFCCC, 1998) most countries agreed that they would limit greenhouse gas emissions. As electricity production accounts for a significant portion of the emissions, much of the burden will fall on this sector. Increased use of renewable energy sources, including hydropower, is one suggested way in which the emissions targets can be met.

Unfortunately, the very fact that renewable energy resources harness the natural climate means that they are at risk from changes in climatic patterns. As such, changes in climate due to higher greenhouse concentrations may frustrate efforts to limit the extent of future climatic changes.

3 CLIMATE IMPACTS

Hydropower is currently the only major renewable energy source contributing to global electricity supply. Given the expectation of a threefold increase in hydropower production over the next century, the continuing significant contribution from hydropower warrants a closer investigation of the potential impacts of changing climate on hydro.

3.1 River Flows

At first glance, rising global precipitation would seem to provide opportunities for increased use of hydroelectricity. Unfortunately, such increases will not occur uniformly over time or space, and many regions are projected to experience significant reductions in precipitation. In addition, the temperature rise will lead to increased evaporation. The combination of changes in precipitation and evaporation will have profound effects on catchment soil moisture levels. The soil provides storage and regulates runoff regimes. Drier soil absorbs more rainfall, tending to reduce the quantity of water available for runoff, while more saturated soils absorb less rainfall increasing the likelihood of flooding.

In river basins that experience significant snowfall, higher temperatures will tend to increase the proportion of wet precipitation. This may increase winter river flows, lead to an earlier spring thaw and reduce summer low flows (Gleick, 1986). Figure 2 shows a hypothetical example of this.

Climate change impacts studies have, in general, relied on rainfall-runoff models to translate changes in precipitation and temperature into altered river flows. GCMs provide information on how climatic variables may change in the future. Unfortunately, each GCM tends to predict a different change in temperature and precipitation, which results in significant and often contradictory differences between resulting river flow impacts. An alternative is to examine basin sensitivity to changing climate, through the application of uniform changes in precipitation and temperature.

A significant body of knowledge exists regarding the impact of climate change on river flows (e.g. Gleick, 1986; Arnell & Reynard, 1996). Many suggest significant sensitivity to climate change.

Reibsame et al. (1995) examined climate impacts on several major rivers. For the Zambezi, GCM sce-

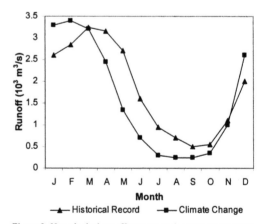

Figure 2. Hypothetical runoff patterns under current and potential climate change scenarios

narios suggested that mean annual runoff may decline by 17% or rise by 18%. The most severe change occurred with the Nile which under one scenario mean flows fell to less than a quarter of their historic level. Overall, Reibsame et al. (1995) note that river basin sensitivity increases with aridity, and this, to some degree, explains the severe fall in Nile flows.

Despite differences between the study techniques used and river basin characteristics, Arnell (1996) drew the following conclusions:
1 Runoff is relatively more sensitive to precipitation change than temperature change.
2 River basins tend to amplify changes in precipitation.

Whilst changes in annual runoff are a useful indicator, often the seasonal changes are more profound. For example, Mimikou et al. (1995) found that for the Mesohora basin in Greece a 20% fall in precipitation accompanied by a 4°C temperature increase resulted in a 35% reduction in annual runoff. However, the impact on summer flows was almost twice as large, and the fall in winter was limited to 16%. This pattern is repeated in many other studies and is a result of changes in soil moisture content.

3.2 Hydroelectric Generation

Hydropower potential is defined by the river flow, and therefore changes in flow due to climate change will alter the energy potential. More importantly, as most hydropower schemes are designed for a particular river flow distribution, plant operation may become non-optimal under altered flow conditions.

The capability of a given hydro installation to generate electricity is limited by its storage and turbine capacities. These place limits on the amount of carry-over storage to allow generation during dry spells, and also the degree to which benefit can be derived from high flows.

A number of studies have examined the impact of climate change on hydropower production (those listed in Table 1 are a representative sample). Published results suggest that the climate sensitivity of energy production is related to the storage available: in general terms the greater the degree of storage the lower the sensitivity. Additionally, turbine capacity

limits the ability of schemes to take advantage of higher flows.

Other than energy volumes, the impact on generation reliability has been examined in a number of studies (e.g. Mimikou & Baltas, 1997). Garr & Fitzharris (1994), among others, relate both hydropower production and energy demand to climatic variables in their examination of how climate change will affect the ability of the electricity supply system to meet demand.

3.3 Revenue and finance

Despite such studies, none published to date has quantified the potential impact on the perceived or actual financial performance of hydro stations.

Hydro is characterised by low operational costs but high capital costs. As a result, the debt repayment period for a hydro scheme is often significantly longer than for fossil-fuelled plant. Despite high fossil-fuel costs, hydro will often be at a disadvantage, and would not be favoured by short-term orientated investors. As with all generation methods, electricity sales revenue is the only way of servicing the capital debt. If reductions in runoff and output were to lead to reductions in revenue, this would adversely affect the return on investment and hence the perceived attractiveness of the plant. Therefore, there is a possibility that potential schemes would not be pursued.

If potential hydro schemes are abandoned or production from existing facilities is limited by runoff changes, then the likely alternative is that fossil-fuelled stations will have to be constructed to cover the deficit. Not only would this require additional capital to be used, but also would probably result in additional carbon emissions, thus exacerbating climate change (Whittington & Gundry, 1998).

Many large hydropower developments in less developed countries have been built with the intention of stimulating economic development. Often, these are internationally financed and repaid in hard currency. Reductions in revenue may make it difficult to repay the debt, severely stressing weak economies, while the shortfall in electricity availability will hamper Governments' development attempts (Whittington & Gundry, 1998).

The magnitude of capital investment required for hydropower installations, together with the increasing penetration of private capital in the industry makes it imperative that project analysis takes account of potential climatic effects.

Table 1. Examples of potential changes in annual hydro generation resulting from changes in temperature and precipitation.

Region/River	Temperature	Precipitation	Production
Nile River*	+4.7°C	+22%	-21%
Indus River*	+4.7°C	+20%	+19%
Colorado River **	+2.0°C	-20%	-49%
New Zealand ***	+2.0°C	+10%	+12%

Notes: * Reibsame et al. (1995), ** Nash & Gleick (1993), *** Garr & Fitzharris (1994).

4 INVESTMENT APPRAISAL

To assess the threat that climate change poses to future hydropower investment, there is a requirement for a robust methodology. The diverse nature of hydropower installations and climatic conditions pre-

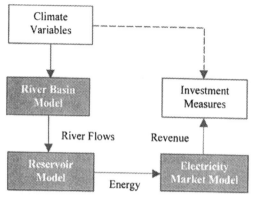

Figure 3. Software tool structure.

cludes any form of accurate regional or global analysis at this stage. Therefore, an analysis on a case by case basis is necessary.

To assess the impact on investment it is necessary to consider the problem from the standpoint of a potential investor. They will be primarily concerned with the impact on a range of investment indicators, and, as such, a methodology derived from traditional hydropower appraisal was devised.

The techniques of hydropower appraisal are long established. However, the continuing reliance on historic flows to indicate future flow conditions is not prudent given the prospect of climate change. Some recent project appraisals have attempted to deal with climate change by uniformly altering river flows. Unfortunately, this practice is inadequate as it fails to take into account the tendency of a river basin to amplify precipitation changes.

The complexity of the task necessitates a software tool, the basic specifications for which are introduced elsewhere (Harrison et al., 1998) and illustrated schematically in Figure 3. The use of a rainfall-runoff model removes the reliance on historic flows by providing a link between climatic variables and river flows. This enables the relationship between climate and financial performance to be examined effectively.

The rainfall-runoff model is calibrated using monthly historic river flow and climate data. Following this, suitable operational, financial and economic data enables simulations to be rapidly carried out.

5 RESULTS

Software has been developed by the authors to meet the required specifications. The software was tested using an actual planned scheme: sample results are presented here. The chosen scheme has limited reservoir storage capacity and is intended to operate as a run-of-river plant. The river flow regime is highly seasonal and is not influenced by snowfall. Basic operational and financial information was extracted from a traditional feasibility study of the scheme. Simulations indicated that the software delivers production estimates and investment measures that are comparable with figures found in the feasibility study.

A sensitivity study was carried out with the model driven by historic precipitation and temperature data uniformly changed to simulate climate change. Results suggested that runoff and energy production are sensitive to rainfall change, and that runoff changes are significantly greater than the precipitation variation. Although storage is limited, production sensitivity is lower than runoff. Energy production is less sensitive to increases in flow as much of the excess flow is spilled.

The assumption of a single energy price means that the investment sensitivity follows a similar pattern to production. Figure 4 shows the response of internal rate of return (IRR) and discounted payback to rainfall variations. IRR is positively related to rainfall, whilst discounted payback period shows the opposite trend. The greater sensitivity to flow reductions can be seen.

Net present value is not shown in Figure 4 as the NPV variations significantly larger. The compounding effect of revenue changes over the project lifetime means that NPV ranges from -200% to 140%.

Although these results are only preliminary, they indicate that the financial performance of the scheme is sensitive to rainfall changes. Furthermore, they imply that in regions that experience reduced rainfall, hydropower could become less competitive. As such, investment in hydropower projects will be less likely, and the ability to limit climate change will be reduced.

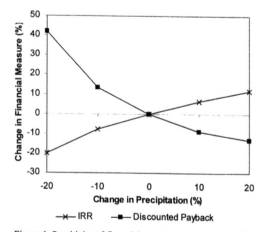

Figure 4. Sensitivity of financial appraisal measures to uniform changes in precipitation

6 CONCLUSIONS

Climatic change is expected to result from the release of significant quantities of man-made emissions of greenhouse gases. One of the key methods of limiting the extent of change is through the use of renewable energy sources, including hydropower. Unfortunately, the reliance of hydropower on climatic conditions means that the changes predicted may affect it adversely. In particular, and given the increasing importance of private capital within the electricity industry, the financial performance of hydro schemes may be damaged. Subsequently, hydropower will be less competitive and alternative, presumably fossil-fuelled schemes will take precedence, reducing our ability to reduce greenhouse gas emissions.

A range of impacts on river flows and hydropower production have been identified, together with a consideration of the potential consequences of failing to take account of climate change when planning hydro schemes. A methodology and associated software tool have been briefly introduced which enable quantification of changes in investment performance as a result of changes in climate. Preliminary results of its use on a planned scheme are presented. The results indicate that investment measures show significant sensitivity to changes in rainfall. This implies that in regions that experience reductions in rainfall, hydropower will become less competitive. Therefore, investment in hydro projects is less likely to occur and our ability to control greenhouse emissions is lessened.

REFERENCES

Arnell, N. 1996. *Global Warming, River Flows and Water Resources*, J. Wiley & Sons Ltd.

Arnell, N.W. & Reynard, N.S. 1996. The effects of climate change due to global warming on river flows in Great Britain, *J. Hydrol.*, 183: 397-424.

Garr, C.E. & Fitzharris, B.B. 1994. Sensitivity of Mountain Runoff and Hydro-Electricity to Changing Climate. In Beniston, M. (ed), *Mountain Environments In Changing Climates*: 366-381. London: Routledge.

Gates, W.L., Rowntree, P.R. & Zeng, Q.-C. 1990. Validation of Climate Models. In Houghton, J.T., Jenkins, G.J. & Ephraums, J.J. (eds), *Climate Change The IPCC Scientific Assessment*: 92-130. Cambridge University Press.

Gleick, P.H. 1986. Methods for evaluating the regional hydrologic impacts of global climatic changes, *J. Hydrol.*, 88: 97-116.

Harrison, G.P., Whittington, H.W. & Gundry, S.W. 1998. Climate change impacts on hydroelectric power, *Proc. 33rd Universities Power Engineering Conference (UPEC '98), Edinburgh, 8-10 September 1998*: 391-394.

Houghton, J.T., Jenkins, G.J. & Ephraums, J.J. (eds) 1990. *Climate Change The IPCC Scientific Assessment*, Intergovernmental Panel on Climate Change, Cambridge University Press.

Mimikou, M.A. & Baltas, E.A. 1997. Climate change impacts on the reliability of hydroelectric energy production, *Hydrol. Sci. J.*, 42 (5): 661-678.

Mimikou, M., Kouvopoulos, Y., Cavadias, G. & Vayianos, N. 1991. Regional hydrological effects of climate change, *J. Hydrol.*, 123: 119-146.

Nakicenovic, N., Grubler, A. & McDonald, A. (eds) 1998. *Global Energy Perspectives*, Cambridge University Press.

Nash, L.L. & Gleick, P.H. 1993. The Colorado River Basin and Climatic Change: *The Sensitivity of Streamflow and Water Supply to Variations in Temperature and Precipitation*, US Environmental Protection Agency, Washington D.C.

Reibsame, W.E., Strzepek, K.M., Wescoat Jr., J.L., Perritt, R., Gaile, G.L., Jacobs, J., Leichenko, R., Magadza, C., Phien, H., Urbiztondo, B.J., Restrepo, P., Rose, W.R., Saleh, M., Ti, L.H., Tucci, C. & Yates, D. 1995. Complex River Basins. In Strzepek, K.M. & Smith, J.B. (eds), *As Climate Changes : International Impacts and Implications*, 57-91. Cambridge University Press.

UNFCCC 1998. *Kyoto Protocol to the United Nations Framework Convention on Climate Change*, UNFCCC.

Whittington, H.W. & Gundry, S.W. 1998. Global Climate Change and Hydroelectric Resources, *IEE Engineering Science & Technology Journal*, 7 (1): 29-34.

Wood, A.W., Lettenmaier, D.P. & Palmer, R.N. 1997. Assessing climate change implications for water resources planning' *Climatic Change*, 37: 203-228.

Hydropower in the New Millennium, Honningsvåg et al (eds), © 2001 Taylor & Francis, ISBN 90 5809 195 3

Hydropower in a wider energy and greenhouse gas perspective

Per Øyvind Herpaasen, Geir Dyrlie, Anders Ruud & Eivind Torblaa
Statkraft SF, Høvik, Norway

ABSTRACT: The text discusses in broad terms, hydropower as an environmentally friendly energy source of energy, focusing on the changes which have taken place within the planning of a modern hydropower project as compared with previous practice. To-day it is usual to give priority to projects with designs which are acceptable from an environmental point of view and from the point of view of local participation as well as having an attractive economic return.

The European Union has set itself ambitious targets as far as the future energy "mix" in Europe is concerned, with a sizeable input of renewable sources in order to meet their obligations under the Kyoto Protocol. As everyone is aware, hydropower is very attractive in that connection with low CO_2 emissions, during the whole length of the project - something which is confirmed in this article.

Statkraft is of the belief that society´s desire to solve the enormous problems confronting climate and energy policy-making will see to it that sound, sensitive hydropower projects will still be able to emerge as part of a general investment in renewable energy sources in Europe. The article examines some of the international activities and arrangements which are in the course of being established in order to promote investment in renewable energy sources in the immediate future.

1 SOME FACTS ABOUT STATKRAFT SF

Statkraft SF is Norway´s largest producer of electric power and its mission is to plan, erect and operate energy units, and to buy and sell energy.

Statkraft´s vision is to be a leading North European energy company, with hydropower as its technical speciality. The company is owned by the Norwegian Government and was established as a government company in 1992.

A government company is operated as is any other company with full administrative and financial freedom, and has a board of management selected on the basis of administrative and financial professionalism.

Statkraft SF has 890 employees, and is full or partial owner of 92 power stations. The company has operational responsibility for 54 of these whilst 38 are operated by others. Altogether Statkraft produces 34 TWh (about 9,000 MW). Statkraft is responsible for about 30 percent of the country's power production and owns about 40 percent of its water reservoir capacity.

The development of the energy markets towards de-regulation and cross-border trading open the way for

new business opportunities. Statkraft is an active participant in on-going restructuring activities, for example through long-term power exchange agreements, acquisition of ownership shares in companies and the establishment of trading activity on the Continent.

To be able to exploit Statkraft's competence within hydropower development, the company participates actively in development projects in growth areas for hydropower outside Europe – in part as

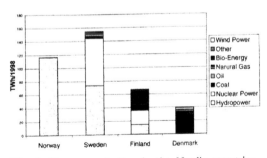

Fig. 1. Power production in the Nordic countries (except Iceland) classified by energy source

project developer and owner, and in part as implementer.

In a market characterised by an ever tighter energy balance, Statkraft wants to utilise its competence in continuing to develop further the hydropower potential in Norway. Increasing emphasis will be put on developing alternative renewable energy on a commercial basis, primarily wind power, but in the long run other alternative energy sources will be exploited. In the North European market Statkraft wishes to use its comparative advantage in the increasing competition that deregulation brings.

2 HYDROPOWER IS ENVIRONMENTALLY FRIENDLY

There has been considerable development within hydropower in the wake of the major developments in the middle of the last century. This applies to technology, costs and development alternatives, and we have become more conscious of the environmental impacts of hydropower developments. Through extensive mapping and research, both before and after construction, we have acquired fresh knowledge. This naturally means that hydropower developments and energy plants are no longer what they used to be. Modern developments no longer need to a "lay waste Nature". The times when riverbeds were left dried out, when vast areas were flooded, and the consequences of this sort of development, are now a thing of the past.

Not least the view of what are acceptable impacts of hydropower development has changed radically in the course of fifty years. This development has even been expressed by the Norwegian Prime Minister in his New Year speech where he said "our goals must be more ambitious when it comes to acceptable impacts of future developments".

Certainly, hydropower is basically a unique source of energy that does not emit greenhouse gases. In operation it is also superior in allowing rapid regulation adjustments which can be adapted to the present consumer pattern. Hydropower is predominant in energy efficiency whereby 85 – 90% of every litre of water passing through a hydropower plant is converted to electricity. Other sources of energy operate with totally different and much lower efficiency rates.

As a major producer of hydropower we are concerned that hydropower gets the environmental status it deserves. Hydropower is a renewable source of energy and has been one of the mainstays of Norway's industrial growth with close on 100% of the energy coming from hydropower. Clearly there are good and bad examples of the environmental impact of hydropower. This is largely a question of when the development has taken place. Modern hydropower plants in Norway are constructed in such a way as to spare the surroundings to the extent possible. Mitigating measures that were inconceivable only 30 – 40 years ago today are a matter of course when it comes to hydropower developments. In this connection we would like to mention the following advantages: 1) Minimum discharge and diversion works adapted to life in and around the watercourse, 2) Genetic banks for preservation and reproduction of local fish types, 3) fish ladders to facilitate the wandering of anadromous fish.

3 LOCALLY BASED PLANNING

To obtain the early participation and influence of other interests, a type of "reverse strategy" is now being implemented in the project development. Up to now it has been the norm to base projects on technical/economic criteria, and then to compensate for damage by way of different measures. Nowadays the emphasis is on the environment and local interests setting out the criteria for the size and design of the hydropower plants. In this way we can reap Nature's bounty. This implies that the need for residual discharge will determine the dimension of the power plants' capacity, especially at times of the year with naturally critically low discharges. This is important in relation to the need for water quality, recipient capacities, aesthetics, landscape appearance, outdoor recreation. Furthermore, one has to look at where the construction best fits in before the plans are launched, so that these may be acceptable from the very outset.

4 HYDROPOWER IN A CLIMATE CONTEXT.

4.1 *CO₂ and hydropower*

A life cycle assessment of hydropower has been made with the aid of the life cycle methodology LCA. As this technique has primarily proved suitable for assessments and comparisons of emissions to water and air in addition to the use of resources, we think that this is a suitable way to produce data on hydropower and CO_2 - emissions in a life-cycle perspective. Table 2 shows emissions of CO_2 from hydropower plants in comparison with other energy sources (SEG 1999). In spite of having used approved methods to arrive at the results, there is a great amount of uncertainty associated with these figures. In the main though we feel that hydropower is a very good environmental choice regarding emissions of GHG.

It is interesting to note that if electricity from Norwegian hydropower plants was to be produced by a coal fired power plant of medium efficiency, ap-

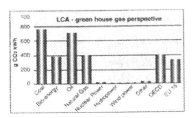

Fig. 2. Illustration of various energy sources in a greenhouse gas perspective

1997

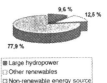

Target for 2010

■ Large hydropower
☐ Other renewables
☐ Non-renewable energy source

Fig. 3 Future share of renewable energy within the EU

proximately 80 million tons of CO_2/annum would have been emitted. (NOU 1998)

Statkraft's hydropower production is mainly based on the storing of water. In total we regulate approximately 120 reservoirs in more than 30 watercourses. We have followed the debate with great interest as to whether the storing of water for the purposes of hydropower production may, in some cases, cause larger emissions than if the production were from fossil fuel. To our knowledge the calculation of hese values is not absolutely accurate, but emissions from hydropower will only be greater where large tracts of tropical forest have been flooded to make way for reservoirs, thereby generating CO_2. Statkraft's view in this matter coincides with that of the conclusions in the report by the World Commission on Dams – Dams and Development "Current understanding of emissions suggests that shallow, warm, tropical dams are more likely to be major GHG emitters than deep, cold boreal dams" (WCD 2000).

4.2 The European energy system – future potential for RES

The European Union has decided that it is desirable to direct energy production in Europe more towards renewable energy sources. To meet the environmental requirements set out in the Kyoto Protocol, there is political agreement within EU that renewable energy is to make an important contribution to fulfilling the members' commitments on climate. Ambitious goals have therefore been proposed in the EU directive to increase the share of RES-E from 337 TWh/annum in 1997 to 675 TWh/annum in 2010, a share of 22.1 percent. The individual member country will have to fix target figures for consumption in accordance with the guidelines set out in the directive. In the directive, RES-E is defined as non-fossil fuels, wind, solar, geothermal, wave, tidal, hydroelectric, and biomass.

For individual countries in the EU various measures for subsidies and quota requirements will be used to stimulate increased use of renewable energy. Various initiatives have also been implemented to

introduce arrangements with certificates for international trade of renewable energy.

5 EXAMPLES OF PLANS FOR NEW ENVIRONMENTALLY FRIENDLY DEVELOPMENTS

Statkraft SF continuously assesses the potential for increasing its production capacity through new developments and through provision for better efficiency, in relation to existing power plants and the construction of new power plants. Small and large measures are examined, considerable emphasis is put on environmental impacts in the development of these measures and emphasis is put on involving other parties and local players early on in the process.

By providing more water to existing developments, it is often possible to avoid establishing new reservoirs since we can make use of the existing ones. In addition new encroachments often come in connection with existing ones so that it is possible, to a greater extent, to avoid incursions in hitherto undisturbed areas. Through reconstruction it is also possible to achieve a better rate of efficiency in the plant and to achieve an increase in production through that. Refurbishments and extensions are therefore in principle less prone to conflict than new developments, and it is possible to derive a lot of energy relative to the drawbacks associated with them.

(a) Example - 1200 to 1500 GWh (300-350 MW)

An example of using the new strategy is to be found in our existing facility at Røssåga which, together with its side and adjoining catchments, represents a considerable energy potential. During its latest administrative processing of the Conservation Plan for Watercourses and the Master Plan for Water Resources the Storting (the Norwegian Parliament) resolved that a licence could be applied for a sizeable development of the area. This related both to projects directly elated to existing plants, and to new developments in the area. Viewed in this light, previous encroachments upon Nature can also be more effectively used than when the plants were built 50 years ago. But there will be some new incursions e.g. in adjoining catchments. By way of dialogue at an early stage and through the involvement of local society and different professional environmental expertise, it is possible to gather knowledge on local values and priorities. This puts the focus on the relevant problems at an early phase in the development. Environmentally adapted residual water flows and different measures can make for good and acceptable conditions even after new regulation. This should be so in order that life in the watercourses with its biological diversity is not diminished and that those who live along the watercourse can still derive great enjoyment out of it. In this way it will be possible to harvest Nature's surplus without too great environmental drawbacks.

(b)   Example - 40 to 70 GWh (8-12 MW)

Another project near the same area, is an example of a measure which local people would like to promote as a purely environmental project. In this case the hydropower potential between two already regulated lakes can be exploited. But in such a way that one of the lakes can eventually emerge with a high and fixed water level, the supply of sediment from the glaciers can be reduced and some old scars on the landscape can be repaired. In this way the water can almost be restored to its original condition, and once more emerge as a true pearl of nature. In this way it is possible to improve conditions around existing regulation works while at the same time getting more energy out of the system.

(c)   Example   65 GWh (45 MW)

An example of the refurbishment and extension of an old power plant is Bjølvo. On the 7th April 2000 Statkraft were granted a licence to construct the new Bjølvo power plant in the Hardanger region. In the new power plant annual production was to be increased in relation to the existing plant by about 65 GWh to 387 GWh. The increase was due to new additional transmissions of power, to reduced flood loss and to better rate of efficiency. This is a good example of a future-oriented exploitation of resources carried out in connection with previous regulation work and existing facilities, without any major environmental problems.

The new power station and related infrastructure is "concealed" within the mountain. This is in contrast with today's power station which is easily visi-

ble with its dominant penstock running down the mountainside. The power station will also feed local industry with water for processing activities, as well as constituting a necessary supplement to the drinking water supplies for the village of Ålvik, providing it with water of good quality.

A comprehensive environmental monitoring programme has also been established for the construction period of about 3 years and a corresponding programme is being established for the operational period in order to monitor the impacts.

Conclusion 1:
There is a whole series of examples of sound environmental projects which will make for achieving the EU target of a good energy "mix". However there is little point in setting limits as to the installed power. On the contrary focus should be placed on the environmental impact of the intrusion.

Conclusion 2:
ENEF measures applied to the existing production capacity which today may be considered as uneconomic, should be regarded as environmentally friendly ones in that they provide new RES capacity.

5.2 Review/renewal of licences

In Norway we are now at the starting point of a period of comprehensive upgrading of existing licensing conditions associated with the operation of the power plants. In the first instance the upgrading shall address itself to the environmental aspects of today's regulation regime. Typical conditions which will be subject to change are new requirements for minimum water flow, changes in the regulation of the reservoirs, measures to improve biotope conditions within the area of regulation and so on. In some cases a fund for local business and an annual contributions to local sports and leisure activities have also been discussed.

As a concrete example we can mention an ongoing case where Statkraft is applying to renew "the licence for the regulation of the Numedal river".

As of today the authorities have not completed the processing of the case, but in order to illustrate the views of the Ministry of Petroleum and Energy, we have extracted the following quotations from the Ministry's comments on the Parliamentary Bill No. 37 on provisions on water flow, physical measures and so forth:

- *"The Ministry of Petroleum and Energy will give priority to setting out provisions in relation to water flow to provide a balanced and responsi-*

ble distribution of the water resources in the watercourse."

- *"The Ministry is of the opinion that it is absolutely necessary to set out provisions on water flow that ensure a viable production of salmon in the Numedal river."*

- *"At the same time the Ministry stresses that the disadvantages in the upper parts of the river course, the reservoir areas, shall be mitigated to the extent possible without endangering the salmon stocks in the watercourse."*

- *"This means that agreed water flows must be within what is reasonable in order to preserve the salmon stocks."*

From a biological point of view it is primarily the provisions on minimum water flow, altered regulation of the reservoir and physical measures that will give a better local environment through increased biological life and reproduction of fish. In addition a proposal has been put forward for contribution of funds (annual payment) to local fishing and outdoor recreation interests and to order Statkraft to establish a hatchery facility and implement extensive studies on the subject of natural science.

For Statkraft the proposed alterations mean an annual reduction in the production of 55 GWh. In today's values the measures will amount to NOK 200-250 million.

Conclusion: Renewal of hydropower licences aims at upgrading older facilities to today's environmental standard. Following renewal, the system can be compared to new renewable energy and should qualify for support at par with other renewable energy.

6 ENVIRONMENTAL CERTIFICATION, ENVIRONMENTAL PRODUCT DECLARATIONS AND CERTIFICATES FOR GREEN ELECTRICITY

6.1 *ISO-14001 as basic Environmental Management System*

Since 1994 Statkraft has continued to establish a system for quality management and in 2001 the company became one of the first to be certified in accordance with the new and process-oriented ISO-9001:2000 standard. This has given us a management tool that is well suited to the handling of environmental questions. On this basis we are preparing environmental certification in accordance with ISO-

14001, which will form the basis of the company's environmental management system. Amongst other things this contributes to a systematic overview of the environmental aspects in every single watercourse. These overviews are valuable in various connections.

Firstly they will form the basis for a general documentation and environmental reporting internally in the company and also externally. In this connection it is also easier to communicate the company's efforts in preventing and reducing negative impacts on the environment and to give the company a clearer and more positive environmental profile.

Secondly such systematic overviews will form the basis for a dialogue with local interest groups in order to arrive at improved solutions in existing facilities, and also together to find acceptable solutions to the development of new production capacity. Openness and the involvement of local interested parties and authorities will be decisive factors for success in the future.

Thirdly, in this way we can establish a system to meet the requirements in a future certification market, at the same time making it possible to make environmental declaration on the whole or parts of our production.

6.2 *The Environmental Product Declaration (ISO 14025)*

Statkraft participates in a working group with, amongst others, Norsk Hydro, Norwegian Electricity Industry Association, Danish Elsam and technological groups in Norway and Denmark (the Nimbus Project), in which the aim is to establish an environmental product declaration common to all products. The project has its basis in the values of the implemented LCI/LCA-studies, and its purpose is to develop a standard for an environmental product declaration, type III (EPD III), which has to be verified/certified by a third party. The final standard will become a Nordic model in which a new international standard for environmental product declarations are tested (ISO 14025). The new ISO standard will then be tested over a period of three years.

An EPD III contains and accounts for an internationally approved set of environmental impacts and data which can be tested. The purpose of the declaration is to reveal the environmental impacts deriving from the production of 1 kWh of power delivered to the client's door.

In this way it is possible for us to provide major purchasers of electric power, reliable documentation, should they wish to make use of environmentally friendly energy e.g. wind or hydropower, in their production. If the consumer is receiving electric power derived from several sources, e.g. a mix of brown coal, atomic power and hydropower, this will be made clear on the declaration.

An EPD III will probably contain information on the company, the company's environmental management systems, the resources used in the construction and operations of the plant, the emissions which have an impact on global warming, which generate acid rain, which impact on the Ozone Layer, on ground ozone, the eutrophication of rivers and the management of wastes from the plant. If the plant is pulled down after it has been used, the use of resources and emissions associated with that operation will also be declared.

Using such environmental product declarations, it will be possible to provide the consumer with a set of differentiated choices, and so the possibility for establishing markets for different types of environmentally friendly energy.

The NIMBUS project will put forward a proposal for a standard model for such an environmental product declaration in the first quarter of 2001.

6.3 *RECS*

RECS (Renewable Energy Certificate System) is one of many initiatives set up by different industrial and political organisations in several European countries and its purpose is to establish a platform for trading in certificates across European national borders. Today RECS has about 50 members from different organisations in the Netherlands, Belgium, Italy, the United Kingdom, Germany, Ireland, France, Austria, Denmark, Norway, Sweden and Finland. A test phase has been established from the 1st January 2001 in which some of the member countries will test the system to see if trading in such certificates across national borders can actually be done despite the different support arrangements and requirements for RES-E in individual countries. The test phase will last for 18 months.

Norway is participating in this test phase because it is important to contribute to the establishment of a credible and effective system of co-operation in certificates. At the same time it is important to be involved in the setting of the ground rules on increased use of RES-E. On the basis of the expertise on hydropower which we in Norway possess, it is important to bring to light the environmental benefits so that the system of certificates will stimulate the further development of hydropower production in the form of new construction and rehabilitation.

7 PUTTING A VALUE ON THE ENVIRONMENTAL IMPACTS OF HYDROPOWER

The costing of environmental impacts has recently gained increasing attention. Government authorities and the bigger purchasers of power on the international market can be potential users of the information which is contained in estimates of environmental costs. The European decision-making authorities are expected to consider environmental costs in their choice of subsidies/quotas/charges arrangements . The method will also possibly be employed in connection with the licensing of new hydropower projects. Statkraft will therefore participate actively in the first attempt by the energy sector to generalise the total environmental costs of hydropower.

The LC//LCA method is directed towards pollution issues and has been developed in relation to traditional industry emitting to the air and water. The special impacts of hydropower are captured to a very limited extent by this method.

Another approach is the so-called value-setting studies. In these it is assumed that all, or at least most of the most important effects and impacts of a planned power project are known. The impacts of the plans are thereafter presented to a representative selection of the population. As a rule one or other representative body from local society is used since it is they who are most affected by the encroachment. Regional, national and international bodies can also be used , depending on the purpose of the investigation.

The environmental costs of the encroachment, for example of power development, are defined as the extent of the damage inflicted by the encroachment on the watercourse after mitigating measures have been implemented. The level of the environmental cost is calculated in monetary units (in this case, Norwegian øre) per kWh. By using a monetary measure, different projects can be compared, so that Norwegian hydropower projects can be compared with projects in other countries and with other sources of energy.

The first stage in the Environmental Cost Project was established by EBL and implemented in 2000. Statkraft SF, BKK and SFE took part in the project.

The purpose of the project was to derive an estimate for the cost expressed in øre/kWh. The starting point for the studies in establishing values was the planned development at Sauda and Øvre Otta. The values from these studies were then transferred (benefit transfer) to the same electricity portfolio as for the LCI/LCA study. The project's results can be used in many contexts, amongst others as a supplement to the LCA study and so will contribute to a more complete picture of the impacts of hydropower.

On the basis of the above, the environmental cost within the Norwegian hydropower portfolio was estimated as:

> About 2.4 øre per kWh
> i.e. US¢ 0.27 per kWh

Corresponding figures exist for other energy sources. A comparison is set out in the table below. However the assumptions behind the figures shown in the table are associated with a degree of uncertainty. Statkraft SF is now working on the extension of the environmental costs project. Amongst other things, the next stage will include the addition of several watercourses/power plants, as well as a new valuation study and a comparison of the environmental costs of Norwegian hydropower versus other hydropower and other energy sources.

7.1 Eco-point/kWh

The LCA method and the environmental cost method can both be used as starting points for estimating so-called "Eco-points" . Eco-points can thus permit a comparison of the results from both approaches. Where an energy source is attributed few eco-points this means that the environmental costs are low. However the estimates of eco-points will vary depending on individual country's environmental policy and its national environmental goals. It is therefore difficult to make direct comparisons between energy sources across national borders. Nonetheless, given the necessary reservations, the figures in the table below do give a rough impression of the differences in environmental costs between eight different energy sources. The eco-point calculations will probably be taken further and as-

Tab. 1 Environmental costs of Norwegian hydropower versus other hydropower and other energy sources.

Energy Cycle	Environmental costs (øre/kWh)
Natural Gas	1.5-47.1
Biomass	2.0
Wind	0.4-2.0
Hydropower	2.4

Tab. 2 Eco-points giving a rough impression of the differences in environmental costs between eight different energy sources

Energy Source	Eco-points (approximate values)
Coal	2 412
Biomass	694
Oil	1 598
Gas	482
Atomic Power	40
Wind	19
Solar	93
Norwegian hydropower	30

sessed in more detail in the next stage of the Environmental Costs Project.

7. REFERENCES

NOU - Official Norwegian Reports 1998:11 *Energi- og kraft balanse mot 2020 (Energy and power towards 2020)*, Oslo: Elanders Publishing

SEG Consult 1999, *LCA for Hydroelectricity in Norway based on STØ's report O.R. 58.98*

WCD - World Commission on Dams 2000 *Dams and development*, UK: Earthscan

Hydropower in the New Millennium, Honningsvåg et al (eds), © 2001 Taylor & Francis, ISBN 90 5809 195 3

Electrification of rural areas in Eastern Himalayas through water wheels

D.Kedia
Industrial Consultants, Guwahati, Assam, India

Amitabh Kedia
Jk Urja, Gurgaon, Haryana, India

ABSTRACT: The Eastern Himalayan region in India comprises of five mountainous states with low levels of per capita power consumption. The more accessible villages in the region have been connected to the grid while remoter ones are being served by diesel generating sets. Both the approaches have their share of problems. The transmission grid is expensive to set up and maintain while transporting diesel to remote sites is expensive. Since most of the villages have a peak load requirement of about 10 - 50 kW, water wheels are found to be the most viable energy source for the region. The traditional water wheel has been modified and upgraded and used in installations at several sites. The paper discusses the viability of water wheels in the Eastern Himalayas, and the various initiatives taken by the Government to promote their usage.

1 INTRODUCTION

The Eastern Himalayas flank the north-eastern region of India, which is drained by the mighty Brahmaputra river originating near the Mansarovar lake in far off western Tibet. The territory is predominantly mountainous and is blessed by nature with copious rainfall. The monsoon clouds rising from the Bay of Bengal are trapped by the horseshoe shaped trough formed by the mountains, compelling them to shed their moisture on the steep slopes. The discharge flows down in torrents to be caught by the Brahmaputra in the valley.

Endowed with such geographical features, it is but natural that the region boasts of the highest rainfall anywhere in the world and has also been identified to possess the largest hydro potential site on the river Siang, which is estimated to be capable of generating 20,000 MW of hydro power at one single location.

2 NORTH EASTERN INDIA

Seven states in the extreme north east of India consist of a separate region that unfortunately is the most underdeveloped in the entire country, due to several factors, primarily geographical. Consequently, the region is far behind the national average, with respect to most economic indicators.

Five out of the seven states are almost entirely mountainous and are very sparsely populated. The village electrification levels in some states are lower than 50% against the Indian average of above 85%. The per capita power consumption (PCC) in the entire region is only 96 kWh against the national level of 320kWh. States like Arunachal Pradesh are still lagging behind with PCC levels of 66 kWh.

The population density varies from the lowest of 10 in Arunachal Pradesh to the highest level of 286 in Assam, which is even higher than the all India level of 273 persons per square kilometer. However, in reality, the interior mountain areas of the hilly states are very sparsely populated with tiny hamlets spread out in distant locations.

3 RURAL ELECTRIFICATION

Some states in the region have already achieved close to 100% village electrification, as per available statistics. However, the percentage population coverage is significantly lower. Till date, the rural electrification program is almost entirely based on a centralized generation system relying on lengthy transmission and distribution networks.

The villages that have been electrified at present, are the ones more approachable. Those remaining are very difficult to approach, at times several kilometers of foot march away from the nearest road. This is usual for the remote communities of Arunachal Pradesh, where accessibility is the prime hurdle. The experience gained so far invariably points to serious problems, both technical and economic in nature, in connecting these locations to grid power.

The cost of installation as well as maintenance of

such long lines in the hostile terrain is extremely high. The connected load as well as the total energy demand, restricted mainly to evening lighting usage, is very poor, and as a result the revenues are inconsequential. The situation is rendered worse due to the imposition of very low rates for sale of power by several state governments, which are at times lower than even the cost of generation.

India in general, and its north-east region in particular, are reeling under a perennial shortage of peak load capacities. During the evening peak hours, the brunt of the power cuts is borne by the poor rural customers who lack political clout. The end result is a situation where these remote communities are usually deprived of their precious share of electricity during the few hours of the day when they need it most.

Whenever the power does flow, the voltage and frequency levels are rarely maintained. When the lines are damaged, which is not very infrequent, the incentives to repair them expeditiously are lacking both at the organizational and worker levels.

In this scenario, the importance of decentralized power generation to meet India's social goal of cent percent rural electrification assumes great significance.

4 ENERGY OPTIONS

The options available to the planners for providing energy to the remote areas are not very many. The difficult terrain, sparse population, and very low load demands render most options expensive to construct and economically unattractive to maintain.

The density of population in the interior areas is very low. An average village consists of less than 100 persons. The peaking load requirements do not exceed 10 to 15 kW for each distant location. As there are no industrial or other activities, there is no demand during the of-peak hours. The total consumption is thus very low and the system load factor is poor.

Coal based thermal schemes (as also nuclear) are not feasible in the hill states as no viable fuel sources are available nearby. A few natural gas based projects have however been installed wherever sources have been discovered in recent years.

Large hydro schemes require huge capital investments, which are beyond the resources of the states. Moreover, the long gestation periods, the rising trend of environmental consciousness and strategic considerations in the border areas, have dissuaded planners in the past to tap the vast hydro potential of the region. However, the Government of India is slowly coming forward to execute selected projects, the generation from which is planned to be evacuated for use mostly outside the region.

The modern non-conventional energy sources like wind and solar power are not likely to make a major impact on the energy planning options for this region in the foreseeable future.

The evacuation of power by long transmission lines to distant consumption centers has severe constraints. The cost of construction in this difficult terrain is very high. The maintenance of the lines is not only expensive but also very erratic due to both natural and human factors. This results in an unstable and unreliable supply to the consumers. The line losses for transmission are very substantial and generally exceed 20 to 30% of the power transmitted. Even if the lines are drawn and maintained with great efforts, the available revenue earnings are negligible, rendering the exercise highly uneconomical.

The seemingly simple and fast maturing alternative available to the administrators is the installation of small diesel generating sets in isolated locations. This option is frequently and rather indiscriminately resorted to, for base load supplies to remote locations or as peaking load substitutes to major towns. However, this is a prohibitively costly mode of power generation. Fuel is air lifted to several sites, making the cost even higher.

5 SMALL HYDRO - THE MOST VIABLE ALTERNATIVE

Small Hydro presents itself as the most viable energy alternative for the remote areas of north-east India, in view of the constraints discussed above. This option provides a tailor made solution for the region, with its hilly region and copious rainfall.

The principal characteristics justifying the suitability of Small Hydro in this area can be outlined as:

- At most locations, economically viable sites can be identified close to clusters of villages. The necessity of long transmission lines is thus substantially reduced, thereby cutting down the project costs and reducing other disadvantages.
- Although the per unit cost of generation by the smaller schemes is higher, the ultimate landed cost at the consumption point after adding the cost of transmission and the line losses, is quite competitive, in fact much cheaper than the other alternatives as per current cost trends.
- The smaller schemes have very low total requirements of funds. The financial requirements are either met from the state's own resources or easy loans are provided by the Rural Electrification Corporation, the Power Finance Corporation, or other such financial lending institutions. The Ministry of Non-conventional Energy Sources, Government of India, has recently come out with a scheme to subsidize upto 50% of the total cost of the project (with a maximum ceiling on the subsidy at Rs. 30 million per installed MW) for the entire north east. Thus a substantial part of

the cost is funded by the Government to encourage development of Small Hydro.

- These projects have a low gestation period, of upto two working seasons extending from 20 to 24 months. Proper advance planning and fast execution can reduce the total time even further.
- Multi purpose schemes, combining power generation with irrigation and water supply benefits, make them economically more attractive.
- Some very remote unapproachable sites glaringly demonstrate the suitability of Small Hydro in this difficult territory. ANINI (3 x 50 kW), MECHUKA (2 x 50 kW) and TUTING (2 x 50 kW) are three such district/circle headquarters towns in the extreme north of Arunachal, where no road connections exist. Sets were airlifted for installation at these stations. Earlier, diesel fuel was carried to these outposts by helicopter or by several days foot march.
- Environmental consciousness and the resultant restraints have been increasingly responsible for the delay in implementation of several major hydro projects in India. Small Hydros, most of which are "run of the river" type of schemes, are environmentally compatible and cause minimal disturbance to the surroundings. Displacement of human population is rarely a problem with these smaller projects. In fact, the Small Hydros help conserve the forest resources by providing alternative electrical energy for the lighting, cooking and heating needs of the village folk.
- These projects, both during construction and after implementation, provide fresh avenues of employment to the local workforce. The civil construction work is normally executed by the local contractors, which saves the drain on the forest resources which will otherwise be caused by influx of outside labor force.

6 HISTORY OF USAGE OF WATER WHEELS IN EASTERN HIMALAYAS

Like the Persian Water Wheel or the vertical shaft Norse Mill used by the Mediterranean civilizations since ancient times to extract the energy of flowing water, the inhabitants in the remote highlands of the Himalayas had learnt long ago to utilize the potential of water dropping from a height. The mechanical power output from these water wheels was used to drive implements to grind wheat and other household cereals. The device is known as CHASKOR in Arunachal Pradesh and is fabricated locally with timber. A vertical wooden shaft is fitted at the bottom with blades made of flat timber planks. The water of streams flowing from a height is directed by means of open flumes at the runner which rotates the shaft. Two circular stones are fixed at the top to serve as a grinding mill.

7 CENSUS OF WATER WHEELS IN ARUNACHAL PRADESH

The Chaskors are mainly located in remote northerly areas of the state of Arunachal Pradesh bordering Tibet.

Arunachal Pradesh is the largest state in the entire region having an area of 84,750 sq. kilometers, but is very sparsely populated. The northern areas are mostly inhabited by Buddhists who have traditionally used hydro energy for rotating their Prayer Wheels.

Under the UNDP-GEF HILLY HYDRO PROJECT program efforts are being made to conduct a census of the existing population of these water wheels. However due to the spread-out nature of the locations, the process is consuming substantial time.

According to rough estimates around 5000 such wheels are operational in the interior areas of the state.

8 STATUS IN OTHER EASTERN HIMALAYAN STATES

According to the surveys made and the latest information collected, no positive reports of existing water mill sites have been received from the states of Assam, Nagaland, Mizoram and Tripura.

In the rural areas of Manipur, a very innovative and interesting design for pounding of paddy is in use. The device is called PANI DHANKI. The unit consists of a lever supported on a fulcrum which has a up and down seesaw motion. One end of the lever has a bucket like construction, where the water falls from the top and fills up the cavity. The other end has a pounding hammer which sits into a pit filled with paddy. The bucket, when filled, increases the weight of the lever on one side and lifts the hammer. This in turn makes the water drain out resulting in a reduction of the weight which makes the hammer side fall into the pit filled with paddy. The cycle repeats itself resulting in pounding the paddy into rice at an approximate speed of around 40 kg per day.

The state of Meghalaya has no traditional water wheels of the type used in the other Himalayan areas. However, a site at Mylliem on the Shillong-Cherrapunjee road has an installation of locally fabricated Pelton wheel, installed around 30 to 40 years back. This unit with about 30 kW output was used to drive a bone-mill but was shut down a decade back due to managerial problems faced by the family owning the installation.

Sikkim is a traditionally Buddhist area where the prayer wheels driven by water were in use since a long time. However it has not yet been possible to locate working units used for any kind of grinding or other end uses.

Similar situation exists in West Bengal where,

although the northern hill districts have similar terrain and potentials, no installations have yet been identified.

In Assam, due to the efforts of the Assam Science Technology and Environment Council, several new sites have been identified, which have power potential varying between 10 kW to 30 kW. It is planned to develop multipurpose installations at some of these sites, with several end use applications, including electrification wherever required.

9 GOVERNMENT INITIATIVES

Since the chaskor or the traditional waterwheel was already a part of the life for the villagers in the remote mountains it was felt that a program should be initiated to provide technological input for their overall improvement. It was decided to incorporate new designs that will help in improving the efficiencies and outputs of these devices. It was also decided to make use of these water wheels as prime movers for generation of electricity.

The North Eastern Council which is the supervising government body in the region, agreed to provide funds to the Rural Works Department of the Government of Arunachal Pradesh for the upgradation of several existing sites in the Bomdilla area of west Kameng district of the state.

10 UNDP-GEF HILLY HYDRO PROJECT

The UNDP-GEF Hilly Hydro Project is an ongoing initiative supported by the World Bank Global Environment Facility and the Government of India to demonstrate and promote the use of small scale hydro power in the 13 Himalayan states of India. One of the objectives of this project is seeking to upgrade and develop 100 water mills in different regions with improved technology that will serve as prototypes for upgrading the remaining water mills in the region.

11 UPGRADED DESIGNS OF WATER WHEEL

A low head Pelton wheel called the Multi Purpose Power Unit (MPPU), fabricated out of sheet steel, has been used as the runner for the modified water wheel. It was mounted on a vertical mild steel shaft that was in turn supported by a ball bearing at the top and a thrust bearing at the bottom. Water was directed at the runner through a closed PVC conduit which had a controlling sluice valve and a nozzle fitted at the runner end. Two circular stones were mounted at the top of the shaft for grinding of grains.

The entire structure was mounted on an angle

iron frame fabricated out of the bolted components for ease of transportation. The frame structure is installed below the traditional house constructed with timber and thatch. Only the grinding stones are visible above the floor level and there is no perceptible change in the superficial looks of the traditional mill reducing the psychological impact on the rural operators, who are mostly women.

Work on devising suitable upgrade designs is being undertaken also by the Alternate Hydro Energy Center at Roorkee and by the International Consultants of the UNDP-GEF project at UK. The Nepal experience has been found to be the great help in the selection of suitable designs. Although the final designs are yet to be decided a decision may be taken to select one of the following:
1 MPPU in either vertical or horizontal configuration, or another improved runner design.
2 Cross Flow Turbines - in either open or closed enclosures.

12 END USE REQUIREMENTS

Although the traditional application of the chaskor in Arunachal is for grinding wheat or maize, the immediate necessity of applying these units for paddy hulling is very essentially felt. The traditional rice eating areas in the north-east could make good use of hullers driven by water wheels. Oil expellers, water pumps, ice-making machines, small sawmills, etc. could also be used with these prime movers.

The generation of electric power in the evening will be acceptable not only in the areas without electricity but also in other remote villages where the quality of power supplied is hopelessly bad. The electrical; energy, when not being used for lighting purposes could be diverted for water heating, space heating or cooking as being experimented particularly in Nepal.

13 SOCIO-ECONOMIC IMPACT

Most of these areas where the water mills are located still follow the traditional barter system, where the miller normally charges one kg of grain for every 20 kg of grain milled. However, a system will have to be devised for the collection of revenue for the usage of electrical energy.

The units being installed currently are planned to be almost entirely funded by the government agencies. However, as the program gathers momentum, it is expected that the demand for the devices will increase and the villagers will come forward of their own asking for remodeling of existing water wheels or the installation of new ones.

It has been suggested to the state government to create a fund that will be disbursed through rural co-operative banks at very low rates of interest. The revenue earned by the individual owners should be sufficient to service the loan and pay back the principal in a short period of time.

14 MNES POLICY

The Ministry of Non-Conventional Energy Sources (MNES), Government of India, is planning to propagate a program to subsidize the installation of water wheels in the hilly states of the Himalayan region.

A site has been selected at LONGPAI in the state of Assam for the installation of a cross flow turbine with several end uses. The site has a head of 12 m with a discharge of 200 LPS. It is proposed to install a cross flow turbine for generating power, rice huller, spices grinding and water pumping.

15 INFRASTRUCTURE REQUIREMENTS

To popularize this program in the remote hilly areas of Eastern Himalayas some important policy remodeling is required to be achieved at the earliest.

15.1 *State policies*

The concerned state Government should formulate and announce their energy policies, particularly with respect to the Renewable Energy technologies.

15.2 *Renewable Energy Agency*

RETs upto a pre decided generation level (around 100 kW or so) should be segregated from the regular Power Department or State Electricity Boards and handed over to separate Renewable energy Agencies which are not yet operational in most of the states. Generation of electrical power upto the aforesaid preset level and its local distribution must be left entirely with the REDA existing or newly created.

15.3 *Publicity*

The general public should be made aware of the advantages of the new technologies and they should be encouraged to avail of the facilities.

15.4 *Manufacturing*

Local skills must be developed to identify, survey, manufacture, install and maintain the equipment at the remote sites.

15.5 *Financing*

A core fund should be created by the concerned State Government, with funds available from various Government of India or other international programs. This fund should be routed through a local bank that has wide spread branches throughout the state, to provide loans to the willing entrepreneurs at subsidized rates of interest.

Capital subsidies should be avoided as far as possible or at least minimized and the public should be encouraged to refund the loans within the committed periods.

16 THE NEPAL EXPERIENCE

Nepal is the only country in the world where significant progress has been made to upgrade water wheel designs and gain experience from their actual performance in the field. Several international agencies, mostly from Europe, have contributed to the process, both with respect to technological inputs a well as financing. This has helped in the development of around ten indigenous manufacturers who are engaged in this field.

The Multi Purpose Power Unit (MPPU) as was installed in Salari, was the first phase of the upgradation. Around early 1980s the CROSS FLOW turbine started picking up and has become very popular. Other innovations like the PELTRIC sets which is a vertical shaft Pelton wheel coupled to an induction motor or standard PELTON turbines are also in use.

Nepal has an estimated population of 25,000 GHATTAs (as the water mills are locally known) in the country. Around 1000 new improved units are already in operation with MPPU, cross flow or other types of turbines.

A majority of the improved units are being used primarily to produce mechanical power for agro processing, particularly grain milling, rice dehusking, and oil expelling. Only around 20% of the installations also generate electricity to be sold locally to the villagers. Many developers have found innovative end uses like, air compressors, ice making machines, paper making plants and even a small conveyor to carry up the mountains.

The financing has been provided mostly by the Agricultural Development Bank of Nepal at an interest rate of between 15 to 17.5%. No government subsidy is available except on electrical generators used in very remote areas where the equipment is to be manually transported as head loads.

However, the experience in Nepal is not entirely happy. The revenue collected is very poor and consequently the rate of loan repayment is unacceptably low with many accounts falling bad every year.

In the north-eastern states, there is a scope for learning a great deal from the Nepal experience encompassing the technical, financial as well social aspects. We can adopt the successful results and avoid getting in to the traps that were encountered by the Nepalese.

17 CONCLUSION

The ambitious programs for the upgradation of the traditional water mill has been taken up by the government agencies to improve the living standards of the rural population living in the most interior settlements. The economy of these areas has no inputs of any other form of energy.

The improved systems are expected to encourage the development of local cottage industries and also provide electricity to the remote areas, most of which are either totally unelectrified or receive very bad and unreliable quality of power supply.

The program can supplement the economic upliftment policy of the government to a great extent when successfully implemented.

Hydropower in the New Millennium, Honningsvåg et al (eds), © 2001 Taylor & Francis, ISBN 90 5809 195 3

Environmental aspects of power generation – thermal v/s hydropower plants

Prof. P. Krishnamachar, Dr. Rajnish Shrivastava & M.S.Chauhan
Maulana Azad College of Technology (A Regional Engineering College), Bhopal-462007, India

ABSTRACT: With the development of society, demand for energy is increasing day by day. With increase in population and settlements and environmental consciousness, stringent conditions are imposed on power generation. The paper analyses environmental impacts of hydropower vis-à-vis thermal and wind power generation in terms of area requirement, cost analysis and impacts based on life cycle assessment.

1 INTRODUCTION

Energy is the basic resource input for every developmental activity. For every 1% increase in GDP, 1.6 % increase in power generation requirement is estimated in India. With exponentially increase in power demand, (9% per year in India)the rate of exploitation of resources has to increase accordingly, to meet the demand. With increase in population and settlements and environmental consciousness, stringent conditions are imposed on power generation. Environmental impacts can be broadly considered on land water and air environment.

Compensation for loss of forest area due to submergence by hydropower projects and environmental pollution by thermal power projects are two salient environmental concerns. In India , hydropower projects are required to compensate for submergence by afforesting three times the submerged area.

Thermal power produced cogenerates Fly-Ash, CO_2, SO_2 and NO_x The Bio- mass of forest is considered to be the best remover of most of these pollutants. The paper analyses the environmental impacts of thermal power generation v/s generation of hydropower and makes a comparative study of remedial measures .The paper quantifies these impacts and stresses the need to consider the cost of the remedial measures in considering the economics of thermal power generation

2 HYDROPOWER PROJECTS

The storage type of hydropower projects submerge land area under reservoir waterspread. It is difficult to apportion the submergence to hydropower in the case of multipurpose plants. Data from typical purely hydropower projects of India, France and USA have been analysed . Table-1 shows the submergence in hectares / MW of these projects. It varies from 0.01 to 16.17. The submergence of run-of-river projects is negligibly small and limited to river bed area.

As power is a product of flow and head, the sub-

Table 1 Submergence by purely hydro power plants of India, France and USA

S. No.	Name of Project	Country	Submergence (Area in Hectares)	Capacity (MW)	Hect/MW
1.	Baira Siul	India	15.20	180	0.08
2.	Salal	India	940.00	690	1.36
3.	Chamera-1	India	975.00	540	1.81
4.	Dhauliganga-1	India	28.68	280	0.10
5.	Rangit	India	19.00	60	0.32
6.	Sawalkat	India	1105.55	600	1.84
7.	Dulhasti	India	85.00	390	0.22
8.	Baglihar	India	98.63	450	0.22
9.	Teesta-III	India	12.00	1200	0.01
10.	Grandual	France	400	400	16.17
11.	Vouglans	France	1600	272	05.88
12.	Lanau	France	158	25	6.32
13.	Mont-Gnis	France	688	364	1.89
14.	Roseland	France	320	501	0.63
15.	Denver	USA	32.36	30	1.07
16.	Carter lake	USA	57	75	0.76
17.	Estes	USA	66	45	1.48
18.	Marshal Ford	USA	11757	1944	6.04
19.	Mormon Elat	USA	384	120	3.2
20.	Ralston	USA	57.4	15.4	3.7

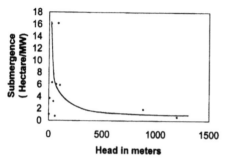

Figure 1. Submergence in hect./MW V/s head in meters of purely hydropower projects.

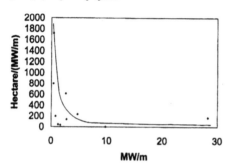

Figure 2. Submergence in hect./(MW/m) v/s power per unit head of purely hydropower projects.

mergence/MW varies widely with head, this trend is illustrated in fig.1.

If power per unit head is considered the plot as shown in fig.2 indicates reduction in submergence in hect./MW/m.

Assuming that the area submerged is predominantly forest area, the submergence is amply required to be compensated by afforestation . In India a minimum of three times the submerged area is required to be afforested thus the effective compensation for submergence would be 0.03 to 48.51 hectares/MW.

It is estimated that the gross submergence in India will be of the order of 0.8 % of the total topographical area while a fourth of it will be effective forest area submerged , when the complete hydropower potential is developed.

3 THERMAL POWER

In India 64,150 MW thermal power produced co-generate annually 290 million tonnes of Co_2, 2.25 million tonnes of So_2,1.71 million tonnes of No_x and 75 million tonnes of Fly-ash .

Vegetation is the best known absorber of the most of the objectionable pollutants co-generated by thermal power plants. During photosynthesis trees and plants absorb Co_2 and give out more O_2 into the

Table-2 Typical removal rates for So_2 and No_x (tons/hect/year)

Pollutants	Deciduous	Conifer
SO_2	0.36	0.12
No_x	0.69	0.24

atmosphere than they breath in . However in the nights and when there is no sunshine, the vegetation does not regenerate oxygen thus the net absorption of Co_2 and conversion into O_2 by vegetation is rendered marginal.

The reservoirs of hydropower projects create water bodies that have positive impact on environment, water bodies are considered to be ultimate absorbers of Co_2 discharged by power plants that burn fossil fuels and the reservoirs on land do play an important role in reducing the Co_2 content though the oceans are the ultimate resort.

Co_2 is considered to be the major contributor to the greenhouse effect and the global warming which is estimated to be 1.5-2.5 ^{o}C by 2020 AD associated with 15-77 cms rise in sea level.

Typical values of rate of absorption of So_2 and No_x by different classes of forest is given in table-2

Hence So_2 produced per MW of thermal power generation requires about 300 hectares conifer or 100 hectares of deciduous forest. Similarly for No_x produced per MW requires 110 and 38 hectares of conifer or deciduous forest respectively for its absorption. Each MW of power generated by thermal power plant is estimated to require 1 hectare of land for disposal of fly ash. Though fly ash is used on acidic lands and in building industry , utilisation of fly ash is marginal in India.

4 OTHER SOURCES OF ENERGY

Wind turbines are becoming more popular for generation of energy. They require unobstructed tracks of land and as well as approach . On an average 24 hectares of area is estimated to be the requirement per MW of wind power generation considering its low power factor. To reduce the level of noise from wind turbine to an environmentally acceptable limit, a distance of 300 m from a 250 KW wind turbine generator is estimated to be required . Area required per MW works out on the above basis to 112 hectares / MW. This may however reduce for a cluster of wind turbine generator.

Solar photo-voltaic panels require on an average 1.7 hectares/MW and typical solar thermal power plant required 3.4 hectares/MW.

5 COMPARATIVE STUDY

The land area requirement per MW generation for different sources of energy discussed above is as shown in the Table-3.

Table-3. Area requirement for different energy options

Source of Energy	Land area requirement hectare/MW (on an average)
Hydropower	0.1 – 4 (submergence) 0.3 – 12 (afforestation)
Thermal power	134 (deciduous) 410 (conifer)
Wind energy	24 112 (Noise criterion)
Solar thermal	3.4
Solar photo-voltaic	1.7

Table-4 Generation cost for different energy options.

Energy Options	Generation cost in Indian Rupees/KWh
Hydro	0.60
Thermal	3.80
Wind	2.50
Solar thermal	5.66
Solar photo-voltaic	11.00

It is clear from the above that the area required even for afforestation at 3 times the submerged area for hydropower plants is much less than the area requirement of thermal and wind energy plants. Though the solar energy generation requires area comparable to that for hydro, solar energy is not yet economically viable for general application

A recent estimate of generation cost of different types of power plants in India is as given in Table-4.

Table-5 Environmental impacts of power generation.

Energy Options	Energy Payback Ratio	Greenhouse gas emission (kteq.Co2 /TWh)	Land requirement kmsq./ TWh/y.	SO2 emission (t So2/ TWh)	Nox emissions (tNOx/ TWh)
Hydro with reservoir	48-260	2-48	2-152	5-60	3-42
Run-of-River	30-267	1-18	0.1	1-25	1-68
Bituminous Coal	7-20	790-1182	4	700-32321	700-5273
Lignite		1147-1272		600-31941	704-4146
Wind	5-39	7-124	24-117	21-87	14-50
Solar Photo voltaic	1-14	13-731	27-45	24-490	16-340

The more rational way of comparison would be from Life cycle assessments (LCA) which includes in itself impact of all the steps involved in creating a product. LCA based on land requirement, emission of greenhouse gasses, No_x, So_2 and energy pay back ratio is as shown in Table-5.

6 CONCLUSION

Land requirement from the afforestation view point of hydro is about 5% of land requirement for thermal and 10% of wind power generation.

Hydropower plants cause marginal ecological changes but no pollution.

The generation cost of hydropower is $1/6^{th}$ of thermal and $1/4^{th}$ of wind power generation.

7 REFERENCES

Naidu, B.S.K. 1999. Environmental aspects of large hydropower projects. *Quantification of environmental impacts of large power projects; Proc. National workshop-cum-seminar,Bhopal,25-26 February1999.*CBIP.

Smith,W.H.1990. *Air pollution and forest: Interaction between air contaminants and forest ecosystems.* Springer-Verlag

--------,2000.IEA hydropower agreement, annex III; *Hydropower and the environment: present context and guidelines for future action.* Vol. III: Main report

Hydropower in the New Millennium, Honningsvåg et al (eds), © 2001 Taylor & Francis, ISBN 90 5809 195 3

Hydrological Computations for Hydropower Projects in Changing Environment

H.V.Lobanova
Russian State Hydrometeorological University, St. Petersburg, Russia

ABSTRACT: Changing environmental, touching upon a subject of hydropower, is connected with an impact of direct man's activity in river channels and on watersheds and with modern climate change. As a result, the observed inflow to hydropower and reservoir is inhomogeneous process and can include three homogeneous components at least: "natural" fluctuations of different time scales, runoff component connected with man's impact and a homogeneous component, connected with modern climate change. On the other side, some homogeneous components can be nonstationary over the time. Suggested methodology of hydrological computations in such changing conditions has been developed as well as new methods, including statistical criteria for an assessment of homogeneity and stationarity, extended for the peculiarity of hydrological information (autocorrelation and non-symmetric distribution), robust empirical-statistical methods for an extraction of homogeneous components from observed data, methods for a determination of the time model's kind and extrapolation, methods for an assessment of random errors of design hydrological values.

1. BACKGROUND

There are two main groups of factors impacted on hydrological events in changing conditions: modern climate change and anthropogenic influence on watersheds and in river channels (different kinds of regulation by dams, water intake and outtake, etc.). The existing approach for the assessment of flood frequency is based on the distribution functions theory and it suitable for homogeneous and stationary conditions (Kritsky & Menkel, 1981; Rozhdestvensky, 1990). The application of this approach in changing conditions is limited by two main reasons:

- non-homogeneity of time series of hydrological characteristics connected with extraordinary outlying observations, different factors of anthropogenic influence and climatic processes of different time scales, including long-term modern climate change;
- non-stationarity of parameters of these time series connected with non-stationarity of anthropogenic factors and climate change.

The third main peculiarity of time series of hydrological characteristics, that is not taken into account usually, is their non-regularity of appearance, that varies from several flood events during a year (storm floods) to some events during century (as mud streams). As a result, the new approach for estimation of flood frequency in changing conditions need to develop.

2. MAIN STAGES OF NEW APPROACH

Algorithm of suggested approach includes the following main stages:
- assessment of extraordinary outlying observations by statistical criteria;
- assessment of man's activity influence and extraction of runoff components connected with anthropogenic impact and restoration of «natural» runoff time series for different input informational conditions;
- extraction of homogeneous different time scale components in «natural» conditions including long-term component of modern climate change;
- determination of time series models of stochastic or deterministic-stochastic kind for extracted runoff components connected with anthropogenic impact;
- determination of time series models for runoff component, connected with modern climate change and for «natural» runoff components;
- extrapolation of deterministic-stochastic models on the basis of auto-regression, scenarios and

expert estimations for future period of water project operation;
- extrapolation of stochastic models by distribution function for future period;
- determination of common computed value in changing conditions as a sum of obtained quantile for «natural» conditions plus additions of extrapolated man's impact and climate change components.

3. METHODS

3.1 Assessment of extraordinary outlying data

Time series of hydrological characteristics could be included extraordinary values, which are outlying data (extraordinary maximums or minimums, as a rule) from the common empirical distribution. The statistical tests are used for assessment of statistical significance of such outlying data. The most popular are Dixon's and Smirnov-Grubbs criterions. Main peculiarities of these and many other criteria are normal distribution and random behavior of the sample which is used for the testing. From other side, the time series of hydrological characteristics have non-symmetric distribution and sometimes – a significant auto-correlation. Therefore the classical tests Dixon's and Smirnov-Grubbs have been extended for the sample with auto-correlation and non-symmetric distribution by Monte-Carlo stochastic modeling. The results are given for example in works: Lobanov & Lobanova (1983), Lobanov (1984). Main conclusion is that the developed criterions increasing the statistics of the tests for non-symmetric distributions and decreasing them for auto-correlated samples. The particular relationships and recommendations are given.

3.2 Extraction of man's impact components

The main factor of man's influence on floods and low flow is seasonal re-distribution of runoff by reservoirs. It leads to a reduction of maximum flood discharges and increasing of low flow. Other man's influence factors, such as water intake and outtake, wood cutting, etc. are not significant for water discharges as usual. The common man's impact component (for case of dam regulation) could be obtained as a difference:

$$\Delta R = R_{real} - R_{nat}, \qquad (1)$$

where ΔR = quantitative assessment of dam influence; R_{nat} is the runoff in dam site in natural conditions; R_{real} is the runoff in dam site in real conditions of operation.

For the synthesis of natural runoff in dam site several methods could be suggested:
- the method of inflow and water balance of reservoir:

$$R_{nat} = R'_{nat} - (P - E - \Delta R_1 - \Delta R_2 + \Delta R_3 \pm \ldots) \qquad (2)$$

where R'_{nat} = "natural" runoff in the site of river falling into the reservoir; P = precipitation onto the surface of reservoir; E = evaporation from free water surface; ΔR_1, ΔR_2, $\Delta R_3, \ldots$ = runoff changes accordingly by the filtration to shores of the reservoir, by the filtration to the bottom of the reservoir, additional underground inflow to the reservoir and other factors;
- the relationships between the runoff in dam site during the natural period and its analogues to do temporary interpolation of these relationships during the hydropower station operation;
- the relationships between the runoff in dam site and in downstream sites, which are under the natural conditions, but only if link degree in the equations characterizes the natural runoff;
- the using of the runoff formation models computed for the natural period of runoff formation. These models can be used for the runoff synthesis during the hydropower station operation period under the condition, that the runoff factors fluctuation regime has been remained as a natural regime;
- difference methods, which determine the natural runoff through a difference between the upper stream site runoff and the dam site runoff in common during the natural period and the station operation period;
- the determination of reservoir influence (effect) using the analysis of common time-series characteristics (mean value, variance, etc.) during two periods, they are: periods before and after the hydropower station construction.

The using one or another of six suggested methods is determined by such parameters as the availability of necessary information or a necessary computation accuracy. Application of the suggested methods for extraction and analysis of flood and low components connected with large dam's regulation for Russian rivers is given in the work (Lobanova, 1999).

3.3 Extraction of climate change components

There are two ways for determination of modern climate change: use of climate scenarios (Jones, 1999) and extraction of long-term climate change component from observed hydrological time series with its future extrapolation in the nearest future (1-2 decades). The first way has a great uncertainty connected with different results obtained by different scenarios and is applied for average (annual, seasonal) meteorological characteristics at least. Therefore the empiric-statistic way has been used together with the scenario assessment. For extraction of long-term climate change component from the observed data, three groups of statistical methods have been used:

- consecutive averaging with periods 10-12 years, that is as a filter of high and middle frequencies of time scale of climate variability;
- different methods of smoothing (Lobanov & Lobanova, 1999);
- truncation method of decomposition (Lobanov, 1995).

Of course extracted long-term climate change component ($Z_{cl.change}$) is not homogeneous and consists of two parts: natural ($Z_{cl.nat}$) and anthropogenic ($Z_{cl.ant}$) changes (Folland, 1996):

$$Z_{cl.change} = Z_{cl.\,nat} + Z_{cl.\,ant} \qquad (3)$$

In the first approximation the analysis and monitoring of climate change can be fulfilled on the basis of the complex process $Z_{cl.change}$. The separation of these two parts is possible by the physical explicable models or by a comparison of data in two different periods: today and before industrial time on the basis of the longest time series (more than 200-300 years).

Two main characteristics of extracted climate change component are obtained: statistical significance (as contribution in the common time series variation) and direction of the tendency in the nearest period. The results obtained by different methods are compared and combined under the same conclusion about significance and direction of climate tendency. Extrapolation of climate component in the nearest period is based on the autoregression model.

3.4 Choice of the time series model

The kind of a model for each extracted component (man's impact and climate change) defines the strategy of further extrapolation and obtains under the following conditions:

$S(Ch) < Ey$ - deterministic (dynamic) model,
$Ey< S(Ch) <100\%$ -Ey   - dynamic-stochastic model,
$S(Ch)>100\%$-Ey   - stochastic model,   (4)

where $S(Ch)$ = standard deviation of the errors in the model; $Ch = f(t)$;, t = time; Ch = informative characteristic; Ey = error of the process (in %).

In the case of deterministic-stochastic model the autoregression is used for the extrapolation and in the case of stochastic one – distribution function.

3.5 Assessment of empirical frequency of non-regular hydrological events

Well-known formula $i/(n+1)$ is suitable for regular random events, where i = rank, n = common number of events. Flood events can be several times in a year and extremes of extremes – several times in century. For calculation of empirical probability (frequency) of distribution of such non-regular events a new formula has been developed:

$$P_i(X_i) = \frac{1}{n+1}[1+\frac{n-1}{n-T_m}(n-\sum_{j=1}^{m-i+1}T_j)] \qquad (5)$$

when $X_1(T_1) > X_2(T_2)> \ldots > X_m(T_m)$ and where
- $T_1, T_2, \ldots$ = duration of hydrological events or time intervals between them;
- m = common number of random events in the time series;
- i = rank number;
- n = size of the sample.

3.6 Extrapolation of stochastic model

The first problem is connected with effective fitting of empirical distribution of hydrological characteristics to chosen analytic approximation (Pirson III, Kritsky-Menkel distribution and other). The most adequate criterion includes two kinds of errors: standard error of fitting for whole distribution ($\sigma\varepsilon$) and maximum error in the zone of small probabilities, which is used for extrapolation (max ε). This way the suggested criterion will be:

$$\sigma\varepsilon + max\ \varepsilon = B\varepsilon \to \min \qquad (6)$$

The second problem connected with setting of designed probability. This probability or frequency is linked with the time interval, that includes period of observation (T_{ob}) and period of future operation of water project (T_{op}). Therefore the designed discharge frequency (P_d) has to include both of two these periods and the formula of the designed probability for will be as follows:

for upper limit:
$$P_d = [1/(T_{ob} +T_{op})] *100\% \qquad (7)$$

for lower limit:

$$P_{low\ for} = [1- 1/ (T_{ob} + T_{for})] *100\% \qquad (8)$$

The last problem of extrapolation by stochastic model connected with determination of design discharge errors or random errors of quantiles. Today the theoretical approach is used but two other empirical methods could be suggested too:
- period of observation divides into two parts: hypothetic period of observation (T'_{ob}) and hypothetic period of operation (T'_{op}) and their ratio has to be equal to the ratio of the real observation and operation periods:

$$T'_{ob} / T'_{op} = T_{ob} / T_{op} \quad \text{and}$$
$$T'_{ob} + T'_{op} = T_{ob} \qquad (9)$$

- observation periods of the longest time series of floods over the world are divided into two parts: observation sample ($T'_{ob} = T_{ob}$) and operation sample ($T'_{op} = T_{op}$) for testing and estimation of quantile errors that are taken the same for the future too;

3.7 Calculation of design discharges in changing conditions

The designed discharge on the final stage is obtained as an algebraic sum of stochastic «natural» component and extrapolated components connected with factors of man's activity and modern climate change with the particular errors:

$$Y_p = (Y_{p\ nat.} + \Delta Y_{i\ nat.}) + (Y_{cl.ant./T=f(P)} + \Delta Y_{cl.ant.}) +$$
$$+ (Y_{ant./T=f(P)} + \Delta Y_{ant.}), \qquad (10)$$

where Y_p = designed discharge of p probability in changing conditions;

$Y_{p\ nat.}$ = designed discharge of p probability in "natural" stationary conditions;

$Y_{cl.ant./T=f(P)}$ = extrapolated component on the period T (period of operation) connected with man's activity;

$Y_{cl.ant./T=f(P)}$ = extrapolated component on the period T (period of operation) connected with modern climate change;

ΔY = random error for each homogeneous component.

4. APPLICATION

4.1 Catastrophic mud floods in the Alatau mountain region

Ten catastrophic mud floods have been chosen on the Karatal River for period of observations 1958-

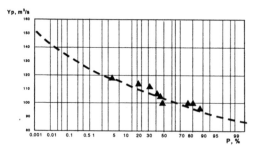

Figure 1. Distribution of catastrophic mud floods on the Karatal River.

1973. Period of existing mud events varied from 1 hour 40 minutes till 3 hours 20 minutes. Interval between floods varied from 10 days till 4 years. Analysis by Dixon and Smirnov-Grabbs criterions allowed to identify the minimum discharge as outlying event and this value was eliminated from the time series. For rest part of sample a formula (5) has been applied for calculation of empirical probability and analytic Kritsky-Menkel distribution has been used for fitting of empirical points. The result of the approximation is shown in Fig.1.

4.2 Catastrophic rainfall floods in the Far East of Russia

Time series of catastrophic rainfall floods exceeded a beginning flooding discharge in 6000 m³/s has been chosen on the Bureya River – site Kamenka. Observation period from 1911 to 1993 includes 94 such floods, interval between floods varied from 10-14 days to 2-3 years. Two approaches have been applied for determination of design floods: distribution of all catastrophic floods and distribution of maximum in each year floods (the existing approach). For each approach two approximations by Pirson III and Kritsky-Menkel distributions have been fulfilled. The results of computations of design floods for several probabilities (P) are given in Table 1. Although the design floods in two approaches are enough close in this case the errors of approximation for the case of all catastrophic maximums was in 3-4 less than for

Table 1 Design floods computed by two approaches for the Bureya River – site Kamenka (m³/s).

P,%	All maximums		Maximum in each year	
	Cs/Cv =12.7	Cs/Cv=15	Cs/Cv=2.7	Cs/Cv=6.0
0.1	17800	21700	17800	22400
0.5	14800	17300	15300	17200
1	13600	15400	14100	15300
5	11000	11200	11400	11400
10	9990	9560	10200	9900

Q, m³/s

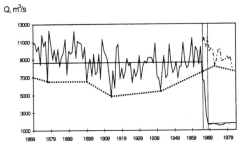

Figure 2. Observed time series of maximum discharges and extracted homogeneous components (the Nile River – Aswan site).

existing approach. Therefore the design floods will have more accuracy.

4.3 Maximum runoff computations in complex changing conditions

Time series of maximum monthly discharges at the Nile River – Aswan site has a observation period since 1869. From the middle of 1960s maximum runoff has been regulated by Aswan dam. Period of the reservoir filling was 3 years only after that the average discharge became 1670 m³/s with variation from 1500 m³/s to 1780 m³/s. For computations in changing conditions man's impact and climate change components have been extracted. «Natural» maximum runoff has been determined by relationships with upper sites for period of hydropower operation. After that the long-term climate change components has been extracted by empirical-statistic methods. The observed time series, «natural» and climate change component are shown in Fig.2.

The scheme of design flood computations is given in Fig.3.

At the first the design discharge with error (ΔY) for natural conditions has been obtained (Y_{pnat}), after that the negative correction on dam's regulation

Y_p, m³/s

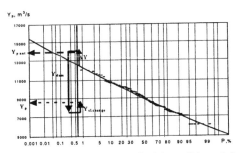

Figure 3. Estimation of design flood in conditions of dam's regulation and climate change impact (the Nile River – Aswan site).

(Y_{dam}) and positive correction on climate change ($Y_{cl\ change}$) have been added and the final result for chosen frequency or future period of operation in 33 years using (10) is given (Y_p).

4.4. Low flow computations under hydropower regulation

River runoff on the Don River has been regulated by hydropower and reservoir in downstream from 1952. Let the new project (water intake) takes place with period of future operation (T_{op}) 1988-2007. The period of observation (T_{ob}) for the Don River – site Razdorskaya was 1881-1987, the period of man's impact (T_{ant}) 1952-1987. Observed time series of low flow is given in Fig.3 with two main periods: "natural" and "anthropogenic".

Restoration of "natural" low flow has been fulfilled on the basis of relationships with non-regulated time series in upper sites using the follow equations:

$$Y_R = 0.572Y_{Kh} + 1.055Y_{KonD} + 39.724 \qquad (11)$$

with $R=0.922$
for 1952, 1954-1955, 1957-1958 and

$$Y_R = 1.612Y_{Kh} + 7.284 \qquad (12)$$

with $R=0.726$ for 1953, 1956, 1959-1987,

where Y_R = low flow at site Razdorskaya;
Y_{Kh} , Y_{KonD} = low flow at upstream sites of the Don River without reservoir impact (site Khovanscy and site Kalach-na-Donu);
R = correlation coefficient.

Low flow component connected with man's impact (Tzimlianskaya hydropower station regulation) has been determined as difference in (1) for period 1952-1987. Relationship between "anthropogenic" component and "natural" low flow is given in the Fig5. It is seen that the increasing of low flow under dam's regulation can be $Y_{p\ ant} = 28$-86m³s⁻¹ for minimum discharges $Y_{nat} < 200$ m³s⁻¹.

Y m³s⁻¹

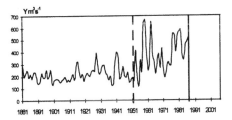

Figure 3. Observed time series of low flow for the Don River – site Razdorskaya

285

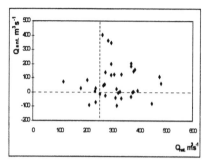

Figure 5. Relationship between anthropogenic component (Q_{ant}) and "natural" (Q_{nat}) low flow.

Determination of low flow components connected with climate variability and climate change has been fulfilled on the basis of model of the decomposition:

$$Y = Y1_{cl\,var} + Y2_{cl\,var} + Y_{cl\,com} \qquad (13)$$

where $Y1_{cl\,var}$ = low flow component connected with interannual climate variability (process of small time scale);

$Y2_{cl\,var}$ = low flow component connected with decadal climate variability (process of middle time scale);

$Y_{cl\,com}$ = low flow component connected with climate change (process of large time scale);

Result of decomposition by truncation method is shown in Fig.6.

In this case the process of interannual scale has average period of fluctuations in 4 years with variation from 2 to 10 years. Process of the decadal climate variability has average period of cycles in 16 years with variations from 9 to 23 years.

Extrapolation of low flow component connected with climate change on the basis of dynamic-stochastic model has been realized by equation:

$$Y_i = 1.898Y_{i-1} - 0.914Y_{i-2} + 2.160 \qquad (14)$$
$$with\ R = 0.998$$

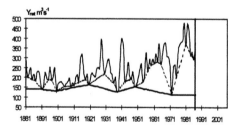

Figure 6. Separated components of climate variability and climate change.

Result of extrapolation on the period 1988-2007 by equation (14) with forecast error is $Y_{p\,cl}$ =91.7 m^3/s^{-1} and by extrapolation of trend gradient is $Y_{p\,cl}$ =81.0m^3/s^{-1}.

Extrapolation of low flow component connected with climate change on the basis of stochastic model has been made for probability:

$$P_d = [1- 1/ (T_{ob} + T_{op})] *100\% = 99.2\% \qquad (15)$$

where T_{ob}=107 years and T_{op}=20 years.

Under the using of joint minimums (climate change and middle-scale climate variability) as 27 random events the following results of approximation and extrapolation by distribution function have been obtained:
criterion of fitting: $B\varepsilon$ = 12.9%
($\sigma\varepsilon$ =2.9%, max ε =10.0%);
extrapolated value: Y_{ext} (P_d)=113 m^{3}s^{-1};
error of extrapolation: ΔY =1 m^{3}s^{-1};
designed value: Y_{pcl}=112 m^{3}s^{-1} .

Under the using of minimums connected with climate change component only in sum of the 9 random events, the common results are as follows:
criterion of fitting: $B\varepsilon$ = 8.0%
($\sigma\varepsilon$ =1.8%, max ε =6.2%);
extrapolated value: Y_{ext} (P_d)=93.8 m^{3}s^{-1};
error of extrapolation: ΔY =20 m^{3}s^{-1};
designed value: Y_{pcl}=73.8 m^{3}s^{-1} .

Computation of designed low flow for management has been fulfilled on the basis of equation:

$$Y_p = Y_{p\,cl} + Y_{p\,ant} \qquad (16)$$

where Y_p = designed low flow in conditions of climate change and man's impact;
$Y_{p\,ant}$ = value of low flow component connected with the most unfavourable man's impact.

Variation of low flow component connected with climate change is in the range:
$Y_{p\,cl}$ =73.8 m^{3}s^{-1}-112 m^{3}s^{-1},
where: $Y_{p\,cl}$ = 73.8 m^{3}s^{-1} is a guaranteed minimum;
$Y_{p\,cl}$ = 92.4 m^{3}s^{-1} is a weighted average minimum (weights are proportional of errors of computed values).

Minimum possible addition of low flow, connected with man's impact $Y_{p\,ant}$ is about 28 m^{3}s^{-1}.(Fig.5).

As a result, the designed low flow in conditions of modern climate change and reservoir regulation will be Y_p =102-120 m^{3}s^{-1} . Real observed minimum Y=121 m^{3}s^{-1} was in 1955.

The results, obtained by developed approach, have

been compared with the results obtained by existing method without taking into account the climate change. Two "natural" time series are used: one with homogeneous "natural" period 1881-1951 and the second with the restored "natural" period: 1881-1987. The results are as follows for P_d =99.2%:

a) Y_{ext} = 126 m^{3}s^{-1} with $B\varepsilon$ = 13.6% ($\sigma\varepsilon$ =2.0%, $max\ \varepsilon$ =11.6%);

b) Y_{ext} = 110 m^{3}s^{-1} with $B\varepsilon$ = 8.3% ($\sigma\varepsilon$ =1.3%, $max\ \varepsilon$ =7.0%);

Maximum obtained error of the design value is ΔY =1 m^{3}s^{-1}. Therefore $Y_{p\ nat}$ = 123 m^{3}s^{-1} in the first case and $Y_{p\ nat}$ = 107 m^{3}s^{-1} in the second case. With correction on reservoir control: Y_p = 151 m^{3}s^{-1} in the first case (for "natural" time series 1881-1951) and Y_p =135 m^{3}s^{-1} for the second case (1881-1987). Unfortunately this values are more than the real observed minimum Y=121 m^{3}s^{-1}, which took place in 1955 under reservoir regulation. Therefore the suggested approach of hydrological computations in conditions of complex impact of man's activity and climate change gives more realistic results than the existing one.

5. CONCLUSION

The following main results have been obtained:
- common strategy of computations has been suggested for low flow management in conditions of climate change and man's impact;
- methods of modeling and extrapolation have been developed for this situation;
- application of the strategy and methods have been given and result is realistic;
- comparison with existing approach has been made and it has been shown that the existing methodology based on stationary principle gives unrealistic result for changing conditions.

References

Folland C. (1996). Current Climate Change: Can We Detect a Human Induced Influence? European Conference on Applied Climatology. Abstract Vol., Norrkoping, Sweden, May 1996, p. 3-4.
Jones, J.A.A. (1999) Climate change and sustainable water resources: placing the threat of global warming in perspective. Special Issue. *Hydrological Sciences Journal*, **44** (4), 541-557.
Kritsky, S.N. & Menkel, M.F (1981) *Hydrological bases of river runoff management*. Science, Moscow, (in Russian).
Lobanova, H.V. (1999) Assessment of Sustainable Floods and Water Resources Management by reservoirs in Russia. *Proc. Int. Conf. Lakes99*, Copenhagen, Denmark.
Lobanov, V.A. & Lobanova, H.V. (1999) Cold Climate Characteristics. In: *Urbane Drainage in Specific Climates*, Vol.2, Unesco.
Lobanov, V.A. & Lobanova, H.V. (1983) Adaptation of Smirnov-Grabbs test for non-symmetric and auto-correlative consequences. *Ref. Journal Geophysics*, 8, 14-28.
Lobanov, V.A. (1983) Assessment of efficiency of the statistical tests for outlying data in the case of the Pirson III distribution. *Ref. Journal Geophysics*, 8, 130-142.
Lobanov, V.A. (1995) Statistical Decision in Changing Natural Conditions. *Proc. Int.Conf. on Statistical and Bayesian Methods in Hydrological Sciences*, Paris.
Rozhdestvensky, A.V. (1990) *Assessment of efficiency of hydrological computations*. Gidrometeoizdat, St.Petersburg, (in Russian).

A contribution to ecological-economical aspects of hydro power plants

H.B.Matthias, E.Doujak & P.Angerer
Institute for Waterpower and Pumps, Vienna University of Technology, Vienna, Austria

ABSTRACT: Using hydropower is more disputed than ever before, because erection and use is always combined with significant interventions in systems of running water with extensive ecological consequences. Necessary developments in the future will have to minimize ecological effects when building and using hydro power plants. Nowadays the protection of nature by law includes also the setting of time limits for allocation of riparian rights and increasing the amount of permissions, which have to be obtained for new hydropower plant projects. The additional costs caused by several ecological aspects lead to rising costs for the whole hydropower plant and can decide the feasibility of such projects. Because of this the additional costs have to be estimated even in the planning-stage of the project. In this paper the ecological problems in the field of hydro power plants, like given amounts of compensation water, treatment of residues of gratings and fish ladders including the upcoming costs will be the main topics.

1 INTRODUCTION

Proceeding from the approaching change in questions of the water cleanliness by the European Union, the matter of ecological guidelines and measures in case of building or refurbishing hydro power plants will become more and more important. This paper should give an impulse and show the possibilities of an ecological assessment of hydro power plants. With the European recommendation for water cleanliness (WRRL) the water policy of the European Union will be defined in an new way and also set on a future oriented basis. This base has a big effect on the water use for electricity production through hydro power plants.

1.1 Central points and aims

Declared aim of these recommendations is the preservation and the improvement of the aquatic environment, but at the same time a protection of the sustained water resource policy. The member countries will be engaged to following general targets:

- Avoidance of a deterioration as well as protection and improvement of the condition of the aquatic ecological systems
- To promote a sustained water resource policy on the basis of a long-term protection of the existing resources
- A contribution to minimize the effects of floods and droughts.

2 AGE STRUCURE OF AUSTRIAN HYDRO POWER PLANTS

2.1 Hydro power plants with a bottleneck capacity below 5 MW (small hydro power plants)

The last statistics of existing hydro power stations is based on the datas of 1994, which were published 1996, but unfortunately a completely list is not available. The quote of production of electricity through hydro power is about of 9%.

Number: 1690 Hydro power plants
Capacity: 601.5 MW
Energy: 3051 GWh

The effective amount of small hydro power plants based on estimations can be shown as follows:

Number: 4000 – 5000 plants
Capacity: 700 MW
Energy: 3500 GWh

that means that nearly 15% of the small hydro power stations in relation to the electricity production has not been acquired yet.

A percentage of nearly 2,24% (or 38 hydro power plants) of the „official" 1690 hydro power plants are storage power plants with a capacity of 36MW (or 6.05%). Also the capacity of these power plants with about 960kW is clearly over the average.

In the field of small hydro the relationship between capacity and the number of plants is like a „quasi-hyperbolic" curve. In a schematic representation it would be possible to say that a lot of

„small" hydro power plants are comparable to a few „bigger" ones. The age and the distribution of the age of these small hydro power plants shows a great possibility to identify the „Replanting"-potential of all these power plants. Replanting in this case means not only to rebuild these power plants on the basis of their service life, but also to take the advantage to optimize the electrical power output by adapting the design layout.

Remarkable for the situation in the field of hydro power is the time where most of the hydro power plants have been founded, the breakdown as a result of the first world war, the recovery in the twenties, the crises between 1930 till 1950 with the second world war, a constant development till 1980 and big boom up to 1990 through promotion of the individual countries. After that time the development and building of new hydro power plants stopped because of economical unattractiveness and public disapproval.

If we take into account that the use of the water for electricity production restricted to 90 years was first announced in 1934 the number of hydro power plants with a residual life time can be estimated. So far as hydro power plants in this age have not been modernized resp. refurbished, or their rights for using water to produce electricity has not been returned to the country, we can calculate with about 50 - 60 hydro power plants with expiring permissions of water use for electricity within the period of 2021 to 2030. A renewal of this permission is unavoidable with the public authority demand of modernizing the hydro power plant which means a lot of new requirements and costs especially in terms of ecological aspects.

2.2 Hydro power plants with a bottleneck capacity above 5 MW

The age structure as well as the distribution of the bottleneck capacity and the yearly work capacity of the hydro power plants in this stage in Austria are shown in the following figure:

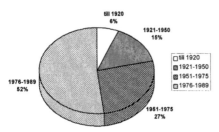

Figure 1. Number of hydro power plants in Austria with a bottleneck capacity above 5MW (Fuhrmann 1991)

Because of the poor data of the age of the hydro power plants (build in the time between 1921 till 1950) an estimation about the licenses of water use for electricity production is not useful resp. possible.

3 COSTS OF A HYDRO POWER PLANT

The costs to build a new or to reconstruct an existing hydro power plant can be divided into two parts : the investment cost (fixed assets) and cost for operation (floating capital).

The fixed plant costs essentially consist of the costs: (a) for the civil work (e.g. powerhouse, weir, water chamber, gates, trash rack and rack cleaning machine), (b) for the mechanical components (e.g. control units and turbines), (c) for the electrical components (e.g. generator, transformer) and (d) other costs (e.g. planing, approval).

As these costs depend to a high degree on the location of the hydro power plant, only a rough estimation of them can be given with high variations from case to case. In some cases the costs for civil work can be about 40 - 50% of the whole costs (Berg et al. 1991). The costs for mechanical equipment (e.g. turbines, gearing, control system) can rise up to 20 - 25% in large hydro power plants in the field of small hydro even a percentage of 30% of the whole costs is possible. The electrical equipment can have a percentage of about 5 - 10% and the costs for steel water construction about 5 - 7% of the investment costs. The rest will be costs for different opportunities like planing, extra charges or interest costs (interest for the investment asset during the building phase ...).

The specific entire investment of newly built hydro power plants in the power range below 1MW are about 63,000 till 165,000 Austrian Shillings per kW (this is about 4578.38 till 11,991€ per kW) (Kaltschmitt & Wiese 1997). With increasing of the installed power the costs will decrease and vary in the range of 10MW capacity between 56,000 and 70,000 Austrian Shillings per kW (this is between 4069.68 and 5087.10€ per kW). In case of small hydro power plants these costs can be much higher.

If only the turbine sets will be changed the costs for this work will be about 7000 Austrian Shillings per kW (this is about 508.71€ per kW) and for small scale hydro power plants about 35,000 Austrian Shillings per kW (this is about 2543.54€ per kW) (Kaltschmitt & Wiese 1997).

For small hydro power plants the following figures show the dependence on the specific total costs K_{ges} and the installed capacity for a statistical middle range. It is clear to see that the specific total costs will decrease with increasing capacity resp. with increasing head of the power plant. In dependence of the individual boundary conditions these curves are only rough estimations and can vary up to 50% in isolated cases.

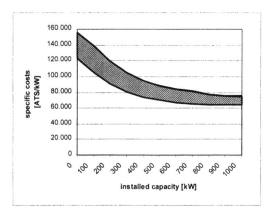

Figure 2. Specific investment costs for building new small hydro power plants depending on the installed capacity (Kaltschmitt & Wiese 1997)

Apart of the nominal capacity the costs depend also on the local available head of the given low head power plant. Hydro power plants with the same nominal capacity have lower investment costs by increasing head.

Independent of that, the costs for ecological measures can vary between 10 - 20% of the total investment costs as reported in (Kaltschmitt & Wiese 1997) and (Hasenleithner 1999).

The following paragraphs show the estimation of costs for the above mentioned parts of a hydro power plant. The given comments are important to understand the highly complex task of calculating the investment and are the base for the calculation of the ecological measures at the end of the paragraph. All formulas are based on mean values of hydro power plants which have been build in the past and are reported in the specific literature.

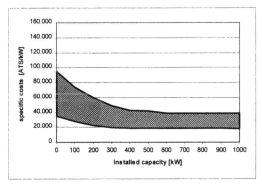

Figure 3. Specific investment costs for refurbishing old small hydro power plants depending on the installed capacity (Kaltschmitt & Wiese 1997).

3.1 Investment costs for a new hydro power plant

A good estimation for the investment costs can be found at (Berg et al. 1991) and depends on the installed capacity P [kW] and the design head h_f [m]. The resulting formula will look like:

$$K = C \cdot \left(\frac{P}{h_f^{0,3}} \right)^y \tag{1}$$

where K = investment costs [ATS]; y = constant []; C = Constant [ATS]; P = capacity of the plant [kW]; and h_f = design head [m].

The estimated costs of the individual kinds of costs are represented mean values by this formula with different constants C and exponent y.

The calculated formulas are only valid for low head power plants with a installed capacity below 2MW and a design head up to 15m. For hydro power plants with higher head the influence of the type of turbine is too big and therefore it is not really easy to give a general formula for this cases.

The following cost specifications are based on the year 1990 and can be projected to a certain year by the use of suitable factors.

Investigating the investment costs regardless whether it is a new building, reconstruction or refurbishing of a existing building, the service life resp. the residual life span of the different parts has to be taken into account. The literature provides different values, but in average it is possible to calculate with the following data refering to (Kaltschmitt & Wiese 1997), (Weiss 1992), (Ländergemeinschaft Wasser 1996):

Table 1. Technical service life of different components.

Component	Micro scale hydro	Small scale hydro	Big scale hydro
El. nominal capacity [kW]	50	500	5.000
Civil structure, service life[a]	60	80	80
Mechanical equipment, service life [a]	30	40	80
Electrical equipment, service life[a]	30	30	40
average full operation hours per year	4500 – 5500	5000 – 6000	5500 – 6500

On the basis of these assumed technical service lives the depreciation of the different components can be calculated and is part of the investigation and cost estimation of hydro power plants.

3.1.1 Civil costs

If a new hydro power plant is build, the exact dimensions of the needed civil work can only be estimated because of different local settings. There-fore the costs for this work vary in a high manner and precise calculations can only be done if the local geological and topographical conditions are known.

Some other influence factors are the knowledge of dimension and frequency of floods or the annual bed load of the river, which can vary these costs significantly. All this factors and conditions apply onto the civil work and therefore also onto the estimated costs K_a and can be roughly estimated with the following formula (Berg et al. 1991):

$$K_a = 300,000 \cdot \left(\frac{P}{h_f^{0,3}} \right)^{0,71} \tag{2}$$

where P = installed capacity; and h_f = design head.

Ecological measures like for example renaturation of dams and so on can be assumed with another 5% of the above calculated investment costs for civil work (Berg et al. 1991).

3.1.2 Mechanical equipment

The mechanical equipment essentially consists of the turbines, turbine control unit, gearing, cooling water supply, drain and ventilating equipment. A gearing is only necessary for lower heads to raise the very low turbine revolution to a higher and economically useful generator revolution. The main parameters for a cost estimation of the turbine sets are the design parameters like the diameter and the head. These design parameters depend on the capacity of the turbine resp. the discharge and the head. Higher head and same discharge means stronger dimensions of the turbine components and therefore higher prices. The estimated costs for the mechanical equipment can be represented as function of capacity and head without the knowledge of the exact design of the turbine. For z turbine sets the formula can look like:

$$K_m = 221,000 \cdot \left(\frac{P}{z \cdot h_f^{0,3}} \right)^{0,85} \cdot (z + 0,1) \tag{3}$$

where P = installed capacity; h_f = design head; and z = number of turbine sets.

The cost differences by using certain turbine types is not included in this estimation. This circumstance can vary the result a little bit.

This fact is only a problem in the field of small scale hydro power because the share of the mechanical equipment is about 20 - 30% of the total costs. In larger hydro power plants this share is lower and therefore the estimation of the total costs does not vary as much as for small hydro power plant.

3.1.3 Electro-technical components

These components mainly consist of the parts generator, transformer, main-switching device, control equipment including automatization and the power production for the plant itself.

The costs can be estimated with the following formula:

$$K_e = 9100 \cdot \left(\frac{P}{h_f^{0,3}} \right)^{0,98} \tag{4}$$

where P = installed capacity; and h_f = design head.

3.1.4 Steel-water construction

The percentage of steel-water construction in the hydro power plant is highly depending on the design of the whole plant and has often no coherence to the capacity. An estimation of the upcoming costs can only be done by including a standardized "base-equipment" consisting of the following elements:

- Trash rake
- Rake cleaner
- Inlet barring
- Outlet barring

The costs for the steel-water construction K_{St} including the above components can approximately be estimated by the following formula (Berg et al. 1991):

$$K_{St} = 16,520 \cdot \left(\frac{P}{h_f^{0,3}} \right)^{0,84} \tag{5}$$

where P = installed capacity; and h_f = design head.

3.1.5 Indirect building costs

These costs include the costs for planning, insurance, geodetically measurements and also ground water observations. As an estimation according to (Berg et al. 1991), these costs will be 8.5% to 20% of the direct building costs, depending on the size of the plant. GORDON (Weiss 1992) has set up an empirical formula for the calculation of the planning costs:

$$K_i = F \cdot C_i \cdot \left(\frac{P}{h_f^{0,3}} \right)^{0,74} \tag{6}$$

where K_i = costs for planning [ATS]; F = exchange rate [ATS/US$]; C_i = cost factor [103US$]; P = installed capacity [MW]; and h_f = design head [m].

The values of the cost factor C:

Table 2. cost factor C – GORDON (Weiss 1992)

		C [103 US$]			
		K1	K2	K3	K4
Small hydro	P < 10MW	4.3	13.0	71.4	721
Hydro power plant	h_f < 350m*	6.4	19.2	105.4	1065
Hydro power plant	h_f < 350m**	10.4	31.1	171.3	1730

K1 Reconaiss. Surveys & Hydrol. Studies,
K2 Pre-feasibility Studies,
K3 Feasibility Studies,
K4 System Planning & Engineering.
* + 30%, if there is a tunnel with a length of over 5km
** tunnel and underground work is already included

3.1.6 Common building costs

These costs include charges for bank ,law , permission and administration. As an estimation according to (Berg et al. 1991), these costs will be 3% to 4% of the direct building costs.

3.1.7 Interest rate of the investment

To determine the interest rate for the investment, the total investment has to be calculated with the related capital-interest rates over the defined reconstruction time. The mean reconstruction time for smaller hydro power plants is between 18 and 30 months.

According to (Weiss 1992), the mean reconstruction time can be calculated with the following formula:

$$i_k = \frac{m}{n} \cdot i_0 \cdot \sum_{j=1}^{m} x_j \cdot \left(1 + \frac{1-j}{m}\right) \tag{7}$$

where k = number of investment parts; i_k = relatively interest rate of the investment part number k; m = number of time increments; n = time increments per year; i_0 = constant annual interest rate for capital investment; and x_j = percentage of the investment costs per time increment.

With the formulas (2) to (7) it is possible to calculate the total costs for the building of a river hydro power plant:

$$GK = (1.03....1.04) \cdot 300,000 \cdot \left(\frac{P}{h_f^{0,3}}\right)^{0,71} +$$

$$+ 221,000 \cdot \left(\frac{P}{z \cdot h_f^{0,3}}\right)^{0,85} \cdot (z+0,1) + 9100 \cdot \left(\frac{P}{h_f^{0,3}}\right)^{0,98} +$$

$$+ 16,520 \cdot \left(\frac{P}{h_f^{0,3}}\right)^{0,84} + F \cdot C \cdot \left(\frac{P}{h_f^{0,3}}\right)^{0,74} +$$

$$+ k \cdot \frac{m}{n} \cdot i_0 \cdot \sum_{j=1}^{m} x_j \cdot \left(1 + \frac{1-j}{m}\right)$$

3.2 Operation costs

The main indicators for the annual maintenance costs are:

- Operation
- Maintenance
- Administration
- Depreciation
- Taxes and charges

3.2.1 Operation, maintenance and management

The operation costs of optimal designed low-maintenance hydro power plants are relatively low. One reason for example is the remote operation and control which effects in a lower percentage of costs for the staff. Variable costs arise for staff, mainte-

nance, administration, transfer to reserve for replacements, waste removal of the trash rake and for insurance. The individual costs vary depending on the local occurrence of different plants.

Especially in older power-plants, the costs for the staff are dominating the fixed operation costs. For that reason, newly built power plants are equipped with a high automatization-level resulting in lower fixed costs. However, on very small hydro power plants, the costs for the staff still cover a high percentage of the operation costs.

Beside these costs also additional cost-factors like operating materials and reparations have to be taken into consideration. An estimation of these additional costs is specified in (Berg et al. 1991).

$$K_{sk} = 11,200 \cdot P^{0,55} \tag{8}$$

where P = installed capacity.

Nonproductive-times because of standstill during maintenance are not covered. Looking at the whole operation time , these expenditures which are specified in (Berg et al. 1991) are at around 1% of the whole investment costs.

In (Kaltschmitt & Wiese 1997), the yearly expenditure for operation is specified with 1% to 4% of the investment-costs. For small hydropower-plants, this percentage usually will be higher than for normal hydro power plants.

3.2.2 Taxes and charges

These costs vary from case to case and for that reason it is very difficult to summarize them in a general formula.

3.3 Influence of the ecological measure on the energy-costs

According to (Weiss 1992), the energy costs are characterising an indicator for the economical efficiency of the plant and can be calculated via the following formula:

$$K_{Ei} = k_I \cdot \frac{P_N}{A_i} \cdot (a+s+v) + f_i + b_s \tag{9}$$

where K_{Ei} = energy-costs [ATS/kWh]; s = mean tax value; v = insurance value, calculated over the operation time of the plant; k_I = investment in correlation to the nominal capacity:

$$k_I = \frac{K_I}{P_N} \tag{10}$$

where K_I = capital investment [ATS]; P_N = nominal capacity [kW]; A_i = annual output [kWh]; and a = annuity-factor:

$$a = \frac{(1+p)^n \cdot p}{(1+p)^n - 1} \tag{11}$$

where p = interest rate [%]; n = number of interest-periods; and f_i = specific fixed costs for the staff and for maintenance:

$$f_i = \frac{K_{Pi} + K_{Ii} + K_{Vi}}{P_N \cdot T_i} \qquad (12)$$

where K_{Pi} = costs for the staff [ATS/a]; K_{Ii} = maintenance-costs [ATS/a]; K_{Vi} = different additional costs [ATS/a]; T_i = actual annual usage[h/a]; and b_s = specific additional operating costs [ATS/kWh].

For storage power plants it is necessary to consider an additional function $P_N = f(Q,H)$ to include the variation of the head. Because of these variations, the sum of the real operation hours T_i per year will be higher than the sum of the calculated operation hours T_i'.

$$T_i' = \frac{A_i}{P_N} \qquad (13)$$

and with

$$T_i' \approx n_{AV} \cdot 8760 \qquad (14)$$

where n_{AV} = capacity factor.
finally

$$\frac{1}{T_i'} = \frac{P_N}{A_i} \approx \frac{1}{n_{AV} \cdot 8760} \qquad (15)$$

Including additionally the costs for environmental and ecological motivated measures, the formula for the specific energy costs results in:

$$K_{Ei,Oko} = \frac{P_N}{A_i} \cdot (a+s) \cdot (k_I + k_{I,Oko}) + k_I \cdot \frac{P_N}{A_i} \cdot v + f_i + b_s$$

$$\qquad (16)$$

where $k_{I,Öko}$ = specific additional investment costs

for ecological balanced measures and environmental measures:

$$k_{I,Oko} = \frac{K_{I,Oko}}{P_N} \qquad (17)$$

where $K_{I,Öko}$ = additional investment costs for ecological balanced measures and environmental measures [ATS].

It has to be taken into consideration, that for the additional investment costs for ecological measures only the annuity and the taxes and usually no insurance-costs have to be included, because normally no additional insurance-contract has to be taken out and also most of the ecological measures in the surrounding of the hydropower plant will not be in the property of the hydro power plant owner.

As an example with the following assumptions (Berg et al. 1991):

investment costs K_I =	ATS 140 Mio
nominal power P_N =	5 MW
annual output A_i =	27 GWh/a
interest rate p =	6%
number of interest periods n =	30 Jahre
mean tax value s =	3.5%
insurance-value over the operation-time v =	2.5%
annual output A_i =	27 GWh/a
interest rate p =	6%
number of interest periods n =	30 Jahre
mean tax value s =	3.5%
insurance-value over the operation-time v =	2.5%
true yearly utilization T_i =	5400 h
specific fixed costs for personal staff and maintenance f_i =	0.1 ATS/kWh
additional specific operation costs b_S =	20% of f_i = 0.02 ATS/kWh

The following diagram shows the energy-costs depending of the ecological balanced measures, calculated with the above values.

With the assumptions shown above, the ecological balanced measures result in additional costs of 0,05 to 0,11ATS/kWh.

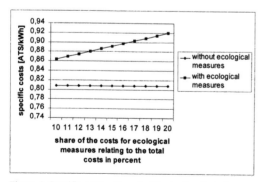

Figure 4. Specific costs with and without of ecological measures.

4 CASE STUDY

Based on the given formulas to estimate the proper costs of building or refurbishing hydro power plants some case studies should be discussed here at this paragraph. All three hydro power plants are existing power plants in Austria and have been refurbished in the last years. Aim of the case study is to identify the main parameters which are necessary to build up the number of equations for the ecological assessment of hydro power plants and of course to evaluate the formulas.

The main parameters for the calculation are the nominal capacity, the annual work, the head as well as the number of turbine sets to estimate the mechanical investment costs. The whole data is summarized in the following table.

Table 3. Summarized data

	HPP 1	HPP 2	HPP 3	HPP 4
P [kW]	201	1630	1900	650
A [kWh]	1,310,000	7,800,000	9,280,000	4,400,000
H [m]	3.18	5.68	6.65	6.45
z []	1	2	2	1

If we put the data of table 3 into the formulas (1) to (7) we will get the following index for the investment, the specific capacity and work costs:

Table 4. Investment, capacity and work costs.

	HPP 1	HPP 2	HPP 3	HPP 4
	costs in 1000ATS	costs in 1000 ATS	costs in 1000 ATS	costs in 1000 ATS
Civil costs	10,124	39,546	42,636	20,038
Mechanical costs	14,928	42,317	46,306	33,801
Electrical costs	1171	7677	8517	3004
Steel water construction costs	1062	5323	5819	2381
Civil common costs	405	1582	1705	802
Interest for civil work	1012	3955	4264	2004
Total sum	28,702	100,399	109,247	62,030
Spec. Capacity costs	142	62	57	95
Spec. Work costs	0.022	0.013	0.012	0.014

The grafical output looks like follow:

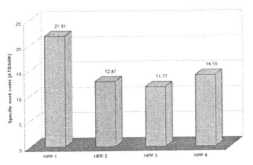

Figure 5. Specific work costs, calculated with the formulas (1) to (7).

As shown in table 4 and figure 5 above, the specific work costs for the HPP 2 to HPP 4 is around 12ATS/kWh to 14ATS/kWh. This range corresponds very good with the specifications given by the HPP-operators. Only HPP1 has higher work costs due to its low nominal capacity. It has to be considered, that the cost saving due to the reuse of

existing components is not included in the calculation of the revitalization project.

To calculate the production costs, the service life of the components has been assumed according to table 1 and the interest rate for the annuity has been set to 6%. Together with annual operation costs (3% of the total investment), the production costs can be calculated (Kaltschmitt & Wiese 1997):

Table 5. calculation of the production costs out of the determined investment costs.

	HPP 1	HPP 2	HPP 3	HPP 4
Annual operation costs 1-4% of inv. [ATS]	861,065	3,011,984	3,277,408	1,860,893
Annuity factors				
Civil costs [ATS/a]	613,267	2,395,378	2,582,564	1,213,750
Mechanical costs [ATS/a]	992,111	2,812,469	3,077,572	2,246,473
Electrical costs [ATS/a]	85,053	557,745	618,785	218,230
Steel water construction costs [ATS/a]	70,575	353,771	386,707	158,277
Civil common costs [ATS/a]	55,024	214,919	231,714	108,901
Interest for civil work [ATS/a]	137,559	537,298	579,285	272,252
Production costs [ATS/kWh]	1.49	0.88	0.81	0.96

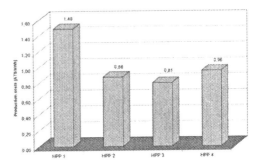

Figure 6.Production costs for the 4 hydro power plants.

Except HPP 1 (low capacity), the production costs vary between 0.80ATS/kWh - 1.00ATS/kWh.

The distribution of the costs for revitalization of the HPPs 1, 2, 3 have been specified by the HPP-owners. Together with the assumption that the investment costs of table 4 are not including costs for the ecological measures and that the costs for fish ladders (FAH) will be 50% of the costs for ecological measures, these costs can be calculated:

Table 6. Estimated costs of the fish ladder system (FAH)

	HPP 3	HPP 2	HPP 1
Investment costs [ATS]	103,784,600	96,383,477	26,405,992
Ecological costs [ATS]	5,462,347	4,015,978	2,296,173
Share FAH (half of the ecol. costs) [ATS]	2,731,174	2,007,989	1,148,087
Costs FAH relating to the head [ATS/m]	410,703	353,519	328,025
Costs FAH relating to the nom. capacity [ATS/kW]	1437	1232	5712
Costs FAH Relating to the annual work [ATS/kWh]	0.29	0.26	0.88

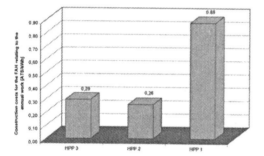

Figure 7. Specific construction costs for fish ladders relating to the annual work (HPP 1, 2, 3)

As seen in figure 7, the specific construction costs for fish ladders relating to the annual work are in coherence with the installed power output and the work capacity. Only the costs for HPP 1 are to high, because of the relatively high total investment costs.

The coherence between the specific construction costs for fish ladders and the nominal head cannot be seen very clear due to the different local situations.

Additionally it has to be remarked, that also if the power output and the work capacity is in coherence with the costs, the geological situation of the HPP will have a more or less influence.

REFERENCES

Kaltschmitt, M. & Wiese, M. 1997. *Erneuerbare Energien – Systemtechnik, Wirtschaftlichkeit, Umweltaspekte.* Wien: Springer Verlag

Fuhrmann, C. 1991. *Ein Beitrag zur Beurteilung über den Revitalisierungsbedarf von alten Wasserkraftanlagen.* TU Wien: Dissertation

Berg W. et al. 1991. *Leitfaden für den Bau von Kleinwasserkraftanlagen.* Baden-Württemberg: Wasserwirtschaftsverband Baden-Württemberg e.V.

Hasenleithner, C. 1999. *Personal informations.* Ennskraft

Weiss, P. 1992. *Ein Beitrag zur Planung und Projektierung von Kleinwasserkraftanlagen.* Rheinisch-Westfälische technische Hochschule Aachen: Dissertation

Länderarbeitsgemeinschaft Wasser, Arbeitsgruppe Kosten-Nutzen-Untersuchungen in der Wasserwirtschaft. 1996. *Wirtschaftlichkeit und Technologie der Wasserkraft.* München

Hydropower in the New Millennium, Honningsvåg et al (eds), © 2001 Taylor & Francis, ISBN 90 5809 195 3

A micro-hydro pilot plant for mechanical pumping

R.Pallabazzer
University of Trento, Trento, Italy

A.Sebbit
Makerere University, Kampala, Uganda

ABSTRACT: This paper deals with a micro-hydro power-plant driven by a pump used as a turbine (PAT), which has been set up and tested in Uganda. The plant, including a gravity water depuration system, supplies clean water to a village by means of a pump directly driven by a PAT working under a waterfall head. As there is no reliable method that allows to calculate the turbine performances from the pump ones, a preliminary laboratory investigation was carried out by testing some commercial pumps as turbines, in order to know the available performances and to make a suitable choice.

The overall efficiency of this mechanical system is much higher than it should be with electric conversion systems; moreover, electric systems should increase the power-plant overall cost and complexity and reduce the reliability. The preliminary tests, the technical solutions and the field tests are presented and discussed.

1 INTRODUCTION

Whereas in the industrialized countries the major energy problem is to increase the amount and the concentration of the energy production, in a large part of the world this is only one aspect of the problem, the other and frequently bigger aspect being the energy distribution, as most human settlements are remote and the energy must be produced on site; the harmonic development of a country requires that this multitude of small remote energy needs, which involve a great many people altogether is not disregarded.

In this case the energy production from renewable sources generally is the most reliable, sustainable and cheap solution. Among these sources the hydropower is the best when available. Usually, the mini and micro-hydropower projects are carried out by small local and/or international organizations that generally have no difficulties as to the civil works but frequently lack the necessary skills for designing the power generating sets (especially as for the water turbines), that therefore are entrusted turnkey to some international manufacturing company. This is the most practical but not always the most reliable and sustainable solution and doesn't facilitate the spreading of micro and mini-hydro systems.

The full hydro project could be better managed by small organizations when resorting to the PAT systems. The use of reverse running pumps working as hydraulic turbines is well known and was rather diffuse some years ago as simple and cheap systems to save energy in industrial plants (HPRT). This happened in the 70s, when the international oil crisis urged industrial countries and companies to save energy. As the situation later got progressively better for these countries, the use of HPRT systems declined. However, the energy crisis remains in progress in a large part of the world, where the problem is to recover energy from natural and free resources rather than from industrial plants.

Usually the water turbines are specially designed and custom-made for a given water head and flow rate, that is for a given site; this means that they are not very cheap nor quick to set up and are suitable only for large energy generation and not very encouraging for small power and autonomous self-managed systems.

On the other hand the hydraulic machines can work in the four quadrants, that is as pump, as turbine and as water brake, depending on the rotational direction, the flow-rate and the load. For the pumps to work as turbines they need only to be fed by a water head and to be free to run reversely, being coupled with a mechanical or electric load. The performances are very good, the efficiency is not very lower and sometimes higher than in direct running, while the *bep* (best efficiency point) head and the flow rate are usually much higher.

Moreover, unlike the water turbines, the pumps of small power are widely produced all around the world, are very cheap and easy and swift to buy. Therefore, for small power and self-managed systems it is more suitable to resort to a pump used as turbine than to buy a specially designed water tι bine.

The PATs can be used either to drive an electric generator or to drive a pump (Figure1). In the last case the aim of the system usually is to convert water energy from high head to high load, as when an aqueduct has to be fed.

However, there is a problem: while the characteristics of the commercial pumps are usually known, those of the pump running as turbine are unknown. They should either be known by measuring them or by means of complicated calculations provided the geometry is known. However both conditions cross out the merits of the PATs and make their use not feasible for people without technical background and facilities. The suitable solution should be to find an approximate criterion to predict the PAT performances by calculating them from the pump ones.

In the past some manufacturers have made this for their pumps, which are all medium-high power machines [see for instance Laux (1979), Buse (1981), Javia (1981) and the 1982 *Worthington Techn. Rev.*], but it has been shown that these methods cannot be used without large errors for different models of pump, particularly in the range of the low-medium specific speeds [this was shown by Williams (1995), Boccazzi et al.(1997a and. 1997b)]. Actually, there is no previsional procedure for small power machines, that is the most frequent case of remote *self*-production of energy. In order to find better previsional correlations for small pumps, some Italian laboratories are carrying out a promising research, whose preliminary results were recently presented by Amelio et al. (2000). However, when a mini or micro-hydro project is intended to be carried out by using PATs, in the meantime, the only practical way is: a) to choose a pump with *bep* head and flow rate about 25% lower than the required values as turbine, and b) to test it; which is possible when some University lab is involved in the project.

The Ssezibwa hydro-power project, presented here, has been carried out by the Makerere University (Uganda) and the University of Trento (Italy), under the financial support of the Italian cooperation.

2 THE SSEZIBWA PROJECT

Uganda is a country very rich in ground waters, both large and small permanent watercourses, that actually is the main local energy resource; the national power network is currently supplied by the Owens Falls dam (180 MW) and its extension at Nalubaale dam (80kW), both on the Victoria Nile in Jinjia. However, the electric network is not spread out

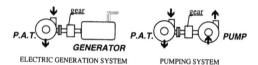

Figure 1. Power uses of the PATs

on all the territory and only 5% of the population is connected; the rural areas that are far from main roads and towns have no power for welfare and productive activities. On the other hand, wherever watercourses are present small hydro-power plants could be profitably set-up with moderate investments and practically no production costs, except for maintenance.

Small hydro-power plants can be easily realized in two ways: using pumps with reverse flow, for waterfall applications, or building cross-flow turbines for river stream applications. Small size centrifugal pumps are very diffused commercially and therefore are very cheap; they can be reliably used in the reverse mode, producing up to 5-10 kW. Cross-flow turbines are not produced commercially but their very simple design allows them to be built locally in a good machine-shop. In both cases the machines have to be investigated from the theoretical and the experimental point of view, in order to determine the operative behavior of the hydro-power system, to select the suitable machine for a given site and to improve it.

The Ssezibwa project was carried out in order to supply water to a rural community in Kayanja, whose population is over 2500. Access to clean water constituted a major problem for the community who needed to fulfil their basic needs of having clean water at a reasonable cost. As a waterfall - Ssezibwa Falls- is located not far away (about 2.5 km) from the village center, it was used for this aim. The river flow rate is large enough both for supplying water to the village and for feeding a hydro-power plant. The project diverted a small fraction of the river flow to drive a turbine, while a smaller flow was diverted to be cleaned and pumped to the village through aqueduct.

As the project was proposed and had to be managed by University staff, it had to be carried out with some restrictive conditions: limited budget, local contractors, sustainability and reliability.

This involved making some choices; the major of which were: a) small size of the power system (micro-hydro power); b) to limit mechanical equipment; therefore, the water treatment was committed to gravity filtration systems without any mechanical assistance; c) to exclude electricity generation and to drive the pump directly by the turbine; d) to use a commercial pump working reversely as turbine.

In spite of or just because of the above limits, the project was defined as demonstrative and hence resulted into a pilot plant.

Figure 2. Ssezibwa Falls, 1995-1999

The project was carried out through five phases: preliminary layout, choice of the mechanical system, executive design, building and setting up, testing. The choice of the power components (pump and PAT) required a laboratory research in order to find the suitable system. A first selection was done on the basis of the required main data: power (1-2 kW) and turbine water head (~16 m); this allowed the choice to be restricted to a few commercial pumps, which had to be tested to know exactly their performances when running reversely.

3 DESCRIPTION OF THE PLANT

Along the Ssezibwa river there is a waterfall (Figure 2) of about 16 m of head, where it was relatively easy to settle the plants for the power and the water cleaning systems. There was no hydrological information available about the river; however, the river is perennial and a low-river flow-rate of about 1800 m^3/h was measured, a small fraction of which (~65 m^3/h) had to be diverted from the upper river through a forebay tank and a short penstock to feed

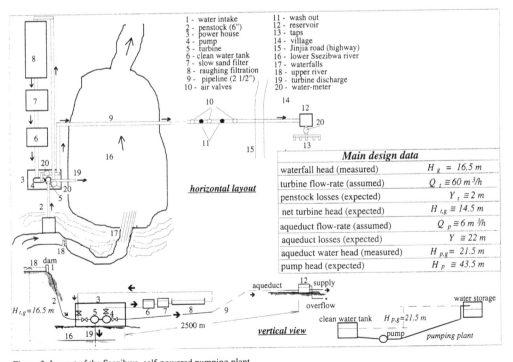

1 - water intake
2 - penstock (6")
3 - power house
4 - pump
5 - turbine
6 - clean water tank
7 - slow sand filter
8 - raughing filtration
9 - pipeline (2 1/2")
10 - air valves
11 - wash out
12 - reservoir
13 - taps
14 - village
15 - Jinjia road (highway)
16 - lower Ssezibwa river
17 - waterfalls
18 - upper river
19 - turbine discharge
20 - water-meter

horizontal layout

vertical view

Main design data	
waterfall head (measured)	$H_g = 16.5\ m$
turbine flow-rate (assumed)	$Q_t \cong 60\ m^3/h$
penstock losses (expected)	$Y_t \cong 2\ m$
net turbine head (expected)	$H_{t,g} \cong 14.5\ m$
aqueduct flow-rate (assumed)	$Q_p \cong 6\ m^3/h$
aqueduct losses (expected)	$Y \cong 22\ m$
aqueduct water head (measured)	$H_{p,g} = 21.5\ m$
pump head (expected)	$H_p \cong 43.5\ m$

Figure 3. Layout of the Ssezibwa self-powered pumping plant

299

both the turbine and the cleaning plant; the seasonal level fluctuations could be regulated by a weir. Figure 3 shows the layout of the plant, together with the main design data.

As the water cleaning system should have to be by gravity filtration, the pump had to work between the clean water tank on the site and a final reservoir in the village, while the turbine had to work between the upper and the lower river.

After surveying and mapping the area, some of the data were measured (waterfall head, aqueduct elevation, pipe lengths), some other were assumed (turbine and pump flow-rates) while the head losses were calculated on the basis of the usual hydraulic methods; therefore they were *expected,* which was a serious mistake, as it will be illustrated later.

On these grounds, the turbine and the pump were selected within the international commercial availability. As a direct mechanical coupling between turbine and pump was assumed in order to simplify transmission saving the gear, the power group should run at the same free equilibrium speed.

4 EXPERIMENTAL INVESTIGATION AND COUPLING PREVISION

Three centrifugal pumps (A, B, C) to be used as turbines were tested and the results are presented in the Figure 4 and the Table. As expected, turbine head and flow rate were higher than for direct rotation; but also the turbine efficiency was higher.

The fact that the turbine *bep* flow-rate is higher than the pump one is quite general for any specific speed and depends on the different shape of the head curves (decreasing for the pump and rising for the turbine) while the internal loss curves are not very different [see Boccazzi et al. (1997b)]; this shifts the optimum toward higher flow rates. Regarding to the improvement of the *bep* efficiency, it can be expected for small PATs; in fact when the pump runs reversely the impeller hydraulic losses usually are

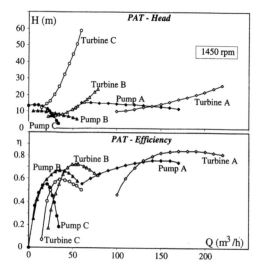

Figure 4. Head and efficiency of the three PATs

lower and this is mostly important in small pumps, as the flow deceleration in high curvature channels imposes strong secondary flow vortices. Moreover, in small machines the increase of flow rate gives a fraction loss coefficient decrease, as the Reynolds number is usually in the transition region. The improvement of the efficiency was confirmed in some other tests on low specific speed pumps [Boccazzi et al. (1997a), Amelio et al. (2000)].

As it can be seen from the Table, the *bep* turbine head of the machine A at rated speed of 1450 rpm is too high for the available water head (~16m); while the *bep* efficiency of the turbine C (0.59) is much lower than that of turbine B (0.73). Therefore, in order to avoid the speed reduction that should be needed with machine A, the machine B was chosen to drive the mechanical system.

The turbine B was then tested at various rated speeds showing a good similarity behavior (Figure 5).

Table: *BEP* data of the PATs, rated speed = 1450 rpm

Pump type	running	A	B	C
Impeller diameter (mm)		219	174	202
Head, H (m)	Pump	12.9	8.5	12
	Turbine	18.8	13.1	25.5
		+46%	+54%	+112%
Flow-rate, Q (m³/h)	Pump	150	35	19
	Turbine	180	55	35
		+20%	+57%	+84%
	Pump	0.76	0.67	0.55
Efficiency, η	Turbine	0.84	0.73	0.59
		+10%	+9%	+7%
Specific speed Ns (metric)	Pump	43.4	28.7	16.3
	Turbine	35.6	26	12.6
		-18%	-9%	-22%

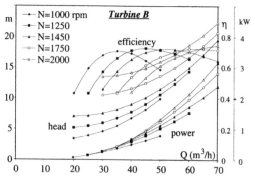

Figure 5. Characteristics at constant speed of the turbine B

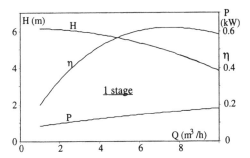

Figure 6. Commercial characteristics of the pump

As regards the pump D of the mechanical system (that had to be a multistage pump because of the high load of the aqueduct), it was not tested as the pump performances were given by the manufacturer as usual (Figure 6).

The choice of the pump was restricted by the effort to find a solution allowing the direct coupling between turbine and pump (that is without gear), that so had to run at the same speed.

On these grounds, it was decided that the power group should be made by the turbine B driving a 12-stage pump D, as shown in Figure 7.

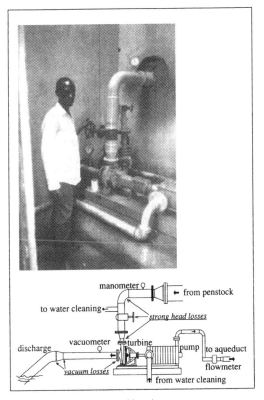

Figure 7. The power system with a view

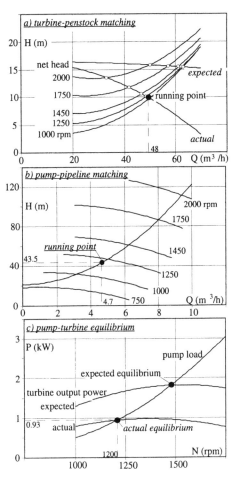

Figure 8. System expected and actual equilibrium

By means of the known performances of the turbine and the pump and of the expected load curves of the penstock and of the aqueduct, the matching conditions at different speeds was determined (Figure 8a and 8b), and from these and the known respective efficiencies the power-speed equilibrium between pump and turbine was defined.

The expected equilibrium parameters are shown in Figure 8c.

5 BUILDING AND SETTING UP OPERATION

Owing to both the needs of saving and the unskilled management, the field works were much less simple than the laboratory and design ones, taking almost two years with many troubles, accidents and stops.

The main problems have been: clearing the area, the hard rock blasting, a river flood (which destroyed a bridge on the river, that had to be rebuilt), the collapse of the roughing filtration tank (due to

301

the fraudulent lack of steel within the concrete), the leakage through the badly waterproofed walls of all the reservoirs, the many thefts of underground pipes (that stopped only when a thief was caught and killed by the enraged people of the village), and especially the manufacture and the setting up of the penstock. Few of these problems were unforeseeable, most were due to the saving needs which forced the management to turn to an unskilled small contractor, which had to provide for the steel works too. The initial budget was 35,000.00 US$, later increased to 40,000 US$ during operations; it was very little and only the almost free labor of the University staff allowed the fulfillment of the project.

Regarding the problems and troubles that were mentioned above, although with more money and delay, all were resolved except one, the penstock.

The penstock was made by cutting, welding, flanging and assembling second-hand old pipes, with internal thick incrustations of plastics; pipe fittings and reduction fittings were very roughly forged; the sealing was inefficient, especially on the depression discharge pipe; moreover, the setting up was made by hand-dragging on the slope (see Figure 9), so that a lot of stones and loam was accumulated within the pipes. This caused the quick stop of the turbine at its first starting up as stones and loam closed the narrow channels of the small impeller. Turbine and penstock were disassembled and cleaned, a grill was fitted in, but the penstock could not be rebuilt and remained essentially the same. The result was: *head losses much higher than expected.*

6 PLANT AND MACHINE TESTING

Finally, the Ssezibwa plant was started in October 1999 and, although its performances are lower than expected, it has been running since then without further troubles and damages (Figure10).

A testing campaign was carried out with the help of the simple instrumentation shown in Figure 7, essentially manometers, flowmeters and tachometer; simulation was obtained by a regulation valve.

It was immediately clear that the shaft speed and the pump delivery were lower than expected. As it was possible to ascertain that the head losses in the aqueduct pipeline were in agreement with the calculation model that was used, that is the load law was correct, the problem had to be the penstock, where much higher losses took place leaving a much lower net water head to the turbine.

On the basis of the measured data, it was possible to calculate the real losses in the penstock. Particularly, high losses take place (see Figure 7) at the turbine inlet (because of wrong diameter of the inlet pipe), at the sharp 90° curve, at the valve, at the turbine outlet (as the turbine is 1.8 m above the river, the discharge pipe should be in depression; however,

Figure 10. Water supplying in the village

Figure 9. The blasted rock and the penstock

the sealing was ineffective and most of the turbine elevation was lost).

The actual turbine net water head is shown in Figure 8a and can be compared with the expected one, while Figure 8c shows the actual pump-turbine equilibrium.

In the end, owing to the unforeseen high head losses in the penstock, the turbine net water head was 10 m instead of 15.5 m and the output power (0.93 kW) was about half the expected one (1.8 kW); so the shaft speed took the equilibrium at about 1200 rpm instead of 1450 rpm; accordingly, the pump delivery was 4.7 m^3/h instead of 6.2 m^3/h.

Fortunately, this delivery was enough to meet the needs of the village, as the depuration capacity of the plant was 30 m^3/day, which can be delivered in 6 ½ hours instead of 5 hours of operative running.

7 CONCLUSIONS

From the scientific point of view the Ssezibwa Project was not noteworthy, but from the technical point of view it gave some useful contribution to people working in the area of the sustainable development. In fact, under many aspects the Ssezibwa Project can be considered positively concluded; that is:

1) the self-powered mechanical system (turbine+pump) proved to be very reliable, as it is running without problems and maintenance since October 1999, in the hard environmental conditions of the equatorial forest and with the only commitment of an unskilled operator;

2) the overall efficiency (available energy/output energy) was very high (~37%), while it should be much lower (~20%) with electric generation; moreover, the mechanical reliability is much higher than the electric one;

3) the use of a commercial pump running as turbine proved to be a low cost and high efficiency solution, allowing a suitable choice among many available models, a quick supplying and a compact mechanical group, without resorting to specialized manufacturers, that is, with only the efforts of University staff;

4) the overall cost (40,000 US$) of the plant was very little, the power group weighing by only 15% upon the budget;

5) the social impact was very significant, especially the linking of the University to the needs of the local people; it was also an interesting training for the University staff which had an opportunity both to learn also from mistakes and to experience the many aspects of the project: prevision, experiment, designing, building, managing and field testing.

REFERENCES

Amelio M., Barbarelli S., D'Amico C., Florio G. 2000: Modello di Calcolo delle prestazioni di pompe centrifughe utilizzate come turbine, *55° Congresso ATI*, Matera, Italia

Boccazzi, A., Gaetani, P., Sala, R. 1997a. Experimental Investigation on Hydraulic Losses of a Centrifugal Pump Operating as Turbine. *ASME 3rd Pumping Machinery Symposium*, Vancouver

Boccazzi A., Pallabazzer R., Sebbit A. 1997b. Pumps as Turbines for Micro-Hydropower. *ISAAE 97*, Jahor Bahru, Malaysia, 266-272

Buse, F. 1981. Using Centrifugal Pumps as Hydraulic Turbines. *Chemical Eng.*, Jan. 1981, 113-117

Javia, R.H. 1981. Pumps as Turbines. *Pump Design Conference*

Laux, C.H. 1979. Multistage reverse running pumps as energy-recovery-turbines in oil supply system. *Sulzer Special Print* KG PU 3

Williams A.A. 1995. Pumps as Turbines. *IT Publ.*, ISBN 1 85339 285 5

"Centrifugal pumps as hydraulic turbines". *Worthington Technical Review* 2100-S1-816-JKG (3-5). 1982

Hydropower in the New Millennium, Honningsvåg et al (eds), © 2001 Taylor & Francis, ISBN 90 5809 195 3

SHP promotion by tradeable certificates – will it work or not?

B.Pelikan
University of Agricultural Sciences, Vienna
Department of Water Management, Hydrology and Hydraulic Engineering

ABSTRACT: During the last decades the promotion of SHP as well as all the other renewables has been done mainly by individual tariff systems. There was little international homogeneity due to the national differences in sources, political lobbying and public acceptance. The ongoing liberalisation process needs new compatible systems meeting both national and international demands and interests. Creating a system of tradeable certificates may be a successful step. It offers both regulation and free market. The position on the scale between the boundaries mentioned is defined by the framework conditions politically fixed. The main keywords in this context are: quotes to be reached, obligations, penalty, reliability and international harmonising. A couple of European countries have already passed the legislative step and have already started the implementation phase. Experiences are rather poor and the therapy is "learning by doing".

1 INTRODUCTION

Speaking about Small Hydropower (SHP) means speaking about renewable energy. Although it seems to be trivial it isn't at all. The reason for a certain distinction respectively separation of SHP is to be found in the general history of hydropower being a traditional source since centuries and having lots of good and bad examples as well as experiences. To most people HP is simply existing but not a subject of interest or promotion. The great majority at least of the European population have not yet realised the advantages, the renewable character and the backbone function within the renewables.

Consequently acting pro SHP means convincing people – to open their eyes and to put SHP in an objective correct position. Even national legislation – for example in Austria – has set signs of division between SHP and all the other renewables. Simply nonsense looking at European goals and policy.

The existence and necessity of promotion of all kinds of RE is part of public as well as political knowledge and acceptance. Only a minority asks for the fundamental reason or if it is some promotion at all. To have a clear view is of great importance. Usually promotion tools serve as a tool of compensation in order to make all the market players competitive within equal and fair conditions. In contradiction to that simple definition the situation on the energy market is not based on equality in cost structure and

completeness of costs. There is simply no fairness. The key, able to solving all economical problems of RE is internalisation of external costs. Once realised at least the most attractive renewable sources will easily be competitive with the non-renewables without any additional support. The aspect described puts a quite different spotlight on all so-called supporting tools for renewables, because they are only compensating manmade disadvantages.

2 BASICS

Recently there is no doubt within the continuously growing European Union, that the contribution of renewable energy sources should be increased significantly and the recent proposal of a directive promoting RE contains the principles and some national targets.

Secondly the need for compensation tools correcting the lack of truth in over all cost calculation is state of knowledge and such tools are essential for RE to survive in a continuously liberalising energy market. Obviously there is some fundamental contradiction between any "supporting" systems and the free market. The idea to implement renewable energy certificates is a certain kind of compromise – an artificial and observed market.

Due to the general features of a certificate system the frame and the conditions provided have key function in positioning the individual result between the boundaries "regulated" and "free". At a first glance that seems to be a disadvantage but instead it is the main chance to establish the system and make it working. To keep this chance all the influencing facts have to be variables for continuous adjustment. From the strategic point of view the implementation process should start from a rather regulated version, watching and evaluating the market reactions and observing whether the function refers to the intentions and the targets can be reached.

Time factor in general is most important, because there is only little resistance against breaks of function. Producers may economically collapse even after some weeks of earning less or no money. Consequences in the system have to be drawn immediately after identifying the problem. The legal framework is decisive in that case and has to robust.

A TGC system is said to be implemented for example in Austria, Sweden, the UK, the Netherlands and Italy. Up to now there is rather no experience whether the certificate system is in principle suitable for renewables or at least for SHP. Running discussions on national and international level have shown some points of delicacy and some "must – conditions" to have a chance – no guarantee.

The main points of discussion are:

Certification and criteria, handling, procedure
Who is obliged
Quotes
Penalty and its "recycling"
Price of certificates and trading phase
Mechanism of flexibility
European harmonising

3 THE PRINCIPLE OF TGCS

The very principle of the system of tradeable certificates is a splitting of the value of energy unit in a physical and an ideological part representing the greenness respectively the sustainability of a unit of renewable generation. The physical value corresponds with the market price of energy – whatever it is. A variable amount depending on world energy market mechanism very little influenced by RE. The ideological part – the certificate – refers to the same unit – the kWh – proving the origin.

Both parts are to be sold by the producer in two divided markets. The physical component is subject of resale – the ideological part needs obligations to create the market respectively the demand. This value may be variable or fixed or something in between. However - both parts of the total value should cover the RE production costs.

4 CERTIFICATION AND CRITERIA, HANDLING, PROCEDURE

From the official European point of view SHP in general is appointed to hand out certificates corresponding to their production. The limitation of SHP is 10 MW. Similar to other RE sources there are no additional conditions to be met.

To make it absolutely clear – the first and singular step is the certification of the production unit – the SHP station. Due to the fact that only energy delivered to the grid is subject of TGC system, grid operators have all information necessary to certify the site. The procedure itself may be carried out by governmental or non-governmental bodies following the criteria *installed capacity* and *feed into the grid*.

In a later stage of developing a certificate system one can imagine distinctions between more than one "green" varying from "dark" to "light" green aiming at an improvement of environmental quality and considering differences between the renewable sources. In an initial phase like now this strategy would destroy the political coherence of all RE.

The second and multiple step is the reliable issuing and handling of the certificates. Again the grid operators may serve easily having primary information. The meter readings are the basis of issuing. Further on it seems to be the best way to count all production at a central body and to make GC electronic documents handled on internet basis. Any producer and any supplier may have an account, offering further opportunities of marketing processing. Obviously the approach to the individual accounts is limited.

5 WHO IS OBLIGED

As said above to make GCs to TGCs a market has to be created by setting obligations. This can be done at any point of the electricity supply chain between production and consumption. The mostly preferred version is the obligation of consumers (polluter pays principle) or the suppliers. Although the obligation of consumers may be useful from a strategic point of view, it is hardly to be handled or executed. Consequently the version most easily to be controlled is the obligation of suppliers.

6 QUOTES

Extremely sensitive is the question of quota and it has to be fixed as a variable value. Additionally this question is closely connected with the borders of system operation. (National or trans-national level).

However it does not make much sense, to create a wide gap between the recently existing quota and the target. Undoubtedly the gap is the driving force increasing the production. But on the other hand there is a relation between the "temperature" of the market and the size of the gap. The bigger the more exited. The overall target is a continuous but responsible increase of RE and not the big boom offering no robustness. In figures the "gap" shouldn't be bigger than roughly 5 - 10% of the target, repeating the possibility to adjust the values easily and short-termed.

7 PENALTY AND ITS "RECYCLING"

Following the target of having a driving force by means of a quota any obligation needs some enforcement. The main aspects are the size of the penalty and how to enforce it. (MITCHEL and ANDERSON, 2000).

Referring to the size two models may be offered, both of them corresponding with the price of the TGC: A fixed amount added or a certain percentage added. Considering the variability of the price of the TGC following market rules it seems to be the more clear solution to add a certain amount fixed in advance by the government. So anybody obliged can do his calculation precisely.

Speaking about the absolute size the additional penalty should ensure that meeting the obligation is cheaper than pay the penalty. The additional amount has to correspond with operational costs, contract costs and administrative costs. Some experiences have shown that about 50% of the price of the TGC may convince the obligatees to buy certificates instead of paying penalty. (Netherlands Green Label System)

If some of them prefer paying penalty for strategic or political reasons the question about making use of the money incoming has to be raised. Best case may be a fund, earmarked for two purposes: buying remaining certificates and supporting the more expensive sites or resources included in a unique certificate system. Within that system of return the question of quotes becomes extraordinary important, because the bigger the "gap" between certificate supply and demand the more money will fill up the fund.

The enforcement of the obligation could be done by the government of the country of the energy supply for both national and foreign supply companies. Both of them have to meet national regulations. Otherwise they will lose their supply permission. That seems to be a theoretical scenery because foreign suppliers will likely try to get quotas of the national

market. One strategy for that aim may be being "greener" than others.

In detail the government may authorise a controlling body for supervision.

8 PRICE OF CERTIFICATES AND TRADING PHASE

In theory the price of certificates will be set by market rules (supply and demand). In practice the price will be influenced by the market structure (individual or pooled), time restrictions, validity and strategic items.

Starting with the market structure SHP is well known for small but many units. For example in Austria there are about 2000 producers, 50% of the being smaller than 100 kW and 90% of them being smaller the 1 MW. Most of the owners are laymen with their another main job.

The counterpart is a limited group of energy supply companies acting professional, having market experience, best information and usually some financial power behind. Easy to be understood that the differences mentioned will discriminate the producers in negotiations and contracting. The inequality could be reduced by building producers groups creating bigger dealing volume. A second measure to ensure fairness is providing a high degree of transparency – a model similar to stock exchange, showing the recent market situation, price movement etc.

For example the Danish system is underpinned by a guaranteed minimum price for TGCs of 0,014 Euro/kWh. That seems to be little, but the penalty in case of not meeting the obligation is almost three times higher at 0,037 Euro/kWh fixing a maximum price for TGCs.

As mentioned above (counting and administrating certificates) even the market place should be a virtual one, connecting market participants. On basis of certificates accounts buying and selling procedures can be carried out and controlled easily. The approach to the "market place" should be limited by accredited participants such as producers, suppliers, big consumers and bodies, representing the groups mentioned.

9 MECHANISM OF FLEXIBILITY (BANKING AND BORROWING)

Time aspects belong - as said in the introduction – to the key decision points of TGC system. Special features of SHP must be considered in that connection.

Banking means "storing" of TGC produced transgressing one redemption period. The only actor of banking is the producer. If obligatees are allowed to bank, the system is severely endangered to be crashed down. Nevertheless the period of banking should be limited according to production features of SHP. Between "wet" and "dry" years the production may differ +/- 30%, means a certificate should have a rolling validity of two years.

Following the idea of virtual marketing it would be easy to mark the TGC with the precise production respectively registration date. After reaching the limitation the GC will be deleted automatically from the account.

The idea of borrowing meets the interests of actors having a shortfall of TGC within one redemption period (a year) and adding the difference to their target in the following redemption period. In general this idea is regarded as unacceptable, but considering production features some extension (maximum half a year) would make sense under the condition, that the TGC market was empty and the shortfall is maximum 8% of the obligation (one month value).

10 EUROPEAN HARMONISING

Although it is not simple to set up a national TGC system, it should be stated that the real benefit of such a system may finally occur in its international (EU or wider) expression. Within that step all the key issues dicussed above have to be shifted on international level. The keyword "harmonising" is to be discussed on the following topics:

Criteria and definitions
National subsidies
Who gets the credit
How to control and finally
How to avoid complicated administration

Concerning criteria and definitions the first steps of European harmonising have already be done by an international group called RECS, starting with a first paper of so-called basic commitments.

More difficult seems to be the variety of national subsidies to be divided into direct and indirect subsidies. Their consideration and finally their withdrawal will be initiate severe resistance.

The question of who gets the credit for TGC – the producing or the consuming country- and some topics connected with is completely unsolved up to now and a really complex subject.

Finally the topics control and administration. It is a well known fact as well as proofed by experience that there is a progressive connection between the size of a system and the administrative body although it is said to be degressive. To fulfil commitments and obligations in time it is unavoidable to create a slim, simple and internet-based structure with short reaction time and effective and direct competencies. From the technical point of view there is no problem to be seen.

But if TGC systems can be set up on national level the problems of enlargement are problems only based on individual national interests. Implementing TGC on international level provides learning that climate change problems have little national but big world wide relevance and – in general - any instrument being effective has to have the same feature as the problem itself.

REFERENCES

MITCHEL C. and ANDERSON Teresa 2000 : "The implications of tradeable Green Certificates for the UK", ETSU Project Number: TGC (K/BD/00218)

SCHAEFFER et al. 2000 : "The implications of tradeable green certificates for the deployment of renewable electricity", ALTENER research project XVII/4.1030/Z/98-037

MOODY et al. 2000 : RECS – Basic Commitment, internal not published paper

BOGNER St. 2000 : "Studie zur Marksituation für Kleinwasserkraftwerks-Zertifikate sowie über die Grundkonzeption eines Modells für den Zertifikatehandel", expertise not published

Water and Environmental conflicts in The Pangani River Basin, Tanzania – Hydropower or Irrigation?

Paul Christen Røhr & Ånund Killingtveit
Department of Hydraulic and Environmental Engineering, NTNU, Trondheim, Norway

George Michael Nderingo
Ministry of Agriculture and Cooperatives, Zonal Irrigation Unit, P.O. Box 1843, Moshi, Tanzania

Peter Elias Kigadye
Tanesco Ltd, P.O. Box 9024, Dar Es Salaam, Tanzania

ABSTRACT: In 1995, the Pangani Falls Hydropower Plant was operational after several years planning and construction. For obtaining maximum production, maximum inflow to the reservoir is preferred. This is in conflict with the irrigation water requirements for the upstream agricultural activities. This paper focuses on the environmental conflict between hydropower production and upstream water users, especially agricultural irrigation using the example from Pangani River Basin. Simulations show the benefit from various use of the water. With present irrigation system, the use of water for hydropower gives a better economic return than agriculture. With some improvement of irrigation efficiency, the economic return will be more even between the users and the consumers of water. It will also contribute to smaller water consumption with present level of agricultural activities in the area. This is coupled with an approach for distributed hydrologic modeling, which take the vegetation, topography and landuse into consideration.

1 BACKGROUND

Agriculture, which occupies 90% of Tanzania's labor force and account for half of the GDP, is given high priority in the ongoing Structural Adjustmentprograme although Tanzanian-Authorities (1999) state that the expected development has not appeared. High attention to agricultural development can therefore also be expected in the future. In the hill slopes of Mt. Kilimanjaro, agriculture has long historical traditions, which have developed some of the oldest and most extensive traditional furrow irrigations systems in Africa (Lein 1998). Expeditions as far back as in the early 1880 give descriptions of the hillside of Mt Kilimanjaro saying that "there where scarcely a ridge without its own irrigation channel". (Johnston 1886). Formal and informal collaboration between the Chaggas of Kilimanjaro based on old traditions managed to develop and maintain these furrow systems with hundreds of farmers for several hundred years. During the recent times in the post-independent Tanzania, the "Ujima" policy emphasized larger public estates to a greater extent and left the control and maintenance of the irrigation system to the public. Foreign aid has made it possible to extend agriculture to the lowland plains where unreliable rainfall conditions make irrigation necessary for securing the crop. Irrigation is based on both surface water from nearby rivers and to a certain extent groundwater yield from natural springs and drilled wells.

For development and rehabilitation of the poor economic structure of Tanzania and to encourage small-scale private industry and trade, electric power is of great importance. Insecure supply and rationing of power cannot be regarded as encouraging. The recent years several hydropower plants have been put into operation and secured the power supply to a certain extent. In connection with the Pangani Falls Redevelopment in the late 1980ies and early 1990ies, the scarcity of water resources in the Pangani Bain was further discovered and water management was emphasized. The Pangani Basin Water Office, PBWO, was established in 1991 to deal with the water resources in the river basin. Presently all water users in the basin, from the big power producer to the simple smallholder, have to apply for a Water Right granted by PBWO for utilizing the water from the rivers in the basin. In this context the idea about a research program on Water Management in Pangani River Basin was developed as a co-operation project between University of Dar es Salaam and the Norwegian University of Science and Technology starting in 1997 and running for four years. The research project focus on changes in landuse and how this inflect the hydrological response in the downstream river shed and with that the Water Management in the river basin. It involves various spheres of study like botany, geography, social science, economy and hydrology for considering different aspects of Water Management. The project involves Ph.D. and master theses in both Tanzania and Nor-

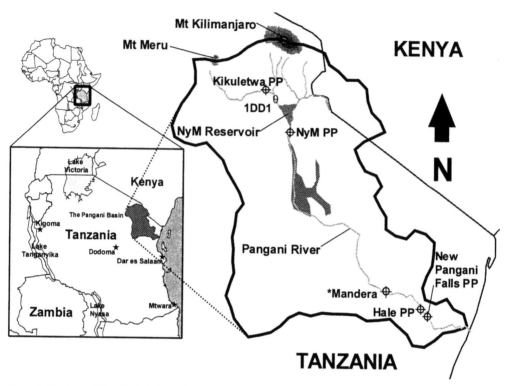

Figure 1. The Pangani River Basin in Tanzania, East-Africa. Reservoirs, Rivers, Power Plants (PP), International Borders and Basin borders are shown. (*Planned Power Plant).

way. It is funded by the Norwegian Agency for Development.

Previously several master theses have been performed focusing on various topics related to water management and the conflicts between water users. Some consideration has also been put on landuse and climatic changes and how they influence the downstream hydrological conditions in the watershed. The following will give some overview of the past and present research activities in the basin related to hydrological modeling, landuse and the contrast between hydropower and agriculture concerning economical return.

2 THE BASIN, ITS WATER AND USERS

The 42,200 km² Pangani River Basin is located northeast in Tanzanian reaching 450 km from the Southern slopes of Mt Kilimanjaro and Mt Meru and Southeast to its outlet in the Indian Ocean near Tanga. See figure 1 for an overview. About 90% of the population of 1.3 million in the upper part of the river basin depends directly or indirectly of agriculture. The highland above approximately 1000 masl receive up to 2000 mm of precipitation annually and

is heavily populated with up to 300 persons/km². The high densities have forced people to migrate from the fertile hillsides of Mt Kilimanjaro down to the more poor lowland plains. The lowland plains receive less than 500 mm of precipitation annually and here irrigation is necessary for securing crops against the uncertain rainfall and subsequent drought. There is one major reservoir, Nyumba ya Mungu, NyM, located in the central part of the river basin, which can store 80-100% of annual runoff. Much of the irrigation based agriculture in the basin are located upstream this reservoir. The groundwater yield in the area is considerable. A few large springs contribute with a yield of more than 20 m³/s, which form a major part of the inflow to NyM during the dry season.

It is estimated that a total area of 29,000 ha is under irrigation of which some are big irrigation schemes of several thousand ha. They utilize surface water from nearby rivers and to a certain extent groundwater. Due to unlined channels, poor on-farm water distribution and no regulation facilities, the irrigation efficiency can be as low as 20-30%.

The main users of water in the basin, which PBWO issues water rights for, are irrigation, domestic, industrial and hydropower production. Due to population increase and irrigation development recent years, the water demand has increased and stressed the scarce water resources in the area.

3 LUMPED HYDROLOGIC MODELLING

Water available for hydropower production is often found by use of hydrologic modeling. If the model is flexible enough, e.g. irrigation can be taken into consideration directly in the model. A lumped hydrologic model for parts of the catchment upstream Nyumba ya Mungu is modeled by Lefstad & Bjørkenes (1997). The objective for their work was to create a model that was able to simulate the water balance in the area and to include routines for calculating soil water storage, spring water yield, water used for the irrigation schemes and to simulate the discharge in the rivers. With reference to the schematic outline of the watershed on figure 1, Lefstad & Bjørkenes (1997) have modeled the catchment above the gauging station 1DD1. Their model is somewhat more detailed than on the figure. It is divided in 7 sub-catchments, which represent the rivers with their catchments upstream. The PINE-modeling system (Rinde 1998) was used for the modeling work. Each sub-catchment is divided on 8 different elevation zones. The rainfall-runoff modeling is based on the concept of the HBV-model (See e.g. Bergstrøm (1992), Killingtveit & Saelthun (1995)). In addition to the two linear tank response routines in the HBV-model, they include two different types of spring response routines, which operate

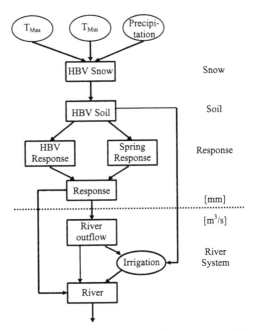

Figure 2. Structure for one sub-catchment in the model by Lefstad & Bjørkenes(1997)

in parallel with the HBV-response. The principal components for one sub-catchment can be seen in figure 2. A water balance calculation is used down to the dotted line where the water is converted to runoff. This is identical for all the sub-catchments

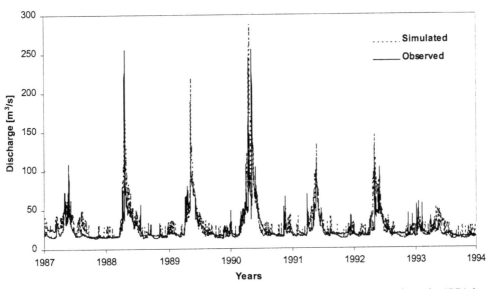

Figure 3. Example of calibration result from a lumped model representing the catchment above gauging station 1DD1 showed on figure 4. Data from Lefstad & Bjørkenes (1997).

with small distinctions. The irrigation is taken care of with irrigation nodes, which receive water from the response routines. Irrigation water requirements are calculated by the pan-evaporation method and crop coefficients for different crops. The simulated discharge at 1DD1 is compared with the observed discharge. The explained variance, R^2-criteria, is used for supporting the evaluation of the calibration. The result of the calibration can be seen on figure 3. The criteria for goodness of fit, gave a value of $R^2=0.34$.

Lefstad & Bjørkenes (1997) use their calibrated model for running simulations with various changes in climate and landuse. The climatic changes tested, are increase/decrease in precipitation, temperature and evaporation. When the precipitation is changed, the simulated discharge shows some changes in the wet season, while the changes are small for the rest of the year. Temperature changes do not give any changes in simulated discharge, but it is pointed out that this might be due to the calculation methods used in the model. Changes in evaporation bring some changes in the simulated discharge. The changes in simulated discharge are proportional with the changes in evaporation. Combinations of climatic changes show the same results as individual changes. Lefstad & Bjørkenes (1997) conclude that their scenarios for climatic change do not show any remarkable reduction or increase of the discharge in the river.

The changes in landuse considered in the simulations are increase in irrigated areas, increased irrigation efficiency and increased population. The increase in population is assumed to give a proportional increase in the irrigation parameters like irrigated area and water abstractions. Increased population leads to increased irrigation areas and increased abstractions of water from the river. This means a greater offtake of water from the river and with that, a reduction of discharge in the river that is considerable, particularly during the dry season. For an annual population growth of 1.5% assumed from 1997 to 2006, the decrease of discharge in the river was found to be 8 m^3/s in the low flow season. Lefstad & Bjørkenes (1997) point out that field investigation of water use for an individual family and field size, will give estimates that are more reliable.

Table 1. Economic return for each m^3 of water for various types of crop.

| Crop | Season | Economic return, TSh/m^3 | | |
		IE = 0.30*	IE = 0.35*	IE = 0.40*
Rice	Dec-May	32	37	42
	Jun-Nov	27	32	37
Maize	Nov-Mar	32	37	42
Beans	Aug-Oct	33	39	44
	Dec-Feb	31	36	41

*IE = Irrigation efficiency

jor irrigation schemes. Only the major irrigation schemes are selected for inclusion in the model, to see how different withdrawal of water from the Pangani River would affect the power production. There are many more minor irrigation schemes, but the selected ones represent a considerable part of the area and they have reliable and systematic data for their monthly water abstractions.

The simulation system is divided in 4 different components:
– The production system, hydropower and irrigation
– The consumer system, electricity and crop
– The regulation reservoir
– The operation strategy

The Water Requirements for different crops are estimated by use of the Penman's formula for calculation of the potential evpotranspiration. With use of a crop water coefficient, the total water demand for the irrigation scheme is calculated for various crops and different alternatives of irrigation efficiency. The economic return for each m^3 of water for various types of crop are shown in table 1. It was computed for the current Irrigation Efficiency, IE, of 0.3 and for two alternatives with improved Irrigation Efficiency of IE = 0.35 and IE = 0.40.

Corresponding to table 1, the Energy Equivalent and the belonging value of the water for hydropower production at different locations along the river can be seen in table 2. From Tanzania Electricity Company, Tanesco, the economic return for power production is put at 0.089USD/kWh. This give the economic return for power production at different locations as shown in column three in table 2. (Exchange rate 1USD=710TSh)

For the upper part of the shed where also the agricultural activities based on irrigation are most in-

4 AGRICULTURE VS HYDROPOWER

A work on the conflict between hydropower and irrigation was done by Nderingo (1999). He considers the economic benefit from various alternatives of irrigation efficiency and compares these with the alternative hydropower production. A simulation model for the Pangani River Basin is prepared including the hydropower plants and some of the ma-

Table 2. Economic return per m^3 of runoff from catchments above various Power Plants in Pangani River Basin.

Power plant	Accumulated Energy Equvalent kWh/m^3	Economic return TSh/m^3
Kikuletwa*	0.64	40
Nyumba ya Mungu	0.60	38
Hale	0.55	35
Pangani Falls	0.42	26

* Not operational for time beeing

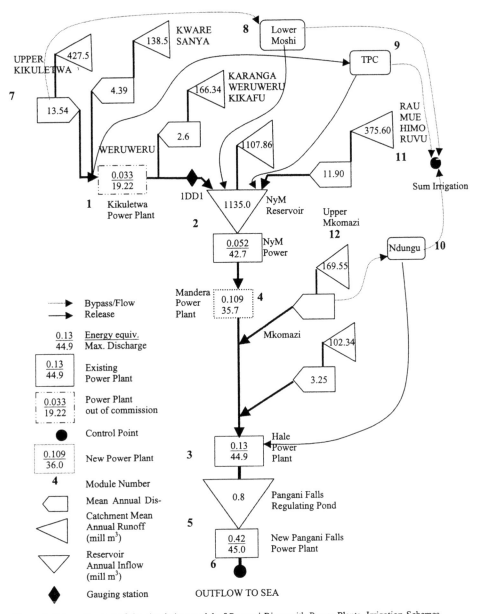

Figure 4. Schematic view of the simulation model of Pangani River with Power Plants, Irrigation Schemes and Reservoirs. See also figure 1 for a scale map. Some modified from Kigadye (2000).

tense, the economic return from power production is about 40 TSh/m^3 of water, which is about 30% more than for crops with irrigation efficiency about 30%. However by improving the irrigation efficiency to 0.35-0.40, crop are more competitive with power production also in the upper part of the river basin. Nderingo (1999) state that considerable investments are necessary for improving the efficiency. He gives the following recommendations:
– Strict enforcement of efficient use of water

– No more abstractions if expected power production are not met
– Conservation of catchments

Many farmers use the water inefficient and abstract much more than necessary for irrigating their farms. This affects both downstream farmers and power production. This can be improved by managerial, educational and technical efforts. The number of abstractions should remain constant and in case of further demand for water, improvement of efficiency should be focused. The granted water rights should be strictly monitored.

Further work with a simulation model for Pangani River is done by Kigadye (2000). He extend and verify the model made by Nderingo (1999) including some planned hydropower schemes and try to uncover whether there have been any decreasing trend in hydropower production due to increased offtake for irrigation. A schematic overview of the simulation model for the Pangani Basin can be seen on figure 4. The model is a continuation and developed version of the model used by Nderingo (1999) in his work.

Kigadye (2000) run simulations and compare them with actual production at the existing power plants in the basin. Due to only 3 years of simultaneous data, the results are somewhat diverging which Kigadye (2000) claim uncertainty about the existing power plants, quality of discharge data and differences in strategy for operation of the power plants.

The simulations done by Kigadye (2000) show that further development of hydropower in the river will increase the value of the water upstream. If planned power schemes are constructed, the energy equivalent and corresponding value of the water if used for hydropower production, will be as shown in table 3.

Compared to the economic return from agricultural production computed by Nderingo (1999), further development of hydropower might make the agricultural production more disadvantageous due to its poor efficiency in use of water. Improved irrigation efficiency will change this relationship.

When considering the trends, the actual power production at Nyumba ya Mungu show some decline through the 1990s, but Kigadye (2000) indicate that the period of simulation are too short for establishing the trend. The decrease in power production at the reservoir corresponds to a decrease in inflow, but no corresponding patterns can be found when analyzing the precipitation data from the area. He suggests that this can be due to unnecessary loss of water into irrigation schemes during flood season.

Table 3. Economic return per m³ of runoff from various catchments in Pangani River Basin if Kikuletwa Power Plant is reconstructed and Mandera Power Plants is constructed. (Energy equivalents from Kigadye (2000). Power prices from Nderingo (1988).

Power plant	Accumulated Energy Equvalent kWh/m³	Economic return TSh/m³
Kikuletwa*	0.75	47
Nyumba ya Mungu	0.71	45
Hale	0.66	42
Mandera**	0.53	33
Pangani Falls	0.42	26

* Not operational at time beeing
** Planned new Power Plant

Ungated offtakes and poor management of offtakes involve uncontrolled offtake of water also during flood season when irrigation is not necessary. The rainfall and runoff pattern was also investigated for a 18 year period from the early 1960s to late 1970s, but no trends can been observed for this period.

Landuse changes may influence the offtake of water and the water consumption in the catchment. An increase in offtake of water from the river will mean decrease in river discharge and with that a smaller potential for power production at downstream power plants. Models that can have landuse as a parameter can be used for predict the changes in river discharge and power production.

5 DISTRIBUTED HYDROLOGIC MODELLING

5.1 Approach for modeling water use

Compared to the previous description of the lumped model approach, a more accurate estimate of the water consumption could be found by use of a distributed hydrologic model. Various landuse and vegetation and their changes can more easily be taken into consideration. An existing modeling tool capable of calculating the runoff from a catchment based on soil conditions, vegetation, topography, landuse and meteorology is described by Rinde (1999). Using this model system, a semi-distributed hydrologic model is under preparation for parts of the catchment above 1DD1 on the system described on figure 4. Three minor sub-catchments have been selected in the hillside of Mt Kilimanjaro. The model will take the changes on landuse into consideration and predict their influence on the runoff from the various catchments.

A thorough understanding of the present situation is essential and this is emphasized in the hydrologic modeling. When satisfactory knowledge and simulations of the known situation is achieved, various changes in landuse and water management policy can be investigated. This can be the previous situation or the predicted future scenarios. The changes in hydrologic conditions in the catchment can then be determined and used as input for hydropower simulations. The influence on hydropower production due to changes in e.g. landuse can then be determined.

In addition to the present modeling tool, emphasize is also put on the description of the groundwater module since this play an important role in the water yield on the lowland plain and contribute with a considerable part of the runoff in the dry season. Many theories about the groundwater recharge have been raised, but few numerical proofs have been presented. Many various areas have been assumed to be the source of recharge for the groundwater aquifers in the area. Water balance calculations for three minor selected catchments in the hillsides of Mt Kilimanjaro can contribute to uncover the contribution to the groundwater recharge. The various component sin the water balance can be sorted out and the groundwater component determined.

5.2 Description of model

The present model calculates the distributed water balance above, on and under the soil surface down to the root zone. Effects of changes in landuse, which concern the vegetation, surface or sub-surface, can be described by the model. A response function receives the outflow from all the area elements and forms a runoff hydrogram from the whole catchment. A figurative description of the model structure in each area element and the response function can be seen in figure 5. The simulated discharge q on figure 5 can be used as input to a simulation model for hydropower production.

The distributed parameters utilized in the model can be seen in table 4. Many of these data can be developed from digital landuse maps, which can give adequate values for the different parameters. The user depending on his accessibility for data can decide the resolution for the model. If your data are very coarse, there is neither cause nor need for making a very detailed grid in your model. On global basis, a 1 × 1 km Digital Terrain Models, DTMs, are available free of charge which could be used for distributed hydrologic modeling. (E.g. http://edcdaac.usgs.gov/). Further division of this into a smaller pixel size, can make it possible to utilize a more detailed description of other parameters even if the DTM are more coarse. The global DTM, which have been used, have a resolution or a grid of 1 × 1 km pixels. Each pixel in the DTM have been

Table 4. Distributed parameters, which are utilized by the hydrological model.

Parameter	Unit	Explanation
Landuse	-	Lake, forest, agriculture, etc.
Elevation	m	Terrain elevation
Vegetation cover	%	Coverage of high vegetation
Vegetation height	m	Average tree height
LAI max	-	Maximum annual Leaf Area
LAI min	-	Minimum Leaf Area Index
LAI low	-	Maximum Leaf Area Index for underscrub
Surface storage	mm	Hollow storage on soil surface
Infiltration capasity	mm/ timestep	Infiltration capacity to the root zone
Field capasity	mm	Field capacity in root zone

divided into 4 new, which give a resolution or a pixel size of 250 × 250 m. This make it possible to describe other parameters as landuse more refined. A spatial resolution of 1 × 1 km for the altitude should be sufficient for taking altitude related effects into consideration. A 250 × 250 m grid for landuse will take the spatial variations in landuse into consideration in a more detailed way than e.g. a 1 × 1 km grid. It is no problem to divide the grid further into smaller pixels, but the disadvantage of a smaller pixel size is the increase in the amount of data, which requires increased calculation time.

The results produced by the model are both distributed and lumped. For the user, it is often of interest to have some knowledge also about the intermediate state variables in the model during the simulation period. E.g. soil saturation, interception and snow storage at a certain time or development through the year. Example of intermediate state values can be seen in table 5. Some examples of the response output from the model can be seen in table 6. The aggregated state variables and response output are given as time series with one value for each time step and is written to a data file if the user specifies this. The data series of runoff from the hydrologic model can then be used as input to the model simulating hydropower production.

The distributed data from the model is exported as a map at an interval given by the user, e.g. daily,

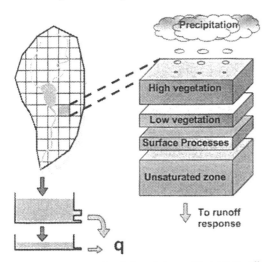

Figure 5. Figurative description of the model structure. All processes above, on and under the soil surface down to the root zone are described distributed.

Table 5. Example of state variables in the distributed hydrological model.

Distributed parameters	Unit
Actual interception in high vegetation	mm
Actual interception in low vegetation	mm
Snow storage	mm
Relative soil saturation	%
Lumped parameters	Unit
Average actual interception in high vegetation	mm
Average actual interception in low vegetation	mm
Average snow storage	mm
Average relative soil saturation	%

315

Table 6. Example of response output from the distributed hydrological model.

Distributed parameters	
Aera distributed precipitation	mm
Aera distributed temperature	oC
Evapotranspiration from high vegetation	mm
Inflow to the soil surface layer	mm
Soil outflow	mm
Surface flow	mm
Lumped parameters	
Aerial precipitation	mm
Aerial temperature	oC
Average evapotranspiration from high vegetation	mm
Average inflow to the soil surface layer	mm
Average soil outflow	mm
Average surface flow	mm
Runoff from the whole catchment	m^3/s

every 7 days or every 30 days. A regular GIS can be used to view the distributed output after each simulation and prepare visual demonstrations of how e.g. soil saturation or evapotranspiration develop over the season.

6 DISCUSSION AND CONCLUSION

The paper describe some of the various modeling work performed with Pangani River Basin as an example. Rainfall-runoff models and hydropower simulation models have been combined for analysis of the optimum water use for both agriculture and hydropower production. At present stage the hydropower production is economically favorable due to low irrigation efficiency in the irrigation industry. An improvement of the rather low irrigation efficiency can make agricultural production more efficient in the future. New hydropower development in the river will increase the power output from each m^3 of water, and make irrigation less favorable. Various changes in landuse may change the evaporation and water consumption and give less water for hydropower production.

The effects of such changes in landuse can be analyzed by use of a distributed hydrologic model. An approach for distributed modeling which have landuse as a parameter is presented in the last part of the paper. With use of a distributed hydrologic model, a more detailed description of the water use can be achieved and coupled with the model for agriculture and hydropower production, this can give a better description of the conditions in the river shed. The landuse can be taken into consideration, possible changes can be modeled, and their consequence can be evaluated before actual changes are performed in the field.

REFERENCES

Bergstrøm, S. (1992). *The HBV-model - Its structure and application* (RH No 4): SMHI.

Johnston, H. H. (1886). *The Kilimanjaro Expedition*. London: Kegan, Paul and Trench.

Kigadye, P. E. (2000). *Hydropower Projects in Pangani River, Tanzania*. Master of Sciene, The Norwegian University of Science and Technology, Trondheim.

Killingtveit, A., & Saelthun, N. R. (1995). *Hydrology*. (1 ed.). (Vol. 7). Trondheim: The Norwegian Institute of Technology, Division of Hydraulic Engineering.

Lefstad, L., & Bjørkenes, A. (1997). *Diploma Thesis in Water Resources Planning. Hydrological studies in The Upper Pangani River, Tanzania*. Diploma Thesis, The Norwegian University of Science and Technology, Trondheim, Norway.

Lein, H. (1998). Traditional versus modern water management systems in Pangani River Basin, Tanzania. In L. de Haan & P. Blaikie (Eds.), *Looking at maps in the dark. Directions for geographical research in land management and sustainable development in rural and urban environments of the Third World* (pp. 52-64). Utrecht/Amsterdam: Royal Dutch Geographical Society, Faculty of Environment Sciences, University of Amsterdam.

Nderingo, G. M. (1999). *Hydropower and irrigation in Pangani River - Conflict or cooperation?* Master of Sciene, The Norwegian University of Science and Technology, Trondheim.

Rinde, T. (1998). *A flexible hydrological modelling system developed using an object-oriented methodology*. Doctoral Thesis, The Norwegian University of Science and Technology, Trondheim.

Rinde, T. (1999). *Landpine, a hydrologic model for simulation of how landuse changes inflect runoff* (Note) In Norwegian: *Landpine. En hydrologisk modell for simulering av arealbruksendringers innvirkning på avrenningsforhold* (Notat). Trondheim: SINTEF Bygg og miljøteknikk.

Tanzanian-Authorities. (1999). Tanzania, Enhanced Structural Adjustment Facility. Policy Framework paper for 1998/99-2000/01. In I. M. Fund (Ed.) (pp. 15). http://www.imf.org/external/np/pfp/1999/tanzania/index.htm : International Monetary Fund.

Hydropower in the New Millennium, Honningsvåg et al (eds), © 2001 Taylor & Francis, ISBN 90 5809 195 3

Increasing energy output from existing hydro power system in Tasmania

Goran Stojmovic
Senior Engineer Dams, Hydro Tasmania, Australia

Joanna Sheedy
Dam Engineer, Hydro Tasmania, Australia

ABSTRACT: With future load demand and entry of the Australian island state of Tasmania into the National Electricity Market, Hydro Tasmania, the state's chief energy producer, is looking to expand the existing network. With the focus on renewable energy sources and Hydro Tasmania's existing electricity network being based on hydro, enhancements to the existing hydro generation network will help meet this required expansion. Preliminary assessment indicates that a minimum of 80 MW-av additional energy to the existing 1,160 MW-av system is potentially viable hydropower potential within Tasmania.

1 INTRODUCTION

The island of Tasmania is a state of Australia and lies about 250km south of the state of Victoria on the Australian mainland. It is surrounded by Bas Straight in the north, the Tasman Sea in the east and the Indian Ocean on the south and wèst. Tasmania is about the size of the Republic of Ireland and island of Hokkaido (Japan). The land area is 68,331 km^2, with the distance from east to west being 315 km and from north to south 286 km. Tasmania has a temperate, maritime climate with four distinct seasons. The total Tasmanian population is 472,000.

Tasmania's World Heritage Area covers about 20% of the state. In addition many national parks and reserves exist outside the World Heritage Area. Almost 36% of Tasmania lies within some form of environmentally protected area. Generally, but with some exceptions outside the World Heritage Area, this protected area is closed to further hydropower development despite containing the largest undeveloped river systems in Tasmania.

2 THE PRESENT STAGE OF HYDRO DEVELOPMENT

The Tasmanian electricity production network is almost entirely based on hydro energy, the resources for which have been developed over a hundred-year period. The present Tasmanian power system is made up of twenty-seven hydropower stations and one thermal power station. The total installed capacity of the system is 2,500 MW with the thermal power station contributing 240 MW of this.

The hydro power stations currently supply about 1160 MW-av (10,000 GWh) of electrical energy annually. The peak load is 1,500 MW and the available storage capability is 1680 MW-av (14,724 GWh).

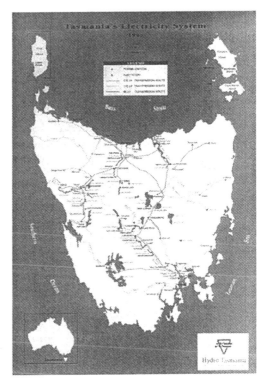

Figure 1. Map of Tasmania

Electricity demand increased by approximately 5% during the past year.

3 FUTURE IMPACTS ON HYDROPOWER DEVELOPMENTS

3.1 Basslink and National Electricity Market

Presently Tasmania's electricity network is isolated from the mainland of Australia. Significant research and planning has gone into the interconnection of Tasmania and Victoria via an undersea cable, known as Basslink. The timing of Basslink is subject to completion of environmental and regulatory approvals, but is anticipated in 2003.

Currently Hydro Tasmania receives a tariff for energy sales that does not vary with the time of day. With the interconnection of Tasmania's electricity system to the mainland of Australia, Hydro Tasmania will be able to participate in the competitive wholesale national electricity market where electricity prices are bid into a pool.

In addition, increased power demand is likely to result from entry into the National Electricity Market.

3.2 Gas Reticulation

Tasmania has a large energy-intensive industrial base, which needs additional sources of energy in order to expand. The introduction of natural gas into the Tasmanian market will provide an additional source of energy and will strengthen the Tasmanian economy. However, the introduction of gas will also introduce competition in electricity generation and put downward pressure on Hydro Tasmania's market share. While this will be offset to some extent by the increase in electricity demand generated by the stimulus that gas would provide to the Tasmanian economy, Hydro Tasmania is in the process of developing strategies to ensure that the adverse impacts on its values and returns are minimised.

Despite uncertainties, Hydro Tasmania supports the introduction of gas and the option of converting the existing thermal power station from oil to gas have been jointly studied with the gas pipeline developer to assist the gas project to proceed.

3.3 Renewable Energy

In December 2000, as part of its greenhouse gas abatement strategy, the Federal Government announced that it would legislate to boost the contribution of the renewable energy industry to Australia's electricity supply by 2% by the year 2010 and introduce a system of tradeable certificates for renewable generation. This legislation will underpin substantial growth in renewable energy generation over the next decade since it will result in a pre-mium, the so-called mandated renewables premium, being applied to the sale of energy that is generated from renewable sources.

For Hydro Tasmania the renewables premium will be applicable to increased sales of energy resulting from enhancements of the existing generation system. (The quantum of the increase is expected to be measured yearly in relation to a baseline that may be set below the average annual energy generation during some period prior to January 2001.)

4 PROSPECTIVE HYDROPOWER DEVELOPMENT

4.1 General

A sharpened focus on renewable energy developments has lead to investigation into opportunities for further hydro developments. With expected growth in power demand both within Tasmania and in mainland Australia, Hydro Tasmania is looking for opportunities to enhance its existing hydro generation system and increase the overall system capacity. In addition, with the imminent introduction of the mandated renewables premium, enhancements to the existing system will achieve greater economic returns.

An initial assessment of remaining hydropower potential within Tasmania identified many opportunities for future developments. The assessment was based on both the collation of previous investigations carried out and recent studies to identify opportunities by examining where the capture of spill, utilization of head, reduction in friction losses in conduit systems, and increased system inflows would be most beneficial. Hydropower potential was assessed in the following categories:

- small hydro developments (0.5 MW – 15 MW)
- catchment diversions
- increased storage capacity (headpond raising)
- increased peaking capacity within existing schemes
- new large developments (15 MW – 350 MW)
- redevelopment of existing, older schemes that require substantial refurbishment
- conduit and penstock cleaning
- turbine runner refits

4.2 Environmental Constraints

As stated previously, Tasmania's World Heritage Area covers about 20% of the state. Developments in this area are non-viable as they cannot presently be constructed due to the World Heritage Properties Conservation Act (passed in 1983 by the Australian Federal Parliament). Furthermore, developments outside the World Heritage Area but within National

Parks and some other reserve areas are also non-viable due to environmental constraints.

Schemes located within Conservation Areas and some other reserves (approximately 10% of Tasmania) and schemes affected by existing irrigation and municipal users would require significant consultation to assess their viability with respect to existing environmental issues. Depending on the type and scale of work and the outcomes of the environmental impact assessment, development of these schemes may be possible. Such schemes have been given a marginal environmental rating.

Even for schemes where no environmental issues are apparent an environmental impact assessment is required prior to approval for development.

4.3 Economic Constraints

For any scheme to proceed it must be shown to be economically viable. The potential economic viability of potential hydropower development is dependent on the expected market prices and requirements. Schemes were assessed on the basis of their unit cost of energy production.

Schemes that are currently economically viable have a unit cost less than the base market price even if Tasmania does not enter the National Electricity Market (less than A\$40/MWhr). These schemes, if they are environmentally acceptable, are the most desirable for Hydro Tasmania.

Schemes may be economically viable if the unit cost of energy production is less than the anticipated energy sale price if Tasmania enters the National Electricity Market and with the introduction of the mandated renewables premium (between A\$40/MWhr and A\$70/MWhr).

Schemes are economically non-viable where the unit cost of energy production exceeds the estimated energy sale price with both entry into the National Electricity Market and the renewables premium (greater than A\$70/MWhr).

5 SMALL HYDRO POTENTIAL

Hydro Tasmania is looking closely at the potential small hydro resources within Tasmania and its part in meeting the future power load forecasts as an environmentally sensitive energy source and economic means of increasing hydropower output from Tasmania. Small hydro developments with installed capacities of between 0.5 MW and 15 MW were considered.

In some cases small hydro installation constitutes only the utilization of riparian release, in other cases it utilises spill or utilizes head loss within existing schemes. Small hydro schemes involving the construction of new major dams were disregarded.

The total small hydro potential identified is 28 MW-av, provided by a total of 15 schemes ranging from 0.3 MW-av to 8 MW-av. Of this some 5.4 MW-av could be generated by 4 schemes which are economically viable under current market conditions and environmentally acceptable subject to the completion of an environmental impact assessment. A total of 12 schemes providing an additional 18 MW-av of energy output have the potential to be economically viable and environmentally acceptable and will be further investigated.

It is likely that other viable small hydro developments within Tasmania may exist since some 200 schemes were eliminated from consideration on the basis of very broad criteria. A reassessment of eliminated options is planned for the near future.

6 CATCHMENT DIVERSIONS

Many catchment diversions were briefly examined, of which several showed potential. To determine which existing schemes to investigate potential for further catchment diversions into the scheme, an assessment was made to determine which schemes gain most benefit from increased inflows.

Two schemes providing 1.7 MW-av are economic and any environmental issues are likely to be resolved. One scheme (0.9 MW-av) involves only the reinstatement of some existing but disused pumps and the other scheme (0.8 MW-av) was previously investigated and designed during the construction of the Anthony Power Scheme near the Tasmanian west coast.

A total of 5 catchment diversion schemes, providing a potential increase in energy output of 4.6 MW-av have the potential to be economically viable and environmentally acceptable and will be further investigated.

Other potentially viable catchment diversion schemes may exist given that a complete study of every catchment within Tasmania has not been completed.

7 INCREASE EXISTING STORAGE CAPACITY

Some storages within Tasmania spill frequently and this spill represents potential energy.

The five storages that were initially assessed for an increase in their existing storage capacity were selected after consideration of the amount, duration and energy value of storage spill. Furthermore, storages where it may be beneficial to capture spill were disregarded if raising of the full supply level leads to flooding of an upstream power station or if spillway gates are already installed.

All five storages examined appear to be economic after preliminary investigation. Two options have some significant environmental issues to address. An estimated total of 13 MW-av could be achieved through raising these five storages. The technical viability of raising has not been addressed in detail but given that dam raising in the order of 1 to 3 metres only is required, no technical issues are immediately apparent with the five storages so far investigated. If further, more detailed assessment of these five storages verifies that increasing the storage capacity is economic then other storages may be considered.

The initial assessment did not include the capture of a power spill from station with multi-storages. In particular, for all storages associated with Tungatinah Power Station, the total power in spill is about 11.6 MW representing about 25% of the long-term average power. By raising just one headpond of this station an energy gain of 2 MW-av would be achieved.

8 INCREASE PEAKING CAPACITY WITHIN EXISTING POWER SCHEMES

Currently, Hydro Tasmania receives a tariff for energy sales that does not vary with the time of day. When Tasmania enters the National Electricity Market, energy prices will be subject to demand and Hydro Tasmania may increase economic returns by increasing its energy generating capacity during peak times. Schemes that are currently run to provide base load generation may be required to provide peak generation on a daily basis.

The potential for reconfiguring Hydro Tasmania's existing power system to provide additional peak generation was broadly examined. Potential opportunities for developing peaking capacity were identified with respect to installation of additional turbines and development of pumped storage schemes between existing reservoirs.

By installing additional turbines in existing schemes, the number of generation hours per day for the scheme can be reduced. This is expected to yield a higher return with regard to peaking premiums since more generation could occur during times when energy prices are higher.

Only a few opportunities were identified. Gordon Power Station (located in south-west Tasmania) has three 150 MW machines installed with additional space provided for a further two machines. Conceptual investigation indicates that the installation of either a fourth machine of 150 MW or two machines, one of 150 MW and the other 35 MW to supply downstream riparian requirements, is warranted. Further investigation of an additional 95 MW power station at Poatina (located in central Tasmania) to enhance the existing station capacity of about 330 MW is also warranted.

The pumped storage schemes investigated recently appear to be technically viable but are not economic even with consideration to the mandated renewables premium.

9 NEW DEVELOPMENTS

In the past, Hydro Tasmania investigated many large power developments. Due to the creation of the World Heritage Area since these investigations, more recent investigation has focused on rivers outside the World Heritage Area and other environmentally sensitive areas. (A basic economic reassessment of previously investigated large developments now located within the World Heritage Area, showed that none were economic.)

Thirteen schemes with installed capacities ranging from 15 MW to 350 MW were broadly assessed. Of these 13 are non-viable due to environmental restrictions. Only one scheme has no obvious environmental issues that would hinder development. This scheme, with an installed capacity of 26 MW (13 MW-av), is also the only currently economic scheme. One other scheme, with an installed capacity of 15 MW (11 MW-av), is likely to be economic with the introduction of the mandated renewables premium and although environmental issues are evident they may not preclude development.

10 REDEVELOPMENT OF EXISTING SCHEMES

Hydro Tasmania owns and maintains several ageing power schemes, three of which – Tarraleah Scheme, Tungatinah Scheme, and Lake Margaret Scheme – require substantial expenditure within the next 40 years as part of their planned refurbishment programs. As a result of previous studies, refurbishment of these schemes was recommended over any redevelopment. Refurbishment of these three schemes has been reexamined in light of current economic constraints and future approach to energy sales and capital expenditure to determine if any redevelopment option has incremental benefit over the planned refurbishment.

Redevelopment of these schemes constitutes major changes to the existing configuration of the scheme such as replacement of power stations and replacement of above ground conduits with tunnels and shafts.

Tarraleah Power Scheme has an installed capacity of about 90 MW with energy output of around 78 MW-av. The scheme was constructed between 1938 and 1951. It consists of headponds, canals, hilltop pipelines and two power stations. The headworks of the lower power station, Tarraleah power station, and the power station itself were the focus of redevelopment options considered. Only one option examined was found to have potential incremental economic benefit over the planned refurbishment if Tasmania enters the National Electricity Market and the renewables premium is introduced. This option replaces the majority of the existing headworks with a new shaft and 3 km tunnel and replaces the existing power station with a new power station containing a single 115 MW machine. Further investigation would be required prior to recommending that this option be pursued.

The Tungatinah Power Scheme was constructed during the 1950s. Run-off is collected in a series of headponds that culminate in the forebay to Tungatinah Power Station. The power station incorporates five vertical axis 26.1 MW Francis turbines. The scheme is operated as one of a series of run-of –river schemes and provides base load power with one or two machines and peak power with the remaining machines. Ten alternative redevelopment options were considered. These options covered items such as a shaft and tunnel to replace penstocks, and a single surface penstock in conjunction with either the existing or up-rated machines. No option was found to have any incremental economic benefit over the planned refurbishment.

Lake Margaret Scheme has unique cultural and heritage value, factors which formed the basis of a prior decision to refurbish the scheme. The upper part of the scheme consists of an 11 metre high concrete gravity diversion dam at Lake Margaret, a single woodstave pipeline, a single penstock and a power station. A lower station, with associated pipeline and penstock, was added in 1932 and subsequently closed in 1995.

Previously examined redevelopment options for Lake Margaret were reviewed. Options incorporated closure of the upper station and construction of a new single large power station, reconstruction of a new lower power station and diversion of Lake Margaret Scheme to two adjacent power schemes. The most economically favourable option appears to be the closure of the upper station and diversion of Lake Margaret to an adjacent scheme, either with or without a power station, providing incremental energy output of 5.8 MW-av and 3.8 MW-av respectively. If, on the basis of consideration of heritage and community expectations, it is recommended that the upper station be retained and the planned refurbishment maintained then the construction of a new lower power station would be economic with the introduction of the renewables premium. The average incremental energy output of this option is about 2.5 MW-av.

11 PENSTOCK CLEANING PROGRAM

To date, Hydro Tasmania does not have in place a regular penstock cleaning program. Substantial increases in electricity generation could be achieved if such a program was implemented. The build up of slime and other algae increases friction losses thereby reducing the net head of water available to the power station. Turbines must be operated at higher speeds to produce the same power output. In most cases this is unachievable and power output reduces.

Due to the cost of the penstock cleaning process, it is generally uneconomic to maintain conduits at minimum slime thickness. A cost benefit analysis was carried out on conduit systems for nine power stations and a regular cleaning period was defined to maximise the benefit of cleaning. The nine stations were selected due to their existing high head losses resulting from long tunnel and penstock lengths and large heads. Cleaning of concrete lined tunnels and unlined tunnels was not considered.

There is a lack of accurate flow and pressure measurements in Hydro Tasmania's conduit systems and it is impossible to accurately determine conduit headloss as a result of slime growth. Based on the simple assumptions made in the preliminary study – constant rate of growth of slime of 1mm per quarter of a year with a minimum achievable roughness of 0.75 mm and a maximum attainable roughness of 6.75 mm – the optimum benefit for each of the nine stations ranged from 0.05 MW-av to 0.5 MW-av with the optimum cleaning interval ranging from 0.25 years to 1.5 years. For the nine stations examined a total additional output of 2.5 MW-av could be economically derived from a regular penstock cleaning program if the optimum cleaning period is adopted.

A field trial in one large system (Tarraleah Hilltop Pipelines) is planned for the near future and will compare the effectiveness of water blasting and conduit dewatering in reducing headloss due to biological growth. Once cleaned, the conduit head losses will be routinely measured to determine actual growth rates and economic analysis redone on the basis o factual cost data and true slime growth rates to determine the optimum cleaning interval for all conduit systems. On the basis of these results, a monitoring and maintenance manual for the long-term control of head loss in Hydro Tasmania's long conduit systems will be prepared.

12 STATION RUNNER UPGRADES

By upgrading turbines at a number of existing power stations, additional energy output from Hydro Tasmania's existing power system can be economically achieved. The maximum energy increase will be realized if the existing generating plant is replaced although this involves large capital cost expenditure. At a considerably lower cost a slightly reduced improvement can be achieved through the redesign and either modification or replacement of critical components, such as turbine runners, guide vanes and water passages. Efficiency improvements of up to 5% may be achieved in some older power stations.

Through the upgrade of 26 turbines at 7 power stations approximately 18 MW-av of additional energy could be economically produced. A further 4.5 MW-av may be economic dependent on the existing plant efficiency, mandated renewables premium and cost of upgrade.

Additional financial benefit from increasing the maximum capacity of generating plant is planned for investigation.

13 CONCLUSIONS

Substantial hydropower potential within Tasmania's existing electricity network that is both economic and environmentally acceptable has been identified in recent studies. The potential identified to date represents a minimum only and it is likely that further potential will be identified for planned further study.

To more completely define hydropower potential in Tasmania a 'hydro atlas' will be prepared. This atlas will provide good graphical summaries of hydropower potential and utilisation for each catchment within Tasmania. Furthermore, approximately 200 sites for hydropower development that were discarded in the initial assessment process without any individual study will be reassessed.

REFERENCES:

Watson, B, March 1992, Report on Hydropower Schemes Investigated Between 1958 and 1991, CER 1059
Titchen, John, 18/11/1991, DRAFT Proposal to Meet the reference Load Projection with Renewable Energy Sources
Scanlon, Andrew, 1995, Water Power
Hydro-Electric Corporation, 1999, 1999 Hydro-Electric Corporation Annual Report
Water Resources Hydrographic Group, October 1994, 1993 Annual Data Review (Includes DPIF Sites)
Hydro-Electric Corporation, 1998, Future Options for Tasmania's Hydropower System
Rawlinsons, 1988, Australian Construction Handbook, Rawlhouse Publishing Pty Ltd, Perth, Western Australia.
Hydro-Electric Corporation, Data Books,
Quinlan, P; Sylvester, M; Stojmirovic, G., May 2000, Pre Feasibility Study System Impact Associated with Conduit Cleaning, Report No. GEN-0412-CR-001
G Stojmirovic, S Haslock, P Southcott, June 2000, Tarraleah No. 2 Canal Mini Hydro Feasibility Study, Report No. COR-0093-CR-002
J Sheedy, June 2000, Monpeelyata Canal Outlet Mini Hydro Pre-Feasibility Study, Report No. COR-0093-CR-004
G Stojmirovic, Tasmanian Remaining Hydro – Mini Hydro Options Report., Report No. COR-0093-CR-001
P Mathers, Tarraleah System Asset Management Study, Executive Report, May 1993
B.Knoop, Tungatinah Power Scheme Re-development Pre-Feasibility Study, Report No. GEN-0061-FR-002
P Mathers, Lake Margaret System Asset Management Study, Executive Report, May 1995
Hydro Electric Corporation, Review of Lake Margaret Scheme, Progress Report No.1, January 1985

Hydropower in the New Millennium, Honningsvåg et al (eds), © 2001 Taylor & Francis, ISBN 90 5809 195 3

National Hydropower Plan (NHP) Study in Vietnam

V.D. Thin
NHP Director, Electricity of Vietnam

E. Skofteland
NHP Senior Adviser, Electricity of Vietnam

G. Lifwenborg
NHP Project Manager

ABSTRACT: Vietnam is presently developing a National Hydropower Plan (NHP) with a time horizon to year 2020. The Study comprises a full investigation of the five main rivers estimated to possess about 3/4 of the country's exploitable hydropower potential. This paper gives a brief presentation of organization and structure of the Study and the methodology used. Based on a carefully selected methodology, and with due consideration to environmental and social concerns - and other users of water resources - the plan will rank projects which should be brought forward first for a license, when justified by power demand forecasts. Based on the results of generation system simulations with hydropower and thermal candidate projects, a ranking of hydropower projects will be made, which, together with appropriate thermal generation units, will ensure power supply until 2020. The Study will also provide Vietnam with a comprehensive assessment of Vietnams potential hydropower resources.

1 INTRODUCTION

The National Hydropower Plan (NHP) Study, Stage 1, in Vietnam was launched in April 1999 and will be concluded in November 2001.

The NHP-Study is jointly financed by the Norwegian Agency for Development Cooperation (NORAD) and the Swedish International Development Cooperation Agency (Sida). It is carried out under the overall supervision of the Client, Electricity of Vietnam (EVN), represented by the Management Board of National Hydropower Plan Study. A Steering Committee has been appointed with members from all relevant Vietnamese agencies, chaired by the Ministry of Planning and Investment. To assist the Client, an institutional cooperation has been established between EVN and the Norwegian Water Resources and Energy Directorate (NVE). A joint venture comprised of SWECO International (leading), Statkraft engineering, and Norplan, is the Consultant selected to carry out the Study.

The ongoing Stage 1 of the National Hydropower Plan comprises the following five priority river basins (see figure 1): Da, Lo-Gam, Ca, Se San, and Dong Nai. The threshold value is fixed at minimum of 30 MW for individual projects.

A Stage 2 is foreseen as a follow-up and completion of the NHP-Study. In this Stage, the aim of providing flood control in combination with hydropower development be given priority. It is also planned to include analyses of the potential of mi-cro/small hydropower in remote mountainous areas not expected to be connected to the main grid in foreseeable future.

2 HYDROPOWER POTENTIAL OF VIETNAM

Vietnam with a population of about 80 million and an area of 330,000 km², is rich in natural resources such as water resources, coal, oil, and gas. In 1999 the total installed capacity available for generation of electricity reached 5,679 MW. Hydropower is presently the dominating generating source with some 3,000 MW installed capacity, or about 53% of the total. Of the remaining, coal and oil count for 22%, and gas and diesel for 25%. The share of hydropower is expected to be reduced in the future, according to EVN's power demand forecast.

The mountainous topography and tropical monsoon climate profoundly affect the quantity and distribution of water in Vietnam. Rainfall is highly uneven causing frequent and often disastrous floods. Mean average rainfall is about 2,000 mm, but most accumulates between May and November. About 70-75 % of the annual flow is generated in 3-4 months, and 20-30 % in one peak month.

It would seem that Vietnam in general has a very favorable water situation, given the amount of rainfall and extensive network of rivers, and in relation to population size. However, several of Vietnam's rivers originate outside the border. More than 90 %

of the Mekong Basin lies outside Vietnam, and 90 % of its flow is generated outside the country. Also about half of the Red River Basin lies outside the country, and about one third of its flow originates in China. Thus, the upstream management of water in many rivers is largely beyond Vietnam's control.

Floods cause extensive damage and loss of lives in Vietnam every year. In the south, the Mekong Delta is largely unprotected. In central Vietnam, the smaller steep rivers are very affected by severe flash floods causing erosion and inundation, and the possibilities for flood control by building dykes are limited. The flood disaster in 1999 left behind 791 people dead. In the Red River Delta in the north an extensive network of dykes has been built. The existence of the Hoa Binh reservoir (9.450 billion m³) in the Da River also plays an important role in the flood control of the Red River. Hoa Binh is a multipurpose project, also with Vietnam's biggest hydropower plant with an installed capacity of 1,920 MW.

It is evident that improved flood control is an urgent matter for Vietnam, and locations were the establishment of reservoirs can serve the functions of both flood control and hydropower generation will be given priority.

Vietnam has a roughly estimated exploitable hydropower potential of 80-100 TWh, with installed capacity of some 18-20,000 MW. The five priority rivers being covered by the NHP-Study account for about 75 % of what is estimated as total exploitable. By including seven more potential hydropower projects from other rivers in the final simulation, the Stage 1 of the NHP-Study will cover about 80 % of the estimated total exploitable hydropower in Vietnam.

After completion of the hydropower plants which are under construction, the installed capacity will increase to about 4,100 MW, or slightly more than 1/5 of the theoretical hydropower potential in Vietnam.

Vietnam also has a considerable potential of micro to small hydropower estimated to 1,600-2,000 MW, of which only about 66 MW has been developed.

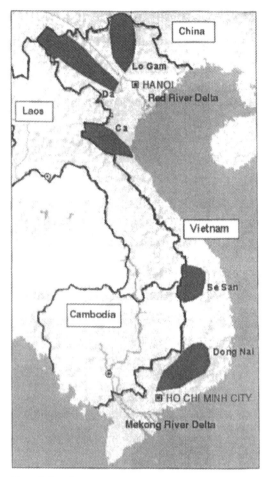

Figure 1. Location map

3 STUDY OBJECTIVES

In most countries, hydropower development has traditionally been based on an individual case by case evaluation of the project's role in the power system. Until the beginning of the 1970s, project economy and power transmission issues were the dominating criteria for developing a project. Thereafter, environmental and social concerns of negative impacts of the hydropower projects grew rapidly, and the need for hydropower master plans emerged. Such master plans aimed at providing assessments of technical/economic and environmental/social aspects to enable a comparison of projects, and should pro-

vide the ranking order in which they should be forwarded for formal licensing procedures according to laws and regulations.

According to the Terms of Reference, the main objective of the NHP-Study is to provide the Vietnamese Government with alternative power system (generation and transmission) development strategies to decide on national long-term power demand. Objectives other than economic power development shall be emphasized, and evaluation of these strategies from viewpoints of economic efficiency, investment requirements, macroeconomic issues, environment and social impacts, will be continuously undertaken by the Government.

The Study should also assist the Government in overall decision-making, taken into consideration regional development priorities.

Another important objective is the development of domestic institutions in Vietnam, analytical tools and staff's skills to the stage that the integrated multi-sector approach to water resources planning will become the norm in Vietnam.

As hydropower is an important generation resource of Vietnam, the Study will also consider alternative power development to such an extent that giving priority to hydropower schemes is possible.

The Study will also attempt to consider the hydropower development within the context of a national and also river basin water resources plans, as prescribed in Vietnam's new Law of Water Resources (1999). This Law is still under implementation, and several institutional and procedural questions remain to be solved.

In the Terms of Reference it is clearly stated that investments and constructions of hydropower projects at various planning stages shall not be delayed or stopped until the NHP-Study is completed.

4 STUDY STRUCTURE AND PARTICIPATION OF LOCAL SUB-COSULTANTS

The Stage 1 of the Study is structured into three main parts:
a) Water resources planning studies
b) Hydropower studies
c) Environmental and social impact studies

The Consultant has established five study teams comprised of key personnel from the joint venture.

The execution of the Study is organized into five consecutive phases:
- Phase I: Inception studies and coarse screening
- Phase II: Field work and primary data collection
- Phase III: Project definition and fine screening
- Phase IV: Analyses of alternative power development strategies

Throughout the Study activities, the Consultant is making maximum use of contracted Vietnamese sub-consultants, all being governmental institutions, selected in cooperation with the Client. The major part of the field work and data collection have been carried out by the following sub-consultants:
- Institute of Water Resources Planning
- Power Engineering Consulting Company No 1
- Power Engineering Consulting Company No 2
- Institute of Geography
- National Institute of Agricultural Planning and Projection
- Institute of Ethnology
- Institute of Energy

5 TRANSFER OF KNOWLEDGE

Being part of Norway's and Sweden's development assistance to Vietnam, issues on transfer of technol-

ogy and institutional strengthening have been given much attention. An institutional cooperation has been established between EVN and NVE, including the stationing of an adviser in EVN for the duration of the Study. It is considered important that EVN be prepared to regularly update the National Hydropower Plan after the ongoing Study is completed.

In the execution of the Study, the Consultant is making maximum use of sub-contracted Vietnamese professional institutions as local consultants, as listed above. A number of workshops and other training activities are organized on various issues. The role of the foreign experts is primarily to provide guidance on the work of the local consultants, both during their scheduled visits, and by the presence of the project manager for longer periods. The main part of the field work and data collection have been done by the local consultants, who are also participating in parts of the assessments. Also members of the Client's staff have participated in the workshops and field work.

Furthermore, several study tours have been organized to Norway and Sweden on a wide range of issues, including visits to hydropower schemes.

6 STAKEHOLDERS' PARTICIPATION

As stakeholders are defined organizations and individuals who in one way or the other will be influenced by the construction and/or the operation of a hydropower project. Stakeholder organizations identified in the Study are 9 ministries, 5 state committees, 22 people's committees, 7 governmental institutes, and 8 Vietnamese non-governmental organizations.

During the NHP-Study, great efforts have been made to involve the various stakeholders as much as possible. This has been a difficult task for several reasons, one reason being that the NHP-Study is a planning exercise only, and the final decision, if any, to actually implement a particular hydropower project may be taken far into the future. This may naturally lead to a less enthusiastic involvement by the stakeholders.

The approach to obtain a dialogue with the stakeholders has been to organize workshops both in Hanoi for the centrally located organizations, and in the provinces. It has proved advantageous to involve the local sub-consultants in the presentations of the projects in the provinces. Keeping the dialog in Vietnamese and not in the English language (to be interpreted) has provided an improved and also more lively involvement by the local participants.

It is the aim of this part of the Study to establish procedures for stakeholders' participation that can be used in Vietnam in future planning of hydropower development.

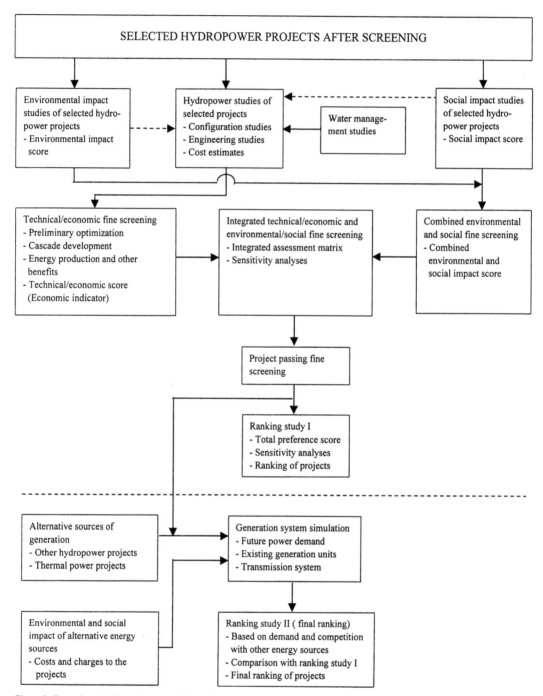

Figure 2. General methodology for phases III and IV.

7 STUDY METHODOLOGY

7.1 General

In Norway a master plan for hydropower development was made in 1981-84 following a number of environmental conflicts connected to planning and construction of hydropower schemes. A total of about 100 TWh had already been developed. The plan covered technical and economic aspects as well as environmental and social impacts. The planning

326

started with a coarse screening where 310 projects with 542 alternatives and a total power potential of nearly 40 TWh were selected for a further fine screening process. The plan concluded with a ranking of the projects in 16 groups and then into 3 categories. Projects in category I (about 11 TWh) were made available for licensing as required to meet the demand forecasts for year 2000, 15 years ahead. Projects in category II (about 7 TWh) were not considered candidates for hydropower development before year 2000, but could be made available on the basis of a new and broad evaluation. Projects in category III (about 20 TWh) were not considered relevant for hydropower development, mainly because of high costs or conflicts with other users, mainly environmental interests. The master plan was endorsed by the Parliament and has since then been revised regularly. It serves as an operational framework for hydropower development in Norway.

In Nepal, the World Bank initiated a Master plan for medium (10-300MW) hydropower projects. The planning was carried out in the period 1995-97. The selection of projects was made from a group of 24, and 7 projects passed to the phase of fine screening and ranking.

In the methodology selected for the NHP-Study in Vietnam, the experiences from both the Norwegian and Nepalese studies have been applied. A considerable number of adjustments have been introduced, however, to adapt to specific Vietnamese conditions, and the methodology is still subject to revisions. Throughout the progress of the Study, the methodology has been discussed with all relevant Vietnamese authorities and stakeholders, and valuable comments have been received from the World Bank, the Asian Development Bank, and NVE through the institutional cooperation with EVN.

It may be noted that when Norway made the master plan for hydropower development in the mid 1980s, a major part of the exploitable potential had already been developed. In Vietnam, however, only about 1/5 of the estimated exploitable potential has by now been developed. It is considered important that Vietnam's National Hydropower Plan be regularly updated. It will then be possible to make use of new and revised data, and also to change the order of priority of the projects in light of new information and changes in users interests.

7.2 Phase 1, coarse screening

The objective of the coarse screening in phase I was to assess the identified hydropower projects in the five river basins based on a number of technical/economic and environmental/social criteria. Other water uses were also taken into account. It was defined as a desk-study, mainly based on existing documentation provided by the Client and others.

The coarse screening of 47 potential hydropower projects in the five priority rivers were completed. Some of these projects are mutually exclusive, meaning that they represent various alternative developments of the same hydropower potential. During this phase, extensive discussions were held between the Client and the Consultant on both methodology and the number of projects to enter the next phases.

The final stage of the coarse screening was the integration of the technical/economic preference scores with combined environmental/social preference scores, presented in an assessment matrix as shown in principle in figure 2. The method selected gives preference score values ranging between 0 and 100. Part of the detailed methodology is that there is no vertical line corresponding to the horizontal "qualifying technical/economic score" line. None of the projects may be ruled out due to environmental/social impacts only, and there is no basis for establishing a defined environmental/social threshold for which projects should be taken out regardless of its technical/economic performance. It should also be noted that projects below the qualifying technical/economic score line do not pass the screening regardless of their environmental/social impacts.

The slope of the slanted lines indicates in principle a defined balance between the technical/economic and the environmental/social merits of a project. Being on the same line, projects 1 and 2 (figure 2) are considered equally attractive from an integrated technical/economic and environmental/social view for the selected slope.

There is no "correct" slope of the line, as there is no "correct" methodology for such assessments. In establishing the slope, special considerations are given to projects close to the coarse screening line. To establish the quantifying border separating the projects, considerations are given to the merits of projects close to the borderline.

It is important to have in mind that the aim of the screening is not a ranking of projects, but a sorting of identified projects into two groups, "passed" or "not passed". The underlying data with its uncertainties, the comparison of non-comparable elements, do not justify using a more sophisticated method to define the coarse screening line.

Sensitivity analyses have been performed to examine the effects on the screening of changing the weighting between key issues, for example the weighting between individual environmental and social parameters, and between economic and noneconomic parameters.

The coarse screening resulted in 22 projects to be further analyzed in the next phases.

7.3 Phase II, Field work

In all five priority basins planning had already been carried out to some extent, but individual plans var-

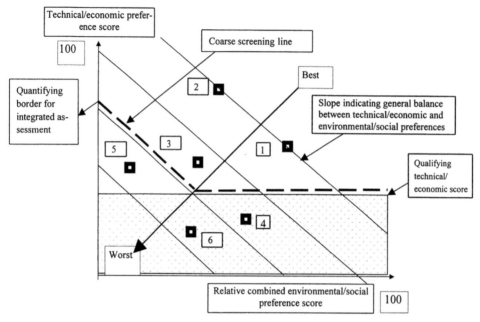

Figure 3. Assessment matrix for the integrated assessment of identified hydropower projects. Projects 1,2,3 are passing the screening, and projects 4,5,6 are not passing the screening.

ied considerably as to the quality of data and the level of planning, some at feasibility level, some at pre-feasibility level and below. Extensive field work and data collection were, therefore, carried out in phase II to fill data gaps in preparation for the fine screening in phase III, and to make a comparison of projects possible.

7.4 Phase III, fine screening

7.4.1 General

The phase III fine screening of the 22 projects (of which some are mutually exclusive) is now going on. The principal objective of phase III is to select the most promising hydropower projects of those identified in phase II. The following sub-studies are part of this phase:

- Integrated river basin planning studies.
- Configuration and layout studies, including cost estimates.
- Environmental and social impact studies.
- Fine screening and ranking

A major challenge will be the integration and inter-action of the water resources planning and the environmental and social impact studies, with the hydro-power studies.

The overall general methodology is given in figure 3. The more detailed methodology is basically the same as in the coarse screening, but it is still be-ing refined. Some important features of the detailed methodology are given in the following.

7.4.2 Water resources planning

A classification of multipurpose projects has been carried out, based on the relative importance of the candidate projects, and in accordance with the most appropriate strategies identified for each of the five river basins. After having developed the strategies for future water allocation in each basin, the most important multi-purpose projects will be included and further examined. The resulting list of multi-purpose projects in each basin will be included in the basin characteristics and will form an input to the optimization of the hydropower projects. The selection of multipurpose projects has in principle been based on the magnitude of reservoir regulation they may provide, based on reservoir volume, regulation percentage (reservoir volume divided by mean annual flow), and the flood control percentage (reservoir volume divided by mean annual flow in the assumed damage areas of floods).

The value of a project is indicated by the following non-hydropower parameters: water supply, flood control, irrigation, salinity control, transportation, and environment. By introducing the Mike Basin model, the water balance in the river basins will be simulated for a period of 35 years using the hydrological series 1961-1996. The water use categories quantified in the process will be water supply, irrigation, and flood control, in addition to hydropower. Fisheries, transport, wastewater management, mitigation of saltwater intrusion in delta areas, and any other interests, will be treated in a descriptive manner only, as they will not affect the flow regime.

7.4.3 Hydropower studies

The aim of the configuration studies is to define in more detail the main parameters of the selected 22 projects, primarily the full supply level, the volume of active storage, and the installed capacity. Secondly, the most appropriate dam type will be determined and the appropriate arrangements for spillway, waterways, power station, etc.

Each basin will be simulated with all existing and planned hydropower projects in each basin included in cascade. Restrictions imposed by other water users as well as environmental and social concerns will be included in the simulations.

Engineering studies are carried out to make more detailed studies of the design of each hydropower project. This will be input to the cost estimates that will be based on the bills of quantities estimated in a cost manual developed for the Study. The cost estimates are based on prices as estimated relevant for international competitive bidding.

To support the optimization studies, indicative incremental costs for variation of installed capacities, reservoir level, and operational level, will be derived.

The costs of environmental and social mitigation will be included in the costs estimate.

7.4.4 Environmental and social impact studies

During phase 2 of the Study, extensive field data were collected for all the passed 22 projects. Based on experience from the coarse screening, the following 10 environmental parameters were selected, partly qualitative and partly quantitative of nature:
1. Agricultural land
2. Water quality
3. Erosion/sedimentation
4. Fish
5. Forestry
6. Flora
7. Fauna
8. Aquatic life
9. Protected areas
10. Fragmentation potential

These parameters are reflecting impacts on the physical environment (nos 1-3), the biological environment (nos 4-8), and the two last parameters reflect possible conflicts due to the geographical location of the project.

The information to be analyzed in the social impact studies is collected by two methods. Firstly by the participatory rural appraisal technique in which the field teams will identify target groups of village inhabitants with special focus on women and key informants. Secondly through interviews with officials at the three administrative levels (province/district/commune). The social parameters applied are given in table 1.

The total number of people to be resettled is a key parameter. A low magnitude will mean few people to be resettled. The importance is high for those with few alternatives to their current life-stile, especially ethnic minorities.

The screening will consider three project phases with reference to potential social impacts:
- Project location
- Construction phase
- Operation phase

The evaluation of both the environmental and social parameters will be made according to both magnitude and importance of the impact according to a scale from 0 (no magnitude/importance) to 4 (high impact/importance). For the total environmental impact, a scale 0-100 will be used.

Mitigation and enhancement measures are identified according to established international principles, when available. The aim is to minimize adverse social impacts and describe enhancement measures towards beneficial impacts of the projects.

7.5 Fine screening and ranking study I

The approach adopted in the fine screening process for the technical/economic evaluation will be based on less parameters than in the coarse screening, as some parameters used in phase 1 will be taken into account in the hydropower simulations. Consequently, parameters related to project size, type, firm power potential, system improvements, and cascade benefits, will be reflected in the results of the simulations.

The potential multi-purpose benefits will be included in the fine screening as added benefits to the respective projects, while water allocations for other uses will be reflected as restrictions for the hydropower generation.

Some of the projects are located in rivers shared by neighboring countries and will be affected by upstream developments. These aspects will be considered in the ranking of projects, but not in the fine screening process.

The aim of the fine screening process is to determine the generation capacity of each project, taking possible cascade operation into consideration, and to calculate the levelised cost-benefit expressed as USc/kWh over the project's estimated lifetime. Following the estimation of levelised cost, the hydropower projects will be grouped according to technical-economic attractiveness in a 0-100 scale. To assess the economic merit of each project and carry

329

Table 1. Parameters applied in social impact analyses

Level	No.	Parameter
I. Regional, river basin area		
	1	People resettled
	2	Host area relations
	3	Ethnicity
II. Catchment area		
	4	Water-related health
	5	Water access and rights
	6	Migration
III. Project area		
Socio-cultural	7	Directly affected people
Socio-cultural	8	Partially affected people
Socio-cultural	9	Ethnic groups and history
Socio-cultural	10	Extension service
Production	11	Land use
Production	12	Farm output
Economic	13	Secure access to food
IV. Downstream area		
	14	Fishery
	15	Water availability and quality

out the screening, a benchmark value for the levelised cost will be established based on an alternative thermal power expansion. Thus, the thermal levelised cost will form the qualifying border for projects passing the fine screening. In principle, hydropower projects superior to thermal alternatives will be left as potential projects for the future.

The combined environmental/social scoring is based on the assessment of 25 parameters as given above (10 mainly environmental, 15 mainly social) assigning them quantitative (magnitude) and qualitative (importance) values. The methodology takes into account detrimental as well as positive impacts, and also mitigation possibilities. The balanced impact score will be transformed into an environmental/social preference score on a 0-100 scale.

Each environmental/social parameter will be assigned a relative weight to express the importance attached to a specific parameter, and the determination of the weighting scales will be a subjective exercise. The opinion of the stakeholders providing the Vietnamese input in this matter will thus be of key importance.

Certain types of impacts can be mitigated or compensated for, but there are always costs involved. A high potential for successful mitigation/compensation means that the project may be considered less detrimental and the negative scoring will be reduced.

The final stage of the fine screening will be the integration of the technical/economic with the environmental/ social preference scores as illustrated by figure 2.

The objective of the ranking study I is to present a total preference score, by adding. As the total preference score will be based on none-comparable values, qualified judgements based on sensitivity analyses must be used to justify the ranking. The ranking study I will not take into account how the individual hydropower projects fits into the power demand forecasts and the overall power system. This will be further analysed in phase IV.

7.6 Phase IV, Alternative power development strategies

The general objective of phase IV is to identify a sequential development of the hydropower projects passing the fine screening in phase III in relation to the power demand forecasts. This also means that the appropriate timing and order of priority will be examined. To enable this analyses, considerations of other alternative sources of power will be required, such as existing and planned thermal power plants. A number of alternative scenarios will be simulated to provide a recommendation on priorities and timing of the projects. In addition to the projects in the five priority rivers, seven potential projects from other river basins will be considered in phase IV.

All projects will be evaluated against each other and ranked according to a national power system approach. The main difference from the fine screening simulations in phase III is that the demand for electricity during the simulation period will be introduced as a governing parameter. It is then not only the generation capacity of the projects that is of interest, but also the timing of the generation in relation to the demand.

Thermal energy (coal, oil, and gas) is considered the only realistic alternative to hydropower. The environmental impacts from thermal power stations will be evaluated to the degree found necessary for comparison with impacts from hydropower stations. These impacts will be evaluated in relation to location, emission to the atmosphere, and release of cooling water.

Based on the results of the generation system simulations with a mixed system of hydropower and thermal candidate projects, a final ranking (Ranking study II) of the hydropower projects will be made, which, together with appropriate thermal generation units will ensure Vietnam's power supply to the year 2020.

8 DATA BASE AND INFORMATION SYSTEM

As part of the Study, a data base and information system has been established. It is a water resources management oriented system, based on Microsoft Access. The system has been established at the premises of the Client and is now operational and being loaded with the various data collected by the Study. All relevant information is geo-referenced, and can thus be presented in GIS-systems. It will be possible to import information directly to river routing models and water allocation models.

It is the intention that the comprehensive data base being established will be made available for all other Vietnamese organisations involved in water resources and environmental planning, as well as obligations to update the data base with new and revised data.

Hydropower in the New Millennium, Honningsvåg et al (eds), © 2001 Taylor & Francis, ISBN 90 5809 195 3

Rehabilitation and improvement of the Túneis hydroelectric undertaking. Hydroelectric and environmental issues

J.Vazquez & L.Gusmão
COBA, Lisbon, Portugal

D.Estrela
EDA, Azores, Portugal

ABSTRACT: The Túneis hydroelectric development is located in the S. Miguel island, Azores Archipelago, and was constructed about 50 years ago. It is the main development of the Ribeira Quente multistage water power development, being owned by EDA – Electricidade dos Açores. COBA has recently completed the design and technical assistance concerning major works in the waterway and the increase of the installed capacity and the system automation. Initially, the actions were expected to include essentially the waterway improvement, the power generation unit replacement and the system remote control. In late 1997 and 1998, the occurrence of heavy rains on the catchment area, favored surface erosion and the occurrence of floods and landslides that have strongly affected all the zone and the waterway elements. This paper describes the major actions undertaken and discusses the methodology followed to achieve the integrated treatment of the different hydroelectric and environmental problems in question.

1 INTRODUCTION

The Túneis Hydropower Development is located in the S. Miguel Island, in the Azores Archipelago, and is one of the four undertakings owned by "EDA – Electricidade dos Açores" within the multistage hydropower development of the Ribeira Quente, taking advantage of the stream heads below level 200, where flows are naturally regulated by the Furnas Lagoon. Here, besides the Túneis undertaking, three other undertakings are in operation, namely the Tambores and Canário, upstream, and Foz da Ribeira Quente, downstream, the powerhouse of this last one being close to the stream mouth.

The Túneis Hydropower Development has come into operation in 1951 and uses the flows from the Quente stream. At the site of the water intake, that is immediately downstream an old hydrometric station, this stream forms a river basin of 33.2 km², integrating the Furnas Lagoon. The annual mean rainfall and flows are 2137 mm and 1.87 m³/s, respectively, for which contributes the regulation made through an underground flow from the Furnas Lagoon, as shown in the curves regarding the average year monthly rainfall and flows (Fig. 1) and in the curve for the annual mean duration of daily mean flows (Fig. 2). These curves clearly show the flows regularity at the water intake site.

For the water intake, located at about level 137, and corresponding to an original nominal flow of 2.0

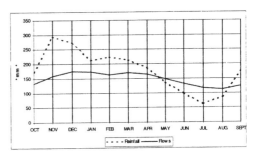

Figure 1. Monthly rainfall and flow curves

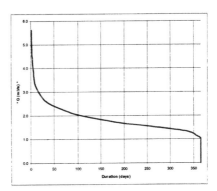

Figure 2. Túneis Dam. Curve of the annual mean duration of daily mean flows

m³/s, a small weir with a 4 m-high and 15 m-long masonry gravity section, and a 12 x 34 m² settling chamber were constructed, incorporating trash racks and several bottom outlets.

The canal is 2320 m long and is located on steep slopes. It has a 2 m-wide rectangular cross section and with a variable height, although not less than 1.40 m, being mainly in masonry construction. Its mean bed slope is of about 0,3‰. Along its development there are 10 bridges, 21 underpasses and 2 overpass culverts, 2 small tunnels, the canal having also 3 gates, 2 bottom outlets and 2 side spillways.

The head pond was partially constructed in masonry and partially in reinforced concrete, it is 80 m long, with variable width and height up to 6.5 m and 3.5 m, respectively, and is provided with a bottom outlet and a spillway. The steel penstock, ranging from 0.95 to 0.80 m diameter and 135 m long, has a net head of 82 m and is supported on masonry massifs with metallic cushions each 7.5 m. The initial penstock had two fixation massifs, one close the head pond and the other near the powerhouse. With the exception of one dilatation joint, existing close the initial massif, the penstock was totally welded.

The hydropower plant, at about level 52, is mainly a stone masonry building, where a 1390 kW nominal power Francis group was installed. In the recent past, the maximum possible power attained and the annual mean power generation observed were about 1200 kW and 6.5 Gwh, respectively.

Taking into account the large hydropower potential of this development and its long age, largely overcoming its design horizon, it was initially foreseen to carry out the upgrading of its conveyance structures, and the replacement of its power-generation group, which already had significant outcome and operation limitations, with an higher capacity group, as well as to develop the power plant automation.

The serious incidents, due to great heavy rains meanwhile arisen from the last third of 1997, made it necessary to carry out some immediate and short-term actions, these being important works and embracing the whole infrastructures of the development. In turn, the created situation has implied the need to review the scope of the general intervention and the construction works assumptions, regarding the observed need to carry out the systematic rehabilitation and strengthening interventions of the whole conveyance structures, in order to prevent new incidents to arise, or, at least, to minimise its effects.

In fact, the present precarious situation of the Quente stream basin and, in particular, of its flood-plain, extremely weaken regarding flood runoffs, does not allow to ensure that works located in its bed and banks are sufficiently reliable. Thus, it was necessary to globally think this problem, as it is the only procedure enabling to develop a set of interventions ensuring the necessary security to these and other infrastructures adjacent to the stream.

It was then with a wider perspective and scope than the initial scenario would allow to foresee that the studies and works for the rehabilitation, strengthening, increase in capacity and automation of the Túneis Hydropower Development were carried out, in view of ensuring the possible reliability conditions to this development – all these actions being duly justified by its technical and economical feasibility.

2 GEOMORPHOLOGIC CONDITIONS

The Quente stream starts at the south versant of the Achada highlands (Tambores stream), above the Furnas Lagoon, and it runs windingly, initially westwards, and from level 200 - slightly upstream the confluence with the Amarela stream -, southwards.

The existence of several volcanic sets in the Quente stream basin, but secondary regarding the volcano on whose crater the Furnas Lagoon is located, turns the overall volcanic morphology extremely complex.

The Quente stream and its tributary watercourses generally run on extremely steep banks, having deeply excavated the valleys and created versants with steep slopes that maintain the youth phases force of the morphogenesis up to levels extremely near the river mouth.

The Furnas basin presents a dense cropping pattern and is covered by projection pyroclastic material, with variable thickness, and some outcrops of trachyte and andesite in some local areas.

The projection materials, which are the predominant formations of the Quente river basin, have a high permeability and particular mechanical characteristics which may sometimes ensure the stability of very steep slopes, but they often undergo some instability process when subjected to percolation and weathering phenomena, sometimes involving significant amounts of mud soil.

These phenomena are intrinsically connected to the Island's hydrologic conditions, with intense rainfall, implying higher runoffs, what is partially explained by the progressive replacement of the trees and shrubs covers with pastoral areas.

Several kinds of geological formations may be found over the interested area, namely debris and pyroclastic formations (projection materials) and eruptive rocks.

The debris formations prevail at the bottom of the valley, represented by an alluvial-flood deposit, and they are formed by blocks and boulders (basaltic and trachytic), covered by different sizes sandy material. Thus, the debris formations mainly interact with the water intake which is at level 137 and with the power plant surrounding area, at about level 52, and in some sections with the water conveyance system – slope deposits and blocks.

The pyroclastic formations interest the most part of the water conveyance system and its composition is extremely varied. The more recent, upper formations are a result of the accumulation of volcanic projections, to a great extent occurred in the historic ages, being, in general, friable. In their composition are included ashes, pumice and chaotic tuffs with a more or less significant conglomeratic nature. Eruptive rocky formations appear in some localised stretches of the canal.

The local morphologic and geotechnical characteristics favoured incidents to arise, which have been observed with more or less frequency along the undertaking's lifetime, and had a particularly serious nature in recent past. Mention is made, namely, and with regard to the Túneis undertaking, to the de-

struction of the weir walls and to the degradation of the structures and infrastructures in some sections of the canal and bridges, to the slippage of large massifs over the canal and head pond – with failure of some sections of the walls-, to the failure of a penstock support and its deviation, the total blocking of the power plant and restitution area, as well as to the destruction of the respective accesses, with flooding and damaging of the power plant and respective equipment (Photos 1 and 2).

3 REHABILITATION AND REINFORCEMENT INTERVENTIONS PREVIOUSLY CARRIED OUT

As a result of those incidents, several interventions had been carried out for the clearance, rehabilitation and reinforcement of the development which, besides having enabled the rapid restart of its exploitation, have allowed to improve significantly the safety of the different structures of the water conveyance system.

The following sections seek to make reference to the main interventions carried out, which, in some cases, have been often repeated due to the succession of incidents caused by the various occurred floods.

3.1 Water intake

i) Removal of the transported sediments and logs of wood that blocked the trash racks and the settling chamber itself; ii) Repair of the trash racks; Removal of part of the transported solid material from the reservoir; iii) Unblocking and resume of the operation of the different bottom outlets of the weir and settling chamber.

3.2 Canal

i) Unblocking and clearance of the canal along a large part of its extension; ii) Cut of the reinforcing beams of the bridges' decks, in order to enable the equipment to pass along the canal, and their further replacement with steel reinforcement bars, bolt with screws to the side walls of the canal; iii) Removal of the large amounts of mass displaced and/or softened that buried the canal in two zones; closure of the canal in these zones, through the execution of a reinforced concrete slab and the heightening of the canal walls, in order to enable the circulation "bob-cat"-type equipment to move inside; iv) Reconstruction of some sections of the canal whose structures have been destroyed and structural improvement or strengthening of some culverts; v) Reconstruction of the bearings and strengthening of the deck of a bridge whose piers had been broken and dis-

Photo 1. Failure of the piers bridge

Photo 2. Landslide that destroyed and buried a channel stretch

placed and the deck's bending was already excessive, with several fissures and water seepage; vi) Strengthening of the foundation footings and of the abutments of several bridges; vii) Unblocking and clearance of the most important culverts of the water conveyance system;

3.3 *Head pond*

i) Unblocking of the head pond and removal from all its surrounding area of the sediment transported and softened due to successive detachments and sliding of earth material, blocks, tree trunks and shrubs; ii) Reconstruction and strengthening of a reinforced concrete wall section; iii) Cleaning and maintenance of the trash rack and associated equipment.

3.4 *Penstock*

i) Creation of a new access way to the penstock, what was often made due to the numerous incidents arisen; ii) Successive cleaning of transported material, tree trunks and shrubs from the small watercourse that crosses the penstock area; iii) Strengthening of the broken support and realignment of the penstock, taking into account that during a second flood there was a new misalignment and the reinforced support and two other contiguous support have broken again. A new reconstruction of the initially broken support was made, then already integrating the final metallic element anticipated for the new penstock saddle, and also including a reinforcement mainly in the higher adjacent support. A new attempt for the realignment was undertaken and was partially successful; iv) Execution of a drainage work for protection of the affected penstock section, through an alignment of reinforced concrete steps, including upstream a reinforced concrete coarse screen to minimise the impacts in the penstock.

3.5 *Power plant*

i) Removal of the deposited material in the streambed and banks, affecting the whole outlet area and both banks – operation that had to be repeated for several times and involving large quantities of material; ii) Construction of a footbridge upstream the power plant that has immediately restored the accessibility to the maintenance and exploitation personnel to the power plant. The steel beams of the deck were destructed by other floods, and their reconstruction was undertaken at a higher level; iii) Construction of an overflow bridge and necessary earthworks in both banks in order to ensure the permanent accessibility to heavy vehicles to the power plant area; iv) Con-

struction of protection walls in both banks in the power plant surrounding area; v) Construction of a gabion weir upstream the power plant, in order to protect it from the impact of large blocks and to allow some flood protection.

As referred, some of these works have been affected owing to new floods and because the upstream floodplain had a large amount of granular material and yet to the execution in its floodplain of provisional accesses - with coarse-grained material - for the protection works of the Ribeira Quente road. These materials are easily transported during floods, thus aggravating the erosion action created by the flows and affecting the downstream infrastructures.

All these actions have been carried out by local contractors with EDA's direct support and control and with the contribution of COBA for the studies and in the follow-up and supervision of the works.

4 UPGRADING, STRENGTHENING AND INCREASE OF THE INSTALLED CAPACITY IN THE DEVELOPMENT

The Túneis undertaking is the most important development in the multistage hydropower development of the Ribeira Quente stream and has an annual mean power generation of more than 60% of the total power generation.

The technical and economic simulation studies carried out concluded that it would be very interesting to increase the nominal flow to 2.3 m^3/s, thus enabling the new turbine to have an installed power capacity of 1600 kW. The annual mean power generated will be 11.4 Gwh.

With these assumptions, the studies carried out enabled to define a set of interventions to implement in the several structures of the development, which have been accomplished in September 2000 and are synthesised in the following sections.

4.1 *Weir, settling chamber and water intake*

For the rehabilitation and upgrading of this area of the Development, the following main interventions were carried out (Photo 3):

i) execution of two small gabion flood protection works, immediately upstream, for erosion and sediments transport control;

ii) interventions for the upgrading and strengthening of the sill, strengthening of the weir and heightening of the side walls, in order to prevent their overflow in case of flood; reconstruction of the bottom slab of the dissipation basin; construction of a wall for protection of the right abutment of the intake weir's basin, and stabilisation and drainage of the adjacent area;

iii) refurbishment of all gates of the weir's bottom outlets, settling chamber and canal's entrance;

iv) strengthening of the toe of the right wing wall of the settling chamber and side spillway, preventing eventual flood erosions;

v) electric supply to the area, lighting of essential points of the works, and installation of pump and expeditious equipment for clearance of the settling chamber and adjacent areas to the water intake;

vi) intervention for the improvement of the flow conditions, immediately upstream and downstream;

vii) creation of vehicles access for the maintenance operations, on the right bank, this extending along the reservoir's peripheral area, and including an overflow bridge, in order to also contribute in the erosion prevention in the banks;

4.2 Canal and associated structures

The interventions carried out, in addition to the numerous actions previously implemented, in view of the complete rehabilitation and upgrading and the increase of the transport capacity in the canal, have essentially been of three types:

i) Outside the canal, corresponding to works that do not specifically interest the canal section and that were possible to carry out without interfering with the exploitation of the Development, namely works for the protection, drainage and reinforcement of the slopes and benches of the canal and for clearance and upgrading of the culverts, and yet for the installation of removable grid panels at the entrance of these culverts. As to the slopes' protection, drainage and reinforcement works, the solutions used were mainly the protection with wire nets and dowels, the covering and strengthening with shotcrete with or without dowels, and the surface and deep drainage.

Photo 3. Intake after upgrading

ii) At the top of the canal, corresponding to the works for closing several sections with a reinforced concrete slab and to the punctual heightening of the walls, for hydraulic reasons, by using concrete pre-cast blocks. These works, though directly involving the canal's structure, did not interfere significantly with its hydraulic section, and have been mostly carried out during the operation of the Development and only with some stops for the transport of material and equipment.

Inside the canal, corresponding to the works for rehabilitation, covering and improving of part of the net hydraulic section of the canal. Due to their nature, these interventions have necessarily implied the stoppage of the Development, and they had to be made when there was a guarantee that no further circulation of equipment able to damage the canal's covering would happen. As to the rehabilitation of the covering affected by the fall of blocks and trees and by the circulation of equipment towards the already carried out urgent works, it was mainly in the canal's bed that interventions had to be made. In these areas and in all other where it was deemed necessary to execute a new bituminous cover, after previous repairing and preparation of the surfaces a three coating bituminous paint was applied, strictly accomplishing the specified procedures for the coating. These actions were also necessary to ensure the increase in the transport capacity to 2,3 m^3/s.

The interventions for the upgrading in the side spillways and refurbishment of bulkhead gates and bottom outlets of the canal are also comprised in the group of actions to undertake with the Development being out of service. As to the side spillways, it was mainly necessary to determine its level for the new exploitation condition, while the interventions in the bulkhead gates and bottom outlets mainly concerned the respective equipment, which have been object of an overall rehabilitation – treated under a specific section of this paper.

4.3 Head pond

Regarding the serious incidents occurred, the instability observed in the confining steep and almost inaccessible slopes, and the sensitivity of this structure, it was deemed convenient to construct a self-supporting frame, leaned against the inner wall of the head pond, operating as a conduit for ensuring the continuous supply to the penstock, even in case of falls of large masses into the inside of the head pond.

This frame interests the whole slope's greater sensitivity area.

Walls, fixing and protection wire nets, and drainage works have been executed at the bottom of the platform and in the slopes.

4.4 Penstock

The interventions anticipated for the penstock and adjacent area in addition to the previously carried out expeditious works were the following:

1 – At the penstock's level:
    i)   correction of the alignment and local deformations and treatment of the inside and outside coating of the whole penstock, with the installation of a manhole;
    ii)   strengthening and upgrading of the massifs with partial demolition and reconstruction of a reinforced concrete section, and replacement of the penstock supports;

2 – At the slope and supporting infrastructures' level:
    i)   installation of a steel stair along the penstock, linking the head pond to the power plant;
    ii)   creation of an upper platform for the access to the head pond, and implementation of two levels for upper drainage and protection of the penstock's upper section. For that purpose, concrete gutters bolt to the massif were foreseen be installed, being the lower gutter used as the final section for the access to the head pond;
    iii)   upgrading and strengthening of the pedestrian access along the slope with punctual restraint and drainage works.

After weighing the prices presented by the contractor and the difficulties in the execution of some of these works, it was decided, already in work's phase, to replace totally the penstock, not to install the steel stair along its layout, and also to simplify the drainage works by only carrying out one drainage level for protection of the penstock's upper section.

These works have shown a high degree of difficulty in their execution and their drainage and protection elements will need frequent maintenance interventions taking into account the erosive nature of the very steep slopes and the relatively high flows drained by the small stream that crosses the penstock alignment.

4.5 Power Plant

The power plant and installed equipment have also been affected by the floods, as well as the whole surrounding area. The situation then created was so serious that it entailed taking immediate and short-term measures in order to prevent large incidents and accidents to arise and also to resume the Development's exploitation.

As referred the interventions carried out were very important and led to a new arrangement of the area, providing it with a new protection weir at the entrance, a footbridge, an overflow bridge for heavy equipment, protection walls in both banks and access from the road to the upstream area of the power plant. These works have been preceded by an exhaustive activity (several times repeated) of removal of blocks and accumulated solid material.

The power plant itself had already been upgraded and extended in a way to be able to receive, in proper conditions, the new group and respective electric equipment (Photo 4).

The activity to carry out in the power plant during this phase, in the scope of the civil construction works, thus mainly comprised the demolition and concreting works inherent to the foundation and support massif for the new group and the execution of gutters for cables and for installation of the electrical boards. It was also necessary to heighten the restitution canal's wall up to the level of the power plant platform, adapting the canal to the new group, and to heighten the downstream stoplog, in order to rise the reservoir level at the restitution. The respective electromechanical equipment is treated under a specific section of this paper.

Regarding the protection of the power plant and adjacent areas from flood flow situations, a gabion weir was also constructed, about 0.5 km upstream, its right abutment being linked to a retaining wall of the Ribeira Quente road.

Immediately downstream the power plant, a small gabion weir was also constructed for the control of the flood flows in another watercourse.

4.6 Hydromechanical Equipment

The interventions in the hydromechanical equipment of the dam and water conveyance structures aimed at the renewal of the equipment, and at its adjustment to the remote control automatic operation, including the installation of an automatic trash rack cleaning machine in the head pond.

Photo 4. Powerhouse after upgrading and penstock scheme

338

There was the concern to assess the resources of the local metallomechanic industry in view of their participation in the manufacture and assembly of the different equipment for the water conveyance system. With this aim, several adequate solutions were studied, all accomplishing the operability objectives imposed on a modern hydropower development. This decision has shown to be adequate as it enables that, in future, during the operation of the development, EDA will have a different capacity in the interventions for the repairs that will be deemed necessary, this besides the interest in diversifying the local industry in view of their participation in the construction and repair of other units.

In the weir and settling chamber four gates of the bottom outlet and water intake have been replaced, whose maximum size is about 2 x 2 m², this having implied the realignment of the trash racks bars of the settling chamber as they were in service for few years and were in a good state of preservation. All these equipment have been endowed with electrical-powered control, with the possibility to work with a remote control from the power plant.

The decks existing along the canal used to close the bottom outlets have been replaced. The criteria used in the solutions proposed for the fixed elements and decks were the easy adaptation to the existing masonries, the introduction of waterproofing elements to prevent the water losses and, finally, the ease of construction. All these gates, whose maximum size does not exceed 1000 mm, have been kept with manual control. There was simultaneously the concern in using the most adequate materials for each purpose, namely the AISI304-type stainless steel in the fixed elements for the settling of the sealing elements and AISI410 bolts for their fixing to the gate.

Due also to the overall changes in the head pond, it was necessary to introduce new equipment and also to replace the existing one. The old trash racks of the penstock have been replaced and an automatic trash rack rake gantry was introduced.

During the design phase, the penstock was completely inspected and ultrasound measurements of all thickness were carried out. As referred, it was initially foreseen its rehabilitation comprising its reinforcement by material deposition in the areas with a thickness lesser than 5 mm and the repair of existing deformations, resulting from the impact of tree trunks.

Given the prices presented in the tender, it was decided in the construction phase to complete replacement of the penstock, its original size having been kept, and remaining with a uniform diameter of 900 mm and a thickness of 8 mm, in order to avoid the installation of reinforcements to prevent the situation of inner void.

The recommended supporting steel cushions were designed in two bolted halfs, linked to the masonries

by rock bolts with nuts, thus ensuring the possibility of disassembling in case of need during future interventions for the penstock's coating. A flow meter and three manholes were intercalated in the penstock, which did not exist in the original penstock, thus not enabling its inner coating.

4.7 Electromechanical Equipment

The intervention in the power plant equipment may be considered as a normal intervention involving the uprating of the power plant capacity by increasing the flow up to the limit of the transport capacity of the existing water conveyance system (from 2.0 m³/s to 2.3 m³/s). In this intervention, all equipment of the power plant from the turbine-alternator group and respective butterfly inlet valve, and the electric equipment have been replaced and, as well, automation equipment, enabling the automatic operation of the development was introduced.

It must be emphasised that during a previous analysis, carried out some years ago in the original group, it has been observed that this group was only led to a maximum capacity of 1050 kW; above this capacity there was an intense cavitation. The hydraulic operating conditions have been assessed and it was concluded that the machine was being put into operation with an excessive suction height. This problem was corrected by the installation of a stop-log downstream, in the tailwater canal, this having enabled the machine to be operated up to a maximum power of 1200 kW. The power limitation (the nominal capacity was 1390 kW) was further due to other problems in the machine, namely in the turbine's shaft, as it was misaligned, this causing an excessive heating in the turbine's bearing, which rise with the increase in power. All these limitations and problems confirmed the need and interest in the replacement of the initial group.

Given the fact that the original building of the power plant was prepared for the installation of a second group, EDA has decided not to make the disassembly of the original group. The modifications introduced in the building for the extension of the rooms for the electrical equipment did not cause significant changes in the perfect integration of the building with the surrounding landscape.

The new turbine-alternator group is composed by a horizontal Francis-type turbine, with a turbine output power of 1650 kW for a maximum discharge flow of 2,30 m³/s under the maximum net head of 81,3 m. The synchronous rotational speed is 1000 rpm. The nominal power of the alternator was fixed in 1600 kW.

As already mentioned, all electric equipment of the power plant has been replaced. Given that it is anticipated its operation in an isolated network, a diesel group was also installed to enable the starting

339

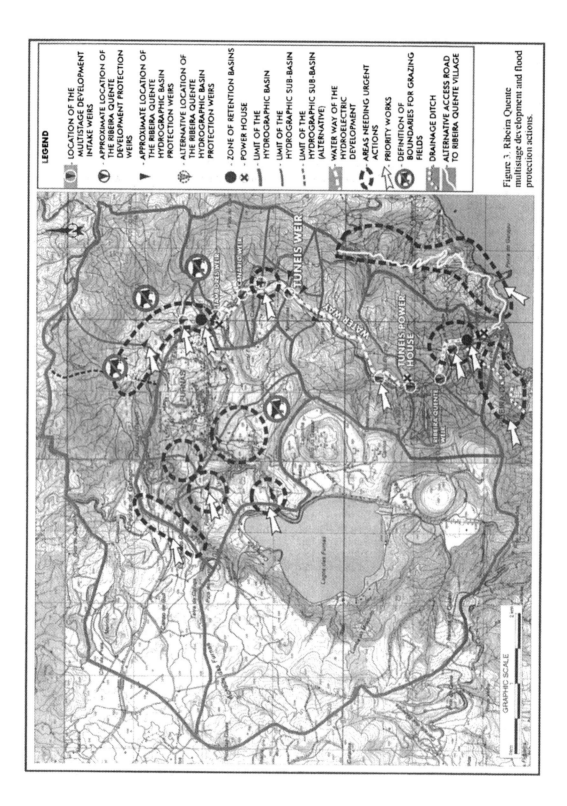

LEGEND

	LOCATION OF THE MULTISTAGE DEVELOPMENT INTAKE WEIRS
	APPROXIMATE LOCATION OF THE RIBEIRA QUENTE DEVELOPMENT PROTECTION WEIRS
	APPROXIMATE LOCATION OF THE RIBEIRA QUENTE HYDROGRAPHIC BASIN PROTECTION WEIRS
	ALTERNATIVE LOCATION OF THE RIBEIRA QUENTE HYDROGRAPHIC BASIN PROTECTION WEIRS
	ZONE OF RETENTION BASINS
	POWER HOUSE
	LIMIT OF THE HYDROGRAPHIC BASIN
	LIMIT OF THE HYDROGRAPHIC SUB-BASIN
	LIMIT OF THE HYDROGRAPHIC SUB-BASIN (ALTERNATIVE)
	WATER WAY OF THE HYDROELECTRIC DEVELOPMENT
	AREAS NEEDING URGENT ACTIONS
	PRIORITY WORKS
	DEFINITION OF BOUNDARIES FOR GRAZING FIELDS
	DRAINAGE DITCH
	ALTERNATIVE ACCESS ROAD TO RIBEIRA QUENTE VILLAGE

Figure 3. Ribeira Quente multistage development and flood protection actions.

340

of the group and the supply to the associated services.

The automation service introduced enables the power plant exploitation in a totally abandoned regime, and it will also remain totally prepared to work under a remote control from EDA's central dispatch power plant.

5 FINAL REMARKS AND CONCLUSIONS

From the studies and works carried out for the Túneis Hydropower Development, we may, in brief, make the following final remarks and conclusions:

The expeditious works previously carried out due to the serious damages caused by the floods were extremely efficient and enabled the rapid recommencement of the Túneis Development exploitation, as well as the creation of improved conditions for the overall major intervention in view of its upgrading, strengthening, increase of installed capacity and remote control. These rehabilitation actions implied a large and direct involvement of the Owner in all the work fronts, even where the technical feasibility of the interventions was questionable. Regarding the nature of these works, the Designer provided the follow-up and assistance to the Supervision team during their execution, adapting the solutions to the specificity of each individual situation and available means.

We may say that the large diversity and complexity of the interventions carried out in a very short term and with extremely adverse meteorological conditions were only possible due to such a direct involvement of all intervening parties, this kind of works being otherwise destined to be unsuccessful.

The final works regarding the strengthening, increase of installed power capacity and automation of the Development were achieved in September 2000.

In this initial period after the works conclusion, the undertaking has been specially surveyed to allow the maximum of knowledge about its behaviour.

The power plant was already put to work at its maximum power and all infrastructures have shown, in general, a satisfactory behaviour. In fact, almost all structures integrating the waterway have properly fulfilled their functions. The hydromechanical and electromechanical equipment have shown a very satisfactory behaviour allowing, in general, easy and efficient exploitation conditions.

Meanwhile some periods of heavy rains and floods occurred that enable to better test and slightly adjust the works executed.

Mainly, it shall be necessary to adjust the control of the water level in the canal of the waterway, to punctually make some additional interventions in some of its slopes and sections, where some localised mass falls occurred again, and to be prepared to periodically execute the removal of sediments from the intake area. Frequent maintenance of the penstock adjacent area will be also needed, taking into account the erosive nature of the soils.

In fact, in this kind of actions related with very erodible and unstable slopes, the methodology to follow must include a successive approach to the situation based in the actual behaviour of the works undertaken and trying to learn with the incidents – in order to avoid too much or too less amount of actions and costs to solve the complex and diversified problems in equation.

As it was referred, regarding the hydrological conditions and morphology of the works implantation area, and besides the remote control installation, it is essential to bear in mind the preoccupation for a frequent inspection maintenance of the water conveyance system in a way to prevent new damages from occurring.

Taking into account the incidents that occurred in the whole Quente stream basin, and the need to prevent new serious situations from arising, as these would certainly affect the Ribeira Quente multistage hydropower developments, particularly that of Túneis, it is indispensable that flood protection works are carried out in the whole basin, at the upstream and downstream of the developments, thus protecting the there existing structures and infrastructures, and to make a set of accessibility and drainage interventions in order to safeguard people's and properties safety in the area. Just as an indication, Figure 3 presents a first lay-out with actions of this nature that are being studied for the basin, and it shall be noticed that for some of the weirs foreseen the hypothesis of hydropower development is being analysed in this and other neighbour basins with the same kind of problems and solutions anticipated.

Development of Hydropower Technology and Design

Sedimentation in the Koka Reservoir, Ethiopia

Michael Abebe H.
Medium Scale Hydropower Study Project, Ministry of Water Resources, Ethiopia

ABSTRACT: Ethiopia is one of the few African countries that have great potential for producing hydroelectricity and irrigation development. The estimated total hydropower potential of Ethiopia is about 30-40GW; of this vast potential, only 1% has been utilized up to now and more than 87% of the country's electricity out put is generated by hydroelectric facilities. Koka is the first large hydroelectric power plant established in Ethiopia. It was commissioned in May/June 1960 followed by Awash II and III cascade power plants and currently all the above three power plants supply about 26% of Ethiopian's power production. According to a recent bathymetric survey, the capacity of Koka Reservoir has been reduced from 1667 Mm^3 in 1959 to 1186 Mm^3 in 1998. The loss on total capacity over 38 years is 481 Mm^3 i.e., 28.9% of the total storage volume. The average annual loss of capacity is 0.74%. In this article it is attempted to discuss the impact, extent and alternative measures to prolong its service life.

1 INTRODUCTION

The Koka hydro power plant is located about 90km southeast of Addis Ababa, the capital city of Ethiopia. Awash and Mojo are the two main rivers which flow to the Koka Reservoir. The Awash River originates from the southwest highlands to the west of Addis Ababa. It flows in the southeast direction up to Melka gorge, where it starts changing its direction to the east and joins the Koka Dam. The Mojo River originates in the northeast highlands of Addis Ababa and flows to the south direction and joins the Koka reservoir.

The area is dominated by bimodal rainfall type with a small peak in April and maximum peak in August Due to the orographic effect, the rainfall pattern in the catchment area increases from east to west and the mean annual rainfall is 1012mm.

Land use in the area is mainly dominated by moderately to intensively cultivated subsistence based cropland, grazing land, settlement areas (Addis Ababa City and other small towns) and eucalyptus trees, shrubs and grass cover some parts of the highland areas. In addition to the swampy area that covers about 140 km^2 of land, there are also small storage reservoirs (Legedadi, Gefersa, and Abasamuel) and small closed lakes basin in the catchment.

The geology of the basin is predominated by sedimentary rocks such as limestone and sandstone. Site investigation in the reservoir area carried out before the dam construction indicates that the area was swampy and alluvial plain, through which the river runs in meanders.

Key Power Plant Data

Catchment area	10747 km^2
Average annual runoff	1602 Mm^3
Average annual flow	51m^3/s
Dam type	Concrete gravity dam
Dam crest level	1593.2masl
HRWL	1590.7masl
LRWL	1580.7masl
Intake level	1570.9masl
Maximum storage	1667 Mm^3
Reservoir area at HRWL	171 km^2
Tunnel length	71.5m
Diameter	5.5m
Capacity	127m^3/s
Penstock (No.)	3
Installed capacity	43MW
Number of units	3
Type of turbine	Francis

2 SEDIMENT YIELD

Sediment yield of a catchment area can be determined by periodic sampling of the river flow to measure sediment concentration for various water discharges. There are a number of ways to

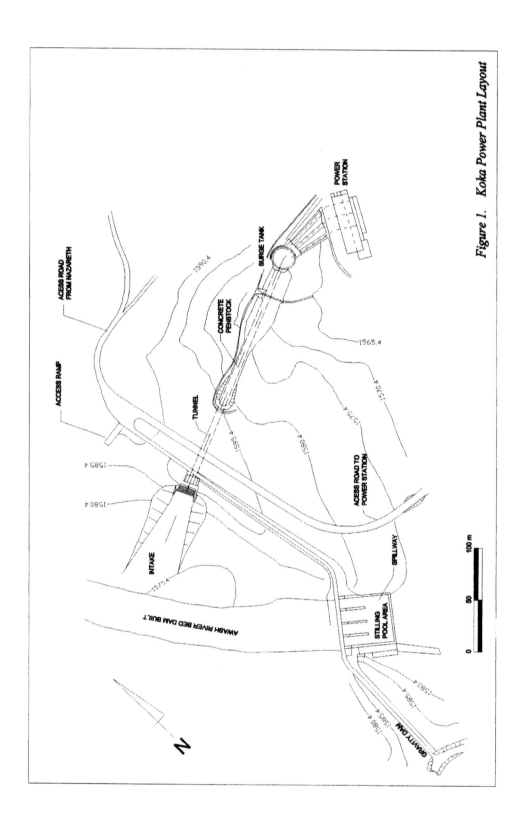

Figure 1.   Koka Power Plant Layout

346

characterize the sediment loads of rivers. Among these, for computation of sediment yield in the design of a water resource project, considerations are usually given to suspended sediment load and bed load.

The conventional method to determine sediment load in a river is to measure the concentration of suspended sediments in the river flow under various flow conditions by use of point-integrated or depth-integrated samples. In Ethiopia, the most usual method to collect suspended sediment yield is the depth- integrated sampler, i.e.; collecting samples at equally spaced stream vertical in the river cross section.

On the other hand, sediment is also carried by the river as bed load material. Although various types of bed load samplers have been designed and tested, there are no simple and reliable measuring techniques available for bed load sampling. The measurement of bed load transport presents many difficulties and results are frequently unreliable. Different researchers revealed that the bed load in a reservoir might vary the suspended load several times, though more commonly it lies between 5-25% range. In Ethiopia bed load is always assumed as a certain percentage of suspended load, which is usually 10-25%, depending upon various factors.

Suspended sediment load transported to the Koka Reservoir was estimated on the basis of sediment sampling carried out at both stations. The sediment concentration and water discharge relation for these stations were developed by the Ministry of Water Resources, Department of Hydrology and these equations are given below.

$$S_A = 55*Q^{1.5}$$

$$S_M = 230*Q^{1.5}$$

Where S_A and S_M are Suspended sediment load transported in Awash and Mojo rivers in ton/day; and Q is river discharge in m³/s.

Similarly, the sediment sampling carried out at Wonji town, down stream of the dam, has the following rating equation and this was adopted to compute the out flow sediment load from the reservoir.

$$S = 200*Q^{0.6}$$

The grain size distribution obtained from suspended sediment sampling carried out in both rivers revealed that clay 55%, Silt 44.5% and Sand 0.5%. In addition to this, samples conducted in one of the Awash River tributaries, upstream of the dam, during pre construction period justified that the sediment load consists of mainly very uniformly graded silt nearly 60% of the total load (Norcosult, 1957). Similarly, the geo-technical investigation carried out in the reservoir area at the same period

also identified alluvial deposit of about 19m thick clay and some small zones of sand and tuff.

Thus, it can be concluded that the dominant sediment particles in the reservoir are clay and silt. Since bed load measurements haven't been carried out either in the rivers nor in the reservoir, the total sediment load (suspended + Bed load) entering the reservoir should be estimated on basis of suspended sediment load. Fan and Morris (1998) suggested that if the stream bed material is dominated by clay and silt without sand, the bed load is <2% of the suspended load. However, as the sediment load of the Koka Reservoir consists of clay, silt, and small amount of sand, the bed load assumed for this study is 10% of the suspended load. In the following table the area considered for the ungauged area excludes the reservoir area at maximum water level i.e., 240km².

Table 1. Sediment inflow to the reservoir (1960-2000).

Sub basin	Area (km²)	Susp. sediment load (ton, 10⁶)	Total inc. 10% BL (ton, 10⁶)	Specific sediment yield (t/km²/yr)
Awash	6599	375	413	1565
Mojo	1425	99	109	1912
Ungauged	2585	152	167	1615
Total	10570	626	689	1630

2.1 Specific weight of deposited sediment

Samples of sediments taken from reservoirs may provide useful information of the density of deposits. The density of deposited material in terms of dry mass per unit volume is used to convert total sediment Samples of sediments taken from reservoirs may inflow to a reservoir from a mass to volume. Basic factors influencing density of sediment deposited in a reservoir are (Lara and Pemberton, 1963),

1. The manner in which the reservoir is operated.
2. The texture and size of deposited sediment particles.
3. The compaction or consolidation of deposited sediments.

Table 2. Types of reservoir operations

Type	Reservoir operation
1	Sediment always submerged or nearly submerged
2	Normally moderate to considerable reservoir draw down
3	Reservoir normally empty
4	River bed sediments

The size of the incoming sediment particles has a significant effect on density. Sediment deposits composed of silt and sand have higher densities than those in which clay predominates. The classification of sediment size according to American Geophysical union, which is applied in the other world, can be applied as follows.

Table 3. Grain size classification

Sediment type	Size range (mm)
Clay	< 0.004
Silt	0.004-0.062
Sand	0.062-2

Once the reservoir type has been selected, the initial density of the sediment deposits can be estimated using the following formula:

$$W_1 = W_c P_c + W_m P_m + W_s P_s$$

Where:

W_1 = Specific weight (kg/m^3)

W_c, W_m, W_s are coefficients of Clay, Silt and Sand of the deposited sediment

P_c, P_m, P_s are percentages of Clay, Silt and Sand of the deposited sediment.

Table 4. Coefficients of clay silt and sand

Reservoir type	W_c	W_m	W_s
1	416	1120	1150
2	561	1140	1150
3	641	1150	1150
4	961	1170	1550

In order to obtain the bulk density of a sediment deposit at the end of a specified consolidation period or the average weight over a time interval, Lane and Koelzer (1953) presented an empirical formula for the density-time relationship. The formula takes into account the grain size of the sediment and the method of operating the reservoir. Thus the density of deposits in reservoirs after a period of reservoir operation is:

$$W_t = W_1 + K \log t$$

Where:

W_t = Specific weight after t years of compaction.

W_1 = Initial specific gravity

K = Constant dependent up on the size analysis of the sediment.

However, part of the sediment will deposit in the reservoir in each of the t years of operation and each year's deposits will have a different compaction time. Miller (1953) developed an approximation of the integral for determining the average density of all sediment deposited in t years of operation as follows. This equation incorporates the assumption that sediment accumulates at a constant rate every year.

$$W_t = W_1 + 0.4343 * K[(t/(t+1))(\ln t) - 1]$$

Thus, based on suspended load measurements, data obtained from site investigation and the grain size distributions were considered to compute the specific weight of Koka Reservoir. These are clay

50%, Silt 40%, Sand 10%, reservoir operation type; Type 2, i.e.; normally moderate to considerable drawdown and t is 40 years.

The specific weight obtained from this computation was 1110kg/m^3 (1.11 t/m^3) and this figure was adopted to convert the inflow sediment mass to volume.

Table 5. Sediment deposited in the reservoir

Total load	Sediment out flow	Deposited		Average trap efficiency
(ton, 10^6)	(ton, 10^6)	(ton, 10^6)	(m^3, 10^6)	%
689	25	664	598	96.4

Based on the above analysis the mean annual specific sediment yield of the basin is1630t/km^2/yr or 1469m^3/ km^2/yr

2.2 Sediment distribution in the reservoir

One of the most common techniques to describe the shape and sediment distribution patterns in the reservoir is based on the logarithmic plots of depth versus storage capacity.

Table 6. Reservoir shape and location of sediment accumulation.

Reservoir Type	Class	m	Location of accumulation
I	Lake	2.5-4.5	Top
II	Flood plain-Foothill	2.5-3.5	Upper middle
III	Hill	1.5-2.5	Lower middle
IV	Gorge	1.0-1.5	Bottom

Here m is the inverse slope of the depth versus capacity curve drawn on a log-log paper.

The value of m for Koka Reservoir is 4.2 which is categorized in reservoir Type I, Lake type, and this illustrates that the dominant sediment accumulation is located on the upstream side of the reservoir.

2.3 Reservoir survey

Sedimentation survey of existing reservoirs has become the more efficient and reliable method for determining the amount, density, and location of sediment deposited in a reservoir. In addition, reservoir survey has vital importance for:

I.   Determining the probable future loss of reservoir capacity.

II.   Correcting the capacity curve periodically to assure more efficient operation.

III. Evaluating the effects of watershed and climatic factors on the rate of sedimentation in order to propose effective and economical measures to reduce the rate of sedimentation.

IV. Planning storage reservoirs in a similar watershed area.

Reservoir surveys at Koka Dam were made five times: in 1969, 1973, 1981, 1988 and 1998, and various estimates have been made of the level of sedimentation in the reservoir. According to the 1998 bathymetric survey the storage capacity of the Koka Reservoir reduced by 481 Mm^3 from its original storage capacity. Based on this result the mean annual silt deposition rate was 12.7 Mm^3 per year i.e., 1201 $m^3/km^2/yr$. The specific sediment yield obtained from the suspended sediment measurement was higher by 18.2% than the above value.

3. REMOVAL OF SEDIMENT FROM KOKA RESERVOIR

Rainfall variability, loss of storage capacity by sediment and increasing demand of electric power made the Koka power plant subjected to power rationing for a period of two-four months in the year 2000. Even though it is still operational, the power plant has frequently facing difficulties to operate efficiently. From recent survey data, the invert level of the intake (1570.9) is below the deposited sediment level. Similarly, the active storage of the reservoir has been filled by sediment.

Thus, in order to pro-long the reservoirs economic lifetime the following possibilities have been assessed.

3.1 Flushing

The efficiency of sediment flushing depends on the topographic position of the reservoir, the capacity and elevation relations, the outlet elevation; the character of deposited sediment, the flushing discharge, the mode of operation, and the environmental effect. Ideally, the reservoir should be almost fully drawdown using large bottom gates, which minimizes both backwater effects and hydraulic detention time.

The Koka Dam has a rectangular section 4x4-m bottom outlet providing a discharge capacity of about 119m^3/s at the highest regulated water level head difference, whereas the mean annual discharge in August is 256m^3/s. On the other hand, there are two cascade power plants down stream of the Koka Dam, Awash II and Awash III and in addition to these power plants, more than 50,000ha of irrigation farmlands are available at down stream, which are using the released water from Koka Dam. Thus, considering the following factors, the possibility of flushing/sluicing at Koka Reservoir is improbable.

1) Inadequate reservoir bottom outlet capacity.
2) The reservoir topography is not suitable for flushing.
3) Environmental effect at downstream: Two run-off river power plants and irrigation systems at down stream could be affected by the release of highly concentrated sediment. Even though it might increase the fertility of the irrigation farm and consequently promote the agricultural production, intakes and canals cannot tolerate heavy sediment load. This would only move the sediment problem from one location to another.
4) Change in channel morphology leads to flooding at down stream Villages and towns.
5) Losing the reservoir water would create loss of power, which consequently leads to severe shortage of power in the country and loss of irrigation water.

3.2 Dredging

Dredging can be an alternative measure to reduce deposited sediment when there is no favorable topography and inadequate low level outlets for sediment flushing. It can be carried out in various methods. In order to carry out mechanical excavation in Koka reservoir, the water stored in the reservoir should be released and this also creates problems discussed in the flushing techniques. In addition to the above problem, mechanical excavation for large storage reservoir is not economical. From recent study at Legedadi Reservoir, the cost of dredging was estimated at 65-75 Eth. Birr/ m^3 (8-9 US $ / m^3).

The other method of dredging is hydraulic dredging (conventional or by gravity). One of the methods is siphoning, which differs from a conventional dredging because of using only gravity with out using a pump. From other projects experience, such type of dredging is efficient in small storage reservoirs. The conventional type hydraulic dredging is carried out using a pump. It releases a mixture of silt and water to down stream. Thus, besides its expensive operation and losing the precious water, it has also environmental problems for downstream users and most unlikely to apply at Koka Reservoir.

3.3 Heightening of the existing dam

In order to attain the lost storage capacity by sedimentation, heightening of the existing dam often practiced in some parts of the world. Heightening of the Aswan dam had carried out two times to increase its storage capacity. Similarly heightening works were conducted on Cienfuens dam (Spain) and Cheufra dam (Algeria).

Raising of an existing dam often involves more problems than building a new dam. This justifies that before adopting this method detail investigation on technical capability, economic viability and environmental effect should be assessed cautiously.

4. CONCLUSION

All sediment management alternatives potentially feasible at each site should be evaluated on the basis of technical, economic and environmental feasibility. As described in the previous sections, unless remedial measures are adopted to sustain the function of the Koka Dam, the reservoir will become virtually useless in the near future.

Because of environmental effect at downstream of the dam and economical loss, abandoning of the reservoir is not recommended. Heightening of the dam by two meter resulted in additional storage capacity of about 430Mm^3. This increases the storage surface area at maximum water level by about 70km^2 and this would results inundation of 5000ha farmlands.

At a preliminary level of investigation the only alternative that seems promising to prolong the reservoir economic lifetime is the raising of Koka Dam. But this also needs detail economic, environmental and technical investigation.

REFERENCES

Grim A., 2000, *Hydrological forecasting for Koka reservoir, in Awash River*, M.Sc. thesis, Department of Hydraulic and Environmental Engineering, the Norwegian University of Science and Technology, Norway

Halcrow, 1989, *Master plan for the development of surface water resources in the Awash basin*, Volume III, Addis Ababa, Ethiopia

Jacobsen,T. 1997, *Sediment problems in reservoirs, Control of sediment deposits*. Doctoral Thesis, IVB Report B2-1997-3. The Norwegian University of Science and Technology (NTNU), Norway

Lane E.W & Koelzer, V.A. 1953 *Density of Sediment Deposited in Reservoir*, Report No.9 of A study of methods used in measurement and analysis of Sediment Loads in Streams, St.Paul United States Engineering District, St.Paul, Minn.

Lara, J.M d Pemberton,E-L., 1963, *Initial unit weight of deposited sediments"* Proceding, federal interagency sedimentation Conference, PP. 818-845, USDA-ARS Misc. publication, 970

Lara, J.M d Pemberton,E-L., 1963, *Initial unit weight of deposited sediments"* Proceeding, federal interagency sedimentation Conference, PP. 818-845, USDA-ARS Misc. publication,970

Miller, C.R., 1953, *Determination of the unit weight of Sediment for use in Sediment Volume computations*, United States Bureau of Reclamation

Morris G.L. & Fan J., 1998, "Reservoir sedimentation Handbook. Design and management of Dams, Reservoirs, and watersheds for sustainable use. McGraw-Hill, New York.

Ministry of Water Resources, Hydrology department, 1999, *Koka reservoir bathymetry survey report* (draft), Addis Ababa, Ethiopia

Norconsult, 1957, *The Koka project, Hydroelectric development in the Awash River, Conditions and specifications*, Oslo, Norway.

Støle H., (1999), *Hand out, Literature for the withdrawal of water from sediment loaded rivers*, department of Hydraulic and Environmental engineering, The Norwegian university of science and Technology, Norway

Tahal & Metaferia Consulting Eng.PLC. (1999), *Bathymetric survey for the Legedadi, and Geffersa reservoirs and Master plan study for Legedadi, Dire and Gefersa catchment areas*, Interim report, (draft), Addis Abeba, Ethiopia

Hydropower in the New Millennium, Honningsvåg et al (eds), © 2001 Taylor & Francis, ISBN 90 5809 195 3

New aspects of research on the prediction of cavitation

T.Baur & J.Köngeter
Institute of Hydraulic Engineering and Water Resources Management (IWW), Aachen University of Technology, Germany

ABSTRACT: Hydraulic modeling of cavitation processes is a demanding task in engineering and research. In the last years the IWW proceeded in the development of a hybrid-model in order to reflect the scale-effects which occur in predicting cavitation inception. The hybrid model treats the cavitation process as two decoupled processes, a hydrodynamic one and a bubble dynamic one. The hydrodynamic part is modeled by a small-scale physical model, whereas the bubble dynamic part is solved mathematically in full scale. In this paper a comprehensive explanation of the latest state of the hybrid model is given which includes the measurement techniques to record time-dependant vortical flow fields, the used algorithms for the calculation of the hydrodynamic pressure and the implementation of the bubble dynamics. Finally some sample results are given, presenting theoretically and experimentally achieved data on cavitation inception in a turbulent shear layer.

1 INTRODUCTION

The design of hydraulic structures for hydropower systems always has to take into account cavitation. From its unwanted effects, such as damage to concrete structures, pumps, turbo-machinery and cavitation noise it is an important phenomenon from the engineering standpoint. Cavitation has been observed to occur mainly in turbulent shear flows when the local fluid pressure drops below the fluid vapour pressure and small bubbles form in the low-pressure cores of fluid vortices. Model tests are usually carried out to estimate the probability of cavitation, especially cavitation inception, in a prototype. Unfortunately, physical small-scale-modeling of cavitation phenomena for a given prototype is still very difficult and often unreliable. The reason for this behavior is the limited understanding of the physical process of cavitation inception, so that the engineer and the researcher still has to deal with scale-effects. The need for a reliable prediction is very strong. Consequently, much effort has been put into the improvement of hydraulic modeling.

2 HYDRAULIC MODELING

2.1 *Physical and mathematical models*

Information on hydrodynamic processes in prototypes can be obtained from tests with different types of models. The most common models are physical or mathematical models.

Physical models are generally small-scale models which have the same geometric properties as the prototype. To build a physical model, a modeling law must exist, which describes the set up of the model. In hydraulic engineering, Reynolds or Froude models are widely used. For the prediction of cavitation phenomenon generally the critical cavitation number σ_i is applied, which is defined according to Thoma (1925) as

$$\sigma_i = \frac{p_{ref} - p_v}{1/2 \cdot \rho \cdot u_{ref}^2} \tag{1}$$

The index i stands for inception. p_{ref} and u_{ref} are the hydrodynamic pressure and flow velocity in the undisturbed inflow, p_v is the vapor pressure and ρ the density of the fluid. While using the results from physical models, one presumes that the determined critical cavitation number describes cavitation inception also in the prototype.

$$\sigma_{i,Model} = \sigma_{i,Protype} \tag{2}$$

Furthermore, the determined critical cavitation number should not be dependent on the chosen reference free stream velocity because the velocity is taken for its definition.

In mathematical models the hydrodynamic process is described by mathematical equations which can be derived from analytical investigation of the

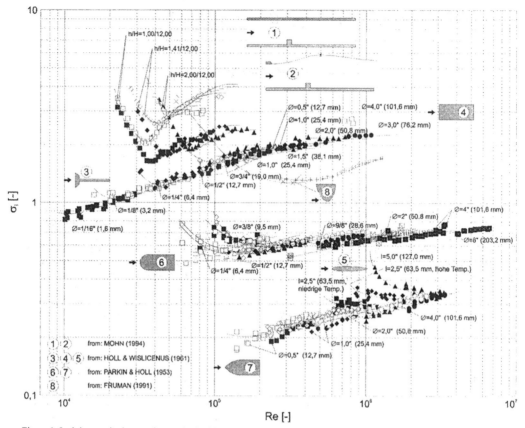

Figure 1. Incipient cavitation number σ_i obtained in cavitation tests for different test bodies from Leucker (1998).

physical background. Mathematical models are more flexible in modeling hydrodynamic flow phenomena. Nevertheless, if a problem should be solved in a reasonable time, simplifications have to be made and thus the model results do not exactly coincide with the prototype. For the investigation of cavitation inception, this second type of hydraulic modeling demands highly sophisticated methods. Due to the improvements of Finite Element (FE) flow simulation programs it might be possible in future to calculate the required parameters with the same accuracy and economic costs as physical models. By then, mathematical models are usually avoided because of the complexity of the required two-phase flow modeling.

2.2 Scale effects

If a physical model is used to predict cavitation phenomena for a prototype (especially cavitation inception), the cavitation process is influenced by several parameters which can strongly differ between model and prototype. Thus, different scale effects can occur (Kermeen & Parkin, 1957), which are caused by two

groups of parameters not taken into account in the definition of the critical cavitation number σ_i in Equation 1.

2.2.1 Water quality
On the one hand, the process of cavitation inception is dependant of the water quality. In this case the water quality is not only represented by the temperature and viscosity, but also by the nuclei spectrum in the flow. The water quality influences the critical fluid pressure p_{crit} at which cavitation starts. If the fluid has no additional tensile strength, then cavitation will occur if the minimum hydrodynamic pressure p_{min} in the flow field is equal to the vapor pressure p_v. Moreover, a high concentration of dissolved gas can increase the critical pressure p_{crit} due to over-saturation of the test fluid. It is necessary to either adjust p_{crit} to the vapor pressure at every measurement of cavitation inception (Yang & Lee, 1999), or to measure it. To directly measure the tensile strength, Keller (1981) developed a vortex-chamber nozzle and replaces the vapor pressure p_v with the measured critical pressure p_{crit} in Equation 1.

From this point of view, there is no evidence about same water qualities in the model and the prototype and consequently water quality scale effects have to be expected. Also time scale effects are induced in a small-scale model because a cavitation nucleus stays not the same time in a zone of low pressure as in the prototype. Since a cavitation nucleus needs some time to respond to a pressure fluctuation (this time depends on the size of the nucleus), the nuclei spectrum in the model has to be adapted. However, the variation of single parameters is often impossible, e.g. is the nuclei spectrum generally influenced by the test facility.

2.2.2 Flow quality

On the other hand the quality of the flow field in terms of turbulence effects the process of cavitation inception (Arndt, 1976). Unsteady turbulent motion, especially the formation of coherent vortical structures, produces pressure fluctuations. A time-dependent vortex can produce a pressure reduction in its core which is much lower than the minimum pressure of the time-averaged pressure field.

Several cavitation inception tests were carried out in recent years where results were obtained at a constant velocity by varying the system pressure and measuring desinent cavitation. A collection of results is presented in Figure 1. It is obvious that the critical cavitation number σ_i is not constant at all for different Reynolds numbers Re. Obviously, there is a dependency between the cavitation number and the free stream velocity u_{ref}. This dependency is called velocity scale effect and clarifies that the general application of the definition in Equation 1 is not even valid for a certain scale. Nevertheless, a good qualitative agreement can be seen in the results. Against small velocities many curves show increasing cavitation numbers σ_i.

The variation of the obstacle size (Mohn, 1994) yields similar dependencies. Although the test bodies have the same shape but different sizes, the cavitation number increases against bigger test body sizes. It can be possible that a hydraulic model does not show the inception of cavitation whereas the prototype is under influence of a well developed state of cavitation. These effects are called size scale effects.

Fundamental work has been done with the empirical formulation of the critical cavitation number for different flow configurations by Eickmann (1992) and Keller & Rott (1999). Numerous measurements of the conditions for the critical cavitation inception with different test body geometries and shapes as well as inflow velocities and pressure conditions were undertaken. The results were obtained under strict control of the fluid characteristics, so that water quality effects could be avoided. The results quantify the scale effects due to the flow quality and show that they exist even when the critical

pressure of the fluid is precisely measured (and applied in Equation 1). Such an experimental approach is very time-consuming because of the reduced flexibility of the model type. The advantages of a mathematical model are missed here. A combination of both models seems to be promising.

3 HYBRID MODELING

Concerning the before mentioned facts, a hybrid model consisting of a physical and a mathematical part can combine the advantages of both types of models. In contrast to the pure physical or mathematical modeling of cavitation inception, the cavitation process in the hybrid-model is divided into two sub processes. Since a cavitating flow is a two-phase-flow, the cavitation process is splitted into two major parts, a liquid and a gaseous one (see Figure 2).

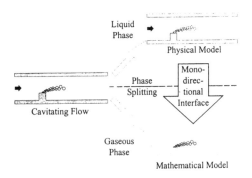

Figure 2. Hybrid model for prediction of cavitation inception in turbulent shear layers.

The liquid phase (or hydrodynamic part) is modeled by a physical model whereas the gaseous phase (or bubble dynamics part) is modeled by a mathematical model. Using the physical part, it is possible to receive exact hydrodynamic quantities, i. e. in this case the flow field which acts on a cavitation nucleus, by saving a lot of computing time. Simulating subsequently the dynamics of a prototype-nucleus in the scale of the prototype with the mathematical part, can avoid scale effects due to water quality. As mentioned before and shown by Eickmann (1992) these effects are very complicated to be avoided in a pure physical model.

Although the splitted process (the two-phase-flow) cannot decomposed physically, it can be modeled separately if the interaction between the processes are exactly defined. An interface connecting the two major parts takes this interaction into account.

The hybrid model was developed at the IWW by Stein (1984) and has been continuously improved (Mohn, 1994; Leucker & Rouvé, 1994). In the following paragraphs the main parts of the model will be explained.

3.1 Physical model

The main interesting features in the hydrodynamic part of a flow configuration are the vortical flow structures. They are the key elements of the turbulent flow and present the critical phenomena for cavitation inception. The main improvement in the latest state of the physical model is the implementation of the Particle Image Velocimetry (PIV) measurement technique. With this technique, it is possible to quantify the unsteady flow velocities in two-dimensional regions of interest (Raffel et al. 1998). The PIV is combined with a high-speed camera to provide a high temporal resolution of the flow field. Details about this high-speed PIV system can be found in Baur & Köngeter (2000). Figure 3 shows the experimental set-up for the investigation of the liquid phase in the model.

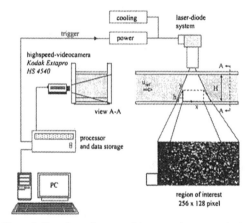

Figure 3. Experimental set-up of the measuring system in the physical model.

The flow field of interest is illuminated by a diode laser system with four high-power laser diodes resulting in an output power of up to 160 W. In a light sheet, seeding particles of same density as the fluid become visible and represent the unsteady flow field. Their motion is recorded with a highspeed-videocamera of type *Kodak Ektapro HS 4540* at a recording rate of 9000 frames/second. The power supply of the laser guarantees pulses of same repetition rates as the camera and short pulse duration for sharp, motion free exposures.

The digital images are processed by a digital cross-correlation technique using interrogation windows of size 64 × 64 pixels. The spatial step width is 16 pixel. After calibration, the PIV-method yields the displacement vectors at the mesh points in a square grid pattern. In combination with the time

difference between two successive images, the unsteady flow velocities in the flow field can be calculated. Several differential quantities can be further estimated in the interface of the hybrid model (Baur & Köngeter, 2000).

The combination of the recording rate of 9000 frames/second and a pulse duration of 25 µs provides high resolution analysis of fast events. Using this measurement technique with time-series PIV, special focus can be put on the time-dependent development of the vortical structures in the flow field. This approach makes time-series PIV with continuous data possible.

To understand the mechanics of turbulence a visualization of vortices is required. Let u, v, w be the velocity components for the x, y, z direction. In the present investigation the Reynolds decomposition (e.g. Adrian et al. 1998)

$$u = \bar{u} + u' \qquad (3)$$

with time averaged velocities $\bar{u}$ and unsteady fluctuations u' is applied to separate the small-scale coherent turbulent motion from the large-scale recirculation zone behind the obstacle. Time-averaging of the total velocity a each grid point yields the large-scale motion. So the translation of vortices caused by the large-scale field is removed and the analysis of the turbulent characteristics easier.

3.2 Interface

The interface connects the physical and mathematical part of the hybrid model and is thus responsible for the relevant data exchange between the physical and mathematical part. It was mentioned before, that the parameters of the liquid phase (e. g. coherent turbulent structures) have a very strong influence on the cavitation nuclei (the gaseous phase). On the other hand the impact of the gaseous phase on the liquid phase can be neglected (assuming that the cavitation bubbles are small). On these reflections, a monodirectional interface has been chosen.

Within the interface the input for the mathematical part is calculated. Like the entire hybrid model the interface is also composed of several independent modules. In the following the modules *scaling-up of velocities*, *pressure calculation* and *nuclei characteristics* are described.

3.2.1 Scaling-up of velocities

In order to provide correct data for the simulation of the prototype-nuclei dynamics, the unsteady velocity fields have to be scaled up into the scale of the prototype. According to the chosen modeling law (generally Froude model) the velocity components in x- and y-direction as well as their co-ordinates are postprocessed. This approach assumes the existence of identical qualitative flow structure in the model and the prototype. Size and velocity scale effects can not taken into account here. Thus, the later presented test-case uses 1:1 as the scale of the physical model.

3.2.2 Pressure calculation

The unsteady pressure field is determined from the measured velocities and its derivatives (given in the Eulerian frame of reference). The velocity derivatives are calculated by finite differences approximation (i.e. central differences of the second order). At the points of the four edgelines of the grid the gradients are calculated with forward or backward differences approximation according to the availability of data.

The calculation of the integral quantity p, which is the unsteady fluid pressure, is carried out, assuming an incompressible flow field with $\partial \rho / \partial t = 0$ and constant viscosity ν, through the numerical integration of the unsteady Navier-Stokes-equations in three-dimensional form and Cartesian coordinates. For the two flow directions x and y in the measurement plane the equations are indicated by Equation 4a and 4b.

x-component:

$$\frac{\partial(\rho u)}{\partial t} + \frac{\partial(\rho u^2)}{\partial x} + \frac{\partial(\rho uv)}{\partial y} + \frac{\partial(\rho uw)}{\partial z} =$$
$$-\frac{\partial p}{\partial x} + \nu\left(\frac{\partial^2(\rho u)}{\partial x^2} + \frac{\partial^2(\rho u)}{\partial y^2} + \frac{\partial^2(\rho u)}{\partial z^2}\right) \quad (4a)$$

y-component:

$$\frac{\partial(\rho v)}{\partial t} + \frac{\partial(\rho uv)}{\partial x} + \frac{\partial(\rho v^2)}{\partial y} + \frac{\partial(\rho vw)}{\partial z} =$$
$$-\frac{\partial p}{\partial y} + \nu\left(\frac{\partial^2(\rho v)}{\partial x^2} + \frac{\partial^2(\rho v)}{\partial y^2} + \frac{\partial^2(\rho v)}{\partial z^2}\right) + \rho g_y \quad (4b)$$

The left side of the Navier-Stokes-equations consists of the temporal and convective acceleration of a fluid particle. On the right side the pressure, the viscous force per unit mass and for the y-component the gravity forces per unit mass is part of the equations. The unsteady approach with the gradients $\partial u / \partial t$ and $\partial v / \partial t$ becomes possible due to the data availability with high temporal resolution.

Additionally, the three-dimensional approach is considered to be important because the spatial behavior of the turbulent structures could be proved to take place in the three spatial coordinates. According to Baur & Köngeter (1998) the flow field under investigation is not clearly two-dimensional but can be described with a three-dimensional topology model.

Due to the unavailability of velocity derivatives in the two-dimensional measurement planes, some simplifications are required. They are described in detail in Baur & Köngeter (1999).

The pressure gradient Δp between two grid points is finally integrated according to

$$\Delta p = \int dp = \int \left[\frac{\partial p(x,y)}{\partial x} dx + \frac{\partial p(x,y)}{\partial y} dy \right] \quad (5)$$

with use of Equation 4a and 4b. The boundary conditions, from which the numerical integration is started, is the top edgeline with the interpolated time-averaged pressure (Baur & Köngeter, 1999). The explicit integration scheme is indicated in Figure 5.

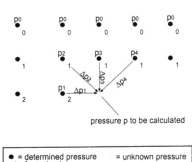

pressure p to be calculated

● = determined pressure   = unknown pressure
i   from step i

Figure 5. Schematic of pressure integration paths with direction from top to bottom.

For each grid point the already calculated neighbor grid points are used as integration paths. The pressure in the investigated grid point is determined as the mean of the so estimated values in order to reduce the effect of uncertainties in the measurements:

$$p = \frac{1}{4}\left[\begin{array}{c} (p_1 + \Delta p_1) + (p_2 + \Delta p_2) \\ + (p_3 + \Delta p_3) + (p_4 + \Delta p_4) \end{array} \right] \quad (6)$$

Additionally, the whole procedure is carried out from the bottom edge-line. The overall result yielding the unsteady pressure field is taken from the distance weighted result of both integration directions.

3.2.3 Determination of nuclei characteristics

Another requirement for the mathematical simulation of the gaseous phase, i.e. the prototype-nuclei, is input information about the radius r_0 and the concentration c_0 (both at initial time t_0). These two parameters are obtained by water quality measurements in the prototype-fluid or have to be estimated from experience.

3.3 Mathematical model

Apart from the investigation of the turbulent flow field including the hydrodynamic pressure distribution, the mathematical simulation of the motion of a cavitation nucleus and its time-dependant dynamic behavior in the prototype is required. The before calculated time-series of the pressure field can be used to determine the pressure versus time curve for a cavitation nucleus. The investigation of this part of the hybrid model focuses on the significance of the flow field for cavitation inception.

Cavitation inception is produced when the nucleus is transported into the region of a critical pressure reduction. The probability of cavitation in the cores of coherent vortices is therefore dependant from the following effects:

1. the size and concentration of cavitation nuclei in the incoming flow,
2. the pressure distribution in the time-dependant flow field,
3. the effects of fluid vortices on the nuclei pathlines.

Using the mathematical model, the interaction between the pathline of a nucleus and its bubble dynamics has to be considered. On the one hand, the major parameter for the time-dependant surrounding pressure is the pathline in the Lagrangian frame of reference. On the other hand, effects the continuously changing radius of the nucleus about its pathline due to lift forces of buoyancy. The interaction requires the successive computation of position and size of the nucleus. The modules of the mathematical part are *nuclei tracking* and *bubble dynamics simulation*.

3.3.1 *Nuclei tracking*

The nuclei tracking is also carried out by mathematical post-processing of the flow velocity data in prototype scale. Similar tracking approaches are carried out by Tio et al. (1993) and Iyer et al. (1999), but only synthetically generated vortical flow fields are used. The improvement of the here presented method lies in the fact that it is based on measured velocity data from the physical model.

The flow velocity of the nuclei is taken to be equal to the velocity of the fluid. Assuming the no-slip condition, shear forces do not effect the phase boundary and the shape of the nucleus remains spherical. According to Isay (1989) produce the lift forces acting on the bubble of maximum radius $r = 0.1$ mm a velocity component v_{lift} in y-direction of:

$$v_{lift} = \frac{2}{9} \frac{g \rho r^2}{\mu} \qquad (7)$$

In Equation 7 μ is the dynamic viscosity. The computation of the pathline has to be carried out assuming two-dimensional motion within the measurement plane.

In every time step of length Δt the radius of the nucleus is considered to be constant, so that its displacement in Cartesian co-ordinates can be derived from every point in the measurement plane by:

$$\Delta x = u \cdot \Delta t \qquad (8a)$$

$$\Delta y = (v + v_{lift}) \cdot \Delta t \qquad (8b)$$

3.3.2 *Bubble dynamics simulation*

The cavitation process of a single spherical cavitation nucleus can then be described by the extended Rayleigh-Plesset equation:

$$r \ddot{r} + \frac{3}{2} \dot{r}^2 = \frac{1}{\rho} \left(p_d + p_{g0} \left(\frac{r_0}{r} \right)^{3\kappa} - p_{fl} - \frac{2\alpha}{r} - 4\mu \frac{\dot{r}}{r} \right) \qquad (9)$$

with $r = r(t)$, $\dot{r} = \dot{r}(t)$, $\ddot{r} = \ddot{r}(t)$, being the actual radius of the bubble and its time derivatives; ρ the density of the fluid; p_v the vapor pressure of the fluid; r_0 the original radius of the nucleus (at initial time t_0), p_{g0} the partial gas pressure (at initial time t_0); $p_{fl} = p_{fl}(t)$ the actual fluid pressure at the phase boundary; κ the adiabatic coefficient ($\kappa = 1.4$ for air), α the surface tension and μ the dynamic viscosity. This differential equation describes the dynamic behavior of the bubble due to inertial and frictional forces assuming constant mass of the included gas. The gas viscosity and the compressibility of the fluid are neglected (Plesset & Prosperetti, 1977). By using the adiabatic coefficient according to Isay (1989), a change of temperature of both phases is regarded, but a temperature interaction neglected.

The initial conditions for the simulations are the parameters described with the index 0, which is the gas partial pressure p_{g0} and the initial radius r_0. The response of these parameters on the time dependant pressure $p_{fl}(t)$ is simulated with a *Matlab* program by using a Runge Kutta method for the solving of the differential equation.

Cavitation inception occurs, when a critical size of the cavitation nucleus in relation to the surrounding fluid pressure is reached. This size is described by the critical radius r_{crit}:

$$r_{crit} = \sqrt{\frac{3 \cdot p_{g0} \cdot r_0^3}{2 \cdot \alpha}} \qquad (10)$$

In this stadium, the dynamical development of the cavitation structure cannot be further simulated with Equation 9. The geometry of complex cavitation zones would have been to be taken into consideration by other equations. Chahine (1995) developed a boundary element method to cover the demands on this field. The present paper focuses on the cavitation inception process so that the simulation with the Rayleigh-Plesset equation (Equation 9) can be stopped after the critical size of a nucleus has been detected.

4.1 Flow configuration and PIV set-up

To demonstrate its performance, the hybrid model is applied for a test-case. In this test case the prototype is presented by the physical model itself. The scale of model and prototype is consequently 1:1. The conditions for cavitation inception are determined in a first step by the hybrid model and later compared with the measurements at the prototype, i.e. the physical model.

The physical part of the hybrid model is represented by the cavitation test rig of the IWW. This test rig is very small (contents of water approx. 1.8 m³) and is operated in recirculation mode. Cavitation inception is produced in the turbulent shear-layer behind a surface-mounted rectangular obstacle of size h = 12 mm under pipe-flow conditions. The flow height and width is H = 84.1 mm and B = 120 mm respectively. In the test-rig flow velocities of 2.0 m/s < u_{ref} < 6.0 m/s are investigated, giving a Reynolds number of 24,000 < Re < 72,000 related to h.

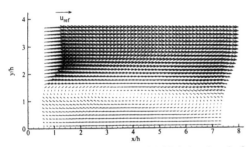

Figure 6. Time-averaged flow field behind the obstacle for Re = 72,000.

The investigated measurement plane is the streamwise wall-normal plane directly behind the obstacle (see also Fig. 3). The seeding particles, the scatter of which is examined and taken as representation of the vector field over a very short time, are polyamide particles with grain size diameter d_m = 28 μm. The camera resolution of 256 × 128 pixel yields at a calibration for a 9.4 cm × 4.7 cm region of interest a pixel size of 0.37 mm. The video system captures up to 5800 single frames and provides a digital data transfer onto the PC. The mean velocities are obtained from time averaging all PIV realizations. The result is given in Figure 6 for Re = 72,000.

The size and concentration of cavitation nuclei in the incoming flow was determined by an optical analysis from Mertens (1988). The recorded spectrum allows for the definition of a characteristic radius r_0 for the simulation of r_0 = 0.05 mm.

4.2 Unsteady flow velocities

Figure 7 shows a typical time-series of instantaneous high-pass filtered velocity fluctuations. The cores of coherent vortical structures are indicated by the circles. The flow pattern shows a roughly circular clockwise-rotating vortical motion and a clear Q2 ejection underneath. Such a vortex pattern corresponds to the physical model of primary and secondary vortices (Bernal & Roshko, 1986, Hussain & Hayakawa, 1987). In the measurement plane a cross-section of the primary vortices and low-speed fluid below is realized. In the wake of vortex A, a second vortex B is generated with same orientation. Vortex C follows with reduced elongation.

The presented sequence of PIV realizations shows the efficiency of this measuring technique. Vortical structures are temporarily and spatially resolved. The velocity data is further used to calculate their effects on the hydrodynamic pressure field.

4.3 Unsteady pressure distribution

The calculation of the hydrodynamic pressure is carried out in the model interface for the sequence shown in Figure 7. The dimensionless pressure coefficient Cp is defined by:

$$C_p = \frac{p - p_{ref}}{1/2 \cdot \rho \cdot u_{ref}^2} \qquad (11)$$

Figure 8 presents the main effect of the coherent vortical structures on the pressure distribution. The dimensionless pressure coefficient reaches values down to C_p = -1.2. It can be verified that the calculated pressure shows a minimum in the vicinity of the vortex cores indicated by the circles. Thus, the conclusion that cavitation inception is likely to start in the center of time-dependent vortices is physically proved.

The purpose of the further investigation is to produce a data base which contains the extreme flow phenomena being critical for the production of cavitation inception. For different Reynolds numbers, a sequence of 5800 PIV realizations (maximum number for camera storage) and total length of T = 0.64 s is processed. By a statistical analysis the minimum pressure produced by a vortex during this time span in the flow field is extracted. The response of cavitation nuclei on these extreme flow events, turning up at a certain time in a certain point of the flow field, is then simulated.

4.4 Response of cavitation nuclei

The response of the cavitation nuclei to the extreme pressure event produced by an extreme coherent vortical motion is described here for the sequence recorded at Re = 48,000. This sequence contains the minimum pressure $C_{p,min}$ = -2,78 which is produced

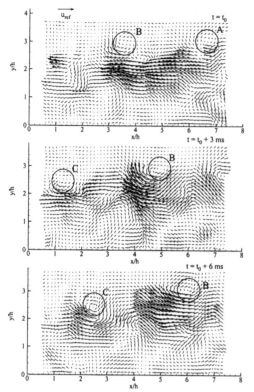

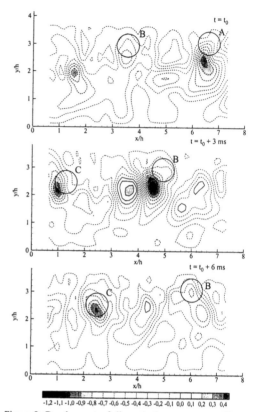

Figure 7. Development of velocity fluctuations (u',v') in the measurement plane, time spacing Δt = 3 ms, primary vortices are indicated by A, B, C.

-1,2 -1,1 -1,0 -0,9 -0,8 -0,7 -0,6 -0,5 -0,4 -0,3 -0,2 -0,1 0,0 0,1 0,2 0,3 0,4

Figure 8. Development of dimensionless pressure coefficient C_p in the measurement plane, time spacing Δt = 3 ms, primary vortices are indicated by A, B, C.

by a primary vortex. The critical start co-ordinates x_0 and y_0 of a pathline going through this temporarily and spatially known minimum pressure are determined by iteration. The result of the nucleus tracking yields not only the pathline but also the time-series of surrounding fluid pressure relative to the system pressure.

According to the bubble dynamics simulation for this pressure time-series, a system pressure in the test rig of p_{sys} = 19,300 N/m² is necessary to initiate a critical radius r_{crit} of the nucleus resulting in a theoretically determined critical cavitation number of σ_i = 2.46. The initial size of the nucleus is r_0 = 0.05 mm. The simulation is carried out also for r_0 = 0.025 mm and r_0 = 0.10 mm respectively.

Figure 9 demonstrates the pathline, the time-series of the surrounding fluid pressure $p_{fl}(t)$ and the bubble radius response at an initial size of r_0 = 0.05 mm.

The calculation beyond the point of critical cavitation inception shows the rebound effect as a consequence of the increasing hydrodynamic pressure as the nucleus has passed the vortex. The application

of Equation 9 is physically not precise but the results, however, correspond to the real behavior of a collapsing bubble.

The simulation is repeated for the sequence recorded at Re = 36,000 where the minimum pressure is $C_{p,min}$ = -3.93. The critical cavitation numbers σ_i are listed in dependency of the simulated initial radius r_0 in Table 1.

Table 1. Critical cavitation number from hybrid modeling.

	r_0 = 0.025 mm	r_0 = 0.050mm	r_0 = 0.100 mm
Re = 36,000	3.45	3.41	3.05
Re = 48,000	2.46	2.55	2.38

4.5 Comparison of cavitation numbers

The physical model of scale 1:1 is used to determine the critical cavitation number of the prototype in an experiment. Taking into account the water quality effect on the process of cavitation inception by applying the method of Yang & Li (1999), the critical cavitation number is determined for different flow velocities u_{ref}.

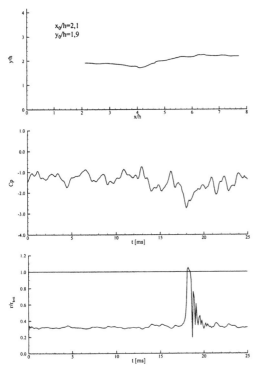

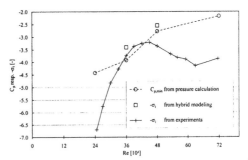

Figure 10. Comparison of theoretically (for $r_0 = 0.05$ mm) and experimentally determined critical cavitation numbers -σ_i with minimum pressure coefficient $C_{p,min}$.

Figure 9. Pathline, critical development of surrounding pressure Cp and radius r/r_{crit} of a cavitation nucleus during the recorded sequence; cavitation inception at $\sigma = 2,46$ after $\Delta t = 18.0$ ms; Re = 48,000.

Figure 10 shows the experimental results and provides a comparison to the results from pressure calculation and hybrid modeling (Table 1). In theory we assume that the test fluid has no additional tensile strength. Using Equation 1 and 11, this indicates that the cavitation number for cavitation inception σ_i equals the minimum pressure coefficient -$C_{p,min}$:

$$\sigma_i = -C_{p,min} \qquad (12)$$

From the simulation of the nucleus response (Fig. 10) a discrepancy between $C_{p,min}$ and -σ_i (from hybrid modeling) can be observed which is characterized by $C_{p,min} < -\sigma_i$ and does not correspond to Equation 12. The reason lies in the fact that the growth of the nucleus with initial radius r_0 has to be considered. In this case, the hydrodynamic pressure p_{crit} which is critical for cavitation inception in this nucleus is lower than the vapor pressure p_v so that in the investigated vortex cavitation occurs.

Comparing the critical cavitation numbers σ_i from the different investigations (hybrid modeling and experiments), it becomes clear that hybrid modeling in the test-case yields absolute values which are smaller than the measured numbers. It is possible that the recorded sequences contain vortices which are not representative for extreme pressure reduc-

tions. The value and the duration of the minimum pressure in the recorded sequences is not able to produce a critical growth of the simulated cavitation nucleus at higher cavitation numbers.

In order to judge about the probability of cavitation inception, the prototype nuclei spectrum has to be known from which the initial parameters for the simulation can be deduced. In such a module, which has not been tested yet, the concentration of prototype-nuclei is combined with the area of critical regions in the inflow giving a number of critical nuclei. The areas are defined by the group of starting coordinates which let the cavitation nuclei go throw a critical growth.

The presented results are taken as an example to demonstrate the actual state of the physical and mathematical parts of the hybrid model. Obviously the data base containing velocity information from the flow field, where cavitation is expected, is the decisive basis for the investigation. The recording of the sequences has to guarantee that extreme flow phenomena are captured. From this point of view, a statistical approach has to be considered in the physical model. Work has to be done on this field in order to improve the procedure.

5 CONCLUSION

Hybrid modeling is applied to predict cavitation inception in a prototype. Due to arising scale effects, pure physical modeling is not reliable enough so that the combination with the mathematical simulation of the bubble dynamics was applied. It is planed that in future investigations, cavitation inception experiments in a small scale physical model can be rejected and replaced through the calculation of the unsteady pressure distribution. Scale effects due to the water quality of the test fluid and prototype fluid are avoided. Additionally, the drawback of subjective decisions in cavitation inception experiments about the occurrence of cavitation bubbles, which arises when optical or acoustical measures for their detection are not available, is not present.

Due to the modular design of the model it can be widely used for different applications. Some of the shown modules are specific to the used test facility, measurement system or flow configuration. These specific modules can be easily modified or replaced to adjust them to the special purpose which is needed for other configurations.

The hybrid modeling guarantees the determination of the critical cavitation numbers taking into account the growth of a cavitation nucleus interacting with measured vortical structures. The mathematical simulation of the nuclei dynamics in the predetermined pressure fields can be carried out in prototype scale. However, work has to be done also on the investigation of the physical reasons of velocity and size scale effects. These latter effects were mentioned in the paper in order to recall their significance but not further considered. Work on the flow quality effects has also been done at the IWW but will be presented elsewhere.

6 ACKNOWLEDGEMENT

The presented investigations are part of a research project which was supported by the German Research Foundation (DFG) with reference number Ko 1573/4.

7 REFERENCES

Adrian, R.J.; Christensen, K. T., Soloff, S. M., Meinhart, C. D., Liu, Z. C. 1998. Decomposition of turbulent fields and visualization of vortices. *Proc. Ninth International Symposium on Applications of Laser Techniques to Fluid Mechanics, Lisbon, Portugal.*

Arndt, R. E. A. 1976. Semiempirical Analysis of Cavitation in the Wake of a Sharp-Edged Disk. *Journal of Fluids Engineering*, Vol. 98: 560-562.

Baur, T. & Köngeter, J. 1998. The Three-dimensional Character of Cavitation Structures in a Turbulent Shear Layer. In H. Brekke et al. (eds), *Proc. XIX IAHR Symposium on Hydraulic Machinery and Cavitation*, Singapore: 490-499.

Baur, T. & Köngeter, J. 1999. PIV with High Temporal Resolution for the Determination of Local Pressure Reductions from Coherent Turbulence Phenomena. In R. Adrian et al. (eds), *Proc. 3rd International Workshop on PIV, Santa Barbara, USA*: 101-106.

Baur, T. & Köngeter, J. 2000. High-speed PIV and the postprocessing of time-series results. *Proceedings of the Euromech 411 Meeting on Application of PIV to Turbulence Measurements, Université de Rouen, France.*

Bernal, L. P. & Roshko, A. 1986. Streamwise vortex structure in plane mixing layers. *Journal of Fluid Mechanics, Vol. 170:* 499-525.

Chahine, G. L. 1995. Bubble Interactions with Vortices. In S. Green (ed.), *Fluid Vortices, Kluwer Academic.*

Eickmann, G. 1992. Maßstabseffekte bei der beginnenden Kavitation - Ihre gesetzmäßige Erfassung unter Berücksichtigung der wesentlichen Einflußgrößen. In T. Strobl (ed.), *Berichte der Versuchsanstalt Obernach und des Lehrstuhls für Wasserbau und Wassermengenwirtschaft der TU München*, Nr. 69.

Fruman, D. H., Dugué, C., Cerutti P. 1991. Tip Vortex Roll-Up and Cavitation. *Cavitation and Multiphase Flow Forum, ASME FED, Vol. 109:* 43-48.

Holl, J. W. & Wislicenus, G. F. 1961. Scale Effects on Cavitation, *Journal of Basic Engineering. Trans. ASME, Series D, Vol. 83:* 385-398.

Hussain, A. K. M. F. & Hayakawa, M. 1987. Eduction of Large-Scale Organized Structures in a Turbulent Plane Wake. *Journal of Fluid Mechanics, Vol.180:* 193-229.

Isay, W. H. 1989. Kavitation. *Schiffahrts-Verlag Hansa.*

Iyer, C. O.; Ceccio, S. L., Tryggvason, G. O. 1999. Capture Of A Bubble By A Concentrated Vortex. *Proc. 3rd ASME/JSME Joint Fluids Engineering Conference, San Francisco, USA.*

Keller, A. P. 1981. Tensile Strength of Liquids. *Proc. of the 5th Intern. Symp. on Water Column Separation*, IAHR Work Group, Obernach, Germany.

Keller, A. P. & Rott, H. K. 1999. Scale Effects on Tip Vortex Cavitation Inception. *Proc. 3rd ASME/JSME Joint Fluids Engineering Conference, San Francisco, USA.*

Kermeen, R. W. & Parkin, B. R. 1957. Incipient Cavitation and Wake Flow Behind Sharp Edged Disks. *Calif. Inst. Technology, Hydrodyn. Lab. Rep. 85-4.*

Leucker, R. & Rouvé, G. 1994.Prediction of Cavitation Inception with a Hybrid-Model. In Cheong, Shankar, Chan, Ng (eds), *Proc. of the 9th Congress of the APD-IAHR, 24.-26. August 1994, Singapore*: 478-485.

Mohn, R. 1994. Zur Modellähnlichkeit des Kavitationsbeginns in abgelöster, turbulenter Strömung. *Mitteilungen des Instituts für Wasserbau und Wasserwirtschaft der RWTH Aachen*, Bd. 92, Academia Verlag, St. Augustin.

Parkin, B. & Holl, J. W. 1953. Incipient-Cavitation Scaling Experiments for Hemispherical and 1.5-Caliber Ogive-Nosed Bodies. *Report No. NOrd 7958-264, Ordnance Research Laboratory*, The Pennsylvania State College.

Plesset, M. S. & Prosperetti, A. 1977. Bubble Dynamics and Cavitation. *Annual Review Fluid Mechanics, Vol. 9:* 145-185.

Raffel, R.; Willert, C. E., Kompenhans, J. 1998. Particle Image Velocimetry - A practical guide. *Springer-Verlag Berlin/Heidelberg.*

Stein, U.. 1984. Zur Untersuchung der Strömungskavitation unter Berücksichtigung von Turbulenz, Wirbelbildung und Blasendynamik. *Mitteilungen des Instituts für Wasserbau und Wasserwirtschaft der RWTH Aachen*, Bd. 43, Academia Verlag, St. Augustin.

Thoma, D. 1925. Die experimentelle Forschung im Wasserkraftfach. *Zeitschrift des Vereins Deutscher Ingenieure, VDI-Verlag GmbH, Berlin*, Band 69.

Tio, K.-K.; Linan, A., Lasheras, J. C., Ganan-Calvo, A. M. 1993. On the Dynamics of Buoyant and Heavy Particles in a Periodic Stuart Vortex Flow. *Journal of Fluid Mechanics, Vol. 254:* 671-699.

Yang, Z. & Li, Q. 1999. On Comparison between Results of Cavitation Inception Tests. *Proc. 3rd ASME/JSME Joint Fluids Engineering Conference, San Francisco, USA.*

Hydropower in the New Millennium, Honningsvåg et al (eds), © 2001 Taylor & Francis, ISBN 90 5809 195 3

Numerical modelling of the intake parts of small hydropower plants

A.Drab, J.Jandora & J.Riha
Water Structures Institute, FCE, Brno University of Technology, Czech Republic

O. Neumayer
AQUATIS, a.s., Brno, Czech Republic

ABSTRACT: The efficiency of hydropower plant is significantly affected by the design of its intake part. To design optimal shape of the intake parts of small hydropower schemes it is possible to use experimental and/or numerical research. Experimental research for small hydropower plants seems to be very expensive while numerical research is much cheaper and less time consuming. Numerical research can even produce a number of variants of different shapes in quite a short time. In the Czech Republic, plenty of small hydropower plants have recently been additionally built at existing weirs and dams. Hence, a special attention must be given to the evaluation of velocity field distribution just behind the turbine inlets. It is useful to use numerical research for these situations. For computation the inlet parts, 2D or 3D model could be used. The numerical simulation is usually performed as steady state viscous and turbulent flow and e.g. k-ε turbulence model can be used. The comparison of 2D and 3D modeling results is carried out to show at which conditions and with what accuracy it is possible to use more simple 2D modeling technique.

1 INTRODUCTION

According to the Czech national standards, the small hydropower plants (HPP) are defined as installations, whose power output does not exceed 5 MW. These hydropower schemes include both micro-sources with installed output in a range of few kW with design discharge about few litres per second and head within hundred metres both quite big schemes with quite great power output and discharge of about 150 m³/s and only few metres of hydraulic head. Especially the latter hydropower schemes have to be designed with special attention to the efficiency of entire hydraulic circuit to minimize the head losses. The very important part of the hydraulic circuit of the low head hydropower schemes is an intake part, which influences significantly velocity distribution in front of the turbine inlet and therefore its efficiency.

As the experimental hydraulic research is relatively expensive and time demanding, the numerical modelling techniques are recently extensively developed and used for the design of intake parts of smaller low-head hydropower plants. The paper deals with the comparison of two numerical methods and approaches based on the application of two different computer codes:

- Finite element 3D and 2D CFD model based on the k-ε method for the turbulent flow solution. For the solution the ANSYS-FLOTRAN computer code was used.

- Finite difference 2D depth integrated model based on curvilinear mesh. For the solution the MIKE21 code was used.

The most of larger "small" low-head hydropower schemes in the Czech Republic are designed at the relatively big rivers close to existing weirs or barrages at the place of old log sluices or lock chambers. Therefore the design is space limited and the turbine inlets must fit more or less the existing river-weir configuration. The uniformity of the flow in front of turbines must be achieved by convenient structural arrangement of the inlet part comprising appropriate shape of division pillar, guiding vanes and an inlet sill. The vortexes and dead zones must be avoided in the inlet flow domain and uniform velocity distribution in front of the turbines must be achieved.

Four practical applications of numerical solution of hydropower plants located on the Vltava and Ohre rivers were carried out before the final turbines inlet design. The chief designer of the scheme was AQUATIS, a.s., a building owner is the Vltava river basin agency Prague (Povodí Vltavy s.p., Povodi Ohre, s.p.). The objects of the solution were following small hydropower plants:
- Libcice – Dolany on the Vltava river in Prague;
- Klecany on the Vltava river;
- Doksany on the Ohre river;
- Vranany on the Vltava river;

In case of the HPP Libcice the numerical computations were compared with the results of experimental laboratory research performed at the Water Structures Institute at the Technical University in Brno.

2 MATHEMATICAL MODEL

2.1 Three-dimensional (3D) model

The 3D turbulent flow model is based on the solution of time-averaged Navier-Stokes equations. The model presented uses the presumption of the steady-state flow, water is assumed to be incompressible and Newtonian. The equations can be expressed as follows (Rodi 1980):

$$\frac{\partial \bar{v}_i}{\partial x_i} = 0 \tag{1}$$

$$\rho \bar{v}_j \frac{\partial \bar{v}_i}{\partial x_j} = -\frac{\partial \bar{p}}{\partial x_i} + \frac{\partial}{\partial x_j}\left[\mu\left(\frac{\partial \bar{v}_i}{\partial x_j} + \frac{\partial \bar{v}_j}{\partial x_i}\right) - \rho \overline{v_i' v_j'}\right], \tag{2}$$

where $\bar{v}, \bar{p}, \rho$ and μ are the mean velocity in the time, mean pressure in the time, fluid density and dynamic viscosity. The last term at the right side of the equation (2) is the contribution of the turbulent motion to the mean flow and is called the Reynolds' stress. In the presented paper, the Reynolds' stress is approximated using the assumption of Boussinesq's eddy viscosity (μ_t), i.e.

$$\tau_{ij}^t = -\rho \overline{v_i' v_j'} = \mu_t\left(\frac{\partial \bar{v}_i}{\partial x_j} + \frac{\partial \bar{v}_j}{\partial x_i}\right) - \frac{2}{3}k\rho\delta_{ij}, \tag{3}$$

where k is the kinetic turbulent energy, ε is the rate of dissipation of turbulent energy and δ_{ij} Kronecker delta. Variable μ_t is expressed using k-ε model (Wilcox 1994):

$$\mu_t = c_\mu \rho \frac{k^2}{\varepsilon}, \tag{4}$$

$$\rho \bar{v}_j \frac{\partial k}{\partial x_j} = \tau_{ij}^t \frac{\partial \bar{v}_i}{\partial x_j} - \rho\varepsilon + \frac{\partial}{\partial x_j}\left[\left(\mu + \frac{\mu_t}{\sigma_k}\right)\frac{\partial k}{\partial x_j}\right], \tag{5}$$

$$\rho \bar{v}_j \frac{\partial \varepsilon}{\partial x_j} = c_{\varepsilon 1}\frac{\varepsilon}{k}\tau_{ij}^t\frac{\partial \bar{v}_i}{\partial x_j} - c_{\varepsilon 2}\rho\frac{\varepsilon^2}{k} + \frac{\partial}{\partial x_j}\left[\left(\mu + \frac{\mu_t}{\sigma_\varepsilon}\right)\frac{\partial \varepsilon}{\partial x_j}\right], \tag{6}$$

Coefficients - c_μ, $c_{\varepsilon 1}$, $c_{\varepsilon 2}$, σ_k, σ_ε - are assumed as constant - Tab.1 (Launder, Spalding 1974).

Table 1. Coefficients of k-ε model.

c_μ	$c_{\varepsilon 1}$	$c_{\varepsilon 2}$	σ_k	σ_ε
0.09	1.44	1.92	1.0	1.3

Boundary conditions express known velocity distribution at the domain inlet (sufficiently far from the area of interest), the wall boundary condition (velocity vector equal to zero) and zero pressure condition at the domain outlet.

2.2 Two-dimensional (2D) approximation

2.2.1 2D approximation in 1m thick layer

The flow equations and boundary conditions are formally the same as those described in the paragraph 2.1. The 2D approximation of this type has to express the character of the flow in 1m thick layer sufficiently far from the inlet bottom and from the free water surface. The inaccuracy of the approximation consists mostly in the non-considering variations in the rivet and turbine inlet bed while variations in water level can be neglected.

2.2.2 2D shallow flow equations

The 2D shallow equations are based on the depth averaged flow velocity (depth integration of Navier-Stokes equations). In the study, the steady state approximation with constant turbine inflow discharge and water level position were used. Shallow water equations are frequently used as mathematical model for water flow in coastal areas, lakes, estuaries, etc. These equations can be obtained by integrating the horizontal momentum equations and the continuity equation over the depth $a(x,y) = h(x,y) - z_b(x,y)$. The result of the integration over depth is (Vreugdenhil, 1994), (MIKE 21, 2000):

$$\frac{\partial p}{\partial x} + \frac{\partial q}{\partial y} = 0 \tag{7}$$

$$\frac{\partial}{\partial x}\left(\frac{p^2}{h}\right) + \frac{\partial}{\partial y}\left(\frac{pq}{h}\right) + ga\frac{\partial h}{\partial x} + c_f\frac{p}{h}\sqrt{\frac{p^2}{h^2} + \frac{q^2}{h^2}} - E\left(\frac{\partial^2 p}{\partial x^2} + \frac{\partial^2 q}{\partial y^2}\right) = 0$$

$$\frac{\partial}{\partial y}\left(\frac{p^2}{h}\right) + \frac{\partial}{\partial x}\left(\frac{pq}{h}\right) + ga\frac{\partial h}{\partial y} + c_f\frac{p}{h}\sqrt{\frac{p^2}{h^2} + \frac{q^2}{h^2}} - E\left(\frac{\partial^2 p}{\partial x^2} + \frac{\partial^2 q}{\partial y^2}\right) = 0$$

where $h(x,y)$ is a position of water level related to a fixed reference level, p is the flux density in the x-direction given by $p = v_{xa} \cdot a$, q is the flux density in the y-direction given by $q = v_{ya} \cdot a$, v_{xa} and v_{ya} are the depth averaged velocity components in x- and y-directions, a is the depth of the flow, g is acceleration due to gravity, c_f is friction coefficient, E is eddy viscosity coefficient and $z_b(x,y)$ bottom level.

3 NUMERICAL SOLUTION

3.1 Three-dimensional (3D) model

The 3D flow equations were solved using the finite element method, for the numerical solution including the post- and pre- processing the ANSYS-FLOTRAN commercial computer code, isoparametric hexahedral 8-nodes FLOTRAN ET 142 finite elements were used.

3.2 Two-dimensional (2D) approximation

3.2.1 2D approximation in 1m thick layer

The 2D flow equations were solved using the finite element method, for the numerical solution including the post- and pre- processing the ANSYS-FLOTRAN commercial computer code, isoparametric quadrilateral 4-nodes FLOTRAN ET 141 finite elements were used.

3.2.2 2D shallow flow (SW) equations

The 2D shallow flow equations were solved using the finite difference scheme, for the numerical solution using curvilinear finite difference mesh the DHI MIKE21 computer code was used.

4 PRACTICAL APPLICATIONS

The numerical analysis was carried out for four small hydropower plants. In all cases the solution was part of the process of designing the inlet part of the hydropower scheme. The shape and dimensions of the flow domain were based on the preliminary design (first estimate) issuing from the engineering approach and experience. The shape of the flow domain were derived from the ground plan, cross and longitudinal sections of the hydropower plant inlet. The upstream boundary was always determined sufficiently far from the hydropower plant, so that the velocity field at the river of approach will not be not influenced by the intake object.

4.1 Hydropower scheme Libcice

Water structure Libcice - Dolany is located on the km 27.20 of the Vltava river approximately 20 km north of Prague. It consists of the movable flap gate weir with three spans and lock chamber located on the right bank navigable canal. The left bank log sluice was liquidated and at its extended space the hydropower structure was designed and is now in operation. At the machine hall two straight-flow Kaplan turbine sets are installed - the runner diameter is $D = 3350$ mm, maximal turbine discharge is 2×80 m³/s, the hydraulic head is varying from 4.25 to 2.5 m, installed output is more than 5 MW.

An extended space on the left bank of the Vltava river, where the hydropower plant is located, was quite narrow (the land owner demands), so the inlet part had to be designed relatively narrow. Therefore the turbine axis span had to be quite small. To avoid high velocities at the inlet sill, it was designed as inclined (see Figure 1). The left boundary of the inlet sill is two times farther from the turbine inlets than the right one; therefore the inlet bed is created by the skew surface. The flow (velocity field) is predominantly three-dimensional. This solution called up the question of the irregular velocity field at the turbine inlets and of the origin of inlet vortexes close to the water level just in front of turbine inlets. To avoid and to minimize these effects, comprehensive experimental and numerical research was carried out at the laboratory of Water Structures Institute at the Brno University of Technology.

The goal of the research was to verify designed shapes of the inlet part and to measure the velocity field at the cross section just behind the turbine inlets (14 m in front of the turbine axis).

The permissible difference (prescribed by the turbine contractor) of the local velocities in particular observation points of measuring cross section from the average velocity was ±10%. Therefore the original design (without inlet piers) was improved by the proposal of baffle piers located at the inlet sill. 9 variants of piers number and shape were analysed both by experimental and numerical modelling. The final ninth variant is shown in Figure 2. Based on results of the research, the turbine inlet was optimised also in terms of re-arranging the upstream nosing of the division wall between the weir and the hydropower plant to avoid vortexes in front of the turbine inlet and to improve the velocity field distribution. The ±10% condition mentioned above had to be fulfilled and moreover the sum of discharge differences at all four quadrants of measuring profile from the regular discharge distribution had to be less than 5%.

For the original and all the other variants solved, three basic operational stages were solved:
1. maximum discharge through both turbines − 2 x 80 m³/s ;
2. flow only through the right turbine with maximum turbine discharge 80 m³/s;
3. flow only through the left turbine with maximum turbine discharge 80 m³/s;
4. flow only through the right turbine with the reduced discharge 60 m³/s;
5. flow only through the left turbine with the reduced discharge 60 m³/s;

For all stages mentioned above, following parameters were observed:
- flow direction at the inlet sill, close to the upstream edge of the partition pillar, guiding vanes, at the space between inlet sill and the turbine inlet;
- forming of the vortexes at the free water level upstream from the turbine inlets;
- flow velocities at selected measuring section downstream of the turbine inlets at the 5 x 5 = 25 points observation net.

At the final ninth variant of the turbine inlet arrangement, following additional flow parameters were observed for all operation stages 1 to 5:
- forming of the vortexes just upstream of the turbine inlets;

- velocity field at the selected measuring section downstream of the turbine inlets at the denser network of 9 x 9 = 81 observation points.

The ground plan scheme of the HPP is shown in Figure 1, the simplified layout of the flow domain for 3D and 2D model is shown in Figure 2.

Figure 1. The ground plan of the Libcice HPP.

For the numerical model, the shape and dimensions of the structure were taken from the project of Libcice hydropower plant. The detailed shape and dimensions of guide vanes and the piers were used in accordance with the experimental model (Hynková et.al 1995). All these data were available as an ACAD files and were transformed to the CFD software.

Boundary conditions were applied as follows. The banks of the Vltava river, the weir flap, partition walls, land piers and guide vanes were assumed to be zero-velocity boundaries. At the inlet cross section on the right side of the flow domain, the known velocity field was introduced. The open turbine inlets were assumed as outlet with zero pressure boundary condition (known water level eventually).

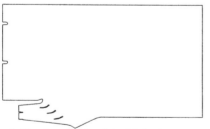

Figure 2. The flow domain of the Libcice HPP with baffle piers.

Model calibration was carried out using the set of velocities measured during the experimental modelling. It can be stated that the sufficient agreement between results of numerical modelling and the values obtained from the experiment was reached. The results of the final and selected discharge scenarios variant can be seen in Figures 3 to 6.

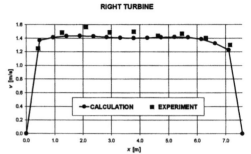

Figure 3. HPP Libcice - comparison of observed and calculated values of total velocities at the inlet of the right turbine when both turbines are opened, total discharge is 160 m³/s.

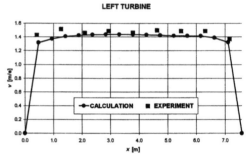

Figure 4. HPP Libcice - comparison of observed and calculated values of total velocities at the inlet of the left turbine when both turbines are opened, total discharge is 160 m³/s.

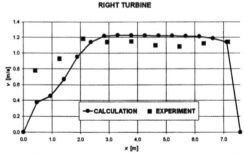

Figure 5. HPP Libcice - comparison of observed and calculated values of total velocities at the inlet of the right turbine when only right turbine is opened, total discharge is 60 m³/s.

Some differences between observed values of velocity vectors and those calculated by 2D numerical model can be explained by the following facts:
- the numerical solution uses (2D) approximation while experiment was performed using the 3D scaled physical model (see Figure 6);
- the numerical solution at the vicinity of division wall between two turbine inlets is influenced by the presumption of sharp edges of the upstream

364

face of the wall, while the experiment used rounded face. These differences are visible especially on the left side of Figure 5 and at the right side of Figure 6 .

LEFT TURBINE

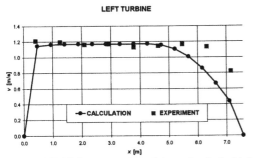

Figure 6. HPP Libcice - comparison of observed and calculated values of velocity distribution at the inlet of the left turbine when only left turbine is opened and total discharge is 60 m³/s.

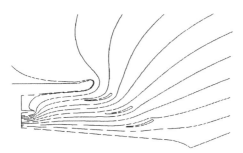

Figure 7. HPP Libcice - flow traces through the inlet part of the left turbine when the only left turbine is opened, total discharge is 60 m³/s - variant with baffle piers.

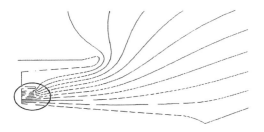

Figure 8. HPP Libcice - flow traces through the inlet part of the left turbine when only left turbine is opened, total discharge is 60 m³/s - variant without baffle piers.

Results of 2D mathematical model are generally corresponding with results of the experimental model, especially in the quality and character of the flow field. Some differences between solutions are

discussed in the previous paragraph. Moreover, the mathematical model enables visualisation of the velocity and flow fields at all critical regions of the entire flow domain for arbitrarily chosen variant of the shape of the flow domain and boundary conditions. Examples of the model output in the form of flow traces and velocity vectors are shown in Figures 7, 8, and 9. The velocities at the turbine inlet cross sections during the flow without baffle piers and with them are compared in Figure 10.

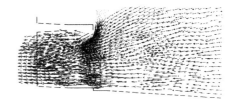

Figure 9. HPP Libcice - velocity field at the inlet part of the left turbine when the only left turbine is opened, total discharge is 60 m³/s - variant without baffle piers.

RIGHT TURBINE

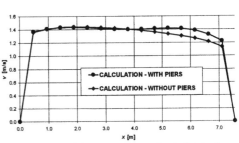

Figure 10. HPP Libcice - comparison of calculated values of total velocity distribution at the inlet of the right turbine during the flow without baffle piers and with them, when both turbines are opened and total discharge is 160 m³/s.

4.2 Hydropower plant Klecany

The water structure Klecany (see Figure 11) is located on the km 36.833 of the Vltava river and consists of the movable flap gate weir, left bank navigation canal with two lock chambers and right bank raft-log sluice. The original structure was built in 1899, the movable weir structure was reconstructed in 1981. In the decade from 1980 to 1990, the log sluice was reconstructed into the small hydropower plant. It consisted of 4 straight-flow Kaplan turbine sets. In 1999 the small hydropower plant was reconstructed, instead of four turbine sets (each 3 m³/s) two turbines with maximal turbine runner 2,300 mm, maximum discharge 2 x 20 m³/s, the total head within the range 1.3 to 2.6 m and total output 1 MW were designed.

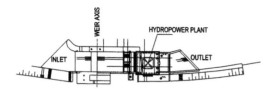

Figure 11. The layout plan of the Klecany HPP.

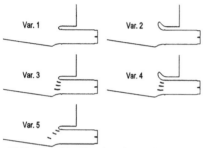

Figure 12. Details of the flow domains at Klecany (5 variants of the inlet shapes of the HPP).

The purpose of the study and numerical modeling was to analyse following inlet shapes (Figure 12):

1 original design (variant 1);
2 improved shape of the upstream nosing of the division wall between the weir and HPP (variant 2);
3 original design completed with 3 baffle piers (variant 3);
4 variant 2 with 3 baffle piers added (variant 4);
5 original design completed with 3 baffle piers with improved shape (variant 5)

Following operation conditions were analyzed:

1 maximum discharge 2 x 20 m³/s through both turbines;
2 discharge 20 m³/s through the right turbine only;
3 discharge 20 m³/s through the left turbine only.

Diagrams of total velocity distribution in horizontal direction at the inlet of the turbines for all variants and maximum discharge through both turbines are shown in Figures 13 and 14. Figures 15 to 17 show comparison of variants 1, 4 and 5 using flow traces when both turbines are opened and total discharge is 2 x 20 m³/s. The best variant seems to be the variant 4, nevertheless the investor chose the variant 5 - the original shape of the upstream nosing of the division wall with 3 baffle piers.

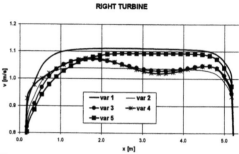

Figure 13. HPP Klecany - the comparison of total velocities at the inlet of the right turbine when both turbines are opened, total discharge is 2 x 20 m³/s.

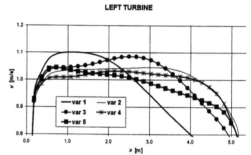

Figure 14. HPP Klecany - the comparison of total velocities at the inlet of the left turbine when both turbines are opened, total discharge is 2 x 20 m³/s.

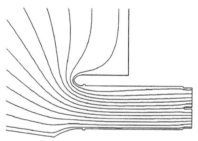

Figure 15. HPP Klecany - variant 1 - flow traces through inlet part of the left turbine when both turbines are opened, total discharge is 2 x 20 m³/s.

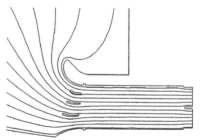

Figure 16. HPP Klecany - variant 4 - flow traces through inlet part of the left turbine when both turbines are opened, total discharge 2 x 20 m³/s.

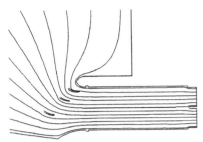

Figure 17. HPP Klecany - variant 5 - flow traces through the inlet part of the left turbine when both turbines are opened, total discharge is 2 x 20 m³/s.

4.3 Hydropower plant Doksany

The hydropower scheme Doksany is located on the km 10.254 of the river Ohre, the right bank tributary of the Elbe river. The hydrotechnical complex consists of the roller-gate weir and short right-bank headrace between the rough trash racks and hydropower house. In the machine hall two straight flow Kaplan turbines with runner diameter D = 1,580 mm, total maximum discharge 30 m³/s, the head varying from 1.5 to 3.5 m and total HPP output 0,8 MW were assumed.

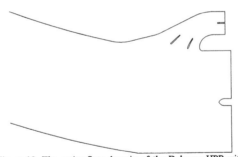

Figure 18. The entire flow domain of the Doksany HPP with baffle piers (Variant 3).

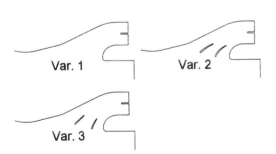

Figure 19. Details of the flow domain (variants of the inlet shapes of the HPP Doksany)

The purpose of the numerical modelling was to assess following inlet shapes (Figure 19):
1 original design (variant 1);
2 two guiding vanes at inlet sill (variant 2);
3 improved shape of guiding vanes (variant 3).

Following operation conditions were analysed:
1 maximum discharge through both turbines: 2 x 15 m³/s;
2 flow only through the right turbine - 15 m³/s;
3 flow only through the left turbine - 15 m³/s.

Results expressed using diagrams of horizontal velocity distribution at the inlet of the turbines for all variants and maximum discharge through both turbines are shown in Figures 20 and 21. Total horizontal velocity distribution for flow (discharge 15 m³/s) through right or left turbine only is shown in Figures 22 and 23. Figures 24 to 26 show comparison of variants 1, 2 and 3 using the flows traces for both turbines opened and total discharge 2 x 15 m³/s. For the final design, variant 3 (inlet part with 2 improved guiding vanes) was chosen as the best solution.

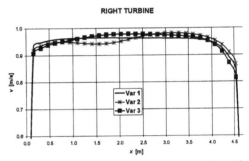

Figure 20. HPP Doksany - comparison of total velocities at the inlet of the right turbine when both turbines are opened, total discharge is 2 x 15 m³/s.

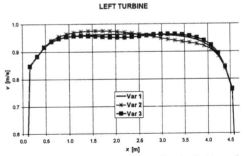

Figure 21. HPP Doksany - comparison of total velocities at the inlet of the left turbine when both turbines are opened, total discharge is 2 x 15 m³/s.

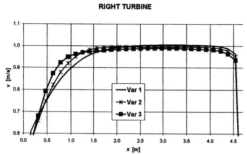

Figure 22. HPP Doksany - comparison of total velocities at the inlet of the right turbine when only right turbine is opened, total discharge is 15 m³/s

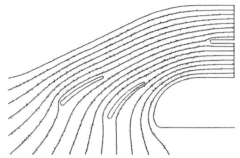

Figure 25. HPP Doksany - flow traces through the inlet part, both turbines are opened, total discharge is 30 m³/s - variant 2.

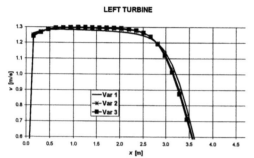

Figure 23. HPP Doksany - comparison of total velocities at the inlet of the right turbine when only right turbine is opened, total discharge is 15 m³/s.

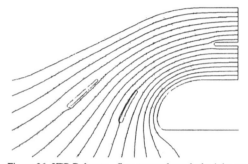

Figure 26. HPP Doksany - flow traces through the inlet part, both turbines are opened, total discharge is 30 m³/s - variant 3.

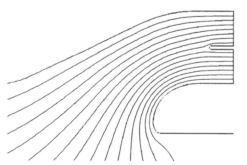

Figure 24. HPP Doksany - flow traces through the inlet part, both turbines are opened, total discharge is 30 m³/s - variant 1.

4.4 Hydropower plant Vranany

The water structure complex in Vranany (Figure 27) is located on the km 11.55 of the Vltava river close to the city of Melnik. It consists of the movable 3-span flap gate weir, raft-log sluice, fish ladder, lock chamber, left bank navigation canal with the lock chamber and right bank raft-log sluice. The designed small hydropower plant is assumed to be located in the old lock-chamber (which is now not used), the inlet canal is created by the improved upstream head of the lock chamber. In the hydropower plant, one straight flow Kaplan turbine with turbine runner D = 3350 mm, maximum discharge Q = 80 m³/s, design head 4.0 m and turbine output 2.7 MW is assumed to be installed.

The purpose of the numerical modelling was to calculate the velocity field at the turbine inlet located at the past navigation lock for the maximum discharge 80 m³/s. In case of the HPP Vranany, the extensive research and comparison of 2D, 2DSW and 3D techniques was carried out. The solution showed (Figures 28 and 29) following results:
- Well developed 3D flow can be observed at the division walls close to the lock chamber heading and just in front of the turbine, where the lock bed declines.
- 2DSW solution fits quite well horizontal velocity distribution both at the lock heading and close the turbine inlet. Along the turbine intake, vertical velocity components are mostly about 100 times

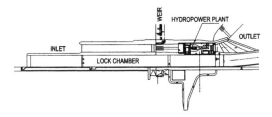

Figure 27. The ground plan of the Vranany HPP.

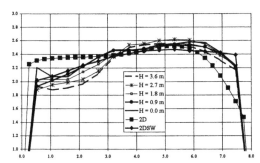

Figure 28. Comparison of total velocities between 2D, 2D SW and 3D calculation at the inlet of the turbine (100 m in front of intake - XS1), total discharge is 80 m³/s.

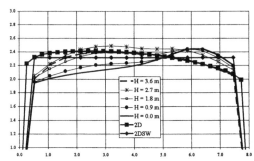

Figure 29. Comparison of total velocities between 2D, 2D SW and 3D calculation at the inlet of the turbine (30 m in front of intake - XS2), total discharge is 80 m³/s.

smaller than horizontal ones, so the 2DSW approximation can give reasonable results. Some inaccuracies occur only close to the "zero flow" boundary at the lock chamber side walls, where the roughness within 2DSW can hardly be applied.
- 2D approximation gives reasonable quantitative results and agreement with 3D model only when the inlet bottom does not vary too much. In case of non-uniform inlet bottom, results of the solution give only qualitative view on the flow field. It can show critical regions where vortices and dead zones could occur, without closer information about the size of flow velocity.

- Both 2D and 2DSW solutions give no information about differences between horizontal velocity components along the vertical (see Figure 29).
- Calculated values of total velocities along the intake part (past navigation lock chamber) show that the length of the intake part is sufficiently long, i.e. the velocity distribution is quite regular (Figures 28, 29 and 30). Along the regular, about 100 m long inlet part, all approximations give very similar results due to approximately constant water depth.

Figure 28 shows velocity distribution at the section located about 100 m in front of the turbine just behind the nose of the division wall where the bottom of the chamber is approximately horizontal. Figure 30 illustrates flow traces through the flow domain solved by 2D model with two analysed cross sections XS1 and XS2.

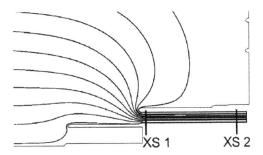

Figure 30 Flow traces through the flow domain, total discharge is 80 m³/s - 2D calculation.

5 CONCLUSIONS

The 3D modeling of turbulent flow gives the best results when compared with 2DSW or 2D model. However the flow domain is sometimes (in case of baffle piers) rather difficult to model and demanding in terms of number of elements in the mesh and computing time. These factors can influence the cost of the research dramatically and in many cases are limiting, especially when solving extensive number of variants of the shape of the inlet combined with various discharge scenarios. As the simplified 2D or 2DSW domains can relatively easily be defined and analyzed using reasonable number of elements, it can be recommended to solve preliminary variants using 2D or 2DSW model. The final variant should be verified using 3D numerical model. In case of bigger hydropower schemes, it is suitable to verify and compare results of numerical modeling by experimental modeling. Another possible approach is to use 3D approximation only in front of the turbine and at another "critical places" and for the remaining flow domain to use 2D or 2D SW approximation. The solution is always site specific and general recommendation dealing with flow domain extent can hardly be given. The sensitivity analysis can strongly be recommended, especially in terms of the flow domain extent selection and computation mesh density design.

6 ACKNOWLEDGEMENT

The research was supported by the research project of the Brno University of Technology MSM 261100006.

REFERENCES

Hynkova, E. & Riha, J. & Jandora, J. & Neumayer, O. 1997. Experimental and Mathematical Modelling of the Flow at the Intake Part of the Hydropower Plant Libcice at Vltava river. International The Wasserbaukolloquium, Dresden 1997.

Hynková, E. et al. 1995. Experimental research of hydropower scheme Libcice - Dolany, Final report, TU of Brno, 1995, In Czech (unpubl.).

Rodi, W. 1980. Turbulence models and their application in hydraulics, International Association for hydraulic research, state-of-the-art paper, Delft, 1980

Launder, B. E. & Spalding, D. B. 1974. The numerical computation of turbulent flows, Computer methods in applied mechanics and engineering 3, 1974, pp. 269-289, North-Holland Publishing Company

MIKE 21. 2000. User Guide, DHI Water & Environment, 2000

Vreugdenhil, C. B. 1994. Numerical methods for shallow water flow, Kluwer Academic Publishers, 1994

Wilcox, D. C. 1994. Turbulence Modelling for CFD, DCW Industries, Inc., 1994

Hydropower in the New Millennium, Honningsvåg et al (eds), © 2001 Taylor & Francis, ISBN 90 5809 195 3

Foundation work at the Penstock Finchaa Power Plant – Ethiopia

Kolbjørn Dønåsen
Senior Hydropower Engineer, MRIF - MNITO, Norconsult AS, Sandvika, Norway

ABSTRACT: Finchaa Hydro Power Plant - Ethiopia - was constructed in 1970 with 100 MW installed capacity. The headrace consists of a tunnel 4,5 km upstream the Surge Chamber and 1450 meter Penstock of which the last 900 meter to the Powerhouse is founded on soil sloping to the Powerhouse.
During a number of years the Penstock showed signs of movement and increased stress, which at the end was considered entering close to the tension strength limit of the Penstock. A test report demonstrated that this was the actual situation and forecast a complete breakdown of the Penstock during the coming few years. The reason for movement of the ground was assumed to be a general creep in the ground on which the penstock was founded. The rehabilitation work consisted of installation of a Double Hinge Joint and installation of prestressed rock anchors.

1 EVALUATION OF PENSTOCK SAFETY

The history of evaluation of the Penstock Safety and remedial measures is long and has involved a number of recognized consulting companies from Europe and America. In 1995 the feasibility studies and conclude on remedial measures. The same group of engineers should also prepare tender documents and invite for international bidding of the construction work.

1.1 *Displacement measurements*

A database has been established by NGI for the displacement measurements of the Penstock. The database includes the monitoring of the Penstock welds and anchorblock reference points since January 1994. The database includes additional measurements of installed inclinometers and load cells from the commissioning after the remedial work was completed at the end of March 1999.

The true maximum heave was close to 180 mm between Anchor Block no 1 and 2. The maximum heave at Anchor Block 1 was about 125 mm. The annual heave and settlement of the Penstock due to swelling and shrinkage in the ground are about +/- 20 mm.

1.2 *Stability analyses*

Two programmes of soil testing were carried out. The purpose of these programmes has been to check and verify the conditions of the slope colluvium underneath Anchor Block 1. One programme included identification tests, oedometer and compaction tests. The other program included also clay mineral tests (XRD) and simple shear strength tests. The result showed that the amount of swelling minerals is limited to 5 - 10 % and the swelling potential is limited. The shear strength of the soil is considerably less than anticipated by the consultant for the design of the powerhouse slopes.

It was concluded that the heaving of the Penstock hardly can be driven by a swelling mechanism. Analyses of movements has, however revealed that there are seasonal movements that may be caused by swelling. Stability analyses of the slope using the shear strength obtained from simple shear tests show that the factor of safety against sliding may be as low as F = 1.1. Together with the fact that slides have occurred immediately north of the Penstock this leads to the conclusion that Anchorblock 1 is founded on a slope with poor stability. The movement of the Penstock and specially of Anchorblock 1 is due to creep movements that occur in plastic soils when the shear stress is approaching the strength. It was assumed that the movement takes place along the shear zone roughly parallel with the original slope. Measurements from the installed inclinometers indicate however that the creep takes place along the zone between soil and rock surface. Stopping the movement can only be achieved by improving the stability of the slope, i.e. reducing the shear stresses and / or increasing the shear strength.

1.3 Evaluation of surface penstock

With assistance from DNV (Det Norske Veritas) the risk for rupture was evaluated. The conclusion was that there were no immediate danger but this was based on a high degree of uncertainty. Under certain conditions it was assumed that with continued displacements at present rate and critical stress calculated to 587 MPa, the working stresses might reach the critical stress before year 2000.

1.4 Evaluation of the embedded penstock

Even though the upper embedded penstock has survived two short dewaterings, the safety margin against buckling is dubious. There is real risk that a dewatering may create a bulging of the lining due to over pressure in the surrounding rock. Consequently temporary measures had to be taken during the dewatering stage.

This stress change sign when gate movement change direction, exposing the transition zone to fatigue.

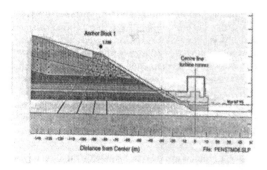

2 THE PROPOSED REMEDIAL MEASURES

2.1 Earth and soft rock excavation

The crest of the slope around Anchor Block 1 was to be trimmed back to lines and slopes aiming to reduce weight and by this increase stability in the Anchor Block area. About 8000 m^3 was excavated and removed from the area and the upper side of the excavated area was supported by a new retaining wall. Boulders of sandstone and occasional shale were broken and removed without blasting. Landscaping was arranged also to allow transport of steel parts

and general access to the different work areas along the Penstock.

2.2 Slope protection

The slope surface along the Penstock was to be protected against rain and erosion. Underneath and immediately to the sides of the Penstock the surface was to be protected by a concrete slab. Elsewhere the slopes were to be protected by turfing. The slab was reinforced and provided with expansion joints every 10 meter.

Stone pitched drainage was arranged on top of the new retaining wall and existing drain trenches were re-established and connected to the old drains.

2.3 Strengthening of anchor block 1

Anchor Block no 1 should be strengthened by grouting and new reinforced concrete structures as the concrete of the Anchor Block showed signs of distress. The concrete was cracked in places e.g. at the downstream end. Construction joints had calcite precipitation, and to secure integrity of the concrete the cracks should be grouted by using a polymer (epoxy) grout.

The new structures were formed as two beams, one on each side, heavily reinforced and suitable to transfer load from the prestressed rock anchors evenly to the entire Anchor Block structure. The integration with the old structure was secured by horizontal prestressed tendons.

2.4 Prestressed ground anchors

Initial design philosophy was to install prestressed ground anchors to increase the effective stresses and thereby the shear capacity of the ground and, if possible, to relocate the Anchor Block to it's original position. The anchors were installed while the Power Plant was in operation and the pre-stressing took place after cutting the Penstock for installation of the Double Hinge Joint and the Bifurcation for the 4th unit in the Powerhouse. The 4th unit is being constructed under a different contract so design of the Bifurcation was done jointly with the other project. Total capacity of the Ground Anchors installed at Anchor Block 1 was 20.000 kN with 4 anchors on each side.

Also the foundation for the Bifurcation and the intermediate foundation for a saddle support was supported by ground anchors.

2.5 Deep drains

In the project area deep drains were installed with the purpose to maintain a low ground water table to increase the general stability of the slopes.

2.6 *Structural steel works*

Relocation of the anchor block

Difficulties were experienced in trying to regain completely the original position. Due to the heave built up over years and the stress the two ends of the cut Penstock did not meet and realignment was necessary Two additional cuts of the Penstock were planned before the stop of Power Plant operation which also proved to be necessary to be able to line up the Penstock for installation of the Double Hinge Joint.

Manhole

A manhole was installed upstream of the Double Hinge Joint to allow access to the inside of the Penstock in this area.

Bifurcation

For practical reasons it was decided to install a Bifurcation to the Penstock under this contract to allow connection of a 4th unit with minimal stop of production when this is to be commissioned.

Drain of embedded part of the Penstock.

Due to problems during the construction of the Power Plant with partial collapse of the embedded part of the Penstock lining, it was assumed highly risky to empty the Penstock without a drain system and monitoring of outside pressure of the embedded part of the Penstock downstream the surge tank. A number of holes were drilled and installation of drain arrangement was prepared but proved to be unnecessary as no outside pressure was detected.

Double hinge joint

The installation of the Double Hinge Joint, which is designed to allow angular movement in both ends, was aiming to allow continued movement in the ground without increasing the stress in the Penstock during the coming years. The angular movement allows heave of 200 mm of the Penstock before a relocation of its foundation is necessary.

The pipe upstream of the Double Hinge Joint is supported by a saddle support designed for easy adjustment of its height in case future movement will necessitate such adjustment.

2.7 *Instrumentation*

The area is closely monitored by a levelling instrument to regularly follow the annual and seasonal

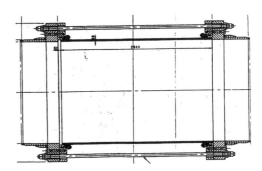

heave of the pipe and the anchor blocks. Ground water level is also regularly monitored.

During construction work 4 of the installed ground anchors in Anchor Block no 1 was equipped with load cells to monitor the prestressed load in the anchors. 3 inclinometers were installed in the area to get further information about the stability and the general creep of the area to compare with the assumptions made during the design stage. Results from the inclinometer readings have confirmed a creep movement between the underlying rock surface and the clay layers immediately above.

3 ADMINISTRATIVE ARRANGEMENT

3.1 *Project management*

The Owner of the Power Plant Ethiopian Electric Power Corporation (EEPCO), organized the Project Management by use of its own staff. The study, design and construction work was funded by the Norwegian funding agency NORAD (Norwegian Agency for Development Cooperation), who assigned NGI (Norwegian Geotechnical Institute) in cooperation with Norconsult International (Civil work and Resident Engineer), Norplan and Statkraft Engineering (Steel works) to finalize the studies and prepare Tender Documents. The Construction Contracts were based on the FIDIC Conditions of Contract and divided in two; the Civil Work and the Steel Work.

3.2 *Construction contracts*

Civil works

The Civil Work Contract was entered into with the Norwegian contractor NOREMCO with base in Tanzania. A project office was established in Ethiopia and the earth and concrete work was done by a local subcontractor Varnero. The ground anchors, the deep drains and injection work were done by the Norwegian subcontractor Entreprenørservice. The work was done during a period of about 10 months after signing of the contract.

Steel works

The Steel Work Contract was entered into with the Austrian Company Voest Alpine, at the same time as the Civil Work Contract. After initial preparations on site the main part of the work had to be done during the planned stop period of 28 days. After very hectic activity involving the Employer's staff, the Civil Work Contractor and first of all the Steel Work Contractor's staff, the Power Plant was put in operation again after 26 days and nights of work.

The high tensile forged steel STE 500 has a thickness of 27 mm at the place of installation and altogether 42 meter of the steel pipe wall was cut and welded, without one single fault during this period, in addition to local transport installation, drain and support work.

4 CONTINUED MONITORING OF THE PENSTOCK

The monitoring program is decided to continue particularly to achieve data for the period immediately after construction work, but also in a long term program for future altering of the foundation of the Penstock if the heave of the part of the Penstock upstream the Double Hinge Joint exceeds the design criteria for the Double Hinge Joint.

It should be commented that the life time of the installed inclinometers is considered to be short while the information gained from the readings has already given significant information about what is presently happening to the slope.

Hydropower in the New Millennium, Honningsvåg et al (eds), © 2001 Taylor & Francis, ISBN 90 5809 195 3

Strain gauge measurements of friction on radial dam gate bearings

Henning Føsker, Halvard Bjørndal & Terje Ellefsrød
Norconsult AS, Sandvika, Norway

Kjell Knutsen
Sira-Kvina kraftselskap, Tonstad, Norway

ABSTRACT: At least five known radial gate failures are caused by trunnion bearing seizure. The radial gate arms are normally designed to withstand bending moments from nominal friction on the bearings. Experience has shown that lack of lubrication and years of deterioration leads to increased friction and even seizure of the bearings. The bending moments produced by bearing friction imposed on the gate arms are beyond the moment capacity, resulting in collapse. Norconsult has developed a method for measuring trunnion bearing friction while the gates are in service. By means of strain gauges attached to the gate arms, the strain caused by the bearing friction is measured during gate movement. The signals from the strain gauges and a gate position transmitter are logged digitally, allowing direct presentation of test results. The method gives objective and precise verification of measuring results, and the risk of subjective and wrong assessments is minimized.

1 BACKGROUND

1.1 *Introduction*

Increased friction and seizure of bearings on dam gates is a growing concern. At least five radial gate breakages in Norway and abroad are caused by bearing failure. Consequently, there is uncertainty among dam owners regarding the state of the bearings on radial gates after several years of service under severe conditions and sparse maintenance.

1.2 *Deterioration of bearings*

Early dam gates (before 1980) are most commonly designed with trunnion bearings using a carbon steel shaft and bronze radial and axial bearings. The shafts are often chromium plated, and the bearings are usually lubricated by a manual grease lubrication system.

Dam gate bearings are static bearings with small rotation and relatively few movements during operation. This operation mode is not suitable for a bearing requiring external lubrication as the lubricant is not distributed on the high pressure sector of the bearing.

When injected to the radial bearing, the lubricant often find its way to the low pressure side, leaving the high pressure load transformation sector unlubricated. The axial bearings are often connected to the same lubricant pipe as the radial bearing. This gives an evenly distributed load, and the grease escapes through the low pressure side of the radial bearing.

The bearing material and opposing surface often represent an unfavorable material combination becoming corroded and contaminated. This is probably the most important factor in deterioration of bearings. When water is present and an electrolyte can occur with the carbon steel shaft as the anode. This leads to corrosion on the bearing surface of the shaft and increased roughness from pitting and corrosion products. Tight bearing radial clearance worsens this effect.

1.3 *Damage mechanisms and consequences*

Radial gate arms are often slender steel structure dimensioned to withstand bending and buckling from the water pressure. Friction forces are often not taken into consideration, or nominal values of friction coefficient are used. Normally a lubricated bronze bearing and carbon steel shaft has a nominal friction coefficient of $\mu = 0.2 - 0.3$

Deterioration of the bearings leads to increased friction, for which the arms are not dimensioned. When the increased friction causes breakage, this is due to both fatigue and instantaneous breakage. In particular, cylindrical bearings with a manual lubrication system combined with an automatic or remotely operated gate are a vulnerable configuration.

All radial gate collapses lead to economic loss due to loss of power production. Excessive discharge results in downstream flooding and lowering of the reservoir, affecting boat traffic and recreational use of the water way.

1.4 Arm design

Exposure to failure due to increased bearing friction depends on the design of the arms. Areas of concern are the transition zones between the trussed beams constituting the arms and the hub accommodating the bearings. As an example we have analyzed the effect of a friction coefficient of $\mu = 0.5$ on the Lundevann dam radial gate arm. At full water head it imposes a bending moment resulting in a bending stress of 7MPa in the large cross-section hub. The bending moment is transferred to the more slender trussed beams, constituting the gate arms. In the transition zone between hub and arms, the stress concentration is especially high in this example. A friction coefficient of $\mu = 0.5$ results in a bending stress in the transition zone of 100MPa as shown in Figure 1. This stress change goes from plus to minus when the gate movement changes direction, exposing the transition zone to fatigue.

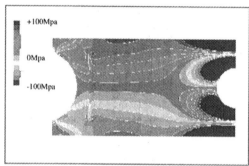

Figure 1. Lundevann dam radial gate arm, hub section. Stress resulting from friction coefficient, $\mu_{\text{BRONZE-STEEL}} = 0.5$

1.5 Trunnion design

Trunnion designs vary with regards to vulnerability for bearing seizure. For the common cantilevered trunnion shaft anchored to the abutment, symptoms of bearing seizure are difficult or impossible to detect by visual inspection. High friction moments can be transferred through the shaft without visible indications of increased bearing friction before the collapse occurs.

The trunnions may include a shaft lock allowing rotation of the shaft in the console, causing the shaft lock to be deformed after a bearing seizure.

2 EVALUATION OF METHODS

2.1 Inspection and maintenance

Inspection and maintenance of gate trunnion bearings often implies dewatering of the dam gate and dismantling of the bearing. Dewatering of the gate can be accomplished by lowering the reservoir. This is, however, both time-consuming and costly if pro-duction is lost. More common is dewatering by installation of stop logs in front of the gate. This is less expensive than lowering the reservoir, but still more expensive than other methods. Another disadvantage is that during the de-commissioning period the discharge capacity of the gate is not available. Dismantling of radial gate trunnion bearings is subsequently not routinely carried out to determine the condition of the bearing. Dewatering and dismantling of the bearing is carried out when and if the symptoms of bearing seizure are evident. On some radial gates total collapse has occurred without prior symptoms.

2.2 Detection of change in maneuvering forces

In theory an increase in lifting force could be detected when bearing friction increases. However, the bearing friction constitutes only a small fraction of the total lifting force. On a typical radial gate with upstream lifting chains, the total lifting force of 156kN is dominated by the gates own weight of 126kN (81%). The remaining 30kN (19%) is friction in rubber seals and trunnion bearings. To separate the two friction forces, the friction on the seals must be based on an assumption regarding the friction coefficient between rubber seals and the embedded stainless steel frame. Assuming $\mu_{\text{RUBBER-STEEL}} = 0.9$, the friction on the seals constitutes 20kN (13%) of the lifting force. The remaining, 10kN (6%) is the nominal bearing friction based on $\mu_{\text{BRONZE-STEEL}} = 0.2$. A variance of this relatively small force, compared to the total lifting force, is complicated to detect and is unreliable since the method is based on assumptions.

2.3 Detecting bending moments in radial gate arms

As a part of a safety revaluation program for dams, a diagnostic technique has been developed for radial gate bearings without dewatering the gate or dismantling the bearings. The method detects friction on the bearings during operation of the gate.

2.4 Measurement technique

The forces acting on the trunnion bearing can be divided into two classes:
□ Perpendicular forces on the bearing surfaces is the result of water pressure, gate weight, operating forces and friction forces from the rubber seals. The resulting force from the water is the dominating the force on the trunnion.
□ Shear forces parallel to the bearing surface due to friction. Without friction ($\mu = 0$), no shear forces will appear. These forces will give bending moments in the gate arms.

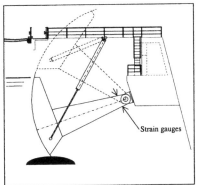

Figure 2. Strain gauges attached to the radial gate arms, close to the bearing.

The shear forces on the bearings can not be measured directly without modifying the gate structure. Instead the mechanical stress variation is measured in the gate arm, near the trunnion bearing, using strain gauges. One strain gauge is attached at the upper side of the gate arm, and one is attached at the lower side of the gate arm as indicated in Figure 2. Each strain gauge measures the surface mechanical stress in parallel to the main stress direction in the gate arm.

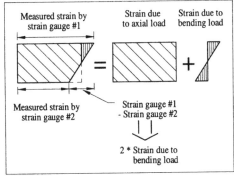

Figure 3. Stress measured by strain gauges.

Superimposed axial and bending forces cause the measured stress indicated in Figure 3. By connecting the strain gauges in a Wheatstone half bridge circuit the two measured stresses are subtracted and the output only indicates active bending moment in the gate arm (ref /1/). By using this circuit, superimposed axial (normal) strain is compensated, and the strain caused by thermal changes is compensated to a high degree.

Figure 4 shows the measured stress in the right gate arm at the Lundevann dam with the original bearing thoroughly lubricated. The gate was opened, almost closed, opened and finally closed. The vertical gate

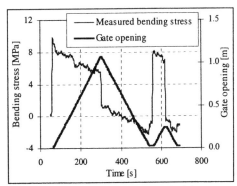

Figure 4. Lundevann dam, measured stress in right gate arm.

opening was measured simultaneously as shown in the figure.

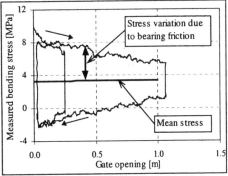

Figure 5. Mean stress variation.

The measured stress caused by bending is shown at the Y-axis and the vertical gate opening at the X-axis, see Figure 5. The figure shows that the measured stress creates a hysteresis curve superimposed on a mean stress variation. The variation in mean bending stress is mainly caused by variation in how the flow forces act on the open gate. These forces are dependent only on the gate opening, not the direction of the movement. The variation of these forces are not used in the analysis and may be ignored. The friction forces in the bearing always act against the direction of the gate motion. When the motion changes direction the measured bending stress, due to friction forces in the bearing, also changes direction. In addition the friction force varies with the gate opening, as the water load on the gate alters.

Dependent on the hoist design, the friction forces from the gate rubber seal have some influence on the total measured friction force. Upstream chain hoist has no influence. In this actual design the rubber seal friction gives less than 5% effect on the measured bending stress when we assume $\mu_{RUBBER-STEEL} = 0.9$

(worst case). This estimation is compensated in the calculations.

The friction torque at the bearing trunnion is calculated based on the measured bending stress due to friction. This calculation is done as a function of moment of inertia in the gate arm at strain gauge location and radial distance from the strain gauges to the bearing centre line.

The water load on the gate is calculated as load from the static water pressure at the wet surface of the gate. The accuracy of this calculated load is high when the gate is closed. At increased gate opening, larger areas are exposed to flow velocity, which decrease the static pressure. Our method will overestimate the load when the gate opening increase, but according to pressure distribution calculation on a gate (ref /2/) this error will be limited to approximately 15 % at 25% gate opening. This calculation gives the main input to the radial load at the bearings transmitted through the gate arms.

Dependent on the design, the gate arm axial force is split into two components, one acting on the axial bearings and the second on the radial bearings. This split is dependent on the gate design. When both radial load and friction torque on the trunnion bearings are known, we can calculate the average friction coefficient on the bearings. Figure 6 shows the friction coefficient dependent on the gate opening and direction of motion.

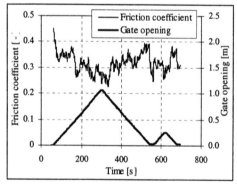

Figure 6. Calculated friction coefficient

3 MEASUREMENTS ACCOMPLISHED

3.1 General

Since the method was developed in 1998, a total of 30 bearings on 15 radial gates in six different dams have been measured. On one of the dams where increased bearing friction was detected, Lundevann Dam (owner: Sira-Kvina Kraftselskap), repeated measurements were carried out .

3.2 Presentation of results

Table 1. Key data of gates measured and results

W x H m	Bearing Shaft	Lubr. system	Friction coefficient		Prior lubri-cation
			Nominal	Measured Left / Right	
15 x 5.2	B/S	Grease	0.2	0.08 / 0.09	Yes
8.0 x 2.8	B/SS	Selflub.	0.1	0.35./.0.25	N/A
20 x 8	B/PS	Selflub.	0.1	0.15 / 0.10	N/A
13 x 4	B/S	Grease	0.2	0.75 / 0.50	No
13 x 4	B/S	Grease	0.2	0.45 / 0.45	No
11.5 x 5	B/S	Grease	0.2	0.15 / 0.14	Yes
11.5 x 5	B/S	Grease	0.2	0.14 / 0.13	Yes
11.5 x 5	B/S	Grease	0.2	0.12 / 0.09	Yes
13 x 4	B/S	Grease	0.2	0.55 / 0.40	Yes
13 x 4	B/S	Grease	0.2	0.22 / 0.35	Yes
12 x 6.3	DU/PS	Selflub.	0.15	0.18 / 0.20	N/A
12 x 6.3	DU/PS	Selflub.	0.15	0.17 / 0.19	N/A
17 x 5	B*/PS	Selflub.	0.1	0.71 / 0.58	N/A
17 x 5	B/PS	Selflub.	0.1	0.43 / 0.42	N/A
13 x 4	C/SS	Selflub.	0.15	0.09 / 0.13	N/A

B = Bronze, S = Carbon steel, PS = Chrome plated carbon steel, SS = Stainless steel, DU = Glacier DU bushing, C = composite. * Oiles 500 w/SL4 lubricant plugs.

3.3 Lundevann Dam

The Lundevann Dam is the reservoir for the Åna-Sira hydropower plant. In the outlet of Lundevann a concrete dam was constructed with two spillways and surface radial gates.

Table 2. Key data for the radial gates on the Lundevann Dam

B x H	: 13.0m x 4.2m
Reservoir level	: El. 48.5
Bottom sill	: El. 44.5
Gate radius	: 7000mm
Old radial bearing diameter	: 250mm
Old axial bearing diameter	: 370mm
Old bearing material:	: Cast steel shaft/bronze bush
New radial bearing diameter	: 216mm
New axial bearing diameter	: 350mm
New bearing material	: Stainless steel/Orkot**

** Trade Mark from Busak + Shamban, non-metallic self-lubricating bearing material

The measurements were carried out on the bearings under three different conditions:
1 Original bearing without prior lubrication
2 Original bearing thoroughly lubricated
3 New bearing with self-lubricating bushing

3.3.1 Measuring the original bearings

The first measurements on the four bearings, two on each gate, were carried out in 1998. Prior to the measurements, the gates had been stationary for one year without maintenance or lubrication. The measurements revealed increased friction values in all bearings. The friction on the bearings of the most frequently used gate No. 1 was especially high. On the first opening movement, the friction coefficient was 0.75, dropping to left 0.60/ right 0.50 in the second opening. The coefficient of friction remained considerably higher than expected for the material properties of the bearings.

In 1999, we repeated the measurements after the bearings had been thoroughly lubricated and moved repeatedly to distribute the grease. The friction coefficient then fell by 25% to 0.45 and 0.35 on the same bearings. Our conclusion was that the bearings had suffered permanent damage and we recommended dewatering and replacement.

3.3.2 *Inspection and replacement of the bearings*

In 2000, the gates were dewatered and the bearings were dismantled. The inspection revealed traces of seizure on the bronze surfaces and seizure and corrosion on the surfaces of the cast carbon steel shaft. It was also evident that the manual lubrication system had not functioned as intended. Partly due to clogged grease channels and partly due to the fact that the grease escapes from the bearing on the low pressure side.

The bearings were completely replaced. New shafts were manufactured in stainless steel, SIS 2387. For radial and axial bearing, a non-metallic material was chosen. Orkot® a brand name from Busak + Shamban. Orkot® TLM Marine grades are non-asbestos composite materials incorporating woven fabric reinforcement and solid lubricants within a thermosetting resin matrix. The manufacturer gives a friction coefficient between 0.10 and 0.15 against stainless steel.

3.3.3 *Measuring the replaced bearings*

After assembly and commissioning of the new bearings, the friction on bearings on gate no. 1 was measured. The result shows that the friction coefficients are within the values given by the manufacturer. The measured friction coefficient is 0.09 and 0.13 for the left and right bearings respectively. The results from the measurements of the new bearings with the known properties and characteristics of the bearing materials is also a verification of the diagnostic method.

3.3.4 *Comparison of the measurements*

Figure 7 shows the results from three of the measurements of the right bearing on gate No. 1 on Lundevann dam.

The initial (static) and dynamic friction of the original unlubricated bearings is almost constant. When lubricated the original bearing has almost the same static friction as unlubricated, while the dynamic friction drops by 25%.

The uneven shape of the graph for the original bearing indicates increased roughness of the bearing

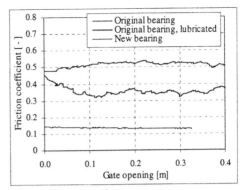

Figure 7. Comparison of measurements

surfaces, while the even shape of the graph for the new bearing indicates smooth bearing surfaces.

4 CONCLUSION

The method of measuring bearing friction by means of strain gauges provides dam owners with a better diagnostic technique. Experience has shown that the method will detect bearing failure at an early stage before the friction moment exceeds the gate arms bending moment capacity. The method is characterized by high reliability and accuracy. The influence from dynamic conditions and the friction on the seals is insignificant. Measurements are carried out without dewatering or decommissioning the gates. The method gives objective and precise verification of measuring results, minimizing the possibility of subjective and wrong assessments.

REFERENCES

Ref /1/ Hoffmann, Karl. *An Introduction to Measurements using Strain Gauges.* 1989, Darmstadt: Hottinger Baldwin Messtechnik GmBh.

Ref /2/ Wickert, G and Scmausser, G. 1971. *Stahlwasserbau.* Berlin: Springer Verlag.

Hydropower in the New Millennium, Honningsvåg et al (eds), © 2001 Taylor & Francis, ISBN 90 5809 195 3

Design of unlined headrace tunnel with 846 m head at Lower Kihansi, Tanzania. Filling Experience

A.Halvorsen & J.A.Roti
NORPLAN AS, Natural Resources Division, Norway

ABSTRACT: : Design of the 2200 m low-gradient, unlined headrace tunnel with a maximum water head of 846 m was critical in regard to ground conditions. Rock stress testing in access tunnel during excavation revealed unsatisfactory stresses assumed caused by extensional tectonism. Directed by an extensive test program, the power house complex was relocated about 700 m to increase overburden and improve rock stresses. A comprehensive program of pre-grouting in the headrace tunnel face was successful for the reduction of permeability and thereby loss of water. Instrumentation for monitoring of the response of the rock mass around the headrace tunnel during filling was mainly based on pore pressure measurement in strategically placed bore holes in the area of the transition between the unlined headrace tunnel and the steel-lined penstock. A period of 4 weeks was used for the careful filling of the tunnel. Initial total leakage out of the pressurized tunnel were 390 l/s, which is reduced to 215 l/s, i.e. in average 0.08 l/s per meter shaft + headrace tunnel 15 months after filling. The outflow rate is still decreasing.

1 PROJECT DEVELOPMENT

Lower Kihansi Hydropower Project in Tanzania is located in the Rufiji basin, some 550 km southwest of the capital Dar-es-Salaam. It is owned by the Tanzania Electric Supply Company (TANESCO).

The project was conceived in the *Rufiji Basin Hydropower Study* of 1984, which suggested that the Kihansi river was the most favorable for future hydro power development in the basin. A feasibility study published in 1990 demonstrated the technical feasibility and economic viability of the project. Based on the study, the World Bank and TANESCO subsequently agreed to incorporate the implementation of the Lower Kihansi project into a sector loan package - the Power VI Project Program.

In December 1991, NORPLAN was selected to carry out a feasibility review of the Lower Kihansi scheme, and if this proved positive, to continue with the final design. Tendering and supervision of construction followed as an extension to NORPLAN's initial contract.

Feasibility review and field investigations were conducted in 1992, followed by the detailed design and preparation of Tender Documents. Considering the ground conditions, NORPLAN found that a deeply sited tunnel system would be advantageous, compared to the shallow tunnel system previously proposed. The design suggested by NORPLAN included an unlined headrace tunnel with maximum head of close to 850 m and underground power house.

The actual construction phase started in July 1994, with the mobilisation of the Chinese contractor SIETCO, who won the bid for preparatory works. In July 1995 the Italian contractor Impregilo SPA, who won the bid for main civil works, mobilized for the underground works. Construction was successfully completed in February 2000, within budget and time schedule.

At present, the power plant at Kihansi contributes with in order of 40% of the total electricity production in Tanzania

2 DESCRIPTION OF PROJECT

Lower Kihansi Hydropower Project includes a 25 m high concrete gravity dam which impounds a small reservoir with a total storage of 1.6M m³. The intake connects to the headrace tunnel via a circular unlined vertical headrace shaft (25 m²), some 500 m deep. The unlined headrace tunnel slopes at a gradient of 1:7. At the downstream end are the stonetrap and the transition section to the steel penstocks. The tunnel is 2200 m long and has a cross-section of 30 m², except in the last downstream 600 m where its cross-section is 37.5 m² to allow for lining, if necessary, in the zone of the highest pressure. No surge shaft or chamber had to be provided due to the relatively short tunnel length and the low water velocities, combined with the use of Pelton turbines.

The tailrace tunnel has a length of 2100 m and a cross-section of 34 m². It slopes gently downstream at an inclination of 1:900 and connects to an 800 m open-cut canal that evacuates the water into the existing Kilombero river system.

The power house cavern is excavated deep into the mountain massif and is 12.6 m wide, 98 m long and 32 m high, with space for possible future installation of 2 additional turbines. The power house is connected to the outside by a 1900 m access tunnel (40 m² in cross-section) and also by a separate cable tunnel which was chosen as an extra security measure for the 220 kV cables passing from the underground power house to the outdoor switchyard.

An overview of the project is shown in Figure 1, showing the waterway as plan and as longitudinal section. A comprehensive description of the project is given by Saidi, F. X., Lindemark, J. and Wilhelm, V. C. (2000).

3 DESIGN OF THE HEADRACE SYSTEM DEPENDENT ON SUFFICIENT ROCK STRESSES

A head of 846 m could be achieved with the location of the tunnel in the escarpment of the Udzungwa mountain range. This escarpment, belonging to the eastern branch of the east African rift system, is formed by large scale block faulting. The rocks are mainly competent gneisses of the Pan African Mozambique Belt, subjected to high grade metamorphism. The degree of faulting and jointing is in general moderate to low. Most pronounced is a joint/fault system oriented perpendicular to the tunnel system with partly high permeability.

An unlined, low-gradient water tunnel with as high head as 846 m head had not been constructed before, anywhere in the world. In 2 other hydropower plants, Tjodan and Nyset-Steggje in Norway higher heads were obtained in 45° inclined shafts, 875 m and 964 m respectively. The design at Kihansi is more vulnerable in regard to leakages, as the high head section in a tunnel with inclination 1:7 is longer than in a 45° shaft, shown schematically in Figure 2.

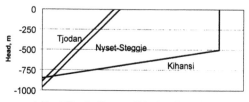

1. Nyset-Steggje, Norway - 964 m head in 45 degree shaft
2. Tjodan, Norway - 875 m head in 45 degree shaft.
3. Kihansi, Tanzania - 846 m head in 1:7 tunnel + vertical shaft

Figure 2: Highest water heads in unlined tunnels / shafts.

With a concept as chosen at Kihansi, detailed knowledge about the ground conditions and adaptation of the design to these are a condition for a successful construction. An important question for the designers was if rock stresses were sufficient. With too low stresses, the confinement of the tunnel could be insufficient, resulting in hydraulic failure. The

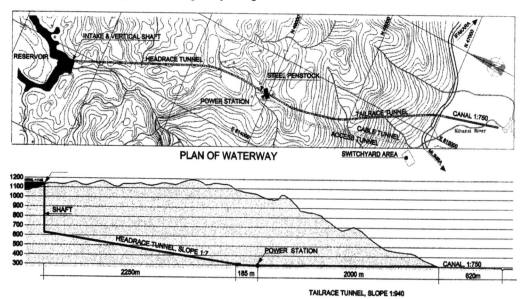

Figure 1: Project overview

primary aim of the field investigations was to ensure that there were sufficient internal stresses in the rock mass for adopting an unlined design for the headrace tunnel, giving large cost savings.

Stress measurements by use of hydraulic fracturing methodology were first conducted in deep, core drilled holes from the surface. After some costly and time-consuming attempts, where test equipment was lost in deep drill holes, a different approach was chosen. Hydro-fracturing tests would be done from short holes drilled from within the tunnel during excavation. If the results from these tests were unsatisfactory, the tunnel layout would have to be modified. That meant that a tentative design had to be presented in Tender documents, based on assumed rock stresses. The contract conditions were written to allow the power house to be sited deeper into the rock massif if necessary, since this would result in larger rock cover and probably improve the rock stress conditions for the critical part of the headrace tunnel. Use of unit price as well as unit time system was the important basis for contractual regulations in case of relocation of the power house.

4 ROCK STRESS MEASUREMENTS

A detailed strategy plan for rock stress measurements and decision making was concluded on in due time. If sufficient rock stresses could be confirmed by testing at 3 different stations, located from Chainage 800 to Chainage 1000 in the Access tunnel, the original, tentative location of power house, at around Chainage 1070, could be maintained.

Hydro-fracturing testing as well as triaxial testing began when 800m of the access tunnel had been excavated. Initial testing gave insufficient minimum principal rock stresses and relocation of the power house seemed unavoidable. The encountered stress pattern was characterized by sub-horizontal minimal principal stresses oriented north - south, close to parallel to the tunnel axis and perpendicular to the main joint orientation. This pattern is assumed to reflect the original stress situation with the low minimum principal stresses explained by extensional tectonism.

When the testing finally was completed at Chainage 2093, totally 19 holes with lengths between 20 m and 140 m had been drilled for hydraulic fracturing / jacking testing in the tunnel. The drilling was done partly from the tunnel face, ahead of the tunnel, partly from niches behind the face. In addition testing was done in 2 deep holes core drilled from terrain above the tunnel. Totally 97 hydro-fracturing tests, 27 hydro-jacking tests and 22 triaxial tests by use of overcoring method were conducted. The testing was done by the SINTEF Rock and Mineral Engineering, Norway, assisted by experts from SOLEXPERTS, Switzerland. During the final stage

of the testing, Dr. Tore Dahlø of SINTEF died as a consequence of a tragic accident in the tunnel.

5 INTERPRETATION OF ROCK STRESS MEASUREMENTS

The measurements were conducted to determine whether the level of minimum stress at the transition between unlined headrace tunnel and steel-lined penstock attained the level of 10 MPa, as required to leave the tunnel unlined. Continued stress measurements by hydro-fracturing / hydro-jacking and by overcoring methods indicated an improved stress situation deeper into the rock massif.

The minimum stress estimates from the instantaneous shut-in pressures (ISIPs) of the hydro-fracturing tests generally showed large scatter and were often less than the estimated pre-disturbance pore pressure. It was suspected that the stress tests were affected by stress alteration around the tunnel due to drainage causing a pore pressure draw-down. To test this, hydro-fracturing and hydro-jacking tests were conducted in two long, horizontal holes drilled ahead of the excavation face into relatively undrained rock, and a third vertical hole drilled behind the face. All were near the critical location where the pressure tunnel enters the powerhouse. The results showed that the minimum stress estimates in the ahead-of-face holes were much higher than those from the behind-face holes, consistent with the hypothesis. A model was developed to explain the coupling between the pore-pressure and stress fields in terms of fracture compliance and poro-elasticity, and was used to correct the stress estimates for the effects of pore pressure draw down around the bore holes during testing. The corrected minimum stress estimates exceeded the 10 MPa limit for an unlined tunnel. Figure 3 shows a compilation of stress measurements at the various test stations in the access tunnel, also indicating how the rock stresses were adjusted for effect of drainage (pore pressure draw down).

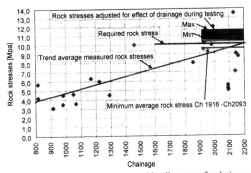

Figure 3: Measured rock stresses, with adjustment for drainage effect

The results of this study are relevant to any situation where stress tests are to be conducted in deep tunnels or excavations. A detailed description of the rock stress measurements and the influence of pore pressure draw down is given by Dahlø, T. et al. (in press).

With this conclusion, the location of the transition between unlined and steel-lined tunnel could be decided on. The power house complex was moved 730 m into the mountain beyond the initial location, and with more than 700m of overburden.

In Table 1 the required and measured minimum principal rock stress for initial as well as as-built design is shown.

Table 1. Required and measured minimum principal rock stresses, σ_3, in initial and as-built design

		Initial design	As-built design
Overburden at penstock, m		600	750
Water pressure / overburden ratio		0.65	0.89
Minimum principal stress, σ_3, MPa	Required	10	10
	Measured	6	≥10
Minimum principal stress / water pressure ratio	Required	1.2	1.2
	Measured	0.7	≥1.2

Table 2. Summary of geological conditions in headrace tunnel

Chainage	Lithology / tectonisation	Average RQD
0 - 825	Mainly massive granitic and dioritic gneiss / low to moderate degree of jointing and faulting.	95
825 - 920	Mainly dolerite and biotite rich gneisses / schistose (thrust faulting?), moderate degree of jointing.	80
920 – 1250	Massive granitic / dioritic gneisses with dolerite and biotite rich interlayers / low degree of jointing, but some schistose zones.	93
1250 – 1520	Granitic gneisses interlayered with meta-diorites and micaceous gneisses / faulted sections and moderate to high degree of jointing..	72
1520 – 2200	Granitic gneisses interlayered with meta-diorites and micaceous gneisses / low to moderate degree of jointing and faulting.	95

6 GEOLOGICAL CONDITIONS IN HEADRACE TUNNEL

A summary of the geological conditions of the headrace tunnel is shown in Table 2. Potential leakage is mainly connected to E-W oriented, sub-vertical joints.

Degree of jointing is reflected in the RQD value, see Table 2 as well as Figure 4. The RQD values are in general low. During excavation the degree of water inflow was low in the lower part of the tunnel, where the RQD values are high.

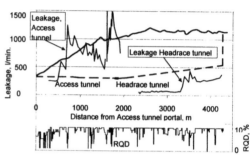

Figure 4. Leakage from access tunnel and headrace tunnel during excavation, measured weekly at portals as tunnel face proceeded

7 TRANSITION BETWEEN THE HEADRACE TUNNEL AND THE DRY TUNNEL SYSTEM

A 120 m long horizontal steel penstock liner from the power station to the headrace tunnel was designed to give an acceptable pore pressure gradient. The sealing between the water filled, high-pressure tunnel and the dry tunnels was taken care of by concrete plugs in the by-pass tunnel and the penstock tunnel, lengths 60 m and 70 m respectively. Figure 5 shows the layout in the transition area, including concrete plugs and grout curtains.

Grouting works were designed and performed ahead of the excavation in the areas of the concrete plugs. In addition, comprehensive grouting of the rock around both plugs was done in 24 m long holes before casting, with use of cement and a grouting pressure of up to 90 bars, and after casting contact grouting between concrete and rock. After water filling post grouting works were done to reduce the leakage encountered at the plug in the Bypass plug area.

8 HYDRO-GEOLOGICAL CONDITIONS

The tunnel is located in a North-South oriented ridge, formed by erosion of the two rivers on both sides of the ridge: Kihansi to the west and Udagaji to the east. A longitudinal profile of the ridge along the tunnel is presented in Figure 1. Figure 5 shows the calculated ground water pressure lines in a cross-section of the ridge, located just upstream of the plugs, prior to tunnel excavation and after completed

384

filling of the tunnel. Estimated flows to the Kihansi and Udagaji Rivers for one set of estimated permeability conditions of the rock. The model indicate an estimated rise of water level above the tunnel in the order of 50 m.

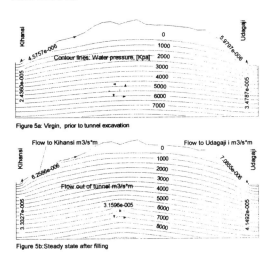

Figure 5a: Virgin, prior to tunnel excavation

Figure 5b: Steady state after filling

Figure 5: Water pressure lines in a cross-section upstream of the plugs prior tunnel excavation and after completed filling.

The hydro-geological conditions of the ridge are decided by:
- A system of sub-vertical E-W-oriented joints across the ridge approximately perpendicular to the tunnel axis. These joints, assumed to be tensional, are perpendicular to the minimum principle stress direction. They are water bearing, partly open / partly filled with weathering materials. Degree of weathering varies and generally these materials are residual soils of a uniform grading in the clayey silt to fine sand fractions. The weathered material is assumed susceptible to internal erosion. Width of the joints varies generally from 0 to 10 mm. In places, however, the thickness of the severely weathered joint may measure up to 100 mm. Parts of the joints, where the weathered materials have been washed away, have a relatively high permeability compared to the rock mass otherwise.
- Distance between the sub-vertical E-W joints is generally in the order of 10-20m. Between the joints, the rock is normally solid without pronounced joints or weaknesses.
- Permeability of the rock in N-S direction at tunnel level is very low and negligible compared to the permeability of the sub-vertical E-W joints.
- The severely weathered zone, of up to 40-50 m thickness, at ground surface above the tunnel has a higher permeability, causing a distribution of

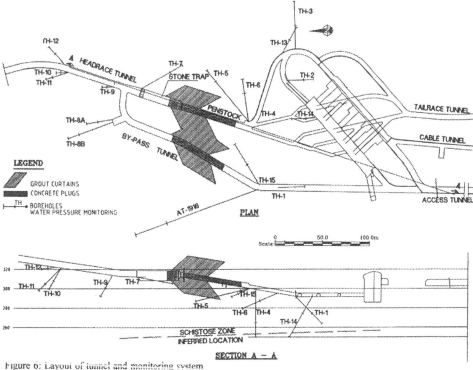

Figure 6: Layout of tunnel and monitoring system

385

ground water from the huge ground water reservoir at the Plateau west of the project area along the ridge. Therefore, some of the tributaries to Kihansi and Udagaji carry water also in the dry season.

- Ground water distributed southwards along the ridge communicates through the E-W joint system into the tributaries, often located along E-W lineaments.

A pronounced system of E-W oriented joints crosses the tunnel system just upstream of the relatively compact section of the rock, in which the penstock and bypass plugs have been located. This joint system crosses Kihansi River at and downstream of the main Kihansi Falls and can be followed over to the Udagaji River. Some of the tributaries to both rivers have water flow all year trough.

The hydro-geological conditions are characterized by a highly anisotrop permeability: high in the joints perpendicular to the tunnel and very low permeability parallel to tunnel axis. This pattern could be observed during excavation of the tunnel: Water pressures up to 6 MPa could be measured in sounding holes crossing E-W joints only 3 m ahead of face. All water entering the tunnel during excavation came from the E-W joint system.

Figure 7 shows estimated permeability of the rock perpendicular to the tunnel axis, calculated on the bases of water pressure tests and or measurements of water passing through soundings systematically drilled ahead of face.

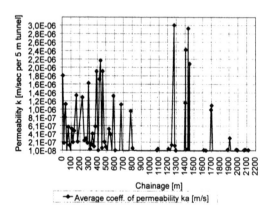

Figure 7: Permeability of rock perpendicular to the tunnel axis, and grout take for grouting at face. Coefficient of permeability k is calculated based on water pressure tests /water inflow in soundings and ahead of face

9 LEAKAGE PREVENTION IN HEADRACE TUNNEL

An estimate of expected water leakage out of the headrace tunnel was made using the permeabilities,

estimated on the basis of the soundings, taking the future net head along the tunnel into account. The estimated leakage at each joint system crossed by the soundings as well an estimated net head curve are shown in Figure 8.

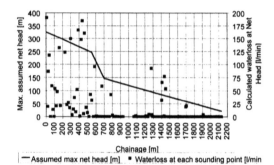

Figure 8: Estimate of water leakage from headrace tunnel, calculated on the basis of the permeability determined at the soundings ahead of face.

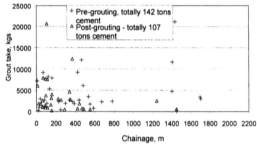

Figure 9: Grout takes in headrace tunnel

Economic analyses were made to obtain parameters for reducing leakage from the headrace tunnel. The net present value of production losses at the Kihansi project was estimated to USD 24 000 /l/s. This high value of the potential water loss entailed that great investments in reduction of permeability were warranted. Pre-grouting should always be performed at face when possible, to reduce permeability in zones in the most cost efficient manner. It was stated that pre-grouting was feasible and efficient when rock permeability, measured as Lugeon value, exceeded 0.5-1 L.

Pre-grouting by use of cement, both ordinary Portland cement and micro-cement, was conducted in sections where unacceptable permeability was detected by water pressure testing. The criteria for grouting was set to the permeability k> 2 E-7 m/s. Usually, sounding was performed with three holes of 18 m length. The conductivity of the rock was determined by water pressure testing of each hole.

If deemed necessary, post grouting was also done from the excavated tunnel in leaking zones. Figure 9 shows the distribution of grout takes along the tunnel.

Various types of sectional linings and local structural measures in addition to grouting were considered and designed for highly permeable sections. Type and length of such measures were selected based on a cost – benefit estimate.

In the lower part of the headrace tunnel, at Chainage 66 – 90 and 111 - 144, sections of concrete linings were applied to reduce potential leakage through some pronounced joint systems.

10 INSTRUMENTATION FOR MONITORING DURING WATER FILLING

Water filling of the headrace system was considered an especially critical operation and the response in the rock mass around the tunnel had to be analysed in detail. A monitoring system included pore pressure measurement in strategically placed bore holes, both upstream and downstream of the concrete plugs. NGI (Geotechnical Institute of Norway) was engaged for the technical design and installation of the instrumentation. Pore pressure measurement was done in totally 17 holes, with lengths from 25 to 140 m. Location of holes is shown in Figure 6. High pressure steel tubes connected each piezometer hole to pressure transducers. The steel tubes were placed in concrete protection in the pressure tunnel and lead through the Bypass concrete plug to the downstream end of the plug. Measures were taken to prevent leakage along the tubes through the plug. All transducers were placed in the dry zones downstream of the Bypass and the Penstock Plugs. No transducers for monitoring piezometer pressures or water pressures were located in the water pressurised tunnel system. DigiQuartz absolute pressure transducers with an accuracy of 0.005% were used. Signals from the transducers were transmitted to 2 automatic data loggers, with connection to computers in a control room in the power house.

Precise recording of the water level in the tunnel or in the shaft was necessary for analysis of loss of water during filling. For this purpose, a DigiQuartz transducer was placed in the lower end of the tunnel and connected to one of the data loggers. In case of failure in this system, and to be able to read the water level in the shaft even more precisely, a high precision level transmitter was brought to the site for use in the shaft.

In addition to the monitoring of water pressures in the various piezometers and in the headrace tunnel, a system to monitor strain in a sectional concrete lining and possible deformation of joints in rock was installed. The system consisted of extensometers placed at the rock / concrete boundary across E - W joints. Furthermore, two extensometers for monitoring possible radial strain in the concrete

lining were installed. The vibrating string extensometers were connected to a AC/DC transformer and to the datalogger downstream of the Bypass Plug by cables, which were lead through the concrete plug in the same way as the steel tubes for pressure monitoring.

A simple, manual system for monitoring inflow of water into the dry part of the tunnel system downstream of the plugs was established. Furthermore, water flow in tributaries and streams to Kihansi and Udagaji rivers were monitored manually daily during and after filling. Water levels in bore holes in rock above the tunnel and close to the dam site were monitored manually on a daily basis.

11 FILLING OF HEADRACE TUNNEL

Filling of the headrace tunnel and the shaft was planned with at least 3 weeks duration to avoid high local hydraulic gradients in the joints close to the tunnel, and to avoid excessive changes in rock stresses around the tunnel. The plan included various stops at certain filling levels in the tunnel and in the shaft in order to monitor net inflow from rock into the tunnel or outflow from the tunnel into the rock at different water pressures. The filling was done using pumps as the construction program at the dam structure did not allow for rising the reservoir level above the threshold of the intake structure in due time.

Filling of the headrace tunnel and intake shaft up to final reservoir level took totally more than one month.

During filling, the pore pressure build-up in the rock mass around the tunnel as well as the water pressure in the tunnel was closely followed by automatic logging of some 20 piezometers installed in drill holes. This monitoring, with automatic readings every 10 minutes, allowed a continuous recording of the filling rate and gave valuable information on stress development around the tunnel. Detailed results of a selection of monitored piezometers during the filling are shown in Figure 10. The monitoring continued for several months after the filling. Results for a period of 4 month after start filling are shown in Figure 11. Locations of the piezometers are shown in Figure 6. The pressures recorded in the diagrams in Figures 10 and 11 are adjusted to the same reference level, i.e. the level of the transducers located in the gallery at the downstream end of the Bypass plug. Consequently, the difference between the recorded water pressure in the tunnel and the pressure monitored at a piezometer, reflects the hydraulic gradient between the respective piezometer and the tunnel.

Based on results from the rock stress measurements, water pressure testing ahead of face, as well as a hydro-geological analysis of ridge, the virgin ground water pressure in rock at the downstream end of the headrace tunnel was estimated to 5.5-6.0 MPa. Before filling, the piezometers showed variable influence of the tunnel on ground water pressures. Piezometer AT1916, located some 120 m away from the pressurized tunnel in a joint system crossing the tunnel system just upstream of the Bypass plug, showed a pressure of approximately 3.0 MPa before filling. Before tunnel excavation crossed that joint system, a pressure of 5.3 MPa was recorded in AT1916. Piezometer TH7, located in the rock at the upstream end of the Penstock plug some 25 m away from the stone trap, showed a pressure of 4 MPa prior to filling.

During the first stages of the filling, as recorded in Figure 10, the piezometers in rock showed only slight pressure increases until the water pressure in the tunnel had exceeded the piezometer pressures by approximately 1.5 MPa, i.e. a water head difference in the order of 150 m. This applies to all piezometers located in rock beside the pressurized tunnel. The distances from the tunnel were between 10 to 120 m. The average hydraulic gradients were correspondingly between 1 to 15.

All piezometers in rock around the pressurized tunnel, the "wet zone", were located in joints crossing the tunnel. Most of these joints yielded water during drilling of soundings ahead of face during tunnel excavation, or showed high permeability during water pressure testing. The joints had a clear communication to the tunnel, and were consequently grouted at face and, in most cases, also post-grouted.

The results of grouting were tested by test-holes and water pressure testing generally at pressures of 9 MPa. The joints generally contained some decomposed material, assumed susceptible to erosion.

At filling levels in the tunnel, corresponding to pressures higher than approximately 1.5 MPa above the piezometer pressure before filling started, the pressures in the "wet zone" piezometers increased at the same rate as the water pressure until the pressure reached the magnitude of virgin ground water of 5.5-6MPa. Thereafter, the piezometer pressures in the "wet zone" equalized more or less with the water pressure inside the tunnel. The pressure differences remained generally between 0 and 0.3 MPa. A detailed study of the pressure development, showed that the increases in most cases did not take place smoothly, but often stepwise. Also pressure decreases with time were observed. An explanation for this development may be found in the erosion susceptible material in the joints: As the pressure and the gradients between the tunnel and the joints outside the grouted zones increased, high local hydraulic gradients across the residual soil increased, leading to erosion of this material. This lead again to sudden equalization of pressures between the tunnel and the piezometer in the corresponding joint. Such sudden increases of pressures, up to 2 MPa, took place within 10 minutes at several occasions. The eroded material was re-located within the joint system, causing a new build up of pressure differences.

The recordings of piezometer TH 9 in Figure 10 shows clearly such pressure development at pres-

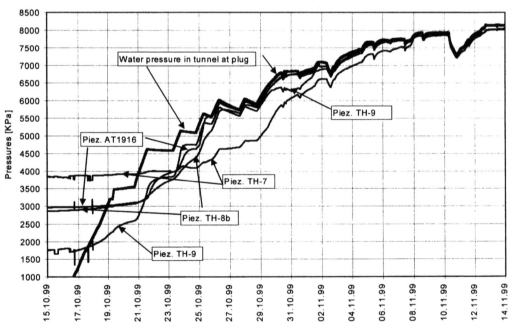

Figure 10: Recorded water pressure in tunnel and in selected piezometers during filling of tunnel/shaft.

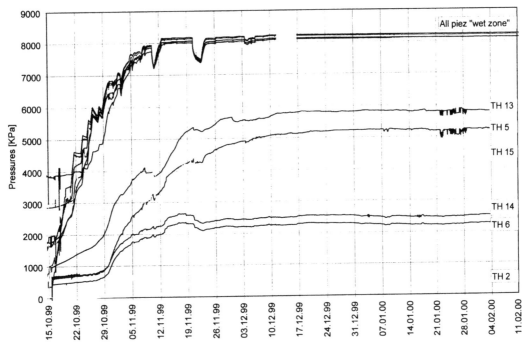

Figure 11: Recordings of piezometers and water pressure until 4 month after start filling.

sures 6.3 to 6.8 MPa. At the latter pressure, piezometer TH 9 again communicated directly with the water in the tunnel.

Pore water pressures recorded in some of the piezometers located in the "dry zone", downstream of the plugs, are shown in Figure 11. A distinct brake in the pressure increase development took place at water pressure 6 MPa in the tunnel, corresponding to the virgin ground water pressure in the rock for all piezometers located in the joints. Piezometer TH 15, is located in rock of very low permeability without any distinct jointing, between the downstream ends of the concrete plugs. This piezometer showed practically no pressure increase until the water pressure in the tunnel reached approximately 7.5 MPa. Then it increased within short time to the same pressure level as the other piezometers located in joints within the "dry zone", i.e. 4 MPa. Thereafter the piezometer showed an unstable behavior with sudden fluctuations between 4.5 and 1.8 MPa. About one month after start filling, the fluctuations came to and end and the further pressure increase followed a development similar to the other "dry zone" piezometers. It has finally stabilized at a pressure of 5 MPa, i.e. in the same order as the others. The permeability in the rock were piezometer TH15 is located, is still very low, in spite of the observed pressure behavior.

The behaviour of the pressure in piezometer TH15 may be explained partly as the consequence of erosion and subsequently collapse and blocking of the residual soil in joints at high local gradients. Because of low permeability, pressure changes within short time spans may be distinct. Similar occurrences could also be seen in other piezometers, supporting the idea about erosion. Natural stress adjustments in the rock when the ground water pressure exceeded by far the virgin water pressure in the rock, could also be an explanation for the sudden pressure changes.

One year after filling of the headrace, pressures in all piezometers have stabilised at reasonable levels.

The monitoring of strains in the sectional concrete lining show that no strain occurred neither radial nor longitudinal in the lining.

During the filling operation, water inflow in the "dry zone" of the tunnel system was followed closely. Several stations for manual monitoring were established. Stations in the eastern area (Penstock plug and eastern tunnels) showed only small inflow, totally around 2 l/s during the first month after filling, decreasing to less than 1 l/s 2 months later.

In the western area, downstream of the bypass plug, the inflow was significantly higher. Total inflow, including some inflow from the access tunnel, shown in Figure 12, was less than 10 l/s when the water head in the tunnel was below 750 m. With higher head, the inflow increased to 50-60 l/s. A major part of this came in the gallery of the bypass plug, or just downstream of the plug. The inflow within the plug area was apparently due to communication between the sub-vertical joint system within the plug and casting joints. Subsequently, a grouting program in the bypass plug gallery was implemented. After this grouting, the inflow in the western area has stabilised at 27 l/s. A second significant water inflow from joints occurred some 60 m downstream of the plug. The leakage way for this water might be up along joints crossing the tunnel upstream of the plugs, through the permeable weathered zone below terrain, and down again along the leaking joints. These joints also give good drainage around the power house. Therefore, no attempt was made to grout these joints.

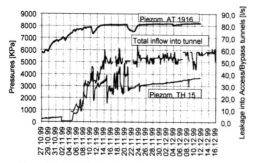

Figure 12: Water inflow into the western "dry zone" tunnel system. Pressure in piezometer AT1916, also representative for the water pressure in the tunnel, is shown for reference.

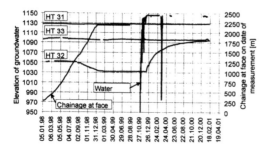

Figure 13: Ground water levels recorded in bore holes above the tunnel from before excavation started to after filling.

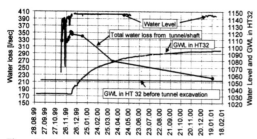

Figure 14: Total outflow from Shaft and Headrace tunnel just after and up to 4 months after filling was completed. Reservoir water level and water level in HT32 are shown for reference.

Before tunnel excavation started, ground water sounding was initiated in 3 bore holes in the ridge above the tunnel. Changes in the water levels are shown in Figure 13. Any influence from tunnel excavation and filling was only observed in one of the bore holes, HT32. Bottom of this hole is located approximately 480m above the headrace tunnel at Chainage 310. The water level in HT32 started to sink about 3 weeks after the tunnel excavation met the permeable zones causing the abrupt increase in inflow in the tunnel, about 900m north of HT32, see Figure 4. The water level started to rise again some 1.5 months after start of filling of tunnel. 15 months after completed filling the water level is about 40m above the virgin water level. The water level has not yet completely stabilised (February 2001).

After completed filling of the headrace tunnel and shaft, the water loss has at some occasions been measured, reading the rate of drop in water level with a high-precision level transmitter when closing the intake gate. As seen from the "Outflow" curve in Figure 14, the total water loss decreased from 390 l/s. just after completed filling to 215 l/s. 15 months later. The loss of water is considered to consist of 3 components:

1 Water inflow in the "dry zone" of the tunnel system, amounting to approx. 30 l/s.
2 Water outflow to terrain, observed as increase in water flow in creeks. The complete flow increase is difficult to monitor. At altogether 8 stations, the increase is estimated to 100-150 l/s.
3 Filling up of ground water reservoir. So far the rise of water table in HT32 is more than 40m. This component of the water loss from the tunnel is temporary, until the new water table has stabilized.

REFERENCES

Saidi, F. X., Lindemark, J. & Wilhelm, V. C. 2000. Kihansi Falls pave the way for power development. *International Water Power & Dam Construction*, Vol. 52, No. 3.

Dahlø, T. , Evans K., Halvorsen, A. & Myrvang, A. In press. Adverse effects of pore-pressure drainage on stress measurements performed in deep tunnels: An example from the Lower Kihansi hydroelectric power project, Tanzania. *International Journal of Rock Mechanics*.

Evans, K., Dahlø, T., Myrvang, A., Roti, J. A. In press. Mechanisms of pore pressure-stress coupling and implications for stress measurements conducted in deep tunnels. *Pure and Applied Geophysics*.

Hydropower in the New Millennium, Honningsvåg et al (eds), © 2001 Taylor & Francis, ISBN 90 5809 195 3

Implementation of SAP R/3 maintenance management system for Lesotho Highlands Water Project (LHWP)

Futho Hoohlo
Lesotho Highlands Development Authority

ABSTRACT

The increase in automaton and resulting high fixed costs of operational systems, as well as an increase in the number of legal and public demands for environmental protection and system safety standards, mean that increasingly, Plant Maintenance (PM) plays a decisive role in the success of a business such as the LHWP. The maintenance of the project incorporates inspections (determining the actual condition of the plant), regular maintenance (maintaining the ideal condition).

In order to achieve this goal, the PM module implementation was in phases: Analysis (establishing required structures and business processes); design (configuration and system testing); and implementation (software loading, final testing and operation).

Introduction

The Lesotho Highlands Water Project, in its progression since 1986, has been one of the World's most ambitious projects. It has resulted in the construction of Katse dam, a world class construction and pride of Lesotho, a network of transfer and delivery tunnels and an underground power station ('Muela Hydro). The purpose being to divert the natural river systems of Lesotho and feed the, ever thirsty, industrial heartland of the Republic of South Africa, the Gauteng province, and as well generate power for Lesotho. This is clearly a multi-million investment whose Operation and Maintenance needs to be of the highest quality and efficiency, hence a need for a high powered and comprehensive Maintenance Management package.

In order to operate and maintain plant as effectively, efficiently and cheaply as possible, with minimal interruptions to production, and maximum availability, Lesotho Highlands Development Authority (LHDA – a body administering and operating the Lesotho Highlands Water Project – LHWP) has realized the need for introducing a powerful, effective and user friendly work management system, for the operation and maintenance of LHWP. LHDA opted for SAP R/3 System (System Application, Products in Data Processing, release 3). The R/3 System provides data processing solutions for all areas of business and the following were implemented: *Plant Maintenance (PM), Materials Management (MM), Human Resources (HR), Financial Accounting (FI) and Controlling (CO).* This paper will be concentrating on the Plant Maintenance module and will look at:

- The four implementation phases;
- Management of operational systems and technical resources using Functional Locations and Equipment (Hardware Breakdown Structure);
- Entering of malfunctions and other direct maintenance requirements as maintenance notifications and forwarding to planning for processing;
- Defining of activities that recur on a regular basis using maintenance and inspection plans and releasing at a scheduled point in time;
- Carrying out of maintenance tasks using maintenance orders;
- Completing of maintenance orders and capturing of history, and
- Successes and problems encountered.
- With the introduction of SAP R/3 PM module to the everyday operations, a lot of red tape and paper work has been reduced by electronically assigning and managing daily activities.

A.   ANALYSIS PHASE

1.   LHDA Plant Maintenance Requirements – Business Processes

It is at this stage that LHDA O & M Branch sat down with the SAP consultants to formulate processes which were applicable to the every day running of the 'Muela PowerStation and Water Transfer as a whole. It should be borne in mind at this stage that although the adopted module in this case was Plant Maintenance (PM), the whole analysis stage was to as well come up with a

System that will be a work management package which was to handle other issues which were not directly related to equipment (Plant) maintenance such as administration.

The maintenance requirements for LHDA were grouped into four main processes that were used to control all maintenance needs and these were namely:

a)   Maintenance Due to Malfunction

This was defined as a breakdown or eminent breakdown of any piece of equipment that would stop normal operation of generation of electricity or delivery of water to the Republic of South Africa.

b)   Maintenance Due to Planned Requirements

This was defined as all maintenance work generated from a pre-planned schedule where inspections, monitoring readings, scheduled services or routines were required.

c)   Maintenance Due to Unplanned Requirements

This was defined as maintenance resulting from scheduled and unscheduled inspections, deviations, modification and/or any maintenance requirement not defined in the above-mentioned processes.

d)   Master Data

This was defined as any data requirements that needed to be captured on the SAP system for future use and included the following:

- Functional Locations[*1]
- Equipment
- Work Centres
- Strategies
- Packages
- Maintenance Item
- Task Lists
- Bill of Materials

It provides a database and skeleton for the operation of this system. The creation or change of Master Data was required for the LHDA Plant when object[*2] were to be installed, build, scrapped or changed during the life cycle of 'Muela Hydropower plant and water transfer system as a whole.

2.   Hardware Breakdown Structure

In order to manage our operational systems and technical resources, functional locations, equipment were used to represent the Hardware Breakdown Structure. These were further subdivided into spare structures using Plant Maintenance Bill of Material. To summarise, the following elements made up the system structuring:

Structuring Rules	-	Structure Indicator
Technical systems	-	Functional Locations
Individual Technical System	-	Piece of Equipment
Spares Structure	-	Equipment/Functional Location Bill of Material (BOM)
Spare	-	Material

Functional Locations, Pieces of Equipment and Bills of Material incorporated several hierarchy levels and enable a flexible means of representing technical structures, without limiting the depth of the structure as indicated in the 'Muela structure in the appendix.

Parallel to this analysis phase there were actual teams set-up within the O & M branch to start collecting all the master data for the whole project. This included establishing and compiling a list of Functional Locations and pieces of equipment from the different contractors in the LHWP and wherever these were not available, actual nameplate data taking of all other equipment's.

3. Maintenance Notifications & Orders

As it was identified in the LHDA maintenance requirements that in addition to regular activities, malfunctions or other exception situations require prompt action as they arise from time to time. To deal with these the Plant Maintenance module contained a comprehensive notification system for creating and processing notification & orders.

Maintenance notifications describe Functional Location affected piece of equipment, and in the LHDA O & M Branch can be raised by any person who requests action to be taken.

Notifications are used not only to initiate maintenance tasks, but also for documenting technical completion confirmations.

A maintenance task would be processed using a maintenance order, in which the type, scope, dates and resources for carrying out the work are described. Orders can be classified as:

   a)   Regular PM Orders

These are released by maintenance schedules at due dates. They are determined in advance by the maintenance schedule in terms of scope and dates and represent a regular load for the PM workshops.

   b)   Planned PM Orders

This would be types of order where maintenance requirements may arise for planned repairs from maintenance notifications or routine inspections.

   c)   Unplanned PM Orders

Sometimes immediate action due to unforeseen machine breakdowns, accidents or other malfunctions has to be taken, and it is under these circumstances that a rush order is created, which requires no additional planning, but it used simply as a trigger for maintenance activities.

4. Maintenance Planning

One of the main functions of maintenance is to ensure a high level of availability of the production system in this case hydropower generation and water transfer. It was therefore recognised by LHDA that preventive maintenance is the ideal way to avoid system breakdowns, which in addition to the basic repair costs often incur far higher subsequent costs as a result of the loss in production. LHDA therefore recognized the need for creating its procedures (Isolation, Switching and Maintenance) to described repetitive activities in which the individual operations, times and resources required are defined. These incorporate the type and sequence of individual maintenance routines and determine functional locations and pieces of equipment at which these activities are to be carried out.

B.   DESIGN, CONSTUCTION AND IMPLENTATION PHASE

Following the analysis phase was now the actual design of the work management package along the discussed and bid down parameters. This stage involved user reps[3] travelling form 'Muela to Johannesburg for three (3) days of the week for sic-seven months and working with the consultants in designing a configuring the system to fit the set guidelines form the analysis phase.

It is at this phase that the interaction of all the modules (Plant Maintenance, Materials Management, Human Resources, Financial Accounting, Controlling and Treasury) took place through integration tests. A typical example would be allocating spares to a work order and after the job has been complete, reconciling the costs for both material and human resources. This would therefore bring together the Financial Accounting, Plant Maintenance, Controlling and Material Management modules.

This phase as well as served as a platform for knowledge transfer from the consultants to the user reps, who are LHDA employees at different branches of the organisation, who would eventually train the end users.

C.   TRAINING

Having gone through these stages, the end product now had to be delivered to the end users in the form of training.

There were four different types of training required by LHDA O & M staff and these would allow them to successfully handle the SAP system.

1.   Windows Training

Since the SAP system is based on the Windows concept, the first section of the training would cover Windows literacy, for those staff without Windows literacy. The concept behind this was to establish a comfort zone, amongst the end users of being able to move between different windows.

It was crucial to identify the correct time to commence with the Windows training as this formed the basis of the rest of the training required. The timing of this training was as well crucial so as not to start too early.

2.   Jump-start Training

This part of the training was to equip the trainees with the essential generic skills to operate SAP and these were:

- Knowing what the SAP system is;

- Being able to log "on" and "off" the SAP system;

- Being able to navigate through the system;

- Being able to run a report and route it to a printer;

This course was required before the detailed modular training could be given. After the completion of this training, it was once again essential that trainees would have the opportunity to practice what they learnt.

3.   Train the Training

As indicated in the paper earlier, the user reps. Worked with the SAP consultants for all the phases of the project, and this was a way of training the very people that were going to deliver training to the end users by way of knowledge transfer. The users reps. Were therefore responsible for the development of appropriate training material (training and user documentation).

4.   User Training

For the Plant Maintenance module, user training would involve all the LHDA O & M staff, since they would all be end users. This meant training in all implemented sections of this module for appreciation and understanding of how the different sections tie together within the Plant Maintenance module, and the rest of the modules implemented by LHDA.

D.   OPERATION OF SYSTEM

Initiation of Notifications (Deviations)

With every Operation and Maintenance of a plat, there need to be some process that is followed, in order to ensure safety of personnel and plant and ensure efficiency and speed in execution of tasks. As well we require capturing any history associated with pieces of equipment as this simplifies future troubleshooting or maintenance.

There is always scheduled periodic maintenance, and in addition to these, there are other unforeseeable situations, which arise, form time to time. These situations require prompt and speedy action, and this Plant Maintenance package contains a comprehensive notification system.

A Notification describes a technical state of a piece of Equipment or Functional Location. They represent a workload that needs to be build into planning for the Operations and Maintenance team. These notifications are not only used to initiate maintenance tasks, but also for documenting technical completions and confirmations, and these will be explained later in the paper.

A notification is initiated by any employee within the Operations and Maintenance team whenever any action needs to be taken on a situation or piece of equipment. It should be borne in mind that within our Operation and
Maintenance activities, this Plant Maintenance package does not only address the technical task required on plant but also is used to facilitated any administrative chores (e.g. Scheduling of meetings). This is used more appropriately as a work management package.

A maintenance notification contains general data on the type, time and name of he person making the notification and information regarding the object concerned, as indicated in the Fig.1 below.

There is a further provision for technical input which contains information regarding the object component affected, the damage and, if known, the cause of damage.

There are four type of notifications that are used, namely:

> Maintenance Request – which is a request for maintenance action on plant which has not been planned or not part of a periodic schedule.

>> Malfunction Report – which is a critical request for maintenance action on plant or otherwise, where there is danger or imminent danger to life or plant and might affect production.

>> Engineering Request – which is a request for modification of plant or procedure.

>> Activity Report – which is more a capture of history on activities that were executed without initiation of the above three notifications (this notification tends not to be used as all work, ideally, should be, and is, done under a work order (to be explained later in the paper) created from any of the other three notification stated and explained above).

Processing of Notifications

After the notification has been created, there is a process that it has to go through before a work order can be created, scheduled and executed. This is to ensure that all the necessary and correct entries have been made and as much information as possible has been entered for making easy the work execution stage. It should be noted at this point that, a notification has a system and user indication which Information on what stage, in the process, the notification is (see appendix). On initiation of a notification, the user status defaults to SUPR

(Supervisor) status. This is the supervisor for the work center that is anticipated to execute the work indicated in the notification, and does all the above-mention information verification. With all information correct and adequate, the supervisor changes status to PLAN (Planner).

For all the notifications that are directly related to production, before the planner can created orders and schedule them have to go through the production meeting for discussion. This is a way of capturing more information which will ease and speed the troubleshooting process and as well a way disseminating information, amongst staff, on the status of the plat.

The system is able to indicate who initiated the notification, through a unique user name which is given to each user, and when it was initiated. Therefore in the case where there is insufficient or unclear information on a notification that has been created, the supervisor, for the work center concerned, as indicated in the process above, changes user status to ORIG (Originator). This indicates to the person who initiated the notification that more information is required.

All these descriptions are captured in the "long text" (editor) of the notification, and indicates who entered the long text.

This however is not a substitute to verbal communication amongst staff on any issue, it is merely a tool that facilitates and captures information for reference, especially in recurring situations.

Creation of a Work Order

Having created the notification and verified that there is correct and sufficient information, there supervisor for the appropriate work center passes it on to the planer (as described above) for the creation of a maintenance work order. Since the planner has the overall planning and scheduling overview of all work that is to be executed, it is only them (planner) who create and schedule orders (as long as the notification created is in PLAN status). However, for a Malfunction Report, whose classification criteria has been explained above, any Supervisor and Planner can create an order from. Malfunctions have the highest priority, as they affect safety and production.

Maintenance tasks are processed using maintenance orders, in which the type, scope, dates ad resources for carrying out the work are described, together with the financial account assignment and settlement and this is linked to the Financial Accounting module which is beyond the scope of this paper. These Maintenance Orders can be classified and distinguished by their degree of planability and criticality to plant into:

Planned PM Order - This would be the type of order where maintenance requirements may arise for planned repairs from maintenance or routine inspections.

Unplanned PM Order - Action, resulting from routine inspections or unforeseen circumstances has to be taken, therefore this type of order is initiated. This is normally created form a Maintenance Request (notification), as explained above.

Malfunction PM Order - Sometimes immediate action due to unforeseen machine breakdowns, accidents or other malfunctions has to be taken, and it is under these circumstances that a rush order is created, which requires no additional planning, but is used simply as a trigger for maintenance activities. These are created form Malfunction Reports (notification).

Engineering PM Order - These are orders for any engineering changes on plant or documentation that need to be carried out. These are created from Engineering Requests (notification). Note that before this order can be created, the notification must pass through an Engineering forum (meeting).

These maintenance orders can either be carried out by the Operations and Maintenance workshops or in some cases by external service companies. Reasons for having maintenance activities processed externally can include cost-effectiveness, specialized knowledge or temporary bottlenecks in internal workshop capacity.

Operations and Sub-Operation in the Order

The scope of the work is described using operations and sub-operations. These represent the individual work steps to be carried out for the maintenance tasks. An operation contains a description, the performing work center and the amount and duration of work. The controlling information determines whether the operation will be processed internally or externally.

Material Required

Material can either be allocated directly to each operation. This can be either stock or non-stock material. Either material reservation (in the case of stock material) or purchase requisitions are subsequently generated for these materials. These material transactions and goods movement are reflected and handled in the Materials Management module also implemented along the Plant Maintenance (PM) module. The details of this Materials Management module are outside the scope of this paper.

Production Resources/Tools (PRT's)

These are any specialized tools or pieces of equipment (e.g. Powerhouse crane, forklift etc.), which can be booked out for a particular job. In addition to materials, each operation in a maintenance order can be allocated one or more resources.

Schedule and Allocation of Work Order

Scheduling

So far, the paper has been discussing the process of creating a notification and an order form it by the planner. All this has to be put into a work plan for execution of the work at a scheduled time with the required resources (human, equipment and material) available.

One of the main functions of maintenance is to ensure a high level of availability of plant in the long term. Preventive maintenance is the ideal way to avoid system breakdown, which in addition to the basic repair costs, often incur far higher subsequent costs as a result of ht loss in production.

In this case, maintenance procedures (task lists) can be created to describe repetitive activities in which the individual operations, times and resources required are defined. These task list contain wither periodic activities or activities which are used selectively. The scheduling rules, for regular activities, are stored as maintenance strategies. These are an indication and control of whether maintenance to be done on a piece of equipment is by calendar (e.g. every six months) or performance based (e.g. circuit breaker to be checked after every hundred operations). A maintenance plan is therefore triggered for order to be executed.

By entering a start date, the maintenance cycle is set in motion. The due maintenance dates and pending tasks are determined on the basis of this date.

Using call intervals, the time period for which you want to generate the maintenance schedules can be easily defined, in other words the date up to which scheduling is to take place.

In this section, planning and scheduling of planned orders has been discussed. Unplanned orders, as explained above, get included in the plan. It should be noted at this point that the frequencies of the plan (work to be done) are weekly, and since unplanned and engineering work orders are created from maintenance and Engineering Requests (notification), they can only be included in the following weeks plan, at the earliest. A planning and engineering meeting is scheduled every Thursday to discuss the following week's work for the planner to be able to run the plan on Friday.

Allocation

After running the work plan, the planner now prints the actual hard copies of the work orders (job cards) which are colour coded for ease of identification and these are distributed to the appropriate work centers for work execution by technicians. The work orders are coded:

Green	-	Planned Order
Yellow	-	Unplanned Order
Blue	-	Engineering Order
Red	-	Malfunction Order

History Capturing, Root Cause Analysis and Completion

Confirmation

At the end of every working day technicians and any person who has been issued with a work order needs to do a confirmation. This is a means of capturing:

- The time that a technician has taken to do a particular job;

- The detail of what was done.

This is to facilitate a number of issues:

- A means of reference for future planning of the same job in terms of time allocation;

- A means of reference for future planning of the same job in terms of technical information;

- A means of capturing any overtime worked. This would be captured in the payroll section of the *Human Resources* module linked to Plant Maintenance, for the purpose of remuneration;

- Assistance on root cause analysis when giving feedback in production meeting;

- Dissemination of information on plant status amongst staff.

The system is able to capture who did the confirmation and any person with access to system can display these confirmations for information. Report writing, by technicians, is one of the very important aspects in maintenance, and this facility, as a daily tool, exposes each and every technician to concise reporting.

Completion

After capturing the times and information on what was done, there is a facility of further entering technical information attached to the particular piece of equipment. This is doe by the supervisor, whose work center executed the work. This is a form of checking and confirming the times entered by the technician in the confirmation and as well ensuring that there is adequate information on what was done, for the purpose of root cause analysis. It should be noted at this point that this is where both the order and the notification are completed.

The supervisor involvement at this stage ensures that work and history capturing is of the highest standard and quality, and facilitates a means of personnel monitoring in terms of the quality and time taken on the job by individual technicians. Reports can therefore be drawn on technicians to monitor performance of individuals.

As well reports can be drawn on any piece of equipment, and technical history be displayed, which facilitates the performance monitoring of any piece of equipment. This is a very powerful facility indeed!

Problems Encountered

The Operation and Maintenance branch had been using a very manual and paper oriented maintenance management system, which whoever was using the system was comfortable with.

As well all known, one of the natural instincts of a human being is to resist change. This was one of the main problems encountered, as very few people are comfortable with venturing into the unknown. Being comfortable with a computer was one of the first problems that had to be achieved with staff.

Getting to understand the new terminology which comes with the new package was another problem that had to be overcome. Some of the terms used had no direct attachment, in term's f the English language, to what was being referred to and below is just some of the examples:

Old Terminology	SAP Terminology
Hardware Breakdown Structure	Equipment Hierarchy
Parts List	Bill Of Material
Procedure	Task List
Deviation	Notification
Department e.g. Electrical, Maintenance, Operation Shift1.	Work center e.g. WWOMMAEL, WOOMOPS1
Plant	Equipment
Resources e.g. Technicians	Capacities
Working Times e.g. Overtime	Capacity Activity Type
Frequency e.g. Weekly, monthly, annually	Package

Configuration incompatibilities (not identified during the design of the package) in this and other modules (Financial Accounting, Materials Management and Payroll) linked to the Plant Maintenance module were one of the most frustrating problems. This meant some transactions (operations) could not be executed and the whole comprehensive process steps that have been outlined above could not be completed. Rooting out these incompatibilities is an ongoing process.

This system has been in use now for twelve months. The "motto" has been:

"No work without a work order"

Through intensive training of staff and continual refresher courses, embracement of this system by the staff has been achieved, and this has addressed the anxieties indicated above having taken ownership of the system, the staff has contributed greatly in shaping up daily maintenance processes using SAP, and improvement of history capturing and knowledge transfer. The result of this has been amazing information sharing and knowledge of plant status at any point in time.

E.   APPENDIX

Functional Location

Element in a technical system (for example, a functional unit in the overall system). Functional location may be structured according to the following criteria:

- Production (production resources and tools)
- Quality Assurance (test equipment)
- Materials Management (serialized materials)

Bill Of Material (BOM)

A complete, formally structured list of the components which make up a product or assemble. The list contains the object number of each component, together with the quality and unit of measure.

Bills of material (BOMs) contain important master data used in many different organizational areas, such as:

- Material requirements planning
- Provision of materials for production
- Product costing

In the SAP R/3 System, you can create the following BOMs:

- Material BOM
- Equipment BOM
- Functional location BOM
- Standard BOM

Work Center

A work center is an organizational unit where a workstep is carried out, producing an output. The work center defines where and by whom an operation is to be carried out. A cost center has a particular available capacity. The activities performed at or by the work center are valuated by charge rates, which are determined by cost centers and activity types. Work centers can be:

- Machines
- People
- Production Line
- Groups of craftsmen

Strategies

These are criteria used in the carrying out of maintenance on pieces of equipment e.g.

1. Performance based strategy
2. Calendar based strategy

Packages

These are frequencies of routine maintenance, e.g. weekly, monthly or yearly

Maintenance Item

Description of the inspection and preventive maintenance activities to be performed on technical objects requiring regular maintenance.

The maintenance objects concerned can be specified directly in the maintenance item or by using an object list.

The necessary activities are described in a task list allocated to the maintenance item.

Task List

This is a step by step procedure when carrying out tasks within Plant Maintenance module.

References

SAP, "R/E System, Plant Maintenance", SAP AG, September 1995

Author

Futho Hoohlo, holds BSc. (Electronic & Electrical) Engineering and MSc. Electrical Power Engineering degrees from the University of Glasgow and University of Strathclyde in Glasgow respectively. He is an Electrical Power Engineer, in the Lesotho Highlands Development Authority, Operations & Maintenance Branch. He is a super user and was joint team leader in the design and implementation of this SAP R/3 Maintenance Management system.

Hydropower in the New Millennium, Honningsvåg et al (eds), © 2001 Taylor & Francis, ISBN 90 5809 195 3

The Groner-membrane: a hydraulic superconductor for rough tunnels?

Tom Jacobsen & Lars Jenssen
Statkraft Grøner AS, Trondheim, Norway

ABSTRACT: Statkraft Grøner has since 1998 developed and patented the Groner-membrane, a technology that can reduce friction losses in unlined tunnels with up to 70% and in lined tunnels with up to 50%. The costs of implementing the technology seems to be significantly lower than known alternatives, such as expansion of the existing tunnel or blasting/drilling of a new parallel tunnel. The technology is environmentally friendly, as additional energy can be produced from existing hydropower plants with hardly any consequences for the environment. The technology is simple. The main component is a flexible pipe, made from a strong PVC/PU coated polyester fabric. The flexible pipe has a circular cross section that fits inside the tunnel walls without being in contact with the rock or the lining. A laboratory experiment was performed in 1999 that verified the potential for reduction of friction loss and indicated that the behavior of the membrane in different situations such as transients, slides, rock fall and damage through puncture was satisfactory. A full scale test is planned in a 12 m^2 tunnel at Håen Hydropower plant south of Trondheim in Norway. A 400 m long flexible pipe with a diameter of 3,0 m will be installed in the tunnel. In addition to its potential for rehabilitation of hydropower tunnels the Groner-membrane may be used in water supply tunnels, to prevent contaminated ground water from mixing with the clean water supplied through the tunnel.

1 INTRODUCTION

A total of 1700 km of hydropower tunnels are in operation in Norway today, and it is assumed that the energy losses due to friction amount to 3-4 TWh (10^9 kWh) annually. Large losses are most common in unlined tunnels constructed before 1970. Furthermore, for many old hydropower plants upgrading that result in increased discharge is not feasible due to the subsequent increase in head loss in the tunnel.

Although unlined tunnels is a Scandinavian specialty, the situation with respect to head losses in tunnels is believed to be similar throughout the world. The capacity of the waterways reduces energy output and limits the possibilities for upgrading. As a consequence valuable energy and peak power output is lost, and the potential of the regulated flow is not fully exploited.

In may 1998 loose thoughts and ideas about reduction of head losses in tunnels materialized into an invention. The basic idea was that if air could be transported in flexible ducts, such as in tunnel ventilation, why not also water? Why not convey the water in a flexible pipe, with a moderate inner pressure, and thus reduce the head loss significantly.

One of the major challenges was to create an inner pressure that is sufficient to create a circular and smooth duct but still low enough to prevent the duct from bursting.

The snowball started rolling. The technology was refined, a patent was applied for and the development work began. A number of companies and institutions were involved in the process, including a company of consulting engineers, a contracting firm, public financing institutions, a membrane supplier and several power plant (-tunnel) owners.

2 THE TECHNOLOGY

2.1 *Principle of the flexible pipe*

The technology is simple. The main component is a flexible pipe, made from a strong PVC/PU coated polyester fabric. The flexible pipe has a circular cross section that fits inside the tunnel walls, and it is therefore not in contact with the rock or the lining. The principle is shown in figure 1 and 2. Although the flexible pipe has a smaller cross sectional area than the tunnel, it is so much smoother that the friction losses may be as low as 30% of that of an un-lined tunnel.

In other words, the technology does nearly the same as a parallel tunnel of equal size.

To control the internal pressure there is a contraction made from steel, at the entrance of the flexible pipe (figure 3). As the flow velocity increases through the contraction, the pressure drops. The low pressure inside the contraction is transmitted to the outside of the flexible pipe through holes in the wall of the contraction. After passing the contraction the flow expands as it enters into the flexible pipe, the velocity decreases and the pressure increases accordingly, while the water outside the pipe maintains the same low pressure as in the contraction. A similar device is installed at the downstream end of the flexible pipe. The result is a constant and controlled pressure difference across the wall of the flexible pipe, even if the pipe is several kilometers long. What makes this solution unique is that the pressure fluctuations due to turbulence always is proportional to the inner pressure.

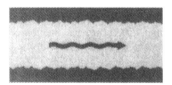

Figure 1: Flow through an unlined tunnel. The uneven surface makes the friction and thus the head loss, much greater than if the tunnel was smooth.

Figure 2: A smooth pipe in the tunnel will reduce the friction loss significantly, even when the cross section of the pipe is smaller than cross section of the tunnel.

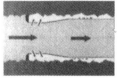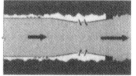

Figure 3: Contractions (e.g. of steel) is installed at each end. Openings, through which water can flow, ensures that an inner pressure is maintained in the pipe. The pipe can therefore be made of a flexible membrane.

3 LABORATORY EXPERIMENT

A laboratory experiment was performed at the Sintef division of civil and environmental engineering in 1999, with the support of Statkraft SF, Statkraft Grøner and the Norwegian research counsel. A 14 m long, 0,4 by 0,4 m "tunnel" was made from acrylic plastic. The flexible pipe was made of thin tent canvas to reproduce the flexibility of a "full scale" membrane. The objectives of the laboratory experiment were as follows:

1. Verify that the technology worked as predicted, i.e. that an inner pressure was created and that a the membrane formed a smooth pipe during steady flow.

2. Investigate the potential for reduction of head loss in unlined tunnels.

3. Investigate how the membrane behaved during different conditions, such as filling, start up and shut down, transients, reverse flow etc.

4. Investigate how the membrane behaved during unwanted situations such as slides, rock fall and puncturing of the membrane.

Figure 4: Laboratory experiment at Sintef. The model was 14 m long, and measured 40 by 40 cm

The experiment was a success. In all ways the membrane behaved as predicted. It was found that the membrane could reduce friction losses in unlined tunnels with 60-70% and that filling, start up and shut down, transients and reverse flow did not introduce problems. The report also concluded: (authors translation from Norwegian)

"Through experiments in a physical model the behaviour of the membrane is studied in different situations such as slides, rock fall and damage through puncture of the membrane. The results show that the membrane behaves satisfactory in the different situations"

Figure 6: Membrane samples at Steinsfoss II headrace tunnel.

5 HYDROPOWER TUNNELS

5.1 *Energy savings*

Reducing the head loss will increase the output from a hydropower plant because a higher net head is available. In addition, the maximum discharge will increase somewhat, and thus reduce losses due to spill of water. A tunnel with maximum flow of $50 m^3/s$ and a head loss of 3,0 m/km tunnel is used as an example. The Groner-membrane will reduce the head loss to 1,0 m/km. If the efficiency of the turbine/generator is 0,80 and the head loss usage time is 3000 hours, 2,4 GWh is saved annually.

5.2 *Upgrading*

Reduced head loss will facilitate upgrading of the turbine/generator. Thus both higher efficiency and higher maximum discharge cam be obtained.

5.3 *Lined tunnels*

The Groner-membrane may also facilitate substantial energy savings in lined tunnels. Calculations show that head losses in a horse shoe shaped concrete lined tunnel with manning number M = 55-60 (n = 0,017 - 0,018) may be reduced with 50%. In lined tunnels the membrane can be very close to the tunnel periphery, and the cross sectional area of the of the flexible pipe can be relatively larger than in an unlined tunnel.

5.4 *Cost savings*

The cost of installation will depend on a number of factors. In addition to the size of the tunnel, the most important factor is the cost of the membrane. The full scale test will provide much information on the membrane quality that is required. However, initial calculations indicate that installation of a Groner-membrane can be done at a significantly lower cost

Figure 5: Senior Researcher Dr. Einar Tesaker studies the inlet to the first contraction.

4 THE MEMBRANE

4.1 *Requirements*

The flexible pipe made of a membrane is the main component of the technology, and a high quality is needed. A PVC or PVC/PUR coated polyester fabric is probably the best choice. Fabric with tear strength up to 180 kN/m or more are available. Such membranes are for example used in "water-bags", in which up to 50 000 m^3 of fresh water is transported at sea.

4.2 *Testing of membranes*

Membranes subjected to flowing water may be subjected to deterioration. In the case of PVC coated textiles, softeners can be washed out. This will reduce the strength and flexibility of the membrane. In March 2000 testing of different membranes started in the headrace tunnel of Steinsfoss II hydro power plant close to Kristiansand in southern Norway. Five samples of four different membrane qualities, a total 20 samples, were mounted on the tunnel wall, exposed to flowing water on both sides. The samples are 0,5 * 1,6 m. Due to the turbulence in the water, the samples were also subjected to continuous flapping or waving.

than construction of an unlined tunnel. A price of a typical unlined tunnel of small cross section in Norway is 10.000 Norwegian kroner, or 1.200 US$.

5.5 *Challenges*

Although many advantages are apparent, there are also a number of challenges. Some keywords are mentioned here:

- Proper fixing to the tunnel wall
- Slides and rock fall
- Sharp rocks and bolts
- Resistance to deterioration
- Resistance to flapping
- Sudden changes in water velocity
- Entrained air
- Reverse flow due to brook intakes
- Remaining water after emptying
- Accessibility

6 WATER SUPPLY TUNNELS

An interesting spin-off of the project is that the Groner-membrane can be used in water supply tunnels. The Groner-membrane will prevent contaminated ground water from mixing with the drinking water. In Norway, this is a problem in a number of tunnels. In Oslo last autumn, 250 000 people had their drinking water contaminated by groundwater that entered a water supply tunnel. In Trondheim, a possible housing area can not be developed because of a water supply tunnel below.

In sewage tunnels, the Groner-membrane can serve its purpose the other way round, by preventing polluted water from leaking and thus harming the environment.

7 FULL SCALE EXPERIMENT

To verify the technology, and to draw the experiences that are necessary before a commercial installation, a full-scale test is planned. Provided that funding is available, the test will be performed during summer/autumn 2001.

7.1 *Location*

An almost ideal location has been found south of Trondheim in Norway, at Håen hydropower plant. The power plant has a four km long outlet tunnel with a cross section of 12 m^2. The maximum discharge is 18 m^3/s. Due to a very rough surface, the head loss at this discharge is close to 3 m/km. Being an outlet tunnel a failure of the flexible pipe will

have small consequences. In addition the access is easy, and the reservoir capacity is relatively large.

The Groner-membrane will be approximately 400 m long and have a diameter of 3,0 – 3,2 m.

7.2 *Scope of full scale test*

The full scale test will give a range of important information. Of great importance is the measured in situ head loss reduction. Head loss in the tunnel will be monitored, both before and after installation. Also, the membrane will be filmed inside during operation, by means of a ROV. (Remotely Operated Vehicle) Further, measurements and filming will be performed during flow changes. There are also a number of practical lessons to learn. Although very much is achieved with proper planning, there are always challenges during installation and operation that will have to be solved.

8 CONCLUSION

Provided the Groner-membrane can reduce head losses as much as the laboratory experiment indicates, and that it can be installed and operated at a competitive cost, the Groner-membrane can significantly reduce the cost of upgrading tunnels and thus hydropower plants.

At this stage the technology is unproven, and there are still many challenges. A full scale test is required before final conclusions can be drawn.

9 REFERENCES

Sintef-report STF 22 F99420, December 1999.

Hydropower in the New Millennium, Honningsvåg et al (eds), © 2001 Taylor & Francis, ISBN 90 5809 195 3

New technologies for dredging of contaminated reservoir sediments

Tom Jacobsen
GTO Hydro AS

ABSTRACT: In the past years the problem of contaminated sediments has drawn an increasing attention. While left undisturbed in sediments these pollutants may cause little damage, but once they are disturbed pollutants and polluted sediment particles may be released into the water column. Appropriate dredging equipment must be able to precision dredge the contaminated layer, minimise resuspension and migration of polluted particles and minimise the water that is added during dredging. In the past eight years a range of sediment handling technologies have been developed in Norway. The Slotted Pipe Sediment Sluicer (SPSS) and the Saxophone Sediment Sluicer (SSS) allows suction of sediment-water mixture into pipelines at high concentrations without causing blocking of the pipeline. Both techniques can be applied without monitoring and no movable parts are required. The GTO ROV dredge and the GTO Subsea dredge have been developed for deburial and dredging of coarse gravel (up 250 mm) The GTO dredges are widely used in offshore operations in the Norwegian and British sector of the North sea. The record depth so far is 540 m. At this operation 1500 tons of gravel were removed in 40 hours by a GTO ROV dredge. The combination of these technologies is a reservoir dredge which can meet the technical and environmental challenges relevant to dredging of contaminated sediments.

1 INTRODUCTION

In the past years the problem of contaminated sediments has drawn an increasing attention. Sediment deposits in harbours, rivers and reservoirs have in many places been found to contain large amounts of toxic agents, such as PCB, PAH, heavy metals etc. While left undisturbed in sediments these pollutants may cause little damage, but once they are disturbed pollutants and polluted sediment particles may be released into the water column. The nature of the problem is different to ordinary reservoir sedimentation, as the volumes are relatively small and are often found in industrialised countries with low sediment transport. Polluted sediments will often have to be treated, ore stored in a confined disposal on shore. Bringing these sediments to the treatment plant or confined disposal site raises several challenges. The sediments must not be re-suspended in the water column. The water that follows the sediments, ore is used for transport of the sediment, must be cleaned. Sediments may have to be dredged from deep reservoirs, and often the size of equipment that can be brought into the reservoir is limited. The sediments may be of variable size, and large items must be expected. Draw down of the reservoir will interrupt the water supply, or it may not be possible

at all. Technology for suction of sediment into pipelines which allow high concentrations combined with technology for subsea dredging of stones may alleviate these problems. Such techniques have been developed in Norway in the past eight years.

2 CONTAMINATED SEDIMENTS

2.1 *Reservoir sediments*

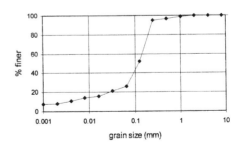

Figure 1: Example of broad grain size distribution. The sample isfrom the Roseires dam in Sudan. (After Pemberton, 1996)

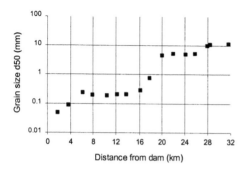

Figure 2: Example of varying grain sizes along a reservoir, from Lake Roxburgh in New Zealand. (Walsh, 1996)

The grain size distribution found in reservoirs will vary from reservoir to reservoir. However, a broad grain size distribution must generally be expected, as shown in the figures 1 and 2.

In addition to the ordinary sediments, large and bulky items, such as trees, roots, stones and waste must be expected.

2.2 Polluting agents

A range of polluting agents can found in sediments. Industry, mining, agriculture transport and sewage are normally the main sources of pollution.

Table 1: Polluting agents

ORGANIC
Hydrocarbons (HC)
Mineral oil
Refined HC
Combustion pr. (NOx)
Synthetic organic compounds
PCB, DDT, KFK, TBT
Plastics
Pesticides
Nutrients (P, N)
Agriculture, sewage

INORGANIC
Heavy metals
Cd, Hg, Pb, Cu, Zn
Radioactive waste
Transurans (Pu, Am, Cm)
Fission products (137Cs)
Others (60Co, 51 Cr)
Industrial waste
ash, acids, mining waste

2.3 Processes

Several processes are important in fixing polluting agents to sediments. Examples are adsorption to Fe-Mn hydroxides and oxides, organic ionic exchange (heavy metals) and adsorption to clay particles and hydrophobic binding to particles. Micro organisms may also play a role. Generally more polluting agents are fixed with high sediment concentrations. Synthetic organic compounds are water repellent, soluble in fat, last long and are easily accumulated in organisms. Clay minerals are chemically active, and therefore important in the ionic exchange process. Heavy metals are found almost exclusively in fine sediments < 20 μ. (Förstner, 1989) At high concentrations (>100 ppm) nearly all heavy metals are fixed to sediments.

3 DREDGING REQUIREMENTS

When contaminated sediments are dredged there are three primary environmental concerns that must be considered before for selecting the appropriate dredging equipment. These include the ability to:

- precision dredge the contaminated layer whether it is thick or thin

- minimise resuspension and migration of polluted particles

- minimise the water that is added during dredging so that the contaminants remain associated with the solid fraction.

More traditional maintenance dredging have been designed using economic and reliability criteria for high production and generally do not satisfy the above criteria. (Young, 1995)

4 NEW TECHNIQUES

In the past eight years a number of sediment handling techniques have been developed at the Norwegian University of Science and Technology, NTNU, (Jacobsen, 1997) and by GTO Subsea AS, a company based in Kristiansund, Norway.

4.1 Slotted Pipe Sediment Sluicer

The Slotted Pipe Sediment Sluicer (SPSS) as a technique developed for suction of sediment into pipelines at concentrations exactly matching the maximum transport capacity of the pipeline. The main feature of the SPSS is that such concentrations can be achieved without the danger of blocking the pipeline. The technology is extremely simple, as there are no moving parts. The SPSS can be described as a pipe with a continuous, longitudinal slot or row of slots along its lower surface. It is fixed close to the original bed / bottom and connected to a suction pipe. The SPSS is operated in two phases, or it can be operated continuously:

1. Sediment is allowed to deposit on top of the slotted pipe until the thickness of the sediment deposit is sufficient for flushing. Because the slots are on the bottom side sediment will not accumulate inside the pipe. Water can thus flow freely through the slotted pipe and out of the outlet pipe.

2. The valve on the outlet pipe is opened, and flushing of sediment starts. Water is drawn through the slots and picks up sediment close to where the slotted pipe emerges from the sediment deposits (the "suction point"). As the sediment is sluiced the suction point moves downstream until all sediment that cover the slotted pipe has been removed.

It can be shown theoretically, and it has also been verified experimentally, that the flow through the slot is concentrated to a short section close to the downstream end of the slot. (In the case where a part of the slot is covered by sediment, this is where the slotted pipe emerges from the sediment.)

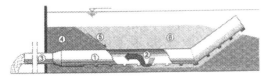

Figure 3: Sediment suction with the SPSS

1: Slotted Pipe   2: Suction point 3: Outlet pipe4: Sediment 5: Sediment sliding down to suction point 6: Removed sediment

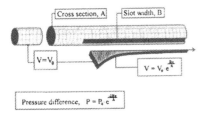

Figure 4: Sketch of flow through infinitely long slot into a pipe. The equations apply to flow of clean water trough the slot.

4.1.1 Field experiment
Field experiments with the Slotted pipe Sediment Sluicer were performed at Jhimruk Hydropower Plant in Nepal in 1994. Jhimruk hydropower plant is a run-of-the-river scheme which exploits the head of approximately 200 m between Jhimruk Khola and Mardi Khola in Western Nepal. The location was ideal for experiments of this kind as they were performed during the testing and commissioning of the power plant. As a consequence the water level could be adjusted on request, allowing both survey and equipment to be installed while the water level was down.

The total length of the slotted pipe was 8,0 m. At the downstream end the slotted pipe was connected to a 45 m long, 125 mm inner diameter flexible pipe with its outlet downstream of the dam. It should be noted that the total head from the water surface to outlet of the pipe was limited to 2,0 m.

The sediment was mostly fine sand and silt, but also included occasional stones up to the size of small potatoes. The median grain size, d_{50}, was 0.12 mm. The sediment was slightly cohesive.

RESULTS
The results of the field experiment can be seen in Figure 7 and is summarised as follows:

– During the two experiments no clogging occurred, and stones small enough to pass the slots were sluiced without any problems.

– The sluicing lasted for 26 minutes and approximately 7 m^3 of sediment deposits were sluiced in both experiments. This gives a capacity of 16 m^3 of sediment deposits per hour.

– The average concentration by volume, C_v, for both experiments was measured at approximately 10 %. This is close to the theoretical maximum concentration as computed by the Durand-Condolios equation. (Vanoni, 1977)

Towards the end of the first experiment the flow of sediment laden water suddenly seized, and (in despair) it was observed that only a small trickle of water came through. The following inspection revealed that the flexible pipe collapsed due to a sharp bend, combined with the external pressure. Even more interesting to observe was that absolutely no sediment remained in the flexible pipe. Thus, what was feared to be blocking of the pipe became a confirmation of the SPSS's ability to handle a special flow situation.

Figure 5: SPSS before the second field experiment. During the experiment the water level was approximately 0.5 m above the sediment.

Figure 6: After the second field experiment in which 7 m3 of sediment were removed in 26 minutes.

4.2 Saxophone Sediment Sluicer

The Saxophone Sediment Sluicer (SSS) is, just as the SPSS, a technique developed for suction of sediment into pipelines at concentrations matching the maximum transport capacity of the pipeline. High concentrations can be achieved without the danger of blocking the pipeline and the technology is extremely simple because there are no moving parts. The SSS consists of a saxophone shaped suction head mounted on a pipeline. It is operated from the water surface. No monitoring is required as it rests on the sediment deposits while dredging is going on. The operation of the SSS can be described as follows:

1. The suction head is placed on the sediment deposits where sediment is to be removed.

2. The valve on the pipeline is opened, and flushing of sediment starts. A sediment water mixture is drawn through the slots and into the pipeline through which it is transported out of the reservoir.

3. When the crater formed has the desired depth, the SSS is shifted to another location and sluicing continues. The procedure is repeated until the desired amount of sediment is removed.

The main feature of the suction head is that it draws water from two places. At the bottom is a row of slots, called the "bottom slots". During normal operation most of the water and all the sediment are drawn through the bottom slots. The upper opening will always be above the sediment deposits. In the event of sediment slides covering the bottom slots, "balancing" water will be drawn from the upper opening and prevent blocking of the pipeline. Due to the lower pressure inside the suction head sediment will still be drawn from the bottom slots, which eventually will be reopened, and normal suction will resume as shown in Figure 8. Because of the length of the inclined bottom part the suction head is not sucked down into soft sediment deposits. Suction only takes place over a part of the inclined bottom part, allowing the suction head to rest on the deposits.

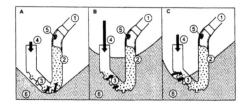

Figure 7: Three modes of operation:

A: Normal operation B: Bottom slots covered with sediment
C: Reopening of bottom slots

1: Outlet pipe   2: Saxophone suction head   3: Bottom slots
4: Flow of water 5: Opening   6: Sediment

4.2.1 Field experiment
As for the Slotted Pipe Sediment Sluicer field experiment were performed at Jhimruk Hydropower Plant in Nepal. The experiments were conducted over a period of three weeks in 1994 and during one week in 1995. The SSS has been operated on 13 days and for a total of 39 hours.

Sediment was sluiced in areas close to the dam. The water was mostly shallow, in most places less than half a metre deep. As the water was too shallow to navigate a raft, the operators had to walk on the deposits while operating the SSS. The SSS was operated as a siphon, that is the pipeline was placed above the dam crest. Priming of the pipeline was necessary before sluicing of sediment could start.

While sediment sluicing took place, the actual head and length of outlet pipe were noted. Samples of sediment water mixture were collected, and concentration measured.

During the experiments, nearly 160 m³ of sediment were removed. Local staff operated the SSS for 64% of time, including all the sluicing performed in 1995.

Figure 9: GTO ROV dredge mounted to a WROV (left) GTO Subsea dredge (right)

Figure 8A local employee demonstrates that only one hour was required to remove 6 m³ sediment. The picture was taken after the water level had been drawn down.

4.3 GTO subsea dredge and GTO ROV dredge

4.3.1 Description

New dredging technology have been developed by GTO Subsea since 1998. (GTO Subsea and GTO Hydro are sister companies) The dredgers have been developed primarily for deburial of subsea structures. Such structures, either they are cables, pipelines or wellheads are (in the North sea) covered by coarse gravel for protection against trawlers. The gravel that is used is crushed rock with d_{max} = 150 mm.

The GTO dredges (patent pending) are unique in that such coarse material can be dredged without blocking of pipelines or destruction of the equipment. Also, the power input is very moderate. For example, the GTO ROV dredge can use the existing interface on standard WROV (Work Remotely Operated Vehicles)

The GTO Subsea dredge can be mounted on belt unit, which will increase maximum capacity, mobility and accuracy of which the suction head can be manoeuvred. Positioning done with DGPS (Differential Global Positioning System) on the surface and acoustic positioning from surface to bottom can be very accurate.

4.3.2 Results

The GTO dredges have performed more than 30 offshore operations in the Norwegian and British sector of the North sea. The deepest operation so far is 540 m below sea level. At this operation 1500 tonnes of gravel were removed in 40 hours by the GTO ROV dredge. The capacity of the GTO Subsea dredge is more than 100 tonnes/hour on continuos dredging of gravel (up to 150- 250 mm maximum grain size) In soft mud the capacity is several times larger.

5 THE COMBINATION: GTO RESERVOIR DREDGE.

5.1 A new dredge

Since 2000, GTO Hydro have developed the GTO Reservoir dredge (patent pending). The Slotted pipe Sediment Sluicer, The Saxophone sediment Sluicer and the GTO ROV or Subsea dredge may be combined into a new type dredge. The dredge will have several advantages, such as:

- Accurate positioning of suction head if desired

- Low resuspension of sediments

- High sediment concentration

- All mechanical equipment can be accessible

- High degree of reliability

- Dredging at great depths

- Dredging of coarse material

- Long distance transport of fine material only

5.2 Description of the GTO reservoir dredge

The two main parts of the GTO Reservoir dredge are the suction unit, which may be a saxophone suction head or GTO ROV or Subsea dredge and the pump unit, where are low water level is created in a barge by pumping of fine material from the barge to the site of disposal.

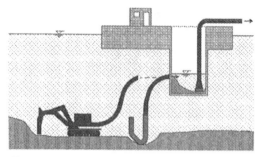

Figure 10: Principle sketch of the GTO reservoir dredge. A GTO dredge on a belt unit or ROW or a Saxophone suction head may be used.

The pump unit may be equipped with a Slotted Pipe Sediment Sluicer to ensure that the sediment concentration delivered to the pipeline exactly matches the transport capacity.

6 CONCLUSION

Reservoir sediments in industrialized areas, in particular where sedimentation rates are moderate, will often be contaminated. Removal of contaminated reservoir sediment puts special requirements to the techniques and equipment which will be used. A range of recently developed sediment handling techniques, combined into the GTO reservoir dredge, can be used to meet these requirements.

REFERENCES

Förstner, Ulrich. 1989. Contaminated sediments Lectures on Environmental aspects. Springer-Verlag

Jacobsen, Tom. 1997. Sediment problems in reservoirs – Control of sediment deposits. Doctoral Thesis, IVB Report B2-1997-3. Department of Hydraulic and Environmental Engineering, The Norwegian University of Science and Technology (NTNU).

Pemberton, Ernest L. (1996) Reservoir sedimentation. Case study - Roseires Dam, Sudan, Africa. Int. Conference on Reservoir Sedimentation, Fort Collins, USA, Sept 9. - 13. 1996.

Vanoni, Vito A. (ed.) 1975. Sedimentation engineering. ASCE Manuals and Reports on Engineering Practice No. 54, New York.

Walsh, Jeremy. 1996. personal communication.

Young R. N. 1995. The fate of Toxic pollutants in Contaminated sediments. In K.R. Demars et al (eds), *Dredging, remediation, and containment of contaminated sediments*: 13-38, Philadelphia, ASTM

Hydropower in the New Millennium, Honningsvåg et al (eds), © 2001 Taylor & Francis, ISBN 90 5809 195 3

Small Lower Head Hydropower Plant Design – Batou Hydropower Plant

Wang Junhong
*Guangdong Provincial Investigation, Design and Research Institute of Water Conservancy and Electric Power
(GPDI), Guangzhou, China*

ABSTRACT: Structures of a small lower water head hydropower plant is not so complicate as large or high water head hydropower plant. But there are various structures all-round in a small one. It is necessary to consider as a large in general structure design, and simplify in some places according to its different from. Batou power plant is a typical type one of that on the river bed. The project has been completed in 1999 and operation well up to now.

1 INSTRUCTION

Batou project is located in Meixian County, the northeast of Guangdong Province, China. Hydraulic structures include power plant, log pass and 8 discharging sluices with each net width of 14m. (Fig. 1) Batou hydropower plant is a typical small lower head project. It is a daily regulation hydropower plant which only 2,500,000m³ of daily regulation capacity. The total generation capacity is 15MW and 3 bulb

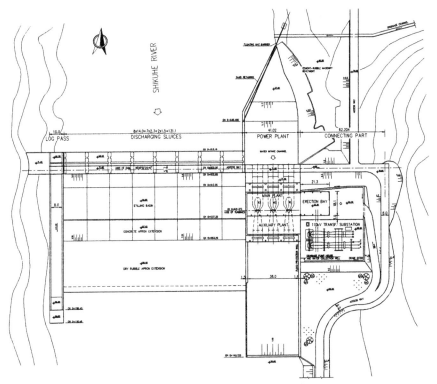

Fig 1. General Layout of Batou Project

turbines installed and 5016 kWh of annual energy out put. The single unit flow is 96.20m³/s and only 7.6m of average water head. The power plant is close to discharging sluices on the left of river bed and the log pass on the right.

Due to the power plant is a component of water re taining structures together with discharging sluices and log way. Not only design consideration according to a common lower head hydropower plant, but also retaining structures should be taken into account. Design criteria for all hydropower plant should be adopted into this project, but it is necessary to take proper optimization somewhere.

2 GEOLOGY

Batou project site is situated upstream 3km of the estuary of the Shikuhe river of Meixian County. Rolling region and undulating topography around the project site, and gradual open topography towards downstream.

The rock mass in site foundation area is mainly a group of fine-coarse quartz sandy stone beds, locally inserted by mudstone, boulder clay, and basite dikes. The strike of rock beds is relatively stable, N70W/NE25° at the left bank, and N35W/NE35° at the right, with a dip to upstream side. Mudstone and clayey siltstone are extremely weak to against weathering events, which the rock beds would be detrited and disintegrated in to graniform material when exposed, and finally, clay and mud would be appeared by the contribution of further weathering. The strength of the weathered rock will be rapidly disappeared when saturated. No regional faults is fund in

site foundation area and the Seismic Intensity is 6 degree in the region.

3 DESIGN PHILOSOPHY

Batou hydropower plant is designed in accordance with Chinese design code SD335-89. Most specifications in the code are applicable to Batou power plant though it is a small lower water head project. The main features of the design consideration are briefly described as follows.

3.1 Generation requirement

The power plant is in the river channel on the left side. Three bulb turbines of GZTF08-WP-365 type are installed with a capacity of 5MW per unit. The power plant is a surface plant on the river bed with a dimension of 62.3m × 17.6m × 33.78m (length × width × max. height. (Fig. 2)

The power plant is consisted of main plant, erection bay and auxiliary plant for its hydropower generation necessary, and 110kV transformer substation as well. Pressure water passageway is at the lower level under each unit with the total length of 57.732m. The generator and turbine runner are installed inside airtight seal bulb body, and the turbine runner center level is EL52.39m. Different structure cross sections from water intake to outlet. The maximum section is rectangular section with 7.348m in width and 8.2m in height, and the minimum section is circular section of diameter 6.86m. There are steel liners both in front and behind of the bulb body in the water passageway, and the other parts are massive concrete structures at all.

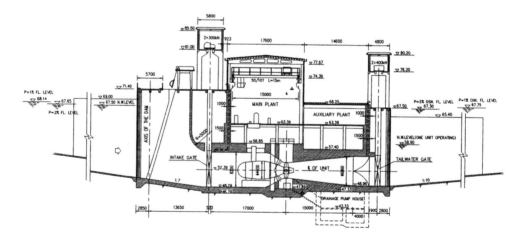

Fig 2. Cross Section of the Power Plant

Under the water passageway at the bottom, there is an operating gallery crossing with the level of EL43.33m. The water collecting well and the pump hose are arranged at the side of #1 unit. Water from the unit service and emergency drainage and seepage of concrete will be collected here and draining away by 4 pumps.

The equipment floor is at EL58.85m, which arrangement by various oil, water, gas pipes, compressed air system and cables, and operating access as well. The upper floor is operating floor with the level of EL63.39m. There are unit speed governors, oil pressure systems and electric instrument installations at this floor.

The erection bay is at the left side of the main plant. The length of the erection bay is 21.3m and the width is 17.6m in accordance with the unit heavy assemble and maintenance. The ground level of erection bay is EL63.39m, which same as the operating floor of the main plant. At left side of the erection bay, there is a plant access gate which linking 110kV transformer substation and access roads outside.

Auxiliary plant is arranged in downstream side of the main plant and close to it. The length is 41m same as the main plant and the width is 15.1m. Auxiliary plant is tiered by two floors. The lower floor is cable floor with the level of EL57.40m and the surface floor is at EL63.39m which same as the operating floor of the main plant. The center control room, high and lower voltage switch rooms, electric laboratory and communication equipment room are arranged at the surface floor.

Both natural ventilation and mechanical ventilation ways are considered under design of the exhaust ventilation system. The upper floor of main plant, such as the operating floor, is natural ventilation due to that on the ground. Mechanical ventilation systems arranged at each lower floors and the bottom gallery because of that always under water and ground surface. Embedded ventilating ducts installed through upstream and downstream concrete structures to the top of the plant, and vent openings arranged at each floors.

3.2 Water retaining requirement

According to the general layout of Batou project, the power plant (main plant) is a part of river barrage together with 8 discharging sluices and log way. Therefore the water retaining structures should be taken into account at the leading structures of the intake of the power plant. The top level of water intake piers is EL71.40m that same as discharging sluice piers by thinking of the water retaining. This top level is designed in accordance with lift water level to gain hydroelectric head and 50-year frequency flood control and 100-year frequency emergency flood protection requirement. At the upstream side of power plant, there are concrete water retaining walls, and better performances of waterproof and seepage prevention should be provided with the walls. There is a sand retaining structure in front of the water intake.

3.3 Flood protection requirement

Structures of the plant upstream are enough to retain flood from upstream. Due to the 100-year frequency flood level is at EL67.75m in the downstream of the power plant, it is higher than the level of EL63.39m which is the operating floor and erection bay floor, even higher than the surface level of the 110kV transformer substation. So that flood protection structures, concrete wall around the plant, will be set up. The wall extending to downstream about 35m along the river bank to protect substation.

At the corner of the transformer substation downstream which back of the wall, there is a water collecting well and a little pump house to draining surface water and rain water around the power plant. The water can be draining off to the river by gravity drain because of lower water level in the river downstream in the most times. In the flood season and rainfall season, the river is high-water level downstream. When the water level is higher than the surface level of the substation, the inverted valve will auto closed to prevent flood down-draught. At this time, surface water inside the walls will be drained off by pumping to protect power plant operation safety.

3.4 Foundation treatment

By the layout of the power plant, only about 20m of the maximum excavation depth and no highwall slope problems. Foundation of the power plant is laid down weakly-weathered rock zone and the foundation rock mass have nice quality and better integrity. Consolidation grouting on the foundation is canceled for the reasons of good foundation rock conditions. Structures concrete were placed on the cleaned-up rock surface immediately.

It is clearly that the rock hydraulic conductivity is relatively poor on the basis of the geological exploration. Commonly, the rock hydraulic conductivity is over 10Lu that above level EL47.0m of strong-weathered or weakly-weathered rock zones and less 10Lu which under EL47.0m. The power plant foundation bottom level is EL45.89m that is in weakly permeable rock. On the other hand, it is less effective from the foundation seepage for only 8.45m of maximum work water head difference. For above reasons, no permeability reducing measures such as grouting is designed at the power plant foundation area. But local foundation treatment is necessary by its geological features during constructing process.

4 STRUCTURE DESIGN

4.1 *Structural joints*

In accordance with the unit equipment arrangement required, the unit spacing is 11.5m and the main plant length is 41m. Thinking of this length is small that can not be resulted in serious structural problems, no structural joints considered between unit structures. There is a structural joint between main plant and erection bay for its different service behavior and foundation structure type and depth.

The main plant upper building and the auxiliary plant are laid over water passageway with combined structures. There is no structural joint between main plant and auxiliary plant. Along the water passageway there is no joint set under the preliminary design stage. But at the beginning of the construction, one structural joint designed in the front of the inlet gate, from the bottom to the top, cross the main plant intake piers. The reason is long structure distance of 57.732m form intake to outlet of the water passageway, which is exceed the definition of the design code although structures always under water, and a little worried about massive structures.

Copper seal and filling material are applied in all structural joints to form a closed water-stop strip along joints.

4.2 *Substructures*

Submerged structures of the power plant mainly consist of foundation bottom plate, water passageway, intake and outlet piers, water retaining wall upstream and flood protection walls around the plant. Foundation bottom plate and water passageway are principal load bearing structures, therefore substructures always are massive concrete structures. The monolithic reinforced concrete and thick shell structure is applied in the water passageway by means of large area of wetted cross section. The thickness of the bottom plate is 1.5m to 2.0m. There are many shaft wells, access ways and openings in submerged structures. It is necessary to consider whether stress concentration will occurred at these places, and structural measures or modifications will be taken into account at the places which stress can not be satisfied by structural calculation.

Water intake and outlet piers are massive concrete structures which thickness is 4.152m and 4.375m of intermediate piers. The upstream wall of the power plant must have functions for water retaining and flood protection. So the concrete wall thickness is 1.5m of that under level EL63.39m and that of 1.0m above to the top EL71.40m. For the flood protection downstream, concrete wall is designed around power plant. The same thickness design consideration as the upstream wall.

It is important that selection of substructures material for its impervious requirement. The waterproof concrete adopted as the concrete grade is not allowed lower than that of C20/W4 (Chinese design code).

4.3 *Upper structures*

The upper structure means that of partial structures of main plant and auxiliary plant above the water passageway below. Under the protection of flood protection walls around the power plant, normal cast-in-situ concrete framed columns and beams construction can be realized. Structures including each level of floors, columns, beams, stairs, crane beams and roof.

Because of less occurrence frequency of flood and high water level on the river, it is unnecessary to retaining high water level outside by concrete walls in the most time, and additional moisture prevention works (e.g. brick walls) inside the wall faces are unnecessary on the flood protection walls. Upstream side columns of the main plant are put on the top of walls directly and main beams of the auxiliary plant connecting with downstream walls directly. As the result of such arrangement, the width of the power plant is shorten 2~3m at least as well as submerged structures. The concrete amount will be decrease.

5 MONOLITHIC STABILITY AND FOUNDATION STRESS

Calculation for the power plant monolithic stability and foundation stress ordinary contain safety against sliding stability along foundation base of the power plant and vertical normal stress of the base. The safety against floatation will be calculated at the case of higher tailwater level.

Three cases are taken into account for monolithic stability and foundation stress analysis of the power plant:

Case 1: Normal operation with the normal water level EL67.50m upstream and EL58.90m water level downstream of only one unit operating.

Case 2: Unit service with the normal water level EL67.50m upstream and mean annual water level EL59.05m downstream, but no water filling in the water passageway.

Case 3: Emergency case with abnormal flood level (P=1%) both upstream of EL68.14m and downstream of EL67.75m. This case calculated for the safety against floatation only.

The equation of shear strength calculation applied to the monolithic stability against sliding as follow:

$$K = \frac{f \Sigma W}{\Sigma P} \tag{1}$$

where K = factor of safety against sliding; f = friction factor of against shear failure between concrete and rock (f=0.55); ΣW = normal component of all loads to the sliding base; and ΣP = tangential component of all loads to the sliding base.

The equation of stability against floatation calculation stated as follow:

$$K_f = \frac{\Sigma W}{U} \tag{2}$$

where K_f = factor of safety against floatation; ΣW = total weight of the calculated unit bay; and U = sum total of the uplift pressure acting to the unit bay.

Structural mechanical method can be applied to calculating vertical normal stress of the foundation base, the equation described as follow:

$$\sigma = \frac{\Sigma W}{A} \pm \frac{\Sigma M_x y}{J_x} \pm \frac{\Sigma M_y x}{J_y} \tag{3}$$

where σ = vertical normal stress of the foundation base; ΣM_x and ΣM_y = sum total of force moment to the axis X and Y of total weight of the calculated unit bay; x and y = distances from calculated point to the axis Y and X; J_x and J_y = inertia to the axis X and Y of the calculated section; and A = section area of the foundation base.

Calculation results from above equation (1) to (3) are detailed as follow Table 1:

Table 1. Calculation results of the power plant monolithic stability and foundation stress

Item	Allow-ance	Case 1 normal	Case 2 service	Case 3 emrgc.
Against sliding K	1.1	3.18	2.67	
Against floatation K_f	1.1	1.92	1.75	1.35
Max. stress of base σ_{max} (MPa)	2.50	0.172	0.156	
Min. stress of base σ_{min} (MPa)	>0	0.123	0.088	

It is clearly that the monolithic stability and foundation stress of the power plant can meet demands of the design code. Calculation results indicate that the power plant is safety under different operating cases.

6 CONCLUSIONS

Small hydropower plant design should be followed general regulations of the design code, such as generator unit arrangement, flood protection, drainage systems, ventilation systems etc. By means of lower water head hydropower plant features, lower water pressures action to the hydraulic structures and no higher excavation slope, relative lower requirement to the structure design and its foundation and slope treatment. Design consideration of structures can be properly simplify in accordance with the power plant layout characteristics. But such structural optimization are limited and determined by calculations, experiences and site conditions.

7 REFERENCES

Wang Junhong, Huang Licai, etc. May 1998. Chapter 5, General Layout and Hydraulic Structures. *Preliminary Design Report for Meixian Batou Hydropower Project*. 5-1 to 5-48, Guangdong Provincial Investigation, Design and Research Institute of Water Conservancy and Electric Power (GPDI), Guangzhou, China

Chen Shurong, ect. April 1998. *Batou Hydropower Project Hydraulic Model Test Report*. Guangdong Provincial Institute of Water Resources and Hydropower Research.

March 1989. *Design Code for Hydropower Station SD335-89*. Ministry of Water Resources, P.R.China.

September 1996. *Design Code for Hydraulic Concrete Structures DL/T5057-1996*. Ministry of Water Resources, P.R.China.

Hydropower in the New Millennium, Honningsvåg et al (eds), © 2001 Taylor & Francis, ISBN 90 5809 195 3

Silica use in RCC Dams – Considerable Cost Savings and Improvement of Quality?

Øystein Lilleland
Norsk Hydro ASA, Hydro Energy

ABSTRACT: Adding silica to the roller compacted concrete (RCC) mix will, under given conditions, save material costs and improve the quality of the fresh and hardened RCC. Laboratory tests and previous experience in Norway confirm these statements, but they have yet to be proved through a full-scale trial and actual construction of an RCC dam. The potential cost savings will vary considerably from site to site and between different countries due to local differences in silica and cement prices. This paper outlines some of the interesting results from the R&D project "RCC for Norwegian conditions", with special focus upon the features of silica-RCC.

1 INTRODUCTION

The era of large dam construction in Norway and northern Europe was more or less over before 1990. The "new" RCC dam concept, which is the fastest growing construction method in the world today, has not been an alternative in Norway due to its short history. Consultants, contractors and others have however been working on several international projects, and have thereby gained knowledge of the construction methods. The successful history of rock-fill and concrete dam construction in Norway represents a general knowledge basis which is assumed to be important in establishing a foundation for the RCC construction method under Norwegian climatic conditions.

In 1996, the Urevatn dam in the southern part of Norway was tendered with two alternatives - rock-fill dam with central asphalt concrete core, and RCC dam. The rock-fill dam alternative was selected for economic reasons.

The experience from the Urevatn dam showed the need for better knowledge of the RCC technology in Norway. This, combined with the plans of Vest-Agder Energy for two new dams at Skjerka, initiated this R&D project (Lilleland 2001). The purpose of the project was:

– to make a design mix with local Skjerka aggregates;
– to investigate freezing-thawing on RCC;
– to investigate the impact of silica use in RCC;

– to make the RCC technology "available" to all interested parties, power companies, consultants, governmental departments and contractors.

This paper focuses on the use of silica in RCC. To the authors' knowledge, silica has, not been used in any RCC dams built so far. In Norway, silica was used with very good results in two big double-curved, concrete arch dams in the 80s.

2 RCC FOR NORWEGIAN CONDITIONS

Of the four main types of RCC (ICOLD 1997), High Paste RCC is found to have the most suitable qualities for Norwegian Conditions. The other types of RCC (Lean RCC, Japanese RCD method and Medium Paste RCC) have not been disqualified, but High Paste RCC is preferable due to climate, regulations set by the Norwegian dam authorities etc. In addition, the development seems to be moving in the direction of more High Paste RCC as the development of the technology matures (Fig. 1).

3 THE POTENTIAL OF SILICA (OR MICRO SILICA)

During the last 10-20 years, silica has become a significant pozzolan in concrete due to its special characteristics. Originally this was a waste product from the ferro-silicon smelting plants/industry, and at the beginning the pozzolan waste at the factories was almost given away.

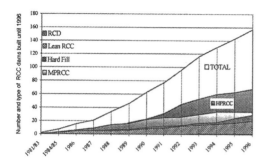

Figure 1. Accumulated number and type of RCC dams built until 1996 (ICOLD 1997).

The particles have a specific surface area of an incredible 20 m^2/g compared to cement particles' area of 0.4 m^2/g. Some of the advantages that have been documented with silica as a pozzolan are:

- Increased workability of the concrete
- Cement content reduced by more than the added amount of silica (2-4)
- More cohesive concrete
- More impermeable concrete

Today silica is acknowledged as a pozzolan and has gained market shares in several segments of the concrete industry. The research on different types of structures so far is with few exceptions positive.

However, this is not the case for RCC dams. To the authors' knowledge, silica has not been used as a pozzolan (together with cement and a pozzolan such as flyash or blast-furnace slag) in any of the dams built so far.

The objections against silica have been that the pozzolan itself will increase the temperature of hydration which is highly undesirable in massive structures like dams. This would be a correct observation if the adding of silica by one part by weight meant the removal of one part of cement. However, this is not the case. It has been proved that one part of silica can entail the removal of two to four parts of cement in a typical concrete mixture without reducing the quality. The hydration temperature should therefore rather be less than that of the original concrete, bearing in mind that the hydration temperature for silica is almost the same or slightly higher than that of Standard Portland Cement (ASTM C150).

The interesting characteristics of silica led Norwegian dam engineers to investigate the impact of adding this pozzolan in two big double-curved concrete arch dams built at the beginning of the 80s. The experiences from these constructions were quite remarkable and formed the foundation for introducing silica as a part of the research program "RCC for Norwegian Conditions." (Lilleland et al. 2001)

3.1 The Førrevatn and Alta dams

The background for introducing silica as a pozzolan in RCC is the experience from two 100-meter high mass concrete arch dams built during the 80s. These dams were built with a thickness of 15 meters (15 meters below HRWL) due to safety measures and regulations. These new regulations made the hydration temperature an important subject of consideration. The adiabatic conditions inside the cross section made it clear that the concrete mixes had to be designed to meet the requirements of low temperature, a certain compressive strength, tensile strain capacity and low permeability. The experience from the first CONDEEP structures in the North Sea (platform foundations of massive concrete) were important in developing concrete for these two dams. During the 70s, extensive laboratory investigations were carried out in order to find a mixture which would meet all the requirements for a dam of this size.

The Concrete Mix designed for the Alta dam in Northern Norway was blended cement with 30% blast-furnace slag produced by a nearby cement factory. The mix produced for the Førrevatn dam was almost identical, but blast-furnace slag was not available and was replaced by low-lime flyash from Denmark.

Table 1. Concrete Mixes designed and used for the Alta and Førrevass dams.

Concrete	Førrevass dam*	Alta dam*
	kg	kg
Cement	110	105
Blast-furnace slag		45
Flyash	40	
Silica	12	10 kg (~7%)
Air-entrainment	1.5-2%	2.5-3.0 liters**)
Plasticizer	2	2
Water content	100	100
Water/cementitious ratio	0.57***)	0.60
Max. aggregate size	120 mm	120 mm

*) From site reports it appears that there are relatively large variations in the water/cementitious ratio among the samples. The table shows the mean values.
**) Measured air content in laboratory at site without aggregates >32 mm.
***) The water/cementitious ratio with the air entrainment included is 0.66.

Data for the concrete mixes which were developed and used for the two dams are shown in Table 1.

The concrete behaved very well during the construction period. The method of construction was laying thick layers in blocks with transverse formwork at the predefined joints in the dam. The concrete was immersion-vibrated which was the main difference from the Japanese RCD method of construction.

418

The focus on reducing the water and cementitious content by using large-sized aggregates was successful. Strict considerations on the aggregate composition (5-6 sizes) prevented problems with segregation. The use of silica was considered (probably correctly) to have a great impact on the lack of segregation. The documentation from these projects was the main reason for facilitating the silica testing in RCC. A paper on the use of silica in the Førrevatn dam was presented at the ICOLD's fifteenth Congress on Large Dams in Lausanne 1985 (Børseth, 1985).

4 TWO RCC PROJECTS IN NORWAY

4.1 The Urevatn dam

The first RCC project in Norway that almost came to fruition was the Urevatn dam in the southern part of the country. The RCC alternative lost in tender competition with a rock-fill alternative in 1996. The explanation of the loss in the competition was a combination of a difficult transport route for plant equipment, strict regulations on concrete, and a volume of 40,000 m³ which was too small at this location. The preparation of the RCC alternative included a laboratory concrete mix program with local aggregates from the site. The optimized design mix showed a compressive strength of 45 MPa. The composition of the mix was as follows:

Portland Cement:	85 kg/m³
High-lime flyash:	115 kg/m³
Water Content:	110 kg/m³
Fine aggregates:	715-765 kg/m³
Coarse aggregates:	1450-1550 kg/m³

4.2 The Skjerka dams

In 1997 a considerably larger project came up where the volume of RCC would be up to 350,000 m³ in two dams at a distance of 1 km. This project is still being considered by the owner (Vest-Agder Energy), but if the project materializes, the owner has decided that RCC will be the only alternative out on tender for one of the dams based on thorough preliminary investigations.

An identical laboratory mix program as for the Urevatn dam was initiated. The laboratory program was designed with reference to ICOLD Draft (ICOLD 1997) on design of mixture proportions. The R&D project was initiated at the same time and carried out in close collaboration with the laboratory program and in the same laboratory (Lahus 1999a).

The aggregates used in the laboratory were fetched from the probable future quarry on site, and were crushed to five sizes at a nearby mobile crushing plant. The maximum aggregate size was 50 mm.

approximately as recommended (ICOLD 1997) for High Paste RCC.

Nine initial mixes with variations in flyash and cement content were tested in the laboratory. The test procedures were as follows (Lahus 1999a):

– Loaded VeBe
– Density of fresh and hardened concrete
– Heat of hydration
– Compressive strength
– Tensile strength
– E-module

The different results are not presented in this paper, but compared with the Urevatn dam, the cementitious content was reduced by 20 kg. The optimum mix was found to have 110 kg of cement and 70 kg of high-lime flyash. The strict adjustments of the aggregate curve and better characteristics of the crushed aggregates are the probable explanations for the difference between the two optimized mixes.

5 THE R&D PROJECT

5.1 The silica laboratory program

In addition to the "standardized testing program", the laboratory part of the R&D project focused on silica as a third pozzolan in the concrete and an extensive testing of freezing-thawing in both facing concrete and RCC with and without silica.

The optimized mixes from the "standard" laboratory program were used as reference mixes when designing the silica-RCC mixes. Three additional mixes with varied silica contents were prepared.

Table 2. RCC mixes with silica.

Mix no. (Cementitious content)	Paste content			
	Water kg	Cement kg	Flyash Kg	Silica kg
1 (180 kg)	115	50	120	10 (5.6%)
2 (174 kg)	95	40	120	14 (8.0%)
3 (167 kg)	95	40	120	7 (4.2%)

Table 3 shows the different tests done on the hardened concrete samples.

Table 3. Laboratory tests - hardened concrete

Mix no.	Tests and moment in time of testing			
	Compressive strength	Perme-ability	E-module	Tensile strength
	Days	Days	Days	Days
1	2, 7, 28	-	-	-
2	2, 7, 28, 93, 182	182	93, 182	93
3	2, 7, 28, 93, 182	182	93, 182	93

The procedure of finding the mix proportioning followed the same procedures as in the "standard" program. The flyash remained fixed and the goal was to investigate if the addition of silica would replace cement with more than a factor of 1.

At first, the fresh concrete was evaluated visually during mixing in the laboratory. Some mixes with very high amounts of silica were tested and were found to be too cohesive. However, all the test mixes behaved very well in the laboratory mixer. One of the most interesting properties of the silica-RCC is the density of fresh and hardened specimens of the theoretical air-free density. The silica-RCC seems to have a higher density than the RCC without silica.

Figure 2 indicates that the density of the silica mixes is 2% higher than mixes without silica. The paste/mortar ratio for the silica mixes 2 and 3 is 0.38-0.385.

The three mixes without silica have a paste/mortar ratio that varies from 0.352 for the Cement-flyash 100-60 mix to 0.385 for the two other mixes. The difference in densities can hardly be explained by the difference in the paste/mortar ratios.

The relationship is shown in Figure 3, which shows that the paste has two fill (air) voids in the fine aggregates, which theoretically have a minimum void ratio of 0.32. The void ratio of the fine aggregates for the RCC mixes with and without silica was 0.35. The paste/mortar ratio should be slightly above this in practice.

The results strongly indicate that the silica has a positive impact on the fresh and hardened density. It is assumed that the silica lubricates the surfaces in the microstructures and the waterfilms between the particles, and also acts as a micro-filler. However, the most important indication is that the silica seems to reduce the need for compressive energy. Full scale, this would mean that the ability to be compacted is better with silica than without silica.

By replacing portions of the cement with silica, with fixed portions of flyash, the mean compressive strength was maintained with increased water/cement ratio. Mean compressive strength after 93 days was 33 MPa for silica-RCC with a water/cement ratio slightly above 2.3. With the same amount of flyash (120 kg), corresponding compressive strengths were obtained for RCC mixes without silica at water/cement ratio of 1.4 (Fig. 4).

The relative cement content, C/C+Fa+Si, compared to the 93-day compressive strength shows that silica mixes have a relative cement content of 0.23-0.24. Corresponding compressive strengths were found for the RCC mixes without silica with relative cement content of 0.42 (Figure 5).

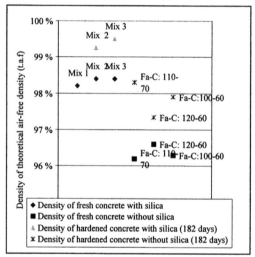

Figure 2. Density of fresh and hardened concrete, with and without silica

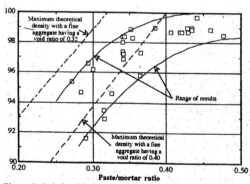

Figure 3. Relationship between in-situ density and paste/mortar ratio from approximately 50 different roller-compacted concretes from 20 projects (ICOLD 1997)

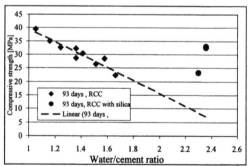

Figure 4. Compressive strength as a function of Water/cement ratio for RCC with and without silica (Lahus 1999b).

We obtain the same or better compressive strength after 93 days with silica mix 2 and 3 than RCC mixes with 120 kg flyash and 60 kg cement or 110 kg flyash and 70 kg cement.

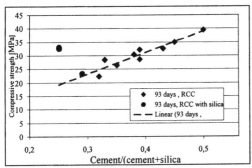

Figure 5. Compressive strength as a function of relative cement content, C/(C+Fa+Si), with and without silica (Lahus 1999b)

These results indicate/confirm that utilization factor should be used on the silica. Mix 2, with a water/cement ratio of 2.4 (95 kg water/40 kg cement) and 14 kg of silica, equals the RCC mixes 120 Fa-60 C or the 110 Fa-70 C. If we disregard a small variation in the flyash content, the water/cement ratio of 1.4 corresponds to a cement content of 68 kg or the 110 Fa-70 C RCC mix according to figure 4. This indicates that 14 kg of silica replaces (68-40) = 28 kg of cement. This gives a utilization factor of k=2 for the silica. The same calculation used for mix 3 with 7 kg silica gives a utilization factor over 4.

The above approach prevails if the utilization factor k is a function of the compressive strength only. However, factors such as tensile strength and permeability also have to be considered to enlarge the picture. The tests on E-modules and tensile strength after 93 days are shown in Figure 6.

tent. The compressive strengths are in the range of 6 MPa (37.4-43.8 MPa) for all the mixes after 182 days. The tensile strength is within a range of 6-8% of the compressive strength for all mixes. It seems that the lower cement content is balanced by the relatively smaller amount of silica with respect to tensile strength. The E-module however, seems to be dependent on the total cementitious content, or even the fact that the silica content has a slightly negative effect on the E-module. The relations here are complex, but the utilization factor does not correspond as clearly as with the compressive strength.

Prior to the permeability tests, it was expected that the silica mixes would be more impermeable than the ordinary RCC mixes. The density of the silica was higher and site reports from the Førrevatn and Alta dams showed impressive impermeabilities ($k=10^{-14}$ m/s), which were partly explained with the use of silica. The high fresh density of the silica mixes strengthened the belief in low permeabilities. However, the permeabilities were higher for the silica mixes than the ordinary RCC mixes. Nevertheless, the permeabilities were fully acceptable and very low for all mixes.

The reason for the unexpected difference between the permeabilities between silica mixes and ordinary RCC has not been found, but there is a suspicion that the relative amount of flyash and cement influence the balance.

Nonetheless, the high fresh and hardened density will probably result in a high density when compacting on full scale.

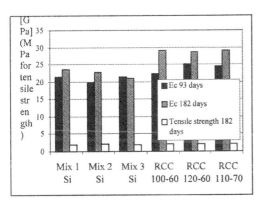

Figure 6. E-module and tensile strength for RCC mixes with and without silica (Lahus 1999b)

The tensile strength after 182 days is more or less the same for silica-RCC and RCC mixes. The E-module after 182 days is lower for silica mixes, which seems to be a result of the lower cement con-

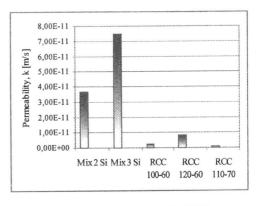

Figure 7. Permeability after 182 days (Lahus 1999b)

The correlation between the in-situ conditions and laboratory tests have to be verified in full-scale trials in order to confirm these suggested correlations. It has been confirmed however that the silica has interesting characteristics for High Paste RCC. The laboratory testing on the temperature of hydration confirms that the silica will decrease the maximum

421

temperature given adiabatic conditions. In simple terms, if we assume that the temperature of hydration increases by 1.4°C/10 kg cement and 7 kg of silica replaces 30 kg of cement, the maximum adiabatic temperature would be 3-3.5 °C lower.

5.2 Freezing-thawing silica-added RCC compared to optimum design mix without RCC

A large part of the scope of the R&D project covered two methods of freezing-thawing investigations. The severe Scandinavian climate represents a challenge to the RCC method and protection of the interior concrete. The High Paste RCC method entails facing concrete, both on the upstream and downstream face of the dam. The development of grout-enriched RCC and other methods produce new interesting aspects on how to create a protective face for the interior concrete.

The R&D project tested facing concretes and RCC with two different methods (with respect to two freezing-thawing phenomena):

1 Scaling at the surfaces
2 Internal cracking

The first effect arises with freezing-thawing cycles at the surfaces of the exposed concrete, which was tested with the "Borås method". Originally developed in Sweden, this method is a candidate for a CEN method.

The second effect is internal cracking as a result of critical saturation within the concrete, and at the same time freezing-thawing cycles that cause higher forces than the tensile strain capacity of the concrete.

The following concrete mixes were tested with respect to freezing thawing:

1 Facing concrete, flyash/cement 0/270 kg, water/cementitious ratio = 0.5
2 Facing concrete, flyash/cement 80/190 kg, water/cementitious ratio = 0.5
3 RCC (interior concrete), flyash/cement 60/120 kg, water/cementitious ratio = 0.53
4 RCC (interior concrete), flyash/cement 60/120 kg, water/cementitious ratio = 0.53 with air entrainment.
5 RCC (interior concrete), flyash/cement/silica 120/40/14 kg, water/cementitious ratio = 0.55 (k=1 for silica)
6 RCC (interior concrete), flyash/cement/silica 120/40/7 kg, water/cementitious ratio = 0.57 (k=1 for silica)

The interesting reports on freezing-thawing (Jacobsen 1999) give a thorough background for the laboratory program. It is not within the scope of this paper to present all of this, but the results are summarized here:

– Facing concrete 1 and 2, with and without flyash is very resistant against freezing-thawing. There is no sign of damage in testing with respect both to scaling and internal cracking. The flyash replacement of cement has no documented effect on the freezing-thawing durability of mix 2.
– As expected, the RCC mix 3 was not durable against freezing-thawing with respect to both test methods, even after 180 days in isolated curing. The requirement of a facing concrete was confirmed.
– Air-entrainment in the RCC increased the resistance against freezing-thawing. The results showed that the air-entrained RCC could replace facing concrete even at locations with severe climatic conditions. The practical problem with the compaction and airpore distribution is however a challenge with the RCC.
– The silica mixes did not stand out compared to the RCC mix without silica, either positively or negatively. The same challenges will emerge with air-entrainment in the silica mixes.

5.2.1 Discussion with respect to freezing-thawing in RCC dams

Grout-enriched RCC against the faces has not been a subject of testing in the project. The interesting aspects of this method could however be tested in a full-scale trial with respect to air-void distribution and subsequent freezing-thawing tests in the laboratory. The savings by not placing facing concrete but an "adjusted" RCC should certainly encourage a closer look at the issue.

The report (Jacobsen 1999) shows that there is a notable difference between the facing concrete and RCC with respect to the thermal coefficient of expansion. Basically this is explained by the relatively larger amount of aggregates in the RCC. With extreme temperature variations and during the hardening period, the interface between the two concretes will have different patterns of movement. I have no knowledge of whether there has been any damage as a consequence of unfortunate variations between facing concrete and RCC (besides cracking of the facing that could be a result of different characteristics between RCC and facing), but generally there is an advantage that mass concrete shows uniform behavior throughout its cross section.

When designing the freezing-thawing laboratory program, we were aware that the freezing-thawing would influence only parts of the dam. Two conditions have to be met in order to make freezing-thawing damages possible; the temperature has to fluctuate between the freezing point (freezing-thawing cycles), and the concrete have to be critically saturated. Without these two factors, the conditions for freezing-thawing do not exist. In principle this limits the exposed areas of a cross section approximately as shown in Figure 8.

A simple first-order calculation shows that the number of freezing-thawing cycles will rapidly decline, moving into the cross section of the dam.

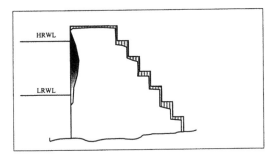

Figure 8 Areas exposed to freezing-thawing in a gravity concrete dam (Lilleland 2001).

6 ASPECTS OF PLANT ENGINEERING

The use of silica as a third pozzolan will entail a more complex system of mixing than with two pozzolans. Modern computerized batching plants should be able to manage adding silica without notable problems. As a positive effect, the silica will probably ensure that the concrete is more workable and that the detrimental segregation will be less problematic.

The silica fumes make the concrete more sticky. Cleaning equipment and possible problems with silica-RCC that sticks to the steel drums of the vibratory rollers may cause problems during RCC placement. The reports from the Førrevatn dam and the Alta dam do not suggest that this has been any problem.

Most of the questions regarding plant engineering and the practical behavior of the silica would be answered by carrying out full-scale trials.

7 COSTS ASPECTS

Using silica as a pozzolan to save 2-4 times cement will obviously reduce the material costs, presupposing that the price of silica is less than 2-4 times the price of cement. This will of course vary between the different sites and continents, where important factors will be the availability of the pozzolan and cement materials (distance to the sources). The addition of silica will of course introduce an extra factor of mixing and batching, which means extra storing, batching and plant costs. However, the costs of adding silica at the plant should not be large. The addition of silica will generate less heat of hydration

in the RCC because the total cementitious content is reduced. This could reduce the need for cooling plant facilities at sites where adding ice would otherwise be required.

Using an example of the silica mix 2 on the Skjerka dam, we can assume a cement price of \$90/ton. With 300,000 m^3 RCC and savings of 30 kg pr/m^3, this would represent savings of \$810,000. Assuming the same price for silica, the total material cost savings would be \$620,000. The extra costs of batching, storing etc. should not be anywhere near the savings in material costs. Furthermore, the effects of better workability and better characteristics would also improve economy. The latter must be tested in-situ before we are able to verify the assumptions.

The Norwegian based company Elkem is the worlds largest supplier of microsilica. They have participated in the R&D project, and is a reccomended source on further commercial and plant engineering information (www.elkem.com).

8 CONCLUSIONS

The performance of silica in the laboratory confirms that there is a utilization factor of k=2-4 for silica compared to cement within specified variations. It seems that the silica content should be within an interval of 4-8% of the cementitious content. More than 10% of the cementitious content will probably cause unwanted effects for the RCC. The balance between flyash, cement and silica in total should also be considered closely. High content of flyash compared to the other cementitious materials seems to influence the hardening of the concrete. These questions still remain.

The promising results of the laboratory testing and the earlier projects with silica use in mass concretes caused us to summarize the silica-RCC this way:

- Higher fresh density and thereby easier compaction and the improved possibility of gaining low permeability in-situ.
- Observed low degree of detrimental segregation in the laboratory, which is promising for the performance in-situ.
- Low permeability, considered as function of the total cementitious content (although higher than the ordinary RCC mixes). The experience from the Førrevatn dam and the Alta dam shows extremely low permeabilities (< 10^{-14} m/s), which is partly explained by the silica.
- Lower heat of hydration as a result of lower cementitious content and the utilization factor of the silica. Silica has the same heat capacity as cement.

REFERENCES

ASTM C150. 1987. *Portland Cement.*

Børseth, Ivar 1985.*Use of silica in the Førrevass dam.* Fifteenth Congress on large dams ICOLD, Lausanne: ICOLD.

ICOLD, Committee on Concrete for Dams. 1997 *State of the Art of Roller Compacted Concrete, (Draft not completed),* ICOLD bulletin 16 May 1997.

Jacobsen, Stefan 1999 *Frostbestandighet valsbetong med høyt flyveaske innhold.* Technical Report, Project No. O 7784. Oslo:Norwegian Building and Research Institute (In Norwegian)

Lahus, Olav 1999a. *Valsebetong og ytterbetong. Ny dam Skjerka.* Technical Report, Project No. O 7784. Oslo: Norwegian Building and Research Institute (In Norwegian).

Lahus, Olav, (1999b). *Valsebetong med silica. Sluttrapport.* Technical Report, Project No. O 7784. Oslo: Norwegian Building and Research Institute (In Norwegian).

Lilleland, Øystein 2001. *Valsebetong for norske forhold.* Oslo: Norwegian Electricity Industry Association (in Norwegian)

Hydropower in the New Millennium, Honningsvåg et al (eds), © 2001 Taylor & Francis, ISBN 90 5809 195 3

Some problems arised with respect to foundation difficulties of an earth dam in Trinidad

R.Maysingh & R.Budhiraja
Ministry of Works and Transport (MOWT), Trinidad & Tobago

C.Stere
Nedeco/Haskoning Co., Nijmegen, Netherlands

A.Popovici
Technical University of Civil Engineering, Bucharest, Romania

ABSTRACT: The Mamoral Dam in Trinidad is a homogeneous earth type of clayey material with 13.50 m maximum height that presents very difficult foundation conditions. The construction of this dam follows to start in the near future. The foundation of the dam consists of sandy, silty and clayey material in soft state for some large area of the dam foundation (Standard Penetration Test, SPT=1...3). The dam body filling material was selected in order to withstand free of cracks to some important foundation settlements, reaching 40 cm maximum value computed by PLAXIS computer code. A special attention was paid to the structure and the foundation of the bottom outlets. The cross profile shape of the structure was selected in order to take over the dam body loads free of dangerous strains in the filling surrounding the structure. The mechanical characteristics of the clayey foundation material will be improved by proper technology.

1 GENERAL DATA

Mamoral dam is part of the Caparo River Flood Control Project in Trinidad. Mamoral reservoir has the function of a temporary buffer, diminishing the peak and delaying the natural storm flood hydrograph. In fact, the Mamoral reservoir is a detention one, located on line Caparo river.

The layout of the Mamoral dam can be seen in Figure 1 (Stere et al 2000). The dam is 515 m long at the crest and 13.50 m maximum high. The surface spillway continued with rapid channel and stilling basin is located at the right bank. It may discharge 14.40 m³/s. The bottom outlets consisting of two reinforced concrete pipes, each of them having 2.00 m nominal diameter are placed in the Caparo river bed. They may discharge 26.70 m³/s.

The dam axis is a rectilinear one in the central zone and has a connection curve of 289 m radius to the right bank. This particular geometry was selected according to local morphology and geology of the dam site, conducting to a minimum volume of the dam body.

According to geotechnical investigations performed in the dam site (Singh et al. 2000), the materials from dam foundation are sedimentary ones, consisting of sandy, silty and clayey materials. The lithological profile, with medium characteristics, considered as representative for Mamoral dam - excepting, the minor river bed - presents the following layers in order from ground surface:

0.00...0.50 m	- Superficial layer - Reddish to Yellowish Brown and Light Gray - Slighty Fissured Silty Clay with traces Fine Roots;
0.50...2.00 m	- Yellowish Brown and Light Gray Slighty Fissured Silty Clay;
2.00...3.50 m	- Yellowish Brown and Gray Silty Clay;
3.50...5.50 m	- Gray Sandy Silty Clay
5.50...9.80 m	- Medium Stiff Dark Silty Clay and Sand

The soft characteristics of the materials from dam foundation, especially of those located near ground surface (Standard Penetration Test, SPT = 1...3), classify this dam in the category of the dams performed in the difficult foundation. The applied solutions within the project - for the dam cross sections, bottom outlets structure and foundation and other appurtenant works - were taken in strict correlation with the geotechnical conditions of the site. Some of them are commented in the present paper.

2 TYPICAL CROSS SECTIONS OF THE DAM

Keeping in mind the richness in clayey material of the dam site proximity, a homogeneous cross section type consisting of clayey material was very recommended. The past experience had shown, the dams built of clayey materials exhibited very good behavior in operation and especially a high resistance against slope failures during earthquakes. Dams constructed of clay soils on clay foundations have withstood extremely strong shaking ranging from

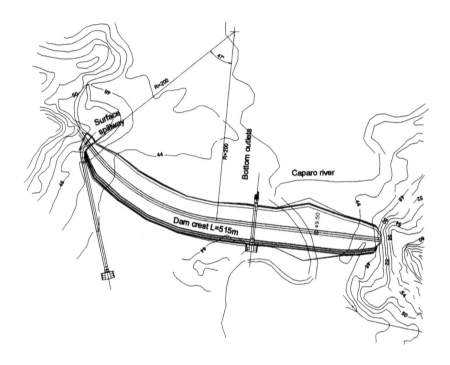

Figure 1. Mamoral Dam. General layout.

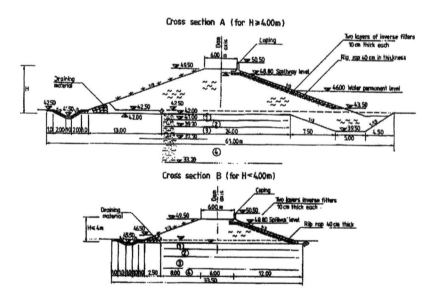

Figure 2. Mamoral Dam. Cross section types.

0.35 to 0.8 g (g-gravity) from a magnitude M=8.25 earthquake with no apparent damage.

In Figure 2 are presented two cross section types applied for Mamoral dam (Popovici 2000).

The profile type A is recommended for H≥4 m (H - dam height). The earth cut-off wall from upstream toe zone has waterproof function and protection against internal erosion. This solution was preferred taking into account that grouting in clayey materials is unusual, rather expensive because of chemical solutions involved and sometimes out of efficiency.

The upstream slope is protected against wave external erosion by a rip-rap layer of 40 cm thick, put on two layers of inverse filters. In the present solution, the rip-rap is laid full length of the upstream slope. Taking into account, the permanent reservoir elevation is 45 maSL, this elevation rising only during floods for short period's time, for esthetical reason the rip-rap may be substituted by green grass between 46.50 and 49.50 maSL elevation. However, the rip-rap will be kept at the connection locations (dam-banks connection, dam-surface spillway connection etc.).

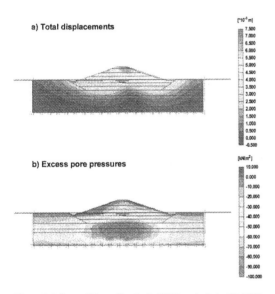

a) Total displacements

b) Excess pore pressures

Figure 3: Mamoral Dam. Results in FEM analysis by PLAXIS computer code: - a) Total displacements after sequence no. 15, - b) Excess pore pressures after sequence no. 15

The downstream slope is protected by green grass.

The profile is drained in the downstream zone by a drainage blanket located at foundation level and by a drainage prism at the downstream toe. Both of them are connected with dam body by inverse filters in order to avoid the risk of the internal erosion. The seepage water from reservoir and raining water are collected in a channel placed at downstream toe

zone, which discharges in canalized zone of Caparo river.

The dam crest with 6 m wide is arranged with two circulation ways. A coping with superior elevation at 50.50 maSL protects dam body against external erosion because of wave overflows.

The profile type B is recommended for H<4 m. Taking into account that the reservoir will be only temporary for short periods time at higher elevations than foundation elevation of this profile, the earth cut-off wall was cancelled.

The profiles follow to be adapted according to local lithological characteristics after excavation works.

Some comprehensive analyses concerning dam profile behaviour during construction and operation have been performed by finite element method using PLAXIS computer code (Wassinsk 2000). The dam settlements during construction and after consolidation, sliding slope safety factors during construction and operation, pseudo-dynamic sliding slope safety factor after an earthquake 0.05 g were main results of these analyses.

The analyses pointed out the necessity to excavate the upper 6 m of soft soil in the river bed zone of the dam and to replace it by the dam body clayey filling material. The dam body filling material was modelled with hyperbolic stress-strain relation (Duncan-Chang) having the following initial main parameters: $\varphi=25^0$, c=15 kPa, E=25.000 kN/m², E_{oed}=40.000 kN/m². The permeability coefficients were $K_x=5\times10^{-4}$ m/day on horizontally and, respectively $K_y=5\times10^{-5}$ m/day on vertically. The dam erection was simulated by 7 layers, each of them having a consolidation time of 30 days.

Some results using PLAXIS computer code are graphically presented in the Figure 3.

In Table 1 are presented the dam slope sliding safety factors after some construction or operation sequences

Table 1

Description of the sequence	Safety factor
Stage No 7, end of the dam erection	1.31
Water permanent level + 46 maSL	1.34
Water extreme level + 48.8 maSL	1.16
Water rapid draw-down from 48.8 maSL to + 46 maSL	1.16
End of consolidation - water permanent level + 46 maSL	1.90
Earthquake 0.05 g - water permanent level + 46 maSL	1.90

3 FILLING MATERIALS FOR DAM BODY

Geotechnical investigations carried out in the Caparo - Mamoral area have pointed out, that in the investigated area all types of materials necessary for dam construction are existing, as follows: clayey material for dam body, natural alluvium (rounded aggregates) for inverse filters, boulders for rip rap and aggregates to be graded for concrete works.

Different qualities of clayey materials, according to presently international practice, may be used as filling in the dam body. However a good quality of the clayey material means economy of the dam volume. The present cross section of the dam was conceived with material having the following main characteristics (see Fig. 4):

Type A: Clayey fill material for the dam body
- Approximate volume: 140.000 m³
- Plasticity index PI = 16...20
- Dry density in the Standard Proctor Tests (SPT) 1.60...1.70 t/m³
- No shrinkage or minimum guaranteed at drying conditions
- No dispersion in water
- No organic material content
- Grading curve recommended can be seen in Figure 4.

Type B: Clayey material for contact zones
- Approximate volume: 5000 m³
- Plasticity index PI = 20...25
- Dry density in the Standard Proctor Tests (SPT) 1.40...1.65 t/m³
- No shrinkage or minimum guaranteed at drying conditions
- No dispersion in water
- No organic material content
- Grading curve recommended can be seen in Figure 4.

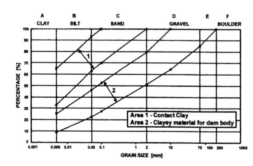

Figure 4. Recommended grading curves of materials in the dam body of Mamoral

The contact zones are considered those presenting high strain or stress gradients. In this case the contact zones are the followings: connection between dam shoulder and right bank, side connections between dam body and surface spillway wall from right bank, side and roof connections between dam body and bottom outlets. Taking into account, the dam foundation is also composed of clayey material, its connection with dam body will be performed with current clayey fill material (type A). This solution will

have positive influence on dam safety, the material type A having higher internal frictional angle than type B one.

Moisture content of clayey materials in natural deposits is recommended to be in the range of Woptimum, such as obtained through SPT. On the contrary, constructive measure (drainage or humidification) will be taken, in order to obtain a filling material having Woptimum+ 2...3% moisture content.

In the case, the natural deposits are not satisfactory concerning clay percent in the filling material, it being higher than those recommended, a current solution applied in this field consists in obtaining a mixture of clayey material with sand + gravel. However, in the present project, because of the small volume of filling, a solution like this is considered too expensive.

Consequently, in case of clayey material characteristics from natural deposits will be very different than the ones above recommended, the dam profile shall be reconsidered accordingly.

A special analysis was performed on the connection of the left bank with dam shoulder because it seemed rather steeper.

The longitudinal profile in the dam axis carried out on left bank has shown the maximum inclination of the dam shoulder is 24⁰. Additionally, the dam cross sections carried out in this zone have usual shape free of geological or morphological special problems.

In the above presented conditions from left bank dam shoulder, the only one constructive measure that is considered necessary in this zone is to include a layer of filling material type B with 0.50 m in thickness on normal direction to surface, along the contact area between left bank and dam shoulder.

4 CONSTRUCTIVE SOLUTION FOR BOTTOM OUTLETS

The cross section of the bottom outlets was settled as two circular pipes of 2.00 m diameter each. One of the pipe is for current operation and second one as discharging reserve (Fig.5).

These two circular voids were covered by a reinforced concrete lining having arch with three centers shape in order to withstand to environmental loads and respectively not to induce dangerous strain or stress state in the dam body surrounding the section (Fig. 5).

The structural analysis of the bottom outlets cross sections in a vertical plane by dam axis performed with PLAXIS computer code has pointed out the following conclusions (Wassinsk 2000):
- the geometrical shape of the bottom outlets section included in the dam body is satisfactory, and was not necessary any correction; some very low horizontal tensile stresses occurs in the dam body

428

at the key section but they are in allowed limits for clay contact (material type B);

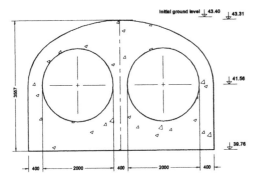

Figure 5. Bottom outlets – current cross section

– the settlements of the foundation are excessive ones in the hypothesis of the natural mechanical characteristics; consequently, the consolidation of the bottom outlets foundation, in order to increase two ... three times the mechanical parameters of the foundation materials resulted to be necessary.

In the longitudinal profile the bottom outlets system comprises in order from upstream to downstream the following components: water intake tower, connection (transition) zone from rectangular to circular section, access bridge from dam crest to water intake, current bottom outlets crossing dam body, stilling basin, risberm (downstream apron) and connection channel to Caparo river.

In the finite element analysis by PLAXIS computer code, the cross section of the bottom outlets was modelled in two variants (Wassinsk 2000):
– as concrete elements,
– as a concrete base with circular tunnel lining elements having a thickness of 40 cm.

The top level of the soil has been taken as + 49.50 maSL to model a cross section under the crest of the dam and respectively + 47.00 maSL to model a current cross section under the dam sloping side. Some results in this analysis are illustrated in Figure 6.

5 THE STRENGTH CAPACITY IMPROVING OF THE BOTTOM OUTLETS FOUNDATION

The necessity for works to improve the strength capacity of the bottom outlets foundation was argued at the points 1 and 4. It is a rather difficult problem because of clayey material from foundation.

In this stage of the project are suggested two technologies to be taken into consideration. The contractor will select one of them according to technical performance of the experimental investigations, cost and experience in this field. These suggested technologies consist in (1) grouting of the sodium silicate and eventually of the calcium - chloride solutions or (2) in repeated blowing of a heavy weight according to a scheme for deep compaction.

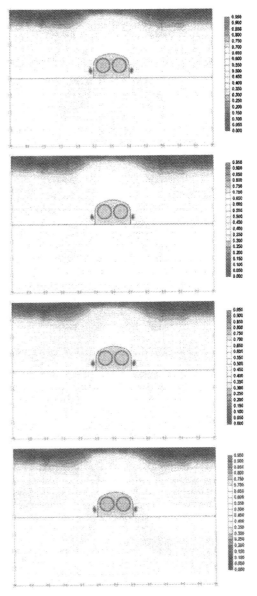

Figure 6 Bottom outlets FEM analysis - relative shear stresses of the four models

The option for one or another technology must be supported by comprehensive tests in site and respective in laboratory.

Grouting of the sodium silicate is a very efficient technology for waterproofing and consolidation of the sand, silt and clayey materials. The sodium silicate solution makes chemical reactions with electrolytes as $CaCl_2$, $Ca(OH)_2$ and s.o. or if they were not in grouted material, they will be introduced from outside (especially calcium-chloride that is grouted after a time interval of sodium silicate grouting).

Usually, the clayey materials grouted with sodium silicate reach to resistance at monoaxial compression of 500 ... 1000 kPa.

The deep compaction by repeated blowing of a heavy weight (tamping crane rammer) is used especially for deposits of clayey materials. The impact generated by tamping crane rammer generates important settlements of the layer because of the collapse of the deposit initial structure and of the mineral particle rearrangement. The blowing of the rammer in successive series on the same location provokes the pore water elimination to adjacent uncompacted deposits where some discontinuous surfaces will appear. In order to allow the pore water excessive pressure dissipation some pauses (breaks) between blowing series are needed.

The compaction degree depends on lithology, grading, and moisture content of the compacted layer. The compaction technology parameters as number of the blowing for each location (usually 5 ... 15), number of phases (usually 2 ... 3) and time duration between phases must be evaluated by in site experiments. A phase is finished when after three successive blowing the settlements are less than 3 mm.

In the present project the parameters computed in the compliance with some usual formulas from literature resulted as follows:
- the depth of the layer to be compacted: h = 4.00 m
- the heavy weight (rammer) from iron (steel) having cylindrical shape with diameter of 1.50 m and height of 1.00 m (total mass M=13.8 t)
- the dropping height of the rammer: H = 5.00 m.

If the works to improve the strength capacity of the bottom outlets foundation would not conduct to satisfactory results, the structural solution of the bottom outlets needs be reconsidered.

6 CONCLUDING REMARKS

The solutions used for Mamoral dam and reservoir are in complete accordance with international standards and the current practice for earth dams carried out on difficult foundation.

It needs to remark, the consolidation phenomenon of the clayey material from body and foundation of the dam will continue a long period time (minimum 2...3 years) after end of the dam erection, because of the very low permeability coefficients. In this period the settlements may be relative important, they need to be surveyed carefully. Strong earthquakes may also provoke some excessive settlements. However, it is considered, the coping of 1 m high (50.50 maSL elevation) ensures a sufficient freeboard against dam overflow.

The risk of the upper zone dam body cracking because of the dry conditions in operation (reservoir permanent elevation 45 maSL versus 49.50 maSL crest elevation) of the dry environment (January - May months in Trinidad) and of the sunstroke is potential possible. However, it is considered very unlikely, the dam body to be break by thermal deep cracks. Instead, superficial cracks are likely to appear but they are not dangerous for the dam safety.

A permanent surveying of the dam behaviour during construction and operation is the key for preventing any negative phenomena.

REFERENCES

Popovici, A. 2000. Report on Final Design of the Mamoral Reservoir Dam. Bucharest - Romania

Singh, B.G. et al. 2000. Mamoral Dam - Report on Geotechnical Investigations Curepe - Trinidad & Tobago

Stere, C et al. 2000. General Report concerning Mamoral Dam and Reservoir Project. Port of Spain - Trinindad & Tobago.

Wassinsk, H. 2000. Mamoral Dam Design - Modelling of Typical Dam Cross - Section. Nijmegen - The Netherland.

Wassinsk, H. 2000. Mamoral Dam Design - Modelling of Bottom Outlet Cross Section. Nijmegen - The Netherland.

Hydropower in the New Millennium, Honningsvåg et al (eds), © 2001 Taylor & Francis, ISBN 90 5809 195 3

Limit condition of distance of force center and rotation center for self-excited vibration of radial gate

Kunihiro Ogihara
Toyo University, Saitama, Japan

Hiroya Emori
Sumitomo Heavy Industry, Tokyo, Japan

Yukihiko Ueda
Ishikawajima harima Heavy Industry, Tokyo, Japan

ABSTRACT: Limit condition of self-excited vibration on radial gate has been analyzed by theoretical model and its condition has certified by model test. The main part of the theoretical analysis is introduced at the conference of Waterpower '99 and the essential part of this is introduced at FIV 2000 (Conference of flow induced vibration 2000). This paper shows that the limit distance of two centers of force and Trunnion pin center is determined by another parameters such as natural frequency of gate system, scale of radial gate and flow condition under the gate.

1 THEORETICAL RESULT

1.1 Introduction

The limit conditions for self-excited vibration arises in radial gate are derived by following situations.

1) The center of hydraulic force acting on the gate surface is placed lower than the center of Trunnion pin. Namely when gate moves downward, the moment force by water pressure must act to turn the gate downward.

2) The instability condition is given by the parameter Pan and gate opening a/H.

3) The parameter Pan is written by a non-dimensional equation, which contains the natural frequency of gate suspension system, damping constant, length of struts, amplitude of gate vibration by disturbance of flow or traffic load and velocity of water flow under the gate.

Results show that the self-excited vibration arises more easy under the following conditions; the smaller gate opening, smaller damping coefficient, higher flow velocity, larger motion under usual flow conditions and lower natural frequency.

1.2 Equation of gate motion

In radial gate operation system, the operation chains or the operation hydraulic rods hang the radial gate and this hang system gives the spring constant in vibration equations. The basic motion of radial gate under the operation system is the rotation motion around the center of Trunnion pins as the first mode of vibration.

So the equation of motion can be written as follows.

$$I\frac{d^2\theta}{dt^2} + R\frac{d\theta}{dt} + kR_0^2\theta = F \qquad \cdots\cdots (1)$$

Here R, k, and I are shown as the inertia of radial gate, damping constant and spring constant respectively. R_0 is the arm length given by the distance between the fixed point of hanged chain and the center of Trunnion pin. F is the external force by water flows under the gate motion. The term θ is the rotation motion of gate.

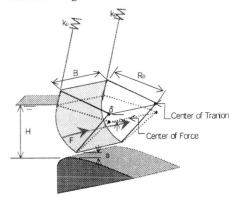

Figure. 1 Relations of centers

1.3 External Force Under Self-Excited Vibration

Usually when the self-excited vibrations arise, the external force of vibration is strongly related to the gate motion. So the term of external force must be dependent on the term θ of rotation motion. External force in Equation (1) is written as a force to make the rotate motion of the gate and its direction of action is same as the gate motion under self-excited vibration. Namely the situation for self-excited vibration is explained as follows.

When the gate moves downward, the water flow is stopped by the decrease of opening under the gate.

This makes the pressure increase and the force by this pressure change push the gate surface.

And this force makes the moment force to turn the gate downward around the Trunnion pin, when the center of force lays downward of Trunnion pin center. For the self-excited vibration arises, the moment force by the gate movement has the same direction of gate motion. If its direction was not same, this force stopped the movement of gate. Therefore the center of curvature of radial gate leaf or skin plate curve must be shifted downward from the center of Trunnion pin. The distance of these two centers is written by δ perpendicular to the force vector from center of Trunnion and H and a are the water depth in upstream side of gate and gate opening respectively.

Discharge from the gate under small opening is written as equation (2).

$$Q = C\sqrt{2g(H-a)}Ba \qquad \cdots\cdots \quad (2)$$

Here C and B are the coefficient of discharge and width of gate. When the gate moves y to downward by the velocity v (=dy/dt), the change of the discharge can be written as follows.

$$dQ = -\frac{dQ}{da}y = -C\sqrt{2g}B\frac{H-\frac{3}{2}a}{\sqrt{H-a}}y \qquad \cdots\cdots \quad (3)$$

Therefore the force derived by momentum equation to apply to the changes of y and v, is written as equation. (4).

$$F = \rho\left|-\frac{dQ}{da}y\right|\frac{dy}{dt}\delta \qquad \cdots\cdots\cdots \quad (4)$$

This gives the external moment force in the basic vibration equation (1). The term y can be written as y=R₀θ and the equation is rewritten by rotation motion θ by applying this relation.

The equation for self-excited vibration is written by combining Equation (5) and Equation (1).

$$F = \left|C\rho\sqrt{2g}B\frac{H-\frac{3}{2}a}{\sqrt{H-a}}R_0\theta\right|R_0\frac{d\theta}{dt}\delta \qquad \cdots \quad (5)$$

Equations for damping constant and angular natural frequency ω_n are given as follows.

$$2\gamma = \frac{R}{I}, \omega_n^2 = \frac{kR_0^2}{I} \qquad \cdots\cdots\cdots\cdots \quad (6)$$

1.4 *Stability Conditions for Self-Excited Vibration*

The second term in the left of Equation (5) takes the energy dissipation and the right term takes energy supply to the vibration system. So the relation of energy in one cycle of vibration in these terms determines the unstable condition. The self-excited vibration occurs usually around the natural vibration system. Now gate motion θ in vibration is taken as small motions as written by Equation (7).

$$\theta = \theta_0 \sin\omega_n t \qquad \cdots\cdots \quad (7)$$

The energy dissipation of the second term of the left in Equation (5) in one cycle of vibration and the energy supply by the external force in one cycle of vibration are derived by integrating the terms in one cycle of vibration.

The condition for unstable is given as Equation (8), such as the supplied energy is larger than the dissipated energy.

$$\frac{\gamma I}{C\rho\sqrt{2gH}BR_0^2\delta\theta_0} < \frac{2}{3\pi}\frac{1-\frac{3}{2}\frac{a}{H}}{\sqrt{1-\frac{a}{H}}} \qquad \cdots\cdots \quad (8)$$

When the value of left term in Equation (8) becomes larger than the right term, a self-excited vibration does not occur. In this equation, the right term is given by only gate ratio a/H and the left has many factors. For the self-excited vibration is occurring, the value of the parameter Pa becomes smaller than the value of the right term in Equation (9).

$$Pa = \frac{hk}{C\rho\omega_n\sqrt{2gH}B\theta_0\delta} \qquad \cdots\cdots \quad (9)$$

1.4 *1.5 Some Consideration on Unstable Condition*

For the parameter Pa to become smaller, the following conditions of each factor are necessary.

- Water depth H is greater and the factors shown of gate scale as the width B and the length of strut R_0 become larger.
- Distance between the center of skin plate curve and the center of Trunnion pin is longer.
- Initial movement of gate is larger or the movement under the hanging condition is larger.
- Damping ratio h is smaller and natural angular frequency is smaller. This latter one means that the natural frequency of gate system is smaller, namely the period of natural vibration is longer.

- Value of moment of inertia I is smaller; this means that gate weight is light.

2 DISTANCE OF TWO CENTERS

2.1 *Distance of two centers under self-excited vibration*

This relation can be rewritten by equations (8) and (9) as follows. Here function F(a, H) is defined as equation (10).

$$F(a,H) = \frac{2}{3\pi} \frac{1 - \frac{3}{2}\frac{a}{H}}{\sqrt{1 - \frac{a}{H}}} \quad \cdots\cdots \quad (10)$$

And parameter Pa is rewritten to three blocks as defined as Cc, Cs, Cr and Cm.

$$Cc = \frac{1}{C\rho}, Cs = \frac{kh}{\omega_n}, Cr = \frac{1}{\sqrt{2gH}BR},$$

$$Cm = \frac{1}{\theta_0} \quad \cdots\cdots \quad (11)$$

Parameter Cc is constant, and Cs, Cr and Cm are related to vibration system, scale of gate and movement of gate in initial condition respectively.
So limit condition of distance of two centers can be derived as following equation.

$$\frac{\delta}{R} > CcCsCrCm\frac{1}{F(a,H)} \quad \cdots\cdots \quad (12)$$

2.2 *Relation of gate opening*

The relation of gate opening and function F(a,H) is given in equation (12). So under the other parameter is given constant, the effect of gate opening is derived as figure 2.
For self- excited vibration occurs, the value of δ/H must larger than the value of curve in figure 2. Namely the upper zone of curve is unstable for self-excited vibration.
When gate opening a/H is larger then 2/3, it is always stable because the function F(a,H) has minus value. And the zone where gate opening is smaller gives more unstable condition, because the value of F(a/H) is smaller.

2.3 *Scale of gate*

The effect of scale of gate is given by parameter Cr and this parameter depends on B, R and H. For the case of the gate width B, gate height H and length of R are same order; this effect on unstable condition is given as figure 3.

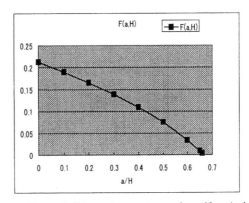

Figure 2 distance two centers under self-excited vibration

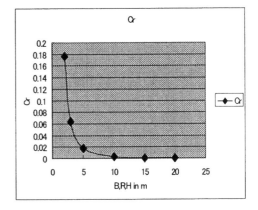

Figure 3 Effect by scale of gate

This relation shows that the scale of gate becomes larger, and value of Cs becomes smaller. This means that the self-excited vibration occurs in larger gate easily.

2.4 *Natural vibration in gate system*

The effects of natural vibration are given by parameter Cs witch is related to natural angular frequency and damping ratio of gate hanging system. In this parameter, spring constant and damping ratio is in numerator and damping ratio is denominator in equation (11).
This relation gives the gate, which has damping constant, is smaller and spring constant is smaller, becomes vibration to be arisen. And when angular frequency is larger, value Cs becomes smaller. This relation is not same as spring constant, but usually natural frequency of radial gate in fields is between 3 to 10 Hz and it does not have changed so much.

2.5 Movement in natural condition

The effect of movement under natural conditions is given by parameter Cm. Movement is derived by waves in upstream, winds from downstream and when the bridge is built with gate operation instruments, traffic road gives movement also.

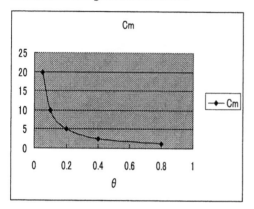

Figure 4. Effect by movement in natural conditions

So parameter Cm becomes smaller when gate movement is larger. This means very important that when the gate does not move, self-excited vibration does not occur. Namely hanging system of gate is not ridged, as wire rope compare to the chain system, movement in natural conditions may be large. So vibration arises in such gate.

3 SOME CONSIDERATION IN FIELD GATE

3.1 Scale of field gate

For checking the limit distance between two centers for actual gate in fields, the following scales of gate is assumed.

Gate parameters:
Radius R_0=14 m, Width B=12 m, Chain length L_1=20m
Moment of Inertia I=4x10^3 ton x m^2

Flow conditions:
Water depth H=15m,
Water velocity from gate $V = \sqrt{2gH} = 17.15 \, m/sec$
Coefficient of discharge C=0.6

Natural frequency and Damping constant:
Natural frequency f=3~18 Hz
Damping constant h=0.01~0.1
Movement in natural condition =0.01~0.1 rad

For calculation of parameters, some values must be calculated such as angular frequency and spring constant.

$\omega_n = 2\pi f = 18.8 \sim 113$ 1/sec
$k = \omega_n^2 I/R^2 = (18.8 \sim 113) \, 4 \times 10^3/14^2$
$= 384 \sim 2306$ ton/sec^2

3.2 Value of parameters

The value of parameters can be calculated by equation (11) by using these values.

$$Cc = \frac{1}{C\rho} = 1/0.6/1 = 1.67 \, cm^3/g$$

$$Cs = \frac{kh}{\omega_n} = \frac{(384 \to 2306)(0.01 \to 0.1)}{(18.8 \to 113)}$$
$$= 0.021 \to 12.3 \, ton/sec$$

$$Cr = \frac{1}{\sqrt{2gH}\,BR} = \frac{1}{17.15 \times 12 \times 14} = 0.000347 \, sec/m^3$$

$$Cm = \frac{1}{\theta_0} = \frac{1}{0.01 \to 0.1} = 100 \to 10 \, 1/rad \quad \cdots\cdots \quad (12)$$

Therefore total values in right hand term without the function F(a,H) is derived as follows.

$CcCsCrCm$

$= 1.67(0.021 \to 12.3)0.000347(100 \to 10)$

$= 0.000121 \to 0.713 \quad \cdots\cdots \quad (13)$

This shows very wide values in these parameters such as the largest value has 7000 times of small value. And the multiplied values of all parameters are non-dimensional value.

3.3 Limit distance of two centers

In figure 2, value of function F(a,H) is from 0.21 to 0.16 in small gate opening, so the limit condition is given by equation (12).
Therefore the relation of limit distance of two centers can be shown in figure 5 by variable values of total parameters.

In this figure, total parameter that is multiply of Cc, Cs, Cr and Cm, is 0.0001 to 1. Usually in radial gate, there is almost small difference of center of force and center of Trunnion pin. The value of it must be less than 20cm in fields, so the value of δ/R is less than 0.01. Therefore the total value of parameter is less than 0.01, can be derived from this figure.

4 CONCLUSION

In field gate, natural frequency is around 5 to 8 Hz and damping ratio is 0.02 to 0.1. So the value of multiplied parameters becomes between 0.001 and

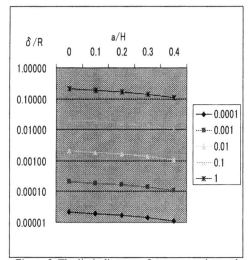

Figure 5. The limit distance of two centers by total parameters

0.01. Therefore the limit distance of two centers δ/R =0.001 to 0.0001 from this figure. And when length of strut is assumed as 5 m, the distance of two centers δ becomes 5 mm to 0.5 mm. In the order of this value, there is the possibility of self-excited vibration in field gate.

For this gate may be supported by soft wire rope system and gate structure may be not so ridged.

REFERENCES

1. Ogihara K, Ueda Y and Emori H, 2000, Flow induced vibration of radial gate under small gate opening, *FIV2000, 7th International Conference Flow Induced Vibrations, 19-22 June 2000*
2. Ogihara K & Ueda Y, 1999, The conditions of self-excited Vibration occurring in Radial Gate, *Waterpower '99*
3. Ogihara K & Ueda Y, 1988 Unsteady condition of self-excited oscillation of roller gate under the submerged flow conditions *Symposium of High dam*, 603-610
4. Ogihara K, Minagawa I, Ueda Y & Ueda T, 1998. Fields test of vibration of draft gate in pump storage station. *Conference of Modeling, Testing & monitoring for Hydro Power plants – III.* 293-300
5. Ogihara K, Nakagawa H & Ueda Y, 1995. Theoretical Analysis on Self-excited vibration of long spans shell roller gate. *1D18, HYDRA 2000, (vol.1)*; 603-608.Thomas Telford, London
6. Ogihara K, Nakagawa H & Ueda Y, 1994. Theoretical analysis on the conditions of self-excited vibration in shell roller gate. *9th APD-IAHR Congress, vol.2*, 125-131
7. Ogihara K, Nakagawa H & Ueda Y, 1992, Self-excited vibration of long span shell roller gate by three-dimensional experimental model *24th IAHR Congress*, D433-D442

Hydropower in the New Millennium, Honningsvåg et al (eds), © 2001 Taylor & Francis, ISBN 90 5809 195 3

3D CFD modelling of water flow in the sand trap of Khimti Hydropower Plant, Nepal

Nils Reidar B.Olsen
The Norwegian University of Science and Technology, Trondheim, Norway

Pravin Raj Aryal
Ministry of Water Resources, Kathmandu, Nepal

ABSTRACT: A three-dimensional numerical model was used to compute the water velocities in a sand trap. The numerical model solved the Navier-Stokes equations on a three-dimensional non-orthogonal finite volume grid. The SIMPLE method was used to compute the pressure, and the k-ε model was used to calculate the turbulence. The computed velocity profiles were compared with the measurements in the physical laboratory model. Good agreement was found using the first order Power-Law Scheme together with Strickler-Mannings roughness coefficient of 75. Higher or lower roughness coefficients caused more deviation between computed and measured velocities. The first-order scheme gave significantly better results than the second-order upstream scheme, which predicted a non-uniform profile with a jet along the bed and the side walls. This may be due to over/undershoots of the second-order scheme.

1 INTRODUCTION

Run of the river hydropower plants in tropical countries often use a sand trap to remove sediments in the inflowing water. The sand trap may carry a considerable cost of the hydropower scheme. It is therefore important that the sand trap has a good hydraulic design. Sediments should be removed to prevent clogging of headrace tunnels/canals, and to reduce wear on hydraulic machinery.

A physical model study has traditionally been used to investigate the hydraulics of the sand trap. Physical models are, however, costly, time-consuming and there are scaling problems for the finer sediments. The alternative is to use a CFD model. In recent years, several studies of numerical modelling of sand traps have been carried out. Olsen and Skoglund (1994) modelled water and sediment flow in a laboratory sand trap. Reasonably good agreement was found for the sediment trap efficiency, although a large recirculation zone was not accurately predicted. Olsen and Kjellesvig (1999) modelled bed changes in a sand trap, and obtained good results compared with measurements in a physical model study. These two studies were carried out on sand traps where the entrance region and the main sand trap were on the same axis. In the current study, the entrance region is skewed at an angle compared to the main direction of the sand trap. This design sometimes have to be used for geological and geometrical reasons specific for the dam site. The sand trap then has to be given a hydraulic design to prevent occurrence of non-uniform flow. In the current study it is investigated how well the CFD model is able to predict the hydraulics for this case.

2 THE CFD MODEL

The numerical model computed the water flow by solving the steady Navier-Stokes equations:

$$U_j \frac{\partial U_i}{\partial x_j} = \frac{\partial}{\partial x_j}(P\delta_{ij} - \overline{u_i u_j}) \qquad (1)$$

U is the time-averaged water velocity, x is a space direction, P is the pressure and u is the time-fluctuations of the velocity around the average, derived from the Reynolds decomposition (Rodi, 1980). The SIMPLE method (Patankar, 1980) was used to solve the pressure term, and the k-ε model (Rodi, 1980) was used to calculate the Reynolds stress term. The convective term on the left side of Eq. 1 can be solved by a number of different algorithms. In the present study two methods have been tested:

1. The Power-Law Scheme
2. The Second Order Upwind Scheme

The Power-Law Scheme is a first-order method. It can often introduce substantial amounts of false diffusion, preventing an accurate prediction of velocity gradients. The second-order upwind method introduces less false diffusion, but it is more unstable and gives longer computational times. Further details on the methods are given by Olsen (1999).

The velocity gradient close to the wall was not resolved in the grid, as this would have required too high number of grid cells. Instead, wall laws were used in the cells close to the wall. The velocity from the center of the cell to the wall was then assumed to follow the formula given below:

$$\frac{U}{U_*} = \frac{1}{\kappa}\ln\left(30\frac{y}{k_s}\right) \quad (2)$$

U is the velocity in the bed cell, U_* is the shear velocity, κ is a constant equal to 0.4, k_s is a roughness parameter and y is the distance from the bed to the center of the bed cell.

The numerical model and the discretization algorithms are further described by Olsen (1999).

3 THE PHYSICAL MODEL STUDY

The Khimti Hydropower Plant is located in western Nepal. An extensive study of the hydraulic performance of the settling basins was carried out by Buthwal Power Company at the hydraulic laboratory of the Institute of Engineering, Tribhuvan University, Kathmandu. This was done in cooperation with the Norwegian University of Science and Technology.

The Khimti Hydropower Scheme includes two parallel sand traps, with different cross-sectional shape. The current study only focuses on the left sand trap, with a hopper-type bed. The length of the physical model was nine meters, of which the entrance region was three meters long and the sand trap itself was six meters. A sketch of the physical model is given in Fig. 1. The width of the sand trap was 0.8 meters and the depth varied from 0.5 to 0.6 meters. The water discharge was set to 17.3 litres/second, giving an average water velocity in the sand trap of 3 cm/s.

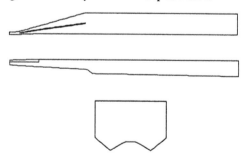

Figure 1. Outline of the sand trap, seen from above (top), longitudinal profile (middle) and a cross-section at the downstream end (lower).

Since the entrance region was slightly skewed compared with the direction of the sand trap, a divide wall was constructed in the entrance region. This should give a more uniform velocity profile in the transverse direction. Another characteristic of the entrance region was that the flow was pressurized for

the first 1.33 meters. The top of the pressurized section was 8 cm lower than the free surface

The water velocity was measured by anchored floats. The floats were dropped at three different sections in the sand trap: upstream section, middle section and downstream section. At each section, five points in the transverse direction was measured. Flow anchors at three different water depths were used for each point in the sections. The floats were timed over a length of 1 meter. The results are given in the following, where they are compared with the CFD model.

4 CFD COMPUTATIONS

The sand trap was modelled using a structured three-dimensional grid. The base grid had 133x19x15 cells in the longitudinal, transverse and vertical direction, respectively. The grid is shown in Fig. 2. The same data used to build and run the physical model study was used as input for the CFD model.

Several computations were done with the two discretization schemes and varying roughness. Best fit to the measured velocities was obtained with the Power-Law Scheme and a Strickler-Manning's coefficient of 90. The resulting velocity field close to the bed in the entrance region is given in Fig. 3. The entrance region is the critical location for formation of skewed flow. The figure shows some recirculation close to the sides. Looking at Fig. 4, showing the same velocity field close to the roof, the profile is more uniform. A longitudinal profile given in Fig. 5, also showing a fairly uniform profile. The recirculation zones are therefore confined to the areas close to the bed and the sides. A cross-sectional view of the velocity vectors is given in Fig. 6. The transverse velocities are an order of magnitude larger in the beginning of the sand trap than in the middle.

Fig. 7 shows transverse profiles of the velocities in the main direction of the flow. Both measured and computed values are shown, and there is fairly good agreement. The velocities are slightly higher at the left side close to the bed. This is shown by both computations and measurements.

Fig. 8 shows vertical profiles of measured and computed velocities. The velocities close to the water surface is higher than close to the bed, corresponding to the classical boundary layer theory. The effect is given for both measured and computed values.

5 DISCUSSION

Some of the input parameters for the CFD model are uncertain. Also, there exist several algorithms for solving the equations for the water flow. To assess the input and the algorithms, a parameter sensitivity test was carried out. One chosen input parameter was the roughness. Strickler-Manning's coefficients of 50 and 105 were tested against the initially chosen value

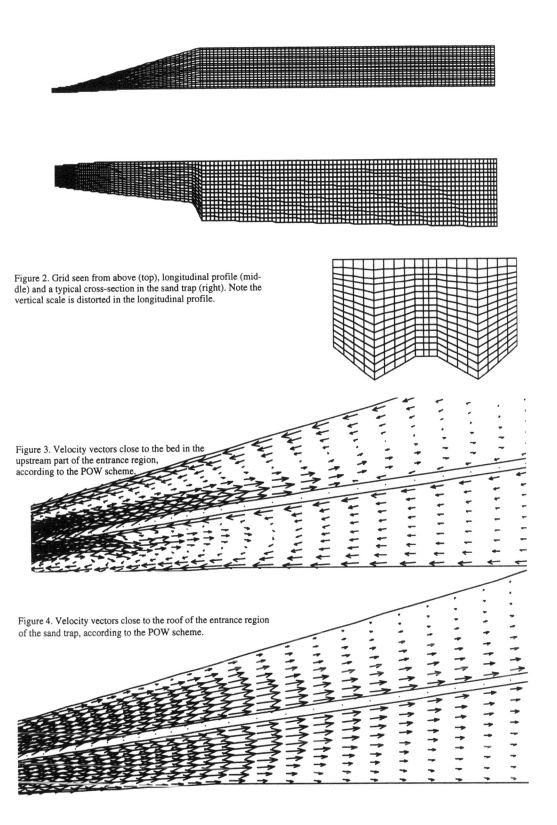

Figure 2. Grid seen from above (top), longitudinal profile (middle) and a typical cross-section in the sand trap (right). Note the vertical scale is distorted in the longitudinal profile.

Figure 3. Velocity vectors close to the bed in the upstream part of the entrance region, according to the POW scheme.

Figure 4. Velocity vectors close to the roof of the entrance region of the sand trap, according to the POW scheme.

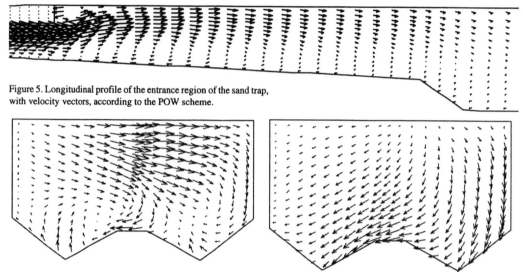

Figure 5. Longitudinal profile of the entrance region of the sand trap, with velocity vectors, according to the POW scheme.

Figure 6. Cross-sections of the sand trap with velocity vectors for the POW scheme. The left cross-section is at the most upstream part of the sand trap, where the average velocity vector is 5 cm/s. The right figure is in the middle of the sand trap, where the average velocity vector is 0.5 cm/s.

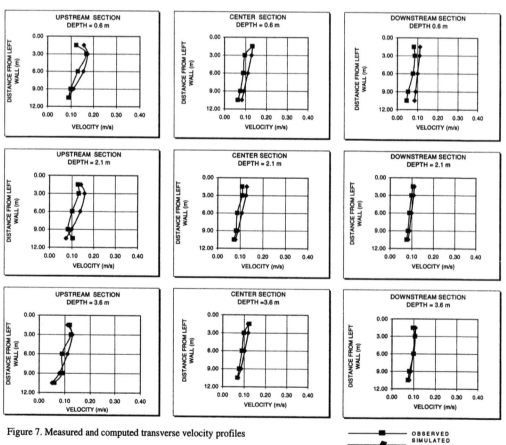

Figure 7. Measured and computed transverse velocity profiles

■ OBSERVED
♦ SIMULATED

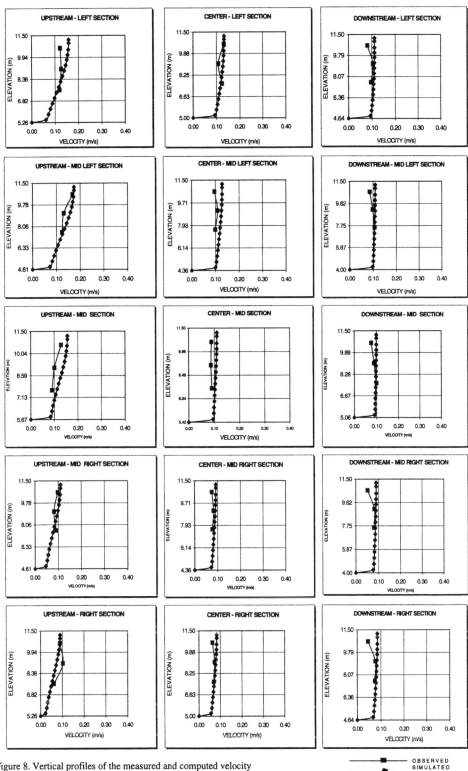

Figure 8. Vertical profiles of the measured and computed velocity

OBSERVED
SIMULATED

441

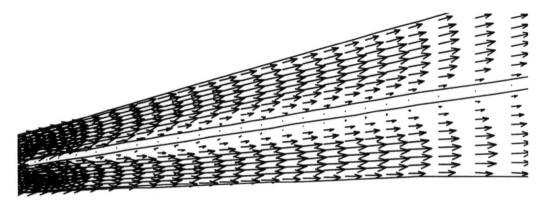

Figure 9. Velocity vector plot close to the bed at the entrance region for the SOU scheme.

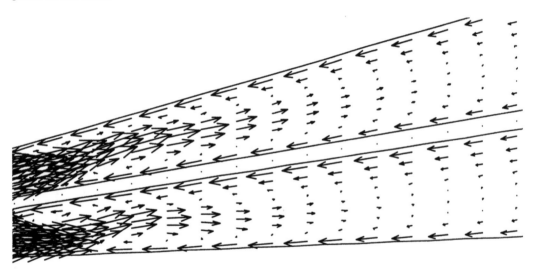

Figure 10. Velocity vector plot close to the roof of the entrance region, for the SOU scheme.

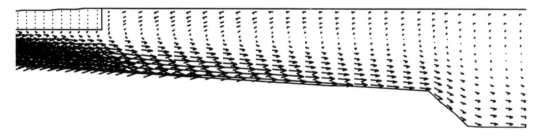

Figure 11. Longitudinal profile of the entrance region of the sand trap with velocity vectors, for the SOU scheme.

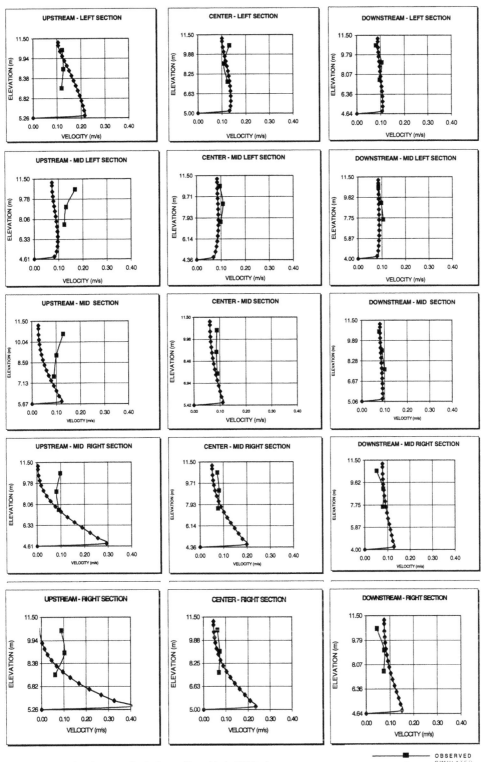

Figure 12. Measured and computed velocity profiles with the SOU scheme.

OBSERVED
SIMULATED

443

of 75. This gave slightly different velocity fields. The correspondence with the measured values was still reasonably good, but not as good as with a Strickler-Manning's coefficient of 75.

Another parameter test was for one of the numerical algorithms of the CFD model. In the k-ε turbulence model, the Boussinesq approximation given below was inserted in Eq. 1:

$$-\overline{u_i u_j} = \nu_T \left(\frac{\partial U_i}{\partial x_j} + \frac{\partial U_j}{\partial x_i} \right) \qquad (3)$$

The variable ν_T is the eddy-viscosity, computed by the k-e turbulence model. The first term on the right side of Eq. 3 is the ordinary diffusive term. The second term is the stress term, which is often negleced. To test this assumption, the stress term was included in one computation. This gave almost identical results as neglecting the term.

The most important algorithm affecting the result for the present case seemed to be the disretization of the convective term. The computations were done with the Second-Order Upwind Scheme (SOU) instead of the Power-Law Scheme (POW). The results are given in Fig. 9-11. Fig. 9 shows the water velocity close to the bed. This can be compared with Fig. 3, showing the same using POW. The velocities are much more uniform using the SOU scheme. Fig. 10 shows the velocity field close to the roof of the entrance region using the SOU scheme. This can be compared with Fig. 4, where the POW scheme was used. The SOU scheme gives a less uniform profile. The reason is shown in Fig. 11, giving a longitudinal profile of the velocity vectors. This can be compared with Fig. 5, showing the same using the POW scheme. The SOU scheme predicts a water current close to the bed, with a recirculation zone at the roof. This is in contradiction with the results from the POW scheme and the measured velocities, as shown in Fig. 12. The reason is probably over/undershoots using the SOU scheme.

6 CONCLUSIONS

The CFD model is able to compute the velocity field in the sand traps of Khimti Hydropower plant reasonably well. The Power-Law Scheme gives results similar to what is measured in the physical model study. The Second-Order Upwind Scheme shows a highly skewed flow pattern, which was not observed in the laboratory. This can be caused by over/undershoots of the scheme. For the present case, the water flow is fairly well aligned with the grid, and therefore the false diffusion introduced by the first-order scheme seems to be small. This is probably why the Power-Law Scheme gives better results than the Second-Order Upwind Scheme.

ACKNOWLEDGEMENTS

We want to thank Hydro Lab Pvt. Ltd., Kathmandu, Nepal for allowing us to use the laboratory data. We also would like to thank Prof. Haakon Støle for his assistance in this project.

REFERENCES

Aryal, P. R. (2000) "CFD Modelling of Flow Patterns in the Sand Traps of Khimti Hydropower Plant", MSc Thesis, Department of Hydraulic and Environmental Engineering, The Norwegian University of Science and Technology.

BPC Hydroconsult (1997) "Khimti 1 Hydropower Project Hydraulic Model Study, Final Report", River Research Laboratory, Kathmandu, Nepal.

Olsen, N. R. B. and Skoglund, M. (1994) "Three-dimensional numerical modeling of water and sediment flow in a sand trap", IAHR Journal of Hydraulic Research, Vol. 32, No. 6.

Olsen, N. R. B. and Kjellesvig, H. M. (1999) "Three-dimensional numerical modelling of bed changes in a sand trap", IAHR Journal of Hydraulic Research, Vol. 37, No. 2.

Olsen, N. R. B. (1999) "Computational Fluid Dynamics in Hydraulic and Sedimentation Engineering", Class notes, Department of Hydraulic and Environmental Engineering, The Norwegian University of Science and Technology. (Can be downloaded from: http://www.bygg.ntnu.no/~nilsol/cfd)

Patankar, S. V. (1980) "Numerical Heat Transfer and Fluid Flow", Taylor and Francis Publishers.

Rodi, W. (1980) "Turbulence models and their application in hydraulics", IAHR state-of-the-art publication

Hydropower in the New Millennium, Honningsvåg et al (eds), © 2001 Taylor & Francis, ISBN 90 5809 195 3

For all future hydro power plants located underground the new high-voltage generator Powerformer™ is expected to be the most supreme alternative

S.Palmer
SWECO International

ABSTRACT: This paper presents a study for the Uri Hydroelectric Power Project in which the existing electro-mechanical installation has been compared with a Powerformer installation. Only those components and structures that have been changed with the Powerformer installation have been studied and the corresponding change of costs has been estimated. For the underground civil works a 22% reduction of the cost for machinery hall, draft tube gallery and appurtenant tunnel systems (headrace and tailrace tunnels not included) has been identified. For the electro-mechanical equipment no substantial change of cost has been identified. For the ventilation system in the underground caverns and tunnels a cost reduction of 16% has been identified. An additional power production of 17 GWh/year due to higher efficiency has been identified. Compared to normal annual production 3200GWh the additional power corresponds to 0.56%. Substantial improvements of environmental factor as well as lower cost for operation and maintenance have been identified. In summary all factors exclusive of higher efficiency show a net present value of some 8.0 MUSD in favor of the Powerformer installation.

1 INTRODUCTION

Since its introduction in the late 19th century the rotating machine for generation of electric power has looked virtually the same. In the traditional design, an increase in the power rating of the generator is achieved by a corresponding increase in the current level while the voltage level usually is fixed, e.g. 13.8 kV. In order to maximize the current carrying capacity, the conductor portion of the armature slot cross-sectional area has to be maximized. The conventional generator design is based on rectangular armature slots and conductor bars and the maximum output voltage is limited to the order of 25-30 kV. Electric power transmission, however, is normally held at voltage levels of several hundred kilovolts. As a result, a power plant based on a conventional generator most often requires a step-up transformer.

By taking a step back to basics, ABB has developed a new class of synchronous machine, Powerformer™ [1,2]. It offers a direct connection to the power transmission network without the need for an intermediate step-up transformer. In contrast to the conventional generator, for any given output power, Powerformer operates at a relatively high voltage and low current. This new high-voltage generator has armature windings with a cylindrical cross-section. In essence, the stator winding of Powerformer is based on proven solid dielectric power cables. The maxi-mum generated output voltage is, therefore, limited only by the state-of-the-art high-voltage cable technology. A number of notable benefits for the plant owner as well as for the network operator come from the power generation directly at the grid voltage. First, Powerformer eliminates the capital costs of the step-up transformer and medium voltage switchgear as well as the corresponding operational costs, e.g. the active and reactive losses.

Figure 1: Powerformer

Therefore, the use of Powerformer can result in efficiency improvements of the total plant in the range of 0.5-2% compared to the conventional generator-transformer system. Second, compared to a conventional power plant, a plant featuring Powerformer comprises a smaller number of components. Consequently, the need for maintenance is reduced at the same time as availability and reliability can be increased. Moreover, an exclusion of the step-up transformer and its auxiliaries makes the layout of a plant implemented with Powerformer more compact than that of a conventional plant [3].

Also, results indicate that Powerformer is more environmentally friendly than an equivalent conventional system [1,2].

Powerformer has a wide application range; it is well suited for both turbo power and hydropower applications [1,2,4]. Also, it finds its application in new installations as well as in refurbishment of existing plants. This paper presents a study for the Uri Hydroelectric Power Project in which a conventional solution has been compared with a Powerformer installation.

2 URI PROJECT

The Uri Hydropower Project is located on the river Jhelum in India, state of Jammu & Kashmir, some 80km west of Srinagar and 850km north-north-west of Delhi. The project utilises a gross head of 256 m and has an installed capacity of 4*120= 480 MW. The power plant generates an average of 3200 GWh per year. The plant designed by SWECO International went into commercial operation in May 1997.

The above-ground works comprise a barrage complex with a 25 m high and 95 m long concrete weir which includes a spillway and undersluices with a total discharge capacity of 2265m^3/s, a canal system with a design capacity of 2*260=520 m^3 /s and a desilting arrangement designed for the continuous removal of sediment coarser than approximately 0.15 mm.

The main components of the underground works consist of a horseshoe-shaped headrace tunnel 10.7 km long and fully concrete lined with a diameter of 8.4 m, two penstocks each 220 m long, steel lined with a diameter of 5 m and underground power house consisting of a machinery hall, transformer hall and a tailrace gate hall, which house four sets of turbines, valves, generators etc. with appertaining pipework and auxiliary equipment, see Figure 2. The powerhouse works also include surge arrangements upstream and downstream of the machinery hall, a cable and ventilation tunnel and a separate access tunnel to the power station. The tailrace tunnel is 2 km long and is of the same dimensions as the headrace tunnel.

The main electrical plant at Uri consists of the generators, transformers and switchgear in the underground power station, control and protection equipment, an outdoor switchyard and diesel units for emergency power. There are four vertical 136 MVA generators in the machinery hall. A flow of 240 m^3 /s is needed to run all four units simultaneously. If the flow in the river drops below this level, power generation continues but with fewer units. The generators have three 50-ton stator sections completely built on site and a rotor weighing 200 tons.

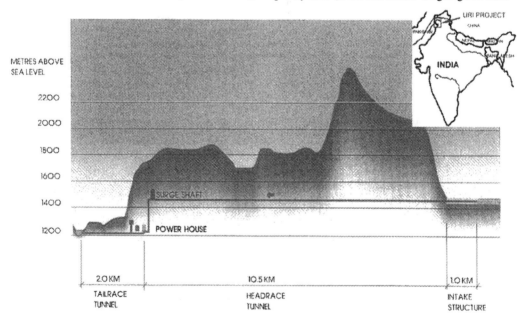

Figure 2: Uri Project, Longitudinal section

446

An encapsulated busbar system carries the power from the generators to single phase transformers in the transformer hall, which is situated in parallel with the machinery hall. There are a total of 13 transformers - three for each generator plus one spare unit. The transformers are directly connected to the gas insulated switchgear positioned above the transformers. However, instead of using busbars to convey the power from the switchgear to an outdoor switchyard, 400 kV oil-filled cables have been used. The cables pass through the lower section of an 800m long horseshoe-shaped cable tunnel (the upper section of which is being used as an air intake for the underground power station). Finally, an overhead double-circuit power line carries the power to the Wagoora substation at Srinagar which is 89 km away.

The general layout drawings of the underground works before and after the modifications due to the Powerformer installation are shown in Figures 3-6.

3 POWERFORMER INSTALLATION

3.1 *Civil Works*

3.1.1 *Summary*
The performed Study [3] has been devoted to analyze the Uri Hydropower Plant with respect to what possible impacts a Powerformer installation solution will have in terms of design and layout changes, investments, benefits, etc. All comparisons in the study have been performed based on available data for the actual Uri Hydropower Project. Alternative general layout drawings have been outlined, and based on these drawings the marginal change of quantities for civil works has been calculated and the marginal cost evaluated. The existing transformer hall and adit tunnel to the same have been deleted and tunnels as entrance tunnel, bus-bar galleries and main access tun-

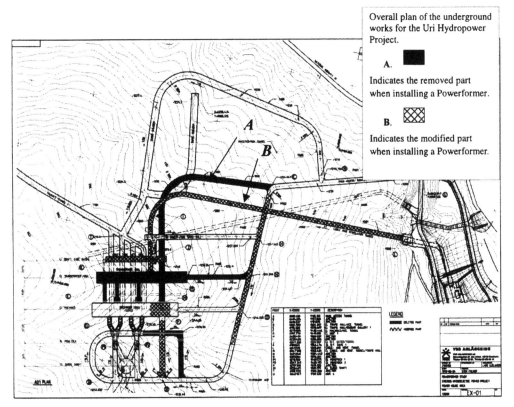

Figure 3: Layout (plan) of underground works before and after a Powerformer installation

447

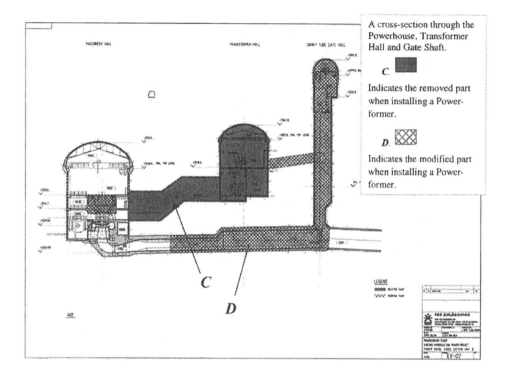

Figure 4: Layout (section) of underground works before and after a Powerformer installation

nel linking the machinery hall and the transformer hall in the actual project layout are not applicable with a Powerformer installation. By deletion of transformer hall and cross-tunnels, the rock stress situation will be reduced considerably in the powerhouse area and consequently eliminates most of the existing rock support. The cost benefit due to the modified civil works for underground caverns and tunnels surrounding or linking the caverns amounts to 22%.

3.1.2 Design changes

By introducing a Powerformer installation some of the existing underground civil structures will not be necessary and some must be redesigned. The following design and layout changes have been made:

- The existing transformer hall with adit tunnel has been deleted. The cave for the draft tube gates has been located closer to the machinery hall.
- Galleries for busbars from generators to transformers have been deleted.
- Main access tunnel has been shortened. The part entering the actual transformer hall from the downstream side for excavation has been deleted.
- Entrance tunnel with ventilation culvert from machinery hall to transformer hall is deleted.
- Chilling plant and ventilation equipment, earlier located in the transformer hall, have been moved

into actual possible storage area in the control building.
- The excavation below the unloading bay area in the machinery hall is extended in order to house the 220 kV gas insulated switchgear (GIS), GIS relay room, auxiliary transformer and potable water plant.
- The cable and ventilation tunnel has been modified, the alignment has been changed, it is shorter compared to the existing tunnel.
- Electrical and mechanical workshops, staff rooms and waste water tank have been located close to the unloading bay area.
- Excitation transformers and auxiliary transformers earlier located in busbar galleries have now been positioned on the generator floor. The encapsulated busbar system carrying the power from generators to single phase transformers has been replaced by a cable system.
- Minor changes for transport and erection of the GIS-equipment have been done.
- The arch ceiling and light steel roof in the machinery hall have been extended to reach pass the new staff building at unloading bay.
- The fire-fighting system could have been reduced in range, but the cost benefits have been neglected in the Study.

- Typically, Powerformer is somewhat larger than the conventional generator. Consequently the valve positions have been moved slightly upstream.
- An auxiliary transformer is introduced to secure power supply during standstill. The transformer is located outdoor at the pothead yard area.

Detailed cost calculations have been performed for the Powerformer alternative compared with the existing layout. All calculations have been based on the contract prices from the actual construction works, and thus are given in a price level of year 1989. However, since the result for the cost estimates is presented in relative figures (percentage of change in investment), the price level (year 1989) for the calculations are of minor importance.

3.2 Electrical Works

3.2.1 Summary
In Uri HEPP the voltage 220 kV has been chosen for the Powerformer, and to give a fair comparison to a conventional alternative a conventional alternative with 220 kV voltage has been set up. The limit for the study [3] is the connection of the transmission line at the pothead yard of Uri HEPP. The study shows that the investment cost of electrical equipment for the Powerformer alternative is to be considered practically the same as the conventional. Only a marginal difference is found, and it is not possible with the uncertainty of the estimations to state if the conventional or the Powerformer alternative is cheaper.

3.2.2 Scope of the study
The study [3] has been limited to equipment located at the underground power house and pothead yard. No considerations are made concerning equipment at spillway, intake area and tailrace or between those, as all such equipment has been assumed not to be de-

pending on the generating equipment. To make a fair comparison between a conventional and a Powerformer 220 kV alternative, a conventional alternative with a transmission voltage of 220 kV has been set up. 220 kV has been chosen for Powerformer as this is a good compromise between optimum voltage for Powerformer itself and a suitable transmission voltage for a transmission line of this capacity and length. The choice is also influenced by the fact that 220 kV is available as a transmission voltage further out in the grid and that Powerformers today are available for this voltage. In a few years time even higher voltages will be available.

3.2.3 Discussions on the Powerformer alternative
By introducing a Powerformer instead of a conventional generator several benefits are reached in the electrical installations:
- Lower losses => higher efficiency
- No block transformers are necessary, less oil underground - less risk of fire and less risk of oil leakage contaminating the environment
- Busducts are replaced by cables, less volume needed
- Less maintenance due to less components and areas
- Higher availability due to less components

No respect is taken to the fact that a Powerformer has a higher capacity to stabilise the network system as it has higher reactive capacity in a short-term respect. A Powerformer has due to design factors a possibility to be reactively overloaded for up to ap-

The study [3] covers three alternatives:
- Existing conventional 400 kV plant
- Hypothetical conventional 220 kV plant
- Hypothetical Powerformer 220 kV plant

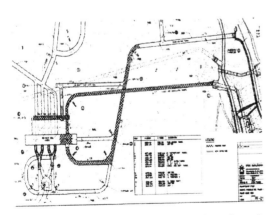
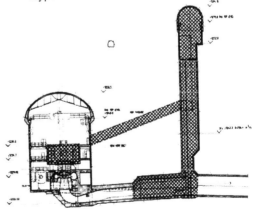

Figure 5 and 6: The underground layout after a Powerformer installation

449

proximately 15 minutes without damage or stability problems. This would in the Uri case be of great interest as the network is fairly weak and at the commissioning showed to be very influent of the operational status at Uri HEPP. This study is not considering the technical or economical value of this, as the limit is the connection of the transmission line. It is recommended to study this further in a continuation of this study [3]. Powerformer is recommended not to have its neutral point directly earthed. If so the harmonics, especially the third tone, will heat the machine considerably and thus reduce the output capacity with as much as up to a third. The system earthing is arranged at the stand alone auxiliary supply transformer at the pothead yard, and in case the reactors are connected, also through those. System earth is assumed to be directly earthed. The system earthing would need to be studied further when including the grid into the study, in respect of voltage rise in case of earth fault.

The weighted total losses, as to the actual operating regime of Uri HEPP, are reduced approximately 0,56% with Powerformer compared to a conventional solution. This is a weighting done with the measured annual variations in the river flow, giving a different number of units being in operation as to the flow. A unit is operated from 50 to 100% of rated load. Below 50% the turbine is inefficient and the unit shut off. When several units are in parallel operation they are sharing the load equally as this gives the best overall efficiency. Consequently the rated need for cooling of air and water circuits is less. The annual additional energy production is approximately 17 GWh. The change in losses is calculated including generator/Powerformer, busducts/cables and unit main transformers, and is done as a weighted value over the actual flow and regulation regime in the river.

An increase in availability is likely to occur. As the number of components at each unit is reduced, the risks of unplanned failures are also reduced. To what extent is very difficult to state as the experience of Powerformer operation is limited. An assumption of 2 less stops à 10 h per year for one unit is estimated as relevant which for this study is assumed to take place at a unit at full output. This will give an increase of 120x20 = 2,4 GWh per year to be added to the increase in annual energy production.

The fact that a Powerformer installation gives less risk of major underground fires due to less amounts of flammable oil has not been evaluated.

The effect of internal faults in the generator compared to Powerformer has not been quantified. It is however our definite opinion that faults are unlikely to occur, and if so those will mainly be phase to ground faults as the winding cables are fully insulated. Compared to a conventional winding the risk of serious faults is considered to be less in Powerformer. If a fault occurs the effect is assumed to be approximately equal, as the higher short circuit voltage for a EHV-winding is balanced by the fact that only phase to ground faults will occur.

Trying to set economical figures to this energy addition gives an interesting result. For Indian circumstances a price of 2.5 INR (Indian Rupies; 1 INR equals 0.024USD) is assumed per kWh. Thus an addition of 17 GWh for higher efficiency and another 2.4 GWh for higher availability gives an annual addition of 48.5 MINR (=1.2 MUSD). Capitalised at an interest rate of 5% for a lifetime of 50 years gives a net present value of 906 MINR (=21.7 MUSD) in favour of the Powerformer alternative. No respect has been taken to less energy losses in ventilation and lighting equipment. The figures are given in table 1 (Benefit of Powerformer).

3.2.4 Cost estimates for electrical equipment investment

The installation costs for electrical equipment for the different alternatives are compared to the actual case, set to 100%. The costs are estimated in today's cost level (1999-03) as the changes in electrical deliveries have altered considerably during the 1990's. The costs are considered as a relative comparison only as actual values are difficult to state accurately.

Table 2: Alternative electrical equipment cost

Alternative	Investment (MUSD)	Percentage
Conventional 400 kV	52	100%
Conventional 220 kV	47.5	91.4%
Powerformer 220 kV	47.1	90.6%

When changing the installations from a conventional solution with generator and unit transformer to Powerformer, the investment cost concerning electri-

Table 1: Benefit of Powerformer

Alternative	Additional energy production (GWh)	Corresponding value (MINR)	Corresponding value (MUSD)
Annual benefit	17+2.4	48.5	1.2
Capitalised at 5% interest rate	n.a.	906	21.7
Capitalised at 10% interest rate	n.a.	529	12.7

cal equipment is marginally altered, see table 2. Many parts of the plant are still the same, giving the same investments for those items. The changes in investment are identified for the following electrical parts:

- Generator → Powerformer
- Busducts → E.H.V. Cables
- Unit main transformers deleted
- Less installations of small power and lighting (less areas underground)
- A stand alone auxiliary supply transformer is added

The investment cost of Powerformer of this type and for this plant is approximately 40% higher than a conventional generator, but one must have in mind that a main unit transformer shall be added for the conventional generator and consequently the difference is less. In this case the cost of Powerformer compared to a generator + transformer is almost equal. This assumption is somewhat conservative as Powerformer is assumed to be less expensive when production is of more serial nature, when methods for construction and erection have become more established. The cost for unit auxiliary supply transformers are assumed to be identical in both alternatives. In this case it seems like the loss reduction is the main benefit of the Powerformer.

3.3 Ventilation Works

By introducing Powerformer the ventilation system of air flow for supply and exhaust air will be reduced when the transformer hall is deleted. Also the demand of heating capacity for the supply air system is reduced. The heat losses from electrical equipment are reduced and the demand of cooling capacity will be lower and the numbers of liquid chillers and air cooling units can be decreased. The demand of ventilation airflow, heating and chilling capacity for the existing plant compared with the Power-former installation is given in table 3.

Table 3: Alternative heating/chilling capacity

PLANT	Airflow Supply/ Exhaust m³/s	Heating Capacity Supply Air kW	Chilling Capacity kW
Existing	32	400	1250
Powerformer	20	250	900

3.4 Operation and Maintenance

3.4.1 Operation

The major benefit of Powerformer is found in the lower total losses for the units. The losses for an operating regime of Uri HEPP type has been found to give an additional annual energy production of 17 GWh corresponding to a higher efficiency of 0.56%. Another 2.4 GWh is assumed to be added from

higher availability due to less components. The higher energy production corresponds to an annual increase in sales of more than 0.5 MUSD. Capitalised the value is 12.6-21.5 MUSD depending on a calculation interest rate of 5-10%. Another major benefit of Powerformer is the fact that tons of flammable oil has been taken out of the underground caves, giving a less risk of major fires.

3.4.2 Maintenance

An assumed reduction in maintenance of 10% is relevant in the Powerformer alternative as the number of major components is less. The components of interest in this matter are the block unit transformers with cooling device, the air insulated busducts, approximately 10% reduced underground areas with lighting, cleaning, etc.

3.5 Environmental Impact

3.5.1 General

Powerformer can be connected to the power grid without a transformer. This new system is therefore more simple and compact than a conventional system. The different size and way of installation will affect the civil works to a certain amount. The changes in design criteria for the civil works and the increased efficiency for Powerformer are described earlier in this report. These changes will also change the environmental impact considerably in some aspects. The main environmental aspects will occur as a consequence from the usage of material and energy. A modern environmental management shall therefore always strive towards solutions and systems, which use a minimum of energy, transportation, materials etc. In this study we have concentrated on the actual changes and how they will affect the environment. In the study [3] the changes in the following areas are considered:

- Energy production
- Energy losses
- Dimensioning of the concrete batching plant
- Dimensioning of the access roads
- Transportation
- Changes in need for materials and chemicals
- Equipment for fire extinguishing

3.5.2 Changes in energy efficiency

Powerformer has less components and the accessibility is estimated to increase and the maintenance to decrease. At present, the increased power production due to accessibility is estimated to 2.4 GWh per year. Due to the improved efficiency, the losses will decrease compared to a conventional system. The reductions of losses for the generator, transformer and conductor busbars are 17 GWh per year.

The reduced losses from Powerformer also change

the design criteria for the cooling and ventilation systems. The reduction of material is not calculated but the energy savings in cooling system, fans and heating of supply air are estimated in a total up to 1,100 MWh per year

The reduction of the cooling system will make it possible to reduce the number of cooling machines and as a consequence, the refrigerating agent will be reduced with 125 kg R134A.

3.5.3 *Dimensioning of the concrete batching plant and access roads*
The need for concrete will be reduced due to the changed design criteria. The reduction is estimated at 4 - 5 %. This will affect the dimensioning of the concrete batching plant and give less impact on the environment.

3.5.4 *Transportation*
The need for external transportation will be changed due to the reduction of the civil works. Less material is excavated, which also reduce the transportation to the site. In the below summary, the reduction of material and transportation are shown:

Change in transportation (Estimated truck capacity 10 m^3):
- Excavation, reductions, 67,000 m^3.
  6,700 trucks, distance 2x1 km   13,400 km
- Concrete, reductions, 14,600 m^3.
  1,460 trucks, distance 2x1 km   2,920 km
- Gravel, reduction, 4,500 m^3.
  450 trucks, distance 2x30 km   27,000 km
- Concrete, reduction, 4,600 tons.
  307 trucks, distance 2x250 km   153,500 km
- Explosives, reduction, 67 tons.
  5 trucks, distance 2x250 km   2,500 km
  Total: 199,320 km

3.5.5 *Changes in need for materials and chemicals*
Equipment for fire extinguishing by carbonic acid for the existing generators will be unchanged with a Powerformer installation, why there will be no changes in the chemicals used.

In Powerformer cross-linked polyethylene is used as insulation instead of epoxy.

The deletion of transformers will reduce the transformer oil in total with 265 tons including a new auxiliary transformer, which still uses approximately 5 tons of transformer oil. The fire fighting system will also be reduced by the deletion of transformers.

The change in voltage will reduce the amount of SF6 -gas within the GIS. The actual amount is not estimated.

The amount of copper and iron will be reduced with approximately 30 %.

The amount of explosives will be reduced with 67 tons.

3.5.6 *Changes in environmental impact*
The above mentioned factors are estimated to be the most important. However, the changes in energy production and losses during the production period are assumed to give the greatest change in environmental impact.

4 CONCLUSION

The study clearly indicates that for hydropower schemes located underground there are substantial benefits in respect of reduced investment costs for the civil works if Powerformer is selected. Other substantial benefits are higher efficiency, lower operation and maintenance costs and improved environmental impact. For all future underground hydropower plants the installation of Powerformer is likely to be the most economical and technical environmentally attractive alternative.

REFERENCES

[1] Mats Leijon et al., "Breaking conventions in electrical power plants", Report 11/37-3, CIGRE General Session, Paris, 1998.
[2] Mats Leijon, "Powerformer™ – a radically new rotating machine", ABB Review, No. 2, 1998, pp. 21-26.
[3] Sten Palmer et al, SWECO International "Powerformer Study – Uri Project, Final Report", 1999.
[4] Mats Kjellberg et al., "Powerformer™ chosen for Swedish combined heat and power plant", pp. 19-23, ABB Review, No. 3, 1999.

Hydropower in the New Millennium, Honningsvåg et al (eds), © 2001 Taylor & Francis, ISBN 90 5809 195 3

Revival & updating of chamera power station

Yogendra Prasad & S.K.Dodeja

SYNOPSIS

Chamera Hydro Electric Project of 540MW (3 x 180) is located in the northern part of the country in Chamba District of Himachal Pradesh State and constructed as Indo-Canadian Joint Venture by NHPC, the biggest hydropower generation company. The Project was commissioned in March, 1994 and generates over 2000MUs annually. In last over 6 years of operation, the Project has seen many challenges including threatening of its survival in the form of disturbances in all the generating units due to tilting of the machines, failure of 400 KV GSU transformers. In fact, all this raised serious apprehensions regarding the performance of generating units & transformers brought from Canada. However, serious studies and diagnosis put in the process further leading to process restoration and revival saved the reputation of Indo-Canadian Joint Venture Project and further motivated to up date the Project with installation of present date generating on-line monitoring system for generating units and establishment of transformers oil testing lab at site itself. Experiences as encountered have provided unique opportunity bringing back the Project to become one of the most important Power Station of the country.

INTRODUCTION

Each Hydroelectric Power plant is different and unique in its characteristics, so are the problems of Hydro units. Faced with major problems, Chamera project has gone for unique solutions. Problems in generating units, GSU transformers etc. required major solutions and preventive measures and having accomplished this challenging task the revival is more significant. Establishment of on line monitoring system for generating units and Transformer oil testing facilities have been steps in the direction of updating and total solution to problems faced by Chamera hydro power station, in last 6 years.

REVIVAL

1. Generating Unit Problems and Solutions

In June 1996, rise in metal and oil temperature of Turbine Guide bearings of unit # 2 caused tripping of the unit. Inspection revealed a concrete piece found wedged between guide vanes 1 & 24 and on removal, an attempt was made to restart the unit but loading beyond 90MW caused high vibrations and abnormal sound from the unit. Re-inspection of the unit revealed Turbine Guide bearing pads damaged (replaced with new bearings). Unit was again synchronized but the abnormal bearing temperatures and noise persisted. Drift of turbine shaft by about 0.25 mm towards 1 O' clock position was observed beyond 90 MW. Rotor was found out of verticality by 0.22 mm/m (acceptable limit 0.06 mm/m). Run out of rotor was 0.14 mm at TGB (acceptable limit 0.1 mm). Involvement of experts of manufacturers of turbine and generator viz. Alsthom Canada and GE Canada was sought, hereafter and rectification works were undertaken in 3 phases.

1.1 PHASE-I

After suggested leveling of thrust cone by grinding to 0.02 mm (earlier out by 0.23mm) and resetting of TGB with gap 0.2 – 0.3 mm, unit was re-synchronized but tripped again due to high TGB temperature, accompanied by abnormal sounds and vibration. Drifting of shaft by 0.25 mm towards 1 O'clock position also continued. Difference between TSD readings of opposite sides of TGB increased to 13.2°C, with metal temperature exceeding 70°C. Unit was kept under observation. Data obtained during these observations was examined by GEC Alsthom. Rotor verticality and shaft run out were found to be disturbed again. Turbine Guide bearing was found damaged again. With activities described on pre-page repeated unit re-synchronized Though bearing temperatures were stable initially, unit again started showing deterioration after a week of operation with same behavioral trends.

In August 1997, high vibration and abnormal sounds were also observed from unit No. 1. TGB oil temperature increased from 60°C to 70°C & hereafter operation of unit No.1 and 2 were continued at 90 MW due to vibration and sound at higher loads.

1.2 PHASE 2

All the experts from Canada & India visited site in Dec'97 to study the data collected, observed physical & investigative alignment & air gap parameters, diagnose for the causes & indicated following factors causing disturbances in the unit.

a. Heavy radial loads on turbine guide bearing more than 40 tons, (for which it was designed) due to blockade of water passage by concrete piece between stay vanes.

b. In adequate grouting below spiral case, stay ring and pit liner areas.

c.   Mis-alignment of shaft and inadequate finish of TGB Journal

d.   Inadequate differential rigidity in the runner removal area & Inclination of head cover.

CIDA experts recommended following restoration plan : -

i)   Re-machining of turbine shaft journal & repair of turbine bearings, matching with new journal diameter with installation of new improved TGB support ring.

ii)   Re-grouting of turbine lower stay ring area and lower pit liner area and erection of extra jacking columns to prevent vertical elastic deflections of turbine head cover and bottom flange of lower pit liner.

iii)   Ultrasonic testing of Generator guide bearing shoes & Re-alignment of shafts.

iv)   Checking of Head Race Tunnel for concrete pieces entry into water conductor system/units.

1.3 PHASE 3

Major restoration works as per the expert's recommendations were undertaken

i)   Shaft journal was machined at site in powerhouse; with surface finish of order 0.8 micron and concentricity of 0.025 mm.

ii)   Turbine bearings were re-babbited at Alsthom Canada and one at BHEL Bhopal simultaneously.

iii)   Turbine guide bearing support for improved strength fabricated at M/s Alsthom-Canada.

iv)   Strain gauges installed at thrust cone to determine the deflection of thrust cone.

v)   6 Nos. column jacks were fabricated and installed with pre-stressed under load of 12-15 tons for support of lower pit liner at runner-removal area.

vi)   Grounding at stay ring and lower pit liner areas with non-shrink grout injected at 0.12 Mpa.

vii)   Thrust bearing support cone flange leveled within 0.05 mm by grinding & re-assembling of unit with verticality, run out, air gaps brought within acceptable limits.

viii)   Inspection of HRT conducted & minor rectifications undertaken.

After completion of unit II works, works of unit I were taken up. Though unit No. 3 continued to operate normally all through this period some inclination of shaft was observed. Accordingly, it was decided to complete renovation works of units 3 also on the same lines.

All the three Units are working most satisfactorily since thereafter.

	Unit I	Unit II	Unit III
Re-commissioning after restoration	29.5.98	15.5.98	11.4.99
Running hours after restoration	10670	12186	8013
Units generated after restoration (MUs)	1085	1693	2107

2. SITE RESTORATION OF 400 KV GSU TRANSFORMERS AT CHAMERA POWER PLANT

A bank of 10 generating transformers of 75 MVA single phase transformer (Federal Poineer – 88/89) were installed in the underground transformer gallery, (commissioned during 1994, and running normally up to 1997). On 30.12.97 during voltage build up, transformers installed in unit 3 R phase failed on inter-turn fault. Spare transformer was installed on unit No.3 R phase.

On 30.3.99, while building up voltage, R Phase transformer installed in Unit No. I failed on inter turn fault, under similar circumstances of failure of unit 3 R phase. Transformer differential protection and Bucholz relay had operated. Further testing revealed low break down voltage of oil, and resistance value of H.T winding higher than that of commissioning value indicating arcing inside transformer. Internal inspection revealed deposition of Carbon all over the bottom frame, bottom and side surfaces of the transformer tank. Water droplets were also observed in the bottom of the tank. No spare transformer was now available.

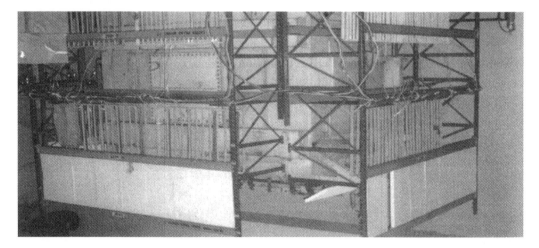

All the manufacturer's experts from the country and from M/s Pauwels Canada were contacted and it was suggested to dispatch the transformers to the manufacturer's premises either in India or Canada for necessary repairs/rectification though site inspections of both the available and failed transformers revealed failure on different windings/coil and cannibalization from these two transformers could make one transfer healthy. Any quantum of interaction with manufacturer was not encouraging. The Canadian manufacturer even intimated "unless transformers are returned to factories and proper repairs are made, the site of repeat failure is too great". In view of period involving in repairs/rectification being over 8 months and monsoon period being nearer, the loss in generation & machine availability in terms of quantity & value being quite substantial a bold decision was taken by NHPC to undertake the repairs at site with some assistance in the form of manpower and supervision from Indian/Canadian manufacturers.

The process of removal of top stampings of coil of Unit # 3 transformer – lifting of damaged coil and healthy coils was undertaken after due testing with further insertion in Unit # 3 transformer with fabrication of special lifting device for handling of coil locally at site. The transformer was re-assembled with connections of the top changer and described testings were also conducted to ensure healthiness of assembled transformer.

1. Turns Ratio Test

Tap No.	Calculated Ratio	Measured Ratio	Phase Deviation
1	17.571	17.572	0.02 c. rad.
2	17.152	17.157	0.02 c. rad.
3	16.734	16.740	0.02 c. rad.
4	16.316	16.323	0.02 c. rad.
5	15.898	15.909	0.02 c. rad.

2. Winding Resistance Measurement: (Hv Winding)

Tap No.	Winding Resistance (Forward)	Winding Resistance (Reverse)	Average
1	0.838	0.840	0.839 *
2	0.815	0.805	0.810 *
3	0.801	0.797	0.799 *
4	0.779	0.776	0.777 *
5	0.768	0.765	0.766 *

3. LV Winding Resistance: 1.95 M *

Removal of foreign particles, carbon particles and moisture from the assembled transformer paused a major engineering challenge, due to lack of vapour phasing facilities at site. Drying of transformer to water content of less than 0.5 % of the weight of paper insulations is the normal practice during manufacturing. Available methods are (a) Circulating hot oil (b) High Vacuum. With site conditions, it was decided to use combinations of both methods.

Using 6 KL vaccum & heating hour plant, circulation of oil was done. Bubble bath was established to flush out carbon particles from inside the coil and assembly, using dry air (at pressure of 5 psi) inducted through oil circulation duct at bottom of the tank. Temperature of oil was kept high to draw maximum moisture. BDV and moisture content were checked on hourly basis circulation of oil continued, so that values of oil at outlet of plant and inside transformer are equal for more than 3 hours.

For creating a heating chamber around the transformer racks, thermocol sheets, Asbestos cloth, and rod heaters were used. On completion of removal of particles, vacuum drying of coil/core assembly using the

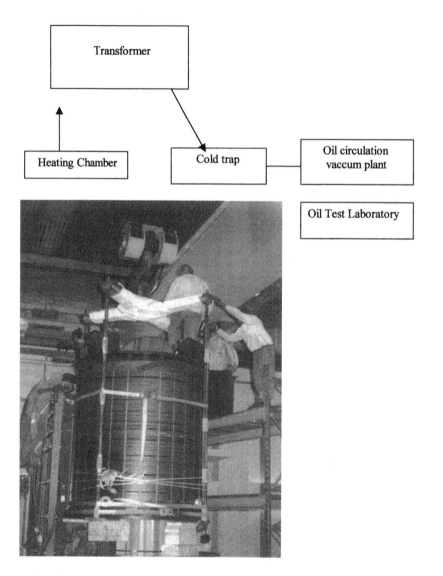

```
┌─────────────────────┐
│    Transformer      │
└─────────────────────┘

┌──────────────────┐   ┌──────────────┐   ┌─────────────────┐
│ Heating Chamber  │   │  Cold trap   │   │ Oil circulation │
└──────────────────┘   └──────────────┘   │  vaccum plant   │
                                          └─────────────────┘

                                          ┌─────────────────────┐
                                          │ Oil Test Laboratory │
                                          └─────────────────────┘
```

tank as the vacuum chamber was carried out. A fine vacuum of 0.2 mm of mercury for 2 days was created. Dry ice trap was fabricated locally and fitted before vacuum pump to take out moisture from system to a level below 15 ml/hr.

Top tank was welded after transformer filled with Nitrogen gas and transformer was shifted to transformer gallery. Conservator, oil pumps, pipelines, coolers, Bucholz relays, etc were assembled with dry air supplied to avoid ingress of moisture. HT pocket, and control wiring etc was completed. Transformer was again vacuum dried, and then filled with new oil, under fine vacuum. Circulation of oil was conducted till BDV, gas content, and moisture content reduced to acceptable values. Other test like winding resistance, turns ratio, capacitance, dissipation factor etc. were carried out.

HV testing was carried out by feeding 13.8 KV at LT side from unit Auxiliary transformer, back charged from D.G. sets. Voltage was fed from D.G. set manually from 0 volt to 415 volts gradually, resulting in LT voltage of 0 to 13.8 KV. Transformer passed voltage test successfully.

HT connections were made, (after vacuum drying of HT pocket) and filled with new oil. Transformer was successfully charged on 16.6.99 and had been operating satisfactorily since than (operated for 7057 hrs till Jan'2001).

This site repair of 400 KV transformer saved loss of $ 25 M for NHPC and also avoided loss of 180 MW of power to the Grid. As per available records, this is the first instance of such repairs of 400 KV transformer at site anywhere not only in the country but in the world.

FURTHER REVIVAL AND UPDATING

3.   *ON-LINE MONITORING SYSTEM FOR HYDRO POWER PLANT*

Subsequent to the repair and restoration works of all the three generating units of Chamera Project the need of constant condition monitoring of the generating units was essential to enable monitor the performance aspects of the generating units on continuous basis. Moreover changing economic scenario linked with availability of the generating units making direct impact on the revenue of the power plant need from change over from periodical maintenance to condition based maintenance was inevitable. On-line monitoring system for generating units was installed during March/April 2000.

3.1   DESCRIPTION OF THE SYSTEM

On-line monitoring system installed at Chamera H.E. project is manufactured by VibroSystM Inc., Canada. It consists of ZOOM 2000 system and AGMS acquisition units (Zoom-Zero outage online monitor, AGMS – Air Gap Monitoring System).

It is a complete data acquisition system for Hydro generators and measures air gap and status parameters such as vibration, pressure and temperature relating to the condition of the rotating generating unit. The system provides alarm notification and external annunciation devices switching on selected dynamic parameters through the use of appropriate relay modules.

AGMS focuses on air gap displacement only and all highly accurate parameters are measured when generators are operated and safety continues alarm monitoring and tripping.

ZOOM system has the capacity to monitor several generators together. It can also be adopted for each generator alone for different parameters to be monitored, but air gap must always be among the parameters to be monitored.

Database from this system is monitored from any far away PC installed with PC anywhere and Vibro System software & links of telephone lines, VSAT, Internet can be used for the same. Presently data from Chamera Power house can be monitored by NHPC. Corporate O&M division, Faridabad with the help of PC installed there over telephone and VSAT.

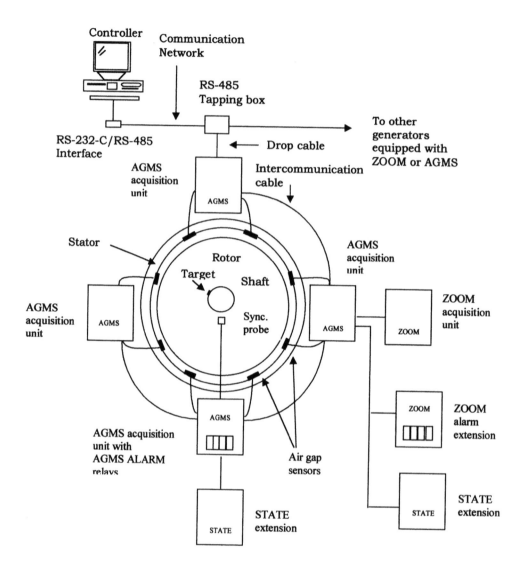

Controller   Communication
             Network

RS-485
Tapping box

RS-232-C/RS-485
Interface

To other
generators
equipped with
ZOOM or AGMS

Drop cable

AGMS
acquisition
unit

AGMS

Intercommunication
cable

Stator

Rotor

AGMS
acquisition
unit

Target

Shaft

AGMS
acquisition
unit

AGMS

Sync.
probe

AGMS

ZOOM

ZOOM
acquisition
unit

AGMS

ZOOM

ZOOM
alarm
extension

AGMS acquisition
unit with
AGMS ALARM
relays

Air gap
sensors

STATE

STATE
extension

STATE

STATE
extension

system layout

3.2 AIR GAP MEASUREMENT

Air gap measurement is the heart of this system. Measurements of all other parameters are correlated with it and depend on it. This principal allows synchronization of all parameters with passing Rotor Poles. Eight nos. of capacitive sensors are installed in each generator for air gap measurement. Four sensors are installed at the top air gap and four are installed at the bottom air gap at 0 degree, 90 degree, 180 degree and 270 degree. These sensors are glued to the stator core. Spacing between the sensors are approximately 90 degrees, symmetrically. Capacitive air gap measurement is immune to oil and carbon particles in the air gap. The polar plot of the air gap at the bottom and the top can be obtained separately on the monitor. The minimum and maximum air gap also can be measured by placing the cursor on the place of measurement on the monitor. The minimum and maximum values of air gap also can be measured. The trend of this measurement for different periods of time, for example today, yesterday, this month, and last month or for a specific period of time can be plotted as a trend graph. This arrangement helps to find out any changes in air gap over a period of time. Changes in air gap normally point out the problem in the bearings or at the pole itself.

3.3 ADDITIONAL PARAMETERS MEASUREMENT

Other additional parameters, which can be measured, are vibration, pressure, temperature, active and reactive power etc.

The parameters available at Chamera power house along with their applications are as given in the table below:

PARAMETER	APPLICATIONS
Vibration	Turbine guide bearing, Upper guide bearing, Thrust bearing
Position	Wicket gates, unit open/close.
Temperature	Stator core, TGB, UGB, Thrust bearing, cooler air, transformers
Active and Reactive power	Generator
Pressure	Spiral casing, draft tube.
Voltage, current	Stator, Rotor (Transducers to be installed)

Additional parameters are acquired by sensors at various systems through 4-20 mA, transducers connected to Zoom and State extensions. Status of these parameters is collected at an interval of 15 minutes at Chamera power plant. Trending of these measurements for any prescribed period of time are available through graphical presentation in the monitor.

3.4 VIBRATION AND DISPLACEMENT MEASUREMENTS

2 nos. non-contact proximity probes are installed at 0 & 90 degree in both in Upper Grade Bearing and the Turbine Guide Bearing for the measurement of vibration at these locations. 3 nos. probes are installed at the bottom of thrust bearing in the axial direction at 0, 20 & 240^0 for monitoring axial vibration of the machine at the thrust bearing.

On-line monitoring system installed in Chamera Project helps to monitor various parameters related to the power plant on real time basis & analysis of the trends helps to reach conclusions regarding the status of the various components of the unit. Monitoring of the parameters like temperature, vibration displacements etc. enable to know the condition of the generating unit. The maintenance works can be planned in advance as per the indications obtained from the monitored values and analyses of the trends of various parameters. Installation of this system has helped reduction of human involvement in the monitoring of various parameters in the power plant and associated errors. Installation of On-line monitoring system has enhanced the condition monitoring of the equipments of power plant.

4. TRANSFORMER OIL TESTING LABORATORY

Failure of 2 Nos. 400 KV 75 MVA GSU transformers at Chamera Project have highlighted the requirement of continuous monitoring of healthiness of the transformers installed. Earlier the tests were arranged from Delhi 600 kms away, where the nearest oil testing lab is located. It was decided to establish transformer oil testing lab at Chamera Power Plant itself. Accordingly, an oil test laboratory has been established at Chamera project and this facility is being utilized for other projects of NHPC in Northern region as well as by other power utilities in Northern India.

4.1 TRANSFORMER FAULT GAS ANALYZER

The TFGA-P200 is a fast micro gas chromatograph manufactured by Morgan & Schaffer, Canada, optimized for rapid analysis of gas samples. In order to measure the dissolved gases in oil, gas sample must be produced in which the gas phase concentrations are related in a known and reproducible way to the dissolved concentrations. With this analyzer a complete gas in oil analysis can be made for Hydrogen, Methane, Carbon Monoxide, Carbon Dioxide, Ethylene, Ethane and Acetylene. Helium gas is used as carrier. The calibration of the equipment is to be done. A chromatogram of the calibration gas is obtained. Gas sample is taken in the syringe, shaked well and put on the equipment. By using the software available concentration of gases like Hydrogen, Methane, Carbon monoxide, Carbon dioxide, Ethylene, Ethane and Acetylene are detected from the chromatogram.

Analysis of interpretation of the test results is also possible with the software oil tran. Database of the samples can be created, and each parameter can be monitored.

Other equipments presently installed at Chamera Oil Test Laboratory are :

1. SERIES AIRE-OMETER – Total amount of dissolved gas in oil.
2. CAPACITANCE AND DECIPESION FACTOR BRIDGE - Accurate measurement of Capacitance and Dissipation factor of electrical insulation.
3. FASTQC TITRATOR – Determine total acid number.
4. COULOMETER – Determine small quantity of moisture.
5. PORTABLE MOISTURE METER - Determine small quantity of moisture.
6. DIELECTRIC TESTER – BDV of insulating oil

With installation of these instruments, health of transformers is checked periodically.

5. CONCLUSION

Continuous and repetitive problems in generating units of Chamera project have forced NHPC to take bold decisions to go for total revival of all three generating units, with review and improvement of design aspect also. After renovation of 3 units, they have been running satisfactorily. On-line monitoring system for generating units has been installed, which monitors the units on a continuous basis. Record of all sorts was made by NHPC by repairing 400 KV transformer at site with minimum time. Addition of transformer oil test lab at Chamera has resulted in continuous monitoring of 400 KV transformers at the project. Revival and updating of Chamera power plant has given stability and reliability to the power plant which works as the back bone of the northern Regional power system, in India. The system of updating is still continuing & studies are in progress for installation of (i) On-line Monitoring System for Transformers (ii) Discharge Measurement through Pen Stocks (iii) Silt PPM measurement through on-line infra-red/optical instruments.

Hydropower in the New Millennium, Honningsvåg et al (eds), © 2001 Taylor & Francis, ISBN 90 5809 195 3

Head loss of inclined fish screen at turbine intake

C.Reuter, K.Rettemeier & J.Köngeter
Institute of Hydraulic Engineering and Water Resources Management Aachen University of Technology, Aachen, Germany

ABSTRACT: For downstream migrating fishes the turbines of hydroelectric power stations are almost insuperable barriers. The mortality due to turbine impact is proved by many investigations (Montén, 1985). A possibility for fish protection is the installation of inclined screens placed in front of the turbine intake to guide the fishes to a suitable bypass. These screens cause a head loss and reduce the production of electricity. For the hydraulic investigation of screens a separation into experimental and numeric investigations is meaningful, since problems in two scales have to be regarded. The paper contains results of the numerical simulation, which were used to define the configurations of the physical experiments. The head loss is determined in physical model tests as a function of the main flow velocity and the angle of attack. A comparison with well-known formulas show, that physical experiments are still required.

1 INTRODUCTION

During their downstream migration fishes have to pass the turbines of hydroelectric power stations in many rivers. The mortality due to turbine impact especially for larger fishes is proved by many investigations (Montén 1985). A wedge wire screens placed in front of the turbine intake is a possibility to guide the fishes to a suitable bypass and to protect them form being injured. The screen constructions can be classified into modular inclined, horizontal angled and vertical screens. Every screen causes a head loss and reduces the production of electricity.

From structural and hydraulic view the effectiveness of the screen must be checked regarding the fish migration and the head losses for the guarantee of an economic use of hydroelectric power. For the hydraulic investigation a separation into experimental and numeric investigations is meaningful, since problems in two scales have to be regarded.

On the one hand the current adjacent to the screen bars has to be examined and the corresponding head losses has to be determined. This can be realized by a physical model. On the other hand only a spacious investigation can supply a sophisticated prediction of the effect of the screen on the entire flow in the area of the turbine inlet, including bypasses. This investigation can be handled more exactly and more efficiently with a numeric simulation due to the higher flexibility and the preventable scale effects. The head loss is determined in the physical model tests as a function of the main flow velocity, and the angle of attack.

The already existing well-known head loss formulas for trash racks often underestimate the actually occurring head losses. The head loss of the selected wedge wire is examined in physical model tests for different flow velocities and angles. The flow velocities for the design flow of 400 m³/s can be derived from available numeric investigations. They serve as reference values for the physical model tests. In the following the numeric model and the simulation results for a sample run of the river power plant are presented. These results are used as boundary conditions for the physical model tests.

2 NUMERICAL EXPERIMENT

2.1 Numerical Method

Computational fluid dynamics provide good tools to predict flow phenomena. The flow phenomena in large-scale geometry are very complex and in most cases fully three-dimensional. This is due to the very irregular boundaries inducing high velocity gradients and a large spectrum of turbulent flow. Therefor the model must be capable of representing the whole eddy spectrum and simulating three-dimensional flow in the irregular boundaries. A three-dimensional 'Large-Eddy' approach, which is based on the Finite-Element method, fulfills these requirements.

For the numerical model calculation a sample run of the river power plant in Germany was used. The three dimensional finite element grid of the intake cove comprises one third of the weir, the separation pier and the four intake chambers of the power plant including the machine casing of the generator. The discretization is limited to the front of the wicket gates. The spatial resolution of the grid must be adapted to the structure of the expected solution of the flow field. Regions with high velocity gradients require small element sizes. Therefor the discretization nearby and inside the intake chambers is finer than in the upstream water bay and in front of the weir. However, a high geometric resolution leads to high node and element numbers. Because of limited computer capacities, it is necessary to develop a compromise between spatial resolution and computational efficiency. This optimization process leads to a grid, which counts 40105 nodes and 34584 elements (Rettemeier et al., 1999). Figure 1 shows an inside view of the grid. The flow direction is towards the intake bellmouth (left hand side). The separation pier with the buffle in front of it is located in the background.

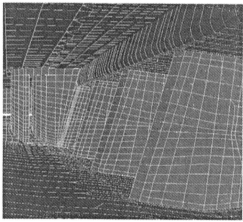

Figure 1. Outline of the discretization (insight)

2.2 Flow Simulation

A great number of simulations have been carried out for comparison of the natural measurements with the numerically modeled flow. For the presented results the inflow conditions has been 400 m³/s. All flow went through the turbines. A selection of the results is presented here describing the general behavior of the current.

Figure 2 indicates the flow on the water surface (plan view 1 - top) and the bottom of the upper water bay (plan view 2 - bottom) at time step t=510 s. The main flow is directed towards the turbines, but cross-flows occur due to the flow around the separa-

tion pier. A vortex shedding zone develops in the range of the separation pier. The cross-flows in the upper water bay are influenced not only by the main flow and the flow around the separation pier but also by the effect of a steeply sloping bottom (Figure 2 – plan view 2 -bottom).

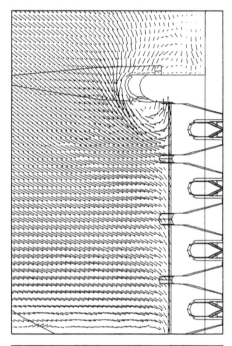

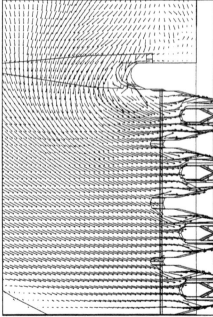

Figure 2. Velocity field at t = 510 s water level (top) and ground level (bottom)

In order to achieve small flow rates transverse to the screen for fish protection, the screen should have a sufficient angle towards the mainstream direction. The geometrical and technical flow conditions at run of the river hydro power plants in favor e.g. an inclined screen. The design of the upper water bay of the sample power plant is schematically represented in Figure 3. The flow field in the example cross section indicates flow velocities around 0.9 m/s with a maximum near the bellmouth around 1.8 m/s.

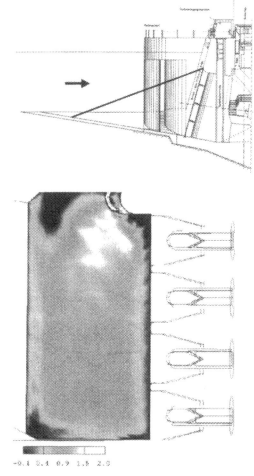

Figure 3. Inclined screen as proposed at a run of the river power plant

These results are considered for the experimental tests. The second iteration will consider the head loss of the inclined screen in the numerical model. The results of the physical model tests can be taken as boundary conditions for a numerical approach and the effect of the screen on the overall flow phenomena can be simulated. Thus the flow phenomena can be observed and the total energy losses can be pre-dicted at the fish screen within the numerical model. Different positions of fish screens as well as varying locations for bypasses can be studied and optimized.

3 SET-UP FOR PHYSICAL EXPERIMENTS

3.1 *General construction*

The experiments were performed in a rectangular flume with a length of 30 m and a width of 0.977 m. The discharge, engendered by two pumps, is limited to $Q_0 = 0.4$ m³/s and measured by a magnetic flow-meter. The water depth and flow velocity throughout the experiments is limited to $h_0 = 0.4$ m and $v_0 = 1.0$ m/s. The declination of the flume is variable. At the outlet of the flume a weir is installed to adjust the water level. The side wall of the flume in the measurement section is transparent.

The wedge wire screen is fixed in a frame. The frame is attached to the bottom of the flume by hinges so the screen declination is infinitely variable in a range of $\alpha=20°$ to $\alpha=90°$.

The frame of the screen and the wall friction of the flume cause a head loss which depends on the flow velocity. In a preliminary investigation this head loss has been determined. In all following experiments this head loss was minimized by adjusting the flume declination.

Figure 4 shows a picture of the experimental set-up in the flume in flow direction and Figure 5 a side view sketch.

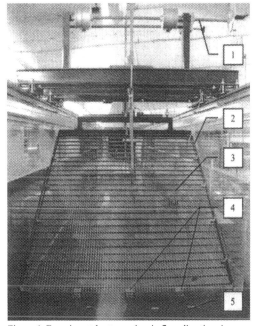

Figure 4. Experimental set-up, view in flow direction, 1 screen declination adjustment, 2 frame, 3 screen, 4 hinges, 5 flume bottom

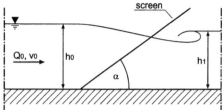

Figure 5. side view sketch of the experimental set-up

3.2 Wedge wire screen

The wedge wire screen consist of rake profiles (1.8×5 mm, rod distance 10 mm) based on orthogonal bearing profiles (2×10 mm, rod distance 50 mm) and is made of stainless steel. This small rod distances are necessary for the eel fish protection. The screen is not scaled and Reynolds similarity is assumed.

3.3 Investigated configurations

The head loss has been determined for 55 configurations. Within all configuration the incoming water level has been adjusted to $h_0 = 0.4$ m and the water level behind the screen h_1 has been measured three times using a manual water level gauge. Eleven screen declinations ($20° \leq \alpha \leq 90°$) and five flow velocities (0.5 m/s $\leq v_0 \leq 0.9$ m/s) have been investigated (compare Fig. 3).

In addition measurements of the velocity profile at the screen have been carried out to get an information about the influence of the screen on the current. For the configuration $v_0 = 0.7$ m/s and $\alpha = 25°$ an ADV-probe (ADV = acoustic-doppler-velocimetry) has been used to determine the 3D velocity vectors in a longitudinal section in the middle of the flume.

Using the same method two cross section in front an behind the screen have been investigated.

4 RESULTS

4.1 Measured head loss

Using the Bernoulli-equation and the continuity the head loss h_v can be calculated with the measured water levels h_1.

$$h_v = h_0 + \frac{v_0^2}{2g} - h_1 - \left(\frac{Q_0}{h_1 \cdot B}\right)^2 \cdot \frac{1}{2g} \qquad (1)$$

where B = breath of the flume; Q_0 = incoming discharge; g = earth acceleration.

Figure 6 shows the determined head loss h_v over the screen declination α. The different curves represent the investigated flow velocities v_0.

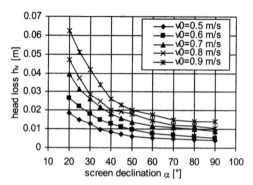

Figure 6. Variation of measured head loss h_v with screen declination α

The head loss increases with increasing flow velocity and decreasing screen declination, especially for declinations $\alpha < 45°$.

4.2 Mathematical formulation

The familiar quadratic correlation between the incoming velocity v_0 and a local head loss h_v is presumed. It can be written as follows

$$h_v = \zeta \cdot \frac{v_0^2}{2g}. \qquad (2)$$

Hence the head loss coefficient ζ is defined as

$$\zeta = \frac{h_v \cdot 2g}{v_0^2} \qquad (3)$$

and can be calculated using the determined values of h_v.

Figure 7 shows the head loss coefficient calculated using equation 3 and the determined values of h_v.

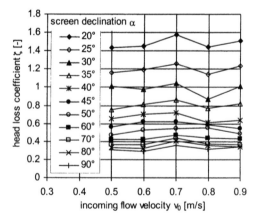

Figure 7. Variation of lead loss coefficient ζ with incoming flow velocity v_0 for all investigated screen declinations α

The head loss coefficient only depends on the screen declination and is approximately constant for every investigated flow velocity v_0. Thus the mean values of the loss coefficients for each screen declination have been calculated and used for every further consideration.

To consider the bearing profile a empirical formula (equ. 4) consisting of two terms for the rake and bearing profiles respectively has been developed. The formula contains shape factors for each profile which have been determined in this experimental investigation.

$$\zeta = \underbrace{\beta_v \cdot \left(\frac{s_v}{b}\right)^{\frac{4}{3}} \cdot \sin^2 \alpha}_{\text{Kirschmer}(1925)} + \beta_h \cdot \frac{s_h \cdot n(h)}{h} \cdot (1 + \cos \alpha) \quad (4)$$

where $\beta_v = 2.377$ (rake profile shape factor); $\beta_h = 0.128$ (bearing profile shape factor); s_v = rake rod breadth; s_h = bearing rod breadth in current f(α); b = vertical rod distance; h = water depth; $n(h)$ = number of rods f(h,α); α = screen declination.

The first term describes the head loss of the vertical rake profiles and has been developed theoretically by Kirschmer (1925). The second term has been empirically developed in this investigation. It is assumed that the total head loss can be calculated by summing the head loss of each profile in the current. The number of profiles in the flow n(h) depends on the screen angle and the water depth.

The correlation between the determined and calculated values of ζ is shown in Figure 8. The dotted line represents the calculated head loss coefficient of the bearing profile and the broken line represents the rake profiles. For small screen declinations the influence of the rake profiles can be neglected while the bearing profiles can be neglected for the vertical screen. The good agreement of calculated and measured head loss is obvious.

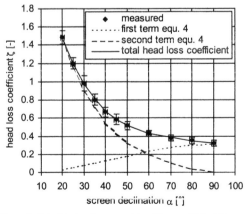

Figure 8. Calculated and measured head loss coefficient ζ

Particularly with regard to the requirements of fish protection small screen declinations are of interest. Therefor the flow component vertical to the screen plain has to be small to preserve the fishes from being pressed on the screen. In order to minimize the head loss, the shape of the bearing profiles have be optimized.

4.3 Velocity profile

The velocity profile contains of 37 points above and underneath the screen. During the determination of the velocity very close to the screen reflections of the acoustic signal of the ADV-probe occur caused by the sharp-edged structure of the screen. These reflections prevent the measurements with this method close to the screen profiles. The blank area of the probe at the water surface is about 6 cm.

Figure 9 shows the measured velocity profile. The velocity vector above the screen are slightly redirected to the water surface while they are redirected to the bottom of the flume underneath the screen. Furthermore an acceleration along the screen can be observed. These flow characteristics are advantageous for fish protection.

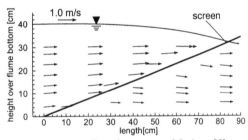

Figure 9. Velocity profile at the screen, v_0=0.7m/s, α=25°

The measurement of the velocity distribution at two cross sections about 2 m in front and behind the screen shows a displacement of the maximum velocities in the profile. While the maximum velocity in front of the screen was found near the surface, it was found near the bottom behind the screen. This proves the redirection of the current downward.

5 COMPARISON WITH FORMER INVESTIGATIONS

5.1 Vertical screen

Table 1 shows calculated head loss coefficients using a selection of references for a vertical screen.

For the vertical screen three of six formulas predict an acceptable head loss (ref. 2/3/4). Two formulas underestimate the loss (ref. 5/6) while one reference (7) overestimates it.

Table 1. Comparison with former investigations, α=90°

	Reference	Head loss coefficient α=90°
1	this investigation	0.32
2	Kirschmer (1925)	0.31
3	Godde (1994)	0.33
4	Yeh & Shrestha (1988)	0.39
5	Fellenius & Lidquist. (1929)	0.10
6	Zimmermann (1969)	0.13
7	Eck (1978)	0.99

These formulas only consider the rake profiles and have been developed for coarse screens (except ref. 4). While the shape of the profiles strongly influence the head loss it is very important to determine the shape factors correctly. In this comparison the shape factors for the investigated screen could only be estimated from the information in the references.

5.2 *Inclined screen*

Kirschmer (1925) performed an experimental investigation for an inclined rake without bearing profiles and found a decreasing head loss for decreasing rake angle (compare Fig. 4, dotted line). For inclined screens the bearing profiles dominate the head loss especially at screen declination smaller α < 45°. Yeh & Shrestha (1988) performed an theoretical investigation neglecting the bearing profiles and found a minimum for the head loss at a declination of α = 80°. In their experimental validation, using a similar screen to this investigation, this minimum was found at a smaller screen angle. The existence of a minimum can not be confirmed in this investigation. Furthermore their formula overestimates the head loss especially for small screen angles.

Comprising it was found that no formula describes the measured head loss at small screen declination sufficiently. To get a reliable information about the head of a wedge wire screen an experimental investigation is still required.

6 CONCLUSIONS

The head loss of an inclined wedge wire fish protection screen has been investigated physically at different declination and flow velocities. An empirical formula is presented in this paper which describes the head loss caused by the screen. The separate consideration of rake and bearing profiles is reasonable. The formula contains shape factors which have been determined for the investigated screen. In future investigations the formula should be validated at screens of different shape.

The measured velocity profiles in longitudinal and cross sections prove the redirection of the current cause by the screen.

The comparison with other investigations only leads to an agreement for the vertical screen. The formulas for inclined screens do not describe the measured head loss sufficiently. Also the minimum of the head loss predicted by Yeh & Shrestha (1988) could not be validated.

While small screen declination are optimal for fish protection, the design of the bearing profiles is very important in order to minimize the head loss.

The boundary conditions for the physical experiments have been determined using a numerical simulation of the complex three-dimensional flow at the turbine intake. The combination of numerical and physical experiments is advantageous, especially because the head loss formula can be used to improve the numerical model and to simulate the screen effected flow and the head loss caused by different screen constructions.

7 REFERENCES

Eck, B. 1978. Technische Strömungsmechanik, Band 1 und 2, Springer-Verlag, Berlin, Heidelberg, New York (in german)

Fellenius, W. & Lidquist, E. 1929. Verluste an Rechen, Hydr. Lab. Practice, ASME, New York

Godde, D. 1994. Experimentelle Untersuchungen zur Anstroemung von Rohrturbinen, Technische Universitat München, Oskar v. Miller- Institut, Heft 75 (in german)

Kirschmer, O. 1925. Untersuchungen über den Gefälleverlust an Rechen, Dissertation, Hydraulisches Inst. d. Techn. Hochschule Muenchen, Mitteilungen Heft 1 (in german)

Montén, E. 1985. Fish and Turbines, Stockholm

Rettemeier, K., Demny, G.; Forkel, C., Köngeter, J., Adler, M. 1999. Einblicke in die Anströmung von Laufwasserkraftanlagen durch Dreidimensionale Numerische Simulationen, Wasserkraft, Heft10, S. 512-518 (in german)

Yeh, H.H. & Shrestha, M. 1988. Free-surface flow through screen, Journal of Hydraulic Engineering, Vol. 115, No. 10, pp. 1371-1385

Zimmermann, J. 1969. Widerstand schräg angeströmter Rechengitter, Dissertation, TH- Karlsruhe, Mitteilung des Theodor-Rehbock-Flußbaulaboratoriums, Heft 157 (in german)

Hydropower in the New Millennium, Honningsvåg et al (eds), © 2001 Taylor & Francis, ISBN 90 5809 195 3

Poles with pins – an innovative element of a hydro-generator

H.-M.Schneider
Hydro Generator Engineering Department
ALSTOM (Switzerland)

A.Fuerst
Hydro Generator Technology Center
ALSTOM Power

ABSTRACT: Engineering hydro generators is traditional plant construction. Nevertheless in detail there are challenging tasks to solve. Some components are mechanically high stressed. One of these parts is the generator pole especially at faulty machine operation like overspeed.

Out of traditional machine design, the solid end plate is known as element to support the rotor winding at the end portion. Some years ago, ALSTOM POWER set a milestone in this area by developing the cylindrical pole coil support pins. In the last decade poles with pin support have been collecting a huge number of operating hours in all kinds of generators.

Now ALSTOM has been developing the pin-pole further. Bent pole coils together with sloped pole tips support the complete pole coil copper perfectly. This designing achieves to carry centrifugal force even without interpolar support structures. Several advantages come along like simple and failure safe construction, which strives for maximum reliability and availability.

1 INTRODUCTION

Hydro generator engineering is traditional machine design task. The mode of operation is well work out but the demand for higher output and efficiency requires constant improvement in detail. Nowadays one main driver for development is the increasing ability of computer simulation. Nevertheless the design can still be improved by new ideas based on experience. This paper is about an improvement in pole design.

2 SOLID AND LAMINATED SALIENT POLES

Hydro generators have salient poles in different design variants. A pole can be a solid component of forged or casted steel, s. Figure 1. Solid poles have high stray losses due to low ohmic resistance in axial direction. Stray losses cause eddy current losses. Solid poles act like conductor rods of a cage damper winding and large currents flow in the pole core during asynchronous operation.

Another pole design is to stack steel sheets mainly to improve electrical properties, s. Figure 4. They are called laminated or stacked poles. Due to this laminar structure the axial ohmic resistance increases. The stray losses and the electric damping effect decrease correspondingly. The stacked pole

Figure 1: Solid pole core

is therefore provided with a damper winding to obtain the required operational stability. Punched or laser cut sheets with a thickness of some millimetres are stacked and pressed together. Axial bolts and screws keep the pole together. The pole ends can have end plates or pins to hold the pole winding at the end region.

3 POLES WITH END PLATES

Traditionally laminated poles are built with end plates. Their function is to hold the pressed sheets together and to support the copper winding against centrifugal forces, s. Figure 2.

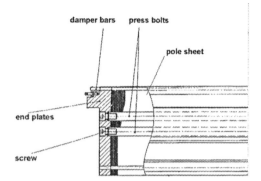

Figure 2: Pole with end plate

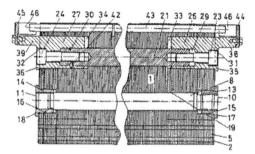

Figure 3: Pole with pins; [1]

The pole end plates have a thickness of several centimetres, depending mainly on the radial load coming from the pole coil. In accordance to the previous paragraph the massive end plate produces pulsation losses. Another disadvantage is their complex geometry that leads to high manufacturing cost for small quantities.

4 POLES WITH PINS

4.1 Description

The pole with pins as shown in Figure 3 and Figure 4 is pressed and hold together by clamping bolts 8 similar to the pole with end plates, s. Figure 2. The nominal diameter of hole 19 in the sheet has the same nominal diameter of the clamping bolt 8. The bolts have female screw thread at both ends. Cylindrical pins 26 and 27 with an eccentric hole are screwed to the clamping bolts 21 by means of socket-head screws 23 and 31. The pin sections 38 and 39 are used for the radial support of the pole coil. Close to the pole surface, damper bars 43 are inserted in the pole and connected by damper plates 44 and 45 to build the damper winding.

4.2 Mechanical stress

The Pole is one of the mechanically highest stressed part of the machine. ALSTOM takes a lot of efforts to proof design of its poles. The stress distribution in pole sheets is complex. Therefore analytical calculation for the stress calculation is not sufficient. Boundary element simulations with BEASY are done to get a high accuracy for the stress simulation. This method shows significant advantages compared with the FEM tools because of high gradients and many small radii. For such an analysis FEM would need a very detailed and time-consuming mesh. Figure 5 shows the stress calculation. The highest stress in pole sheet appear in the area of the claw and in the corner of pole shoe.

The pin diameter depends on the mechanical stress and vary according to the pole shoe height.

Figure 4: Laminated pole with pins

4.3 Electrical aspects

The poles generate the magnetic field of the rotor. The magnetic flux depends on the pole design. In the past poles with pins had a tendency to have a pole shoe height higher than with an end plate. With today's improved simulation facilities it is possible to build poles with similar dimensions for both designs. In case of equal pole shoe height the pole with pins has the advantage of less pulsation losses on the pole shoe surface.

Figure 5: Stress in pole sheet calculated by BEM

4.4 References for pin-poles

In the last decade poles with pins have been collecting a huge number of operating hours in all

Figure 6: Stress calculated by BEM for a pin and geometry model

kinds of generators. Reference plants are shown in table 1. During this long time in operation poles with pins have never shown any failure.

Table 1 Plants using poles with pins

Year of Design	Plant	Number/Diam. of Pins
1987	Pan Jia Kou	2x80 mm
1989	Alberschwende	2x102 mm
1990	Martina	2x90 mm
1991	Currilinque	2x100 mm
1992	Amsteg	5x130 mm
1993	Bhumibol	2x92 mm
1994	Obervermunt	2x100 mm
1995	Nasa Langley	4x94 mm
1998	Three Gorges	3x94 mm
1999	Manantali	2x92 mm

5 NEW DESIGN: POLES WITH BENT POLE COIL AND SLOPED TIPS

Now ALSTOM has been developing the pin-pole further. Conventional poles, as shown in Figure 4, often need an interpolar support for the side component of the centrifugal force.
Figure 7: Forces on a normal pole coil

Figure 7 and Figure 8 explain the forces to the pole coil. To eliminate the interpolar support the tangential component of the centrifugal force has to be neglected. This side component increases for low pole-numbers. In fact nearly all generators need an interpolar support between the pole coils, like that shown in Figure 9.

Poles with bent pole coils, as shown in Figure 11, avoid an interpolar support. The centrifugal forces are taken directly by the pole body. The copper can be supported without interpolar sup-

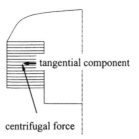

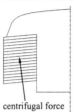

Figure 7: Forces on a normal pole coil

Figure 8: Forces on a bent pole coil; there is no tangential component anymore

port. Furthermore, the new pole design can be build with a constant pole core cross section. This means lower excitation looses in the pole in comparison to trapezoidal cross-section.

The pin pole with bent coil will be applied soon. Figure 11 illustrates a possible design realization. The pole core has a rectangular cross-section. The pole tips and the position of the pins follow the radius of the bent pole coil.

Figure 9: Interpolar support between pole coils

Concluding the benefits of the new pole design:

- Negligible tangential force to the pole coil.
- Compared to pole with trapezoidal cross-section improved magnetical properties.
- Less components and an easier way of assembling and disassembling of the pole.

Figure 10: Rotor with conventional pin pole

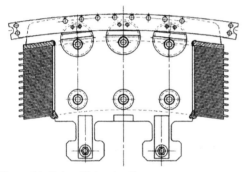

Figure 11: Pole with bent coil

- Better ventilation thereby improved cooling of the pole coil.
- Simple and failure safe construction which strives for maximum reliability and availability.

6 CONCLUSION

The main design principles of pin-poles have been presented. Theoretical and practical aspects of this well approved design are discussed.

The new concept of bent pole coils with sloped pole tips together with pole pins has been presented. It is found that this design has improved magnetical and mechanical properties.

7 REFERENCE

[1] European Patent EP 0 339 271 B1

Hydropower in the New Millennium, Honningsvåg et al (eds), © 2001 Taylor & Francis, ISBN 90 5809 195 3

Theoretical and Practical Aspects of the Design of Desander Outlet Structures

J.Schramm, C.Schweim & J.Köngeter
Institute of Hydraulic Engineering and Water Resources Management (IWW), Aachen University of Technology (RWTH-Aachen), Kreuzherrenstrasse, 52056 Aachen, Germany

C.Jokiel
Lahmeyer International GmbH, Friedberger Str. 173, 61118 Bad Vilbel, Germany

ABSTRACT: Due to geological, geometrically or economical reasons outlet structures are often designed in a way that the flow direction in the headrace tunnel is chosen to be orthogonal to the flow direction in the desander leading to a very compact structure. Investigations that were carried out at a physical model of a desander showed, that this compact design can not guarantee equal flow distribution in the desander and therefor has to be optimized. In the paper we will show a method to easily calculate the resulting velocities in the desander and various ways to optimize the design to achieve equal flow patterns.

1 INTRODUCTION

To ensure carefree operation of the hydro power plant's turbines at rivers with high sediment loads usually a desander serves to clean the water before it flows through the headrace tunnel to the turbines. To achieve best cleaning results one has to assure that the discharge through the desander is equally distributed over the cross section and that the flow velocity will not exceed the design velocity.

In this paper we will show that one of the main influences of the velocity distribution in a desander is the construction of the outlet structure to the headrace tunnel. Due to geological, geometrically or economical reasons desander outlet structures are often designed in a very compact manner. In this case compact means that the flow direction in the headrace tunnel is orthogonal to the flow direction in the desander. Following, we will present results of investigations carried out at a physical model of a desander build at the IWW with the outlet structure designed as mentioned above.

The paper consists of two parts. The first part starts with investigations concerning the flow patterns of the original layout of the settling basin. Two different operating level were taken into consideration. Based on the results of the initial investigations the observed flow phenomena is investigated theoretically. After that, modifications to the original design and its influence on the flow patterns in the desander will be discussed. Final tests proved the suitability of the modifications concerning the improved design of the outlet of the settling basin.

In the second part we will present an alternative design of the desander outlet structure. For this alternative design the suitability will only be proven on theoretical bases.

2 PART I: INVESTIGATIONS OF THE PHYSICAL MODEL

2.1 *Physical model*

The investigations presented in this paper were carried out at a physical model of a desander which belongs to a hydro power plant. The model was built to investigate and optimise the flow conditions in the settling basin in order to guarantee smooth and uniform flow conditions (the velocity is not to exceed 0.30 m/s) in the basin being a prerequisite for optimum removal of suspended sediment. The model constructed at the IWW covered the spillway gates, the intake, the sediment sluice and the desanding basins including the intake, the conveying channels, the equalising racks and the outlet to the headrace tunnel of the power plant (see Fig. 1).

Hydraulic model tests are applied to investigate complex flow processes which cannot be determined by a theoretical approach alone. The hydromechanical phenomena taking place are therefore studied in a scale model. Such a scale model is a reduced representation of the prototype. To transfer the observed processes and measured data correctly from the model to the prototype, the model has to be hydromechanical, i.e. geometrically, kinematically and dynamically similar to the prototype. In the physical model Froudes similarity law is applied as we are dealing with free surface flow. According to the demands of the model tests and the spatial possibilities at the laboratory of the IWW, the length scale was chosen to be 1:25. As the secondary currents have a main influence on the investigated processes the model was built geometrically undistorted ($L_r = H_r$).

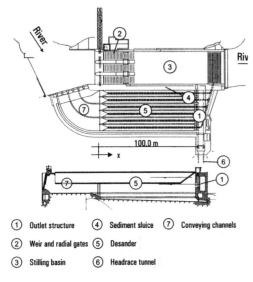

$\circled{1}$	Outlet structure	$\circled{4}$	Sediment sluice	$\circled{7}$	Conveying channels
$\circled{2}$	Weir and radial gates	$\circled{5}$	Desander		
$\circled{3}$	Stilling basin	$\circled{6}$	Headrace tunnel		

Figure 1. Object of investigation

2.2 Investigation

The power plant is going to be operated at two different operating levels. Therefore, to investigate the flow patterns in the settling basin tests for two states of operation were carried out. The first operating level, hereafter referred to as the minimum water level, is set to 612.0 m asl with a river discharge of $Q_R = 400$ m³/s. The second, hereafter referred to as the maximum water level, is set to 618.50 m asl with a river discharge of $Q_R = 100$ m³/s. The discharge through the headrace tunnel is set to $Q_P = 80$ m³/s for both operating level. In case of the minimum operating level $Q_S = 60$ m³/s will run through the sediment sluice. The initial investigations comprise tracer tests and velocity measurements for both the minimum and the maximum reservoir water level at different water levels at five cross sections (0.00 m, 12.5 m, 37.5 m, 62.5 m and 87.5 m in flow direction).

2.2.1 Initial investigation: minimum water level
For the hydraulic tests the river discharge was adjusted to $Q_R = 400$ m³/s. The power discharge and the discharge through the sediment sluice were adjusted to the calculated loads ($Q_P = 80$ m³/s, $Q_S = 60$ m³/s).

First, tracer tests with dye injections in all three chambers were carried out. The tracer front was investigated at different times (Fig. 2). An unequal

velocity distribution with high velocities in the in stream direction right chamber (3) and low velocities in the middle (2) and left chamber (1) could be observed.

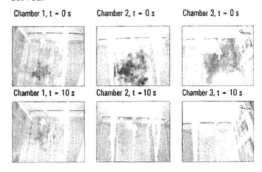

Figure 2. Tracer tests in all three chambers in front of the outlet structure for t = 0 seconds and t = 10 seconds

Next velocity measurements were carried out. The velocities were measured at five cross sections at an elevation of 607.50 m asl. Figure 3 illustrates the velocity distribution for minimum water level. The velocity in the outer right chamber at the outlet structure of the settling basin reaches a maximum of 0.63 m/s exceeding the design velocity by 110 %. The lowest velocity in the settling basin was measured in the left chamber with a velocity of 0.03 m/s. The velocity decreases from right to left in every chamber as well as over the whole cross section. These general flow pattern is only disturbed at the inlet to the settling basin due to secondary currents.

Velocity distribution

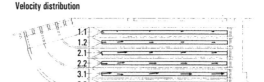

reference vector, design velocity 30 cm/s: ⟶

Figure 3. Velocity distribution for minimum water level (Q_R= 400 m³/s, water level 612.00 m asl)

2.2.2 Initial investigation: maximum water level
For the hydraulic tests the river discharge was set to $Q_R = 100$ m³/s. The power discharge was set to $Q_P = 80$ m³/s and there is no discharge through the sediment sluice.

Again, tracer tests were carried out. The results are identical to the results of the investigation of the minimum operating level (Fig. 2) and therefore will not be displayed here. The only difference are back currents that could be visually observed in chamber 1.1, 1.2 and in chamber 2.1. The back currents seemed to be stronger than the currents investigated at an operating level of 612.00 m asl. A dominating stream with high velocities along the right wall of the settling basin could be observed.

To conclude the preliminary investigations velocity measurements were carried out for the maximum operating level. The results are shown in Figure 4. The velocities were measured at five cross sections at an elevation of 607.50 m asl. Again the velocity distribution is non-uniform but the measured velocities do not exceed the design velocity of 0.30 m/s. Back currents with a maximum velocity of 0.11 m/s are measured at the left side (chamber 1 and 2) of cross section 0+000.00 m.

Velocity distribution

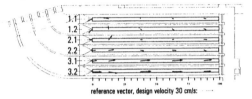

reference vector, design velocity 30 cm/s:

Figure 4. Velocity distribution for maximum water level ($Q_R = 100$ m³/s, water level 618.50 m asl)

2.3 *Theoretical investigation of the flow phenomenon at the outlet structure*

In outlet structures the difference of energy heads between the free water surface and the full pipe flow through the headrace tunnel will be decomposed. In general the turbine water is withdrawn along the flow direction in the settling basin.

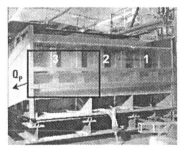

Figure 5. Inclined water surface in the outlet structure

Due to local topographical and geological reasons the design of the outlet structure was chosen to withdraw the water orthogonal to the flow direction. The energy head of the free water surface of the settling basin is aligned in the cross section of the outlet structure. As a consequence an inclined water surface in the outlet structure develops (see Fig. 5).

The shortest way for the water to overcome the potential difference is through the right chamber of the settling basin. To show the influence of the inclined water surface on the distribution of water withdrawal of the three chambers the following calculation is carried out.

2.4 *Calculation of discharge distribution*

In general the discharge of a diving overflow can be calculated with a formula for free overflow (e.g. Poleni's formula). The influence of the downstream water level is taken into account by a coefficient σ_d. Hence, the discharge of the diving overflow can be calculated by the following equation:

$$Q_d = \sigma_d \cdot Q_{Poleni} = \sigma_d \cdot \frac{2}{3} \cdot \mu \cdot b \cdot \sqrt{2g} \cdot h^{\frac{3}{2}} \quad (1)$$

Here, Q is the discharge, b is the width of the weir, h is the overflow height and μ is the coefficient of discharge. The coefficient σ_d mainly depends on the relation of h/h_u (see Fig. 6)

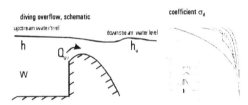

Figure 6. Diving overflow (Bollrich (1992))

The influence of the downstream water level is given by $h_u = 0.7$ to 0.85 h. By increasing the downstream water level the discharge of the diving overflow decreases. In section A-A (Fig. 7) the discharge behind the outlet opening can be calculated to (with v being the velocity):

$$Q = v_u \cdot A \quad (2)$$

The area is given by the formula (with w is the height of the weir)

$$A = b \cdot (w + h_u) \quad (3)$$

On assumption of a frictionless flow the downstream water level h_u is defined to the difference between upstream water level and velocity loss:

$$h = h_u + \frac{v_u^2}{2g} \Leftrightarrow h_u = h - \frac{v_u^2}{2g} \qquad (4)$$

Finally the area can be calculated to:

$$A = b \cdot (w + h - \frac{v_u^2}{2g}) \qquad (5)$$

Under precondition of the design parameter the discharge through the outlet structure can be combined to the following equation:

$$Q = v_u \cdot A$$

$$= v_u \cdot b \cdot (w + h - \frac{v_u^2}{2g}) \qquad (6)$$

$$= (w + h) \cdot b \cdot v_u - \frac{b}{2g} \cdot v_u^3$$

$$\Rightarrow Q = 37.05 v_u - 0.2905 v_u^3 \qquad (7)$$

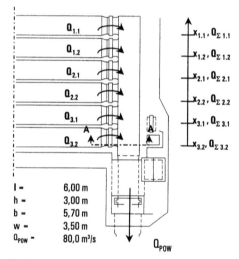

I –	6,00 m
h –	3,00 m
b –	5,70 m
w –	3,50 m
Q_{POW} –	80,0 m³/s

Section a-a:

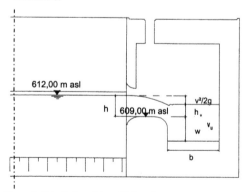

Figure 7. Calculation of the withdrawal

The discharge on location $x_{3.2}$ (see Fig. 7) is equal to the power discharge of $Q_P = 80$ m³/s. By the above mentioned Equation 7 the velocity amounts to

$$Q_{X3.2} = Q_{POW} = 80 \, m^3/s$$

$$= 37.05 v_u - 0.2905 v_u^3 \qquad (8)$$

$$\Leftrightarrow v_u = 2.248 \, m/s$$

With $v_u = 2.248$ m/s and Equation 4 the downstream water level is calculated to

$$h_u = h - v_u^2/2g$$

$$= 3.00 - 2.248^2/19.62 \qquad (9)$$

$$= 2.74 \, m$$

The diagram in Figure 6 reveals the coefficient σ_d:

$$h_u/h = 2.74/3.00 = 0.9141$$

$$\Rightarrow \sigma_d \approx 0.91 \qquad (10)$$

According to Equation 1 and with $\mu = 0.5$ (wide crown) the discharge can be determined to

$$Q_{X3.2} = 0.91 \cdot 46.03 = 41.89 \, m^3/s \qquad (11)$$

The withdrawal of water through the outlet opening 3.2 is 41.89 m³/s. At position 3.1 the discharge is reduced to:

$$Q_{X3.1} = 80 \, m^3/s - Q_{X3.2} = 38.11 \, m^3/s \qquad (12)$$

The velocity, the downstream water level and the coefficient σ_d are calculated to:
Velocity:

$$Q_{X3.1} = 37.05 v_u - 0.2905 v_u^3$$

$$\Leftrightarrow v_u = 1.037 \, m/s \qquad (13)$$

Downstream water level:

$$h_u = h - \frac{v_u^2}{2g} = 3.0 - \frac{1.037^2}{19.62} = 2.94 \, m \qquad (14)$$

Coefficient σ_d:

$$hu/h = 2.94/3.00 = 0.9817$$

$$\Rightarrow \sigma_{uv} \approx 0.40 \qquad (15)$$

Consequently the withdrawn water out of the outlet opening 3.1 is

$$Q_{X3.1} = 0.40 \cdot 46.03 = 18.41 \, m^3/s \qquad (16)$$

The same procedure is applied to the openings $x_{2.2}$ – $x_{1.1}$ with results as given in Table 1.

The discharge which is withdrawn out of both openings of chamber 3 amounts 60.3 m³/s and corresponds to 73 % of the power discharge. According to the law of continuity the velocity in chamber 3 of the settling basin is $v_m = 0.45$ m/s. The calculated velocity v_m fits the interval of measured data (0.29 to 0.51 m/s, see Fig. 3 and Fig. 4).

Table 1. Compilation of calculated discharge through outlet structure openings.

	Q m³/s	v_u m/s	h_u m	h_u/h -	σ_{uv} -	Q_{uv} m³/s
$x_{3.2}$	80.0	2.248	2.742	0.9141	0.91	41.8
$x_{3.1}$	38.11	1.037	2.945	0.9817	0.40	18.4
$x_{2.2}$	19.70	0.533	2.985	0.9952	0.30	13.8
$x_{2.1}$	5.89	0.159	2.999	0.9996	0.30	5.8
$x_{1.2}$	-	-	-	-	-	-
$x_{1.1}$	-	-	-	-	-	-

Because of the increasing downstream water level the withdrawal out of chamber 2 decreases ($Q = 19.70$ m³/s, $v_m = 0.15$ m/s). The velocities are measured between 0.11 and 0.18 m/s. The left discharge of 5.89 m³/s is totally withdrawn through the opening 2.1.

According to the calculation no water is withdrawn out of chamber 1 which corresponds well to the experiments (tracer tests).

To illustrate the investigated phenomenon of varying discharged a detailed velocity profile in front of the outlet structure was measured. Figure 8 sums up the results of the measurements carried out for the minimum operating level.

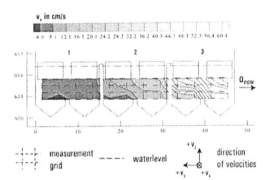

Figure 8. Velocity distribution in front of outlet structure at cross section 0^{+100} m. Operating level set to 612.0 m asl., discharge Q_R is set to 400.0 m³/s.

So fare, all investigations with the original design of the desander have shown a velocity distribution that decreases from the right wall with velocities of about 0.55 m/s to the left wall with very low velocities of about 0.05 m/s. The same flow characteristics have been investigated for either operating level. Therefore, to attain a uniform velocity distribution in the settling basin a modification of the outlet of the settling basin seemed to be indispensable.

2.5 Modification of original design

The unequal flow patterns are due to the fact that the outlet structure was build orthogonal to the flow axis of the desander. To overcome the difference be-

tween the pressure head of the desander and the headrace tunnel, the water will take the shortest way from the settling basin to the headrace tunnel. Almost 75 % of the power discharge are supplied by the outer right chamber. The middle chamber supplies 17% and the left chamber only about 8%. Applying the law of continuity it is obvious that the unequal discharges lead to the unequal velocity distribution in the settling basin. Therefore, to attain the desired equal velocity distribution in the desander the outlet structure of the settling basin had to be modified in a way that every chamber supplies equal discharges to the power discharge.

2.5.1 Modification of the desander outlet

Various tests equalizing the discharges at the outlet of the stilling basin were carried out to attain uniform flow patterns in the settling basin. Equal discharges were achieved by reducing the heights of outlet openings of the desander. This was done using trash boards. The result of this optimization process is illustrated in Figure 9.

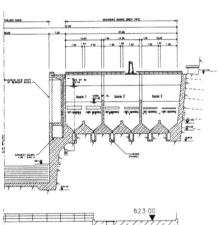

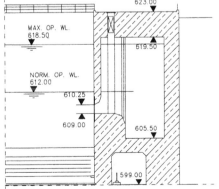

Figure 9. View at modified openings from the settling basin towards the outlet structure(above). Cross sectional view at modified openings of the outlet structure (below).

Because of the unequal potential decline between each opening and the headrace tunnel every opening has a different height (see Table 2).

Table 2. Sizes of the improved outlet openings.

| | | left opening | | right opening | |
		elevation m asl	height m	elevation m asl	height m
Basin 1	crest	609.00	1.25	609.00	1.25
	top	610.25		610.20	
basin 2	crest	609.00	0.83	609.00	0.65
	top	609.83		609.65	
basin 3	crest	609.00	0.45	609.00	0.25
	top	609.45		609.25	

With the improved outlet openings the flow field in the settling basin is divided in two parts. The first part is mainly influenced by the conveying channels and the second part is influenced by the outlet structure of the desander. In the latter the velocity distribution meets the design criterion and has not to be further optimized. In the first part the flow patterns are not yet satisfying with regard to an equal velocity distribution. The peak velocities reaches about 0.50 m/s at some areas which is considered to be to high. Due to the bends in the conveying channels (see Fig. 1), the water is accelerated to the right side of each chamber which leads to high velocities in that regions. To overcome this problem flow equalizing devices were studied to reduce the high velocities below the critical velocity and to equalize the velocity distribution.

2.5.2 Flow equalizing devices

After an equal velocity distribution at the outlet of the settling basin could be assured, measures were taken to equalize the velocities at the inlet of the settling basin thus achieving smooth flow patterns in the whole of the settling basin. With no equalizing devices the velocity distribution is distorted along the cross section (see Fig. 11a). To overcome this problem, calming racks with three different spacing were build and assembled at the beginning of the settling basin (Fig. 10). The vortices induced by the calming racks causes energy dissipation which results in smooth velocity downstream of the racks.

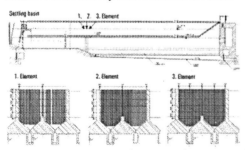

Figure 10. Position of calming racks

The calming racks are constructed with an upstream facing angle. Both the bar spacing and the spacing of the racks is reduced in downstream direction. According to the results of preliminary tests with perforated plates and values given by various literature (e.g. BOUVARD, 1992,) the bar spacing is set to 250 mm for the first, 200 mm for the second and 150 mm for the third element.

In order to determine the optimum spacing of the racks in flow direction further investigation were carried out. The final measures are shown in Figure 11b.

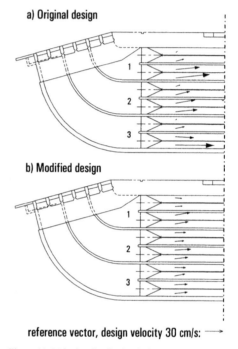

reference vector, design velocity 30 cm/s: ⟶

Figure 11. Velocity distribution before (a) and after (b) installation of calming racks.

2.6 Final investigations of the improved design of the desander

Finally the improved design of the desander was investigated. In five cross sections velocity measurements were performed to verify the flow condition of the improved design. At each cross section velocities in four respectively seven different elevations (depending on the water level) were measured. In each chamber four points of measurement (for each elevation) were established.

In this paper we will only present the results for the minimum water level. Figure 12 illustrates the results of the velocity measurements carried out at the original (Fig. 12a) and the improved design (Fig. 12b) of the desander for the minimum operating level. For the improved design smooth and uniform flow conditions are ensured. The velocities amounts about 0.15 m/s (behind the calming racks) to 0.25 m/s which is obviously below the design velocity.

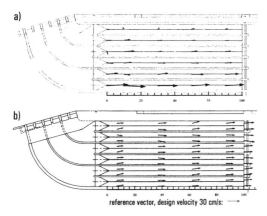

a)

b)

reference vector, design velocity 30 cm/s: ⟶

Figure 12. Comparison of velocity distribution of original (a) and modified (b) design for minimum operating level (612.0 m asl)

The velocity distribution in the different elevation is satisfying as well. The measurements prove that the combination of the modifications made at the outlet structure and at the inlet to the settling basin (calming racks) guarantee smooth and uniform flow conditions in the settling basin

3 PART II: ALTERNATIVE DESIGN

The investigation presented above showed that with the outlet structure designed in a compact manner equal velocity distribution in the settling basin can only be archived with major modifications resulting in high velocities at the small openings of the outlet structure.

To avoid this and to guarantee an equal velocity distribution the best way of designing an outlet structure would be to equalize the length of the outlet structure for every chamber between the settling basin and the headrace tunnel. Doing so the total loss of energy would be equally divided leading to an equal discharge. Unfortunately, that will lead to, in a geometrically sense, large (or long) structure.

Next we will present an alternative design that will provide equal discharge through each chamber without the described negative effects. Different to Part I the suitability of the design will only be proven theoretically.

3.1 Geometry

Due to the fact that there is a variety of literature available (Idelchik, 1986; Richter, 1971) regarding

investigations of flow in converging wyes the following design of the converging of the three headrace tunnel segments was chosen (Fig. 13).

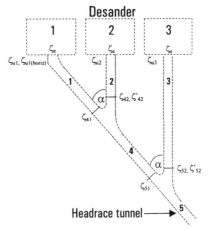

Figure 13. Design of the converging of the three headrace tunnel segments

The starting configuration for the numerically supported optimization of the three headrace tunnel segments is listed in Tab. 3. The diameter and wall roughness were set all equal. Angels α between segments are 45.0°.

Table 3. Starting configuration

	Diameter m	Length m	Manning coefficient $m^{1/3}/s$
Segment 1	5.40	90.0	0.0166
Segment 2	5.40	74.5	0.0166
Segment 3	5.40	109.5	0.0166
Segment 4	5.40	50.0	0.0166
Segment 5	5.40	50	0.0166

As we will show later (chapter 3.3.1), with this configuration applied the discharge through segment 1, 2 and 3 will not be equal causing a in flow direction decreasing velocity distribution from left to right in the settling basin.

3.2 Calculation

To optimize the outlet of the desander in a way that the same discharge flows through the segments 1, 2 and 3 the energy dissipation at the inlet to the tunnels, at bends, at wyes and due to wall roughness has to be considered. The geometry that will meet this criteria is obtained by adjusting the diameter of each section individually in a way that the total loss of energy in all three segments become equal. The adjustment of the pipes' diameter is done in an iterative manner.

The total amount of energy loss Δh (in meter) for each segment is calculated by summing up the local energy losses and the energy loss due to friction of the wall and can be determined from Bernoulli's formula:

$$\Delta h_E = \frac{v^2}{2g}\left(\sum \zeta_{local} + \lambda \frac{l}{d}\right) \tag{17}$$

The resistance coefficients ζ and λ are determined as follows.

3.2.1 Resistance coefficient of inlets
Following Idelchik (1975) the resistance coefficient at the entrance into tubes ζ_{Inflow} in the case of circular cross sections and the velocity in front of the inlet w_∞ considered to be zero can be determined from Weisbach's formula

$$\zeta_{Inflow} = 0.5 + 0.3 \cdot \cos\beta + 0.2 \cdot \cos^2\zeta \tag{18}$$

where $\beta = 90°$.

3.2.2 Resistance coefficient of bends
With $Re > 2 \cdot 10^5$ and the ratio r/d, with r being the radius of the bend and d the diameter of the tunnel, the resistance coefficient becomes (Idelchik (1986))

$$\zeta_{Bend} = \frac{0.21}{\sqrt{r/d}} \cdot 0.6 \tag{19}$$

3.2.3 Resistance coefficient of wyes
Following Idelchik (1986), a wye is characterized by the branching angel α and the ratio of the cross-sectional areas of its branches. When two streams moving in the same direction but with different velocities merge, turbulent mixing of streams occurs. This is accompanied by non-recoverable total pressure losses. In the course of this mixing, momentum exchange takes place between the particles in the medium moving with different velocities. This exchange favors equalization of the flow velocity field. In this case, the jet with higher velocity loses a part of its kinetic energy by transmitting it to the slower moving jet.

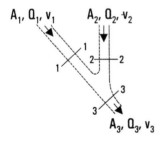

Figure 14. Determination of resistance coefficient of converging wyes

The resistance coefficient of converging wyes can be calculated by corrected formulas by Levin (1940) and Taliev (1952). For a straight passage the resistance coefficient ζ_{13} is given by (see Idelchik (1986)):

$$\zeta_{13} = 1 - \left(1 - \frac{Q_2}{Q_3}\right)^2 - 1.412 \cdot \frac{A_3}{A_2}\left(\frac{Q_2}{Q_3}\right)^2 \tag{20}$$

with ζ, Q and A according to Figure 14. For the side branch the passage the resistance coefficient ζ_{23} can be calculated to (Idelchik (1986)):

$$\zeta_{23} = R \cdot \left[\begin{array}{c} 1 + \left(\dfrac{Q_2 A_3}{Q_3 A_2}\right)^2 - 2 \cdot \left(1 - \dfrac{Q_2}{Q_3}\right)^2 \\[2ex] -1.41 \cdot \dfrac{A_3}{A_2}\left(\dfrac{Q_2}{Q_3}\right)^2 \end{array} \right] \tag{21}$$

with ζ, Q and A according to Figure 14. The value R is given in Table 4.

Table 4: Values of R as given by Idelchik (1986)

A2/A3	≤ 0.35	> 0.35	> 0.35
Q2/Q3	0.0 – 1.0	≤ 0.4	> 0.4
R	1.0	$0.9 \cdot (1 - Q_2/Q_3)$	0.55

3.2.4 Energy loss due to wall roughness
For stabilized flow in the region of purely turbulent flow, the friction coefficient λ can be determined from Prandtl-Nikuradse formula

$$\frac{1}{\sqrt{\lambda}} = 2.0 \cdot \lg\left(\frac{3.71}{k/d}\right) \tag{22}$$

where k is the equivalent uniform roughness of the wall and d is the diameter of the tunnel.

3.3 Results

3.3.1 Equal diameter
To demonstrate the effect of equal diameters for all three segments on the discharge distribution the following calculation was carried out. The location of the calculated coefficients are shown in Figure 13. The results for the discharge distribution are displayed in Table 5. The calculated resistance coefficients are displayed in Table 6.

Table 5. Resulting discharge distribution with equal diameters

	Diameter m	Length m	Discharge m³/s	velocity m/s
Segment 1	5.40	90.0	16.366	0.715
Segment 2	5.40	74.5	21.055	0.919
Segment 3	5.40	109.5	42.579	1.859
Segment 4	5.40	50.0	37.421	1.634
Segment 5	5.40	50.0	80.000	3.493

Table 6. Resulting resistance coefficients (equal diameters)

Inlet	Wyes	Bends
ξ_{e1}=0.5	ξ_{41} = 0.3624	ξ_{u1} = 0.0756
ξ_{e2}=0.5	ξ'_{41}= 0.4876	ξ_{u2} = 0.0756
ξ_{e3}=0.5	ξ_{42} = 0.2682	ξ_{u3} = 0.0756
	ξ_{51} = 0.3818	$\xi_{u1(hor.)}$ = 0.0756
	ξ'_{51}= 0.4463	
	ξ_{52} = 0.2454	

As expected, the discharge through segment 1, 2 and 3 is not equal and will cause the unequal velocity distribution in the settling basin.

3.3.2 *Equal discharge*
Next, the diameter of the segments were changed in a way that equal discharge is provided for all three segments. Table 7 provides the calculated diameters. The resulting resistance coefficients are given in Table 8.

Table 7. Resulting diameters with equal discharge

	Diameter m	Length m	Discharge m³/s	velocity m/s
Segment 1	5.40	90.0	26.667	1.164
Segment 2	4.92	74.5	26.667	1.401
Segment 3	3.95	109.5	26.667	2.181
Segment 4	5.40	50.0	53.333	2.329
Segment 5	5.40	50.0	80.000	3.493

Table 8. Resulting resistance coefficients (equal discharge)

Inlet	Wyes	Bends
ξ_{e1}=0.5	ξ_{41} = 0.3258	ξ_{u1} = 0.0756
ξ_{e2}=0.5	ξ'_{41}= 0.4379	ξ_{u2} = 0.0722
ξ_{e3}=0.5	ξ_{42} = 0.2408	ξ_{u3} = 0.0646
	ξ_{51} = 0.2621	$\xi_{u1(hor.)}$ = 0.0586
	ξ'_{51}= 0.2076	
	ξ_{52} = 0.1245	

Equal discharge in all three segments can be provided using the above calculated diameters. Additional calculations were carried out to show the influence of different and more manageable diameters on the discharge distribution (Table 9).

Table 9: Difference in discharge for different diameters

	Segment 1,4 &5	Segment 2	Segment3
Diameter m	5.40	4.922	3.945
Discharge m³/s	26.66	26.66	26.66
Difference %	0.0	0.0	0.0
Diameter m	5.40	5.000	4.000
Discharge m³/s	25.829	26.917	27.254
Difference %	-3.2	1.1	2.2
Diameter m	5.40	4.900	4.000
Discharge m³/s	26.513	26.199	27.288
Difference %	-0.6	-1.7	1.0

4 CONCLUSIONS

In the first part of this paper we presented investigations carried out at a physical model of a desander. The outlet structure of the desander is designed in a way that the flow direction in the settling basin is orthogonal to the flow direction in the headrace tunnel. The investigations showed, that only with modifications on the design of the outlet structure an equal velocity distribution in the settling basin could be achieved. The unequal velocity distribution caused by the original design of the outlet structure is due to the fact, that the potential difference between the pressure head of the settling basin and the headrace tunnel is overcome by taking almost 73% of the total discharge from the basin of the desander in the immediate vicinity of the inlet to the headrace tunnel. In the paper we could show a way to calculate the discharge distribution of the desander basins caused by the outlet structure.

In the second part of the paper we investigated theoretically an alternative design of the outlet structure. Three pipes were connected on one side to the end of the settling basin and on the other side to the headrace tunnel. As we could show the diameter of the pipes can be adjusted in a way that the energy dissipation in all segments of the structure are equal leading to an equal discharge in the pipes which is responsible for an equal velocity distribution in the settling basin. The calculation for the optimization process was carried out using Bernoulli's formula.

REFERENCES

Bollrich, G., Preissler, G.: "Technische Hydromechanik", Verlag für Bauwesen, Berlin,1992

Idelchik, I. E., (1986): "Handbook of Hydraulic Resistance, Second Edition", Hemisphere Publishing Corporation

Levin, S. R. (1940): "Resistance of wyes of outlet air pipelines, Otoplenie. Ventil., no. 10/11, 5-10

Richter, H. (1971): "Rohrhydraulik", 5. Neubearbeitete Auflage. Berlin/Heidelberg/New York: Springer Verlag

Taliev, V. N. (1952): "Calculation of Local Resistance of Wyes", Gosstroiizdat, 35 pp.

Bouvard, M. (1992): "Mobile Barrages and Intakes on Sediment Transporting Rivers", Monograph Series: Fluvial Hydraulics, IAHR, International Association for Hydraulic Research

Hydropower in the New Millennium, Honningsvåg et al (eds), © 2001 Taylor & Francis, ISBN 90 5809 195 3

Hydropower potential and underground construction risks in Nepal Himalayan region

Gyanendra Lal Shrestha
Acting Director, Butwal power company Ltd, Lalitpur, Nepal

ABSTRACT: Nepal holds a major share of Himalayan range. It consists of high mountains with snow throughout the year. It has been a perennial source of water. Many rivers start from this snow fed catchment. Topography has multiplied the hydropower potential in Nepal. Because of the big variation in elevation from North to South, rivers run in higher gradient with very high head for a relatively shorter length. Diversion of a river from one valley to other through a tunnel has further amplified the feasibility of hydropower project. So far very small amount of water resources has been harnessed for hydropower generation.

Nepal Himal consists of various types of rocks and geological structures. Thus the underground construction works through these domains varies in terms of excavation, support system requirement and stability. The critical issues with regard to the underground works are geotechnical. An understanding of the ground, through which the excavation will be carried out, will be essential. Hence investigations are necessary to clarify the type and magnitude of the risks and how they can be reduced or managed.

Experiences have shown that same type and scale of ground investigation is not suitable to all the hydropower project sites. Site specific modification would be necessary in determining support for the underground construction.

1 HYDROPOWER POTENTIAL

Tectonic movement has given Nepal a major share of Himalayan range. It has resulted a series of high mountains with snow throughout the year. It has been a very good perennial source of water. Many rivers and rivulets start from this snow fed catchment. Topography has multiplied the hydropower potential in Nepal. Because of the big variation in elevation from North to South, rivers run in higher gradient with very high head for a relatively shorter length of headrace conveyance. Diversion of a river from one valley to other through a tunnel has further amplified the feasibility of hydropower project. These features can be observed in the Figure 1.1. With its 6000 rivers and rivulets with 224-billion m³ discharge per year, Nepal is believed to have 43000 MW economically feasible hydropower potential. See Table 1.1.

Very small amount of water resources has been harnessed for hydropower generation. Total installed capacity of hydropower plants in the country is 335 MW only. Some projects are under construction. Those are in the range and it will add 275 MW power to the national grid.

Table 1.1 Some of the main identified hydropower projects

S. No.	Name	Capacity (MW)
1	Seti (West)	750.00
2	Arun 3	402.00
3	Budhigandaki	600.00
4	Kaligandaki no. 2	660.00
5	Lower Arun	308.00
6	Upper Arun	335.00
7	Karnali (Chisapani)	10,800.00
8	Upper Karnali	300.00
9	Pancheshwar	6,480.00
10	Tamur/Mewa	100.00
11	Dudhkoshi (Storage)	300.00
12	Upper Marsyangadi	121.00
13	Andhikhola (Storage)	176.00
14	Sapta Gandaki	225.00
15	Kankai	60.00
16	Sunkoshi Diversion to Kamala	700.00
17	Sunkoshi high dam	200.00

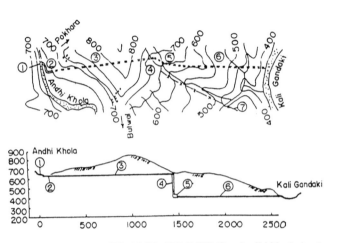

SITE LAYOUT AND PROFILE   Andhi khola hydropower project

1   DAM
2   DESILTING BASIN
3   HEADRACE TUNNEL
4   SHAFT
5   POWERHOUSE
6   TAILRACE TUNNEL

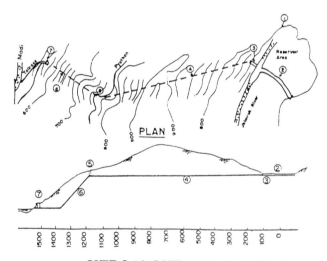

SITE LAYOUT AND PROFILE   Jhimruk hydropower project

1 RIVER TRAINING	5. SURGE SHAFT
2 DAM	6. INCLINED SHAFT
3 DESILTING BASIN	7. POWERHOUSE
4 TAILRACE TUNNEL	

Figure 1.1: Typical layouts of hydropower projects in Nepal

2 CHALLENGES IN HYDROPOWER DEVELOPMENT

Some of the factors, which are standing on the way as hindrances to the development of the hydropower sector in Nepal, include:

- Lack of finance: Historically the government has relied on foreign funding for generation, transmission and distribution of hydropower. Nepali private sector has recently started.
- Seasonal variation in discharge. Monsoon season contribution is 80% of the annual flow.
- Lack of infrastructure: Nepal needs to have some infrastructures within the country so that cheap and reliable power can be produced.
- Limitation in water resources uses. It should be integrated with agriculture, industries and recreation.
- Lack of production of hardware, construction material and construction equipment within the country. So it has little contribution to the local employment.
- Small transmission and distribution network that prevents more peoples and places to have access to electricity. So hydropower market is not extended yet.
- Power market and tariff: As the electricity rates are high it prohibits the Nepali industrialists to consume more power, on the otherhand, because the consumption is low the power market remains small and discourages other promoters to come into the scene.
- Lack of agreement on the basic policy and to form a vision for the hydropower sector for domestic use and power export.
- Fragile and complex geology

3 HIMALAYAN GEOLOGY

The Himalayas are formed by collision of the Indian plate and Tibetan plate. The collision was started in the early Tertiary period (Molnar, 1984). The rate of convergence between the two plates is 41-61 mm/year (Minster and Jordan 1978). As a result several thrust faults have developed. These thrust faults are parallel to general Himalayan trend. There are also several thrust and strike-slip faults, which have northwest or northeast trends. The first shear thrust deformation took place about 20-25 million years ago.

The main orographic and tectonic zones of the Himalayas from south to north are generally devisable into four primary tectonic areas, which largely correspond to the physiographic divisions of the country. All these zones are separated from each other by distinct tectonic elements (Thrusts faults) namely, Himalayan Fontal Thrust (HFT), Main Boundary Thrust (MBT) and Main Central Thrust (MCT). Figure 3.1 represents a schematic geological cross section across the country showing the main geological elements.

Figure 3.1 Schematic tectonic cross section through central Nepal.

4 GEOLOGICAL RISKS

Tectonic movement along with all its geological elements has disturbed the geology in Himalayan region. It has not only caused the fragile geology but also resulted with the series of different types of rocks and geological structures. Thus the underground construction work through these domain varies in terms of excavation, support system requirements, stability and potential for rock burst and squeezing in the underground openings.

Wide disturbed zones are associated with the faults. Minor faults and shearing are common places. Even in good rocks, intense jointing can be observed. Weathering is expected to be deep and extensive, especially under the saddle and this is the area where the cover over the underground structures is low, and in areas where the rocks have been disturbed by faulting and shearing. In addition, majors stress relief joint systems can be expected close to the deep valleys. Underground construction will be very sensitive to these adverse geological conditions.

Lining will be required to prevent leakage into highly permeable ground across the saddle areas where the water table is likely to be below tunnel elevation. Heavy support may be needed in highly stressed ground resulting from high cover or tectonic forces. Ground water inflows may be encountered along faults and highly permeable rocks and could prove troublesome where the head is high or pumping is required in downgrade headings. It can be considered that the risk of displacement along an active fault during the life of the project is low and less of a potential problem than the existing ground disturbance created by the major inactive faults that cross the alignment and which could cause difficulties in excavation.

Detailed geotechnical and geological investigations should be carried out before the design for the main project is started.

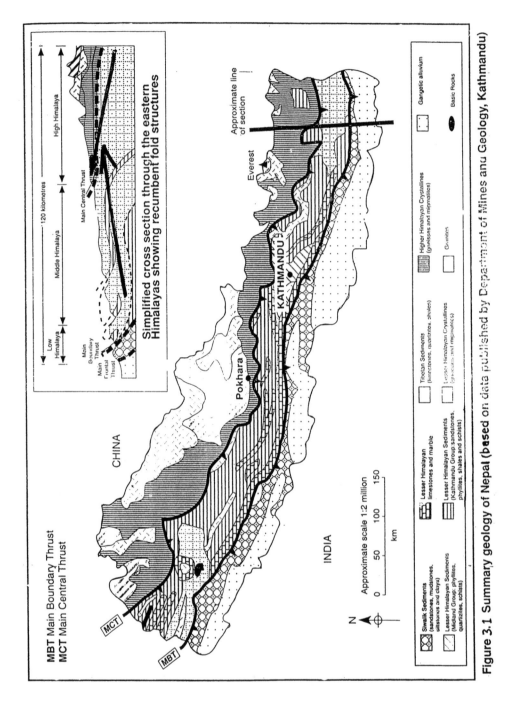

MBT Main Boundary Thrust
MCT Main Central Thrust

Simplified cross section through the eastern Himalayas showing recumbent fold structures

Figure 3.1 Summary geology of Nepal (based on data published by Department of Mines and Geology, Kathmandu)

Legend:
- Siwalik Sediments (sandstones, mudstones, siltstones and clays)
- Lesser Himalayan Sediments (Midland Group: phyllites, quartzites, schists)
- Lesser Himalayan limestones and marble
- Lesser Himalayan Sediments (Kathmandu Group: sandstones, phyllites, shales and schists)
- Tibetan Sediments (limestones, quartzites, shales)
- Lesser Himalayan Crystallines (gneisses and migmatites)
- Higher Himalayan Crystallines (gneisses and migmatites)
- Granites
- Gangetic alluvium
- Basic Rocks

Approximate scale 1:2 million
0   50   100   150
km

5 CHOOSING SITE INVESTIGATION METHODS

Tunnel, underground cavern and shaft are common underground structures in hydropower projects. The critical issues with regard to these underground works are geotechnical. An understanding of the ground through which the excavation will be carried out and areas through which the portal will be developed will be essential. The geotechnical issues carry the greatest risks in construction. Hence investigations are necessary to clarify the type and magnitude of the risks and how they can be reduced or managed. It is essential that all the components of the underground construction for the project should be equally strong and of uniform standard.

The investigation includes geological mapping, geophysical survey, core drilling, test tunnelling, in-situ testing and laboratory testing.

5.1 If good outcrops are available at project area geological mapping will be very useful. Geophysical test needs to be correlated with the core drilling, so that it provides inexpensive and quicker sub surface information. For the underground works at shallow depth and for a small budget geological data can be obtained by mapping and geophysical tests.

5.2 The standard approach to underground investigation is to drill testholes. In the case of the good rock condition, recovery of the samples from the testholes will be good. For the area of simple geology, the data will be representative of a large area and worth drilling the test holes.

5.3 In the case of the crushed rock, recovery of the samples from the testholes will be minimal. For the area of the high ground cover, drillholes would be deep and hence expensive. Although they could provide good site-specific data at tunnel elevation, that might only be representative of short tunnel lengths because of the complexity of the geology. In this condition test tunnels is considered to be a valuable means of investigation because it provides a first-hand look at the ground, potentially at tunnel elevation, but also allows the testing of different techniques of excavation and support and permits pre-construction of tunnel access. The test adit is also driven, to investigate the rock conditions considered to be representative of some of the poorer ground along the tunnel route and to gain early access to a crucial section of the main tunnel line.

5.4 In tunnelling projects where data is difficult or expensive to obtain during the feasibility or design stages, it is now common to undertake investigations concurrently with excavation, for example by drilling ahead of the face to anticipate ground conditions or to use modern technique on the tunnel face. Thus it is not always essential that the critical issues are fully resolved prior to commencing construction, providing that they have been recognised, and can be dealt with, and that the requisite data can be obtained in a timely way to minimise the risks during construction and operation. However this approach must also be reflected in the way that the construction contracts are written to allow flexibility in design and a degree of shared risk between the owner and contractor.

6 CONSTRUCTION RISKS

In Himalayan region, rock quality changes in a short section of underground construction. Drill and blast method adapts better in such condition. Degree of mechanisation depends on the cross-section of the excavation and scale of the project.

The principal objective of the rock support is design of efficient and economical support for underground excavation. Rock strength around the excavated space is utilised to support itself. None of the available classification systems provides ability to design the correct support for underground construction for this region. Therefore site-specific modifications are usually required as compliments to the established support design methods. The applied support needs to be monitored specially in the critical areas. As per the observation additional support may be required.

Where there is a possibility of encountering water if is important to specify that a probe hole should be drilled ahead of the face. Typically, such a probe hole should extend 2-3 times tunnel diameter ahead of the face at all times. A geotechnical engineer should very carefully monitor the drilling. Penetration rate and the quantity and colour of the water return should be recorded on probe-hole log. Sudden change in the penetration rate will indicate the presence of the hard or soft zones, and deviation from normal water quantity and colour may indicate water-bearing fault or fracture zone. This probing is particularly important if a major fault zone that acts as a water barrier is to be traversed. This type of problem can only be solved satisfactorily if there has been sufficient advanced planning, by both the engineer and the contractor, to ensure that an agreed course of actin as been mapped out and that appropriate equipment has been mobilised before the fault is exposed.

7 CASE STUDIES

Some of the underground construction risks and problems encountered are highlighted in the following case studies.

7.1 *Melamchi Test Tunnel*

Melamchi diversion scheme is a raw water transfer project, which includes an intake on the Melamchi River, 40-km northeast of Kathmandu. Water will be diverted through a 27 km long headrace tunnel to the treatment plant the northern edge of the Kathmandu valley near Sundarijal. Adual purpose test/access tunnel was excavated under the saddle area with complex geology at Patibhanjyang. The objectives of this tunnel construction were to investigate actual tunnelling conditions and obtain data regarding costs and progress rates.

Excavation of this tunnel revealed that the actual rocks along the tunnel alignment are poorer (weaker, more sheared) than anticipated and budgeted for. See Table 7.1.

The poorer rock required additional support. A smooth tunnel shape was not possible in this poorer rock, and therefore both the amount of material to be

Table 7.1. Comparison between anticipated and encountered rock conditions in the test tunnel

Rock support type	Budget estimate (m)	Actual conditions (m)
Type I	0	0
Type II	200	0
Type III	160	10
Type IV	40	187
Type V	0	149

mucked out and the amount of support to be installed increased.

Because of the undulation of the subsurface rock layers, quartzite was encountered at the crown at some places. Otherwise the tunnel was drivern mostly in a horizontal transition zone between upper layer of micaceous schist and the lower layer of quartzite. As the tunnel was constructed in the very badly sheared micaceous schist layer, it caused overbreaks at crown at some places where seepage was prominent. Before tunnelling through these very weak rocks, spilling was used to support the crown and drain holes were provided. Then the excavation was carried out without using any explosives. Another reason causing overbreak was the very short stand-up time for providing support in the very weak area.

7.2 Khimti Hydropower Project

Khimti 1 Hydropower Project site is located approximately 100 km due east of Kathmandu. It utilises a gross head of 684 m in the Khimti River between the intake, at elevation 1270 m, and tailwater in the Tamakoshi River. Total length of the waterways including headrace and tailrace tunnel is 10 km. The low-pressure headrace tunnel of 7.6-km length has a cross-sectional area of 11 m^2.

From the downstream end of the headrace tunnel the water is conducted to the powerhouse through a 45° inclined pressure shaft, steel lined and concrete embedded. The powerhouse is located underground, with an access tunnel of 870 m length and an installation of five equal horizontal pelton units with a total capacity of 60 MW. Plan and profile of the project layout is given in Figure 6.2.

The geotechnical investigation for the project included geological mapping, geophysical tests, testhole drilling and laboratory testing of rock samples. For investigation of sub-surface rock conditions, three diamond core testholes were drilled at the top shaft area, the most critical location regarding waterways design. Drilling in the vicinity of the underground powerhouse, which is more than 300 m below a steep hillside, was not economically justified. The permeability tests were carried out in the testholes by the constant head method and by the falling head method. Different lab tests were performed on fresh rock samples selected from rock cores.

The rock within the area belongs to the Melung augen gneiss schuppe zone. The rocks are mainly augen gneiss, but also layers of more banded gneiss occur. Near the base of Midland thrust the rocks consist of mylonite. The southern part of the tailrace tunnel will be close to this mylonite.

The augen gneiss is blocky to massive, hard and compact. The augen structure is formed by the porphyroblasts of feldspar. The augen gneiss is foliated almost parallel to the Midland. Thrust dipping 40-50° northward at the powerhouse area decreasing to 20-30° at the intake area. The banded gneiss is massive to blocky, hard and compact and is characterised by alternation of light and dark minerals. Light minerals are normally quartz and feldspar, and dark minerals are biotite.

The depth and degree of rock weathering varies in the area as judged from surface observations and diamond core drill-holes. The weathering is assumed to

Table 7.2 Some major problems encountered during the construction

S. No.	Problem	Reason	Remedy
1	Alignment change		
1.1	Adit 4: Old Adit tunnel, chainage 0-50 m. Horizontal. Occurred in Feb. 1995 6 x 7 x 15 m3	Tunnel alignment was parallel to foliation plane of highly weathered very weak chloride schist band and colluvium deposit. Mild squeezing was also observed though there was no high overburden. Walls were buckled. Sinkhole formed in February 1995.	Abandoned. New adit was driven through the Gneiss rock.
1.2	Old upper pressure shaft: Chainage 0-186 m (45° inclined). Occurred in April 1997	Presence of fault. The Blocky gneiss rock fractured with clay bands. Presence of ground water through the fractured rock. 2 l/s Limitation in construction method as it is 45° inclined section. Rock type changes form Augen to Granitic Gneiss.	Abandoned. Solved by moving the tunnel further 50 m into the hill.
1.3	Adit 3 junction to downstream: Horizontal. Chainage 12-26m. Occurred in October 1998 14 x 12 x 16 m3	Failure occurred for not providing prescribed support (Type IV rock) Moderately weathered, heavily jointed & fractured Augen gneiss with clay gouge. Recommended construction method of short pull length and spiling use was not followed. Ground water 10 l/m.	New alignment from Chainage 8 m. The recommendations were followed.
1.4	Adit 2 junction to downstream: Chainage	Poor rock: type $\underline{V}$ Tunnel alignment parallel to the foliation plane of the highly weathered and decomposed gray	Realigned from 230 m. Tunnelled with re-

490

S. No.	Problem	Reason	Remedy
	230-260 m. Horizontal. Occurred in August 1997 30 x 12 x 15 m3.	sericite chlorite schist with >20° dipping. Water pressure developed above the crown in impermeable schist of pinching & bulging shape. Presence of ground water 7 l/m. No support within standup time. Initially semi-mechanised construction method was used instead of recommended construction method and support types with short pull length and spiling.	duced pull length (i.e. 1.5 to 2 m) and sets of spilling in 30 cm spacing.
1.5	Adit 1 junction to downstream: Horizontal. Chainage 330-390 m. Occurred in September 1997. 60 x 15 x 20 m3.	Poor rock: type $\underline{V}$ Tunnel alignment parallel to the foliation plane of the highly weathered and decomposed gray sericite chlorite schist with >20° dipping. Water pressure developed above the crown in impermeable schist of pinching & bulging shape. Presence of ground water 8 l/m. No support within stand-up time. Initially semi-mechanised construction method was used instead of recommended construction method and support types with short pull length and spiling.	Realigned from 328 m. Tunnelled with reduced pull length (i.e. 1.5 to 2m) and sets of spilling in 30 cm spacing.
2	Overbreak		
2.1	Upper pressure shaft. Chainage 125-140 m. Occurred in December 1998. 15 x 10 x 15 m3	This area was under the influence of a fault. Sheared, fractured blocky gneiss with clay band. Hydrostatic pressure built up during upper overbreak, which was not released during the sudden flow. Insufficient support which was destroyed during upper overbreak. Rocktype changes form Augen to Granitic Gneiss.	It was yet to be stabilised
2.2	Upper pressure shaft: Chainage	Area was under the influence of fault. Fractured and sheared blocky gneiss with clay	Solved by: Grouting and spiling in 20 cm

S. No.	Problem	Reason	Remedy
	352-364 m Occurred in July 1998. 9 x 12 x 20 m3.	bands. Presence of ground water 8 l/m. Limitation in construction method at this 45° inclined section. Excavation was not carried out from the top to the bottom, which was recommended.	spacing. 1 m excavation by cutting from top to bottom and shotcreting. Providing reinforced ribs of shotcrete.

go deeper into the ground at the top or near the top of the ridges than down at the rivers.

In the beginning, investigation was not carried out concurrently with excavation. Pre-requisite data were not obtained in a timely way to minimise the risks during construction. Some major problems were encountered during the initial phase of the construction of the project. See Table 7.2.

Remedying wrongly designed works or adopting alternative construction methods during construction have been expensive than the cost of the original site investigation. However, system of investigation concurrently with excavation was established by the middle of the construction period. Investigation probe hole was drilled ahead of the face to anticipate ground condition and peizometer tubes were installed for the representative stretch of the excavated tunnel. Experiences gained during the initial phase and information received from the investigation on the tunnel face helped to minimise the risks during the construction.

8 CONCLUSIONS

Permanent snow on the Himalayan range has given Nepal a perennial source of water. Topography with series of mountains and gorges has amplified the hydropower potential of the country. However, the tectonic movements, has disturbed the rockmass and hence resulted with complex geology. So it poses threats to the underground construction works in the Himalayan region.

Detailed geotechnical and geological investigations should be carried out before the design for the main project is started. Experiences have shown that same type and scale of ground investigation is not suitable to all the hydropower project sites. Testhole drilling is the best method where it represents a large area of the site. Test tunnel method is a suitable method for a large hydropower project. Where data is difficult or expensive for the feasibility stage, investigation can be undertaken concurrently with excavation, in a timely way to minimise the risks during construction.

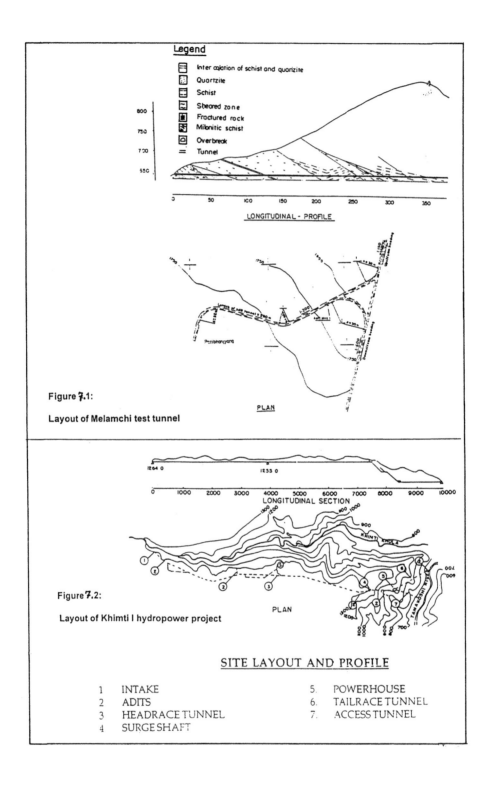

Inter calation of schist and quartzite
Quartzite
Schist
Sheared zone
Fractured rock
Milonitic schist
Overbreak
Tunnel

LONGITUDINAL - PROFILE

PLAN

Figure 7.1:

Layout of Melamchi test tunnel

LONGITUDINAL SECTION

PLAN

Figure 7.2:

Layout of Khimti I hydropower project

SITE LAYOUT AND PROFILE

1	INTAKE	5.	POWERHOUSE
2	ADITS	6.	TAILRACE TUNNEL
3	HEADRACE TUNNEL	7.	ACCESS TUNNEL
4	SURGE SHAFT		

Some site-specific modification would be necessary in determining support for the underground construction. There should be capabilities to provide support within a short stand up time so that the support will be most effective and prevent potential overbreaks. Monitoring needs to be done to observe the performance and adequacy of the support provided.

REFERENCES

Butwal power company Ltd 1996. Melamchi diversion scheme: Bankable feasibility study report.
Khimti services consortium 2000. Khimti I hydropower project: project completion report.
Nepal electricity authority 2000. F/Y 1999/00 a year in review.
Shrestha G L and Thanju R 1997. Tunnelling through the sheared zone in Likhu adit presented in 2nd international course in small hydro development.
Shrestha G L and et al 1999. Rock support design for Khimti hydropower project presented in international symposium organised by NGS and IAEG.

493

Hydropower in the New Millennium, Honningsvåg et al (eds), © 2001 Taylor & Francis, ISBN 90 5809 195 3

Geographical Inforniation System of hydropower installations

T.E.Skaugen

Statkraft Grøner AS, Oslo, Norway

ABSTRACT: Statkraft Grøner AS is developing state of the art Geographical Information System (GIS) for the hydropower schemes in Norwegian watersheds, which are operated by Norway's largest hydropower developer, Statkraft SF. The system is dealing with all types of installations and schemes at different scales, such as gatehouses, fish hatcheries, weir, office buildings, dams, hydropower stations, tributary sub watersheds and hydropower network schemes. The work includes developing a code set based on existing national (and international) code systems and symbol sets. When the GIS is established, the hydropower developer can achieve synergy effects and more effectively operate the watersheds. The GIS can merge the information of all the installations in other databases with the GIS system. Pictures and real-time video recordings can be accessed from the maps, as well as precipitation-runoff models, runoff- and climate stations, and other important information, if needed.

1 INTRODUCTION

The presented work is based upon a project financed by Statkraft SF whitch is going on and will be completed within desember 2001. The project started as a pilot in august 2000 with establishing a Geographical Information System (GIS) of installations owned by Statkraft SF in the two Norwegian municipalities Vinje and Hemnes. The aim of the project is to establish a GIS of all the installations owned by Statkraft SF in Norwegian watersheds.

A GIS connects geospatial information in a digital map. The system is sketched in Figure 1.

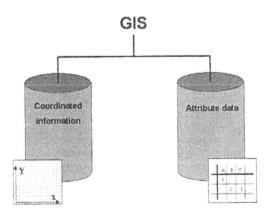

Figure 1. The figure shows a sketch of a GIS. The user is connected to the digital map and the database at the same time.

The use of a GIS to combine information with digital maps is not uncommon today. The problem however has been expensive software and hardware. Digital maps and attribute information, and of course qualified experts, is not readily available.

For the hydropower profession, which deals with watersheds, physical installations in the watershed and information connected to these installations, an established GIS may be useful for the following reasons:

- to store and connect useful information from the watershed
- to search for information from the watershed
- to create sketches or thematic maps of watersheds with installations
- to combine information of installations with environmental information or with new hydropower plans
- to make decisions concerning environmental schemes in the watershed such as weirs or fish hatcheries
- as a tool for holistic watershed modelling

2 METHOD

Maps can be represented in a GIS in two ways; as vector data or as raster data. The raster representation is a set of equally sized cells located by coordinates. Each cell is independently addressed with the value of an attribute. The vector representation maintains three main geographical entities; points, lines and areas. Points are similar to cells except they do not cover area, lines and areas are sets of interconnected coordinates that can be linked to given attributes. As long as the vector data and the raster data are stored in identical coordinatesystem, they can easily be combined in a GIS system.

To create a GIS for hydropower installations, the items are stored as vector data.

2.1 Vector data

Vector data are stored in a chosen coordinatesystem. Information connected to the features is stored in a database. The features can be stored in files separated only by topology, but can also be separated by themes (such as dams, gatehouse, roads and regulated zones around reservoirs).

A unique hydropower *identification code* for all the hydropower installations, when establishing a GIS, must be established for addressing the features. This code must be registered in the database connected to the topology file as attribute data. A *hydropower symbol code*, separated by the identification code for hydropower installations, must be developed to make the system unique for the users.

The quality in coordinates of the registered items is of great importance. Therefore it is useful to develop a quality code for the registered hydropower installations in the GIS. Further information concerning the details of GIS can be found in Burrough (1986).

2.2 Software program

There are different kinds of software tools for the development of GIS. The software, which is used in this project, is the Environmental Systems Research Institute Inc (ESRI) software program ArcView and ArcExplorer. The files created from ArcView can be converted to almost any file format of interest.

3 ESTABLISHING A GIS FOR STATKRAFT SF'S INSTALLATIONS IN NORWAY

A GIS of all the installations in Norwegian watersheds owned by Statkraft SF is to be established. The work is separated into three stages, development of identification- and symbol codes, registration of installations and creating of thematic map routine (section 2). The system is handling all types of installations and schemes at different scales, such as gatehouses, fish hatcheries, weirs, office buildings, dams, hydropower stations, tributary sub watersheds and hydropower network schemes.

3.1 Creating identification codes for every installation in the GIS

When developing a GIS, it is important to establish a unique identification code for every single installation in the GIS. Information connected to only one specific weir or one specific gatehouse can then be stored. The need for the identification code is sketched in Figure 2.

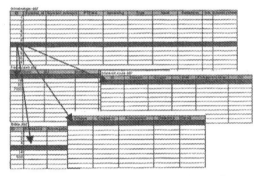

Figure 2. The figure shows how information from different tables can be connected to the features in a GIS with a unique identification code. As well as a unique installation identification code, the need of a unique identification code for each feature is needed.

3.2 Creating general hydropower identification codes for hydropower installations

A unique general hydropower *identification code* for all the hydropower installations must be established for addressing the features. The code sets will become a part of the Systematic Organisation of Spatial Information (SOSI- standard). The SOSI standard is a Norwegian standard, and is developed by Statens Kartverk in Norway. The first version is from 1987. Version 3.1 is now in use but the SOSI standard is under continuous revision. The work is developed in close relationship to international standardization rules and models, especially CENT/TC 287 and ISO/TC 211. The code set is needed to address the features registered in the GIS. The aim is that this codeset can be developed as a national (and international) standard code for hydropower installations. The SOSI standard is divided into featuregroups such as buildings, transportation, topographical forms and technical installations. The hydropower installations embrace all these feature groups. As an example, some features with its hydropower code are presented in Table 1. The codeset is not yet finished, it will however be completed within summer this year.

Table 1. The table presents an example of a identification code list of hydropower installations. The list is based on the Norwegian SOSI standard and is not yet completed. The standard will be completed within the summer of 2001.

Innstallation	SOSI code
Dam	6501
Weir	3205
Office building	5000
Hydropower station	8160
Radiomast	8604
Portal building	5000
Fish hatcheries	6641
Pump station	8264

3.3 Creating a hydropower symbol code for hydropower installations

A unique hydropower *symbol code* for all the hydropower installations must be established for recognise the features in the map. The symbols are based on existing symbols used for the same purpose. The need for a standard is urgent because no such standard exist today. The work with a symbol set will be completed within summer this year.

3.4 Creating a quality code for hydropower installations

A unique hydropower *quality code* for all the hydropower installations should be established for checking the quality of the features. When the GIS is operating, there are no other way to control how exact the coordinats of the installations (features) are. The quality code set is not yet established, but the features will be registered with a code for unknown quality unless the quality of the coordinates is known. The work with upgrading the quality can then be done continuously.

3.5 Registration of all Statkraft SF's installations in Norway in a GIS

There are different ways of organizing the registration of all installations in Norway owned by Stakraft SFs. There are lots of information in analog maps and reports, it is however very time consuming to find all the information needed. To simplify the work, the registration has been based on people who works in Statkraft SF (and retired personnel) who knows the installations very well. Their knowledge of the watersheds and the location of the installations are of invaluable importance. The GIS operator organizes meetings with these people. During the meetings the Statkraft SF personal informs the GIS operator where to put the installations. After a few days with intensive work, most of the installations in the actual watershed(s) are registered with coordinates and information. An example in how to use the GIS is presented in Figure 3.

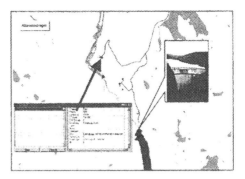

Figure 3. The figure shows a clip of a watershed in Norway. The illustration of popup-information is shown in the figure.

3.6 Creating thematic maps and combining information

When the GIS is operating, thematic maps can be created. Other digital information, such as pictures or digital air photos can be combined with the data. There are also possible to develop routines to make thematic maps so that the user can create thematic maps of any themes of interest.

4 DISCUSSION AND CONCLUSION

When all the data sets are stored in the GIS, they can be accessed from the maps. The information in the GIS can be combined with other geographically referenced information such as flood maps, dambreak data, river restoration (such as minimum streamflow, weirs and pools) and of course area planning data. Information can also be combined with population sets, and other environmental characteristics of the watershed. Results from river basin models, such as Mike Basin and IRAS (Integrated River and Aquifer System) and hydraulic models, can also easily be combined with the database.

When the GIS is established, the hydropower developer can achieve synergy effects and more effectively operate the watersheds. The GIS can merge the information of all the installations in other databases with the GIS system. Pictures and real-time video recordings can be accessed from the maps, as well as precipitation-runoff models, runoff- and climate stations, and other important information, if needed.

The hydropower profession will have profit of a GIS in the communication with municipalities and other institutions in the watershed. As discussed above, thematic maps are easily created with a GIS and they are easy to understand.

The potentials of getting the optimal solutions in different processes and project the Hydropower Company are involved in are also present with a GIS. The GIS is perfect as a tool for creating new plans for the watershed and the time used on planning will decrease. The access to correct information is easy with a GIS. And the information can easily be combined with other digital maps and information.

Internationally, from the hydropower developer's point of view, is increasingly focused on integrated watershed management approach. This implies a closer cooperation with the authorities, various stakeholders and the local society. Better tools for better management of steadily increasing and integrated data sets is therefore of crucial importance.

5 REFERENCES

Burrough, P.A. 1986. Principles of geographical Information Systems for Land Resources Assessment. Clarendon Press Oxford

Hydropower in the New Millennium, Honningsvåg et al (eds), © 2001 Taylor & Francis, ISBN 90 5809 195 3

Powerformer™: Insulation co-ordination and its impact on plant design

E.G.Sörensen
ALSTOM Power Generation AB, Västerås, Sweden

ABSTRACT: Insulation co-ordination of winding system and plant design of system solutions with Power-former™, the brand name of a cable wounded high-voltage generator that offer a direct connection to a high voltage network, is described. The paper deals briefly with parameters taken into account when designing the winding system of a high voltage generator. In comparison with a conventional generator, normally galvanic separated from the high voltage network via a step-up transformer, the winding system of a high voltage generator will be subjected to transient voltage levels that are distinctive characteristics of the principle of system earthing of the network concerned. The number and the duration of prospective single line to earth fault in non-effectively earthed network are also to be considered. Lightning stroke density, typical lightning parameters such as peak current, summer and winter conditions typical of the area concerned as well as the distance between the switchyard and the powerhouse are also essential parameters in the design work on the winding system concerned. The design of the insulation co-ordination of the winding system, herein the procedure of selection and location of surge arresters, selection of cables and the final decision making concerning a totally insulated winding system or not is described. The design of the winding system might have an essential impact on the physical size of Powerformer and the total space needed for the installation of the generation unit concerned. An example illustrating the technical benefits of a totally insulated cable winding system with incorporated surge arresters located close to Powerformer is elaborated.

1 INTRODUCTION

The windings of Powerformer consist in principle of continuous, insulated cables similar to conventional solid extruded cables. The induced voltage in Power-former increases gradually along the stator winding from the neutral side to the line side of the stator core. Therefore the cable used for the stator winding will experience different electric stresses along the length of the winding. It is, therefore, feasible to use a thinner insulation of the first turns of the winding and thereafter increase the insulation thickness. One way of obtaining this is to use a predefined number of different cable dimensions per phase, i.e. a step-wise increase in the insulation thickness, refer to Figure 1. This type of stepped insulation facilitates the optimization of the volume of the laminated stator core. The different cables are joined together outside the active part of the machine. The line side cables wounded in the stator core are connected to external cables for connection to the terminals of a high-voltage unit circuit breaker normally located in a high voltage switchyard of the power plant concerned.

In this paper the term "cable system for Power-former" is defined to include the cables wounded in

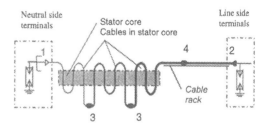

Figure 1. Draft of the cable system for Powerformer. The thickness of the lines illustrates the different insulation levels of the cables incorporated in the cable system.

Explanations:
1. Cable terminations with surge arrester located in the neutral point equipment.
2. Cable terminations with surge arresters located in the high voltage switchyard.
3. Cable joints placed and fixed in a metal bracket located outside the stator core
4. Cable joints for interconnection of the "internal" cables and the "external" cables.

the stator core, the external cables, cable accessories and surge arresters connected to the neutral side terminals and the line side terminals of the cables.

2 DESIGN SEQUENCE

Electromagnetical design of a synchronous machine type Powerformer is an iterative procedure. In addition to generator ratings provided by a prospective client, power frequency voltage rises caused by either persisting or transient fault conditions as well as travelling waves caused by e.g. lightning strokes are to be taken into account in the design work, refer to Figure 2.

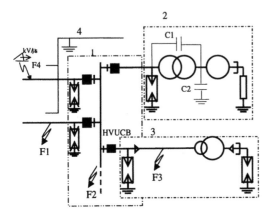

Figure 2. Single line draft illustrating a power plant with a conventional generation unit and high voltage generator type Powerformer. Typical prospective faults have been indicated in the figure.
Explanations:
1: High voltage switchyard with two outgoing lines.
  - HVUCB: High voltage unit circuit breaker
2: Conventional generation unit. C1 and C2 illustrate the capacitive coupling across the step-up transformer.
3. Powerformer.
4: Lightning line protection above the station and the closest part of outgoing lines is illustrated.
F1: Prospective single line-to-earth fault in the network.
F2: Prospective single line-to-earth fault in the switchyard.
F3: Prospective single line-to-earth fault on the line side of Powerformer.
F4: Lightning stroke in the network.

2.1 *Step 1. Transient reactance value*

The voltage level on the terminals of a synchronous machine rises, in case of a load rejection, almost instantly to a value corresponding to the voltage behind the direct-axis transient reactance. As prospective power frequency voltage levels are predominant parameters in the initial design work on Powerformer a typical value is to be estimated and consequently the first step in the calculation sequence is to estimate a typical value of the direct-axis transient reactance of the synchronous machine concerned. Rated power in MVA and rated voltage in kV of the machine concerned in combination with engineering experiences gained from previous de-

sign works on high voltage generators type Powerformer are the main parameters used for this estimation.

2.2 *Step 2. Power frequency and medium frequency transient voltage levels*

The maximum voltage levels on the terminals of Powerformer at prospective faults are to be considered. The following prospective faults have an essential impact on the design of the cable system concerned.

- Single line-to-earth fault in the cable between the high voltage unit circuit breaker and Powerformer, followed by a load rejection.
- Single line-to-earth fault at maximum terminal voltage and unsynchronized Powerformer i.e. Powerformer is not connected to a network.
- Single line-to-earth fault at maximum terminal voltage and Powerformer connected to a non-effectively earthed network.

2.3 *Step 3. Surge arresters*

Per definition the surge arresters located at the cable terminations are an integrated part of the cable system of Powerformer and consequently they must be able to withstand the voltage stresses caused by prospective faults in the network concerned. In addition to the voltage stresses caused by temporary faults in the network the surge arresters must also be able to withstand the stresses caused by e.g. single line-to-earth fault occurring on one of the line side terminals of Powerformer during voltage build-up or prior to synchronising. The temporary overvoltage values, TOV, of the surge arresters are to be matched against the prospective maximum power frequency voltages occurring at the terminals of Powerformer.

Design rules for surge arresters connected to the high voltage side of step-up or step-down transformers or connected to outgoing lines to limit overvoltages from entering the switchyard, might differ from the above mentioned design rules applied for arresters incorporated in the cable system. In an effectively grounded network the voltage rise caused by a prospective single line-to-earth fault depends on the earthing degree of the network concerned. Consequently the temporary overvoltage value, TOV, and the protection level of surge arresters installed in an effectively grounded network will be lower than the corresponding levels of the arresters incorporated in the cable system of Powerformer.

2.4 *Step 4. Travelling waves*

Travelling waves caused by switching operations, lightning strokes or other faults in the network concerned can reach and propagate directly into the cable system of Powerformer.

In the case of a lightning hits a phase conductor the steep discharge current will meet the wave impedance of the line in both directions and the current will give rise to voltage waves that move towards the ends of the line concerned. The overvoltages generated in the cable system of Powerformer by a propagating voltage wave caused by e.g. a lightning stroke in a phase conductor of an overhead line depends on several parameters, such as:

- The value of a prospective lightning discharge current for the region concerned.
- The protecting level of prospective surge arresters connected to the high voltage busbar in the switchyard.
- The protection level of the line side surge arrester incorporated in the cable system of Powerformer.
- The distance between the line side surge arrester of the cable system and prospective surge arresters connected to the busbar in the vicinity of the high voltage unit circuit breaker of Powerformer.
- The distance between the line side surge arrester incorporated in the cable system and Powerformer.

Prospective switching transients generated by opening the high-voltage unit circuit breaker will be limited by the line side surge arresters of the cable system of Powerformer.

2.5 Step 5. Selection of preliminary cable insulation levels

The above-estimated prospective voltage levels, refer to step 2 and 4, are the inputs for the selection of preliminary insulation levels and ratings of the cable system of Powerformer. A safety margin of approximately 20% will be added to the estimated voltage levels.

2.6 Step 6. Design of Powerformer

Electromechanical design of Powerformer will be based on the preliminary cable ratings estimated under step 5.

2.7 Step 7. Verification of the estimated direct-axis transient reactance value

The calculated direct-axis transient reactance value is to be compared with the corresponding estimated value used for the estimation of power frequency voltage transients, refer to step 1. If the results differ with more than a few percentages then one or more calculations, refer to the calculation sequence described in step 1 to 6, is to be performed.

2.8 Step 8. Final overvoltage analysi

After the completion of the electrical and mechanical design of Powerformer, data are available for dynamic simulation work. The magnetic fields and their interaction with the conductors of the winding are simulated with the aid of FEM program.

- Power frequency voltage levels can be simulated by use of traditional generator models.
- Fast transient voltage phenomena propagating through the cable system can be simulated by use of special models of Powerformer.

The total design sequence is illustrated in Figure 3.

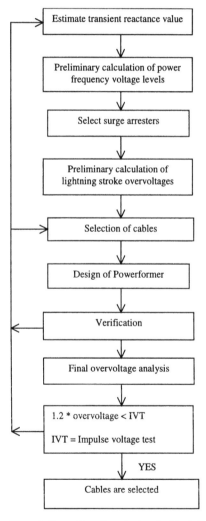

Figure 3. Overvoltage dimensioning of cables in Powerformer.

3 ESTIMATION PROCEDURE

3.1 Estimation of the power frequency voltage rise caused by a load rejection

A fast change of armature current may be due to the application or removal of a fault or to swinging of the generator with respect to the power system or to a load rejection of the machine concerned. During such a change, the flux linkage of the field winding remains substantially constant and a new fictitious internal armature voltage, E_q' (Kimbark 1956) will be defined, refer to Figure 4. The voltage E'= Eq' is generally known as the voltage behind the transient impedance or the voltage behind the direct-axis transient reactance of the machine.

A prospective instantaneous voltage rise on the terminals of a salient-pole machine following a load rejection depends partly on the direct-axis transient reactance of the machine and on the load. In the case of a temporary reactive overload the voltage rise will be larger than if the machine was producing active power only. The maximum voltage level and the subsequent decaying of a prospective overvoltage depends on several parameters such as the load or the power factor, the type of excitation system, the speed increase of the rotor and the time constant of the field winding. As a first rough estimate a prospective maxim voltage rise following a load rejection is estimated not to exceed a value corresponding to x_d', i.e. the machine is loaded with 1 pu reactive load.

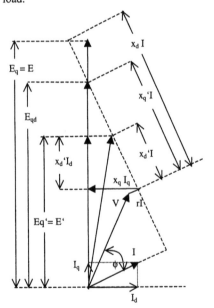

Figure 4. Vector diagram of salient-pole synchronous machine (xq'= xq). Voltage vectors E, Egd and E' all lie on the quadrature axis, both in the transient state and the steady state.

The maximum power frequency voltage level at the neutral point of Powerformer in the case of a prospective single line-to-ground fault at a line side terminal is estimated from the below equation 1, refer to Figure 5.

$$U_{neutral} = \left[1 + \frac{\Delta U_h}{U_h}\right] \cdot \left(1 + X_{du}'\right) \cdot U_{ph} \qquad (1)$$

Where U_h = rated line-to-line voltage; U_{ph} = rated line-to-earth voltage; ΔU_h = the max positive deviation from rated value; and X_{du}' = the direct-axis transient reactance.

The corresponding voltage level on the healthy line side terminals is the above-mentioned value multiplied by √3.

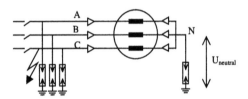

Figure 5. Illustration of a prospective single line-to-ground fault on the line side terminal of Powerformer.

3.2 Selection of surge arresters

The surge arresters incorporated in the cable system must be able to withstand the voltage stresses caused by the above mentioned power frequency voltage levels. In general, surge arresters are not used to protect equipment against temporary overvoltage, TOV, as this would require an enormous numbers of parallel columns of blocks. If Powerformer is to be connected to an effectively earthed network then the TOV_{10s} value of the neutral point arrester is selected to be equal to or above the above estimated voltage level $U_{neutral}$, refer to equation 1. In case of a non-effectively earthed network with permitted fault duration in a range from 10 s to several hours the continuous operation voltage of the surge arresters are also to be taken into account in the selection procedure. The corresponding TOV_{10s} of the line side arresters of the cable system is the TOV_{10s} neutral point value multiplied by √3 plus an additional small safety margin to secure current selectivity between the line side and the neutral side arresters.

The above estimated TOV_{10s} values are the input parameters for the selection of the arresters incorporated in the cable system. Lightning and switching protection levels can be read from the characteristics normally stated in data sheets of prospective arresters.

3.3 Simple model for estimation of lightning overvoltage

When lightning strikes an object the voltage that it develops is a function of the lightning discharge current multiplied by the impedance of the system. For a first estimation of the overvoltages generated in the cable system of Powerformer by a propagating voltage wave a simple generator model based on the surge impedance's of the cable system is used, refer to Figure 6.

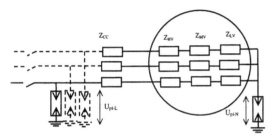

Figure 6. Illustration of Powerformer surge impedance model.
Abbreviations:
- Z_{CC}: Surge impedance of the external cable connecting Powerformer with the high voltage switchyard.
- Z_{HV}: Surge impedance of the internal line side high voltage cable.
- Z_{MV}: Surge impedance of the internal intermediate high voltage cable.
- Z_{LV}: Surge impedance of the internal neutral side high voltage cable.
- U_{pl-L}: The residual voltage of the line side surge arrester.
- U_{pl-N}: The residual voltage of the neutral side surge arrester.

Knowing the surge impedances of the cables the reflection factor, ρ, for each cable can be calculated and subsequently the reflected and the total voltage at each connection point can be defined.

At the connection point for interconnection of the external cable and the internal line side cable, the following expression can be derived, refer to equation 2 and 3.

$$v_1 = v_f + v_{b1} = v_f (1 + \rho_1) \qquad (2)$$

where

$$\rho_1 = \frac{Z_{HV} - Z_{cc}}{Z_{HV} + Z_{cc}} \qquad (3)$$

Where Z_{HV} = Zurge impedance of the internal line side cable; and Z_{CC} = Zurge impedance of the external cable.

The propagating voltage wave, V_f, will be reflected by Z_{HV} and the voltage V_1 can be calculated, refer to Figure 7 and the above equation 2.

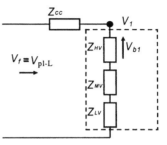

Figure 7. Illustration of the principle for estimated overvoltage calculations at the line side cable entrance in the stator core.

The overvoltage at the point for interconnection of the external cables and the internal line side cables can be determined by equation 4 or 5.

$$U_{rp} = U_{pl} + 2 \cdot S \cdot T \quad \text{for } U_{pl} \geq 2 \cdot S \cdot T \qquad (4)$$

$$U_{rp} = 2 \cdot U_{pl} \quad \text{for } U_{pl} < 2 \cdot S \cdot T \qquad (5)$$

where U_{pl} = Residual voltage (lightning impulse protective level) of the line side arrester (kV); S = Steepness of the impinging surge (kV/s); and T = Travel time of the lightning surge determined as, refer to equation 6.

$$T = \frac{l}{v} \qquad (6)$$

Where $l = l_1 + l_2$ = Distance from the arrester; and v = Velocity of the voltage wave.

However, it might be worth mentioning that the voltage stresses on the terminals of a conventional generator caused by a prospective lightning stroke depends on the capacitive coupling across the step-up transformer, C1, and the capacitive coupling between the busbar and earth, C2, refer to Figure 2. The voltage ratio of the transformer does not reduce the voltage level of a prospective voltage surge.

The length of the cable between the line side surge arrester and Powerformer has a major effect on the overvoltage level at the connection point for interconnection of the external cable and the internal line side cable, refer to Figure 8.

When calculating overvoltages it is also important to take into account the residual voltage of other surge arresters in the vicinity of the line side surge arresters incorporated in the cable system of Powerformer, refer to Figure 2. In some applications it might be feasible to install an additional set of surge

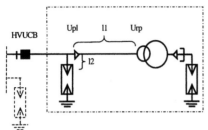

Figure 8. Surge arresters protecting Powerformer.
Abbreviations
- l1: Length of the external cable.
- l2: Distance between cable termination and surge arrester.
- HVUCB: High voltage unit circuit breaker.

arresters on the busbar side of the high voltage unit circuit breaker. The arresters on the busbar side, with a lower residual voltage protection level than that of the line side arresters of the cable system, will protect against incoming voltage surges caused by lightning strokes and breaker operations. The line side arresters of the cable system protect against switching transients caused by the high voltage unit circuit breaker, refer to Figure 8.

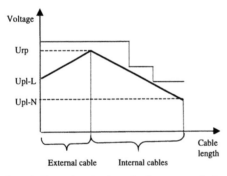

Figure 9. Illustration of estimated voltage stresses in the cable system of Powerformer.
Abbreviations:
- Upl-N: Residual voltage i.e. lightning impulse protective level of the neutral point arrester.
- Upl-L: Residual voltage i.e. lightning impulse protective level of the line side arrester.
- Urp: Voltage level at the point for interconnection of cables.

Experiences from calculations on cable systems indicate that the surge impedance of the external cable is 10 – 20 times smaller than the surge impedance of the internal line side cable of the cable system, refer to Figure 6. The line side cable has the thickest insulation of the internal cables and consequently it has the lowest capacitance and the largest surge impedance. Hence it follows that it gives the highest reflection to a propagating voltage wave and estimated voltage stresses of the total cable system can be derived from two straight lines, refer to

Figure 9. The step-ladder in the figure indicates the estimated voltage impulse test levels as a prospective cable system must be able to withstand.

3.4 *Estimation of lightning overvoltage levels*

The values of prospective lightning stroke currents range from a few hundred amps to approximately 250 kA. Lightning flashes can be divided into two main groups "Upward flashes" and "Downward flashes". Each of the two groups contains "Positive flashes" and "Negative Flashes". The distribution between positive and negative flashes may vary seasonally over the year for different regions of the world or even within different regions of a country.

Downwards negative flashes are considered to be the most important discharge process for practical engineering systems. The possibility of a positive flash incidence has been estimated to about 10 percentages.

For a thorough review of properties and parameters of lightning flashes is referred to the literature (Berger & Andersson & Kröninger 1975). Reference can also be made to international standards e.g. IEC (IEC 71-1, IEC 71-2) regarding lightning parameters useful for protection of systems in engineering applications.

Typical peak lightning stroke parameters and the number of prospective lightning strokes per year for the region concerned are the main parameters to be used in the calculation procedure.

An approach, which may be useful in engineering applications is to, if information concerning lightning currents is unavailable for the region concerned, take advantage of the statistic material collected in e.g. IEC standards (IEC 71-1 , IEC 71-2 & IEC 600099-5). Table 1 illustrates some statistic values from extensive measurements stated in a CIGRE study report (Andersson & Eriksson 1980)

Data have been analyzed from several parts of the world, and the resultant cumulative distribution of peak current amplitude has a median value of about 34 kA.

Table 1. Probability of lightning currents.

Lightning current	Probability
4 kA	98 %
20 kA	80 %
90 kA	5 %

Utilizing the principle illustrated in Figure 9 the maximum voltage stress level caused by an incoming propagating voltage wave can be estimated. Figure 10 illustrates how the estimated maximum value increases with the length of the external cable.

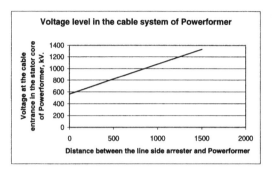

Figure 10. Illustration of voltage levels at the connection point for interconnection of the external cable and the internal line side cable.

From the above figure one can conclude that it is an advantage to install one set of surge arresters close to Powerformer. Experiences from several calculations on prospective installations with Powerformer indicate that the increase of the insulation level of the cables which is needed to meet prospective voltage stresses due to a long external cable may influence on the weight of the stator core with up to approximately 10%.

3.5 Estimation of power frequency and medium frequency voltage levels

Networks that have a large percentage of all transformers with neutrals connected directly to earth with no impedance intentionally inserted in the neutral and where X0 < 3X1, and R0 < X1 are classified as effectively earthed network. Systems operating in this classification have a maximum transient line-to-earth voltage on healthy phases no exceeding two times the operation voltage and a line-to-earth sustained voltage not exceeding 140 % of the operation voltage concerned (Petersson 1951).

In network systems earthed through one or more reactances the fundamental-frequency phase-to-earth voltage will not exceed normal line-to-line voltage, and the neutral-to-earth will not exceed normal line-to-earth voltage. Systems with neutrals earthed through reactances will have maximum transient voltages-to-earth on the healthy phases not exceeding 2.73 times normal. The transient voltage-to-earth at the neutral will not exceed 1.67 times normal line-to-earth voltage (Petersson 1951).

The neutral point of Powerformer is open, refer to Figure 2. After voltage build-up and prior to synchronization Powerformer can be regarded as a small unearthed system comprising a generator and a three phase symmetric capacitive load, refer to Figure 11.

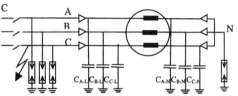

Figure 11. Powerformer illustrated as a small unearthed system. The distributed cable capacitance of each cable has been divided and lumped into two fixed capacitance's, one at the line side and one at the neutral side of Powerformer.

A prospective single line-to-earth fault will rapidly change the electric charges of the faulty phase line side capacitor, but the charges of the healthy phases cannot be altered unless currents are set up in the windings. Medium frequency transient currents will be set up to redistribute the charges. If the prospective single line-to-earth fault appears at voltage maximum, the peak voltage on the healthy phases will not exceed 2.5 times normal line-to-earth voltage (Willheim & Waters 1956). The peak voltage-to-earth at the neutral will not exceed 1.67 times normal line-to-earth voltage (Willheim & Waters 1956). A record of a voltage oscillation at the neutral point in case of a prospective single line-to-earth fault prior to synchronization is illustrated in Figure 12.

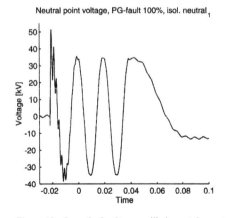

Figure 12. Record of voltage oscillations at the neutral point of Powerformer Porjus U9, 11 MVA 45 kV.

3.6 Cable withstand capability

Knowing the following parameters
- Ratings of Powerformer.
- Philosophy of the system earthing of the network concerned.
- Max continuous operation voltage.
- Max temporary power frequency voltage level at a prospective earth fault in the network and the coherent fault duration time (hours/year).
- Max transient voltage levels at a prospective single-line-to-earth fault in the network concerned.
- Max transient voltage levels at a prospective single line-to-earth fault on the line side of Powerformer prior to synchronization.
- Min levels for voltage impulse tests.

adequate cables for the design work on Powerformer can be selected, refer to section 2.6 Step 6 Design of Powerformer.

4 FINAL OVERVOLTAGE CALCULATION

Electrical transient simulations studies in power systems involves a frequency range from zero to about 50 MHz. The high frequency oscillation range are determined by the surge impedance and the travelling times of connecting lines and by the inductance's and capacitance's involved. Model representations, which are valid throughout the complete frequency range are not possible for all network components. The introduction of high voltage generators type Powerformer have focused on the need for improved models for simulation of fast transient phenomena in cable wounded machines. Lumped circuit models and FEM models for cable wounded machines are described in the literature (Holmberg 2000). Some results from a simulation work on a

prospective installation with Powerformer are shown in Figure 13, Figure 14, and Figure 16.

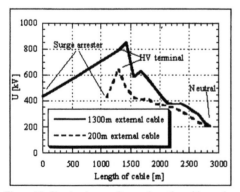

Figure 13 .Simulated overvoltages along the cables and surge arresters that constitute the cable system of Powerformer. Note the peak difference between the curves.

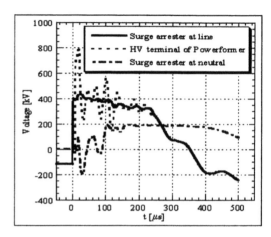

Figure 14. Simulated overvoltages at the end terminals of the cable system and at a point in the internal line side cable.

The experience from simulation works and comparison of results from simulations based on advanced models and results from a simplified cable model indicates almost the same voltage level at the internal line side cable of Powerformer as far as a current wave as specified for the arresters is considered. At the neutral point of Powerformer the simplified calculation model gives a few percentages higher value than the advanced dynamic simulation models, refer to Figure 15.

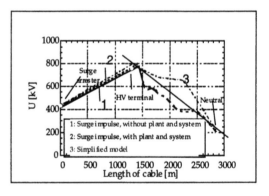

Figure 15. Simulated and calculated overvoltages along the cable system of Powerformer. The result from the simple calculation model gives a somewhat conservative overvoltage value for the internal cables of the total cable system for Powerformer.

5 FINAL DESIGN OF CABLE SYSTEM FOR POWERFORMER

The final design of the cable system might have an essential impact on the physical size of Powerformer and the total space needed for the installation of the

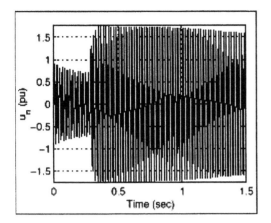

Figure 16. Simulation of the voltage in the neutral point of Powerformer caused by a single line-to-earth fault, fault type F3, followed by a load rejection, refer to Figure 2.

generation unit concerned. If the length of the external cable exceeds a few hundred meters, say 400-500 meters, a cable system with two sets of surge arresters on the line side of Powerformer may be feasible for the total Powerformer installation, refer to Figure 17. The results from use of simulation tools based on models with high accuracy and inputs parameters based on improved statistic material with respect to lightning parameters will influence on the decision making concerning the number of arrester sets which are feasible to be installed on the line side of Powerformer.

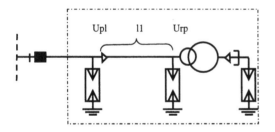

Figure 17. Cable system with two sets of surge arresters on the line side of Powerformer. One of the two sets is located close to Powerformer.

The design solution of the connection point for interconnection of the internal cable, the external cable and the surge arrester located close to Powerformer may vary from installation to installation and the design may range from open air insulated installations to a totally insulated design.

For a refurbishment installation in an underground hydro plant an open-air installation may be feasible as it might be installed in the cell previous occupied by the step-up transformer.

For a new underground hydro plant will an installation with Powerformer, to a larger extent, influence on the civil work and a design solution with a minimum need for space may be preferred.

An under ground powerhouse consists normally of a machinery hall, transformer hall, bus-bar galleries and a tailrace gate hall. The plant design also includes cable and ventilation tunnels and one or more access tunnels to the multi-cavern system constituting the powerhouse. The civil work may also include a certain amount of rock-supports to compensate for internal stresses in the surrounding rock.

A prospective installation with Powerformer reduces the need for space for transformers and busbar galleries and the elimination of one or more step-up transformers automatically reduces the amount of inflammable oil in the plant. Hence it follows that the need of rock-supports and the capacity and the dimensions of the fire protection system and the ventilation tunnels could be reduced.

Less inflammable material and the elimination of the busbar galleries will also increase the personal safety in an underground powerhouse.

Less need for underground galleries and less transformer oil have also a feasible impact on the environment.

Increased efficiency and the consequently improved figures for loss evaluation due to the reduced size of the stator core can also be foreseen.

6 CONCLUSION

The final design of the cable system has an essential impact on the physical size of Powerformer.

Powerformer with cable system has a positive impact on the total plant design, focusing on reduced need for space, increased personal safety and improved environmental impact.

REFERENCES

Anderson R.B. & Eriksson A.J. 1980. Lightning parameters for engineering applications. CIGRE Study Committee 33 Report. Electra, No. 69: 65-102.

Berger K., Anderson B.B. & Kröninger H. 1975. Parameters of lightning flashes. Electra, No.41:23-37.

Holmberg P. 2000. Modelling the Transient Responce of Windings, Laminated Steel Cores and Electromagnetic Power devices by Means of Lumped Circuits. Comprehensive Summaries of Uppsala dissertations from the Faculty of Scince and Technology 592. 97 pp. Uppsala. ISBN 91-554-4877-1. Uppsala: Uppsala University, Tryck & Medier.

507

IEC71-1 1996. Insulation coordination- Part1:Definitions, principles and rules", CEI/IEC 71-2:1996

IEC71-2 1996. Insulation coordination- Part2: Application guide", CEI/IEC 71-2:1996.

IEC 600099-5 1998. Surge arrester- part 5: Selection and application recommendations.

Kimbark E.W. 1956. Power System Stability volume III Synchronous Machines. New York: John Wiley & Sons.

Petersson H.A. 1951. Transients in Power Systems. New York: John Wiley & Sons, Inc., London: Chapman & Hall, LTD.

Willheim R. & Waters M. 1956. Neutral grounding in High-Voltage Transmission.. New York- Amsterdam- London- Princeton: Elsevier Publishing.

Hydropower in the New Millennium, Honningsvåg et al (eds), © 2001 Taylor & Francis, ISBN 90 5809 195 3

Modified design of the power intake at the 660 MW Deriner Dam Project

E.Tanriverdi
Dams and HEPP Division, DSI, Ankara, Turkey

M.Balissat
Stucky Consulting Engineers Ltd., Renens, Switzerland

J.L.Boillat
Laboratory of Hydraulic Constructions, Swiss Federal Institute of Technology, Lausanne, Switzerland

ABSTRACT: The original design for the power intake in front of the 257 m high Deriner arch dam called for a 67 m high concrete tower placed on top of a 143 m deep vertical shaft. Seismic and hydraulic considerations led to a major change in the initial concept of the intake. The shaft was moved away from the dam and a new intake with an inclined inlet structure was designed. Hydraulic model tests were performed to assess that despite the predominantly vertical flow pattern no vortex condition would develop even at exceptionally low reservoir levels.

1 INTRODUCTION

The Deriner Dam Project constitutes the major scheme in the development of the Çoruh River Basin in the north-eastern part of Turkey. Its construction started 1998 with site installation and road diversion works. Beside the Deriner Project there are presently (2001) two other schemes under construction (Borçka, Muratli), while a third one (Yusufeli) has been considered for realisation.

The Deriner Project is located just 5 km upstream of the provincial capital of Artvin. It consists of a 257 m high double curvature arch dam that closes the V-shaped valley and an underground power-house with four units totalling 660 MW. The gross hydraulic head varies between 167 and 212 m.

The geology consists mainly of hard crystalline rock (grano-diorite) crossed by numerous diabase dykes. The site is characterised by a relatively deep fracturing of the rock and by the presence of lateral erosion gorges filled with slope debris (gullies).

The Deriner scheme is completed by two side spillways with frontal overflow sections and circular tunnel chutes. They are dimensioned to pass a flood of 2'250 m³/s, largely in excess of the 100-year flood (1'800 m³/s). For releasing extreme floods, a series of intermediate level outlets have been foreseen in the dam. Their capacity, added to the side spillways, is such that 9'050 m³/s, corresponding to the probable maximum flood (PMF), can be passed safely (Figure 1).

2 REVIEW OF THE POWER INTAKE LAYOUT

2.1 Original layout of headrace

At the previous design stage (final design) it was attempted to have a compact scheme with a minimum length of underground openings and tunnels.

The headrace was conceived as
- a 67 m high free standing hexagonal tower of reinforced concrete with 6 inlets at its base (El. 336), dimensioned for a total discharge of 360 m³/s. The location of the tower was chosen as close as possible to the dam
- a 143 m deep vertical shaft with concrete lining (inner diameter 10 m)
- a 155 m long steel lined stretch of the tunnel used in a first stage for river diversion purposes
- four penstocks leading to the valve chamber and the powerhouse (Figure 2).

Arch dam

Overflow spillway
(2 x 1125 m³/s)

Power Intake   Orifice spillways (8 x 850 m³/s)

Figure 1. View of the Deriner scheme model

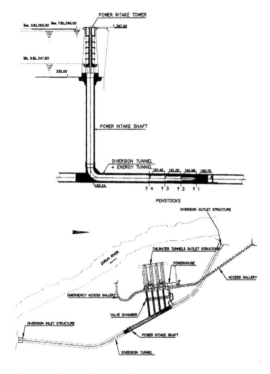

Figure 2. Original layout of the headrace

At the beginning of the construction stage it was decided to
- completely separate the headrace from the diversion tunnel, thus reducing the river diversion operation to a single stage and allowing for more flexibility in the construction schedule, and to
- locate the headrace tunnel at the same level as the valves and turbines axis (El. 180), thus avoiding unnecessary vertical bends in the underground penstocks.

It was furthermore agreed with the Owner to check both the location and the type of structure adopted for the power intake.

2.2 Review of the dynamic behaviour of the intake tower

A reassessment of the seismic conditions at the site indicated that a peak ground acceleration (PGA) $\alpha = 0.2$ g should be assumed for a return period of 475 years, as specified for this type of structure by the Turkish Building Code (Ministry of Public Works and Settlement, Turkish Government, 1998). This value is substantially higher than $\alpha = 0.1$ g assumed previously.

A so-called response spectra analysis (RSA) was then conducted to reassess the dynamic response of the structure to the earthquake. The design spectrum was taken from the Eurocode (Eurocode 8, 1997) and a comparison was made with the spectrum of the Swiss standards (SIA Norm 160, 1989).

The total response of the tower was considered by combining the response of the two horizontal components of ground motion. The vertical component was neglected.

Two extreme load cases were considered:
- earthquake with empty reservoir
- earthquake with full reservoir (max. operating level El. 392)

For the second load case the effect on the tower of surrounding and inside water was taken into account by adding corresponding masses to the lumped mass model (Goyal & Chopra, 1989).

Of particular interest is the overturning moment induced by the earthquake at the base of the tower (see Table 1). Calculations were conducted for both 0.2 and 0.1 g and the results were compared with the previous pseudo-static analysis.

Table 1. Calculated overturning moments at the base of the tower

Type of analysis	PGA	Reservoir	Overturning Moment
RSA	0.2 g	empty **full**	1'040 MNm **1'504 MNm**
RSA	0.1 g	empty full	520 MNm 752 MNm
Pseudo-static	0.1 g	empty **full**	528 MNm **976 MNm**

For the determining condition of full reservoir it can be seen that the overturning moment has been increased by 54%.

To counteract this effect and ensure that no tension condition prevails at the foundation interface, it would be necessary to apply substantial anchoring forces at this level. The increase compared with the previous analysis is in the order of +150%.

Table 2. Required anchoring forces at foundation level (full reservoir)

Type of analysis	PGA	Tensile stress in rock (w/o anchor)	Total anchoring force required
RSA	0.2 g	273 kN/m^2	197.7 MN
RSA	0.1 g	37 kN/m^2	26.5 MN
Pseudo-static	0.1 g	107 kN/m^2	77.3 MN

The additional anchoring requirement (120 MN) should be realised by some 30 additional prestressed anchors of the 400 t class. Such a procedure was felt problematic due to the lack of space and of the difficulty in properly staggering the release of the anchor forces in the rock foundation.

Also a substantial vertical prestressing of the tower itself would be necessary that was not accounted for originally. For these reasons it was decided to search for a completely different solution that offers also better inflow conditions for the intake itself.

2.3 Modified solution with lateral intake

The new power intake takes advantage of the sharply inclined abutment and of the excavation pattern dictated by the dam foundation. Intake structure and shaft are located 40 m upstream of the defunct tower and closer by some 10 m to the dam axis that runs almost parallel to the river course.

The new layout requires less excavation than the previous solution. Furthermore it allows to shift also the headrace tunnel axis towards the river and to compensate the extra-length of the headrace tunnel by shorter underground penstocks (Figure 3).

The 50° inclined inlet consists of 3 major openings measuring 6 by 25 m, followed by a vertical converging stretch connected to the power shaft.

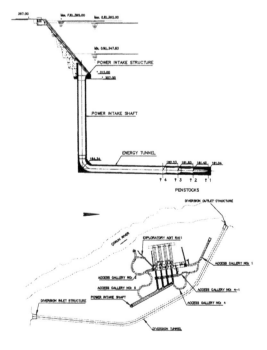

Figure 3. Layout and section of modified headrace

Special care has been given to the concrete forms in order to avoid negative hydraulic pressures and to minimise the head losses in the intake.

The intake is provided with a removable trashrack and a set of stoplogs to be inserted behind the trashrack that allow to empty and inspect the headrace system.

The structure is completed by a gantry crane at the top (El. 397) for picking up and moving trashrack elements and stoplogs (Figure 4).

Considering that the new intake has an essentially vertical inflow pattern and that the units should be operated without problem also at the minimum storage level (El. 347.8), it was decided to evaluate both analytically and physically the possible formation of vortices.

The second task led to the construction and testing of an hydraulic model that is described later on.

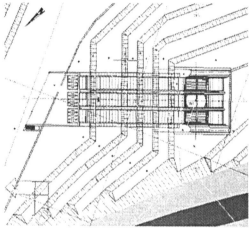

Figure 4a. Layout of modified intake structure

3 INVESTIGATION OF VORTICES

3.1 General

A fluid motion in which the streamlines develop like concentric circles is called a *vortex*. Vortex extending from the free surface down to an intake can lead to air entrainment that can damage hydraulic and mechanical structures. In order to limit head losses and to prevent air entrainment in the intake structure vortex formation should be avoided.

According to rotational energy, vortex can be classified in three types (Graf & Altinakar, 1995 and Sinniger & Hager, 1989):
- rotational or forced vortex
- non-rotational or free vortex
- combined vortex or Rankine vortex

A simply *forced* vortex corresponds to the motion

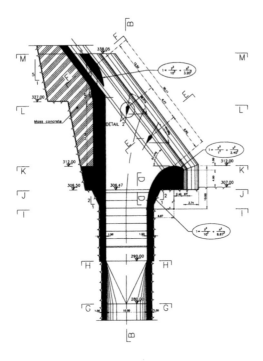

Figure 4b. Vertical section of modified intake structure

of an homogeneous liquid partly filling a cylindrical cylinder of radius R that is rotated at constant angular velocity ω. The free surface obtained is a paraboloid of revolution. The motion of a forced vortex is rotational and it is generated by the transmission of tangential shear stresses (Figure 5).

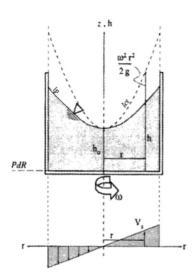

Figure 5. Forced vortex

In the *free* vortex the total head is constant throughout the fluid. Simply this is what happens if the drain in the tank bottom is instantaneously closed and the fluid motion is ultimately dissipated through viscous action (Figure 6).

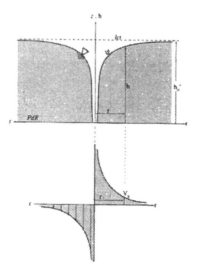

Figure 6. Free vortex

The *combined* vortex flow consists of assuming that a relatively small central portion of fluid is rotating as a highly viscous solid body (*forced* vortex) and that it is combined with a non-viscous *free* vortex region extending radially outwards. The two flow regions are usually matched by a discontinuity in the tangential velocity and by an inflexion point at the water surface profile (Figure 7).

In the central region of the vortex the fluid is assumed to rotate so that the tangential velocity V_t varies linearly with the radius r:

$V_t = \omega \cdot r$

where ω = angular velocity in radians per unit time.

In the free vortex region the velocity varies inversely with the radius and directly with the circulation.

3.2 *Flow behavior at intakes*

In most cases the flow towards a power intake is not uniform. Often the flow is divided up into two flow regions, namely

- primary flow region: the flow from this region moves directly into the intake
- secondary flow region: there is little or no flow to the intake from this region (dead water zone).

It is therefore obvious that very different velocity

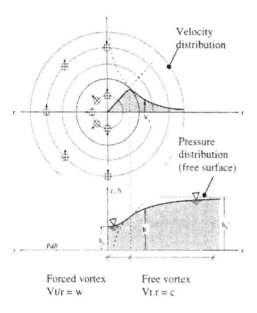

Figure 7. Combined vortex

Forced vortex   Free vortex
$V_t/r = w$ $V_t.r = c$

distributions are occurring in these flow regions. An impulse is generated by the friction at the boundary of the two flow regions that induces a motion in the secondary flow region. If the influence of the primary flow region on the secondary zone is strong enough a vortex can form.

3.3 Classification of vortices

For the present investigation a qualitative classification of the vortices was established according to following criteria:

a) Intensity (Figure 8)

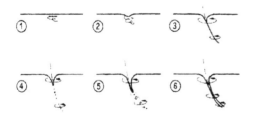

Figure 8. Intensity of surface vortices

1) coherent surface swirl
2) surface dimple / coherent swirl at surface
3) dye core to intake with coherent swirl through-

out water column
4) vortex pulling floating trash, but without air entrainment
5) vortex pulling air bubbles to intake
6) full air core to intake

b) Position and path
The position of the vortex is located in reference to an horizontal square grid (e.g. on model at 10 cm pattern). The observed path of the vortex can be marked with an arrow symbol: ↔

c) Size
● large vortex i.e. Ø > 4 cm on the model
◑ medium vortex Ø 3-4 cm on the model
○ small vortex Ø < 2 cm on the model

d) Time dependency of vortex strength
Distinction can be made between durable vortex, corresponding to more than 50% of observation time, and short time vortex, observed during less than 50% of the time, for given inflow conditions.

3.4 Critical depth of submersion

The critical submergence is generally taken as that submergence at which strong air entraining vortices begin to form. In order to avoid air entrainment by free surface vortices and to minimise swirl entrainment by subsurface vortices an intake must be adequately submerged. Therefore the design of intakes requires assessing with some degree of confidence the required depth of submergence.
The characteristic data of the Deriner power intake are:

Nominal discharge	Q	= 360 m³/s
Shaft diameter	D_p	= 10.00 m
Lowest operating level	W_{Lmin}	= 347.8 m.a.s.l.
Head at start of circular section H_P = 58.83 m		
Shaft section	A_P	= 78.54 m²
Froude number	Fr	= 0.46
Surface tension	σ	= 7.49E-02 N/m²
Kinematic viscosity	υ	= 1.0E-06 m²/sec
Specific weight of water	ρ	= 1'000 kg/m³
Gravity acceleration	g	= 9.81 m/sec²
Asymmetrical approach	c	= 0.72
Symmetrical approach	c_1	= 0.54

The most appropriate formula for determining the depth of submergence in the present case is the one by Jain (Jain et al., 1978), since it refers to intakes with vertical inflow:

$$(h/d)_{cr} = 88 \ Fr^{0.139}(c/\sqrt{g} \ d^{3/2})^{0.723} \qquad (1)$$

Where the Froude number is defined by:

$$Fr = \frac{v}{\sqrt{g \cdot d}} \qquad (2)$$

and d = equivalent diameter of the section.

Depending upon the reference level selected following submergence depths can be calculated:

a) considering the critical submergence from the beginning of the circular section
 - asymmetrical approach: h_{cr} = 22.5 m
                         W_{Lcrit}=302.5 m a.s.l.
 - symmetrical approach: h_{cr} = 18.3 m
                         W_{Lcrit}=298.3 m a.s.l.

b) considering the critical submergence from the intake section at floor level in front of the intake
 - asymmetrical approach: h_{cr} = 18.4 m
                         W_{Lcrit}=325.4 m a.s.l.
 - symmetrical approach: h_{cr} = 14.9 m
                         W_{Lcrit}=321.9 m a.s.l.

Compared with the minimum operating level (347.8 m a.s.l.) the calculation with Jain's formula suggests that there will be no detrimental vortex formation.

4 HYDRAULIC MODEL

4.1 Hydraulic similarity criteria

The dynamic and kinematic similarity of open channel models is, in general, adequately described by the Froude number, i.e. by the ratio of the inertial and gravitational forces. The interpretation of the measurements was based on the assumption that this ratio is the same in the physical model and in the prototype.

Froude similarity criteria can be written as:

$$F_m^2 = \frac{v_m^2}{g \cdot L_m} = \frac{v_p^2}{g \cdot L_p} = F_p^2 \qquad (3)$$

where: F   : Froude number   [-]
         v   : characteristic flow velocity [m/s]
         L   : characteristic length of the wet
             section   [m]
         g   : gravity acceleration   [m/s^2]
         m   : model reference
         p   : prototype reference

According to Froude similarity the scale factors of the most important parameters become:

Table 3. Scale factors of the main physical parameters

Physical value	Scale relation	Scale factor
Length [m]	$L_p/L_m = \lambda$	40
Velocity [m/s]	$v_p/v_m = \lambda^{1/2}$	6.32
Discharge [m^3/s]	$Q_p/Q_m = \lambda^{5/2}$	10119
Froude number [-]	F_p/F_m	1
Reynolds number [-]	$R_p/R_m = \lambda^{3/2}$	252.9

4.2 Scale effects

A scale effect exists if the model behaviour does not satisfy the Froude number as unique similarity criterion.

Concerning vortices different authors put in evidence the necessity of respecting additional similarity conditions:

a) Influence of Reynolds number
 According to a study on vortex flow through orifices (Daggett and Keulegan, 1974) viscous effect can be neglected for
 Re = $Q/(v \cdot d) > 2.5 \cdot 10^4$
 where Q = discharge
         v = kinematic viscosity of the liquid
         d = diameter of the orifice
 Other authors (Anwar et al., 1978, and Anwar & Amphlett, 1980) noted that the vortex formation in horizontal and vertical intakes is unaffected by the reduction scale when the radial Reynolds number satisfies
 Rr = $Q/(v \cdot h) > 2.0 \cdot 10^4$
 where h = submergence of the intake (reference axis or level)

b) Criterion by Jain et al.
 Jain et al. (1978) recommend that the model should be built as per Froude number similarity to any desired scale. They developed a correction factor that enables converting results of critical submergence obtained in the model to the prototype if the Reynolds numbers in the model and in the prototype are known.
 According to Jain et al. the correction factor becomes unity at high Reynolds number, or, more precisely, when
 $(g^{0.5} \cdot d^{1.5})/v > 5.0 \cdot 10^4$

c) Influence of Weber number
 By experimenting with liquids of different surface tensions Jain et al. showed furthermore that vortex formation in case of vertical intake is unaffected by the so called "intake" Weber number (We) as long as

We = $Q/A \cdot \sqrt{((\rho \cdot d)/\sigma)} > 11.0$

Other experiments (Anwar et al., 1978) indicated that vortex formation is also independent of the so called Weber number (Wr) if

Wr = $Q^2 \cdot \rho \cdot h / A^2 \cdot \sigma > 10^4$

where σ = surface tension

Applying the parameters of the Deriner power intake (see 3.4) it was concluded that a model scale of 1:40 has to be used for satisfying all the conditions mentioned above. The determining condition in this case is given by the Weber number Wr.

4.3 Laboratory model and testing

The hydraulic model at 1:40 scale is represented in Figures 9 to 11. It was build-up at the LCH laboratory hall of the Swiss Federal Institute of Technology and placed in a closed basin of dimensions 3.0 x 2.1 x 1.1 m (W x L x H). A valve at the entrance of the basin controls the input discharge. Another gate at the end of the flow circuit controls the outflow. When both discharges are equal the system is in steady conditions and ready for testing. Since this steady condition is difficult to achieve, reference measurements were taken throughout the tests.

Testing was conducted for the combination of the following conditions:

- Reservoir levels at 345 m.a.s.l. (corresponding to min. operating level 347.8 m.a.s.l.), 340 m.a.s.l. and 335 m.a.s.l.
- Discharges of 320 m³/s (corresponding to turbine capacity at min. operating level), 360 m³/s (max. capacity of plant) and 400m³/s.

Each test was performed in a sequence of lowering and then raising the water level. Vortex characteristics were observed during a period of 30 minutes.

4.4 Results on vortices tests

The testing results are summarised in Table 4.

From these series of tests it can be concluded that there is no critical situation regarding the vortex formation, not only for the *normal* case (reservoir level at 345 m.a.s.l., Q = 320 m³/s), but also for all other cases. Even with a 25% higher discharge than the design one and for a reservoir level 10 m below min. operating level, no serious vortex formation could be observed.

The new design of the power intake can be considered as very favourable for avoiding the problem of vortex.

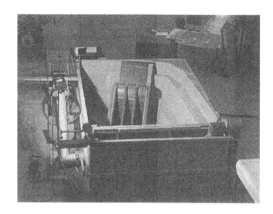

Figure 9. General view of model set-up in the laboratory

Figure 10. Front view of the intake model

Figure 11. Vertical view of the intake model

515

Table 4. Experimental results on vortices formation

Discharge Q (m³/s)	Water level W_L (masl)	Type (1)	Size (1)	Persistence (% of time)
320	345.00	Type 1	o	< 50%
320	340.00	Type 1-2	o	< 50%
320	335.00	Type 2-3	◑	< 50%
		Type 2-3	◑	< 50%
360	340.00	Type 1	o	< 50%
360	335.00	Type 1 +	◑	< 50%
		Type 2-3	◑	< 50%
400	345.00	Type 1	o	< 50%
400	340.00	Type 2-3	◑	> 50%
		Type 1	o	< 50%
400	335.00	Type 2-3	◑	> 50%
		Type 2-3	◑	< 50%

(1) Note: According to 3.3 classification of vortices

4.5 Pressure measurements

A series of *static* and *dynamic* pressure measurements were performed at min. operating level (347.8 m.a.s.l.) and design discharge of 320 m³/s in 14 points distributed along the intake walls between El. 312 (inlet sill) and El. 252 (circular shaft).

The total head loss was found to be 19 cm at the lowest point (no 14). Since the wall roughness of the hydraulic model is smoother than the prototype, the head losses in the prototype will be somewhat higher by 10%. Furthermore the trashrack to be installed in the prototype has not been modelled in the present investigation. It will induce some further head losses in the prototype.

Dynamic pressure measurements were performed using a membrane probe with an accuracy of 0.1 mm at a sampling frequency of 100 Hz. The data were analysed statistically to provide detailed information about pressure fluctuations.

It was found that the standard deviation, expressed in m of hydraulic head of the prototype, was varying between 0.05 and 0.37 (with one exceptional value at 0.75). Furthermore the energy spectrum of the signal was computed for each reading sequence of 5 sec. This spectral analysis revealed a forced vibration induced by the supply pump of the model at a frequency of about 10 Hz. No other vibration perturbation could be observed at any measurement point.

It can be concluded that no negative pressure has to be expected along the walls of the intake structure. Therefore any risk of cavitation can be excluded.

REFERENCES

Anwar H.O. & Amphlett M.B. 1980. Vortices of vertically inverted intake. *IAHR Journal of Hydraulic Research, Vol. 18 No2.*
Anwar H.O., Weller A. & Amphlett M.B. 1978. Similarity of free vortex at horizontal intake. *IAHR Journal of Hydraulic Research, Vol. 16 No 2, pp. 95-105.*
Daggett L.L. & Keulegan G.H. 1974. Similitude in free surface vortex formations. *ASCE Journal of the Hydraulics Division 100, HY11.*
Eurocode 8 1997. *Conception et dimensionnement des structures pour la résistance aux séismes.* Collection Bâtiments.
Goyal A.& Chopra A.K. 1989. Earthquake Analysis and Response of Intake Outlet Towers. *Report No UCB/EERC-89/04.*
Graf W.H. & Altinakar M.S. 1995. *Hydrodynamique.* Traité de Génie Civil / Vol. 14, Swiss Federal Institute of Technology, Lausanne.
Jain A.K., Ranga Raju K.G. & Garde R.J. 1978. Vortex formation at vertical pipe intakes. *ASCE Journal of the Hydraulics Division 104, HY10.*
Ministry of Public works and Settlement, Turkish Government 1998. *Specifications for Structures to be Built in Disaster Areas, Part III / Earthquake Disaster Prevention: Seismic Design of Build*ings.
SIA Norm 160, 1989. *Actions sur les structures porteuses.*
Sinniger R. & Hager W.H. 1989. *Constructions hydrauliques/ Ecoulements stationnaires.* Traité de Génie Civil / Vol. 15, Swiss Federal Institute of Technology, Lausanne.

Hydropower in the New Millennium, Honningsvåg et al (eds), © 2001 Taylor & Francis, ISBN 90 5809 195 3

More power from the Mississippi River: upgrading the Ford Hydroelectric Station

G.M.Waldow
HDR Engineering, Minneapolis, MN, USA

B.J.Bystrom
Ford Motor Company, St. Paul, MN, USA

J.H.Rohlf
Ford Motor Company, Dearborn, MI, USA

ABSTRACT: The Ford Hydroelectric Project was constructed in 1924 to provide economical power for a vehicle assembly plant. After many decades, electrical demand had increased and the original units were unable to generate at rated capacity. Ford initiated a multi-year upgrade and refurbishment program that included new Francis runners and other improvements. This program was creatively cost-justified and successfully implemented in the industrial sector. Many challenges were encountered, but the investment has been rewarded with increased capacity and energy production. The lessons learned during this program will be of interest to others who are planning to rehabilitate or upgrade older hydropower facilities.

1 INTRODUCTION

Many hydroelectric projects have been developed by industrial interests to provide low cost energy for manufacturing purposes. For example, the pulp and paper industry has been particularly well served through the years by its substantial investment in hydropower.

Regardless of routine maintenance policy, all hydroelectric generation facilities eventually become so badly worn that they require major maintenance and rehabilitation. These programs typically involve lengthy outages, complete disassembly of the units, thorough inspection and evaluation of all plant and equipment, and repair or replacement of worn components. The usual objective of a rehabilitation program is to restore civil works and generating equipment to original specifications and performance levels. A well-planned program can extend project operating life for many years.

When implementing a major rehabilitation program, it is often possible to upgrade older generating equipment to improve efficiency and increase power production. Due to industry advancements in recent years, there may be a range of upgrade options available at a given site. To fully optimize facility upgrade potential, the cost of each reasonable option should be evaluated against the benefits which would accrue to the owner over the extended life of the facility. Water rights, hydrology, existing power contracts, regulatory permit issues, alternative power sources and other factors may also have to be considered in the evaluation.

2 CAPITAL-BUDGETING

Hydroelectric improvement programs are relatively costly and normally cannot be accomplished using annual maintenance budgets. In most manufacturing industries, hydropower improvements are classified as capital projects and must therefore compete with all other significant projects for limited investment funds. Individual companies utilize many different methods for evaluating proposed capital improvement projects and making investment decisions. Many corporations have established formal capital-budgeting procedures to rank proposed projects and select for funding only those which are most beneficial to the company's interests.

Virtually every company has a minimum-attractive-rate-of-return (MARR) below which it will not invest. Most companies also have a practical upper limit on the total magnitude of capital investments in a given period. These criteria form the general boundaries for the capital-budgeting process. Some firms also require recovery of their capital investments within a limited time period.

Detailed discussion of capital-budgeting methods is beyond the scope of this paper. However, the fundamental process involves evaluating proposed project costs and benefits using discounted cash flow techniques. These may include internal-rate-of-return, net-present-value or profitability index methods. Appropriate adjustments are made for taxes and depreciation and the projects are ranked in descending order of economic attractiveness. Those projects which do not meet the firm's threshold feasibility criteria are eliminated from further consideration. The remaining projects are then approved for funding in order of rank until a budgetary limit is reached.

There are other considerations which companies may use to change rankings or selectively combine projects for funding. Some of these considerations include management intuition, corporate growth policies, economic conditions, perceived investment risk, and rationing of capital.

3 COST-JUSTIFICATION AT FORD

Ford Motor Company employs a modified internal-rate-of-return (IRR) technique to evaluate proposed capital projects. Corporate accounting determines a time-adjusted-rate-of-return (TARR) for each project depending on its specific characteristics. The TARR enhances the basic IRR technique with respect to time-sensitive variables, taxes, depreciation, labor rates, geography, and other factors which may apply to individual projects.

Proposed projects may be formulated by any authorized "sponsor" in the company and submitted to corporate management for evaluation and funding approval. Sponsor documentation for funding requests must clearly describe the proposed project and explain why it would be a desirable investment for the company. The proposed "cost-justification" argument must be supported by appropriate technical analyses and credible estimates of annual costs and benefits.

Ford maintains a proprietary investment analysis program on its wide area network. This allows individual sponsors to enter spreadsheet data and rapidly determine the TARR for their candidate projects. Accounting manages the program and frequently updates the many internal variables on which TARR calculations are based. This keeps the analysis program current for everyone and makes cost-justification less complex for the many individual project sponsors within the company.

The minimum TARR required to obtain project funding ranges from 20 to 50 percent at Ford. In effect, this means that most projects need to recover their investment costs in four years or less. Even with the company's formidable size and financial strength, there are established limits on the total dollar value of qualified projects that can be funded during a fiscal year.

These relatively demanding evaluation criteria represented a significant challenge when attempting to cost-justify and fund the rather extensive hydroelectric improvement program described in the following sections.

4 THE FORD HYDROELECTRIC PROJECT

Ford Motor Company has owned and operated a hydroelectric station on the Mississippi River at St. Paul, Minnesota for over 75 years. The project was constructed at an existing navigation dam in 1924 to provide power for a new automotive assembly plant overlooking the river. The hydroelectric facility was sized to economically utilize available head and river flow to produce sufficient energy for assembly operations. As a strong advocate for hydroelectric power, Mr. Henry Ford Sr. was personally involved in the site selection and general design of his company's St. Paul facilities.

A 175 m long fixed-crest concrete spillway connects the powerhouse with navigation locks at the opposite end of the dam. The dam functions as a fixed weir to maintain minimum depths in the upstream navigation pool. A 0.6 m high flashboard system extends across the spillway. Eight small sluice gates were provided to pass water through the spillway. Federal Lock and Dam No. 1 is owned, operated and maintained by the U.S. Army Corps of Engineers. However, Ford is responsible for operation and maintenance of three spillway sluice gates and the flashboard system. The hydroelectric station is operated in a run-of-river mode using water that would otherwise spill over the dam.

The concrete and masonry powerhouse was constructed with four vertical Francis turbines. Each turbine was rated for a maximum output of 3650 kW at a head of 10.4 m. The directly coupled generators have a nameplate rating of 4500 kVA at a 0.95 power factor. Power is generated at 13,800 V and fed directly to the assembly plant through cable tunnels. There are no external substations or transmission lines.

The station is manually operated and staffed 24 hours a day, seven days a week. Ford personnel perform all routine maintenance. Specialty contractors are hired for major maintenance, replacements, and improvements.

The hydroelectric station has historically been a reliable source of low cost electricity for Ford's Twin Cities Assembly Plant. The local electric utility purchases all excess hydroelectric generation and provides supplemental power as required. Annual production of nonpolluting energy averaged nearly 79 million kilowatt-hours between 1925 and 1992.

The Ford Hydroelectric Project is licensed by and operated under the jurisdiction of the Federal Energy Regulatory Commission (FERC). This agency licenses nearly all non-federal hydroelectric projects in the United States, including non-federal generating facilities located at federal dams.

Figure 1 is a view of the government spillway and Ford powerhouse from above the navigation locks. The vehicle assembly plant is visible in the background. Figure 2 is a view of the station generator floor.

5 HYDROELECTRIC IMPROVEMENT PROGRAM

A major expansion of the vehicle assembly plant in 1985 resulted in Ford's peak electrical demand exceeding its hydroelectric generating capacity for the first time. In addition, the hydroelectric plant had

Figure 1. The Ford Hydroelectric Station.

Figure 2. The Ford Station Generator Room.

been operating for over 60 years and performance of the original units was measurably deteriorated. The cost of purchasing external energy was significant. Engineering studies were initiated to identify economical options that would increase internal electrical generating capacity. Rehabilitating and upgrading the hydroelectric station was determined to be the best option available.

The hydroelectric rehabilitation concept evaluated several alternatives including: rehabilitation of the existing generating equipment to achieve original performance levels; installation of modern Francis runners in either two or four of the existing units; removal of one existing Francis unit and replacing it with a new vertical Kaplan unit; and installation of a fifth generating unit in the small, but vacant, original hydraulic exciter bay.

Comparative benefit cost analysis indicated that installation of new runners would be the most attractive option. The existing generators had been rewound with additional copper and improved insulation in the late 1960's. With minor modifications, they were believed capable of accepting up to 4850 kW from the turbines. With this value as an upper limit for turbine output, other items such as main shaft integrity and draft tube limitations were analyzed. Enlarging water passage geometry was explored and found unacceptable due to high cost, construction issues associated with structural modification, and FERC license complications. Any improvement project would have to work with the existing structure of the powerhouse.

5.1 Phase 1 Improvements

Based on detailed engineering analysis and further input from turbine vendors, it became evident that upgrading even two units to 4850 kW was cost prohibitive. In addition to the draft tube limitations, there remained some doubt as to whether the generators could be reliably operated beyond their rated capacity. It was therefore decided to limit turbine output to 4305 kW. Upgrading only two of the runners was all that could be initially cost-justified. This was due to the rigorous economic evaluation criteria and the impact of declining flow availability for each additional upgraded unit. In very rough terms, the upgrade project needed to pay for itself in just three years.

Phase 1 of the improvement program therefore emerged as replacement of two of the original cast iron runners with 4305 kW stainless steel runners. It was hoped that replacement of the two remaining original runners could be cost-justified in the future.

5.2 Regulatory Issues

Ford is assessed substantial annual charges by the federal government to pay for FERC administration costs and for headwater benefits provided by the Corps of Engineers' dam. The amount of these annual charges is partially based on installed station capacity. The FERC defines station capacity as the sum of generator nameplate capacities. As such, it was desirable not to exceed the "official" installed station generating capacity specified in the FERC license.

For reasons not known today, the original turbine full gate capacity was substantially less than the direct-coupled generator capacity. In addition, the turbine nameplate capacity was based on the most efficient gate setting. The proposed 4305 kW replacement runners would not cause the original generator nameplate capacity to be exceeded. Thus there would be no recognized increase in station capacity with respect to the project license

The proposed project improvements would increase station hydraulic capacity by 17 m^3/sec. This additional flow is available nearly 40 percent of the time and has historically been spilled over the dam. Utilization of this "surplus" water during high flow conditions presented no significant environmental impact.

Since the proposed unit modifications involved no new capacity and were not substantial in scope or probable impact, FERC ruled that the license amendment process could consist of a single consultation stage. In addition, agency consultation was to focus exclusively on proposed changes and new or incremental impacts; the process was not to address or reopen issues related to the existing licensed facility. These two factors had a major impact in compressing the overall project implementation schedule.

The primary consultation issues involved convincing resource management agency staff that the proposed runner changes would have absolutely no impact during low flow conditions and that the project would continue to operate in a run-of-river mode. A non-capacity license amendment was subsequently issued in a timely manner by the FERC.

5.3 Vendor Selection

A specification package was prepared for competitive tendering. The base scope of work included the design, fabrication and complete installation of two 4305 kW stainless steel Francis runners. A dewatered inspection of one of the units was conducted during the pre-tender meeting. Since other parts of the machines could also be in need of attention, venders were requested to submit an itemized price list for replacing various turbine components.

All tenders originally opened exceeded the available project funding. Ford then requested two of the venders to submit cost reduction proposals. Each vender approached the cost reduction process differently; however both lowered their unit output guarantees. Items such as daily clean-up, crane service for shipping, and miscellaneous painting were deleted from the scope of work or assumed by Ford. Normally the low qualified tender receives the award. However in this case, with independent approaches to reduce cost, the final selection decision was based on a ratio of guaranteed unit output versus total unit cost. The contract was ultimately awarded to Voith Hydro.

5.4 Unforeseen Problems

The Phase 1 project was cost-justified and funded solely on the basis of turbine runner upgrades. In addition to new runners, the contracted scope of work included turning the wicket gate stems, line boring the head cover and bottom ring, and installing new greaseless bushings for the wicket gates. Unfortunately, there was little money held in reserve to address contingencies.

The wicket gates themselves offered the biggest surprise. Clearances found in the lower bushings were 2.54 mm; original design clearances were 0.05 mm. Also, the gate stems had extensive sub-surface cracks, many of which were halfway through the stem. The hollow gates also had cracking on the vertical surfaces of the leaf along the sealing areas. Close comparison with the original design drawings revealed that the gates had actually been cast upside down. The two masses of material to support the stems were reversed, which placed the larger mass at the bottom for the shorter bottom stem, and the smaller mass at the top of the gate leaf. The best available option was to cut off the old gate stems and replace them with new stainless steel stems. The added cost for this work approached 40 percent of the original contract amount. In addition, the project was delayed for several weeks until more funding was authorized.

The single set of station intake bulkheads would have to be committed for several months to rehabilitation work on the first unit. As a precaution, the other three units were sequentially dewatered, inspected and returned to service prior to setting the bulkhead panels and disassembling the first unit. However, the fact that the station had only one set of upstream bulkheads nearly proved devastating.

After the first unit was disassembled and the head cover sent out for machining, a problem arose on one of the other units. The wicket gates could not be closed. This had never happened before in the history of the project. A "clicking" sound could be heard through the head cover, indicating something was probably lodged against the runner. Attempts to identify which wicket gate was stuck resulted in six of the sixteen gate links breaking in rapid succession. Six wicket gates had rotated over-center and flow into the unit could not be stopped because the bulkheads could not be removed from the disassembled unit. The problem unit was on-line at 95 percent gate and performing fine. However there was real concern that the unit would somehow trip off-line and go to runaway. To further complicate the situation, control-wiring changes from years past had eliminated the ability to stop excitation by opening a switch. This meant a unit going to runaway would still have excitation.

The solution to stopping water flow was to fabricate and install a new set of bulkheads. The two bulkhead panels are 4.3 m wide and 6.7 m high. Design, fabrication and installation of the bulkheads were accomplished in just sixty hours. To overcome installation problems during full flow, the new bulkheads were made in eight sections (four per bay) and then installed one at a time. Each section was only 1.7 m high, allowing the water pressure to equalize somewhat before sliding each subsequent section into position. The top sections required the use of hydraulic jacks to lower them into final position.

With flow reduced to extensive leakage, the unit went into a motoring condition. Sand bags and cinders were then used to further reduce leakage. The undamaged wicket gates were closed and the unit finally stopped rotating. The "clicking" sound ceased when the unit stopped turning. Eventually the leakage was controlled and the unit was dewatered. Inspection revealed that a 0.15 m diameter log had somehow lodged between two buckets and had been rotating against the wickets. Naturally this damaged unit was scheduled for immediate rehabilitation and upgrade.

Another surprise was the status of unit alignment. In both units, the stator had to be moved to bring the system into final alignment. There was no evidence of shifting, only speculation about the original installation. Non-symmetrical bearing wear had been observed on one of the units during maintenance.

5.5 *Phase 1 Results*

Generator output of each original unit was carefully measured at different gate settings using existing plant instrumentation. The deteriorated condition of the original units was obvious; performance was well below original commissioning levels. Similar measurements were then made with each new runner in place. Since Ford was interested in relative (as opposed to absolute) unit improvements, more sophisticated testing was not required.

Measured full gate kW output increased by over 40 percent with the new runners and reworked wicket gates. Unit efficiency also increased significantly throughout the normal operating range of the machines. Predicted performance objectives had been achieved. Figure 3 illustrates comparative performance, at available test head, for the first unit upgraded.

strated that the incremental cost for fabrication and installation of two improved runners could be justified by the incremental value of increased energy production. This conclusion was possible only because all costs for mobilization, disassembly, rehabilitation, new wicket gates, and reassembly would be paid for under the contingency account. Ford management accepted a blended cost account approach and authorized the necessary capital funding for the replacement runners.

5.7 *Phase 2 Results*

Rehabilitation and upgrade work was successfully completed on the second pair of units during 1994-95. Before and after performance testing again confirmed predicted unit output improvements. Figure 4 illustrates the measured increases in total plant output achieved with Phase 1 and Phase 2 improvements.

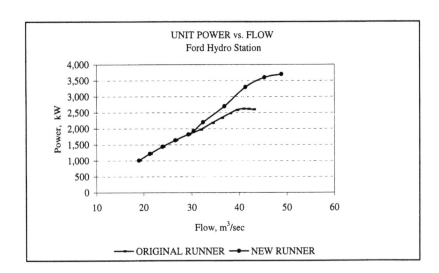

Figure 3. Measured increase in single unit output.

5.6 *Phase 2 Improvements*

Because of the cracked wicket gate stems discovered in the first two units and the potential for catastrophic turbine failure, it was highly advisable to tear down the second pair of units for inspection and probable replacement of the gates. These non-optional repair and rehabilitation costs would be funded from a corporate contingency account.

Based on the successful performance results obtained in Phase 1, there was strong incentive to similarly upgrade the two remaining original units with stainless steel runners. Updated analyses demon-

Unfortunately, Ford's improvement efforts have not been completely successful. During the wicket gate rebuilding process, Teflon coated greaseless bushings were installed using original machine design clearance specifications. Within 18 months, the rebuilt gates started to seize. The problem was due to the Teflon surfaces absorbing water and swelling over time. Contingency funds were used to change out the failed bushings. The replacement bushings were designed with numerous Teflon plugs embedded in their surfaces to provide lubrication and limit potential swelling problems.

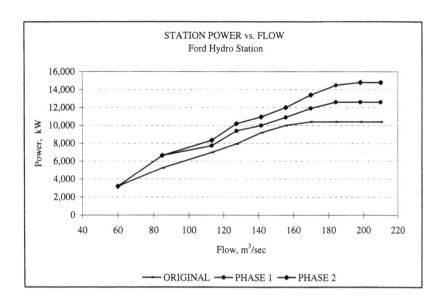

STATION POWER vs. FLOW
Ford Hydro Station

Figure 4. Measured increase in total station power by phase.

5.8 *Phase 3 Improvements*

The original hydroelectric project was constructed with a submerged weir across the tailrace channel. The apparent purpose of the weir was to assure draft tube submergence during low tailwater conditions. The structure was composed of stacked 0.5 m³ concrete blocks. Minimum project tailwater level had been increased slightly in the early 1930's by reconstruction of a downstream navigation dam. However the submerged weir remained in place and negatively impacted net generating head at the project for nearly 70 years.

A 1993 engineering study indicated that removing the weir would measurably increase average annual generation. Analysis showed that lowering the local tailwater and thus increasing net head would not cause cavitation problems in the turbines. Unfortunately, weir removal could not be cost-justified as a stand-alone improvement project because of difficult site access and the high cost to mobilize heavy construction equipment. This constraint caused the weir removal project to be temporarily shelved.

Ford is responsible for maintaining a 0.6 m high flashboard system on the 175 m long concrete dam. The bottom-hinged steel flashboard sections would typically collapse once or more each year when overtopped during flood events or from ice in winter. They could only be raised again when water level dropped below the spillway crest. This dangerous job had to be accomplished by Ford personnel working from small boats. Depending on flow conditions in the river, it was often months, and occasionally over a year, before the boards could be raised again. Loss of the 0.6 m flashboard head had significant energy production impact on machines rated for a net head of only 10.4 m.

Engineering studies showed that an inflatable flashboard (gate) system that could be lowered and quickly raised following a high flow or ice passage event would modestly improve annual generation. An inflatable flashboard system would require only minor concrete modifications and would not impact navigation, spillway capacity, or dam safety. These factors were critical for the concept to be accepted by the Corps of Engineers. Air and condensate piping could be installed in a passageway through the spillway. The small blowers and PLC-based automatic control system would require little space in the powerhouse. In addition to being cost-justified by increased annual generation, the proposed flashboard system would eliminate a hazardous task for Ford maintenance personnel.

Further analysis determined that installation of the inflatable gate system and removal of the weir could be combined and cost-justified as a single head enhancement package. It should be noted that this was not a case of "beating the system" by bundling two projects together in order to get an uneconomical one funded. The required mobilization of a heavy crane and construction barge for work on the dam represented a legitimate change in circumstance that was sufficient to justify weir removal during the same construction contract.

5.9 *Phase 3 Results*

The entire Phase 3 head enhancement package was approved for funding and implemented during 1995-96. Installation was complicated by high river flows during construction. However, the program was a complete success. Annual project generation has increased by approximately five percent as a result of these improvements. Figure 5 shows the flashboard system holding back the partially frozen Mississippi River. A large tree is harmlessly depressing a small segment onto the concrete spillway.

Another current improvement involves excavation and stabilization of the rock bluff above the powerhouse. This work was required by the FERC as a safety measure. Approximately 2150 m^3 of undermined limestone and sandstone base will be removed. Following construction, an approved planting program will be implemented to prevent erosion and preserve the scenic value of the site.

A significant non-structural effort underway is the preparation of an application for a new FERC li-

Figure 5. Project spillway with inflatable flashboard system.

5.10 *Phase 4 Improvements*

The current phase of the Ford hydroelectric improvement program includes installation of an automated trash rake in the intake bay. This rail-mounted system was recently added without compromising the architectural features of the historic powerhouse.

A second project involves the installation of a fully automated project monitoring and control system. The near-term objective of the system is to optimize the overall generating efficiency of the facility. Improved instrumentation with real time monitoring of generator and environmental parameters will enable implementation of predictive maintenance planning. A longer-term objective is to reduce station labor costs by having operators on site only one shift per day. Ford has selected a system vendor and is working to finalize the system design specifications. The system will interconnect with the government locks to better coordinate run-of-river station operation with infrequent navigation lock activities.

cense for the facility. This complex multi-year process involves broad consultation with numerous resource management agencies, special interest groups, and the general public. The applicant is required to explain project operations, financials, environmental impacts, and to justify why it should continue to operate the facility. The applicant is also obliged to perform any reasonable studies the resource agencies determine they need to evaluate the hydro station's impact.

Ford has agreed to perform a one-year fish entrainment and mortality study at the powerhouse. This effort will involve special netting equipment and procedures. Evaluation of captured fish will determine the impact of the turbines on different species and sizes of fish. Following review of the study findings, Ford will work with the resource agencies to determine appropriate mitigation measures.

The station is once again experiencing sticking

and binding of the wicket gates on all units. Investigation has revealed that several hydroelectric stations in the region have had similar problems. It appears that local water chemistry and particulate material can result in swelling of greaseless bushings and clearance cementation.

The problem has been resolved at other hydroelectric stations by increasing clearances to avoid particulate retention and allow for bushings to swell. Ford will soon initiate line boring of the wicket gate bushings to open the tolerances from 0.25 mm to 0.75 mm

Phase 4 projects are being funded with a combination of capital project funds and Ford's internal equivalent of generation revenues.

5.11 Future Improvements

It is difficult to envision an end to all desired and required improvements. For example, at some future date Ford still hopes to cost-justify and install solid-state exciters on the units.

Ford is certain that several needs or required enhancements will be identified during the current relicensing process. Potential projects could include additional public safety measures such as warning signs, lighting, and audible alarms warning of tailwater flow changes. Several resource management agencies have indicated that they expect Ford to construct or finance various recreation enhancements, interpretive displays, and public access improvements on public lands surrounding the Corps of Engineers' navigation impoundment. These issues are being further clarified in discussions with the agencies.

Depending on the results of the turbine entrainment and mortality study, structural changes or fish diversion systems may be required at the station intake. Costs for these items could be prohibitive since they may also reduce energy production.

Somehow funding for these and other unidentified future improvements will have to be found. Managing the Ford Hydroelectric Project will never become boring.

6 CONCLUSIONS

Ford's experience demonstrates what can be accomplished when an ageing hydroelectric facility is upgraded. Performance improvements can be dramatic, but so can the unexpected problems that are encountered.

Substantial turbine performance improvements are possible with modern design and manufacturing technology. Turbine vendors can reliably estimate incremental generation improvements. These estimates can then be used for project analysis and justification.

Owners undertaking hydroelectric project rehabilitation should be prepared for unforeseen problems. They should have contingency procedures and funds in place to react quickly and efficiently. Projects of this nature are rarely completed without a few surprises. The obstacles should not prevent industrial owners from pursuing hydropower upgrades. The economic and environmental benefits obtainable by optimizing this renewable energy resource far outweigh the challenges presented.

The comprehensive Ford hydroelectric improvement program was cost-justified based on a combination of benefits including: purchased power displacement, increased power sales, project life extension, worker safety benefits, and reduced maintenance costs. In addition, it took creative account utilization, strategic bundling of subprojects, and selective scheduling to get the program funded and implemented. The creative methods used and the successful results obtained in Ford's program may be of value to others who are planning to rehabilitate and upgrade ageing hydroelectric facilities.

REFERENCES

Heitger, L.E. & Matulich, S. 1980. *Managerial Accounting.* USA: McGraw-Hill.

Rohlf, J.H. & Waldow, G.M. 1993. Turbine Improvements at the Ford Hydroelectric Project. *Proc. Waterpower 93,* Nashville, 9-13 August 1993. New York: ASCE, Vol. 1. 720-727.

Rohlf, J.H. & Waldow, G.M. 1997 Cost-Justifying Project Improvements in the Industrial Sector *Proc. Waterpower 97,* Atlanta, 5-8 August 1997. New York: ASCE, Vol. III. 1650-1658.

Tarquin, A.J. & Blank, L.T. 1976. *Engineering Economy.* USA: McGraw-Hill.

Hydropower in the New Millennium, Honningsvåg et al (eds), © 2001 Taylor & Francis, ISBN 90 5809 195 3

Stochastic boundary element computation of the reliability of gravity dam

Ming Zhang & Qinggao Wu
Department of Hydraulic Engineering, Tsinghua University, Beijing 100084, P.R. China

ABSTRACT: Based on the method of obtaining the partial derivatives of basic random variables, stochastic boundary integral equations are established, and practically, the method is applied to resolve the engineering problems of concrete gravity dams, considering the random factors including material parameters of dam body, water level of upper reaches, anti-slide friction coefficient of dam base, etc. Numerical examples show that the stochastic boundary element method applied in the paper, to compute the reliability degree of grand volume constructions such as concrete gravity dam, comparatively has the advantage of inputting fewer data and its computational results are more precise.

1 INTRODUCTION

In the process of construction reliability analysis, in particular, to resolve the engineering problems of complex or large volume constructions, random stress-strain field and reliability degree must be computed by numerical computation methods. Stochastic finite element method (Gautam, D. 1992) and stochastic boundary element method (Burczynski, T. & Skrzypczyk J. 1999, Kaljevic, I., Saigal, S. 1993, Nakagiri, S., Suzuki, K. & Hisada, T. 1983, Weidong Wen, Wei Chen & Deping Gao. 1995, Yongjian Ren, Aimin Jiang & Haojiang Ding. 1993)are powerful resorts for the reliability computation of large volume constructions. In comparison with finite element method, boundary element method has unique advantages in the resolution of some special engineering problems, such as boundary value problems, infinite or semi-infinite field problems and singularity problems, etc. Hence, in the paper, a kind of stochastic boundary element method is presented in the basis of two-dimension elastic stochastic boundary integral equations to resolve the engineering problems.

2 COMPUTATION OF STRUCTURAL RELIABILITY

In the analysis of structural reliability, a set of basic random variables $X = (x_1, x_2, \cdots, x_n)^T$ are assumed, and let the failure function of considered structure be $g(X)$, so limit state equation can be written as $g(X) = g(R, S) = g(x_1, x_2, \cdots, x_n) = 0$, where the resistance R and the action S will be in terms of these basic random variables. In general, R is a explicit function of basic random variables. But, to average large volume structures, S depends on the transformation relation between structure load and load effect. It may be obtained by numerical computation of load effect.

An arbitrary set of basic random variables X can be transferred to be a set of independent standard normal variables Y, so the failure function $g(X)$ can be written in the standard normalized system $G(Y)$. The design check point Y^* which has the shortest distance from the origin to the failure surface $G(Y) = 0$ is determined by the method of superposition, then reliability index is conveniently calculated using distance formula $\beta = \sqrt{Y^{*T}Y^*}$ and the probability of failure can be obtained using formula $P_f = 1 - \Phi(\beta)$. Superposition formula used is

$$Y_{i+1} = \left(Y_i \alpha_i + \frac{\nabla G(Y_i)}{|\nabla G(Y_i)|} \right) \alpha_i \tag{1}$$

where $\nabla G(Y)$ is a gradient vector, α is the unit vector in the direction of minus gradient. It is obvious that the calculation of $\nabla G(Y)$ is crucial in the determination of Y^*. And the bulk of the calculation of $\nabla G(Y)$ will be to compute $J_s = \dfrac{\partial S}{\partial X}$. Load effect S of structural computation generally is stresses. in plane problem, $S = \sigma = \{\sigma_x, \sigma_y, \sigma_{xy}\}^{\mathrm{T}}$.

3 STOCHASTIC BOUNDARY INTEGRAL EQUATIONS

The computation of reliability index by first-order second-moment (FOSM) (Yongjian Ren, Aimin Jiang & Haojiang Ding. 1993) method mostly relies on the responding values and their partial derivatives at the design checking point, so stochastic boundary integration equations can be obtained by computing partial derivatives of basic random variables. Deterministic boundary integration equations of tow-dimensional elasticity is an arbitrary set of basic random variables X can be transferred to be a set of independent standard normal variables Y, so the failure function $g(X)$ can be written in the standard normalized system $G(Y)$. The design check point Y^* which has the shortest distance from the origin to the failure surface $G(Y)=0$ is determined by the method of superposition, then reliability index is conveniently calculated using distance formula $\beta = \sqrt{Y^{*\mathrm{T}}Y^*}$ and the probability of failure can be obtained using formula $P_f = 1 - \Phi(\beta)$. Superposition formula used is

$$c^i u^i_l + \int_\Gamma \dot{p}_{lk} u_k \mathrm{d}\Gamma = \int_\Gamma u^*_{lk} p_k \mathrm{d}\Gamma + \int_\Omega u^*_{lk} f_k \mathrm{d}\Omega \qquad (2)$$

where u_k and p_k are stochastic the displacement and the traction on Γ respectively.

In two-dimensional plain stress problems, body integral $\int_\Omega u^*_{lk} f_k \mathrm{d}\Omega$ in the equations can be substituted by boundary integral $\int_\Gamma P_l \mathrm{d}\Gamma$, in which

$$P_l = \frac{(1+v)r}{4\pi E}\left(2\ln\frac{1}{r} - 1\right)\left(b_l n_m r_{,m} - \frac{n_l b_m r_{,m}}{2(1-v)}\right) \qquad (3)$$

Then, equation (2) can be shown that

$$c^i u^i_l + \int_\Gamma \dot{p}_l\, u\ \mathrm{d}\Gamma = \int_\Gamma u^*_l\ p\ \mathrm{d}\Gamma + \int_\Gamma P_l \mathrm{d}\Gamma \qquad (4)$$

Stochastic boundary integral equations are given by computing partial derivatives of above equation with respect to Y_j (k=1, 2, ..., n) respectively. Those equations are

$$C^i \frac{\partial u^i_l}{\partial Y_j} + \int_\Gamma \dot{p}^*_{lk} \frac{\partial u_k}{\partial Y_j} \mathrm{d}\Gamma = \int_\Gamma u^*_{lk} \frac{\partial p_k}{\partial Y_j} \mathrm{d}\Gamma + \int_\Gamma \frac{\partial u^*_{lk}}{\partial Y_j} p_k \mathrm{d}\Gamma +$$

$$\int_\Gamma u^*_{lk} p_k \mathrm{d}\left(\frac{\partial \Gamma}{\partial Y_j}\right) + \int_\Gamma \frac{\partial P_l}{\partial Y_j} \mathrm{d}\Gamma + \qquad (5)$$

$$\int_\Gamma P_l \mathrm{d}\left(\frac{\partial \Gamma}{\partial Y_j}\right) - \int_\Gamma \frac{\partial p^*_{lk}}{\partial Y_j} u_k \mathrm{d}\Gamma -$$

$$\int_\Gamma \dot{p}^*_{lk} u_k \mathrm{d}\left(\frac{\partial \Gamma}{\partial Y_j}\right)$$

Similarly, Jacobean matrix J_s of stress of interior point can be obtained:

$$J_s = \frac{\partial \sigma_{ij}}{\partial Y_l} = \int_\Gamma \left(\frac{\partial D_{kij}}{\partial Y_l} p_k + D_{kij}\frac{\partial p_k}{\partial Y_l}\right)\mathrm{d}\Gamma -$$

$$\int_\Gamma \left(\frac{\partial S_{kij}}{\partial Y_l} u_k + S_{kij}\frac{\partial u_k}{\partial Y_l}\right)\mathrm{d}\Gamma + \qquad (i{=}1,2,...n)\quad(6)$$

$$\int_\Omega \left(\frac{\partial D_{kij}}{\partial Y_l} f_k + D_{kij}\frac{\partial f_k}{\partial Y_l}\right)\mathrm{d}\Omega$$

where $\partial D_{kij}/\partial X_l$, $\partial S_{kij}/\partial X_l$ and $\partial f_k/\partial X_l$ all are explicit functions of X which are easily computed; and u_k, p_k, $\partial u_k/\partial Y_l$, $\partial p_k/\partial Y_l$ can be calculated using basic equations of stochastic boundary element method. J_s of boundary point can be obtained by solve following equations which are

$$\begin{cases} \dfrac{\partial n_k}{\partial Y_j}\sigma_{lk} + n_k \dfrac{\partial \sigma_{lk}}{\partial Y_j} = \dfrac{\partial p_l}{\partial Y_j} \\[2mm] \dfrac{\partial}{\partial Y_j}\left(\dfrac{\partial u_k}{\partial x_m}\cdot\dfrac{\partial x_m}{\partial \eta_a}\right) = \dfrac{\partial}{\partial Y_j}\left(\dfrac{\partial u_k}{\partial \eta_a}\right) \\[2mm] \dfrac{\partial \sigma_{lk}}{\partial Y_j} - \dfrac{\partial}{\partial Y_j}\left[\dfrac{2Gv}{(1-2v)}\dfrac{\partial u_i}{\partial x_m}\delta_{im}\right] - \dfrac{\partial}{\partial Y_j}\left[G\left(\dfrac{\partial u_i}{\partial x_k}+\dfrac{\partial u_k}{\partial x_l}\right)\right] = 0 \end{cases} \qquad (7)$$

where j=1,2,..., n; a donates the element of number a and $\partial u_k/\partial Y_j$, $\partial p_L/\partial Y_j$ can be known first by solve basic equations of stochastic boundary element method. If J_s is known, reliability index of stress at arbitrary point will be computed finally.

4 THE APPLICATION OF SBEM IN THE RELIABILITY ANALYSIS OF GRAVITY DAM

Every material parameters of gravity dam such as elastic modulus, Poisson ratio, and density vary randomly, and its body force and any kind of load action on its boundary are random variables. The variation of stress at dam heel is so severe that the value of reliability degree at dam heel disturbs intensely. To obtain a precise value, in the paper, the reliability degree of resistance of tension at the point a bit away from dam heel is presented as the reliability degree to valuate the resistance of tension at dam heel.

Considering a typical concrete gravity dam whose height H is 110m, slope of lower reaches is 1:0.8, and water depth of upper reaches H_1 is 100m, the acting loads include the gravity force of the body and pressures of the waters of upper reaches. Every random variable distributes normally, and the statistical values of these parameters are tabulated as follow (to the convenience of the computation, it is assumed that the material parameters of fundament is as the same as those of the dam) (see Table 1).

Table1 Random variable parameters.

	Mean value	Standard deviation
Water depth of upper reaches H_1 (m)	100	6.3
Weight density γ (kN·m^{-3})	23.52	0.4704
Elastic modulus E (MPa)	1.96×10^4	1.96×10^3
Poisson's ratio μ	0.16	3.2×10^{-3}
Compress strength σ_c (MPa)	22.8732	4.57464
Tension strength σ_t (MPa)	1.8571	0.37142
Friction coefficient f'	1.0	0.2
Cohesion c' (MPa)	0.98	0.294

In the program, two-order elements, which can result in more precise resolution, is used. The mesh of the body and the foundation of the dam is shown as following figures which have 91 nodes and 46 elements altogether.

The problems of the resistance to tension at the dam heel, the resistance to compression at the dam toe and the anti-slide stability at the interface of the dam and the foundation are analyzed.

Using the parameters listed in Table 1, reliability indices of the resistance to tension and compression obtained in the computation are respectively $\beta_t = 4.527$ and $\beta_c = 4.754$, while the stability reliability index is $\beta_s = 4.308$. It can be seen that the computational results tallies with the practical engineering situation.

It can be known from the figures that the coefficient of variation of each parameter has influence on the stability reliability. Figure 2 shows that the coefficients of variation of anti-shearing friction coefficient f' and cohesion c' greatly influence the stable safety of gravity dam. Comparatively, from Figure 3, the influence of elastic modulus E, weight density γ and water level of upper reaches H_1 are trivial, which can be ignored. These conclusions also

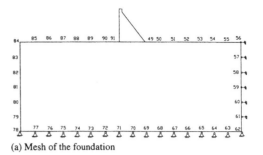

(a) Mesh of the foundation

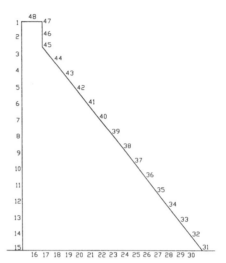

(b) Mesh of the dam body
Figure 1 Boundary element mesh

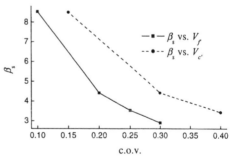

Figure 2 Relations of reliability index β_s with coefficients of variation f' and c'

coincide with practical engineering situations of gravity dam.

5 CONCLUSIONS

In the paper, two-dimensional elastic stochastic boundary integral equations are obtained by computing partial derivatives, and applied in the

529

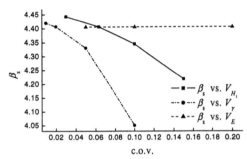

Figure 3 Relations of reliability index β_s with coefficients of variation H_1, γ and E

reliability analysis of intensity and stability of concrete gravity dam. To great volume constructions such as concrete gravity dam, the computation of reliability index of some known danger points or sliding surfaces by stochastic boundary element method, has the obvious advantages that fewer data are inputted and computational results are more precise. Numerical example shows that the method is effective in the reliability problem of practical engineering.

6 REFERENCES

Burczynski, T. & Skrzypczyk J. 1999. Theoretical and computational aspects of the stochastic boundary element method. *Computer Methods in Applied Mechanics and Engineering* 168(1-4): 321-344
Gautam, D. 1992. Stochastic finite and boundary elements. *Proceedings of Engineering Mechanics, Publ by ASCE, May 24-27 1992: 932-935*
Kaljevic, I., Saigal, S. 1993. Stochastic boundary elements in elastostatics. *Computer Methods in Applied Mechanics and Engineering* 109(3-4): 259-280
Nakagiri, S., Suzuki, K. & Hisada, T. 1983. Stochastic boundary element method applied to stress analysis. *Proc. 5th. Int. Conf. On Boundary Elements, Hiroshima, Japan, 1983*
Weidong Wen, Wei Chen & Deping Gao. 1995. Two-dimensional elastic stochastic boundary element and reliability analysis. *Journal of Applied Mechanics* 12(1):8-11
Yongjian Ren, Aimin Jiang & Haojiang Ding. 1993. Stochastic boundary element method in elasticity. *Acta Mechanica Sinica/Lixue Xuebao* 9(4):320-328

Hydropower in the New Millennium, Honningsvåg et al (eds), © 2001 Swets & Zeitlinger, Lisse, ISBN 90 5809 195 3

The repairing technique for the slab of stilling pool of Ankang hydropower station

Sun Zhiheng
China Institute of Water Resources and Hydropower Research, Beijing, China

Zhanglei
China Institute of Water Resources and Hydropower Research, Beijing, China

ABSTRACT: The Ankang Hydropower Station is located at t he upper reach of Hanjiang River in China, and mainly consists of a 128m high concrete gravity dam, a power house and navigation structures and so on. The slab of its stilling pool is made up of more than 7m thick concrete with a strength grade R_{150} which is covered by a 1m thick layer of anti-erosion concrete with a strength grade R_{300}. Due to the poor construction quality, it was found during operation that the anti-erosion concrete layer was separated from the base concrete, which would cause the deterioration and disability of this upper layer. Therefore the plan of using prestressing anchors to strengthen the slab and repairing the defect concrete was put forward. As for the defect repairing, a variety of materials such as modified epoxy mortar, polymers cement mortar polysulfide sealant. GB flexible plate and GBW swelling bars were utilized and suitable repairing methods were adopted. This work included the erosion on the spill way surface, cracks on the slab of stilling pool, and the expansion joints of the slab of stilling pool. After the flood period, the repairing effect was found to be satisfying.

1 INTRODUCTION

The Ankang Hydropower Station is Located at the upper reach of Hanjiang River of China, and mainly consists of a concrete dam, a power house and navigation structures. This project is of a Large Size with its water-holding structure belonging to a class I one of which the design flood peak and spillway design flood are 36700 m^3/s and 45000 m^3/s respectively. The design flood peak for its energy-dissipating structure is 28100 m^3/s.

The construction of the stilling pool began in 1985 and was finished in 1989.After a operation period of more than 10 years, it was found in 1996 during a routine inspection that some defects such as cracks and staggered joint on the 1m thick R_{300} surface concrete slabs. Then the stilling pool was strengthened by using ordinary anchor bars in the same year. But during another check in the end of 1999, some of these anchor bars were found to be pull out of the concrete surface. So a conclusion was drawn through the field investigation that the concrete Slabs of the pool were likely to be lifted, and it would cause the potential disability of this anti-erosion concrete layer. Taking into

consideration that this station played an important regulating role in the Shanxi Electric Network, it was impossible to take a total re-construction for the cushion pool, So the only remedial measures were the strengthening and repairing of these surface concrete slabs.

2 REHABLITATION PLAN

The rehabilitation of the pool slabs was divided into two parts, strengthening and repairing. Strengthening was to use prestressed anchor bars to enhance the integrity of these surface slabs with the base concrete, while the main defects for repairing were focused on the scour holes on the ogee section of spillway and the cracks and some expansion joints of the cushion pool. Details of repairing procedure are described below.

2.1 Repairing for scour holes

The repairing material for doing this job is required to have a strong bondage with the old concrete and excellent properties of erosion and abrasion resistance.

For the scour holes with depth less than 5cm only

polymer cement mortar was used, but when these depth exceeded 5cm, polymer modified concrete was first placed at the bottom followed by a 3cm thick polymer cement mortar layer on the top. The polymer cement mortar and the polymer-modified concrete are

composite materials made by mixing polymer latex's with the ordinary cement mortar and concrete respectively. The hardening process of these materials can be described as follows. With the forming of rigid space structure by cement hydration products, and due to the cement hydration and water loss, the latex are dehydrated. Then the latex particles adhere to each other, and then under the capillary force membranes are formed to fill in the intervals between the crystal phases, which will finally lead to a space net structure of the polymer phase. The introduction of the polymer phase greatly enhances the density and adhesive capacity of the cement matrix, and simultaneously decreased its brittleness. As a result, these polymer modified materials will have a much better flexibility and erosion resistance, and can be used as the thin layer surface repairing materials for hydraulic structures under severe conditions.

The main property of the polymer cement mortar and polymer modified concrete are listed in table I.

Table I   Main properties of PCM and PMC

Material	Compressive Strength (MPa)	Fracture Strength (MPa)	Tensile Strength (MPa)	Bonding strength With concrete (MPa)
PCM	≥35	≥8.0	≥5.0	≥2.0
PMC	≥40	≥9.0	≥5.0	≥2.0

2.2 Repairing for vertical and transverse expansion joints.

The repaired expansion joints are required to have a suitable capacity to accommodate the joint displacement and a good erosion resistance. Mean while, the waterstop structure inside should be well protected. The schematic-repairing plan is shown in figure I.

It can be found from the figure above that two waterstops are adopted, GB flexible plate and polysnlfide sealant. The GB plate has an excellent air-tightness, good bonding strength, strong self-adhesive capacity and flexibility. Not only can it be bond directly to the upper epoxy cement mortar without adhesives under specific conditions, but also it can deform simultaneously with the lower

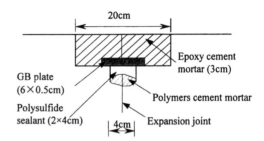

Figure I. schematic figure of expansion joints repairing

polysulfide sealant. This plate acts as the first or the top water-stop. The polysulfide sealant is a two-composition hardening waterproof material, and has a good aging resistance and excellent water resistance, elasticity, adhesion and volume restore capacity after withdrawing compression. Its physical properties can it maintained when subjected to a continuously contracting expanding, vibrating and temperature varying environment. This sealant is used as the second or bottom waterstop. The epoxy cement mortar has high compressive, tensile and bonding strength and erosion resistance, and it is coated on the joint surface for protection purpose. When all the repairing work is done, the epoxy cement mortar is to be cut in the middle to fit for a better displacement accommodation.

Table 2.   Properties of GB plate.

Property	Testing item		Control Index	Tested
Tensile capacity	Normal Temperature	Fracture elongation ratio	≥800	1278
	−30℃	Fracture elongation ratio	≥800	1040
Density(g/cm³)			≥1.15	1.22
Environmental Protection	Rubber product		no poison, no pollution	

Table 3 Properties of polysulfide sealant

Item	Index	Tested
Surface drying time(hr)	≤24	20
Max. Tensile strength(mpa)	≥0.3	0.57
Max. Elongation ration(%)	≥200	338
Drooping(mm)	≤3	1
Poisonous or not	no	no

The properties of the waterproof materials are shown in table 2 and 3.

2.3 Repairing for cracks

This work is to control the seepage through the cracks. The sealing plan is shown in figure 2.

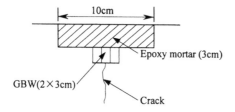

Figure 2. Schematic figure of repairing for cracks

Figure 2 shows that the main waterstop structure of repaired cracks is GBW swelling bar, which is a kind of rubber based material. Made compositely by rubber, inorganic/ organic water absorbing materials, high viscosity resins and others under specify producing process. Having elasticity and plasticity to some extent, GBW can expand in volume when

Table 4. Properties of GBW swelling bar

Item	Index
Max. expanding ration under water(%)	150°C~250°C (adjustable)
Bonding strength with concrete	Exceed its own tensile strength
Water load resistance(mpa)	≥2.0MPa
High Temperature Tolerance 80°C,200h,75°C declination	No fluidity
Low Temperature Tolerance -25°C,200h,bend	No crackles No rigidity
Density g/cm³	1.10~1.30

meeting seepage water and thus pond the cracks to provide impermeability. The surface epoxy cement mortar is to bear the water pressures from the top and the bottom of slabs and provide erosion resistance.

3 REMARKS

Repairing the defects is a painstaking and complex work. What should be done first is to make out a suitable repairing plan according to characteristics of defects, including selecting appropriate repairing materials and construction procedures. But these are only a half success. Another important factor is whether a professional contraction group is available. Before working, a proper contraction organization plan is indispensable, and during working, an expertise personnel is necessary since the repairing procedures should be adjusted in accordance with the field situations to guarantee the plan to be fulfilled. The repairing quality for the defects of stilling pool slabs of Ankang is good and the effect was found to be satisfying after the examining of the flood period. Appropriate repairing materials and construction procedures. But these are only a half success. Another important factor is whether a professional contraction group is available. Before working, a proper contraction organization plan is indispensable, and during working, an expertise personnel is necessary since the repairing procedures should be adjusted in accordance with the field situations to guarantee the plan to be fulfilled.

The repairing quality for the defects of stilling pool slabs of Ankang is good and the effect was found to be satisfying after the examining of the flood period.

Author index